W0234648

Wireless Communication and Networks

UPENA DALAL

Associate Professor
Department of Electronics Engineering
Sardar Vallabhbhai National Institute of Technology
Surat

OXFORD
UNIVERSITY PRESS

OXFORD
UNIVERSITY PRESS

Oxford University Press is a department of the University of Oxford.
It furthers the University's objective of excellence in research, scholarship,
and education by publishing worldwide. Oxford is a registered trade mark of
Oxford University Press in the UK and in certain other countries.

Published in India by
Oxford University Press
YMCA Library Building, 1 Jai Singh Road, New Delhi 110001, India

© Oxford University Press 2015

The moral rights of the author/s have been asserted.

First published in 2015

All rights reserved. No part of this publication may be reproduced, stored in
a retrieval system, or transmitted, in any form or by any means, without the
prior permission in writing of Oxford University Press, or as expressly permitted
by law, by licence, or under terms agreed with the appropriate reprographics
rights organization. Enquiries concerning reproduction outside the scope of the
above should be sent to the Rights Department, Oxford University Press, at the
address above.

You must not circulate this work in any other form
and you must impose this same condition on any acquirer.

ISBN-13: 978-0-19-809888-1
ISBN-10: 0-19-809888-X

Typeset in Times New Roman
by Cameo Corporate Services Limited, Chennai
Printed in India by Magic International (P) Ltd, Greater Noida

Third-party website addresses mentioned in this book are provided
by Oxford University Press in good faith and for information only.
Oxford University Press disclaims any responsibility for the material contained therein.

Dedicated to my husband Devang
and my daughters Parima and Jahnavee

Preface

A communication system is used for transmitting information or data from one point to another. Earlier communication systems, such as telegraphy and telephony, were wired systems, which transmitted information through wires. A major breakthrough in the field came with the advent of wireless technology, which uses radio waves to transmit data, as opposed to wires, to carry signals over the communication path. Wireless communication deals with the principles, techniques, and analytical tools underlying wireless systems, other emerging systems, and standards in the field.

Wireless communication has become the fastest growing segment of the telecommunications industry and has led to exciting technological advances over the last few decades. This has radically changed the way people communicate across the world. Initially, wireless communication was mainly used in military applications. With the commercialization of wireless systems and resource-sharing networks, communicating with people over mobile phones even on the move has become easy, and access to social media and applications provided by mobile service providers has made it much more convenient. In the future, the telecommunications industry is set to become all-wireless with an 'anywhere, anytime, and for anybody' communications scenario. Users will have a single and unique identification number—universal telecommunication number (UTN)—which would enable users to receive as well as make calls from any terminal on any network. The stage is set for 4G, the fourth generation of mobile telephones, which in addition to the usual services would also enable streaming multimedia, HDTV content, digital video broadcasting (DVB), and ultra-broadband Internet access. Mobile companies are already aiming for 5G technology, which has been visualized as the convergence of network access technologies.

ABOUT THE BOOK

This book is primarily designed for undergraduate students of electronics and communications engineering as well as computer engineering, and is suitable for courses on mobile communication, wireless communication, and mobile networks. Basic knowledge of the concepts of communication, signal processing, and probability theory is assumed to be a pre-requisite. A lucid approach, both in terms of language and content, has been adopted throughout the text. Beginning with the fundamental concepts of wireless communication, the book comprehensively covers the various aspects of wireless systems.

As the majority of wireless communication systems today are completely digital, this text focuses only on all the aspects of digital communication in the context of wireless channels, and analog methods have been completely omitted from the text.

The book is divided into four parts which represent the four important aspects of practical wireless systems—Wireless Communication Prerequisites (Chapters 1 and 2), Wireless Channels and Modelling (Chapters 3 and 4), Wireless Communication Techniques (Chapters 5–9), and Wireless Networks (Chapters 10–12). Each chapter begins with a theme and key topics, and gradually explores concepts through detailed explanations and illustrations. A large variety of solved examples have been added to elucidate the application of the theory covered in each chapter. Review questions, multiple-choice questions, and numerical exercises add value to the rich content of the book. The book also contains appendices on the additional topics associated with the subject.

CONTENTS AND COVERAGE

Chapter 1 is the introductory chapter, which describes the basic terminology associated with wireless communication in the present scenario and trends in wireless systems. It discusses various types of wireless systems in terms of major advancements identified in different generations. The chapter helps to develop a basic understanding of the subject, so that concepts in later chapters can be understood easily.

Chapter 2 is related to the infrastructure development of cell-based wireless communication in multi-user environments. An understanding of cell theory is necessary for deciding the size of the cell, locating the transmitter in a cell, and splitting the cell to cover a higher population density. Frequency reuse is the key concept to utilize the available channels efficiently, but it leads to co-channel interference. By utilizing cellular theory, all these problems can be solved. The chapter briefly discusses traffic engineering as well.

Chapter 3 describes radio propagation over a wireless channel. Starting from the free-space propagation model, different types of long-distance radio propagations are discussed in the chapter. Path loss model and multipath effect are also explained; these are necessary to understand the behaviour of the channel in certain frequency ranges. It also delves into the different types of fading effects (such as delay spread and Doppler effect), which are very common in the multipath environment.

Chapter 4 covers the different channel models represented in their mathematical forms. The chapter discusses popular channel models such as the Rayleigh model, the Rician model, and Nakagami, which are all characterized by their probability density functions (PDFs). It also covers popular urban models such as the Okumura and Hata models.

Chapter 5 mainly deals with the concept of source coding and waveform coding. Most real-time signals are analog in nature. Beginning from the digitization of analog signals, further processing must be applied to the source signal to compress or convert it into a standard format. The chapter describes the analog-to-digital conversion process, as well as the errors which result from the conversion, for example, aliasing. Digital transmission formats, special voice coders for low bit rate signals, and data compression methods are also discussed in this chapter.

Chapter 6 describes error handling over a noisy channel. As the wireless channel is more susceptible to noise and multipath effects, error-correcting codes are required. Hence, in this chapter, we describe most of the error-correcting schemes with their error-correction capabilities. The chapter also demonstrates the latest developments, such as Turbo codes, which are increasingly becoming popular as they approach Shannon's limit for bit error rate (BER) performance.

Chapter 7 helps in the understanding of all the basic single- and multi-carrier digital modulation schemes along with their mathematical representations, block diagrams, constellation diagrams, and other important parameters. The chapter comprises conventional methods such as Amplitude shift keying (ASK), frequency shift keying (FSK), binary phase shift keying (BPSK), M-PSK, and quadrature amplitude modulation (QAM) as well as the modified versions of the conventional modulation schemes, such as differential PSK (DPSK), offset keyed quadrature PSK (OKQPSK), minimum shift keying (MSK), Gaussian MSK (GMSK), and M-FSK. Finally, the chapter explains spread spectrum modulation (SSM) and orthogonal frequency division multiplexing schemes, which are especially suitable for the 3G and 4G systems, and elucidates how these techniques are superior to conventional digital modulation techniques.

Chapter 8 illustrates the diversity techniques, equalization methods, and channel estimation to mitigate channel effects. Most of these techniques are important at the receiver's end and help improve the quality of signal reception. An understanding of these concepts is very important because, as a result of these techniques, phase ambiguity due to multipath, frequency-dependent effects, or fading effects can be considerably reduced at the receiver side, and BER performance can be improved. Multiple input, multiple output (MIMO), the latest diversity-based technique, which is based on spatial diversity, is also covered in this chapter.

Until Chapter 8, all the basic theories and fundamentals for establishing a single wireless digital link are described. From Chapter 9 onwards, the focus shifts to the multi-user system environment.

Chapter 9 is related to multiple access techniques. There are numerous ways in which multiple users are allowed to access the available wireless channels on a sharing basis, so that all the users can communicate successfully without any partiality and without interference from one another. This chapter throws light on some of the schemes for multi-user environments in which an individual user's information is transmitted independently, such as FDMA, TDMA, CDMA, OFDMA, and space division multiple access (SDMA). For packet radio systems, random access schemes such as ALOHA, slotted ALOHA, and carrier sense multiple access with collision detection (CSMA/CD) are used for sharing packets over a channel, rather than complete information transmission at a time.

Chapter 10 summarizes the concepts of conventional networking and its applications in a wireless networking environment. Starting with the OSI reference model and layered concept of protocol design, the chapter discusses TCP/IP protocol. It also describes the basic constraints of networking and gives some basic solutions, such as MAC scenario, routing protocols, and transport scenario, along with their applications, and highlights the importance of mobile computing.

The last two chapters provide an introduction to all the existing wireless digital systems, which have been developed on the basis of certain standards and protocols.

Chapter 11 describes the infrastructure-based/cell-based networks which are established permanently and support mobility, such as GSM, CDMA, UMTS, WLL, and LTE.

Chapter 12 describes special categories of wireless systems like ad-hoc networks (e.g., bluetooth), ad-hoc networks with the support of cellular concept, and networks mainly designed for data access or transfer (e.g., Wi-fi and WiMAX). The chapter also expounds on Zigbee, which is a special protocol for the wireless sensor network, and UWB, which is used for ultra-high speed indoor communication.

Appendices A to *F* deal with linear systems theory, algebra for the linear system, probability theory, DSP fundamentals applied to OFDM processing, satellite communication aspects, and Erlang and Poisson traffic tables.

ONLINE RESOURCES

The online resource centre provides resources for faculty and students. The following resources are available for faculty and students using this text:

For Faculty
Solutions manual
PowerPoint presentations
For Students
MATLAB codes

ACKNOWLEDGEMENTS

First and foremost, I thank God. I sincerely thank the senior teachers of my department, retired Prof. Mrs Nila Desai and retired Prof. K.U. Joshi, and my colleagues Mrs Jigisha Patel and Mrs Shweta Shah for their encouragement and valuable support. I am grateful to my husband Devang for his continuous motivation and support. It has been a great pleasure and honour to be associated with Oxford University Press, India. I express my deep gratitude to the entire editorial team as well as the production department of OUP India for publishing this book with a high degree of precision and accuracy. Every effort has been made to produce an error-free text; however, I would be grateful if readers point out any unintended errors or discrepancies. Suggestions for improving the presentation and contents of the book can be sent to the publisher through their website www.oup.co.in or to the author at upena_dalal@yahoo.com.

Upena Dalal

Features of

Theme of the Chapter

k mainly deals with the concepts of wireless digital communication. Though it is assumed that readers of th
familiar with the basic theory of communication, many required concepts are revised as a ready reference
tudents must be familiar with wireless communication systems—both conventional and latest. This chapte
ith a brief revision of the basics of communication and moves on to discuss wireless systems. It explains
digital wireless link with all the necessary blocks that form the basis for wireless systems. The chapter als
transmission rate, channel capacity, bandwidth, and signal-to-noise ratio parameters deciding the perfo
this link along with the types of signals useful for communication theory. It further explores the need fo
e of the best developments in wireless communications, which is possible only if the standards used toda
ess systems are known. Evolution of a system is linked with the previous systems, and the new system i
by analysing the problems of the previous systems and eliminating them. Hence, it is also necessary to kno
opment scenario of the first to fourth generation systems. Once this background is provided and student
lying these from the root level of the wireless link, considering each and every stage of the wireless lin
t of the theory and its application to the system can be correlated and the best solutions can be identified fo
here, anytime communication scenario.

Theme of the Chapter

Provides a glimpse of the topics that the readers
are going to read and understand from a chapter

Antenna size and
carrier follow inverse
relationship; hence,
for portable sizes of
antenna, very high
frequencies are used in
wireless systems.

In wireless communi-
cation, the final form of
transmission is always
analog, irrespective of
whether the modulation
is analog or digital.

Sidebar

Captures important statements to facilitate easy
grasp and quick recap

Note: It is observed that an analog signal consumes less spectrum compared to its digital counterpart and hence requires lesser
bandwidth, because the digital counterpart is the result of sharp transitions.

Note: Due to amplitude compression, logarithmic increase in quantization noise throughout the dynamic range of
a sampled signal will keep the SQNR constant throughout this dynamic range.

Note: For 4 kHz voice digitization, the standard word size used is 8 bits. If an input analog signal is sampled 8000
times/s and each sample is given a code word that is 8 bits long, then the maximum transmission bit rate for
telephony systems using PCM will be 64,000 bits/s.

Note

Highlights important statements so that readers
don't miss them while reading

WIRED MEDIA AS TRANSMISSION LINE

cting wires become a transmission line? It is when the capacitance between the wires and
tributed instead of lumped. This begins to happen when the wire approaches the dimensior
frequency f are related by $\lambda = v/f$). At sufficiently high frequencies, when the length of the
vices is in the order of the wavelength or larger, the voltages and currents between thes
avel back and forth on the wires. Hence, a signal sent out by one device propagates as a
d the wave is reflected unless the receiving device is properly terminated or matched. If th
an interfere with the incident wave, making communication unreliable or even impossible.
mportant when networking computers, printers, and other peripherals, which must be pr
ansmission lines are used to carry the signal from the transmitter front end to the antenna

Box

Throughout each chapter, boxes provide a brief
description of some key concepts and their significance

This transient response
really does have a broad
spectrum

But windowing the data
makes it look more like a
single tone

Fig. 8.11   The result of applying a window function without proper thought

Illustrations

Important topics have been well-supported with
suitable illustrations to allow easy visualization
of difficult concepts. The book contains close to
350 self-explanatory illustrations.

the Book

Examples

Every chapter contains plenty of solved examples to demonstrate the applicability of the concepts discussed.

Example 5.2 The following are the readings for the measurement of quantization error in five consecutive samples. The number of quantization levels in the dynamic range of 2 V is eight.

Sample 1: 1.2 V
Sample 2: 1 V
Sample 3: 0.95 V
Sample 4: 1.41 V
Sample 5: 1.65 V

Find the quantization error in terms of its mean square value.

Solution If the dynamic range is 2 V, then the smallest step size will be 2/8 = 0.25 V.

Hence, the eight quantization l
0.5, 0.75, 1.0, 1.25, 1.50, 1.75, and 2
The measured samples will be a
ing quantization values:

Sample 1: 1.25 V   quantization
Sample 2: 1.0 V   quantization
Sample 3: 1.0 V   quantization
Sample 4: 1.5 V   quantization
Sample 5: 1.75 V   quantization
Mean square error
= $[(0.05)^2 + 0 + (0.05)^2 + ($
= $[0.0025 + 0.0025 + 0.00$
= $0.0231/5 = 0.00462$
Root mean square error = 0.0678

6.4 CHANNEL CODING AND TRADE-OFF

Error-correcting codes can be regarde
Figure 6.5 compares two curves depic
noise density ratio (E_b/N_o) (which is th
a digital signal representing the signal
scheme without coding; the other repr
Though channel coding is incorpora

Coverage

The book covers topics related to wireless communications as well as wireless networks. Topics such as source coding, channel coding, multiple access, modulation techniques, spread spectrum, diversity, and equalization are covered in great detail.

Modular Organization

Mutually exclusive chapters combined with their modular organization allows flexibility to the instructors, so that they can pick chapters as per their respective syllabi requirements.

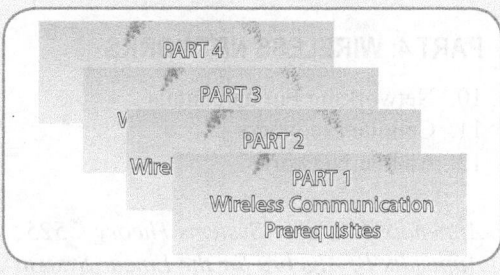

Exercises

EXERCISES

Multiple-choice Questions

3.1 The electric field of an EM wave at a point in free space is in the positive *Y* direction and the magnetic field is in the positive *X* direction. The direction of power flow will be in the
(a) positive *X* direction   (c) positive *Z* direction
(b) positive *Y* direction   (d) negative *Z* direction

3.2 In isotropic propagation, if f_c is the critical frequency and θ is the angle of incidence at the ionosphere measured with respect to the normal, the MUF equals
(a) $f_c \cos\theta$ (c) $f_c \sec\theta$
(b) $\sqrt{f_c} \cos\theta$ (d) $\sqrt{f_c} \sec\theta$

3.3 The skin distance is

3.7 The virtual height of an ionospheric layer is
(a) more than the height a wave actually reach
(b) less than the height a wave actually reache
(c) the same as the height a wave actually reac
(d) none of these

3.8 The ionosphere roughly extends from
(a) 50 km to several earth radii
(b) 50 km to 80 km
(c) 50 km to 400 km
(d) 50 km to 150 km

3.9 The total noise of a satellite earth station rece system consists of
(a) sky noise
(b) parametric amplifier noise
(c) antenna and feeder noise

Exercises

Has a rich set of end-chapter exercises with close to 250 review questions, more than 200 Multiple-choice Questions, and 100 unsolved problems.

Brief Contents

Detailed Contents

List of Symbols

$\alpha(t)$ = Rayleigh amplitudes of the multipath signals

δ = handoff margin

ε = permittivity of free space (capacitance per unit length measured in farads/meter)

ε_0 = dielectric constant of vacuum

ε_r = relative dielectric constant of the earth

$\phi(t)$ = different phases due to different delays of multipath signals

Φ = phase of noise vector

λ = wavelength (of carrier normally)

λ = call arrival rate

μ = permeability of free space (inductance per unit length measured in henries/meter)

μ = mean number of call arrivals

η = aperture efficiency

η_{fdma} = spectral efficiency of FDMA scheme

$\eta_{fdma\text{-}tdma}$ = spectral efficiency of FDMA-TDMA scheme

η_{tdma} = spectral efficiency of TDMA scheme

σ = standard deviation

σ_g = conductivity of ground

σ_x = standard deviation from the mean value

θ = phase variable

θ = directions variable

θ_{err} = phase error

θ_{max} = maximum allowable phase margin

τ = time variable for autocorrelation function

τ_{max} = maximum delay spread on multipath channel

τ_1 = delay between received multipath signals

ω = angular frequency of a signal

ω_c = angular carrier frequency

a's = filter coefficients of an FIR filter

a_n = value of nth chip in PN code

A = traffic in erlangs

A_0 = peak amplitude

A_e = effective receiving aperture area of antenna

A_m = amplitude of the signalling element

A_r = actual receiving aperture area of antenna

B = spread bandwidth

B = busy period in ISMA scheme

B = grade of service

B_c = coherence bandwidth

B_d = Doppler spread

b_g = number of bits in each guard interval

B_j = interference signal bandwidth

B_m = message signal bandwidth

b_{OH} = overhead bits in TDMA frame

b_p = number of bits in each slot preamble

b_r = number of overhead bits per reference burst

b_T = total bits in TDMA frame

B_s = average bits per symbol

B_s = PN signal bandwidth

c = average number of calls in period of observation

c_n = amplitude of nth reflected component

$c(t)$ = PN sequence (subscript t or r represents transmitter PN sequence and received PN sequence)

C = channel capacity in bits/s

C = total effective number of duplex channels available in cell area

$C_{auto}(k)$ = autocorrelation function for speech for kth sample

d = distance between isotropic source and receiver antenna

d = distance between two antenna elements

d_1 = initial period of the busy period in ISMA

d_2 = processing delay within busy period after packet duration in ISMA

d_f = far-field distance

d_{min} = minimum weight

d_0 = close-in distance

$d(U,V)$ = hamming distance between the codes U and V

A_t = largest physical dimension of the transmitter antenna

D = frequency reuse distance

$e(n)$ = estimation or equalization error signal in discrete form

E = electric field strength w.r.t. transmitter

E_b = bit energy

E_{ms} = mean square error

E_o = field strength while propagation in free space

E_s = signal energy dissipated in time T

f = frequency variable

f_1 = spacing between consecutive hopping frequencies in FHSS

f_1, f_2, etc. = subcarrier frequencies in OFDM

$f_1, ..., f_N$ = frequency groups within a cluster f_1 for cell 1 and so on

f_c = carrier frequency

f_d = Doppler spread

f_{max} = highest frequency content of analog in

f_o = centre frequency of the PN signal spectrum

f_s = sampling frequency or Nyquist frequency

$f(t)$ = basic pulse function

$f(x,y)$ = image function (2D)

Δf = carrier spacing between orthogonal subcarriers in OFDM

F = flux density

F = fade margin

$F(x)$ = companding function to compress x integer value

G = offered traffic

G_r = gain of receiving antenna

G_t = gain of transmitting antenna

$G(\theta)$ = gain of antenna at angle θ

$R_S(f)$ = PSD function

H = average call holding time

h_R = receive antenna height

h_T = transmit antenna height

$h(n)$ = channel impulse response (discrete form)

$h(t)$ = channel impulse response

$h(x)$ = polynomial for LFSR design

$h_e(n)$ = estimated channel impulse response in discrete form

$H(\omega)$ = a channel impulse response in frequency domain

H = complex N × M matrix representing the MIMO channel impulse response, subscripts LOS, NLOS represents corresponding components

H or $H(x)$ = entropy (in outcome x)

$i(t)$ = current as a function of time t (variable current)

$i(t)$ = interference signal

$I(t)$ or I = in-phase component of the received modulated (complex) signal

I = mean value of exponentially distributed Poisson arrivals

I_i = interference power of ith cell

I_{int} = total number of interfering cochannel cells

I_o = interference density (per bit)

$I(x_i$ or $x_j)$ = information content in outcome x_i or x_j

J = total interference power in CDMA

k = input bits applied to channel coder

k = ratio of message bit duration to chip duration (hops per message bit)

k = number of channels within a group

k = occupied trunks

K = frequency multiplier in FHSS

K = Rician factor

K = number of users in DS-CDMA

K = variable for smart antenna elements in general

l = number of mobile terminals

L = dimension of vector quantization, number of samples in a block or vector

L = maximum resolvable paths for CDMA

L = total number of duplex channels available to the operator

L = number of multipath components received at each antenna element from 1 mobiles

m = shape factor or gamma distribution (Nakagami model)

$m(t)$ = transmitted message signal

$m'(t)$ = received message signal

M = number of quantization levels, number of signalling elements

M = number of hopping frequencies (in FHSS), number of hopping time slots (in THSS)

M = number of times cluster is repeated

M = a number of receiving antennas for MIMO

$M(f)$ = spectrum of message signal

n = number of bits per quantization level (i.e. number of bits per sample)

n = number of bits/symbol

n = output bits from channel coder

n = number of users/slots per channel in TDMA

n = path loss exponent

N = total noise power

N = maximum number of significant reflected components to model the channel

N = number of users in CDMA

N = one period comprising number of samples, IFFT bin size in OFDM

N = number of trunks

N = cluster size in cell theory

N = number of transmitting antennas for MIMO

N_c = number of chips in a full period of PN code

N_c = number of subcarriers in OFDM bandwidth

N_c = total number of carriers in FDM scheme

N_t = number of data slots per TDMA frame

N_o = noise power spectral density

N_r = number of reference bursts per TDMA frame

N_s = total number of slots in TDMA frame

N_u = number of users supported in FDM

N_{us} = number of users per sector

p = probability of occurrence of information in binary symmetric source

p = number of shift registers in an ML sequence generator

p = persistency

p_p = prime number to generate twin prime sequences

P = power (subscript differentiating 's' signal or 'n' noise power)

P_B = bit error probability

P_o = power radiated by test antenna

P_r = received power at the receiver front end

$Pr_{handoff}$ = power level at which handoff is made

$Pr_{min\ usable}$ = minimum usable power for acceptable voice quality

P_t = (isotropic source) transmitted power

PG = processing gain

$P(\theta)$ = power radiated in a direction with angle θ

$P(t)$ = power as a function of time t (variable power)

$p(x)$ = probability function

$P(x_i$ or $x_j)$ = probability of outcome x_i or x_j

$P(Y/S^{(m)})$ = probability of maximum likelihood with mth sequence when Y sequence is received

$\overline{\overline{P}}$ = local mean power

$\overline{P}$ = area mean power

q = integer deciding the q-r sequences and Hall sequences

q_e = quantization error

$Q(t)$ or Q = quadrature component of the received modulated (complex) signal

r = integer deciding the sequences

Δr = resolution, step size in quantizer

$r(n)$ = received signal (discrete form)

$r(t)$ = received time varying signal

R = resistance

R = data rate or information rate in bits/s

R = cell radius

R_b = bit rate in bits/s

R_c = ground reflection coefficient

R_d = data redundancy

R_{max} = maximum cross correlation

R_T, R_R = correlation matrices of transmitter and receiver, respectively

$R(D)$ = rate of vector quantization in bits per sample (for given distortion D)

$R_c(\tau)$ = autocorrelation function for PN codes

s = number of sectors in a cell

s_1, s_2, etc. = symbols, split from OFDM symbol (frame) to assign the carrier

$s(t)$ = transmitted time varying signal

$s(t)$ = spread baseband at the transmitter

$s'(t)$ = spread baseband at the receiver

S = average signal power

$S(t)$ = frequency spectrum (signal in frequency domain)

$s(n)$ = transmitted signal (discrete form)

$s(k)$ = kth speech sample

S_k = kth frequency sample in the speech spectra

$S^{(m)}$ = mth possible sequence

$s_k(t)$ = kth subcarrier in multicarrier system

$S_q(t)$ = quantized speech signal

t = time variable

t_c = error-correcting capability of the code

t_{chip} = chip duration

t_d = error-detecting capability of the code

t_{di} = delay over the ith channel path in case of multipath reception for SSM system

t_m = message bit duration for spread spectrum system

T or T_o = period of a periodic signal (time for one cycle)

T = traffic observation duration

T_b = One bit interval (pulse duration)

T_{code} = period of PN code

T_f = frame duration containing all time slots

T_g = guard interval between consecutive OFDM symbols

T_m = maximum excess delay over channel

T_{mc} = (multicarrier system) symbol period

T_o = coherent time

T_{rms} = RMS delay spread

T_s = OFDM symbol period

T_{sc} = (single carrier system) symbol period

T^k = shift operator

ΔT_s = sampling interval

v = velocity of mobile

v = velocity of light/velocity of electromagnetic waves

v_f = voice activity factor

V = voltage (subscript differentiate with 's' signal or 'n' noise voltage)

$V(t)$ = voltage as a function of time t (with appropriate identity subscript)

w = variable for channel taps or weights

$w(n)$ = white noise (discrete form)

$w(t)$ = white Gaussian noise varying with time

W = bandwidth

$W_{channel}$ = available bandwidth

W_{guard} = guard band

W_{signal} = per user bandwidth

$W(U)$ = Hamming weight for code word U or V

x = actual number of call arrivals

Y = received sequence

z = number of standard deviations

z = threshold level (or noise threshold level)

PART 1
Wireless Communication Prerequisites

Signal processing and basic principles of communication are the basis for wireless communication, and cellular theory forms the basis for wireless systems. Part 1 of this book consists of two chapters that provide the initial spark by introducing the basic terminologies and describing the existing wireless systems all over the world, along with their development phases called *generations and major features*.

In this part, the readers will also be introduced to the basic necessity of user's mobility for "anytime anywhere seamless connectivity", and that is cellular structure and associated issues.

This way, the section creates a bigger picture of the matters to be touched upon for the development of the complete wireless scenario.

1 Fundamentals and Present Scenario

Theme of the Chapter

This book mainly deals with the concepts of wireless digital communication. Though it is assumed that readers of the book are familiar with the basic theory of communication, many required concepts are revised as a ready reference. Today's students must be familiar with wireless communication systems—both conventional and latest. This chapter begins with a brief revision of the basics of communication and moves on to discuss wireless systems. It explains a complete digital wireless link with all the necessary blocks that form the basis for wireless systems. The chapter also discusses transmission rate, channel capacity, bandwidth, and signal-to-noise ratio parameters deciding the performance of this link along with the types of signals useful for communication theory. It further explores the need for and scope of the best developments in wireless communications, which is possible only if the standards used today for wireless systems are known. Evolution of a system is linked with the previous systems, and the new system is designed by analysing the problems of the previous systems and eliminating them. Hence, it is also necessary to know the development scenario of the first to fourth generation systems. Once this background is provided and students start studying these from the root level of the wireless link, considering each and every stage of the wireless link, every part of the theory and its application to the system can be correlated and the best solutions can be identified for the *anywhere, anytime* communication scenario.

Key Topics

- Fundamental terms of communication
- Wireless communication link model
- Bandwidth and signal-to-noise ratio
- Types of signals

- Types of communication systems
- Wired versus wireless communication
- Types of wireless systems
- Existing technologies and requirements
- Evolution of wireless systems

- First- to fourth-generation wireless systems
- Licensed and unlicensed band communication
- Spectrum policies

1.1 FUNDAMENTAL TERMS OF COMMUNICATION

Wireless communication is a diverse field and its study requires a basic knowledge of many other fields. The overall model of the learning system for wireless communication is shown in Fig. 1.1.

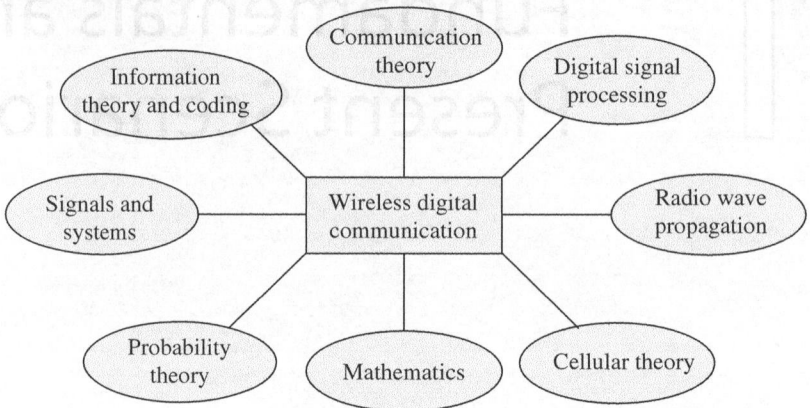

Fig. 1.1 Interdisciplinary learning model for wireless communication

Readers may be aware about many of these fields and may even be familiar with the basic theory of communications. However, in this chapter, we will brush up on all the fundamentals before we get into the details of wireless communication. This section will recapitulate the most frequently used terms in this subject.

Information Communication systems convey messages that originate from information sources. The information may be analog or digital, and accordingly, the communication system can be classified as an analog or a digital system. The sine wave is the fundamental analog information signal. A pure sine wave can be represented by three parameters—peak amplitude (A_0), frequency (f), and phase (θ)—in the form

$$s(t) = A_0 \sin(\omega t + \theta)$$

where $\omega = 2\pi f$, the angular frequency. The analog information may be voice or video (or real-time signals).

Information is to be transmitted by a sender and is to be received and interpreted by a receiver. If the information is in analog form, a conversion is required to process and transmit it into digital form. The smallest unit representing the digital form is a bit, which is a pulse. Digital information may be converted into words, groups of words (frames), code symbols, or any other prearranged units of bits. When no interpretation is applied, these units are called *data*, which may be a raw bit stream. When they are received and interpreted at the other end, they become *information*, which is conveyed.

For binary digital systems, the data or information transmission rate is measured in bits per second. If additional bits are added (for special purpose) to the required data, the efficiency of information transmission reduces. It must be understood that no real information is conveyed by a redundant message, but redundancy is not wasteful under all conditions, especially where error handling is concerned (which will be discussed shortly). In short, a set of information or data with respect to time is the time domain *input signal* for a system, whose frequency contents can be observed in the frequency domain by observing the spectrum. Information theory and the mathematical aspects of measurement of information are discussed in Chapter 5.

Antenna size and carrier follow inverse relationship; hence, for portable sizes of antenna, very high frequencies are used in wireless systems.

The transmitter and receiver systems are connected through a channel. These systems process the input signal in various ways to ensure proper communication. One of the important processes is *modulation*, with which the term *carrier* is associated.

Modulation This is the process by which a signal is transformed into waveforms that are compatible with the characteristics of the channel. Modulation may be of two types: analog and digital. In analog modulation, analog signal is modulated by a carrier while digital modulation is the process by which pulses are modulated into the required digital form or modulated by a carrier. These modulated waveforms usually take the form of shaped pulses [ideally the shape of a sinc function, which is $\sin(x)/x$, in the frequency domain]. However, in the case of *digital band-pass modulation*, the shaped pulses modulate a sinusoid called a carrier wave, or simply a carrier. For radio transmission, the carrier is converted into an electromagnetic (EM) field through an antenna for propagation to the desired destination.

Carrier The transmission of EM fields through space is accomplished using antennas. The size of the antenna depends upon the wavelength λ and the application. The antennas used for cellular telephones are typically small. Wavelength and frequency are related as $c = f\lambda$, where c is the speed of the EM wave in free space. Thus, antenna dimensions indirectly decide the frequency an antenna can transmit. A very large antenna would be required for sending a baseband signal of a very low frequency. To transmit a 3 kHz signal or voice signal through space, without carrier wave modulation, an antenna that spans 15 miles would be required. If the baseband information is first modulated on a high-frequency carrier (e.g., 900 MHz), then it would require an antenna with a diameter of only about 8 cm. Hence, for all portable applications, radio frequency (RF) conversion is necessary.

Another advantage of modulation with a carrier is the multi-user environment. If more than one signal or user utilizes a single channel, modulation with different carriers or the same carrier may be used to separate the different signals (these techniques are explained in Chapter 9). The reception will be based on the tuning of the carriers. Systematic allocation of frequency bands is possible due to the fixed bandwidth and dedicated allocation of carriers. Some modulations can be used to minimize the effects of interference. Such modulation schemes require a transmission bandwidth that is much greater than the minimum bandwidth that would be required by the message (wideband communication). Bandwidth concepts are discussed in Section 1.3.

Transmitter A transmitter performs various functions to make a source signal suitable for transmission. Examples of such functions are converting a non-electrical form of signal into an electrical signal, restricting the range of frequencies, compressing the amplitude ranges, and modulating the signal as per requirements. Not much processing is required in *baseband communication* or *carrierless communication*, such as the local loop wire telephony, as the mouthpiece of the handset gives analog electrical signals that can be directly transmitted for short distances on the wired lines. However, in long-distance communication, a transmitter is required to process, possibly encode, and to modulate the incoming information to make it suitable for transmission over the desired channel and subsequent reception. This is known as *broadband communication*. Eventually, in this type of transmitter, the information modulates the carrier, that is, the information is systematically superimposed on a comparatively high-frequency sine wave. RF upconversion may be followed by the modulator stage, especially for a wireless link, and then the power amplifier stage completes the transmitter part. The signal becomes ready for transmission through an antenna.

> In wireless communication, the final form of transmission is always analog, irrespective of whether the modulation is analog or digital.

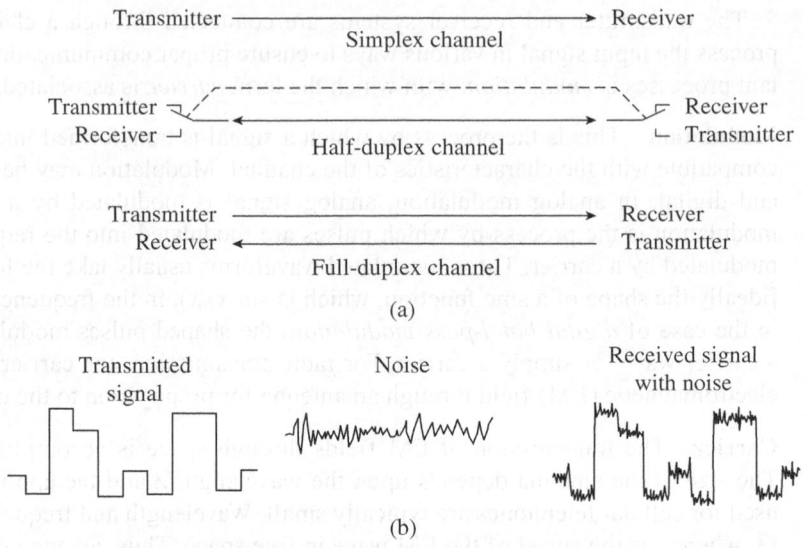

Fig. 1.2 (a) Channel types (b) Addition of noise to signal

Channel It should be noted that the term *channel* is often used to refer to the frequency range allocated to a particular service for transmission, such as a television channel (the allowable carrier bandwidth with modulation); however, in general, a channel is a medium through which a signal propagates towards its receiver. As shown in Fig. 1.2(a), channels may be of three types: *simplex*, *half duplex*, and *full duplex*. Simplex channel implies one way communication, half duplex implies bi-directions communication on a sharing basis one at a time, while full duplex channel implies simultaneous bi-directional communication. *Noise* and *interference* are the most serious problems associated with a channel. It is inevitable that a signal will deteriorate during the processes of transmission, propagation, and reception because of some distortion in the system or because of the introduction of noise. Noise is unwanted energy (usually of random nature) present in a transmission system due to a variety of causes. Since noise will be received together with the signal, as shown in Fig. 1.2(b), it places a limitation on the transmission system as a whole. When noise is severe, it may mask a given signal so much that the signal becomes unintelligible and therefore useless.

Though noise may interfere with a signal at any point in the communication system, its effect will be maximum when the signal is weak. Hence, the most noticeable noise is that in the channel or at the input to the receiver. Correspondingly, when the signal is strong, the noise effects are less. This is defined using the parameter *signal power to noise power ratio* or *signal-to-noise ratio* (SNR). Better the SNR, stronger the signal in the presence of noise. The different types of noise are discussed in Chapter 3.

Receiver There are many varieties of receivers in communication systems since the exact form of a particular receiver is influenced by the opposite tasks to that of the transmitter and many other requirements. Among these requirements are the modulation scheme used, the operating frequency and its range, error-handling tasks, and the type of output device required, which in turn depends on the destination of the intelligence received. Most of the wireless receivers are of superheterodyne type with the intermediate frequency (IF) stage and then the local oscillator and mixer stage for final RF upconversion. Receivers vary in complexity from a very simple crystal receiver, with headphones, to a far more complex *rake receiver*, explained in the chapter 7 of modulation techniques.

As already stated, the purpose of a receiver and the form of its output influence its construction. The output of a receiver may be fed to a loudspeaker, video display unit, radar display, television picture tube, pen recorder, or computer. In each instance, different arrangements must be made, each affecting the receiver design.

1.2 GENERAL MODEL FOR WIRELESS DIGITAL COMMUNICATION LINK

A study of wireless digital communication involves the in-depth study of the whole point-to-point link, covering the fundamentals of each block of the link. This section provides an introduction to the blocks and their importance. The blocks will be explained in detail in subsequent chapters. Figure 1.3 provides a simplified block diagram of a digital communications link. A transmitter begins and ends with an analog signal (except the readily stored or generated digital base). The signal that comes out as multimedia information is analog in nature, which should be first converted into the digital form. Initially, wireless communication was used only for voice communication, but now any signal can be communicated. In the case of video communication, a huge storage capacity and high speed of communication are required, and hence, source encoding for compression of the database is necessary. Here, standard methods may be used to compress the data, and the stored files with standard extensions, such as .jpg, .avi, .mp3, .gif, .tif, and .dat, can be made available for transmission. The basic communication model, as shown in Fig. 1.3, is a *systematic assemblage* of the *forward path* and the *reverse path*.

Source coding/decoding stage   The first step is to convert a continuous analog signal into a discrete or digital bit stream. This process is called *digitization*. The next step is to add information coding for data compression. The information to be transmitted from the source may be human-originated (speech) or machine-originated (data or image). The source encoder with compression eliminates the *inherent redundancy* in the information (thus compressing) to maximize the transmission rate, and the encrypter ensures secrecy of data. The encryption process is described in Chapter 5 of the source coding stage.

> In general, channel coding aspects need more attention in wireless communication, whereas line coding is important in wired communication.

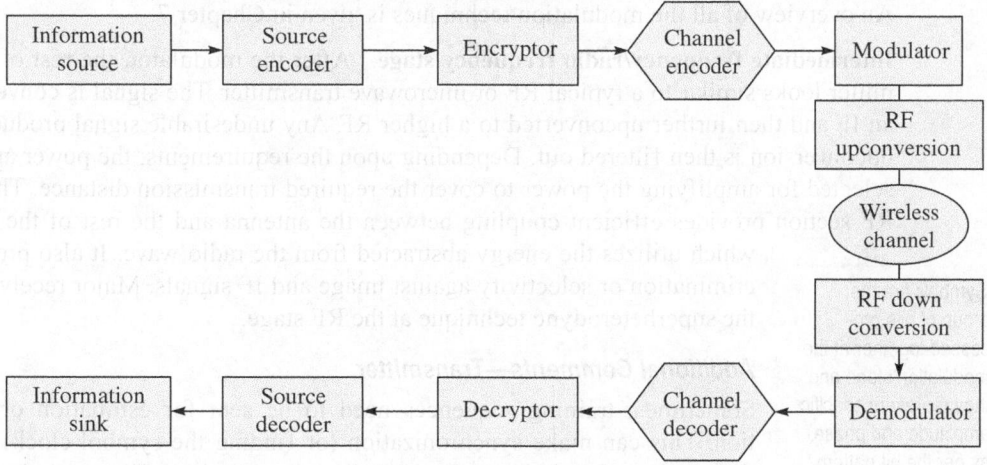

Fig. 1.3   Basic model of wireless digital communication link

Channel coding/decoding stage   Data must be protected against perturbations introduced by the noisy channel, which could lead to misinterpretation of the transmitted message at the receiving end. Line coding techniques are used for inserting systematic amplitude variations, power levels, and synchronization points, whereas channel coding techniques are used to insert structured sequences. Both techniques help in combating channel errors. Data can be protected through the following error control strategies:

Forward error correction   It uses error-correcting codes that are derived mathematically and inserted systematically at the transmitting end, and are able to correct errors at the receiving end.

Automatic repeat request   It uses error-detecting codes with no capabilities of correcting them, (except single bit error correction techniques) but with strategies of retransmissions (sending the same data again) when erroneous data is received.

The channel coding stage systematically adds extra bits to the input data stream, even after the removal of the redundant bits by the source coders to balance the reliable transmission capabilities of the channel. Though sending the extra bits involves extra cost in terms of bandwidth utilization and speed, these bits are used for error correction to enhance the quality of reception. The channel decoder decodes the data in such a way that the effects of noise and interference in the communication channel are minimized. These techniques are discussed in depth in Chapter 6.

Modulator/Demodulator stage   The modulation method to be used should be selected based on the channel characteristics. Channel-related issues and the corrections for channel effects are discussed in Chapters 3 and 8. Corrections should take place before demodulation to reduce the probability of errors. The output from the channel coder is fed into the modulator. Since the modulator deals with complex modulation techniques, there are independent I (in-phase) and Q (quadrature-phase) components in the radio; half of the information can be sent on I and the other half on Q. This is one reason why digital radios work well with this type of digital signals. The I and Q components are separate and orthogonal. The modulator block generates a signal suitable for the transmission channel. The blocks in the reverse path do the opposite of those in the forward path. Modulation techniques are basically divided into three types: *pulse modulation*, *carrier modulation*, and *spread spectrum techniques*. The latest modulation technique based on multicarrier transmission, which eliminates most of the problems of wireless channel, is orthogonal frequency division multiplexing (OFDM). An overview of all the modulation techniques is given in Chapter 7.

Intermediate frequency/radio frequency stage   After the modulator, the rest of the transmitter looks similar to a typical RF or microwave transmitter. The signal is converted up to an IF and then further upconverted to a higher RF. Any undesirable signal produced by the upconversion is then filtered out. Depending upon the requirements, the power amplifier is selected for amplifying the power to cover the required transmission distance. The receiver RF section provides efficient coupling between the antenna and the rest of the hardware, which utilizes the energy abstracted from the radio wave. It also provides discrimination or selectivity against image and IF signals. Major receivers follow the superheterodyne technique at the RF stage.

Symbols are the group of bits processed together at the modulation stage and they represent specific amplitude and phase as per the bit pattern.

Additional Comments—Transmitter

Sometimes, training sequences need to be sent for estimation or equalization. This can make synchronization (or finding the symbol clock) easier for the receiver. Symbols are processed in synchronism. The symbol clock is an essential part of the link and represents the frequency and exact timing of the transmission of the individual symbols. At the symbol clock transitions, the transmitted carrier is at the correct *I/Q* (or magnitude/phase) value to represent a specific symbol. Then the values (*I/Q*) of the transmitted carrier are changed to represent another symbol. The interval between these two is the symbol clock period. The reciprocal of this is the symbol clock frequency. The symbol phase is correct when the symbol clock is aligned with the optimum instant(s) to detect the symbols. One essential step after channel coding in the transmitter is filtering, which is required for good bandwidth efficiency. Without filtering, signals would have very fast transitions between states and therefore, very wide frequency spectra—much wider than is needed for the purpose of sending information. A single filter can be shown for simplicity in the block diagram, but in reality, there are two filters, one each for the *I* and *Q* channels. This creates a compact and spectrally efficient signal that can be placed on a carrier. Many times, pulse shaping and windowing techniques of digital signal processing (DSP) make the communication efficient.

Additional Comments—Receiver

The desired receiver characteristics or issues are as follows:

Sensitivity This is expressed in terms of the voltage that must be applied to the receiver input to give a standard output.

Selectivity This characteristic determines the extent to which the receiver is capable of distinguishing between the desired signal and the signal of other frequencies.

Fidelity This represents the variation of the output with the modulation frequency, when the output load impedance is a resistance. At the lower modulation frequencies, it is determined by the low-frequency characteristics of the audio frequency amplifier. At the higher modulation frequencies, the fidelity is affected by the high-frequency characteristics of the audio frequency amplifier.

Noise figure This is a measure of the extent to which the noise appearing in the receiver output in the absence of a signal is greater than the noise that would be present, if the receiver was a perfect one from the point of view of generating the minimum possible noise. It determines the smallest power that may be received without being drowned out by the noise.

Learning about the *wireless medium* is essential to understand the reasoning behind the specific designs for wireless communication protocols or systems. In particular, the design of the physical and medium access protocols is highly affected by the behaviour of the channel that varies substantially in different indoor and outdoor areas. The diversity and complexity of transmission techniques in wireless communications are far more complex than those of wired communications.

The incoming RF signal is first downconverted to IF and demodulated. The ability to demodulate the signal is hampered by factors including atmospheric noise, competing signals, and signal strength variations. The concept of demodulation is explained in Chapter 7. Generally, demodulation involves the following stages:

Automatic gain control and power control are the important aspects of transceivers and require closed-loop systems.

- Carrier frequency recovery (carrier lock)
- Symbol clock recovery (symbol lock)
- Signal decomposition to *I* and *Q* components
- Determination of *I* and *Q* values for each symbol (slicing)
- Decoding and de-interleaving

- Expansion to original bit stream
- Digital-to-analog conversion, if required

Carrier and symbol clock recovery is a complex issue in the receiver. Both the symbol clock frequency and phase (or timing) must be correct in the receiver to successfully demodulate the bits and recover the transmitted information. Offset in frequency or phase will lead to unsuccessful demodulation. Usually, the frequency of a symbol clock is fixed, and both the transmitter and receiver accurately know this frequency. The difficulty is to get them aligned in phase or timing. A variety of techniques is available and most systems employ two or more such techniques. If the signal amplitude varies during modulation, a receiver can measure the variations. The transmitter can send a specific synchronization signal or a predetermined bit sequence such as 10101010101010 to train the receiver's clock. In systems with a pulsed carrier, the symbol clock can be aligned with the power turn-on of the carrier. In the transmitter, it is known where the RF carrier and digital data clock are because they are being generated inside the transmitter itself, whereas in the receiver, this is not known. The receiver can approximate where the carrier is, but has no information about the symbol clock phase or timing. Creating the carrier and symbol clock recovery algorithms is a difficult task in receiver design. This task can be made easier by the channel coding performed in the transmitter.

Mobile telephony, mobile internet services, and wireless local area networks (WLANs) are a few applications that are based on protocol. The lowermost layer of the protocol stack is the physical layer, which is the wireless link along with the standard specifications. These are explained in part 4 of the book (Chapters 10, 11 and 12).

Some of the useful signal processing aspects observed in the wireless link are Fourier series and Fourier transforms of the various functions (observing the signal in the time and frequency domains), sampling theorem, filters, correlation, convolution, and windowing. Various properties of Fourier transforms are applied at various stages, and these fundamentals can be revised by self-study.

1.3 BANDWIDTH

A signal may have one or more frequency content, which can be represented in the frequency domain. Information, which may be in the form of analog or digital signals, can be represented in the time domain (amplitude versus time plot) and the frequency domain (amplitude versus frequency plot, also called the spectrum). A digital signal is the representation of a signal with discrete values at discrete time. It is produced by the sampling of a continuous envelope of information and will carry discrete, well-defined amplitude levels. Binary coded data is one typical case of a digital system; it takes only two values of amplitude levels, one each for logic 0 and logic 1. It will carry the amplitudes decided for logic 0 and 1. When an analog or a digital time domain signal is converted into a frequency domain signal, the significant frequency components of the spectrum decide the bandwidth. Practically, the signal is band-limited by applying certain techniques to meet certain requirements.

There is no universally satisfying definition for the term *bandwidth*, which is used in the following circumstances:

- It is used to characterize a signal, which can be the input signal or the baseband or broadband to be transmitted. Correspondingly, this is called the signal or transmission bandwidth.
- A channel allocated to the user to allow the transmission of maximum frequency content (allowable range of frequencies) is called channel bandwidth. It decides the capacity of the transmission. Channel bandwidth may be decided by the service provider.

Note: It is observed that an analog signal consumes less spectrum compared to its digital counterpart and hence requires lesser bandwidth, because the digital counterpart is the result of sharp transitions.

- While designing wireless system hardware, including transmitter and receiver, the frequency response of the hardware stages must be such that the total system bandwidth supports the channel bandwidth (or the hardware frequency response must be set accordingly).

Bit rate, symbol rate, and baud rate   Digital data transfer is measured in bits per second, as mentioned earlier, or in symbols per second units. When the number of bits is represented together at the modulator front end, the bit rate is converted into the symbol rate. To understand and compare the efficiencies of different modulation schemes, it is important to first understand the difference between the bit and symbol rates. The transmission bandwidth due to digital modulation techniques depends on the symbol rate, and not on the bit rate (refer to digital modulation schemes discussed in Chapter 7). The bit rate is the frequency of a system bit stream.

$$\text{Symbol rate} = \frac{\text{Bit rate}}{\text{Number of bits transmitted with each symbol}}$$

Each symbol represents M finite states and k bits of information, where

$$k = \log_2 M \tag{1.1}$$

The symbol rate is measured in symbols per second.

The baud rate refers to the signalling rate at which the data is sent through a channel and is measured in electrical transitions per second. It is the reciprocal of the duration of the shortest signalling element. If there is one signal transition per bit, then the bit rate and the baud rate are identical. If two electrical transitions are required for each bit, as in the case of return-to-zero (RZ), then at a rate of 9600 baud, only 4800 bits per second can be conveyed (refer to Section 7.2 of Chapter 7 for further discussion on signalling). The baud rate decides the bandwidth as it decides the highest frequency occurred.

> The spectrum of a signal is the collective representation of all its frequency components along with their amplitude weights.

Example 1.1   Let the symbols be represented by 4, 8, and 16 modulo values at the front end of a modulator. What will be the symbol rates in all the cases if the bit stream is of 256 Mbps?

Solution   A modulator modulates symbols rather than bits. For 4 modulo values, 2 bits/symbol are taken in

for modulation. For 8 modulo values, 3 bits/symbol are taken in, and for 16 modulo values, 4 bits/symbol are taken in. Hence,

for case 1, Symbol rate = 256 Mbps/2 = 128 Mbps
for case II, Symbol rate = 256 Mbps/3 = 85.33 Mbps
for case III, Symbol rate = 256 Mbps/4 = 64 Mbps

Bandwidth of signal and system   A system can be as simple as a low-pass filter or an amplifier or as complicated as an entire satellite communication link. Bandwidth, when referring to a system or a device, usually means the ability to pass, amplify, or somehow process a band of frequencies. However, bandwidth of significant energy for a signal can be subjective. For example, the speech signal bandwidth of maximum energy could be specified as the range between 100 Hz and 6000 Hz, whereas the bandwidth of significant energy for

Note: The line coded signal decides the transmission channel bandwidth in the case of baseband communication, whereas the modulated signal decides the bandwidth in the case of broadband communication.

telephone quality speech could be specified as between 100 Hz and 3000 Hz. The bandwidth for a system is usually defined between the 3 dB points (at higher and lower cut-off frequencies) assuming 0 dB point as the maximum gain, when the system gain is plotted against the range of frequencies.

Pulse degradations are dependent upon the rate of transmission, channel bandwidth, SNR condition, and channel delay.

In strict technical terms, there is no need to differentiate between analog and digital signals, because we just need to look at the spectral content of each signal, the extent of which determines the bandwidth. Typical analog signals, because of their smooth variations, usually have a finite bandwidth, whereas digital signals, due to their discrete nature, usually have unlimited bandwidth. However, it is useful to specify a finite bandwidth for digital signals. To find the most appropriate bandwidth for a digital signal, it is necessary to know the range of frequencies that contains the significant energy of the signal.

We can now make a simple but important observation. When the available bandwidth of a transmission system (medium) is equal to or larger than the bandwidth of a signal that is to be transmitted over the system, and also the actual transmitted signal frequency contents at an instant of time is less than the maximum frequency allowed by the medium, that is,

$$W_{signal} \leq W_{channel} \tag{1.2}$$

then, the entire information content of the signal can be recovered at the receiving end. Conversely, when the transmission system bandwidth is less than the signal bandwidth, some degradation of the signals always occurs because of the loss of frequency components due to its lack of capacity to transfer those frequencies.

Pulse transmission over channel Let us consider digital signals and the bandwidth requirements for pulse transmission. We have to distinguish between the case of an exact reproduction at the receiving end of a transmitted square pulse (which represents a binary digit 1) and a distorted reproduction. An exact reproduction would require a transmission channel with ideally infinite bandwidth, as an ideal square pulse has infinite bandwidth due to extremely high frequency content to retain its sharpness. However, if we only need to detect that a pulse has been sent, we can get by with a finite channel bandwidth. For example, if we were to calculate the effect of an ideal low-pass filter on a square pulse, we would find the output to be a distorted pulse that resembles the original pulse better and better with increasing bandwidth W of the filter. The channel acts as a low-pass filter. Hence, higher harmonic losses are certain. In addition, attenuation also occurs. Ideally, the bandwidth of a binary digital signal with the baud rate same as the bit rate will always be half of that of its bit rate. This is because the consecutive 1 and 0 bits will establish the worst-case condition for transitions, which will decide the highest frequency content, making one cycle of frequency and bandwidths being represented in terms of frequencies normally.

The variation in the bit rate of a channel with fixed bandwidth $W_{channel}$ generates different situations, as shown in Fig. 1.4, because a change in the bit rate will vary the signal bandwidth. Here, the bit rate and signal bandwidth are related mathematically as

$$W_{signal} = 1/2T_b \tag{1.3}$$

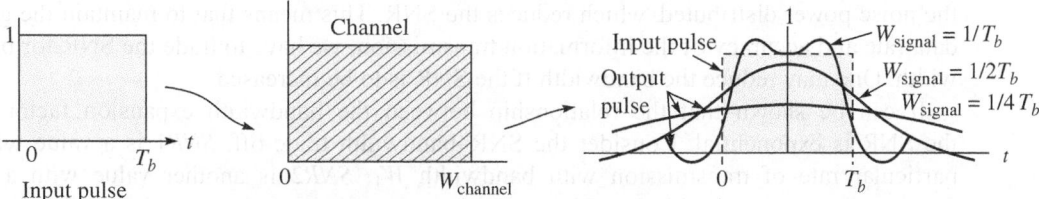

Fig. 1.4 Response of an ideal pulse transmitted through a channel according to the bit rate over a fixed channel bandwidth

SNR–bandwidth trade-off: Transmission rate increases at the cost of reduced SNR due to the noise distributed over the wide bandwidth and vice versa.

where T_b is the 1-bit interval (time duration). For many purposes, this bandwidth yields a resolution with an acceptable error rate.

Since a wireless transmission channel with multipath effects has band-pass characteristics similar to that of a low-pass filter, a pulse propagating over the channel will be affected by the spreading of the pulse. The reasons are explained in chapter 3. As the bit interval becomes narrower, more errors are likely to occur. However, an advantage of digital transmission is that the message is preserved. In analog transmission, the signal becomes irreversibly distorted due to the addition of noise. In contrast, in digital transmission, even though the individual pulses become badly distorted during propagation, as long as the distorted signal that is received can be identified with the presence or absence of a pulse, the original message is preserved. There are some techniques for regenerating the digital signal with the help of a pre-decided threshold level.

Signal-to-noise ratio and channel bandwidth The amount of information that a channel can carry reliably depends on the bandwidth of the channel and the magnitude of the noise present in the channel. The amount of noise present in any channel limits the number of distinct amplitude levels that a signal propagating may have. For example, if a varying analog signal has a maximum level of 10 V and the noise level is 5 V, the signal may have only two levels. On the other hand, if the noise level is only 1 mV, the same signal can be divided into approximately $10\,\text{V}/1\,\text{mV} = 10^4$ levels. Figure 1.2(b) illustrates how noise that has been added during transmission can degrade the signal and hence, its resolution at the receiving ends.

The SNR is the standard measure of the noise level in a system. It can be measured at different stages in the wireless link. It is the ratio of power P_s to noise power P_n. Since power is proportional to voltage squared, we can express SNR as

$$SNR = \frac{P_s}{P_n} = \left(\frac{V_s}{V_n}\right)^2 \tag{1.4}$$

where V_s is the signal voltage and V_n is the noise voltage (because of their multitude of random amplitudes, noise voltages are typically given as rms voltages). SNR is usually expressed in decibels (dB).

$$SNR_{dB} = 10\log_{10} SNR = 20\log_{10}\left(\frac{V_s}{V_n}\right) \tag{1.5}$$

Signal power plays a very important role in successful communication. On increasing the signal power, the effect of channel noise reduces and the signal is received more accurately. A larger SNR allows for a longer distance of transmission. An important feature of signal power is that the SNR and bandwidth are exchangeable. Higher the bandwidth, more will be the noise power distributed, which reduces the SNR. This means that to maintain the given data rate and accuracy of the information transmission, we have to trade the SNR for bandwidth. One may reduce the bandwidth if the SNR is to be increased.

It can be shown that the relationship between the bandwidth expansion factor and the SNR is exponential. Consider the SNR–bandwidth trade-off. *SNR*1 is a value with a particular rate of transmission with bandwidth W_1. *SNR*2 is another value with a different rate and bandwidth W_2. Then, for the same channel capacity, it can be derived that

$$SNR2 \approx SNR1^{W_1/W_2} \tag{1.6a}$$

Thus, if $W_2 = 2W_1$, then $SNR2 \approx SNR1^{1/2}$ \hfill (1.6b)

That is, *SNR*2 is the square root of *SNR*1.

Example 1.2 Compare the SNR requirements for 1 bit/symbol and 2 bits/symbol transmission systems that have a bit rate of 1 Mbps.

Solution The visualization of this example will be better if we treat one symbol block as one pulse, because the symbol rate decides the transmission bandwidth.

Considering first nulls in the sync shaped frequency domain response of a pulse:

For 1 bit/symbol transmission bandwidth W_1, *baud rate* = 1 Mbps (because it takes in 1 bit per symbol).

For 2 bit/symbol transmission bandwidth W_2, baud rate = 0.5 Mbps. Hence, $W_1/W_2 = 2$.

Now, $SNR2 \approx SNR1^{W_1/W_2} \Rightarrow SNR2 \approx SNR1^2$

Thus, theoretically, the second scheme requires a higher value of SNR compared to the first scheme for the same bit rate to be transmitted.

Note for 1 bit/symbol transmissions: We know that for square signals, the spectrum contains odd harmonics of the fundamental, which here equals $1/2T_b$. Thus, the signal's bandwidth is infinite. In practical terms, we use 90 per cent power bandwidth to assess the effective range of frequencies consumed by the signal. The first and third harmonics contain that fraction of the total power, meaning that the effective bandwidth of our baseband signal is $3/2T_b$ or, expressing this quantity in terms of the data rate, $3R_b/2$. Thus, a digital communications signal requires more bandwidth than the data rate: a 1 Mbps baseband system requires a bandwidth of at least 1.5 MHz. However, bandwidth also depends upon the adopted line coding scheme.

Shannon's and Nyquist's equations Channel capacity is decided by the transmission bandwidth and SNR condition, and the relationship is given by Shannon and Nyquist from their independent research. In 1948, Dr Claude Shannon of Bell Telephone Laboratories published a groundbreaking work entitled *The Mathematical Theory of Communication*, in which he described the development of communication systems that transmit data effectively with limits on the exchange of the SNR and bandwidth. The limitations imposed on communication by the transmission with zero errors. We can consider the channel as a pipe through which we send information. Shannon worked on the channel capacity and found the equation for the band-limited signal to be transmitted over additive white Gaussian noise (AWGN) channel as follows:

> Channel capacity is the maximum amount of data that can be pumped through the channel in a fixed period of time and can be measured in terms of bits per second.

$$C = W_{channel} \, \log_2(1 + SNR) \tag{1.7a}$$

Nyquist had given another formula:

$$C = 2W_{channel} \log_2 M \tag{1.7b}$$

Here, $W_{channel}$ is the channel bandwidth in hertz and *SNR* is the power ratio in general, in which S is the signal power and N is the noise power in watts. Equation (1.7) gives the maximum possible data transmission when 1 bit/symbol is transmitted; $M = 2^k$ are the signalling levels. If more bits per symbols are being transmitted, then the maximum rate of transmission of information in symbols per second is C_s, and for k bits/symbols, we can say $C = kC_s$. Combining Nyquist and Shannon relationships, $k = C$/symbol rate, where C is the maximum bit rate capacity, k is the number of bits per signalling element (symbol), and symbol rate is two times the bandwidth of the signal according to Nyquist relation. There may be 2^k different possible bit combinations to send in the form of symbols.

There is a parameter related to the SNR that is more convenient for determining digital data rates and error rates. It is the ratio of the signal energy per bit to the noise power density (noise power per hertz), E_b/N_o. Consider a signal that contains binary digital data transmitted at a certain bit rate R. Recalling 1 watt = 1 J/s, the energy per bit in a signal is given by

$$E_b = ST_b$$

where S is the signal power and T_b is the time required to send 1 bit. The data rate is just $R = 1/T_b$. For thermal noise,

$$E_b/N_o = S/KTR$$

Example 1.3 A standard 4 kHz telephone channel has an SNR of 25 dB at the input to the receiver. Calculate its information-carrying capacity. In addition, find the capacity of the channel if its bandwidth is doubled while the transmitted signal power remains constant.

Solution   SNR = antilog (25/10) = 316
Capacity of the channel in the first case

$$C = 4000\log_2 (1 + 316) = 33.233 \text{ kbps}$$

If the SNR is 316, it means that when the signal power is 316 mW, the noise power is 1 mW. Now, the bandwidth is doubled with no change in the signal power, effectively, the noise power is doubled due to twice the bandwidth. Hence, the SNR drops to half the original value.

Capacity in the second case

$$C = 8000\log_2 (1 + 316/2) = 58.503 \text{ kbps}$$

Thus, the capacity of the channel has increased.

1.4 TYPES OF SIGNALS

Appropriate signal processing can be applied in the transmitter, as well as receiver, if and only if we know the type of the signal. If we are aware of the nature of the signal, we can treat it in the time or frequency domain and can identify the changes applied. In addition, we can decide the approach to deal with the system and performance parameters of the system.

1.4.1 Analog and Digital Signals

Signals are classified in terms of the nature of amplitude. Normally, they are represented in the time domain.

Let us define the analog and digital signals once again in terms of DSP. A signal whose amplitude takes all the values in the specified range over the measuring interval or time, and is continuous in time is called an *analog signal*. Here, the signal can take an infinite number of values, and precision is dependent upon the resolution of the system. If the signal amplitude takes a finite number of values and not all, it is called a *digital signal*. Binary is a special case of digital signal and takes only two values, one each for logic 0 and 1.

1.4.2 Continuous-time and Discrete-time Signals

Signals are classified on the basis of time as continuous-time and discrete-time signals, and are represented in the time domain. A continuous-time signal is specified for every value of time, whatever precise time can be resolved, whereas a discrete-time signal is specified with the gap of measuring instants.

The following are the mathematical representations for signals with peak value A_0:

$$s(t) = A_0 \sin \omega t \quad \text{(continuous-time signal with time variable } t)$$

$$s(n) = A_0 \sin \omega n/N \quad \text{(discrete-time signal with index variable } n)$$

where N is a period of n samples.

A discrete-time signal is represented at discrete instants of time with its natural value or quantized value. The time variable is not continuous and hence, a discrete-time signal can be represented as a sequence of numbers.

From these two types, four different signal categories can be formed:
(a) Continuous-time analog signal (real-time signals)
(b) Continuous-time digital signal (square wave representing a binary signal)
(c) Discrete-time analog signal (with natural value of samples)
(d) Discrete-time digital signal (with quantized value of samples)

A discrete-time signal is represented as a sequence $s(n)$, where n can take in a set of values in the integer range $-\infty$ to $+\infty$. In most cases, the discrete-time signal $s(n)$ is obtained by sampling a continuous-time signal $s(t)$ at periodic interval ΔT_s. So, we can write $s(n) = s(t)|_{t=n\Delta Ts}$.

A *discrete-time system* is one that accepts a set of sequences $s_i[n]$ (*i* stands for the *i*th sequence) and produces a set of sequences $r_j[n]$ as output.

1.4.3 Periodic and Aperiodic Signals

A signal is said to be periodic for some positive constant T_0 (or N for a discrete signal), that is, a fixed interval, if it satisfies the following conditions:

$$s(t) = s(t + T_0) \text{ for all } t \quad \text{(continuous time)} \tag{1.8a}$$

$$s(n) = s(n + N) \text{ for all } n \quad \text{(discrete time)} \tag{1.8b}$$

The smallest value of T_0 that satisfies this condition is called a *period* in terms of time unit. It is obvious that $s(t)$ will remain the same when it is shifted in time by one period. A periodic signal

Note: For the class of periodic signals, decomposition in sinusoidal components is called a *Fourier series*, whereas for the class of finite energy signal (aperiodic), it is called a *Fourier transform*.

Voice signal is a continuous-time analog, aperiodic, random, and energy signal.

must start at $-\infty$ and continue forever. Moreover, it can be generated by repeating the signal $s(t)$ with the period T_0 infinite number of times. The instants from and to which instants of time the period is measured is immaterial due to periodicity; the shape of $s(t)$ during that period must repeat itself an infinite number of times. The signal that occurs for a finite duration of time is called an *aperiodic signal*. Here, the shape of $s(t)$ is not repeated an infinite number of times. It is a time-limited non-repetitive signal.

Most signals of practical interest can be decomposed into a sum of sinusoidal signal components. The signals in the time domain and the corresponding frequency domain equivalents are given in Table 1.1.

Table 1.1   Time and frequency domain signal equivalents

Time domain signal	Frequency domain equivalent
Continuous time, periodic	Discrete spectrum, aperiodic
Continuous time, aperiodic	Continuous spectrum, aperiodic
Discrete time, periodic	Discrete spectrum, periodic
Discrete time, aperiodic	Continuous spectrum, periodic

1.4.4 Deterministic and Probabilistic Signals

A signal can be classified as deterministic if there is no uncertainty with respect to its value at any instant of time. Probabilistic signals, also known as random or non-deterministic signals, cannot be predicted, that is, there is some degree of uncertainty. Deterministic signals can be represented with a mathematical expression, which will be unique. Random signals are generated from random or stochastic processes.

Random functions of time are often referred to as stochastic signals. A stochastic signal may be continuous or discrete in time and may have continuous-valued or discrete-valued amplitudes. Stochastic processes are classes of signals whose fluctuations in time are partially or completely random; examples of such signals are speech, music, image, time-varying channel response, noise, and video. Stochastic signals are completely described in terms of the probability model and theory, but can also be characterized with relatively simple statistics, such as the mean or statistical averages, correlation, and power spectrum. They must deal with the ensemble averages, variance, probability distribution function (PDF), cumulative distribution function (CDF), and so on. Readers can refer to any book on statistical signal modelling to explore these topics further.

1.4.5 Energy and Power Signals

Power is related to signal voltage or current. Here, the continuous-time analog signal is considered, and therefore, we have to deal with integrations in the subsequent formulas. Power signal can be defined as

$$P(t) = \frac{V^2(t)}{R} \quad \text{or} \quad P(t) = i^2(t) \times R \qquad (1.9)$$

where R is the resistance across which power is measured.

In a communication system, power is often represented in the normalized form, assuming $R = 1\,\Omega$, though the resistance may have another value in the actual circuit. The actual value of power is obtained by denormalizing the normalized power value. Conventionally, irrespective of whether the signal is of the voltage or current waveform, the normalization convention for power allows us to express the instantaneous power as

$$P(t) = s^2(t) \tag{1.10}$$

Energy dissipated in the time interval $-T/2$ to $T/2$ of a signal with instantaneous power is measured by the following expression:

$$E_S = \int_{\frac{-T}{2}}^{\frac{T}{2}} s^2(t)dt \tag{1.11}$$

The average power dissipated by the signal during the same interval is

$$P_{av} = \frac{E_s}{T} = \frac{1}{T} \int_{\frac{-T}{2}}^{\frac{T}{2}} s^2(t)dt \tag{1.12}$$

The performance of the communication link depends on the energy of the received signal. Higher the energy, more accurate the signal detection. At the same time, power is the rate at which the energy is delivered. This is necessary because voltages, currents, or EM field intensities are related to powers and they need to be designed as per power requirements. The signal $s(t)$ can be converted into the discrete form by sampling, and samples can be written as $s(n)$, where n is the index value. All these formulas can be rewritten by replacing the integration with summation and $s(t)$ with $s(n)$. Similarly, the changes can be applied to energy and power signals as well.

This fundamental knowledge can be used to differentiate between energy and power signals. While analysing the signals, it is often desirable to deal with the waveform energy E_s. We can classify $s(t)$ as an energy signal, if and only if, it has finite but non-zero energy for all time, that is, when $T \to \infty$.

$$E_s = \int_{-\infty}^{\infty} s^2(t)dt \quad \text{(continuous-time signal)} \tag{1.13a}$$

$$E_s = \sum_{-\infty}^{\infty} |s(n)|^2 \quad \text{(discrete-time signal)} \tag{1.13b}$$

In the real world, transmitted signals have finite energy $(0 < E_s < \infty)$. A finite energy signal has zero average power. However, in order to describe periodic signals, which by definition exist for all time and thus have infinite energy, these are called power signals. Even random signals having infinite energy are power signals. If E_s is infinite, the average power P_s may be either finite or infinite. A signal is defined as a power signal only if it has finite but non-zero power for all time t.

$$P_s = \lim_{T \to \infty} \frac{1}{T} \int_{\frac{-T}{2}}^{\frac{T}{2}} s^2(t)dt \tag{1.14}$$

To study random signals, mathematical models based on the PDF and exhibiting their behaviour are used.

For signal $s(n) = Ae^{j\omega_n}$ has average power A^2.

The classification of energy and power signals is mutually exclusive. An energy signal has finite energy but zero average power (e.g., deterministic and aperiodic signals) and can be generated in a laboratory. A power signal has finite

> Ramp signal is neither the energy signal nor the power signal.

average power but infinite energy (e.g., periodic and probabilistic signals). It is impossible to generate a true power signal in practice, because such a signal has infinite duration and infinite energy.

From the theory of linear systems, Parseval's theorem states that the Fourier transform preserves energy and power. However, the energy (or power) in the complex envelope is not equal to the corresponding energy (or power) in the corresponding band-pass signal.

1.5 TYPES OF COMMUNICATION SYSTEMS

There are two possible options in many scenarios while dealing with communications between two hardware ends—a transmitter and a receiver.

- The input (or baseband) signals may be analog or digital
- The channels may be wired (guided) or wireless (unguided)
- The transmissions may be analog or digital
- The number of bits sent at a time may be serial (one bit at a time) or parallel (more bits at a time, i.e., symbols)
- The communication may be baseband or passband (general terms for broadband and wideband)
- The mode of communication may be synchronous or asynchronous
- The information may be real time or non-real time (stored data)
- The direction of transmission may be unidirectional or bidirectional

Out of the two possibilities, only one can exist at a time. To have a combination of both possibilities, either conversions or convergence in the system is required. As there are two possibilities in the input signals and two possibilities in the transmissions, according to the binary theory, four combinations of communication systems are possible. The systems may be analysed by using a qualitative approach first and then a quantitative approach. Moreover, we must analyse the ideal system and then the actual system, with noise.

In general, communication systems can be of four different types: analog input–analog transmission, analog input–digital transmission, digital input–digital transmission, and digital input–analog transmission. The different types of systems and the corresponding modulation schemes are described here for a proper visualization.

Analog input–analog transmission Wireless communication commercially started with amplitude modulation (AM) radio broadcasting in the range 550–1600 kHz. Thereafter, frequency modulation (FM) transmissions also started commercially in the range 88–108 MHz. In both these systems, the input was in the analog form of audio signal. These broadcast systems still exist. When analog television standards were framed, quadrature AM was selected for video information and FM for audio information for combined audio and video transmission, resulting in vestigial sideband communication. These standards are still followed to maintain compatibility with the older televisions and follow the very high frequency (VHF) and ultrahigh frequency (UHF) ranges. In local loops of wired telephone lines, the analog baseband signal is transmitted without modification in the signal.

> The PCM scheme serves various stages in the communication link—analog-to-digital converter in the source coder as well as modulator in the baseband communication link.

In the near future, commercial systems based on the analog input–analog transmission may become obsolete. The transient period of revolution has already started with digital broadcast systems employing A-D-A conversion stages with the digital audio broadcasting (DAB) and digital video broadcasting (DVB) standards. High definition radio (HD radio) and digital radio mondiale (DRM) systems have also come up. All these systems follow the OFDM modulation scheme, which is suitable for long-distance communication, and hence for broadcasting.

Analog input–digital transmission The pulse code modulation (PCM) scheme, which exists for analog-to-digital conversion (ADC), is considered in the source coding stage of the wireless communication link, though it is the method for analog input–digital transmission. Thus, PCM forms the basis for the source coding stage of the wireless link for real-time input signals such as voice, image, and video. It is discussed in detail in chapter 5. Digital transmission in its baseband form is suitable only for transmissions on the wired lines. To achieve this, ADC is required, which can be achieved through the PCM scheme. PCM signals of 64 kbps bit rate are transmitted over the telephone trunk lines or over the integrated services digital network (ISDN) or broadband ISDN (B-ISDN) channels. Another method for analog input–digital transmission is delta modulation (DM), but because of its practical limitations related to slope overload and sampling rate, it is not standardized in commercial systems. PCM signals can also be converted into frames for transmissions over wired links of computer networks. Differential pulse code modulation (DPCM) and adaptive DPCM (ADPCM) are the modified and bandwidth-efficient versions of PCM.

Digital input–digital transmission When it is necessary to send digital information in its baseband form, the binary form of transmission may not always be suitable, as it may not be compatible with the transmission channel. In addition, the binary form of transmission adds a DC voltage level to the final transmission, which takes more energy in the signal. Therefore, it is required to convert the form of transmission by changing the bit representation format or voltage levels for shaping the signal power, and also incorporating the synchronization points in the signal. In short, the signal can be shaped as per the desired spectrum characteristics for digital baseband communication. *Non-return-to-zero* (NRZ), RZ, Manchester, differential Manchester, and bipolar are some methods that have a final digital form of transmission. These methods are normally suitable for wired line or computer networks; however, they are incorporated in wireless links as well. These methods are also called *digital signalling* as they are a suitable form for ISDN lines. It is also called *line coding*. Line coding can be applied to the digital baseband in wireless communication before the modulation stage. Refer to Chapter 7 for further discussion on this topic.

Digital input–analog transmission This type of transmission is mainly used in the systems that use a modem (modulator-demodulator), either over wired lines or wireless links. Here, the modulation scheme converts the input digital signal into the analog form using the carrier wave. The final wireless communication is always in the analog form. If wireless transmission can be used and the carrier frequency after modulation does not fall in the RF range, it is necessary to use an RF upconversion. If wired communication is used, only a data modem can be used without upconversion. Amplitude shift keying (ASK), frequency shift keying (FSK), M-ary phase shift keying (M-PSK), M-ary quadrature amplitude modulation (M-QAM), minimum shift keying (MSK), spread spectrum modulation (SSM), and OFDM fall into this category. The details of these modulation schemes are provided in Chapter 7.

1.6 WIRED VERSUS WIRELESS MEDIA

The existing systems are not all wireless; a few are wired. The fundamentals of both types of media are described here, which will answer questions regarding the differences between the two systems and the kind of conversions required for the converged system.

The electrical signals in an open wire line, such as a twisted pair, travel at the velocity of light, which is determined by the expression

$$v = 1/sqrt(\varepsilon\mu)$$

(1.15)

Analog input–digital transmission and digital input–digital transmission techniques are cascaded in practice to achieve the required form of transmission signal.

where ε and μ are the permittivity of free space (capacitance per unit length measured in farads/metre) and the permeability of free space (inductance per unit length measured in henries/metre), respectively. In free space, $v = 3 \times 10^8$ m/s, given that $\varepsilon = 9.854 \times 10^{-12}$ F/m and $\mu = 4\pi \times 10^{-7}$ H/m. The signal travels as an EM wave just outside the wires (radiation). It differs from a free space EM wave (such as the one launched by a television, radio, or mobile antenna, which spreads out in all directions) only in that it is bound to and guided by the wires of the transmission line.

> *Note*: Metallic wired media follows the conduction theory and undergoes radiation losses, whereas fibre and wireless media follow the theory of dielectric material as per their natures and do not have radiations.

The following wired media are mainly popular:
(a) *Twisted pair* wirelines, unshielded twisted pair (UTP), and shielded twisted pair (STP), for conventional landline telephone systems, 10Base-T Ethernet cabling, and so on
(b) *Coaxial cable* for closed circuit televisions (CCTV) and cable television network, Ethernet 10Base2, 10Base5 cabling, and so on and transmission lines
(c) *Optical fibres* for long-distance communications, B-ISDN, fibre distributed data interface (FDDI), local area network (LAN), synchronous optical network (SONET), and so on

Twisted pair and coaxial cables provide a reliable, guided link that conduct an electrical signal associated with the transmission of information from one fixed terminal to another. The wires act as filters (due to lumped resistance and capacitance) that limit the maximum transmitted data rate of the channel because of band-limiting frequency response characteristics. A twisted pair wire line can typically support a 250 kbps bit rate, whereas a coaxial cable may typically support 300 Mbps. The signal passing through a wire radiates EM waves outside the wire to some extent, which can cause interference to the nearby radio signals or to other wired transmissions as a noise. These characteristics may differ from one wired medium to another. Laying additional cables in general can double the bandwidth of the wired medium.

An optical fibre is a dielectric guided medium that passes the data through itself as light waves. The carrier frequency range is of the order of 10^{14} Hz. Ideally, optical fibres have infinite bandwidth, but in practice, due to the limitations of sources and detectors and the dispersion effect, the bit rate up to Tbps (terabits per second) is achieved over high-grade

WIRED MEDIA AS TRANSMISSION LINE

When do two connecting wires become a transmission line? It is when the capacitance between the wires and the inductance of the wires acts as *distributed* instead of *lumped*. This begins to happen when the wire approaches the dimensions of a wavelength (wavelength λ and frequency f are related by $\lambda = v/f$). At sufficiently high frequencies, when the length of the connecting wires between any two devices is in the order of the wavelength or larger, the voltages and currents between these two devices act as waves that can travel back and forth on the wires. Hence, a signal sent out by one device propagates as a wave towards the receiving device and the wave is reflected unless the receiving device is properly terminated or matched. If there is a mismatch, the reflected wave can interfere with the incident wave, making communication unreliable or even impossible. Proper termination of the wired link is important when networking computers, printers, and other peripherals, which must be properly matched to avoid reflections. Transmission lines are used to carry the signal from the transmitter front end to the antenna site.

> Dispersion effect is due to group delay of the multiple EM waves of the same light source propagating through the fibre and results in pulse spreading. The similar effect in a wireless channel is delay spread.

optical fibres. Optical fibres exhibit *pulse spreading effect* due to dispersion and hence, bit errors may occur. A dielectric medium allows more than one frequency to pass through it and this is the case in optical fibres in the form of *wavelength division multiplexing* (WDM). A wireless medium (which is also dielectric in nature) supports more than one frequency at a time. All links undergo the effect of white noise.

Compared to wired media, the wireless medium is unreliable; though ideally infinite, it has a low bandwidth, effectively due to the delay spread and intersymbol interference (ISI) effects. However, it supports mobility due to its *tetherless* nature. Different signals through wired media are physically conducted through different wires, but all wireless transmissions share the same medium—air—in the form of unguided EM waves released through an antenna of supporting bandwidth. Thus, it is the frequency of operation and the legality of access to the band that differentiates the variety of wireless services. Wireless networks operate in the following bands:

- 1 GHz–Cellular
- 2.4 GHz–Personal communication systems (PCS)
- 5 GHz–WLANs
- 28–60 GHz–Local multipoint distribution service (LMDS) and point to point (P2P) base station connections
- 300 GHz–Satellite ranges, infrared (IR) frequencies for optical line of sight (LoS) communication or laser communication

These bands are either licensed, such as the cellular bands, or unlicensed, such as the industrial, scientific, and medical (ISM) bands or U-NII bands used for PCS.

PROBLEM OF ELECTROMAGNETIC WAVE PENETRATION THROUGH DIFFERENT MATERIALS

As the frequency of operation and data rates increase, two simultaneous problems arise—the hardware implementation cost increases and the ability of a radio signal to penetrate walls decreases. For frequencies up to a few gigahertz, the signal penetrates through walls, allowing indoor applications with minimal wireless infrastructure inside a building. At higher frequencies, a signal generated outdoors does not penetrate into buildings and a signal generated indoors stays confined to a room. This phenomenon imposes restrictions on the selection of a suitable band for wireless application, though electronic cost has become less significant with time. Concrete and building structures are partially transparent to wavelengths of microwave range and attenuate the signal, and hence, the signal is weaker inside the buildings than outside. However, radio waves and microwaves cannot penetrate a lift (which is essentially a metal box), because the metal is a conductor and EM radiation can penetrate only a small distance into a conductor.

Capacity enhancement Wired media provide an easy means to increase capacity; we can use more wires, as and when required, if it is affordable. In contrast, in the case of the wireless medium, bandwidth is a limited resource, which imposes severe restrictions on the effective capacity. Limited bands are available for operation, and it is not possible to obtain new bands or duplicate the medium to accommodate more number of users in a system. Therefore, researchers have developed numerous techniques to increase the capacity of wireless systems to support more users with a fixed bandwidth. One such method for wireless cellular systems is *frequency reuse*, which is comparable to laying new wires in wired systems. If two cells are at a sufficient distance, then there will be no interference, even when the same frequency is used for communication in these two cells. The theory is explained in

Chapter 2. One may even reduce the size of the cells to overcome the demand of the population. In a wireless system, reducing the size of the cells by half allows twice as many users as in one cell. However, reducing the size of the cells increases the cost and complexity of the infrastructure that interconnects the cells. Multiplexing and multiple access schemes also help to accommodate more users. Capacity issues with multiple users are discussed in detail in Chapters 2 and 9 for various technologies implemented over cellular infrastructure and wireless scenario.

The capacity can be highly improved using smart antenna systems. Single input, multiple output (SIMO), multiple input, single output (MISO), and multiple input, multiple output (MIMO) systems are explained in Chapter 8. Compared to single input, single output systems, capacity increment by 300–400 per cent is possible in cellular environments with such techniques through exploiting the concepts of diversity and multipath, and then combining them. Even OFDM can support multiple users with multicarrier communication in the cellular environment.

1.7 TYPES OF WIRELESS SYSTEMS

There are three types of wireless communication systems:
(a) *Wireless broadcast systems*: The user is always at the receiver end.
(b) *Wireless networks*: Multiple users can exchange data independently being a transmitter or a receiver and share the common resources as per requirement.
(c) *Wireless navigation systems*: This is required for location-based services with the help of the global positioning system (GPS).

Modulation schemes are selected according to the suitability of the system. Wireless link requirement and protocol structures are also different.

Wireless broadcast systems These kinds of systems do not require the cellular structure or device identification numbers (except some special systems with encrypted data). Transmissions occur through a single transmitter and are of sufficiently high-power amplification. Within the predefined range, anybody can receive transmissions with the help of a receiver. These communications are mainly based on frequency tuning. Examples of such systems are AM/FM radio, television, direct-to-home (DTH), DAB, DVB, and mobile television systems.

Wireless networks These types of systems are mainly based on cellular infrastructure or ad hoc connections (forming two different types of wireless networks). Examples include mobile telephone networks, WLANs, and metropolitan area networks (MANs) for broadband access, and wireless sensor networks (in distributed configuration), which are based on cell support. For cell-based systems, at least one transceiver per cell is required, in the form of either a base station or an access point. They are low-power transmitters when compared to broadcast systems. The transmitters (or transceivers) of different cells may be interlinked to form a path between the destination and source devices. These communications are based on frequency tuning plus identification number or address. Ad hoc networks do not always require a cellular infrastructure as they are self-configurable networks. Examples of such systems include Wi-fi, Bluetooth, WiMAX, and wireless sensor networks (in centralized configuration). These networks are discussed in Chapters 11 and 12.

Conversational cellular networks supporting data services as well have to get licensed frequency bands as they are managed by service providers, whereas ad hoc networks are self-configurable and use ISM band.

Wireless navigation systems These services are used for various applications, such as providing turn–by-turn voice-based or onscreen driving directions, automatic rerouting in case of a missed turn, real-time traffic monitoring and upgrade, alerting to slow down, and locating and navigating restaurants, Wi-fi hotspots, and maps. It is a self-correcting closed-loop system working on mobile devices. Navigation services are supported by wireless internet services.

1.8 CELLULAR NETWORKS

Cell is the basic region with a base station tower and a transceiver set having radiating power for the coverage of the basic region. A set of frequencies is allocated to the cell for communication. Multiple cells together form a cellular network. There are three types of cellular networks: cellular voice networks, cellular data networks, and cellular satellite networks. Voice networks are for conversational services, data networks are for internet access through wireless broadband services, and satellite networks are for international support to the other two networks, navigation—GPS, and so on. All the three networks support mobility. Cellular design based on some theoretical aspects is very useful in practice today, without which the existing land mobile communication would be near impossible. It forms the basis for cellular telephony. Cellular networks enable calls to be routed to and from mobile phones, even when their users are moving from one cell to another. They also enable other essential operations such as access to the network, billing, and security. To support such varied operations, a cellular network comprises many elements, each having its own function to perform.

The most important part of a cellular network is the base station with antennas and its associated equipment. To provide seamless connectivity, the system needs to have elements of central control. It also needs to link in with the public-switched telephone network (PSTN) to enable calls to be made to and from the wire-based phones, or between the networks served by different service providers.

Cellular division of an area is very useful to manage coverage, mobile device location, and handover of the services from one cell to another to have seamless connectivity during pedestrian or vehicular mobility.

Various cellular systems are available, such as the global system for mobile communication (GSM) and universal mobile telecommunication system (UMTS), and each system has its own cellular standards. For example, GSM has its own well-defined structure with which the manufacturers' products can be standardized, whereas UMTS has its own structure, standards, and protocols. Despite the differences between the different cellular systems, the basic concepts are very similar. Cellular basics and various cellular networks are explained in Chapter 2.

1.9 EXISTING TECHNOLOGIES

There is an increasing demand for broadband or wideband wireless communication systems because of the need for high-speed communications (mobile internet, wireless video transmissions etc.). At the same time, the telecommunications industry faces the problem of providing telephone services to rural areas, where the customer base is small, but the cost of installing a wired phone network is very high. One method of reducing the high infrastructure cost due to a wired system is to use a fixed wireless radio network. The disadvantage with this is that to enable the rural and urban areas to communicate, large cell sizes are required for obtaining sufficient coverage. It results in problems caused by the large signal path loss and long delay times in multipath signal propagation due to long distances. If we design more number of cells for the rural area, it would be inefficient and expensive due to the low population density. Hence, a modulation technique that covers a longer distance while eliminating the problems of a wireless channel should be introduced in the system.

Other aspects that researchers are currently working on include multipath delay compensation, speed of communication or high bit rate communication, and efficient use of available spectrum for accommodating more users and applications.

Leading Techniques of Modern Era

Several techniques play a leading role in the modernization of digital phone systems, land mobile communication, and wireless internet, with the aim of improving cell capacity, multipath immunity, security, and flexibility. Modern techniques include wideband code division multiple access (WCDMA). The latest development is the emergence of the multicarrier modulation (MCM) or multiple access technique, namely OFDM or orthogonal FDMA (OFDMA). Both these techniques could be applied to provide a fixed wireless system for rural areas. However, each technique has different properties, making it more suited for specific applications. The combinations of both these schemes are also considered to overcome the limitations and to exploit the advantages of both the systems.

The *WCDMA* technique combines two major phone technologies: code division multiple access (CDMA) and GSM. There are several key advantages of WCDMA, some of which are as follows:

- Each transmitter is assigned an identification code; hence, data from multiple transmitters can be carried over the same frequency in the same geographical area.
- It uses power control and adjusts the strength of the signal, eliminating the problem of far-off users being dominated by near users with higher signal strength.
- It is more suitable for densely populated regions and capacity enhancement as compared to CDMA.

The *OFDM* technique is for multi-user access and allows many users to simultaneously transmit in an allocated band by subdividing the available bandwidth into many narrow bandwidth carriers (described in Chapter 7). Information is allocated to several carriers in which the data is to be transmitted, so that the bits on each subcarrier are much longer, drastically reducing the ISI. Thus, it provides the concept of multicarrier modulation (multiple carriers for one digital baseband signal) rather than the conventional single-carrier modulation. The transmission is generated in such a way that the carriers used are orthogonal to one another and non-interfering with each other, thus allowing them to be packed together much closer than in standard frequency division multiplexing (FDM). This leads to OFDM providing a high spectral efficiency.

Broadcast Technologies

The main broadcast technologies are DAB and DVB, which are based on OFDM that forms the single-frequency network concept. Therefore, high-speed, high-quality communication has now become possible. Most of the applications are audio- and video-based entertainment; however, some data services are also supported.

Digital audio broadcasting is a digital radio broadcasting standard that is designed to replace the analog FM and AM radio transmissions. The development of terrestrial DAB (T-DAB) was carried out in the EUREKA 147 consortium formed by broadcasting companies, network operators, consumer electronics industries, and research institutes. The development started officially in 1987, and in 1995, the European Telecommunication Standard Institute (ETSI) standardized DAB. European Telecommunication Standard (ETS) 300–401 became the first standard to include OFDM. In 1997, the second edition of ETS 300–401 was

Approximately 80 per cent of the world's cellular systems are based on GSM technology; most of the remaining 20 per cent are based on CDMA technology.

The bands that are allocated for public DAB services are abbreviated as terrestrial DAB (T-DAB).

released, and the commercial employment of DAB started in 1998. Later, DAB included satellite as well as hybrid satellite or terrestrial broadcasting options. DAB is more robust against noise and multipath fading. It is based on wide-bandwidth broadcast technology and single-frequency network concept; that is, all the transmitters use the same transmission frequency with a very large coverage area.

Technically, there are two main ways of delivering mobile television in today's scenario: via two-way cellular network and via one-way dedicated broadcast network. Some examples of mobile television technologies include DVB-H, satellite digital multimedia broadcast (S-DMB), T-DMB, TDTV (based on TD-CDMA technology from IPWireless), China mobile multimedia broadcasting (CMMB), 1seg (one segment), which is based on Japan's integrated service digital broadcasting (ISDB-T), MediaFLO, general packet radio service (GPRS), and third generation (3G). DVB is a set of standards that defines digital broadcasting using existing satellite cable and terrestrial infrastructures. The DVB project consists of over 220 organizations in more than 29 countries worldwide. DVB standards are published by the Joint Technical Committee (JTC) of the ETSI, European Committee for Electro technical standardization (CENELEC), and European Broadcasting Union (EBU). DVB mostly uses moving picture experts group (MPEG) standards for the compression of audio and video signals. On the basis of distribution, there are four different standards:

- DVB-S is based on satellites
- DVB-C is based on the cable network in houses
- DVB-T is based on terrestrial transmission
- DVB-H is for audio/video streaming (H stands for hand-held) to broadcast television content to mobile devices such as personal digital assistants (PDAs) and mobile phones

Cellular Technologies

Let us have a look at some of the cellular technologies.

GSM and upgradations Currently, the GSM technology is being applied to wireless telephone systems even in rural areas. GSM900, GSM1800, and GSM1900 are the three main specifications of this technology. GSM uses frequency division multiple access (FDMA) and time division multiple access (TDMA) with frequency reuse, which has limited frequency channels to communicate. Since GSM has a high symbol rate, it leads to problems with multipath, causing ISI. Hence, there was a need for a scheme that has no ISI effects at high-speed communications. Enhanced data rate for GSM evolution (EDGE) was introduced for higher bit rate solution. Many service providers compete with each other in providing the maximum possible coverage for mobile telephony. They also try to introduce advanced services to the subscribers in order to acquire the market. Hence, EDGE technology with its high-speed support received a good response and made GSM very popular in parallel data service support.

General packet radio service is the protocol by which packet radio is made possible, and hence data services are added in the GSM system with minor modifications in the infrastructure. It is designed to have wireless web access through mobile telephony service providers.

CDMA and upgradations In CDMA systems, all users transmit in the same frequency band using specialized separate orthogonal codes as a basis of channelization (discussed in Chapter 9). The transmitted information is spread over the spectrum by multiplying it with a wide-bandwidth pseudo-random sequence. Both the base station and the mobile station know these random codes, which are used to modulate the data sent, allowing it to descramble the received signal. The use of CDMA technology started in 1990 with the IS-95 standard, which then developed to IS-95A and IS-95B with further improvements in the voice quality, bit rate, and data services. The next development was CDMA2000. It is now a challenge to cover the global wireless communication using CDMA techniques, and hence, International Mobile Telecommunications-2000 (IMT-2000) has taken up the UMTS project. Using WCDMA, standards are developed for the system even for indoor and outdoor communication. A CDMA high data rate system has been developed by Qualcomm, now called 3G 1X EV-DO, which has improved throughput and made significant enhancements in the downlink structure of CDMA2000.

Long-term evolution (LTE)   This new revolutionary technology is partially commercialized. It is based on subcarrier block transmissions using OFDMA in the downlink and single-carrier FDMA (SC-FDMA) in the uplink transmissions. The research work is still going on in the LTE standard. It is an emerging high-speed wireless technology, described as the fourth-generation (4G) technology, which is based on cellular division.

All these cellular technologies are discussed in Chapter 11.

Ad hoc Networks

Ad hoc networks are the data networks established temporarily without using any infrastructure. However, they take cellular division support for some configurations in LAN and MAN. The number of users in such systems may be limited. Mostly, ad hoc networks are established for personal use or for use within a limited domain, such as an office or a plant. Due to the temporary nature of these networks and their use in personal domain for communication among personal devices, they use ISM band frequencies as their carrier frequency. ISM bands are explained in Section 1.11. Bluetooth, ultra-wideband (UWB), and ZigBee IEEE 802.15.4 (wireless sensor network) are some of the protocols for an ad hoc scenario. Wi-fi IEEE a/b/g/n is an ad hoc network with multiple configurations. The Wi-fi configuration based on the access point is similar to that of a cell because the access point acts as a base station and it has its own coverage area.

Ad hoc networks dealing with internet access follow internet protocol (IP)-based protocols. These networks allow mobility. A central challenge in the design of mobile ad hoc networks is the development of dynamic routing protocols that can efficiently find routes between two communication nodes. A mobile ad hoc networking (MANET) working group has been formed within the internet engineering task force (IETF) to develop a routing framework for IP-based protocols in ad hoc networks. Another challenge is the proper design of medium access control (MAC) protocols for multihop ad hoc networks. WiMAX IEEE 802.16x can be considered as an ad hoc network with multihop. Currently, OFDM is used as a physical layer standard in IEEE 802.11a/g/n and 802.16x protocols, HIPERLAN protocols, and so on. IEEE 802.16x are the protocols for IP-based metropolitan area broadband access networks.

Concept of Convergence in Personal Networking and Broadband Access

The aforementioned technologies can help maintain wireless connections with mobility and ensure that information is made available whenever the user requires it. However, the nature of resource and information sharing differs according to user requirements, and hence, a convergence of the technologies is required.

Convergence leads to heterogeneous networks.

Newly designed mobile devices can support many technologies in one device along with conventional mobile telephony services. The wireless technologies that are coexisting with second-generation (2G) GSM include UWB, Wi-fi, Bluetooth, and various 3G technologies, such as WCDMA and wireless access protocol (WAP). These technologies are working synergistically to meet the unique needs of the users. Apart from this, many systems require interworking among them to pass on the data to the appropriate destination.

Some examples of the convergence are as follows:

- Multiple WLANs can be connected to a WiMAX tower
- Sensor network can collect data through data aggregation techniques and send the collected data through Wi-fi to far-distance sites using broadband services
- A GSM operator can provide faster Internet services with speedy access protocols such as WAP and high speed packet access (HSPA)

A typical example of convergence using an integrated network scenario is shown in Fig. 1.5.

All these technologies and their development phases are categorized in *generations* as per the similarity in the system capabilities, bit rate support, and so on.

Table 1.2 summarizes some of the present wireless digital communication-based systems that are already in practice. Table 1.3 gives a comparison chart for the existing and upcoming technologies for wireless networking.

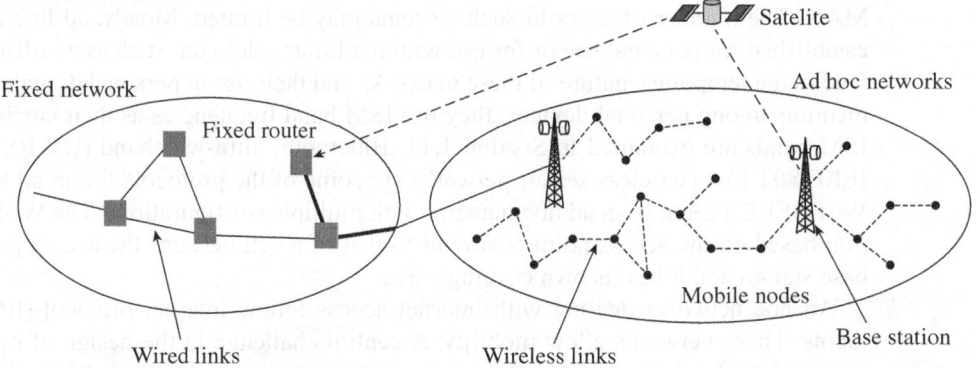

Fig. 1.5 Heterogeneous networks—integrated network combining different kinds of wireless and wired networks

Table 1.2 Summary of applications based on existing wireless digital communication

Application	Existing standard or technology used
Mobile telephony (digital cellular telephony)	GSM, CDMA (IS-95 to CDMA 2000), WCDMA-UMTS
WLAN/MAM/WAN	IEEE 802.11(Wi-fi), 802.16(WiMAX), etc.
Personal area communication	Bluetooth
Digital audio broadcast, HD Radio, DRM	DAB
Digital video broadcast, DTH through satellite	DVB
Mobile satellite communication, global communication	Iridium, UMTS, GPS
Mobile internet access	GPRS, Mobile IPv6, WAP, LTE
Wireless local loops	DECT, CorDECT, CDMA, GSM
Mobile ad hoc networks	All WLAN/WMAN standards and Bluetooth, sensor N/w

Table 1.3 Comparison of most-recent wireless networking technologies

	EDGE	CDMA 2000/1 x EV-DO	Bluetooth	Wi-fi	Wi-fi	Wi-fi	WiMAX	WiMAX	WCDMA/UMTS	UWB	LTE
Standard	2.5G	3G	802.15.1	802.11a	802.11b	802.11g	802.16d	802.16e	3G	802.15.3a	4G
Usage	WWAN	WWAN	WPAN	WLAN	WLAN	WLAN	WMAN Fixed	WMAN Portable	WWAN	WPAN	WMAN/WWAN
Throughput	Up to 384 Kbps	Up to 2.4 Mbps (typical 300–600 kbps)	Up to 720 kbps	Up to 54 Mbps	Up to 11 Mbps	Up to 54 Mbps	Up to 75Mbps (20 MHz BW)	Up to 30Mbps (10 MHz BW)	Up to 2Mbps (Up to 10 Mbps with HSDPA technology)	110–480Mbps	Typically 2–20 Mbps (RB throughput) up to 101.8 for 20 MHz carrier and 162.9 for 2 × 2 MIMO
Range	Typically 1–5 miles	Typically 1–5 miles	Up to 30 feet	Up to 300 feet	Up to 300 feet	Up to 300 feet	Typically 4–6 miles	Typically 1–3 miles	Typically 1–5 miles	Up to 30 feet	Typically 3–18 miles
Frequency	1900 MHz	400, 800, 900, 1700, 1800, 1900, 2100 MHz	2.4 GHz	5 GHz	2.4 GHz	2.4 GHz	Sub 11 GHz	2–6 GHz	1800, 1900, 2100 MHz	7.5 GHz	Multiple bands such as 700, 800, 900, 1700, 1800 MHz 2.1, 2.6 GHz etc. (Different bands for FDD and TDD modes)*

*TDD–Time Division Duplex
FDD–Frequency Division Duplex

1.10 EVOLUTION OF WIRELESS SYSTEMS

In general, the communication link requires a transmitter, a channel, and a receiver to transfer data. Here, the real-time signals and data must be modified in accordance with the channel characteristics and in a suitable detectable format, so that they can be communicated reliably through the media. Wired or wireless media can be chosen for transmission, but at the transmitter and receiver ends, a large amount of signal processing is required; hence, hardware designs need to pay more attention to the portability of the devices and should ensure good quality of reception at the same time. In the present scenario, we have a combination of systems that may have wireless infrastructure with an extensive wired support. However, the future scenario is going to be *wireless everywhere* providing the facility of mobility to the user. Hence, the following points need to be taken into consideration:

Mobility and speed of communication There is a trade off in the systems between the mobility of the user and the speed of communication achieved, as shown in Figure 1.6(b). Figure 1.6(a) represents the mobility versus data rate for various systems. It is an approximate and relative representation. It can be seen that the mobility and bit rate are increasing with the generations. The following can be observed from the figure:

- GSM provides the best mobility but very low data rate support, whereas EDGE achieves higher bit rate but compromises vehicular mobility.

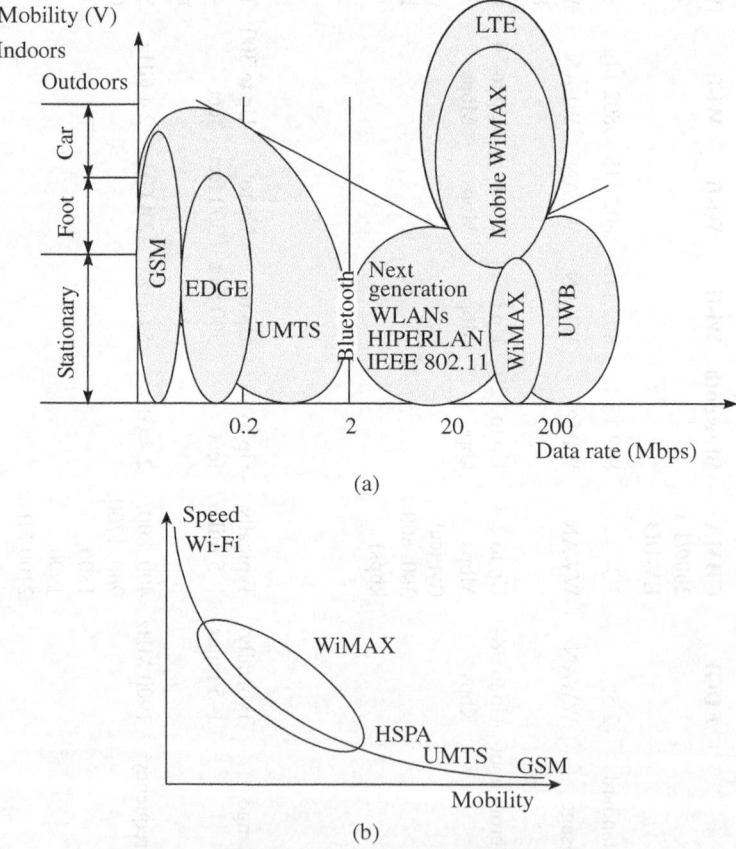

(a)

(b)

Fig. 1.6 Mobility versus data rate for various systems

- Considerable mobility is achieved with UMTS with a little compromise on the bit rate.
- IEEE 802.16e mobile wireless broadband access system and LTE are found to have vehicular mobility with a higher bit rate.
- Bluetooth and Wi-fi do not require high mobility conditions, as they are small area networks and are mostly operative in personal domains. Such low-power systems perform well in terms of data rate.
- UWB gives the highest data rate.

> The cellular infrastructure for UMTS and LTE follow almost similar architecture to that of GSM with the required upgradations in the interfaces and system components.

Wireless communication versus mobile communication There exists a very thin line of difference between wireless communication and mobile communication. Basically, in wireless communication, the focus is on the main link (transmitter + channel + receiver) and its fundamentals for communication, including various blocks of processing the information signal described in Chapters 5 to 8. Here, it is necessary to know the various methods of modifying the data or real signals, modulation schemes, channel characteristics, receiving methods, and so on. Cellular theory provides the systematic platform to have the infrastructure for developing wireless communication links for multiple users without interference. Using the cell concept, users can be identified uniquely even in the mobility mode. In mobile communications, the main focus is on cell-based wireless multi-user telecommunication systems, for which standards and protocols are developed. Here, the user is assumed to be either in steady or in mobility mode.

Growth in hardware Wireless communications were initially developed for military purpose. Gradually, the development in computers, DSP, and chip technology enabled rapid progress in the development of portable, sophisticated wireless units, such as mobile phones as well as laptops and palmtops based on Centrino technology. DSP has become indispensable for existing wireless systems. Today's wireless communication systems are mostly based on processors, VLSI/ASIC/FPGA chips, microstrip RF circuits, and PC interface. Figure 1.7 shows that faster DSP processors (compared in terms of multi-instructions per second—MIPS) are incorporated in systems to support higher bit rate. MIPS is the measure to compute the speed of a DSP processor.

> Wireless communication aims for an optimized wireless link whereas mobile communication aims for an optimized mobile system including architecture and protocols.

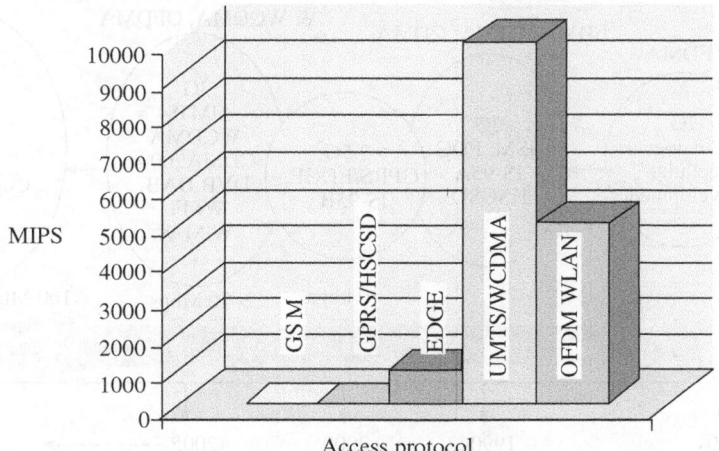

Fig. 1.7 Processing power requirement for wireless protocols and standards according to complexity of hardware

> Frequency planning is required to serve millions of users in terms of different services without interference and using the limited spectrum efficiently.

Frequency planning This is necessary to have frequency planning for various wireless systems to coexist. Wireless channel is an unguided dielectric media and hence, the frequency ranges it can support are ideally infinite. Still, due to many reasons, the full available spectrum cannot be utilized. The RF and the above range utilized for wireless communication are systematically shared; different ranges are used for different applications. Various frequency ranges from the satellites provide global coverage to the cellular system, covering 50–70 km. In contrast, LANs and personal area networks (PANs) provide a maximum range of a few to hundred metres. Hence, the carrier frequency requirement also varies. If the systems are to coexist, they would obtain a crowded frequency spectrum, since there are many factors that want their share of limited frequency resource. Therefore, it is extremely important to use spectrally efficient signal strategies. The current trend to achieve high spectral efficiency is to use adaptivity on all four dimensions: time, frequency, power, and phase. The cellular theory in Chapter 2 and the multiplexing and multiple access techniques in Chapter 9 provide the best techniques of frequency planning.

> *Note*: In short, the requirements of wireless communication include high speed/high bit rate, high spectrally efficiency, zero ISI/ICI, convergence, anywhere and anytime, global coverage, multimedia support, wireless, and digital communication systems.

Latest techniques such as WCDMA, OFDM, Hybrid OFDM, and MIMO will fulfil most of these requirements. Moreover, new approaches, such as software defined radio and cognitive radio, are coming up with a fixed set of hardware (Processor, FPGA, etc.) but with programmable software support to perform signal processing tasks, providing options such as different channel coding or different modulation scheme selection.

1.10.1 First- to Fourth-generation Wireless Systems

There is no specific measure to calculate the years of generations in wireless communication. Rather, the generations are measured on the basis of the considerable innovations in the standards and applications. Analog systems are considered as the start-up and hence they are known as the first-generation (1G) systems. The systems of other generations are illustrated in Fig. 1.8.

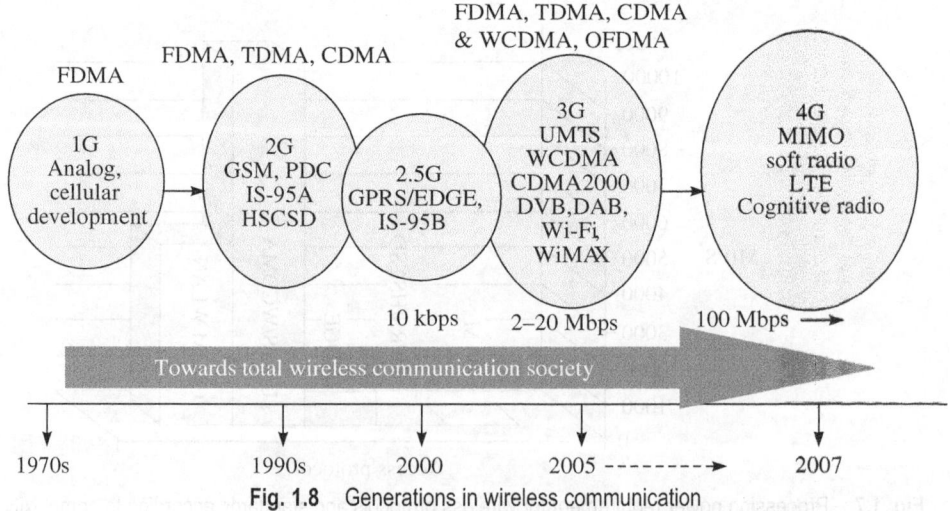

Fig. 1.8 Generations in wireless communication

As mentioned, it is very difficult to distinguish the systems on the basis of generations. For simplicity, complete analog systems mainly dealing with audio (except television with analog video) are classified as 1G systems, including analog mobile phone systems (AMPS). Partially analog and digital are classified as 2G systems. In these systems, audio and images were able to communicate, and the bit rate was very low, about 10–50 kbps. Fully digital systems with audio, image, and video are classified as 3G systems. There was a tremendous rise in the bit rate, of the order of 2–20 Mbps and even up to 54 Mbps in Wi-fi and WiMAX. In 4G systems, high-speed, fully digital, anywhere, anytime, and converged wireless communication is expected with total multimedia. The expected bit rate may reach up to 100 Mbps or more in wireless environment. With evolution in WiMAX standards, UWB, and LTE, development in the 4G systems have started.

> 4G systems are targeted with the bit rate of the order of 100 Mbps and more and that is possible due to multicarrier technique.

The following can be stated as the major differences in the generations:

1G—Cell structure, analog communication
2G—Cell structure, digital communication, convolution coding, power control
3G—Hierarchical cell structure, turbo coding, Hybrid Automatic Repeat Request (HARQ)
4G—Smart antenna, adaptive systems over above scenario

Why does a wireless channel face the problems of high bit rate? The channel faces the problem of delay spread due to multipath fading, meaning that the channels are time disper-sive; this is discussed in detail in Chapter 3. Spreading results in merging of two consecutive pulses. If the bit rate is too high, the bit duration is low; hence, due to the merging, it is very difficult to identify the two separate pulses. This limits the bit rate of the system. Higher-order M-PSK, diversity mitigation techniques such as MIMO, or multicarrier techniques such as OFDM can eliminate the problem of higher bit rate.

The 2G technology for mobile communication originated during the 1990s, before which the conventional telephony based on wired lines was being used. A few military wireless applications and AM, FM, television, radar, and satellite communication systems were the only wireless systems implemented and known to the people. The revolution started with two new systems: the Internet based on wired lines and the cellular-based GSM that depend on wireless channels mainly for voice communication. In 2000, data transmission in the GSM was enhanced, resulting in GPRS, which could use any number of time slots among the total eight slots available for sending data. The technology exists with a data rate of 14.4–64 kbps. Another high-speed data enhancement was made in GSM, called EDGE, in which the modulation scheme is changed from *Gaussian minimum shift keying* (GMSK) to 8-PSK and the transmission data rate can be up to 500 kbps. The GSM system initially was focused on voice services with circuit switching, whereas the current 2.5G technology is focused on circuit-switched voice service and packet-switched data services.

The major challenges before the implementation of 3G were as follows:

- There was slow production of mobile phones and services.
- Wireless Internet for exponentially growing users was difficult to implement until IPv6 was implemented. (Refer to any book on computer networks for IPv6, which is the proto-col for IP layer and includes IP addresses for mobile networks as well.)
- Global roaming with a single number as proposed was yet to be standardized.
- Low-cost flexible mobile devices with all desirable features were yet to evolve.

All these challenges were overcome by the scientists and engineers. The 3G systems were successfully developed, solving major problems. Now, we are into the 4G technologies, mov-ing towards the fifth generation.

The 3G technology is optimally focused on using a single interface number and an advanced core network.

AIMS OF 3G SYSTEMS

- Anywhere and anytime mobile communication with low-cost and flexible hand-held devices
- Wireless data access, particularly with wireless Internet connection, which was motivated by the exponential growth of Internet access
- High data rate of 2 Mbps or more compared to the previous 2G systems offering 10–50 kbps
- High-speed multimedia or broadband services causing shift from voice-oriented services to Internet access (both data and voice), video, graphics, and other multimedia services
- Global roaming support and global communication
- Use of spectrum around 2 GHz and higher whereas spectrum allocation for 2G was 800/900 MHz

---

The 2G technology offered a quiet satisfactory voice communication, but with growing data traffic, the 3G technology has mainly targeted data services, particularly the Internet traffic. The main service component of the 3G technology is quality and reliable data traffic. The journey from 2G to 3G proceeded with an intermediate halt on 2.5G, which provides reliable services with minimal investment. The UMTS is a typical 3G system that uses WCDMA technology as mentioned previously and has the following aims:

- Data services up to 2 Mbps in rural or urban environment
- Voice over a packet-switched IP-based network
- Good spectral efficiency and low delay
- Complete mobility to the user
- Typical applications :
  □ Speech—teleconferencing and voice mail
  □ Message—short message service, email, etc.
  □ Switched data—low-speed LAN, Internet, etc.
  □ Medium multimedia—e-commerce, LAN, and Internet public messaging
  □ High multimedia—video clips, online shopping, and fast LAN and Internet
- High interactive multimedia, for example, video telephony and video conferencing

Some important UMTS applications and their requirements are listed in Table 1.4.

Table 1.4 Important UMTS applications and their requirements

Applications or services	Data rate required	Quality of service required	Time critical data
Messaging (email, etc.)	Low (1–10 kbps)	High	No
Voice	Low (4–20 kbps)	Low (BER < 1e-3)	Yes
Web browsing	As high as possible (>10–100 kbps)	High (BER < 1e-9)	Depends on the material; generally not time critical
Videoconferencing	High (100 kbps–2 Mbps)	Medium	Yes
Video surveillance	Medium (50–300 kbps)	Medium	No
High-quality audio	High (100–300 kbps)	Medium	Yes
Database access	High (>30 kbps)	Very High	No

1.10.2 Beyond Third Generation

During the past 20 years, wireless networks have evolved from the analog, single-medium (voice), and low data rate (few kbps) systems to the digital, multimedia, and high data rate (10–100 Mbps) systems of today.

The International Telecommunication Union (ITU) in July 2003 had made the following requirements for a 4G system:

- At a standstill condition, the transmission data rate should be 1 Gbps.
- At a moving condition, the transmission data rate should be 100 Mbps.

With these high-speed data systems, it is possible to provide users many advanced applications, such as video streaming. A potential 4G system could be used in the family of OFDM, because OFDM can have a transmission data rate of 54–70 Mbps, which is much higher than what a CDMA system can provide. A comprehensive, integrated broadband mobile communication will step forward into all-mobile 4G service and communication. The 4G technology is developed to provide high-speed transmission, next-generation Internet support (IPv6, VOIP, and mobile IP), high capacity, seamless integrated services and coverage, utilization of higher frequency, low mobile cost, efficient spectrum use, quality of service and end-to-end IP system. In short, the 4G requirements are as follows:

- High-speed data communication
- Best quality voice
- Multimedia on mobile
- LAN and intranet or Internet on mobile

1.11 LICENSED AND UNLICENSED BANDS FOR EXISTING WIRELESS SYSTEMS

Wireless channel is shared by a number of users, and the frequency ranges are provided systematically to the users, services, or applications for reliable communication (refer to Chapters 2 and 9). A few frequencies are allocated to the cellular mobile operators, such as Airtel, Hutch, or Idea, who pay heavy charges for using the allocated ranges. Even satellite channels are paid channels because of this reason. Mobile operators cannot invest in huge private infrastructure, such as satellites; moreover, they have to follow government rules. Hence, they have to get the licensed bands for communication. Mobile communications based on GSM and CDMA are made over licensed bands.

Presently, some technologies limited to the user's area without the need for huge or global infrastructure are developed. Some applications of these technologies are PAN, based on Bluetooth, UWB, and WLAN, based on Wi-fi, which are small area communication systems. The frequency range of operation is 2.4–5.6 GHz. Actually, these bands are international bands for scientists and medical officers. As the systems are not concerned with other such systems at far distances, independent communication is possible. For example, in Bluetooth applications, one device with Bluetooth support will search for other active Bluetooth devices within an area of 10 metres. The list of devices will be displayed on the screen and the required device can be selected from the list for communication. Even if any other Bluetooth device is active beyond this range, it will not be listed or connected with the device. These communications are called *unlicensed band communications*. Since they are based on spread spectrum or OFDM technology, secure communication is possible. In spread spectrum techniques, orthogonal codes are present, whereas in OFDM, orthogonal carriers are present.

Ad hoc networks are operated in the unlicensed band, whereas infrastructure-based cellular network operators need to pay for the licensed frequencies.

1.11.1 Spectral Policies

There is a rapidly increasing growth of wireless services as well as development of new technologies. Consequently, the demands on the use of the RF spectrum are rapidly increasing for both the federal government and non-federal users. The spectrum is heavily occupied in the 0.8–11 GHz range for land mobile systems and 3–30 GHz for television and satellite ranges. An approximate representation is given in Fig. 1.9. The spectrum is shared among many service providers. The services include defence and military applications too, and therefore, the spectrum must be managed with certain policies. There is a continuous revision in the spectrum management policies to satisfy domestic and international uses to cope with the latest development and usage scenario. The US, the UK, and many other countries have their own body to manage such concerns and to take specific actions to improve the spectrum management. Policies vary from country to country.

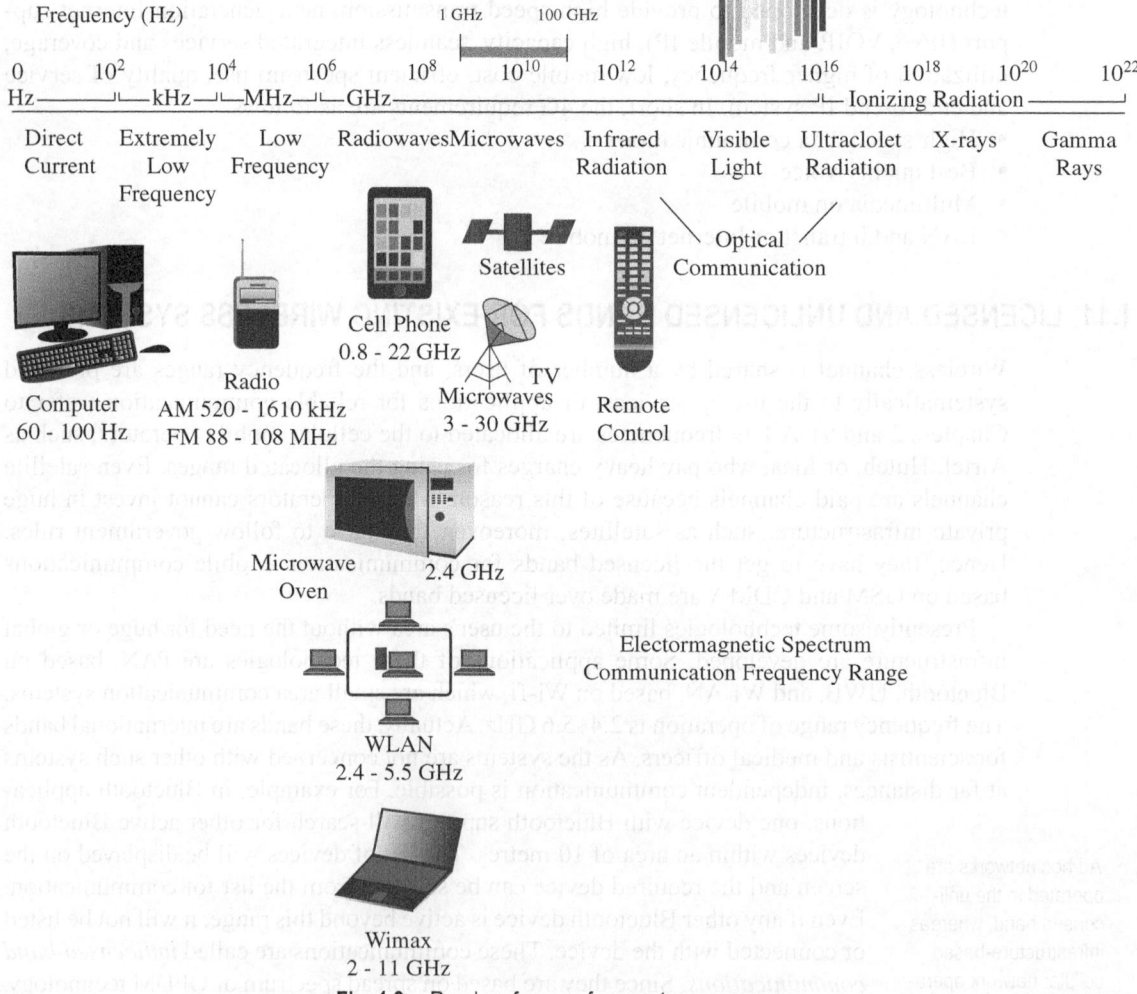

Fig. 1.9 Ready reference for spectrum occupancy

MORE SOLVED EXAMPLES

Example 1.4 Identify the type of signal (energy or power) shown in Fig. 1.10 and calculate the suitable measure.

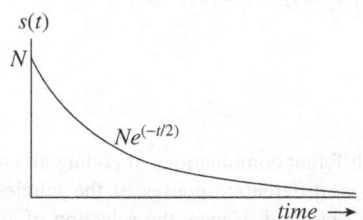

Fig. 1.10 Exponential function for Example 1.4

Solution The signal shown in Fig. 1.10 is an exponentially decaying signal that approaches zero as t approaches infinity. It is not a periodic signal. Hence, it is an energy signal. The suitable measure is energy E_s. From Eq. (1.13a), for a continuous signal

$$E_s = \int_{-\infty}^{\infty} s^2(t)dt = \int_0^{\infty}(Ne^{-t/2})^2 dt = \int_0^{\infty}N^2e^{-t}dt = N^2$$

(*Note*: Readers can try to identify from various other functions whether it is a power signal or an energy signal.)

Example 1.5 Show that the frequency spectra of the square wave shown in Fig. 1.11, is sinc shaped. [Hint: sinc function is of the form sin(x)/x.]

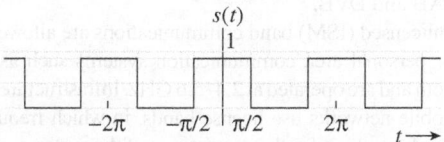

Fig. 1.11 Square wave for Example 1.5

Solution It is better to represent the square wave $s(t)$ in its exponential Fourier series form to get the frequency domain coefficients and the symmetrical form of the spectra.

The signal $s(t)$ can be written in its exponential series form as

$$s(t) = \sum_{n=-\infty}^{\infty} c_n e^{jn2\pi f_0 t}$$

where $f_0 = 1/T_0$, and T_0 is the duration 2π.

$$C_0 = \frac{1}{T_0}\int_{T_0} s(t)dt = \frac{1}{2}$$

$$c_n = \frac{1}{T_0}\int_{T_0} s(t)e^{-jn2\pi f_0 t}dt \quad n \neq 0$$

$$= \frac{1}{T_0}\int_{-T_0/4}^{T_0/4} e^{-jn2\pi f_0 t}dt$$

$$= \frac{1}{-jn2\pi f_0 T_0}[e^{-jn2\pi f_0 T_0/4} - e^{jn2\pi f_0 T_0/4}]$$

Rearranging the terms as per the definition of sine wave in terms of the exponential form and substituting $f_0 T_0 = 1$, we get

$$c_n = \frac{1}{n\pi}\sin\left(\frac{n\pi}{2}\right)$$

This is the mathematical representation of a sinc function.
Now, for $n = 1$, $c_n = \frac{1}{\pi}\sin\left(\frac{\pi}{2}\right) = \frac{1}{\pi}$

for $n = 2$, $c_n = \frac{1}{2\pi}\sin\left(\frac{2\pi}{2}\right) = \frac{1}{2\pi}\sin\pi = 0$

and so on. On plotting the discrete components and their values and joining them, the since shape is obtained (Fig. 1.12).

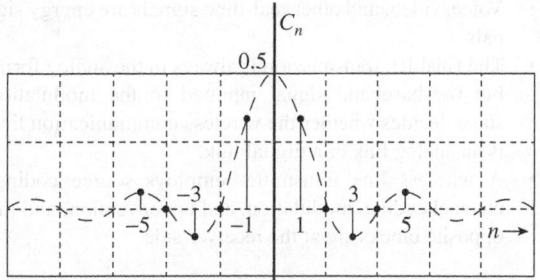

Fig. 1.12 Result of Example 1.5

Example 1.6 Find the Fourier transform of the unit impulse signal $\delta(t)$.

Solution $s(t) = \delta(t)$, where $\delta(t)$ occurs at a time instant $t = 0$

or $s(f) = \int_{-\infty}^{\infty} \delta(t)e^{-j2\pi ft}dt = e^{-j2\pi f \cdot 0} = 1$

or $\delta(t) \Longleftrightarrow 1$

It occupies the whole spectrum. This property can be observed in AWGN noise (Chapter 4).

(*Note*: The various useful Fourier transforms for communication systems are for rectangular pulse function, sinc function, and sinusoidal wave. Readers may go through them as a self-study.)

Example 1.7 Establish the relationship of E_b/N_o with the channel capacity.

Solution As per Shannon–Hartley theorem,
Channel capacity $C = W\log_2(1 + S/N)$
Signal power = Bit energy $E_b \times$ bit rate C
Noise power = Noise spectral density $N_o \times$ bandwidth W
Substituting the values, we get

$$E_b/N_o = [2^{C/W} - 1]/(C/W)$$

SUMMARY

- The bit rate defines the rate at which information is passed, whereas the signalling rate defines the baud rate. The symbol rate is the number of symbols per second; each symbol represents n bits and has M signal states, where $M = 2^n$. This is called M-ary signalling. Baud is synonymous to symbols per second or pulse per second.
- Using the transforms, any time domain signal can be analysed into its frequency components. For every signal, the signal defines the spectrum and the spectrum defines the signal; that is, they are unique and opposite conversions ideally (especially for linear systems) but may not be so practically.
- Bandwidth gives important information about useful frequency components.
- The SNR and bandwidth are exchangeable and are to be balanced always to decide the channel capacity.
- Voice, video, and other real-time signals are energy signals.
- The final RF transmission is always in the analog form, but the baseband signal inputted to the modulation stage decides whether the wireless communication link is an analog link or a digital link.
- A wireless link transmitter employs source coding, channel coding, modulation, and upconversion, and the opposite blocks are at the receiver side.

- With different combinations of coding and modulation schemes, different responses of the wireless systems can be observed. Hence, the selection of an optimum set-up of the protocols and standards is a matter of balancing the requirements.
- Line coding is applied to digital baseband for obtaining the desired spectral characteristics.
- OFDM and CDMA are the important modulation techniques for the latest wireless systems and for next-generation networks.
- GSM is the first digital wireless system, which was then upgraded to EDGE and is supported by the GPRS packet radio protocol.
- UMTS targets worldwide mobile communication with a unique user number.
- WPAN, WLAN, and WMAN are three major networks with different sizes and are based on the IP protocol.
- LTE is a 4G network based on OFDMA.
- The major systems in the broadcast technologies are DAB and DVB.
- Unlicensed (ISM) band communications are allowed only for personal area communication systems such as Bluetooth and are operated at 2.4–5.6 GHz. Infrastructure-based mobile networks use license bands, in which frequencies are planned out for the coexistence of the systems.

EXERCISES

Multiple-choice Questions

1.1 If the transmission bandwidth is W and the available channel bandwidth is $W_{channel}$, what should be the condition that will allow fruitful reception?
 (a) $W = W_{channel}$ (c) $W > W_{channel}$
 (b) $W < W_{channel}$ (d) All of these

1.2 If the bit rate of a data is 1 Mbps, what should be the bandwidth occupied by the rectangular wave?
 (a) 1 MHz   (c) 0.5 MHz
 (b) 0.1 MHz   (d) 2 MHz

1.3 Real audio/video signal is a/an
 (a) energy signal   (c) deterministic signal

 (b) power signal   (d) periodic signal

1.4 Unit ramp signal is
 (a) an energy signal   (c) a periodic signal
 (b) a power signal   (d) none of these

1.5 Which of the following measures cannot be effective in reducing noise?
 (a) Decrease in signalling rate
 (b) Increase in channel bandwidth
 (c) Increase in transmitter power
 (d) Use of redundancy

1.6 The channel capacity C of a band-limited Gaussian channel is defined as
(a) $W_{channel}\log_2(1 + SNR)$
(b) $(1/W_{channel})\log_2(1 + SNR)$
(c) $W_{channel}\log_2(SNR)$
(d) $(1/W_{channel})\log_2(SNR)$

1.7 In communication receivers, fidelity is provided by the
(a) mixer stage   (c) IF stage
(b) audio stage   (d) detector stage

1.8 If a receiver has poor IF selectivity, it will, therefore, also have poor
(a) sensitivity   (c) diversity reception
(b) double spotting   (d) blocking

1.9 Noise figure is used as a figure of merit of a/an
(a) oscillator   (c) amplifier
(b) modulator   (d) isolator

1.10 The selectivity of most receivers is determined largely by the
(a) sensitivity
(b) antenna direction
(c) characteristics of IF section
(d) all of these

1.11 Which one of the following is not a useful quantity for comparing the noise performance of receivers?
(a) Noise figure
(b) Equivalent noise resistance
(c) Input noise voltage
(d) Noise temperature

1.12 Which of the following communication systems is mainly suitable for wireless digital communication?
(a) Analog input–analog transmission
(b) Analog input–digital transmission
(c) Digital input–digital transmission
(d) Digital input–analog transmission

1.13 Which of the following is the scheme for creating a digital database of real signals?
(a) Pulse code modulation
(b) Manchester coding
(c) Binary conversion
(d) Pulse amplitude modulation

1.14 Which of the following systems is a 3G system?
(a) Analog cellular system
(b) EDGE
(c) FM
(d) UMTS

1.15 The capacity of a wireline system can be increased by
(a) TDMA
(b) random access
(c) increasing the number of wires
(d) all of these

1.16 The protocol for a Wi-fi system is
(a) IEEE 802.16d   (c) IEEE 802.11a
(b) IEEE 802.15.3   (d) IEEE 802.15.1

1.17 Which of the following is a system in which long haul communication is involved?
(a) Mobile satellite communication system
(b) GSM system
(c) WiMAX system
(d) Bluetooth system

1.18 The systems that utilizes the ISM band for communication are
(a) GPRS and EDGE
(b) Bluetooth and Wi-fi
(c) GPRS and Bluetooth
(d) Bluetooth and WiMAX

Review Questions

1.1 How are the communication systems classified in general?

1.2 How are the wireless systems classified? State the major changes in the classified wireless systems.

1.3 Presently, what are the systems in which partly wired links and partly wireless communication are incorporated? Can you find the types of cables used in different wired systems?

1.4 Prepare a list of all existing communication systems used in everyday life. Out of these, find which are wired and which are wireless and then prepare a list of the existing wireless systems and the associated standards along with their modulation schemes, bit rate, frequency range of communication, special features, and so on.

1.5 Write short notes on the following terms:
(a) Information   (b) Transmitter
(c) Types of channels   (d) Types of noise
(e) Receiver   (f) Modulation
(g) Carrier   (h) Bandwidth
(i) SNR

1.6 The bandwidth of a channel is 250 KHz. What kind of information signals can be transmitted over it? Why should the system bandwidth be higher than the signal bandwidth?

1.7 What are the various commercial ranges for various wireless applications? Some commercial ranges are used for multi-applications. Which factors are considered to derive reliable communication in these situations?

1.8 With reference to Fig. 1.3, find the theoretical range of bit interval for which the bit occurrence can be detected and establish the relation with the system bandwidth.

1.9 What is the relationship between the fundamental frequency and the period of a signal?

1.10 Shannon and Nyquist formulas of channel capacity place an upper limit on the bit rate of a channel. Are they related? How?

1.11 What are the key factors that affect the channel capacity? Explain how the capacity is affected.

1.12 Explain the SNR–bandwidth trade-off.

1.13 Prove that the relationship between the SNR and the bandwidth expansion factor is non-linear.

1.14 Are the signal spectrum and the signal bandwidth the same? Why?

1.15 List out the various types of signals for communication described in the chapter and draw their waveforms. In which category will the audio, image, and video signals fall?

1.16 Find the Fourier transforms of the well-known functions square, triangular, exponential, and ramp.

1.17 Identify the wireless devices that incorporate various modern processors.

1.18 Represent an EM wave equation with its amplitude, frequency, and phase, assuming that the wave is travelling in any one direction.

1.19 When will a signal be a scalar or a vector? How can scalars and vectors be represented in mathematical form?

1.20 Compare AM, FM, and PM techniques of modulation. What are the drawbacks of these techniques that are eliminated using digital modulation techniques?

1.21 Why is line coding more important for wired line communication?

1.22 Why is the receiver a critical part of a complete wireless link?

1.23 Differentiate between the following terms:
(a) Analog and digital EM signals
(b) Analog and digital communication systems
(c) Guided and unguided media

1.24 List out the requirements of 4G, and from the analysis of the existing standards, find the points at which we are lacking.

or

Which are the areas that should be concentrated upon by the scientists and engineers to have a reliable *anywhere, anytime* communication scenario?

1.25 Develop the requirements of a wireless digital communication transmitter and a receiver in the form of blocks and link them to form a basic link diagram.

1.26 List the basic requirements of UMTS and LTE systems.

1.27 Compare wired and wireless communication and find why a higher bit rate is a problem in the wireless link but not in the wired link. When does a wired link have the problem of a higher bit rate?

1.28 How can we increase the user accommodation capacity on wired and wireless links?

1.29 How do licensed and unlicensed band communications differ?

1.30 Discuss the major changes that took place in the communication systems from the first to the fourth generations in general. Also, discuss separately the changes in the 1G to 4G wireless systems.

1.31 How can you say that wireless digital communication exhibits interdisciplinary approach?

Numerical Problems

1.1 If the bit rate is to be maintained at 10 Mbps, what modifications should be made in a system to cope with SNR variations between 10 dB and 20 dB?

1.2 If square pulses, each of duration 0.05 μs, are to be transmitted at a carrier frequency 100 MHz, what will be the shape of the spectrum? According to this spectrum, find the following:
(a) Null to null (significant energy) bandwidth
(b) Fractional power containment bandwidth
(c) Bounded power spectral density
(d) Absolute bandwidth

Hint: *Fractional power containment bandwidth*: According to Federal Communications Commission (FCC) rules, the occupied bandwidth is the band that levels exactly 0.5 per cent of the signal power above the upper band limit and exactly 0.5 per cent of the signal power below the lower band limit. Thus, 99 per cent of the signal power is inside the occupied band.

Bounded power spectral density: Typical attenuation level might be 35 dB or 50 dB.

Absolute bandwidth: It is the interval between the frequencies beyond which the spectrum is zero. However, for all realizable waveforms, absolute bandwidth is infinite.

1.3 The energies of signals $g_1(t)$ and $g_2(t)$ are E_{g1} and E_{g2}, respectively.
(a) Show that, in general, the energy of signal $g_1(t) + g_2(t)$ is not $E_{g1} + E_{g2}$.
(b) Under what condition is the energy of $g_1(t) + g_2(t)$ equal to $E_{g1} + E_{g2}$?
(c) Can the energy of signal $g_1(t) + g_2(t)$ be zero? If so, under what condition(s) will it happen?

1.4 Determine the energy spectral density of the square pulse $s(t) = rect(t/T)$, where $rect(t/T)$ equals 1 for

$-T/2 \leq t \leq T/2$ and equals 0 elsewhere. Calculate the normalized energy E_s in the pulse.

1.5 The input x and output y of a certain non-linear channel are related as $y = x + 0.22x^3$. The input signal $x(t)$ is a sum of two modulated signals as follows:

$$x(t) = x_1(t)\cos\omega_1 t + x_2(t)\cos\omega_2 t$$

where $X_1(\omega)$ and $X_2(\omega)$ are shown in Fig. 1.13.

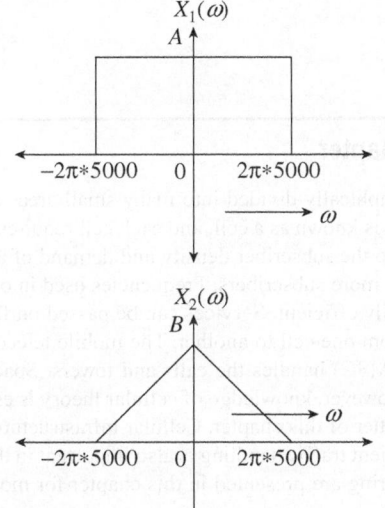

Fig. 1.13 Spectra for Problem 5

Here $\omega_1 = 2\pi(100 \times 10^3)$ and $\omega_2 = 2\pi(110 \times 10^3)$
(a) Sketch the spectra of the input signal $x(t)$ and the output signal $y(t)$.

(b) Can the signals $x_1(t)$ and $x_2(t)$ be recovered without distortion and interference from the output $y(t)$?

1.6 Show that an arbitrary function $s(n)$ can be represented by the sum of an even function $s_e(n)$ and an odd function $s_o(n)$.

$$s(n) = s_e(n) + s_o(n)$$

1.7 In a multilevel signalling, if the number of discrete signal or voltage levels is 8 in a modem and the bandwidth is 4 kHz, find the channel capacity. If the data rate is increased by increasing the number of signalling elements, for a given bandwidth, what will be the expected changes? Comment on it.

1.8 The bandwidth of a channel is 2 MHz and the SNR is 25 dB. Using Shannon's formula, find the channel capacity. If we assume that we can achieve this limit based on Nyquist's formula, find the number of signalling levels required.

1.9 A system with digital signalling is operated at 4800 bits per second. If the signal element encodes a 4-bit word, what is the minimum required bandwidth?

1.10 For the signal shown in Fig. 1.14, find the type of signal and suitable measure to analyze it.

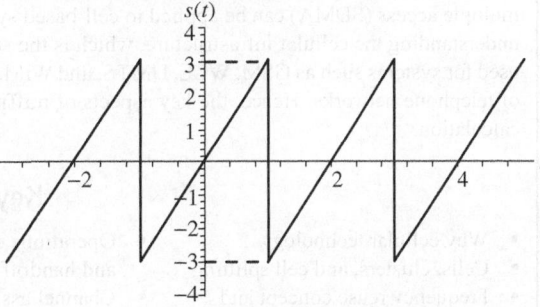

Fig. 1.14 Signal for Problem 10

2 Cellular Theory

Theme of the Chapter

The total coverage area of a mobile telephone network is geographically divided into many small areas and hence, such networks require appropriate infrastructure. Each small area is known as a cell, and each cell requires a tower of its own. The cellular concept allows cells to be sized according to the subscriber density and demand of a particular area. As the population grows, cells can be split to accommodate more subscribers. Frequencies used in one cell can be reused in other cells. This makes the whole technique spectrally efficient. Services can be passed on from cell to cell to maintain uninterrupted conversation as the user moves from one cell to another. The mobile telecommunication switching office (MTSO) or the mobile switching centre (MSC) handles the calls and towers. Space division multiple access (SDMA) can be applied to cell-based systems. However, knowledge of cellular theory is essential for understanding the cellular infrastructure, which is the subject matter of this chapter. Cellular infrastructure is mainly used for systems such as GSM, WLL, UMTS, and WiMAX. Efficient traffic handling is also important in the working of telephone networks. Hence, the key aspects of traffic engineering are presented in this chapter for mobile traffic calculation.

Key Topics

- Why cellular technology
- Cells, clusters, and cell splitting
- Frequency reuse concept and reuse distance calculation
- Cellular system components
- Antennas for base station

- Operations of cellular systems and handoff
- Channel assignment—fixed and dynamic
- Cellular interferences— co-channel and adjacent channel

- Sectorization
- Mobile traffic calculation
- Spectrum efficiency of cellular systems
- Location management

2.1 INTRODUCTION

The concept of trunking (resource sharing) between two central telephone exchanges was used in conventional landline telephone systems. A trunked radio system is a complex, centrally controlled, full duplex radio system that allows sharing of limited radio frequency (RF) channels among a large group of users. It uses control or signalling channels as well. If mobility is to be provided to subscribers, then wireless links and radio trunking are required in the telephone system; these may be systematically provided to the users by developing the required type of infrastructure. The concept of cellular technology encompasses various aspects of

Note: Mean opinion score is a numerical measure of the quality of voice or video, which is an average of the subjective evaluations of a group of listeners. A value 4–4.5 (on a scale of 1—bad to 5—excellent) is referred to as the toll quality value for an analog circuit-switched Public switched telephone network (PSTN) and is normally maintained for long-distance calls.

> Frequency reuse is not allowed within a cluster of planned size, but the same set of frequencies can be reused in different clusters.

such infrastructure, for instance, division of area, frequency management or allocations, and call handling. On the basis of such infrastructure, various mobile telephone systems have been developed, which are discussed in Chapter 11. In short, the modern trunked radio system is a circuit-switched (dedicated connectivity-based) system with the support of packet switching (bursty data traffic) over a cellular infrastructure. The term *cell phone* is often used by the public when referring to a wireless mobile phone. The cellular approach was proposed and developed predominantly by the Bell system in the US in the early 1970s. The following were considered as the initial requirements:

- Large subscriber capacity
- Efficient use of available spectrum
- Nationwide coverage
- Adaptability to traffic density
- Telephone service to both vehicle and portable user terminals
- Toll quality, which is measured by mean opinion score
- Affordability, which could eventually make it a mass-market service

To satisfy these requirements, a simple hypothetical system should have the following two main components:

Central station   This is common for many subscribers and includes the switching equipment and an RF transmitter and receiver.

Mobile telephone   Every subscriber must have a mobile telephone that includes a microphone, a speaker, dialling facility, a radio transmitter, and a receiver.

Apart from these, an infrastructure is required to support user mobility, as well as to cover different areas with seamless connectivity.

Today's cellular technology fulfils all these requirements. It enables coverage of all regions of a country as well as international connections, and efficiently uses the available limited frequency spectrum.

All mobile operators use the available radio spectrum to provide services. Spectrum is a scarce resource and is allocated systematically according to certain policies. It is shared by

---

BEFORE THE CELLULAR AGE...

Before the advent of cellular technology, capacity was enhanced through the division of frequencies, resulting in an increase in the number of channels. However, this reduced the total bandwidth available to each user, affecting the quality of service.

For example, if one million subscribers are to be accommodated, then two million radio channels are required, because it is necessary to have one channel each for uplink and downlink.

Let us now calculate the bandwidth. This will depend on the nature of modulation. Let us assume frequency modulation. For speech, f_{max} is about 4 kHz. If we assume a frequency deviation of 12 kHz, then the bandwidth required is $2(12 + 4) = 32$ kHz. For simplicity, let us round it to 30 kHz bandwidth per channel. Then the total bandwidth required is 2 million $\times$ 30 kHz = 60,000 MHz. This is a very large requirement and is practically impossible.

Cellular technology, on the other hand, requires approximately 50–75 MHz bandwidth to accommodate the number of subscribers requesting the services in today's scenario. The subsequent topics in this chapter will clarify how the cellular technology is so bandwidth efficient.

A GSM system is based on cellular infrastructure and uses a combination of FDMA and TDMA along with frequency reuse that is incorporated within SDMA.

a number of industries, including the broadcasting, mobile communications, and military.

In addition to frequency division multiple access (FDMA), **cellular technology** also uses time division multiple access (TDMA) or a combination of both to increase capacity. However, division of frequency or time alone would not be sufficient to accommodate a large number of subscribers. For doing so, the system requires additional features.

Therefore, cellular technology allows for the division of geographical areas, rather than just frequencies or time. Geographical division requires assignment of channels along with its systematic reuse. This geographical reuse of radio channels is known as *frequency reuse*, and it leads to a more efficient use of the radio spectrum. Refer Chapter 11 for more on the Global system of communication (GSM) system, which is based on frequency reuse. Multiple access schemes are discussed in Chapter 9.

To explain cellular systems in general, it is necessary to discuss the basics of cellular systems, their performance criteria, the uniqueness of the mobile radio environment, the operation of the cellular systems, the reduction of channel interferences, handoffs, and so on. Basically, there are two types of cellular systems: *analog* or *digital circuit-switched* system and *packet-switched* system. These two types of systems are discussed in Section 2.3.

2.2 CELLULAR INFRASTRUCTURE

In modern cellular telephony, rural and urban regions are divided into areas according to specific provisioning guidelines or standard protocols. Deployment parameters, such as amount of cell splitting and cell sizes, are determined experimentally in the cellular system architecture. Provisioning for each region is planned according to an engineering plan, which includes cells, clusters, frequency reuse, and handovers. Today, software tools are also available to plan and design cellular architecture.

2.2.1 Cells, Clusters, and Cell Splitting

A *cell* is the basic geographic unit of a cellular system. When cluster maps are to be drawn, depiction of cells is a challenging task. A circle seems to be the natural choice because an omnidirectional antenna with a circular radiation pattern can be placed at the centre of the cell. However, if a large area is to be filled with circles, there may be either overlaps or gaps in between. Hence, cells are chosen to be of a hexagonal shape, which is the most suitable.

The term *cellular* comes from the honeycomb (hexagonal) shape of the areas into which a coverage region is divided theoretically. Cell-wise, one base station provides transmission over a small geographic area. Cell sizes vary depending on the landscape (topographic locations such as mountains, valleys, and plains). Due to the constraints imposed by natural terrain and man-made structures, the cells are not perfect hexagons.

A group of cells form a *cluster*, the size of which is denoted as *N*. Figure 2.1 illustrates a seven-cell cluster. Each cluster is allocated a set of frequencies.

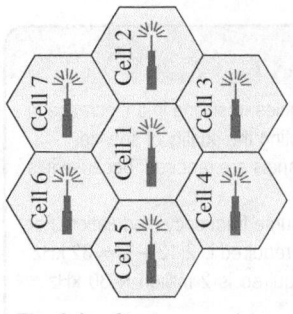

Fig. 2.1 Cluster comprising seven cells (*N* = 7)

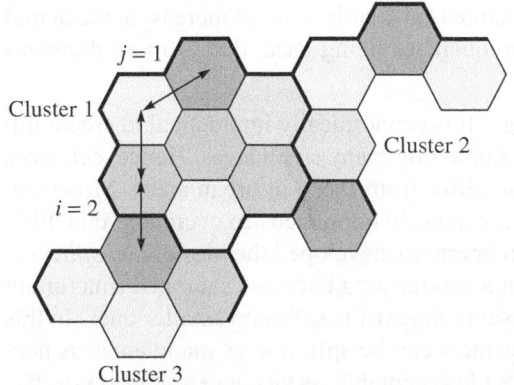

Fig. 2.2 Co-channel cell search in seven-cell cluster with $i = 2$ and $j = 1$ (cluster represented by thick outlines)

Restrictions on Value of N

If we arbitrarily choose any cluster size, we may not be able to cover the entire area using clusters without leaving gaps. To ensure that there are no gaps, N is restricted by the following equation:

$$N = i^2 + ij + j^2 \tag{2.1}$$

For $i = 1, j = 1 :: N = 3$
For $i = 2, j = 1 :: N = 7$
For $i = 2, j = 2 :: N = 12$
For $i = 3, j = 2 :: N = 19$

This can be proved by the use of the geometry of hexagonal cells shown in Fig. 2.2, which gives the significance of i and j. Possible cluster sizes are obtained using Eq. (2.1) only with non-negative integers i and j. The integers i and j determine the relative locations of co-channel cells, as shown in Fig. 2.2.

Example 2.1 Show that the area of a hexagonal cell is $2.598R^2$, where R is the radius of the hexagon.

Solution A hexagon is made up of six triangles as shown in Fig. 2.3.

From the geometry of the figure, $a = R\cos\theta$

or $\quad a/R = \cos 60° = 1/2 \Rightarrow a = R/2$

In addition, $\quad b^2 = R^2 - a^2 = R^2 - R^2/4 = 3R^2/4$

$\Rightarrow b = \sqrt{3/4}R$

Now, $\quad$ area $= 6 \times$ area of triangle $= 6 \times \frac{1}{2}(2ab) = 6ab$

$= 6 \times R/2 \times \sqrt{3/4}R = 2.598R^2$

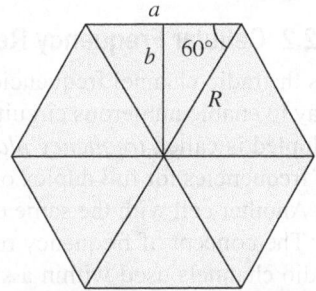

Fig. 2.3 Figure for Example 2.1

A relationship exists between the cluster design equation and frequency reuse as well as reuse distance. However, according to Fig. 2.2, to find the nearest co-channel cell, one must carry out the following:

- Move i (here $i = 2$) cells along any chain of hexagons (here vertically up)
- Turn 60° clockwise and move j (here $j = 1$) cells

The approached cell is the co-channel cell of the host cell from which the moves began.

Cell size The size of a cell depends on the density of subscribers in an area. For instance, the capacity of a network in a densely populated area can be improved by reducing the size of the cells or by increasing the number of cells along with installing low-power base stations. This will effectively increase the number of channels in that area (Fig. 2.4). On the basis of their size, cells may be categorized as *macrocells*, *microcells*, *picocells*, and *femtocells*.

Smaller cells use low-power transceivers to cover smaller regions; hence, frequency reuse can also be increased without interferences. Moreover, smaller cells effectively increase the

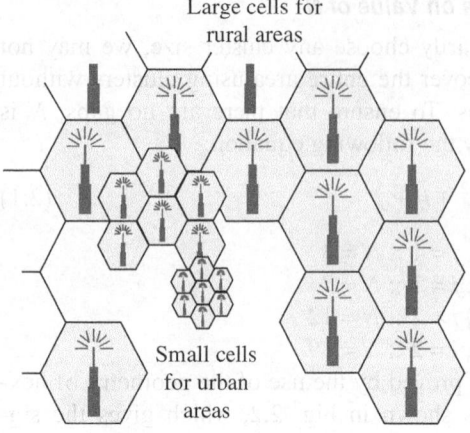

Large cells for rural areas

Small cells for urban areas

Fig. 2.4 Cell splitting concept

number of channels available without increasing the actual number of frequencies being used, due to more divisions of area.

Cell splitting It is economically impractical to create full systems encompassing many small areas. Hence, cell sizes in rural areas differ from those in urban areas. Moreover, urban areas are densely populated. To overcome this difficulty, system operators developed the idea of *cell splitting*. As and when a service area becomes saturated (maximum number of users), the area is split into smaller ones. In this way, urban centres can be split into as many areas as necessary to provide acceptable service levels in heavy-traffic regions, whereas less-expensive bigger cells can be used to cover remote rural regions. Figure 2.4 shows the division of an area into small cells.

> *Note:* Macrocells are for rural areas, microcells are for urban areas, picocells are for local areas such as streets, and femtocells are typically designed for use in a home or small business, which improves coverage and capacity indoors.

2.2.2 Cellular Frequency Reuse

As the radio channel frequencies were available for mobile systems, engineers had to find a way to enable numerous circuit switching-based conversations simultaneously. The solution adopted is called *frequency planning* or *frequency reuse*. A radio channel consists of a pair of frequencies for full duplex operation. Each frequency has its coverage within a cell radius R. Another cell with the same coverage but at a distance D can use the same frequency.

The concept of frequency reuse is based on assigning to each cluster the same group of radio channels used within a small geographic area. A set of N different frequency groups $\{f1, ..., fN\}$ is used for each cluster of N adjacent cells and shared among the cells almost equally. The set of frequencies assigned to a cell is completely different from that assigned to the neighbouring cells. The coverage area of the cells is called the *footprint*. The footprint is limited by a boundary such that the same group of channels, can be used in different cells that are far enough away from each other, so that their frequencies do not interfere. Fig. 2.5 represents the frequency reuse concept without interference.

In Fig. 2.5(b), cells with the same number have the same set of frequencies. Here, if the number of available frequency sets is seven, the frequency reuse factor is 1/7. This means that each cell uses one-seventh of the available cellular channel for reuse. In addition, it is clear from Figs 2.5(a) and (b) that with increase in cluster size, the co-channel cell distance increases.

> If frequency reuse is not properly designed, serious interference may occur among the same frequencies. This type of interference is called *co-channel interference*.

The concepts of clustering and frequency reuse and the effectiveness of frequency reuse can be understood in a better way with the help of the following example:

Let L = total number of duplex channels available for reuse (i.e., frequencies per cluster)

k = number of duplex channels allocated to each cell of a cluster ($k < L$)

N = cluster size (in which there are N cells)

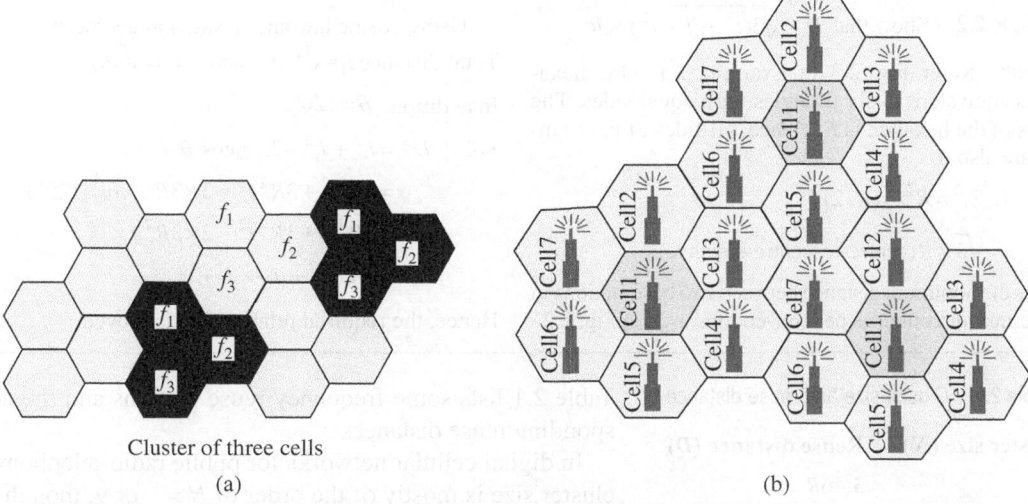

Fig. 2.5 Frequency reuse plan (a) $N = 3$ (black cells show clusters) and (b) $N = 7$ in three clusters without interference

M = number of times the cluster is repeated
C = total effective number of duplex channels available in the area

Thus, $L = k \times N$ (2.2)

$C = M \times L = MkN$ (2.3)

If a service provider has been allocated 21 duplex radio channels and if it uses 1000 clusters, then the effective number of channels will be 21,000. Using demand assignment and assuming statistical results, as many as two lakh customers can be served by a single service provider. If TDMA is also accommodated and single slot per user is assigned, then arbitrarily for eight slots per frame per channel, almost 16 lakh subscribers can be served. Thus, the effective number of channels can be increased by using the techniques efficiently.

Note that this customer base is distributed over a wide area (for example, an entire state); thus, larger the value of M, larger will be the capacity, which means that the cluster size has to be small. However, this will reduce the distance between the cells having the same set of frequencies and hence, will increase the interference. Therefore, there has to be a balance between the requirements of larger capacity and smaller interference. The total bandwidth for the system is N times the bandwidth occupied by a single cell.

Reuse Distance Calculations

The closest distance D between the centres of two cells using the same frequency (in different clusters) is determined by the choice of the cluster size N and the layout of the cell cluster. This distance is called the *frequency reuse distance*. The reuse distance D, normalized to the size of each hexagon, is given as follows:

$$D = \sqrt{3N} \times R$$ (2.4)

Example 2.2 Show that $D = \sqrt{3(i^2 + ij + j^2)} \times R$.

Solution Refer Fig. 2.3 in Example 2.1. One hexagon is equivalent to six triangles with equal sides. The radius of the hexagon is R. Hence, all sides of the triangles are also R.

So, $\quad (i/2)^2 = R^2 - (R/2)^2$

or $\quad i = \sqrt{3} \times R$ (centre-to-centre distance)

This is the minimum distance between two base stations in the vertical direction, as per conventions used in Fig. 2.2.

Using cosine law and hexagonal geometry,

Total distance $i_1 = \sqrt{3}Ri$ and $\quad j_1 = \sqrt{3}Rj$

In addition, $\theta = 120°$.

So, $\quad D^2 = i_1^2 + j_1^2 - 2i_1 j_1 \cos \theta$

$= 3R^2 i^2 + 3R^2 j^2 - 2 \times 3R^2 ij \cos (120°)$

$= 3R^2 i^2 + 3R^2 j^2 - 2 \times 3R^2 ij \times \frac{1}{2}$

$= 3R^2 (i^2 + ij + j^2)$

Hence, the required relationship is proved.

Table 2.1 Cluster size and reuse distance

Cluster size (N)	Reuse distance (D)
4	3.46R
7	4.6R
12	6R
19	7.55R

Table 2.1 lists some frequency reuse patterns and the corresponding reuse distances.

In digital cellular networks for public radio-telephony, the cluster size is mostly of the order of $N = 7$ or 9, though with special techniques, such as diversity reception, smaller reuse distances can be used. GSM can also work with $N = 3$ or 4. Cellular CDMA systems can use $N = 1$ and the same frequency is used in all cells.

2.2.3 Real-world Cells

As mentioned earlier, due to geographical constraints, in practice, it is not feasible to plan cells that are exact hexagons. Computer methods are used for optimized planning of base station locations and cell RF frequencies. Path loss and link budgets are computed from the terrain features and antenna data. These determine the coverage of each base station and interference to other cells. Figure 2.6 (a) is a partial screenshot of a typical cellular planning tool that shows the irregular terrains, and Fig. 2.6(b) is a screenshot of Ericsson TEMS.

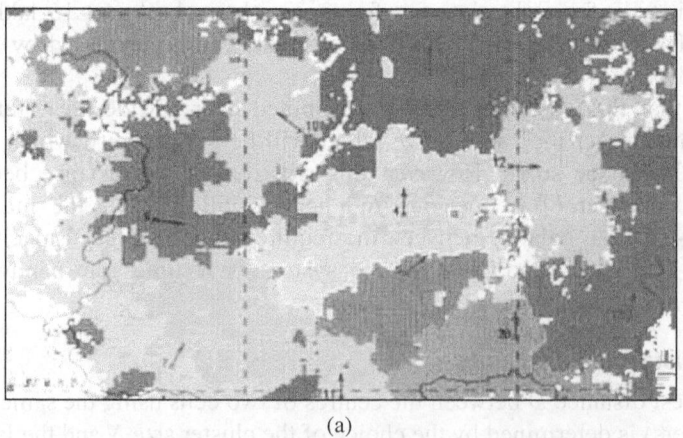

(a)

Fig. 2.6 Cellular planning by simulators (a) Screenshot of a typical tool

Note: Radio frequency and cellular planning software tools with various features have been developed by various companies such as Ericsson TEMS, Motorola Netplan, Agilent ADS, iBWave, and Global Mapper.

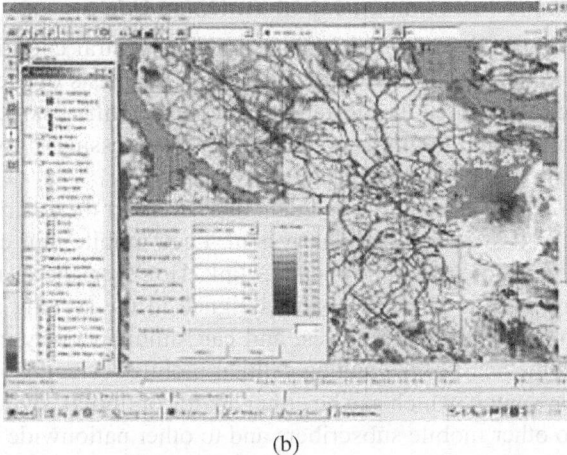

(b)

Fig. 2.6   Cellular planning by simulators (b) Screenshot of Ericsson TEMS

2.3 CELLULAR SYSTEM COMPONENTS

Along with the cell structure, some system components are required for the functioning of a cellular system. A cellular system offers mobile and portable telephone stations the same service provided by the fixed stations over conventional wired loops. It has the capacity to serve tens of thousands of subscribers in a major metropolitan area. A cellular communications system consists of the following major components, which work together to provide mobile service to subscribers.

2.3.1 Analog Circuit-switched Cellular System

An analog circuit-switched cellular system has three subsystems.

Mobile Unit

A mobile unit is also called a *mobile subscriber unit* (MSU). An MSU is a control unit consisting of a battery, a transceiver, and an antenna system that transmits and receives radio transmissions to and from a cell site. MSUs are generally of the following three types; however, many models in each of these three types are now obsolete:

(a) Mobile telephone (typical transmit power is 4.0 watts)

(b) Portable (typical transmit power is 0.6 watts)

(c) Transportable (typical transmit power is 1.6 watts)

The mobile telephone is installed in the trunk of a car, and the handset is installed in a location convenient to the driver. Portable and transportable telephones are hand-held and can be used anywhere, but their use is limited by the charge life of the internal battery.

Cell Site

Cell site is also known as *base station*. It is used to refer to the physical location of the radio equipment that provides coverage within a cell in which the transceiver is installed. The cell site provides the interface between the MTSO and the mobile units. The hardware located at a cell site or base station includes the control unit, interface equipment, RF transmitters and receivers, antennas, power plant, and data terminals.

Advanced Mobile Phone System (AMPS) was the analog mobile phone systems standard developed by Bell Labs and introduced in 1978, which was then discontinued by most of the companies and has been replaced by Digital AMPS.

Mobile Telecommunication Switching Office

The processor of the MTSO provides central coordination and cellular administration for all cell sites. The MTSO consists of the cellular processor and the cellular switch. It interfaces with the zone offices of the telephone company, performs field monitoring, controls call processing, provides operation and maintenance, and handles billing activities.

Radio and high-speed data links connect these three subsystems of the analog circuit-switched cellular system. Each mobile unit can use only one channel at a time for its communication link. However, the channel is not fixed; it can be any one in the entire band assigned to the serving area, with each site having multichannel capabilities, and can simultaneously connect many mobile units. The MTSO is the heart of the analog cellular mobile system.

The cellular switch, which can be either analog or digital, switches calls to connect mobile subscribers to other mobile subscribers and to other nationwide telephone networks. It uses voice trunks similar to telephone company interoffice voice trunks. It also contains data links providing supervision links between the processor and the switch and between the cell sites and the processor. The radio link carries voice and signalling between the mobile unit and the cell site. High-speed data links cannot be transmitted over standard telephone trunks; therefore, they must use either microwave links or T-carriers (wire lines), both of which can carry voice as well as data between the cell site and the MTSO.

2.3.2 Digital Circuit-switched Cellular System

A digital circuit-switched cellular system has four subsystems.

Mobile Station

A mobile station (MS) consists of two parts: the mobile equipment (with battery) and the subscriber identity module (SIM), which contains all the subscriber-specific data stored at the MS side.

Base Station or Base Transceiver Station

Besides having the same function as the analog base station or base transceiver station (BTS), this has the *transcoder/rate adapter unit* (TRAU), which carries out coding and decoding as well as rate adaptation in case the data rate varies.

Base Station Controller

The base station controller (BSC) is a new element in the digital system. It performs the radio resource (RR) management for the cells under its control. In addition to managing the databases of all the sites, it also handles power management, time and frequency synchronization, and frequency reallocation among BTSs. Moreover, it controls handovers from BTS to BTS (except in the case of an inter-BSC handover in which case the control is in part the responsibility of the anchor MSC).

Switching Subsystems

The MSC is the main element that coordinates the set-up of calls between two MS users, as well as between MS and public-switched telephone network (PSTN—landline) users. Gateway mobile switching centre (GMSC) handles the mobile-to-landline or landline-to-mobile calls. Visitor location register (VLR), home location register (HLR), authentication

> The infrastructural hierarchy followed by the cellular systems is MS → BTS → BSC → MSC, where multiple MSs are handled by a BTS, multiple BTSs by a BSC, multiple BSCs by an MSC, and multiple MSCs are networked.

centre (AUC), equipment identity register (EIR), and operation and maintenance centre (OMC) are the other elements in the switching subsystem. The signal transfer point (STP) handles network routing by establishing the route to the HLR for a specific mobile user. This simplifies network management, because only the routing tables in the STP need to be updated as the system grows and the MSC does not have to maintain full routing tables to all MSCs. STP and MSC or STP and PSTN are connected through SS7 signalling.

These elements will be dealt with in detail in Chapter 11 while discussing the GSM system. A typical diagram of a digital circuit-switched cellular system is shown in Fig. 2.7.

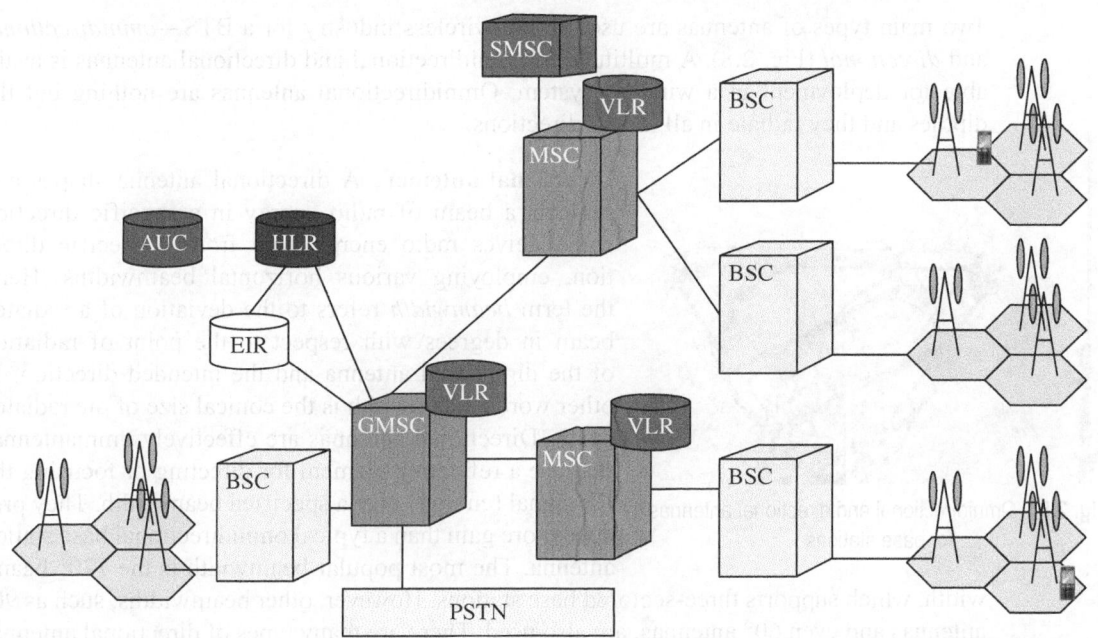

Fig. 2.7　Digital circuit-switched system hierarchy

2.3.3 Packet-switched Cellular System

A packet-switched cellular system is a system typically designed and developed for internet access. It interworks with a digital circuit-switched cellular system and uses the same infrastructure. It has six elements: MS (user equipment), base station, radio network controller (RNC), service support node (SSN), gateway support node (GSN), and charging gateway function (CGF). All these elements and their functions will be discussed in Chapter 11 along with the Global packet radio service (GPRS) system (Section 11.2).

2.4　ANTENNAS FOR CELLULAR SYSTEMS

The antenna elements that comprise a collinear array antenna will be longer for the antennas designed and used for lower frequencies and shorter for the antennas designed and used for higher frequencies. This length in the antenna elements correlates directly to the fact

The dimension of an antenna is inversely proportional to the frequency of the signal.

that lower frequencies transmit longer radio wavelengths and higher frequencies transmit shorter wavelengths. These differences are directly attributed to the properties of radio physics. Thus, an antenna used for low frequencies will be physically longer than that used for high frequencies, if the gain assigned to both is the same. For example, if a wireless carrier deploys a 9 dB antenna that operates at 850 MHz and a 9 dB antenna that operates at 1900 MHz, the 850 MHz antenna will be physically longer due to the longer wavelength inherent with the RF emitted at 850 MHz.

Today, 0 dB (unity antennas) or 3 dB gain antennas are frequently used in urban areas. These antennas are mounted on the sides of buildings or at street level and are used to cover very small areas to support enhancer or microcell deployments.

2.4.1 Antennas for Base Stations

Two main types of antennas are used in the wireless industry for a BTS—*omnidirectional* and *directional* (Fig. 2.8). A multitude of omnidirectional and directional antennas is available for deployment in a wireless system. Omnidirectional antennas are nothing but the dipoles and they radiate in all lateral directions.

Fig. 2.8   Omnidirectional and directional antennas for base stations

Directional antenna   A directional antenna shapes and projects a beam of radio energy in a specific direction and receives radio energy only from a specific direction, employing various horizontal beamwidths. Here, the term *beamwidth* refers to the deviation of a radiated beam in degrees with respect to the point of radiation of the directional antenna and the intended direction. In other words, beamwidth is the conical size of the radiated beam. Directional antennas are effectively omniantennas that use a reflecting element for directing or focusing the RF signal (energy) over a specified beamwidth. They produce more gain than a typical omnidirectional base station antenna. The most popular beamwidth is the 120° beamwidth, which supports three-sectored base stations. However, other beamwidths, such as 90° antennas and even 60° antennas, are also used. There are many types of directional antennas used by wireless carriers—log periodic, Yagi, phased-array, and panel antennas.

Downtilt antenna   At cell sites with a very high tower and a high-gain antenna, coverage shadows may be formed near the tower. To compensate for coverage shadows, antenna manufacturers have developed electrical downtilt antennas and mechanical downtilt kits specifically for the wireless industry. The radiation pattern of a downtilt antenna is electrically or mechanically tilted downwards at a specified number of degrees. Downtilting of antennas decreases distance coverage horizontally, but increases signal coverage closer to the cell site. Omniantennas can be downtilted only electrically, which is accomplished during the manufacturing process of the antenna by adjusting the phasing of the RF signal fed to the collinear antenna elements. Electrical downtilting is the way in which a specific antenna is manufactured, similar to how antennas are manufactured with specific gains assigned to them. Directional antennas can be downtilted either electrically or mechanically. Mechanical downtilting is accomplished by manually adjusting the antennas so that they tilt towards the ground.

An omnidirectional antenna on a tower radiates equally in 360°, whereas a directional antenna is for directional radiation with certain radiation angle.

ADVANTAGE OF USING DOWNTILT ANTENNA

An advantage of using downtilt antenna is that it prevents the occurrence of far-field effect. Far-field effect occurs when the radio coverage projected from site *A* completely overwhelms the intended coverage area of site *B* or other nearby sites. Site *A* may transmit and receive into site *B* or other sites, theoretically leaving these sites unused. This would not only be inefficient, but also result in a huge waste of equipment and frequency resources at cell site *B* and the other nearby base stations. The deployment of a downtilt antenna at cell site *A* would ensure that the intended radio coverage from site *A* stays within its designated coverage boundary. Far-field effect is discussed in Chapter 3 while explaining the concepts of long-distance propagation.

The following are some of the common causes of far-field effect:

- RF power level is too high at the base station.
- The tower is too high at the base station or the base station transmit antenna is too high on the tower.
- The antenna gain is too high at the base station, exceeding its intended coverage area.

> Downtilt antennas are used to increase the coverage near the cell site and reduce the impact of far-field effect in wireless networks.

Although mechanical downtilting is less expensive than electrical downtilting, it distorts (expands) the side lobes of the radiation pattern and might lead to interference with adjacent sectors. A common place to install a downtilt antenna is at a cell site that is on a very tall tower, on a hill, or near a large body of water.

Base station antennas are much more sophisticated and use a much wider variety of designs than mobile phone antennas. This is because base station antennas are required to have a higher degree of gain, ordinarily between 6 and 12 dB for omnidirectional antennas and between 4 and 18 dB for directional antennas. In some cases, 0 dB gain omniantennas are used to support microcell deployments.

A base station antenna is chosen according to the situation depending upon many factors, such as the following:

- Size of the area to be covered
- Configurations of the neighbouring cell sites
- Type of antenna used—omnidirectional or directional
- Antenna's beamwidth in case of a directional antenna
- Allotted RF spectrum the antenna can utilize

2.4.2 Adverse Effects of Base Stations

Base transceiver stations use RF radiation to transmit and receive data. Antennas on cell towers transmit in the frequency range of 869–890 MHz (CDMA), 935–960 MHz (GSM), 1805–1880 MHz (GSM1800), and 2110–2170 (3G). Hence, there is widespread concern about the possible health impacts of the radiation emissions from these stations. Several studies have been conducted to measure such health effects. There is some disagreement among the scientific community about whether such radiation causes negative health effects. The World Health Organization (WHO) fact sheet on the BTS and wireless technologies states, 'There is no convincing scientific evidence that the weak RF signals from the base stations and wireless networks cause adverse health effects.'

Most of the RF fields spread from a high base station antenna like a beam of light from a lighthouse with a shadow area directly below the base station. The RF field strength reduces gradually with distance, as shown in Fig. 2.9. It can also be seen that the radiation has no effect at just below the antenna, but has an effect after 50–250 m from the antenna.

It is concluded by many researchers that long-term exposure to radiation at unsafe levels can cause negative health impacts, such as increased risk of lung and skin cancers, headache, sleep disorder, and sterility in men.

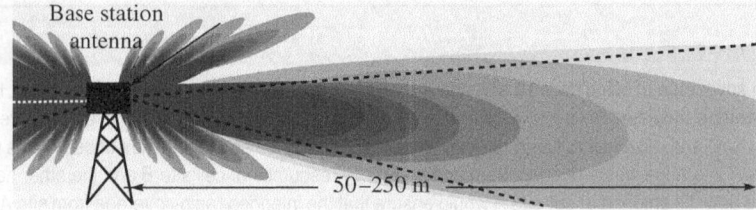

Fig. 2.9 Radiation pattern from base station (Lighter shades of grey indicate decreasing power strength)

The probable adverse effects of the radiation from the base stations are a major concern, especially among residents in urban areas. Independent monitoring of RF exposure levels around mobile phone base stations has been conducted for many years under the control of government agencies. With the number of mobile phone users constantly on the rise, the need for more base stations to cover wider ranges of the world is also increasing. Now, the problem is the number of stations that can safely be located in the same area without causing negative health effects. Research has been carried out on this in various countries, and some restrictions have been made over the range of RF emissions allowed and the number of BTSs that can operate in a single area.

2.4.3 Antennas for Mobile Radio Frequency Front End

Earlier, mobile phones used dipole antennas targeting omnidirectional reception of signals. However, microwave patch antennas are commonly used now in mobile terminals due to their many attractive features such as simple structure, low production cost, lightweight, and robustness. Moreover, dual frequency antenna elements are required as today's standard mobile terminals operate in two frequency bands, for example, GSM900/GSM1800. It is desirable to have more universal telephones that operate in multiple systems; hence, frequency tunable patch antennas are also designed.

Currently, PIFA (planar inverted-F antenna—a shorted patch) wideband antennas are used widely and sometimes fractal antennas are also used. PIFA is the modified version of the quarter-wavelength patch antenna. As the patch is shorted at the end, the current at the end of the patch antenna is no longer forced to be zero. This results in the same current–voltage distribution as that of a half-wave patch antenna. The antenna is resonant at a quarter-wavelength, thus reducing the space needed on the telephone; moreover, it typically has good specific absorption rate (SAR) properties. This antenna resembles an inverted *F* and hence the name PIFA. It has a low profile and an omnidirectional pattern.

2.5 OPERATIONS OF CELLULAR SYSTEMS

The operations of cellular systems can be divided into five different categories, namely mobile-originated call, mobile unit initialization to receive a call, network-originated or landline-originated call, call termination, and handoff procedure.

2.5.1 Mobile-originated Call

There are two possibilities in a mobile-originated call: *mobile-to-mobile call* and *mobile-to-landline call*. A typical diagram of a mobile-originated call in a GSM system is given in Fig. 2.10.

The following are the steps describing the minimum procedure to handle mobile-originated calls:

- Mobile-originated calls are routed to the BTS first and then to the BSC.
- The BSC forwards this call to the MSC.
- The MSC authenticates and routes the call to the called subscriber as per the dialled digits.
- If the called subscriber is in mobility, the process to receive the call on his mobile device is described in Section 2.5.2.

The user places the called number into an originating register in the mobile unit and pushes the *send* button. A request for service is sent on a selected set-up channel obtained from a self-location scheme. The cell site or base station receives it and, in directional cell sites (or sectors in case of sectorization—Section 2.8), selects the best directive antenna for the voice channel to use. At the same time, the cell site sends a request to the MTSO/MSC via a high-speed data link. The MTSO/MSC selects an appropriate voice channel for the call, and the cell site acts on it through the best directive antenna to link the mobile unit. The MTSO/MSC also connects the landline telephone through the telephone company zone office (local exchange).

As an example, the call set-up sequences for mobile-originated calls in a GSM system are given in Fig. 11.3 in Chapter 11.

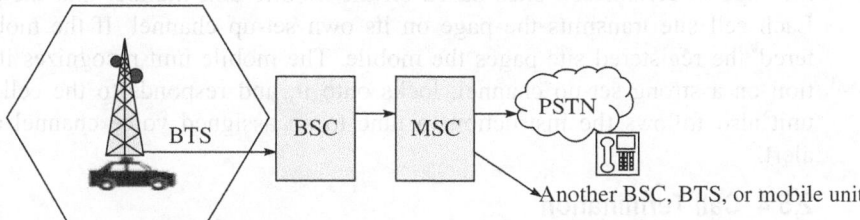

Fig. 2.10   Blocks showing mobile-originated call propagation

2.5.2 Mobile-terminated Calls

There are two possibilities: *mobile-to-mobile call* and *landline-to-mobile call*. A typical diagram for a mobile-terminated call in the GSM system is given in Fig. 2.11.

The following are the steps that describe the minimum procedure to handle mobile-terminated calls.

- Mobile-terminated calls come to the MSC first, where the HLR/VLR enquiry is carried out, and as per the information, the MS is paged in the suitable BSC.

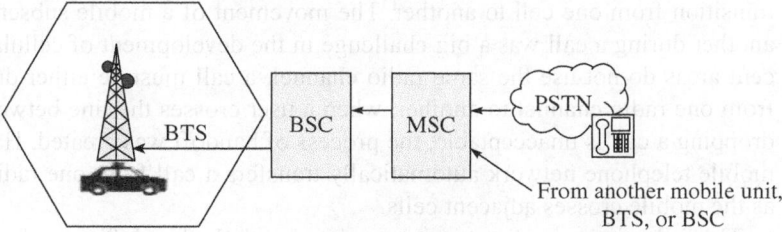

Fig. 2.11   Blocks representing mobile-terminated call flow

- The BSC forwards this page to all BTSs where the actual paging is done.
- After a BTS gets a response from the mobile, it allocates a channel for this call.
- On ending the call, the BTS informs the BSC and MSC.

When a user activates the receiver of a mobile unit, the receiver scans the set-up channels. It then selects the strongest channel frequency and locks on for a certain time. As each site is assigned a different set-up channel, the nearest cell site is selected by locking onto the strongest set-up channel. This self-location scheme is used in the idle stage and is user independent. A major advantage is that it eliminates the load on the transmission at the cell site for locating the mobile unit. The disadvantage of the self-location scheme is that the location information of idle mobile units does not appear at each cell site. Therefore, when a call initiates from a landline to a mobile unit, the paging process is longer. Since a large percentage of calls originate at the mobile unit, the use of self-location schemes is justified. After a specific period (delay), the self-location procedure is repeated.

2.5.3 Network-originated or Landline-originated Call

A feature called *registration* is used for landline-originated calls. A landline user dials a mobile unit number. The telephone company zone office recognizes that the call is for a mobile number and forwards it to the MTSO/MSC. The MTSO/MSC sends a paging message to certain cell sites based on the mobile unit number and the search algorithm. Each cell site transmits the page on its own set-up channel. If the mobile unit is registered, the registered site pages the mobile. The mobile unit recognizes its own identification on a strong set-up channel, locks onto it, and responds to the cell site. The mobile unit also follows the instruction to tune to an assigned voice channel and initiates user alert.

2.5.4 Call Termination

When a mobile user turns off the transmitter, a particular signalling tone is transmitted to the cell site and the voice channel is freed by both the sides. The mobile unit resumes monitoring pages through the strongest set-up channel.

2.5.5 Handoff Procedure

Handoff is also known as *handover*. It can be categorized as *hard handover*, *soft handover*, and *softer handover*.

Hard Handover

Hard handover occurs when the assigned channel frequency undergoes a change during the transition from one cell to another. The movement of a mobile subscriber from one cell to another during a call was a big challenge in the development of cellular networks. As adjacent areas do not use the same radio channel, a call must be either dropped or transferred from one radio channel to another, when a user crosses the line between adjacent cells. As dropping a call is unacceptable, the process of handoff was created. Handoff occurs when a mobile telephone network automatically transfers a call from one radio channel to another as the mobile crosses adjacent cells.

The following are a few terms associated with the handoff procedure:

Handoff probability   It is the probability that a handoff is executed before call termination.

Rate of handover   It is the number of handovers per unit time.

Interruption duration   It is the duration of time during a handover procedure in which a mobile is not connected to any base station.

Handoff delay   It is the distance between the point at which the handoff should occur and the point at which it does occur.

Probability of unsuccessful handoff   It is the probability that a handoff is executed while the reception conditions are inadequate.

Handoff blocking probability   It is the probability that a handover cannot be completed successfully.

The scenario of a handover is shown in Fig. 2.12(a). Here, the handover procedure is between base stations *A* and *B*, and the mobile is moving towards *B* from *A* crossing the

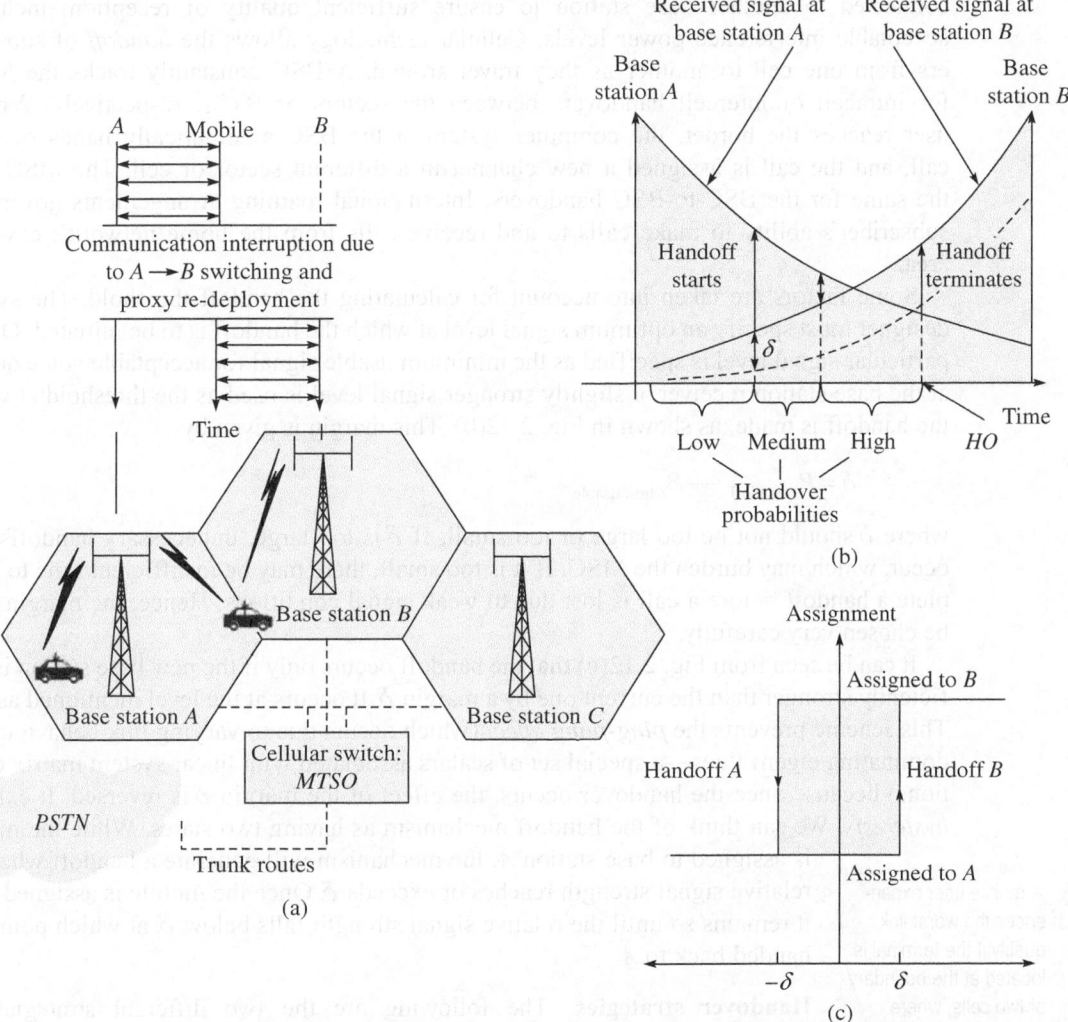

Fig. 2.12   Handoff (a) Cellular infrastructure with base stations and handover concept (b) Received signal strength variations and timing during handoff (c) Hysteresis exhibited during handoff

cell area of A. Before the handover, the lines with arrows represent the ongoing communication between base station A and the mobile. The interruption part is the time zone for the handover procedure. It does not break the conversation but switches from one channel to another. Proactive handover is made. After the handover, the communication continues between the mobile and base station B, shown again by the lines with arrows.

Each mobile uses a separate temporary radio channel to talk to the cell site. The cell site talks to many mobiles simultaneously, using one channel per mobile. As radio energy dissipates over distance, mobiles must stay near the base station to maintain communication.

When a mobile terminal moves outside the coverage area of its base station, network management is assumed to respond appropriately. The mobile should be immediately connected to another base station to ensure sufficient quality of reception, including acceptable interference power levels. Cellular technology allows the *handoff* of subscribers from one cell to another as they travel around. A BSC constantly tracks the MSUs for intracell or intercell handovers between the sectors or BTSs, respectively. When a user reaches the border, the computer system at the BSC automatically hands over the call, and the call is assigned a new channel in a different sector or cell. The MSC does the same for the BSC-to-BSC handovers. International roaming arrangements govern the subscriber's ability to make calls to and receive calls from the home network's coverage area.

Some factors are taken into account for calculating the handoff threshold. The system designer must specify an optimum signal level at which the handoff is to be initiated. Once a particular signal level is specified as the minimum usable signal for acceptable voice quality at the base station receiver, a slightly stronger signal level is used as the threshold at which the handoff is made, as shown in Fig. 2.12(b). This margin is given by

$$\delta = P_{r\,\text{handoff}} - P_{r\,\text{min usable}} \tag{2.5}$$

where δ should not be too large or too small. If δ is too large, unnecessary handoffs may occur, which may burden the MSC. If it is too small, there may be insufficient time to complete a handoff before a call is lost due to weak signal conditions. Hence, the margin must be chosen very carefully.

It can be seen from Fig. 2.12(c) that the handoff occurs only if the new base station is sufficiently stronger than the current one by a margin δ. It occurs at the level mentioned as *HO*. This scheme prevents the *ping-pong effect* (which occurs due to varying link behaviour and dominating eigenvalues—a special set of scalars associated with linear system matrix equations) because once the handover occurs, the effect of the margin δ is reversed. It exhibits *hysteresis*. We can think of the handoff mechanism as having two states. While the mobile is assigned to base station A, the mechanism will generate a handoff when the relative signal strength reaches or exceeds δ. Once the mobile is assigned to B, it remains so until the relative signal strength falls below δ, at which point it is handed back to A.

A mobile user experiences the worst link quality if the terminal is located at the boundary of two cells, where the distances to the base stations are the longest.

Handover strategies The following are the two different strategies for handover:
(a) Centralized methods (used in GSM)
(b) Decentralized methods [used in DECT (WLL)]

Handovers are mostly used for providing uninterrupted services along with improvement of the received signal power or the carrier to interference ratio (*C/I ratio*) of a particular link. It can also be used to free the spectrum resources in a cell that is saturated with traffic. In such a case, handovers become a part of the *dynamic frequency assignment* (DFA) scheme.

Handoffs can be initialized either by measurements done at the base stations or by measurements done by the mobile. Thus, there are two methods of handoff.

Mobile-assisted handoff The mobile measures the power received from the surrounding base stations and continuously reports the results of the measurements to the serving base station. A handoff is initiated when the power received from the base station of a neighbouring cell begins to exceed that from the current base station by a certain level or a certain period of time. This method is currently being used in the mobile systems.

Base station-assisted handoff In the first-generation systems, the strength measurements are made by the base stations and supervised by the MTSO. Here, the base station measures the signals from the mobiles served by it, as well as from the mobiles in the neighbouring cells and reports to the MTSO. The MTSO decides whether a handoff is necessary as well as who needs it. Here, the load of the MTSO is more compared to the mobile-assisted method, and hence handoffs are slower.

Fading effect and handoff Prediction techniques are very often used in a handover. A handoff decision is based on the expected future value of the received signal strength. Another important issue during handoffs is that signal levels change not only due to moving away from the base station (i.e., increased distance), but also due to the relative movement of nearby objects. This is because reflections from these objects also determine the signal level. This variation in the signal level from time to time is called *fading*, which is discussed in Chapter 3. Hence, handoffs may occur even when they are not needed. To avoid this situation, handoffs are based on running average rather than the instantaneous values.

Queuing of handoff request is possible because there is a finite time interval between the time at which the received signal level drops below the handoff threshold and the time at which the call may be terminated due to insufficient signal level. The queue size and the delay time depend upon the traffic condition. Thus, priority-based handover may be possible. However, queuing does not guarantee a zero probability of forced termination.

Soft Handover

We have so far described the hard handoff. However, in systems based on spread spectrum (CDMA), mobiles use the same channel frequency in every cell. Here, the term *handoff* does not mean a physical change in the assigned channel, but rather means that a different base station handles the radio communication task. The MSC simultaneously evaluates the signals received from neighbouring base stations and decides which version of the user's signal is best at any moment. This technique exploits the space diversity provided by the different physical locations of the base stations and allows the MSC to make a soft decision regarding the version of the user's signal to pass. This ability to select between the instantaneous received signals from a variety of base stations is called *soft handover*. The technique is discussed in detail in Chapter 11.

Queuing of handoff request is a method to decrease the probability of forced termination of a mobile call due to lack of available channels.

Example 2.3 A mobile is moving along a straight line from base station *BS*1 to base station *BS*2, as shown in Fig. 2.13. The distance between the base stations is 2200 m. For simplicity, small-scale fading is neglected. The received power $P_{r,i}(d_i)$ in dBm at base station i (here i = 1, 2 for the two base stations) from the MS is modelled as a function of distance d_0 on the reverse link. All distances are in metres.

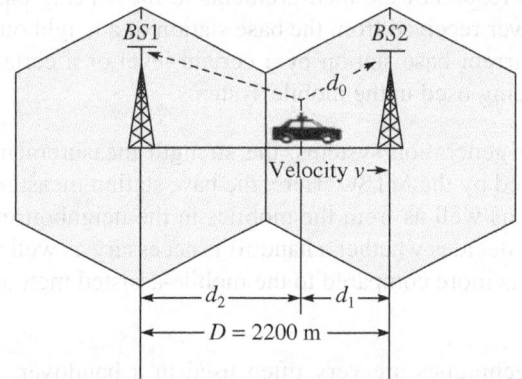

Fig. 2.13 Two base station scenario for handoff calculations

$$P_{r,i}(d_i) = P_0 - 10n\log_{10}\left(\frac{d_i}{d_0}\right)\text{dBm}$$

Given that P_0 = received power at base station *BS*2 at distance d_0 from mobile antenna = 0 dBm at the time of handoff, d_0 = 1 m, path loss exponent = 3, and minimum usable $P_{r\min usable}$ = −87 dBm. The threshold level used by the switch for handoff initialization is $P_{r handoff}$. The mobile is currently connected to *BS*1 and is moving towards *BS*2, allowing handoff procedure, and there is no call loss during the handoff. In addition, the antenna heights are negligible compared to the distance between the mobile and the base station. The handoff time is 4.5 s and the velocity of the MS is 100 km/h. Determine the minimum required margin for handoff δ and the effect of the margin on the performance of the cellular systems.

Solution We have v = 100 km/h = 100 × 1000/3600 = 1000/36 m/s

Distance at which the handoff starts

$$d_1 = \text{Velocity/handoff time}$$
$$= (1000 \times 4.5)/36 = 125\,\text{m}$$

We have $P_{r,i}(d_i) = P_0 - 10n\log_{10}\left(\dfrac{d_i}{d_0}\right)$ dBm.

Here, $P_0 = 0\,\text{dBm}, \dfrac{d_i}{d_0} = 125/1 = 125$, and path loss exponent $n = 3$.

Therefore, the second term in the received power equation will be 62.9 dBm.

$$P_{r,2}(d_2) = P_{r\,handoff} = 0 - 62.9 = -62.9\,\text{dBm and}$$
$$P_{r\min usable} = -87\,\text{dBm}$$

Now, $\delta = P_{r\,handoff} - P_{r\min usable}$
$$= -62.9\,\text{dBm} - (-87\text{dBm}) = 24.1\,\text{dBm}$$

For minimum usable power, the distance from *BS*2 is

$$87/3 \times 10 = \log_{10}\left(\frac{d_i}{d_0}\right) \Rightarrow \frac{d_i}{d_0} = 794.3\,\text{m}$$

The handover procedure will have a margin of 669.3 m, which is large compared to the total distance of 2200 m.

Comments: The received power when the handoff starts is stronger than the minimum usable power; thus, successful handover can be completed. The margin should not be too large or too small. If it is too large, unnecessary handoff burdens the MSC, and if it is too small, there may be insufficient time to complete a handoff before a call is lost due to weak signal condition. A larger margin is beneficial when queuing is used due to unavailability of a channel at *BS*2.

2.6 CHANNEL ASSIGNMENT

We have already discussed the concepts of cell, cluster, and frequency reuse in this chapter. A set of frequencies is to be allocated to each cell, and hence, a cell has that many channels to assign as per the demand. In heavy traffic conditions, all active subscribers in a cell may use up all channels, and there may be a demand for extra channels. When channel reassignment is to be made in dynamic conditions such as a handover, strategies are to be developed such that even if shortage of channels occurs due to heavy traffic, a forceful call drop

Dynamic channel allocation strategy gives the best traffic handling procedure but at the cost of fast processing, huge storage, and computational load, whereas fixed allocation strategy exhibits the worst traffic handling in peak traffic hours.

may not occur. The aim of channel assignment is to efficiently use the available radio spectrum and to provide the best possible service, so that a subscriber does not lose any call due to unavailability of channels. The different ways of channel assignment are fixed channel assignment, borrowing strategy, and dynamic channel assignment.

Fixed Channel Assignment

In the fixed channel assignment method, each cell is allocated a predetermined set of voice channels. Only the unused channels in a particular cell can serve any call attempt within that cell. If all the channels are occupied, then the call will be blocked and the subscribers will not get any service.

Example 2.4 A total of 36 MHz of bandwidth is allocated to a particular frequency division duplex cellular telephone system, which uses two 25 kHz simplex channels to provide full duplex voice and control channels. Compute the number of channels available per cell if the system uses (a) 7-cell reuse and (b) 12-cell reuse.

Solution The total bandwidth given is 36 MHz.

Channel bandwidth = 25 kHz × 2 simplex channels

$$= 50 \text{ kHz per duplex channel}$$

Total available channels within 36 MHz range = 36 MHz /50 kHz = 720 channels

(a) If $N = 7$, the total number of channels available per cell is 720/7 = 102.85 (rounded to 102 channels).

(b) If $N = 12$, the total number of channels available per cell is 720/12 = 60.

Borrowing Strategy

In the borrowing strategy method, a cell is allowed to borrow a channel from a neighbouring cell, if all of its own channels are occupied. The MSC supervises the borrowing procedure and ensures that the borrowing of the channel does not disrupt or interfere with any of the calls in progress in the donor cell.

Dynamic Channel Assignment

In the dynamic channel assignment method, voice channels are not allocated permanently in any of the cell. When a call request is made, the serving base station requests a channel from the MSC, which then allocates a channel to the requested cell. While allocating the channel, the switch follows an algorithm that takes into account factors such as the traffic statistics and the likelihood of future blocking within the cell, frequencies already used, reuse distance of the channel, and cost functions. The MSC allocates a given frequency only if that frequency is not currently used in that cell or any other cell within the minimum reuse distance to avoid *co-channel interference*.

One of the advantages of the dynamic channel assignment strategy is that the likelihood of call blocking reduces. This increases the trunking capacity of the system because all the channels that are available to the service provider are accessible to all the cells. Thus, it increases *channel utilization*.

The disadvantage is that the MSC must be fast and capable of collecting real-time data on channel occupancy, traffic distribution, and radio signal strength indications of all channels on a continuous basis.

2.7 CELLULAR INTERFERENCES

Two major types of cellular interferences are *co-channel interference* and *adjacent channel interference*. Co-channel interference cannot be eliminated by increasing the carrier power of a transmitter. This is because an increase in the carrier transmit power increases the interference with the neighbouring co-channel cells. Frequency reuse distance calculation is critical for managing interference.

Signal-to-interference Ratio for Co-channel Interference

The co-channel cells surrounding a particular cell exist in a circular (or rather hexagonal) pattern. There will be maximum interference from the nearest circle of cells, which are also called *first-tier cells* and are always six in number (Fig. 2.14). Thus, major interference is from the first tier. Depending upon the cluster size, the size of the tier will also change. However, for a larger cluster size, there is very less or negligible co-channel interference from the cells of the second and higher number tiers. In general, if i_{int} is the total number of co-channel cells that create interference for the cell in the centre, then the signal-to-interference ratio (SIR) for a mobile receiver that monitors a forward channel can be expressed by

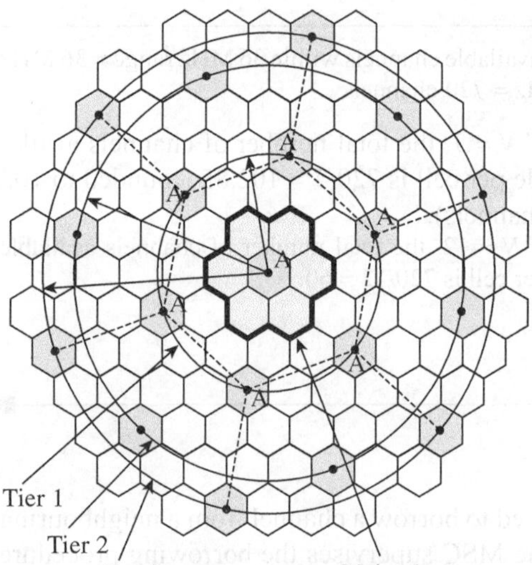

$$\frac{S}{I} = \frac{S}{\sum\limits_{i=1}^{i_{int}} I_i} \tag{2.6}$$

Tier 1
Tier 2
Tier 3   Main cluster

Fig. 2.14 First, second, and third tier co-channel interfering cells for omnidirectional transmissions for cluster size $N = 7$

where S is the desired signal power from the desired base station and I_i is the interference power of the base station of the ith interfering co-channel cell (neglecting noise). If the signal levels of the co-channel cells are known, the SIR for the forward link can be found using this equation.

For calculating the received power after propagation loss, Eq. (3.14) given in Chapter 3 can be used, in which instead of the square of the distance ratio, the nth power is considered, where n is the path loss exponent. In mobile systems, a signal will rarely reach the receiver by the *line-of-sight path*. There will be a scattering and absorption loss due to obstacles. Hence, the empirical value of n ranges from two to five. A typical value assumed for cellular systems is $n = 4$ (mostly assumed). Let us consider the first tier of the six interfering cells. Let all the interfering base stations be equidistant from the desired base station and this distance be equal to the distance D between the cell centres (this assumption is valid because $R \ll D$) and let the path loss exponent be the same throughout the coverage area. Then, Eq. (2.6) can be written in terms of the radius of the cell R, the distance of the interfering base stations to the desired mobile receiver D_i, D/R ratio, and cluster size N.

> *D/R* ratio is the co-channel reuse ratio that helps in maintaining the *signal-to-interference* level to reduce or eliminate co-channel interference.

$$\frac{S}{I} = \frac{R^{-n}}{\sum\limits_{i=1}^{6} (D_i)^{-n}} \tag{2.7}$$

Note: The path loss exponent n represents the link quality for various channel environments. Some typical values are as follows: Free space: $n = 2$; Urban cellular: $n = 2.7$–3.5; Shadowed urban cellular: $n = 3$–5; Obstructed in building: $n = 4$–6; Obstructed in factory: $n = 2$–3; In building line of sight: $n = 1.6$–1.8

We can write $\quad S = k \times R^{-n}$

where k is the proportionality constant, which depends on the power radiated by the base station at the centre. I represents the power radiated by the other base stations in the first tier, which are six in number.

Hence, $\quad I = 6k \times D^{-n}$

Using Eqs (2.4) and (2.7),

$$\frac{S}{I} = \frac{(D/R)^n}{6} = \frac{\left(\sqrt{3N}\right)^n}{6} \quad \text{or} \quad \frac{1}{6}(3N)^{n/2} \tag{2.8}$$

From Eq. (2.8), it is clear that N should be designed critically for the desired SIR. If the SIR is to be at least 18 dB or more, the cluster size must be seven, assuming the path loss exponent n is four. It should be noted that these equations are derived assuming hexagonal geometry. Hence, these results may differ in practice.

The target levels of the SIR for various wireless technologies are as follows:

- *AMPS*: 17–18 dB
- *GSM*: 12–13 dB
- *CDMA*: No need for well-maintained SIR (due to $N = 1$ basis of the system and the fact that it is a noise-based system)

Example 2.5 The desired SIR for a cell-based system is 15 dB. Considering only first-tier interfering cells, find the frequency reuse factor for maximum capacity. Assume path loss exponent $n = 3$.

Solution We have to find the minimum value of N that satisfies the required SIR. We shall do it by trial and error. Let us first assume $N = 7$.

Thus $\quad \left(\dfrac{S}{I}\right) = \dfrac{\left(\sqrt{3N}\right)^n}{6} = (1/6) \times (3 \times 7)^{3/2} = 16.04$

or $\quad \left(\dfrac{S}{I}\right) \text{dB} = 10\log(16.04) = 12.05\,\text{dB}$

Since this value is less than the desired value, we will next calculate it with $N = 12$.

We can show that for $N = 12$, SIR = 15.56 dB. Hence, $N = 12$ can be used.

Signal-to-interference Ratio for Adjacent Channel Interference

If there is interference from the signals that are adjacent in frequency to the desired signal, it is called *adjacent channel interference*. It results from not having enough separation between the adjacent channels or having an imperfect receiver filter that allows nearby frequencies to enter into the passband (as shown in Fig. 2.15b). This problem becomes serious when an adjacent or interfering channel user transmits in a range that is very close to a subscriber's receiver, while the receiver attempts to receive a base station on the desired channel. The scenario is shown in Fig. 2.15(a). The base station in the Figure is in communication with mobile B, and mobile A is the interfering mobile. Mobile B is at a distance that is 20 times the distance of mobile A from the base station. Assuming that both mobiles transmit the same power, the SIR picked up by the base station receiver will approximately be

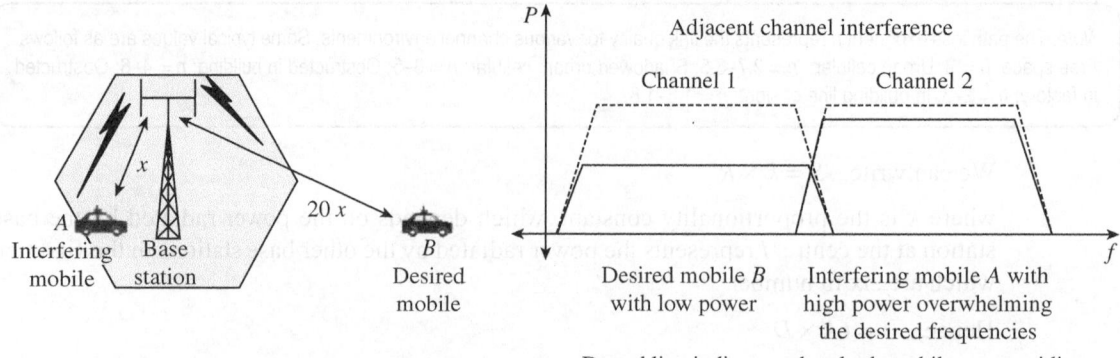

Fig. 2.15 Adjacent channel interference (a) Interference created by near–far problem (b) Interference shown on frequency axis with relative powers of both channels for mobiles A and B

> Adjacent channel interference can be minimized through careful filtering at IF stages and careful channel assignments.

$$\left(\frac{S}{I}\right) = (20)^{-n} \tag{2.9}$$

If the path loss exponent n is assumed to be four, the SIR will become −52 dB.

In other words, *near–far effect* occurs when a mobile near a base station transmits on a channel that is close to the one being used by a weak mobile at a far distance. The base station may have difficulty in discriminating the desired mobile user from the nearer adjacent channel mobile user. If the frequency reuse factor is large or N is small, the separation between the adjacent channels may not be sufficient to keep the adjacent channel interference level within tolerable limits.

Since each cell is given only a fraction of the available channels, a cell need not be assigned channels that are all adjacent in frequency. Adjacent channel interference can be reduced considerably by maintaining a large frequency separation between each channel in a given cell. Sequential allocation of channels in a cell should be avoided; instead, distant channels can form a group of channels per cell.

2.8 SECTORIZATION

More co-channel cells were added to the wireless system over time. Hence, it became necessary to develop a means to increase the system (traffic-handling) capacity without constantly having to split cells, especially because cell splitting is an expensive process. The industry developed a way to migrate from an omniantenna configuration at cell base stations and started sectorizing base stations to obtain more capacity from each base station deployment. Sectorized base stations are created by subdividing an omnicell into sectors that are covered using directional antennas mounted in the same base station location (i.e., the same tower or rooftop). A directional antenna is shown in Fig. 2.8.

Operationally, each sector is treated as a different cell, the range of which is greater than an omnicell. Directional antennas always produce more gain than omniantennas. All subcell directional antennas supporting each sector are located at the same base station. Similarly, all radio equipment for each subcell, or sector, is housed in the same base station. To sectorize a cell, a horizontal, equilateral platform resembling a triangle is deployed on a tower. Each side of the platform is called a *face*. Three, four, or six directional antennas are installed

If the BTSs are located at the centre of the cell, then it is called a centre-excited cell, whereas if the BTSs are on three of the six cell vertices, then it is called an edge-excited cell.

on the platform, depending on the number of sectors. The directional antennas propagate different frequencies or channels assigned within each respective face. Sectorization facilitates wireless engineering and operations in the following ways:

- It minimizes or eliminates co-channel interference.
- It optimizes the frequency reuse plan. This is facilitated through another concept known as the *front-to-back ratio*.
- At a minimum, it triples the capacity of any given coverage area when compared to the capacity offered by deploying omniantennas. Most wireless carriers in the US usually deploy three or, in some cases, four sectors per cell site. In some parts of the world, six-sectored base stations are used.
- In cases where three-sectored base stations are deployed, the directional antennas mounted in each sector will have a beamwidth of 120°. In this scenario, from a graphical viewpoint, sectorization takes a circle (representing an omni base station) and converts it into a three-section pie chart.
- In cases where four-sectored sites are deployed, directional antennas having a beamwidth of 90° are used.
- In cases where six-sectored base stations are deployed, directional antennas having a beamwidth of 60° are used.

The objective behind implementing sectorized base stations is to support 360° coverage from a single location. The number of sectors will dictate the beamwidth of the directional antennas used within each sector. Typically, directional antennas have a beamwidth of 65°–85°. Each sector has its own assignment of radio channels and its own channel set or sets. Each sector also has its own control channels and will handover calls to its adjacent sectors that are housed on the same tower, rooftop, water tank, and so on. Figure 2.16 shows a typical illustration of sectorization in a cell.

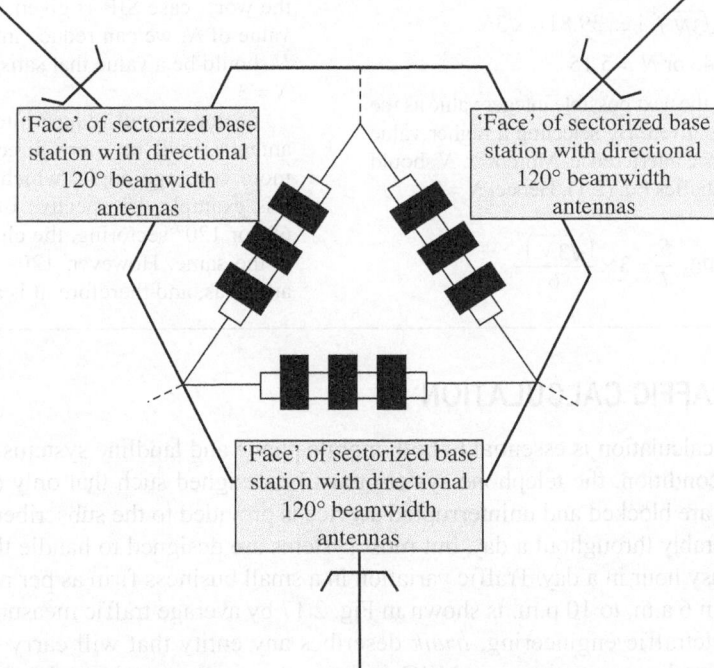

Fig. 2.16 Sectorization in a cell with 120° beamwidth directional antennas

Front-to-back ratio This is defined as the ratio of the forward gain of a given cell's sector (based on the placement of the directional antenna) to the gain 180° to the rear of the co-channel sectors. As directional antennas are used in sectorized cells, RF engineers must take into account the front-to-back ratio to reduce the possibility of co-channel interference caused by the back lobe of the antenna propagation. The back lobe is an RF that is projected 180° to the rear of the antenna's front. Antennas with high front-to-back ratios are sought by engineers and purchasing managers to reduce co-channel interference in a sectorized environment. Manufacturers assign the front-to-back ratio to each directional antenna, based on the electrical, frequency, and gain characteristics of that particular antenna.

Example 2.6 A cellular service provider decides to use a TDMA scheme that can tolerate an SIR of 16 dB in the worst case. Find the optimum value of cluster size N in case of (a) an omnidirectional antenna, (b) 120° sectoring, and (c) 60° sectoring. What will be the advantage of sectoring? Which sectoring will be better, 60° or 120°? Assume path loss exponent $n = 4$.

Solution We have $\left(\dfrac{S}{I}\right) dB = 16 \, dB = 10 \log\left(\dfrac{S}{I}\right)$

So, $\left(\dfrac{S}{I}\right) = 39.81, n = 4$

(a) For omnidirectional antenna, $\dfrac{S}{I} = \dfrac{\left(\sqrt{3N}\right)^n}{6}$

or $39.81 \times 6 = \left(\sqrt{3N}\right)^4$ i.e., $39.81 = \sqrt{3N}$

∴ $3N = 15.45$ or $N = 5.15$

We have to select the next possible integer value as the worst-case SIR is given. By selecting a higher value of N, we can reduce interference. Moreover, N should be a value that satisfies Eq. (2.1). Hence, $N = 7$.

(b) For 120° sectoring, $\dfrac{S}{I} = 3 \times \dfrac{\left(\sqrt{3N}\right)^n}{6}$

or $39.81 \times 2 = \left(\sqrt{3N}\right)^4$ or $2.98 = \sqrt{3N}$

or $3N = 8.9$ or $N = 2.967$

We have to select the next possible integer value as the worst-case SIR is given. By selecting a higher value of N, we can reduce interference. Moreover, N should be a value that satisfies Eq. (2.1). Hence, $N = 3$.

(c) For 60° sectoring, $\dfrac{S}{I} = 6 \times \dfrac{\left(\sqrt{3N}\right)^n}{6}$

or $39.81 = \left(\sqrt{3N}\right)^4$ or $2.51 = \left(\sqrt{3N}\right)$

or $3N = 6.3$ or $N = 2.1$

We have to select the next possible integer value as the worst-case SIR is given. By selecting a higher value of N, we can reduce interference. Moreover, N should be a value that satisfies Eq. (2.1). Hence, $N = 3$.

Sectoring will reduce interference because the antennas will be more directional and will have more coverage angle, which reduces the area. In this example, irrespective of whether we opt for 60° or 120° sectoring, the cluster size requirement is the same. However, 120° sectoring saves three antennas, and therefore, it is a better choice.

2.9 MOBILE TRAFFIC CALCULATION

Traffic calculation is essential to both mobile phone and landline systems. On the basis of the traffic condition, the telephone system must be designed such that only a minimum number of calls are blocked and uninterrupted service is provided to the subscribers. The traffic varies considerably throughout a day, but most systems are designed to handle the traffic during the peak busy hour in a day. Traffic variation in a small business firm as per normal activities in a day from 6 a.m. to 10 p.m. is shown in Fig. 2.17 by average traffic measured every hour.

In teletraffic engineering, *trunk* describes any entity that will carry one call. The number of trunks connecting one MSC with another is the number of voice pairs used in the

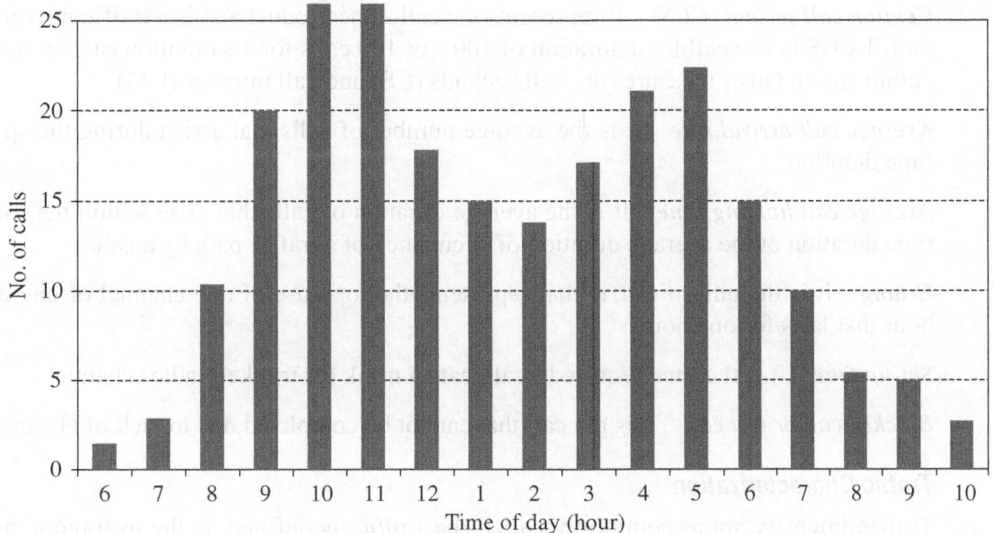

Fig. 2.17 Typical hour-by-hour traffic variation

connection; hence, it is important to determine the number of trunks required between the MSCs. Trunking efficiency refers to the decrease in call blocking and increase in coverage without call drops, at the cost of management overhead.

The following is the list of some definitions important for traffic engineering.

Call blocking It is the non-availability of any trunk or call propagation path.

Busy hour or peak busy hour It is the 60-min interval in which the traffic or the number of call attempts is the highest in a day. It usually varies from day to day or over a number of days.

Time-consistent busy hour It is the one hour period starting at the same time each day for which the average traffic volume or the number of call attempts is the highest over the days under consideration.

Call completion rate It is the ratio of the number of successful calls to the number of call attempts. It can be represented in percentage.

Busy hour call attempts It is the number of call attempts in the busy hour and it decides the network capacity.

Average busy hour calls (completed) = Busy hour call attempts × call completion rate

> Similar to conventional trunk lines between telephone exchanges, cellular systems have radio trunks between MSCs to carry the calls.

Busy hour calling rate It is the average number of calls originated by a subscriber during the busy hour or the call intensity per traffic path during the busy hour.

Day to busy hour traffic ratio It is the ratio of busy hour calling rate to the average calling rate for the day. It indicates how much of the day's total traffic is carried during the busy hour.

Centum call second (CCS)   It represents the call–time product and is a traffic measurement unit. 1 CCS is one call for a duration of 100 s or 100 calls for 1 s duration each or any other combination. Other measures are call seconds (CS) and call minutes (CM).

Average call arrival rate   It is the average number of calls that arrive during the specified time duration.

Average call holding time   It is the average duration of calls that arise within the specified time duration or the average duration of occupancy of a traffic path by a call.

Erlang   It is the unit of traffic that represents the total use of one channel or one call per hour that lasts for one hour.

Set-up time   It is the time required to allocate a trunk (or trunked radio) channel.

Blocked call or lost call   It is the call that cannot be completed due to lack of channels.

Traffic Characterization

Traffic intensity, more commonly called the *traffic*, is defined as the average number of calls in progress. Although this is a dimensionless quantity, the unit of traffic has been given a name—erlang (abbreviated as E). It is named after A.K. Erlang, a Danish mathematician, who published in 1914 and 1917 the first basic results on the number of subscribers that can be served with a given number of channels at a required *quality of service* (QoS), known as the blocking probability. In a group of channels, the average number of calls in progress depends on both the number of calls that arrive and their duration. The duration of a call is often called its *holding time*. As per the definition given earlier, one erlang of traffic can result from one trunk being busy all the time (here, one hour reference duration), from each of the two trunks being busy for half of the time, from each of three trunks being busy for one-third of the time, and so on. Averaged over time, one erlang of telephone traffic occupies exactly one channel. However, the arrival and closure of telephone calls are random processes. As time elapses, one erlang of traffic may occupy zero, one, or multiple channels. It should be noted that the definition of the unit erlang does not say anything about how the traffic behaves statistically about this average. Thus, from the definition of erlang, the traffic carried by a group of trunks is given by

$$A = \frac{ch}{T} \qquad (2.10)$$

where A is the traffic in erlangs, c is the average number of call arrivals during time T, and h is the average call holding time.

From Eq. (2.10), if $T = h$ then $A = c$. Moreover, 1 erlang = 36 CCS = 3600 CS = 60 CM. In short, one erlang of traffic can be generated, for instance, by either of the following:

- One call of infinite duration
- A random process of many calls arriving and closing such that the average number of active calls is one

From this discussion, we can say that $A \leq 1$ for a single trunk or channel.

Example 2.7 The average duration of call for a user is 2 min. If the average number of requests for call per hour is six, find the traffic intensity in erlangs. If the total number of users is 100, find the total traffic.

Solution Let T = average duration of call
R = average number of call requests per unit time
Then, the traffic intensity of a single user is

$$A = T \times R = 2 \times 6/60 = 0.2 \, \text{erlangs}$$

For the total number of users, total traffic intensity = $0.2 \times 100 = 20$ erlangs

This can be interpreted as follows: If 20 channels can handle the total traffic in the worst case, it means there is one call per hour in a fully occupied condition. If more than 20 channels are used, the call will never be blocked.

■

2.9.1 Call Handling and Grade of Service

In a telephone system, a finite number of N channels are available. New calls are assigned a channel until all the channels are full. When all the channels become occupied, the system cannot accept further calls. This state is known as *congestion*. The arrival of a new call can then be handled by either of the following two methods:

(a) Blocked (for a lost call system based on circuit switching)
(b) Queued (for a delayed system based on message or packet switching)

The result of congestion is that the traffic actually carried by the system is less than the traffic offered to it. We may, therefore, write

$$\text{Traffic carried} = \text{Traffic offered} - \text{Traffic lost} \qquad (2.11)$$

The proportion of calls lost or delayed due to congestion is used to measure the service quality, called the *grade of service* (GOS). The GOS B for a lost call system may be defined as

$$B = \text{Number of calls lost/Number of calls offered} \qquad (2.12a)$$

or $\quad B = \text{Traffic lost/Traffic offered}$

$\quad\quad = \text{Probability that a call will be lost due to congestion}$ $\qquad (2.12b)$

The GOS may also be defined as the proportion of time for which congestion exists or the probability of congestion. Thus, if a traffic of A erlangs is offered to a group of trunks or channels having a GOS B, the traffic lost is AB and the traffic carried is $A(1 - B)$ erlangs. Larger the GOS value, worse will be the service provided. The GOS is normally specified for the traffic at the busy hour, as it is the worst-case condition and checks the maximum capacity. At other times, it is always better.

The traffic in erlangs is equal to the mean number of calls arriving during a period equal to the mean duration of the call.

The way in which channels are grouped can substantially alter the number of users handled by a trunked system. Sectoring changes the trunking efficiency. The GOS based on the probability of blocked calls gives the Erlang B formula, whereas the GOS based on the probability of delaying calls up to a certain time gives the Erlang C formula. These are discussed in Sections 2.9.3 and 2.9.4. Erlang tables are given in appendix F.

Example 2.8 During a busy hour, 1300 calls were offered to a group of trunks and 6 calls were lost. The average call duration is 2.5 min. Find the (a) traffic offered, (b) traffic carried, (c) traffic lost, (d) GOS, and (e) total duration of period of congestion.

Solution

(a) $A = \dfrac{ch}{T} = 1300 \times 2.5/60 = 54.17\,\text{E} =$ offered traffic

(b) Six calls are lost. Hence

$1300 - 6 = 1294$ calls are successful

or $1294 \times 2.5/60 = 53.917\,\text{E} =$ carried traffic

(c) $6 \times 2.5/60 = 0.25\,\text{E} =$ lost traffic

(d) $B = 6/1300 = 0.0046$

(e) The total duration of period of congestion can be found from the GOS. The entire traffic is measured on a 1 h basis. So, the fraction of the 1 h under congestion is found as follows:

$0.0046 \times 3600 = 16.56\,\text{s}$

2.9.2 Mathematical Modelling of Traffic

A simple mathematical model is based on the following assumptions:

- Pure chance traffic
- Statistical equilibrium

> *Note*: Mobile traffic calculations are rightly possible if and only if call arrivals follow a Poisson distribution, holding time follows a negative exponential distribution, and busy hour is considered with the statistical equilibrium condition and constant channel state probabilities.

Pure Chance Traffic

If the traffic is pure chance, all call arrivals and call terminations are independent random events. Actually, calls are made by individual users but not at random. However, the total traffic generated by a large number of users is observed to behave as if the calls were generated at random. If call arrivals are independent random events, their occurrence is not affected by previous calls. The traffic is, therefore, sometimes called *memoryless traffic*. It also implies that the number of sources generating calls is very large. If the number of sources is small and many of them are already busy, then the rate at which new calls can be generated is less than what it would be if all the sources were free. Here, call arrival or call attempts, trunks, and so on are counted with integer numbers, and hence, discrete states-based or chained process-based modelling can be applied for mathematical calculations.

A commonly used model for random, mutually independent message (here, call) arrivals is the *Poisson process*. The Poisson distribution can be obtained by evaluating the following assumptions for arrivals during an infinitesimal short period of time δt:

- The probability that one arrival occurs between t and $t + \delta t$ is λt, where λ is a constant, independent of time t and independent of arrivals in earlier intervals. Here, λ is the arrival rate, which is expressed as the average number of arrivals during a unit of time.

> Trunking efficiency is a measure of the number of users that can be offered a particular GOS with a particular configuration of fixed channels.

- The number of arrivals in non-overlapping intervals is statistically independent.
- The probability of two or more arrivals happening during δt is negligible compared to the probability of zero or one arrival.
- The distribution of the number of arrivals in a time interval of t to $t + T$ is independent of the starting time t. T is the interval between call arrivals or the interval between two random events.

The Poisson process is a stochastic process that counts the number of events and the time that these events occur in a given time interval.

The probability of the number of call arrivals in a given time has a Poisson distribution given by

$$P(x) = \frac{\mu^x}{x!} e^{-\mu} \tag{2.13}$$

where x is the number of call arrivals in time T and μ is the mean number of call arrivals in time T. Here, μ is λt. It can be shown that these intervals have a negative exponential distribution.

$$P(T \geq t) = e^{-t/T_{\text{mean}}} \tag{2.14}$$

where T_{mean} is the mean interval between call arrivals, which is equal to h, the holding time. Now, call arrival and call termination are also independent random events. Then, T can also be the interval between these two random events, which also has a negative exponential distribution as given in Eq. (2.14), because most of the calls are short and only few are long. The probability of no arrivals during the period of duration T is

$$P(T) = e^{-\lambda T} \tag{2.15}$$

where $P(T)$ is the probability density function of the duration between two arrivals.

Statistical Equilibrium

According to the statistical equilibrium assumption, the generation of traffic is a stationary random process; that is, the probabilities do not change during the period being considered. Consequently, the mean number of calls in progress remains constant. This condition is satisfied during the busy hour, and it must be remembered that it is the busy hour GOS that needs to be determined. Statistical equilibrium is not obtained immediately before the busy hour, when the calling rate is increasing, or at the end of the busy hour, when the calling rate is falling.

Example 2.9 In a company, on average, one call arrives every 5 min. During a period of 10 min, what is the probability that (a) no calls arrive, (b) one call arrives, (c) two calls arrive, and (d) more than two calls arrive? If the average call duration is 2 min and a call has lasted for 4 min, what is the probability that the call will last for at least another 4 min?

Solution We have $P(x) = \dfrac{\mu^x}{x!} e^{-\mu}$

where $x =$ number of call arrivals in time T

$\mu =$ mean number of call arrivals in time $T = 2$ because during the period of observation the maximum possibility of calls is only two as a call arises every 5 min

(a) The probability that no calls arrive in the duration of 10 min is

$$P(0) = \frac{2^0}{0!} e^{-2} = e^{-2} = 0.135$$

(b) The probability that one call arrives in the duration of 10 min is

$$P(1) = \frac{2^1}{1!} e^{-2} = 0.27$$

(c) The probability that two calls arrive is

$$P(2) = \frac{2^2}{2!} e^{-2} = 0.27$$

(d) The probability that more than two calls arrive is

$$P(> 2) = 1 - P(0) - P(1) - P(2)$$
$$= 1 - 0.135 - 0.27 - 0.27 = 0.325$$

Now, the probability of a call lasting for 8 min when the average call duration is 2 min is

$$P(T \geq t) = e^{-t/T_{\text{mean}}}$$

Here, T_{mean} is the average call duration $= 2$ min

So, $P(8 \geq 4) = e^{-4/2} = e^{-2} = 0.135$

Note: Traffic measurement or calculation errors may occur due to improper traffic modelling or computer processing, database errors, interpretation errors, or statistical errors.

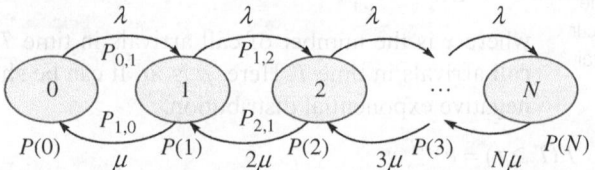

Fig. 2.18 Markov model for a number of occupied channels in a network wit *N* channels

Markov model New calls arrive according to the *Poisson* process at a rate of λ calls per unit of time. Calls have an exponential (memoryless) duration with mean $1/\mu$. The number of active calls is a Markov process. Such a process is called a simple Markov chain and is represented in Fig. 2.18. Here, $P(1)$, $P(2)$, ..., $P(N)$ are the *state probabilities*. $P_{0,1}$, $P_{1,2}$, and so on are the probabilities of a state increment. Similarly, there are also decrement probabilities. These are the *conditional probabilities*, also called the *transition probabilities*. If there is statistical equilibrium, these probabilities do not change and the process is said to be a regular Markov chain.

2.9.3 Erlang B Formula

We can find from the Poisson distribution that

$$P(x) = \frac{A^x}{x!}e^{-A}$$

(2.16)

where $P(0)=e^{-A}$

In Eq. (2.16), $P(x)$ is the probability of x calls in progress. Here, x can have any value between zero and infinity and the sum of their probabilities must be unity. Thus, if call arrivals have a Poisson distribution, so does the number of calls in progress. This requires an infinite number of trunks to carry the calls. If the number of trunks available is finite, then some calls can be lost or delayed and the distribution is no longer Poissonian. The distribution that then occurs is derived using Erlang's formula.

Erlang determined the GOS (i.e., the loss probability) of a lost call system having N trunks when the offered traffic is A. The solution was obtained on the basis of the following assumptions.:

- Pure chance traffic
- Statistical equilibrium
- Full availability
- Loss of calls encountering congestion

The probability of congestion, or of a lost call, or the GOS B for the full availability group of N trunks and the offered traffic A erlangs is given by

Erlang B formula is for a lost call system (circuit switched), whereas Erlang C formula is for a delayed system with the provision of queuing (packet switched).

$$B = \frac{A^N/N!}{\displaystyle\sum_{k=0}^{N}A^k/k!}$$

(2.17)

This is also called the Erlang B formula.

Example 2.10 How many users can be supported for 1% blocking probability if the number of channels is 20? Assume that each user generates 0.1 erlangs of traffic. Find the peak hour traffic with the average duration of call being 2 min. The offered traffic intensities for a number of channels are as follows:

Number of channels	A
2	0.153
4	0.869
5	1.36
20	12

Solution When the number of available channels is 20, $A = 12$.

Each user generates 0.1 erlangs of traffic.

So, the number of users supported = 12/0.1 = 120

Peak hour traffic in terms of calls = number of calls in peak hour

When the traffic load is 1 erlang, the number of calls in 1 h is T/h.

Thus, the number of total calls = $A \times T/h = 12 \times 60/2$ = 360 calls in 1 h

2.9.4 Erlang C Formula

In an Erlang C telephone system, N channels are available. New calls are assigned a channel until all channels are full. When all the channels are occupied, a new call is queued until it can be served. This is in contrast to an Erlang B system in which new calls are blocked.

The assumptions made here are the same as in an Erlang B system, except the fourth one, where the calls that encounter congestion enter a queue and are stored there until a server becomes free. Here, the second assumption, that is, statistical equilibrium, implies that $A \leq N$. If $A \geq N$, calls are entering the system at a greater rate than they leave. As a result, the length of the queue will continually increase towards infinity. This is not statistical equilibrium. The probability of delay formula or Erlang's delay formula can be given as follows:

$$P_D = \frac{A^N}{A^N + N!\left(1 - \dfrac{A}{N}\right)\displaystyle\sum_{k=0}^{N-1}\dfrac{A^k}{k!}} \tag{2.18}$$

This is also called the Erlang C formula.

2.10 SPECTRUM EFFICIENCY OF CELLULAR SYSTEMS

In most cellular systems, each base station can carry more than one telephone call in its cell. Let

k = number of channels per cell
N = cluster size
$W_{channel}$ = total bandwidth for the cellular net
W_{signal} = occupied bandwidth per channel

Then $\quad W_{channel} = kNW_{signal}$ $\tag{2.19}$

The spectrum efficiency S_E of a cellular net can be defined as the carried traffic per cell A_c, expressed in erlangs, divided by the bandwidth of the total system $W_{channel}$ and the area of the cell S_u. So

$$S_E = \frac{A_c}{kW_{signal}NS_u} \tag{2.20}$$

NUMBER OF CUSTOMERS IN A CELLULAR SYSTEM

When we design a system, the traffic condition in the area during a busy hour is one of the parameters that help to determine both the size of different cells and the number of channels therein. This is because the channels are divided among the cells of a cluster. Users share the channels on a demand assignment basis. More the traffic, more will be the reuse of the channels.

The maximum number of calls per hour per cell is driven by the traffic conditions at each particular cell. After the maximum number of frequency channels per cell has been allocated in each cell, the maximum number of calls per hour can be taken care of in each cell. Add the maximum number of calls per hour in each cell (Q_i) over all the cells of a cluster to get the total number of calls per hour using the set of frequencies. If we assume that 65 per cent telephones are active during the busy hour, then the number of customers in a cluster can be estimated by dividing the total number of calls per hour per cluster by 0.65.

> Spectrum efficiency of a cellular system decreases with larger cluster sizes.

Here, A_c is mostly computed from the Erlang B formula, with A_c equal to the attempted traffic multiplied by the probability of success (= 1 − blocking probability). Spectrum efficiency is expressed in erlang/MHz/km^2.

System performance, for instance, expressed in terms of the outage probability or the bit error rate experienced by the user, improves with increasing reuse distance, and so it improves with cluster size. On the contrary, traffic handling improves with decreasing cluster size. Hence, achievement of high system performance and efficient use of the radio spectrum are conflicting objectives for a network designer.

2.11 LOCATION MANAGEMENT

The study of location management aims at tracking a subscriber (with an active mobile unit) on move and at reducing the overhead incurred in locating that subscriber in a cellular environment, though he/she moves continuously from cell to cell. Actually, mobility management in cellular systems consists of two components: handover management (which was discussed earlier) and location management. Location management deals with two issues: location update and paging/call delivery. The objective is to locate mobile terminals with the main purpose of paging to deliver incoming calls to the corresponding base station at a reasonable cost. Location updates are periodically performed; the mobile notifies the network of its new access point and the network database stores the new position or registration area.

In literature, location update schemes are classified into two main groups:

Static or global scheme   Location update is triggered based on the topology of the network.

Dynamic or local scheme   A mobile sends a location update message according to the time elapsed (time-based method), the number of cells visited (movement-based method), or the distance in terms of cells travelled (distance-based method).

In the movement-based method, a mobile keeps count of the number of cells visited. A location update is performed when this count exceeds a predetermined threshold. Similarly in the time-based methods, the total time of visiting a cell is recorded. In distance based methods how many total cells are travelled, with how much distance is recorded.

The following are a few terms related to location management:

Centre cell   It is the cell where the last location update occurred.

Residing area   It is the area in which the mobile unit can be located.

Polling cycle   It is the process performed by the network when a call arrives at a mobile terminal. The network sends a polling signal to the target cell in the residing area and waits for the response.

In cellular data networks, location management is very useful for position-based routing protocols. Network topology plays a very important role in location management. In this case, the networks may be homogeneous or heterogeneous. Moreover, various mobility models are used for location management. Call arrival probability is also crucial when evaluating the performance of a location management scheme.

MORE SOLVED EXAMPLES

Example 2.11 For a reuse factor of seven, consider a cellular system with total 395 channels in the cluster. For uniform traffic, an average call holding time of 100 s and a blocking probability of 2%, find the number of calls per BTS per hour. Also, find the SIR, assuming omnidirectional antenna and first-tier cells of co-channel interference. Assume path loss exponent $n = 4$.

Solution For $N = 7$, the number of channels per BTS = $395/7 = 56.43 \approx 56$

From the Erlang B traffic table for 56 channels with 2% blocking probability, the traffic load is 45.88 erlangs. Therefore, the carried load A is $(1 - 0.02) \times 45.88 = 44.96$ erlangs.

For $T = 1$ h = 3600 s and $h = 100$ s (given), from Eq. (2.10),

$$c = AT/h = 44.96 \times 3600/100 \approx 1619 \text{ calls per BTS/h}$$

Now, $\quad \text{SIR} = \dfrac{\left(\sqrt{3N}\right)^n}{6} = \dfrac{\left(\sqrt{3 \times 7}\right)^4}{6} = 73.5 = 18.66 \text{ dB}$

Example 2.12 Consider a land–mobile system with the following data:

Subscriber attempting calls = 45,000, out of which 35% are mobile–landline users and 65% are mobile–mobile users

Average call holding time for mobile–landline users = 100 s

Average call holding time for mobile–mobile users = 130 s

Average calls per hour for mobile–landline users = 3

Average calls per hour for mobile–mobile users = 5

One MSC can handle 2000 erlangs traffic.

Do we need multiple MSCs? If yes, then how many?

Solution Mobile–landline case:

Traffic per hour = $ch/T = 3 \times 100/3600 = 0.0833$ erlangs

Number of users = $45,000 \times 0.35 = 15,750$

Total traffic = $15,750 \times 0.0833 = 1312$ erlangs

Mobile–mobile case:

Traffic per hour = $ch/T = 5 \times 130/3600 = 0.1806$ erlangs

Number of users = $45,000 \times 0.65 = 29,250$

Total traffic = $29,250 \times 0.1806 = 5282.55$ erlangs

Overall traffic = $1312 + 5282.55 = 6594.55$ erlangs

Now, one MSC can handle 2000 erlangs traffic; hence, four MSCs are required to handle the total traffic.

Example 2.13 If there are 60 radio channels for a BTS to handle the total traffic and the average call holding time is 120 s, how many calls are handled in this cell with a GOS of 3%.

Solution From the Erlang B traffic table for 60 channels and GOS = 0.03, the traffic is 51.57 erlangs.

So, the carried load = $(1 - 0.03) \times 51.57 = 50.022$ erlangs

Number of calls that can be handled = $AT/h = 50.022 \times 3600/120 \approx 1500$

Example 2.14 For a cellular design, the parameters are as follows:

Voice channels to handle the traffic = 780

Reuse factor = 7

Cell area = 1 km^2

Total area to be covered = 100 km^2

Find the number of clusters, number of BTSs, number of channels per cell, and total system capacity in terms of effective number of channels.

Solution The total number of cells is 100 (calculated from the details of area) and hence this many BTSs are required.

Reuse factor is seven; hence, the number of clusters is $100/7 = 14.28 \approx 14$

Number of channels per cell = $780/7 = 111.42 \approx 111$

Effective number of channels in this system = $14 \times 780 = 10,920$

SUMMARY

- The smallest area into which the total area is divided is called a cell and a group of cells form a cluster.
- In a cellular system, cell-wise base stations are provided with transceiver facility to communicate with the mobiles in that area.
- Theoretically, a cell has a hexagonal shape.
- Cell splitting is the further division of a cell area to accommodate more traffic.
- A group of channels is assigned to a cell, which is not repeated in the same cluster but repeated in different clusters in the same manner.
- Frequency reuse is used in cellular systems to efficiently use the limited available spectrum.
- Cellular systems may be of two types: (a) analog or digital circuit switched and (b) packet switched. The components of the two systems differ slightly but the concept remains the same.
- The MTSO or MSC is the central switch to handle the calls and the channel assignments.

- A handoff procedure is necessary during transition from one cell area to another for maintaining uninterrupted conversation.
- Channel assignment may be of two types: (a) fixed and (b) dynamic. Borrowing strategies can be used sometimes.
- Cellular interferences are of two types: co-channel and adjacent channel.
- Antennas for the base station may be omnidirectional or directional. Directional antennas are used for sectorization.
- Microstrip patch antennas are used for mobiles.
- Sectorization increases the traffic-handling capacity.
- The unit of traffic is erlang.
- Mobile traffic calculation can be done by considering the traffic with a Poisson distribution. The Markov chain model can be used for the calculation of Erlang B formula for a lost call system.
- Erlang C formula can be used for a delayed system.
- Continuous location management is very important to trace the location of a mobile for completing a call request.

EXERCISES

Multiple-choice Questions

2.1 Cellular theory is applicable to
  (a) WLL   (c) mobile satellite
  (b) GSM   (d) all of these

2.2 Cluster is a
  (a) set of reuse frequencies
  (b) group of channels in a cell
  (c) group of cells using the same frequencies
  (d) group of cells using different frequencies without reuse

2.3 A group of channels assigned to a cell
  (a) is repeated in the same cluster
  (b) is repeated in different clusters in a different manner
  (c) is repeated in different clusters in the same manner
  (d) varies cell by cell

2.4 Which of the following multiple access schemes can be accommodated by cellular technology-based networks?
  (a) FDMA   (c) SDMA
  (b) TDMA   (d) All of these

2.5 If the cluster size is increased
  (a) the interchannel interference reduces
  (b) the transmission power increases
  (c) the co-channel interference reduces
  (d) the number of reuse frequency increases

2.6 Hard handover means
  (a) handing over the call to another mobile
  (b) changing over the communication channel
  (c) increase in power while moving to another base station
  (d) handing over the call to an MSC

2.7 Cell splitting is done to
  (a) accommodate more traffic
  (b) accommodate more area
  (c) save power
  (d) increase frequency reuse

2.8 In CDMA-based cellular networks, the near–far effect may appear due to
  (a) distant users
  (b) imperfect orthogonality between codes
  (c) interfering signals
  (d) orthogonal codes

2.9 Sectorization means
  (a) dividing clusters into sectors
  (b) dividing the channels
  (c) replacing omnidirectional antennas by directional antennas
  (d) cell splitting

2.10 Which of the following angle is suitable for sectorization?
  (a) 45°   (b) 60°   (c) 100°   (d) 80°

2.11 The number of handoffs will increase in case of
   (a) micro-cellular structure
   (b) cell splitting
   (c) sectorization
   (d) (b) and (c)

2.12 Interference effects in cellular systems are a result of
   (a) the distance between the areas
   (b) the height of the antennas
   (c) the ratio of the distance between the areas to the transmitted power of the areas
   (d) power of the transmitters

2.13 Larger cells are more useful in
   (a) lightly populated urban areas
   (b) rural areas
   (c) densely populated urban areas
   (d) hilly areas

2.14 Rayleigh fading occurs when there is
   (a) an LOS component present
   (b) no LOS component present
   (c) intercell interference
   (d) none of these

2.15 Frequency reuse can be maximized by
   (a) increasing the size of cells
   (b) decreasing the size of cells
   (c) increasing the size of clusters
   (d) increasing the number of users

2.16 Which of the following statements is true?
   (a) In TDMA, a digital air interface standard has twice the capacity of an analog interface.
   (b) In practice, cells are always hexagonal in shape.
   (c) More the handoff margin, more the benefit.
   (d) By increasing the sectors, co-channel interference can be reduced.

Review Questions

2.1 What do you mean by a cell? Why is a cell theoretically assumed to have a hexagonal shape?

2.2 Comment on the relationship between the size of a cluster and the frequency reuse distance.

2.3 How is the reuse distance decided theoretically?

2.4 'Cellular communication increases spectral efficiency but spectral efficiency also depends upon the cluster size.' Justify.

2.5 What are the parameters that will be affected by the size of a cluster?

2.6 What will happen if the power of a cell site transmission is detected in an adjacent cell? What will be the situation of the mobile user at the boundary of the cell?

2.7 Why is the value of path loss exponent more than two?

2.8 How are the mobile-originated and landline-originated calls handled?

2.9 What do you mean by channel assignment? Why is this process required?

2.10 Discuss the constraints of dynamic channel assignment.

2.11 'Directional antennas produce more gain than omnidirectional antennas.' Justify. How is this feature utilized in cellular communication?

2.12 Why is sectorization a critical issue though it looks simple due to directional antennas?

2.13 What are the differences between the outdoor and indoor propagation issues of a cellular system?

2.14 Explain why operators of cellular networks in densely populated areas prefer to put their base stations in the valleys rather than on top of mountains. How does this differ from the situation in rural areas?

2.15 Compare traffic handling in landline telephone and mobile phone systems.

2.16 What is the difference between the Erlang B and C formulas?

2.17 Define Ergodic, Poisson, and Markov processes from the basic theory of random processes and try to correlate them with telephone traffic.

2.18 Analyse the trade-off between sectoring and trunking efficiency loss. Compare the same for 120° and 60° sectoring.

2.19 List the advantages of digital cellular technology in terms of user capacity and performance.

Numerical Problems

2.1 Explain the relative merits and demerits of large and small cluster sizes in cellular systems. Prove that
$$D/R = (3N)^{0.5}$$

2.2 Find the appropriate cluster size N for a cellular system if the SIR requirement is 15 dB. Here, I corresponds to co-channel interference. Assume path loss factor $n = 3$. Find your answer for an omnidirectional antenna and a 120° sectored antenna.

2.3 For the different cluster sizes $N = 1, 3, 4, 7, 12$, and so on, find the SIR in dB for all the cases. Assume that all cells have equal radii and the base stations have equal power and are located at the centre of each cell. Comment your result in the form of some conclusions. What will be the effect of the first tier, second tier, and so on?

2.4 Show that the frequency reuse factor for a cellular system is given by k/S, where k is the average number of channels per cell and S is the total number of channels available to the service provider.

2.5 If 24 MHz of total spectrum is allocated to a duplex wireless cellular system and each simplex channel has 25 kHz RF bandwidth, find the number of duplex channels and the total number of channels per cell site if $N = 4$ cell reuse is used.

2.6 For a cell, the number of available channels is 30. Assume average call length of 2 min. Also, the average number of calls per hour per user is one. Find the capacity loss when going from an omnidirectional to a 60° sectored antenna. The blocking probability desired is 1%. A section of the blocking table is given as follows (the entries show the offered load):

Number of channels	A
5	1.361
10	4.462
20	12.03
30	20.34
40	27.38
51	44.2

2.7 On average, during a busy hour, a business firm makes 100 outgoing calls with an average duration of 3 min. It receives 120 calls, which are incoming at an average duration of 2.5 min. Find the outgoing traffic, incoming traffic, and total traffic in erlangs.

2.8 Show that the probability of delaying a call is

$$P_D = \frac{A^N}{A^N + N!\left(1 - \dfrac{A}{N}\right)\displaystyle\sum_{k=0}^{N-1}\dfrac{A^k}{k!}}$$

where A is the offered traffic expressed in erlangs (λ/μ).

2.9 A total of 25 MHz of bandwidth is allocated to a particular frequency division duplex (FDD—with different uplink and downlink frequencies) cellular telephone system that uses two 30 kHz simplex channels to provide full duplex voice and control channels. Assume that each mobile phone user generates 0.2 erlangs of traffic. Find the number of channels in each cell for a four-cell frequency reuse system. If each cell is to offer capacity, that is 90% of perfect scheduling, find the maximum number of users that can be supported per cell for an omnidirectional antenna case. What is the blocking probability of the

system using the Erlang B formula when maximum users are available in the user pool?

(**Hint:** Assume that each channel can carry 1 erlang of traffic.)

2.10 The cluster size is of seven cells with the blocking probability P_r being 1% and the average call holding time being 2 min. Find the traffic capacity loss in percentage due to trunking for 51 channels when migrating from omnidirectional antennas to 120° and 60° sectoring. Assume that blocked calls are cleared and the average per user call rate λ is one per hour. Use the table provided in Problem 6 for total traffic intensity A. Comment on your results.

2.11 Assume that there are six co-channel cells in the first tier and all of them are at the same distance from the mobile. An SIR of 12 dB is needed for satisfactory forward channel performance in a cellular system. What should be the optimum frequency reuse factor and cluster size if the path loss exponent is (a) $n = 3$ and (b) $n = 4$.

2.12 A 20 MHz total spectrum is allocated for a duplex wireless cellular system and each simplex channel has 25 kHz RF bandwidth. Find (a) the number of duplex channels and (b) the total number of channels per cell site if $N = 4$ cell reuse is used.

2.13 The frequency bands are 825–845 MHz for a mobile unit transmission (reverse link) and 870–890 MHz for a base station transmission (forward link). A duplex circuit consists of one 30 kHz channel in each direction. The systems are distinguished by reuse factors, which are 4, 7, 12, and 19, respectively.
   (a) Find the number of simultaneous communications that can be supported by a single cell in each system.
   (b) Suppose that in each of the systems the cluster of cells is repeated 16 times, find the number of simultaneous communications that can be supported by each system.
   (c) What is the area covered in cells by each system?
   (d) Suppose the cell size is the same in all four systems and a fixed area of 100 cells is covered by each system. Find the number of simultaneous communications that can be supported by each system.

2.14 A mobile user got 500 min free talktime from the service provider. He has utilized 40% of the total usage in mobile–landline calls and the remaining in mobile–mobile calls. How much traffic is created by him if the average holding time for landline calls is 120 s and that for mobile-to-mobile calls is 150 s?

PART 2
Wireless Channels and Modelling

The basic hurdle in the development of a wireless communication link is the wireless channel. Compared to other guided media such as a copper wire and an optical fiber, the wireless channel is an unguided dielectric media, and is more complicated due to its unpredictable nature. The channel has different features as per frequency of communication, distance, and surrounding objects. Modelling of channel is required for analysis or synthesis of various links and their evaluation.

This part of the book has two chapters describing the basic characteristics of a wireless channel, impairments, and mitigation techniques to deal with and thereafter, channel modeling using a mathematical base.

3 Radio Propagation Over Wireless Channel

Theme of the Chapter

Radio propagation is highly dependent upon factors such as the site or location of communication, and the distance between the transmitter and the receiver. The performance of a system can vary significantly depending on the terrain, type of propagation, weather, frequency of operation, velocity of mobile terminal, interference sources, and other dynamic factors. In the same vein, radio propagation differs through various atmospheric layers. Channel considerations too differ for long- and short-distance communication. This chapter describes radio wave propagation along with the channel effects such as noise, attenuation, large- and small-scale fading, multipath, delay and Doppler spread, shadowing, and outage. Using the free space propagation model, the chapter shows the mathematical formulation to calculate various channel parameters.

Key Topics

- Radio propagation fundamentals
- Radio propagation in atmospheric layers
- Radio communication cases
- Free space propagation model
- Ground wave, ionospheric, and tropospheric propagations

- Various channel noises and losses
- Multipath fading effects
- Delay spread: small-scale fading and large-scale fading
- Coherence bandwidth: flat fading and frequency-selective fading

- Doppler effect: fast fading and slow fading
- Shadowing
- Outage, outage probability, and fading margin

3.1 WIRELESS CHANNEL AND RADIO PROPAGATION FUNDAMENTALS

A wireless channel is a dielectric unguided medium that can be analysed in the following different ways:

The theory of electromagnetic radiation was proposed by the British physicist James Clerk Maxwell in 1857 and finalized in 1873.

- By using the fundamentals of electromagnetic (EM) wave theory, such as phase velocity, phase propagation constant, amplitude, frequency, and phase
- By using ray theory, refractive index dependence, reflection, refraction, and diffraction of EM waves (rays)
- By using the fundamentals of digital signal processing (DSP) such as channel transfer function, spectrum, channel impulse response (CIR), and convolution with transmitting signal

The channel can act as a low-pass or band-pass filter in certain conditions. The channel characteristic is of a random nature (channel processes are assumed to be stochastic processes) and depends upon the situation; hence, probability theory and the concept of probability distribution function (PDF) can be applied to wireless channels. The PDF gives the behavioural model for the channel.

3.1.1 Radio Waves

A radio signal exists in the form of an EM wave over a wireless channel, which is radiated through an antenna. EM waves are the energy propagated through free space at the velocity of light (3×10^8 m/s). The energy level decreases with distance, and it can have both vertical and horizontal components.

As shown in Fig. 3.1, EM waves have both electric and magnetic field components, which are inseparable and orthogonal to each other. The planes of the fields are at right angles to each other and to the direction in which the wave is travelling.

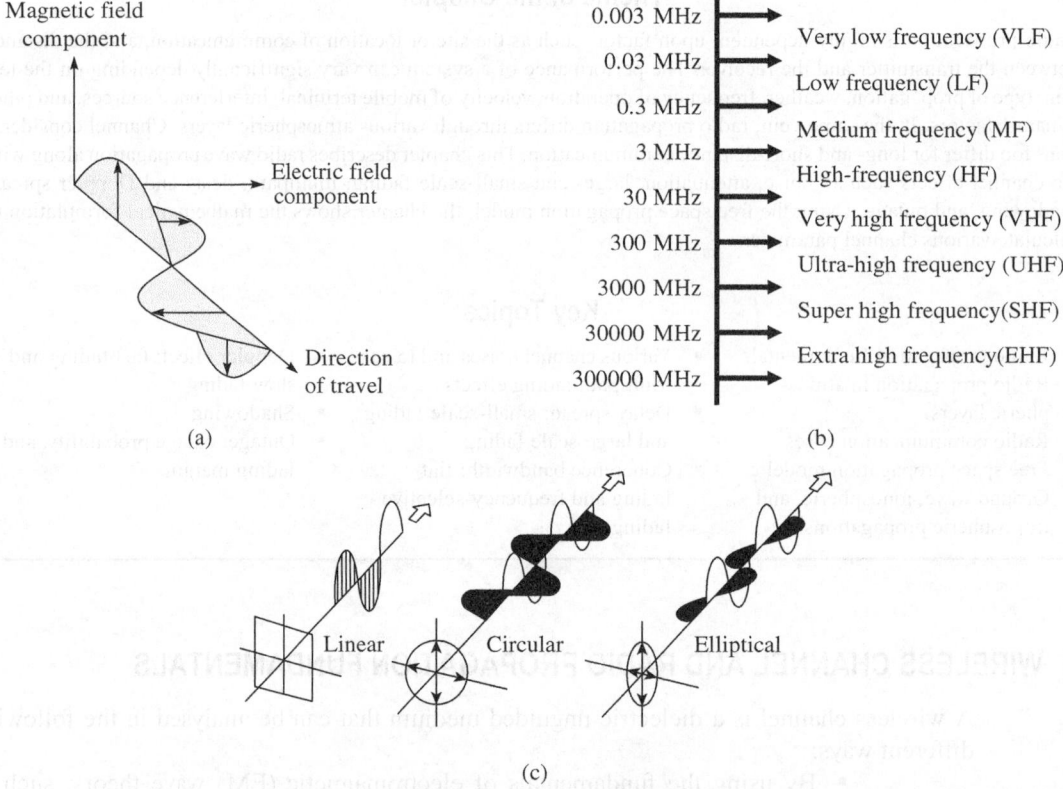

Fig. 3.1 Electromagnetic wave (a) Nature of electromagnetic wave (b) Radio frequency ranges (c) Linear, circular, and elliptical polarized waves

The electric component of the wave results from the voltage changes that occur as the antenna element is excited by the alternating waveform. The lines of force in the electric field run along the same axis as the antenna but spread out in the form of a typical lobe as they move away from it. This electric field is measured in terms of the change of potential over a given distance, for example, volts per metre, and this is known as the *field strength*.

Note: Laser communication by optical rays and infrared communication can be categorized under special cases of wireless communication. These systems require optical device-based solid-state radiators instead of conventional metallic antenna-based radiators.

This measure is often used in measuring the intensity of an EM wave at a particular point.

The second component, that is, the magnetic field, is at right angles to the electric field and hence, at right angles to the plane of the antenna. It is generated because of the current flow in the antenna.

Figure 3.1(b) shows the radio frequency (RF) ranges of the EM spectrum that are suitable for wireless communication.

Polarization of Electromagnetic Waves

Polarization of an EM wave indicates the plane in which it is vibrating. It often has a significant effect on the way in which the wave propagates. While it is important to match the polarization of the transmitting and receiving antennas, the choice of polarization is also important for signal propagation and diversity mitigation. As an EM wave consists of an electric and a magnetic field vibrating at right angles to each other, it is necessary to adopt a convention to determine the polarization of the signal. The plane of the electric field is used for this purpose.

Linear polarization   It includes vertical and horizontal polarizations, which are the most straightforward forms. Here, the wave can be thought of as vibrating in one plane.

Circular polarization   It can be visualized by imagining a signal propagating from a rotating antenna. The tip of the *electric field vector* can be seen to trace a helix or corkscrew as it travels away from the antenna. Circular polarization can be either right-handed or left-handed, depending upon the direction of rotation as seen from the transmitting antenna.

Elliptical polarization   It occurs when there is a combination of both linear and circular polarizations. Examples of each type are given in Figure 3.1(c).

3.1.2 Basic Propagation Mechanisms

The propagation mechanisms followed by a radio wave are discussed in this section. These mechanisms can be better understood if the radio wave is treated as a ray.

Line of sight   The line-of-sight (LOS) propagation is the wave propagation in which the EM ray follows a straight line from the transmitter to the receiver. It is shown in Fig. 3.2 as a direct ray.

Non-line of sight   The non-line-of-sight (NLOS) propagation mechanism is based on and is the resultant of the following mechanisms:

Reflection   This occurs when the propagating wave impinges on an object that is larger than its wavelength. Examples of such objects are the surface of the earth, buildings, and walls.

The received signal is the resultant of the LOS/NLOS signal components along with the effect of fading, shadowing, change of polarization, and attenuation.

Diffraction   This occurs when the radio path between the transmitter and the receiver is obstructed by a surface with sharp irregular edges, which results in the waves bending around the obstacle. Radio signals are diffracted by the boundaries of obstructions, thus preventing total shadowing of the signals behind hills and buildings. However, the amount of diffraction is dependent on the RF used. *Diffraction is more with low-frequency (LF) signals than with high-frequency (HF)*

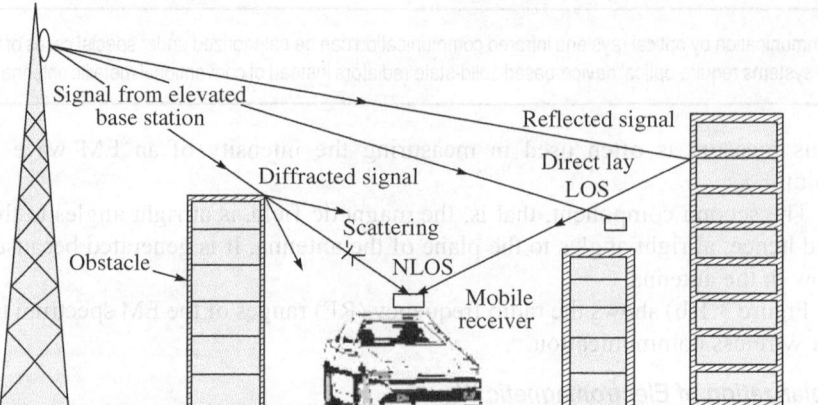

Fig. 3.2 LOS reception by direct ray and NLOS reception by reflected, diffracted, and scattered rays

signals. Thus, HF signals, especially ultra-high frequencies (UHFs), and microwave signals require LOS for adequate signal strength.

Scattering This occurs when the propagating wave is obstructed by objects that are smaller than its wavelength. Examples of such objects are lamp posts, foliage, street signs, and particles in the air.

Refraction Due to variations in the refractive index of the atmospheric layers, the EM wave bends (in the cases other than satellite communication).

All these effects except refraction are shown in Fig. 3.2, an urban scenario.

Propagation Channel Effects

Propagation of radio waves undergoes some environmental effects and effects due to surroundings. Hence, they are deteriorated or reduced in strength at a longer distance.

Attenuation It is the drop in the signal power when it is being transmitted from one point to another. It is caused by the transmission path length, obstructions in the signal path, and multipath effect. This effect can be observed along with LOS or NLOS transmissions, small-scale and large-scale fading, slow and fast fading, or flat and frequency-selective fading, where a drop in signal strength is observed with distance. Any object that obstructs the signal ray passing from the transmitter to the receiver can cause attenuation.

Fading As there are obstacles and reflectors in the wireless propagation channel, the transmitted signal arrives at the receiver from various directions over multiple paths. Such a phenomenon is called *multipath*. *Fading* is the result of multipath in which the signal strength varies continuously with respect to distance and with time from the transmitter to the receiver along with the attenuation. At the receiver, a vector addition of all the multipath-delayed components of the received signal is done. Large- and small-scale fading are explained in Section 3.8.

Wooden objects, glass, bricks, and so on have different attenuation values when an EM ray penetrates through them; hence, attenuation will not be uniform.

Shadowing This occurs whenever there is an obstruction between the transmitter and the receiver, and it can be observed in long-distance as well as short-distance communication. It is generally caused by buildings and hills and is the most important environmental attenuation factor. Shadowing is very severe in heavily built-up areas. Hills can cause a major problem due to the large shadow they produce. Ground wave propagation, discussed in Section 3.4, is affected by shadowing. To overcome this problem, transmitters are usually elevated as high as possible to minimize the number of obstructions. In Fig. 3.2, the car can be considered to be in the shadow portion.

> *Note*: The effects of the aforementioned propagation mechanisms will become significant based on the distance between the transmitter and the receiver and the number of objects in the surroundings. Depending upon the propagation effects, two different forms of channels can be categorized for analysis purpose—time-invariant non-dispersive and time-variant dispersive.

Types of Channels

Wireless communication systems exist in a variety of forms and consequently different channel characteristics are to be considered. However, the following are the two main differentiations:

Time-invariant non-dispersive channel Microwave point-to-point links or satellite communication links are established for very large distance communication where LOS is established or the signal penetrates the atmospheric layers correspondingly and hence, there are no significant reflections or multipaths. These effects are negligible as the distance is very large and the direct signal is very strong. A majority of these links are considered as pure LOS. These are typically designed with slim margins, as the channel will vary little over time, with the exception of possibly increased loss from rain (absorption), scattering due to solid particles or Faraday rotation, or simply attenuation with distance. The dynamic range required at the receiver will consequently be small. Hence, to analyse such a system and link power budget estimation, one can start with the simple free space propagation model and then gradually add the losses in it for the actual scenario. This scenario can be modelled as a time-invariant non-dispersive channel as there is one strong LOS component under consideration and negligible multipath; hence, there is no dispersion and no signal variation with time.

Time-variant dispersive channel Multipath links are usually described by the resultant of both the following:

- *Line-of-sight component* The direct path between the transmitter and the receiver
- *Non-line-of-sight components* The path due to the reflection from reflectors, scatterers, and diffractors

Mobile communication systems operating in the very high frequency (VHF) and UHF bands are typically based on multipath and, therefore, consist of the sum of signals from the

IMPORTANT RELATIONSHIPS FOR A DIELECTRIC MEDIUM	
$E = E_0\sin(wt - bz)$	Representation of an EM wave in terms of its peak amplitude E_0, angular frequency, and phase (z represents the direction of propagation axis in the three-coordinate system, whereas the x-axis and y-axis are to represent the electric and magnetic field components, respectively)
$w = 2pf$	Angular frequency
$l = c/f$	Wavelength
$n = c/v_{em}$	Refractive index of the medium
$k = 2p/l$	Media propagation constant (1 for free space)
$b = nk$	Wave propagation constant
c	Speed of light in free space
v_{em}	Velocity of an EM wave in the dielectric medium

different paths. Reflection, refraction, and diffraction are more pronounced in these systems. This scenario is slightly complex. Potentially, this will lead to multipath fading effect, which results from considerable amplitude variations with time and/or distance. Physical objects will also cause *shadowing*, resulting in a loss of signal, for instance, a vehicle entering a tunnel or travelling behind a building. Additionally, due to the varying proximity between two nodes, a considerably larger dynamic range will be required at the receiver together with an increased link budget margin.

From the foregoing discussion, it is clear that these two communication systems will require different properties from a modulation scheme to transmit the data.

3.1.3 Radio Propagation in Atmospheric Layers

The way the radio signals propagate from a transmitter to a receiver is of great importance while planning a radio system. This is governed to a great degree by the regions of the atmosphere through which they pass. Without the action of the atmosphere, it would not be possible for radio signals to travel across the globe on the short-wave bands or travel greater than the LOS distance at higher frequencies. In view of the importance of the atmosphere, an overview of its make-up is given here. The atmosphere can be split into different layers according to their properties. As different aspects of science look at different properties, there is no single nomenclature for the layers.

The lowest layer is the *troposphere*, which extends from the earth's surface to a height of 10 km. At altitudes between 10 km and 50 km is the *stratosphere*, which contains the ozone layer at a height of about 20 km. Next is the *mesosphere*, extending at altitudes from 50 km to 80 km. Above this is the *thermosphere*, where temperatures rise dramatically. Overlapping the mesosphere and the thermosphere is the *ionosphere*, which again contains four sub-layers.

The layers that are essential for long-distance communication are briefly discussed here (refer to Fig. 3.3).

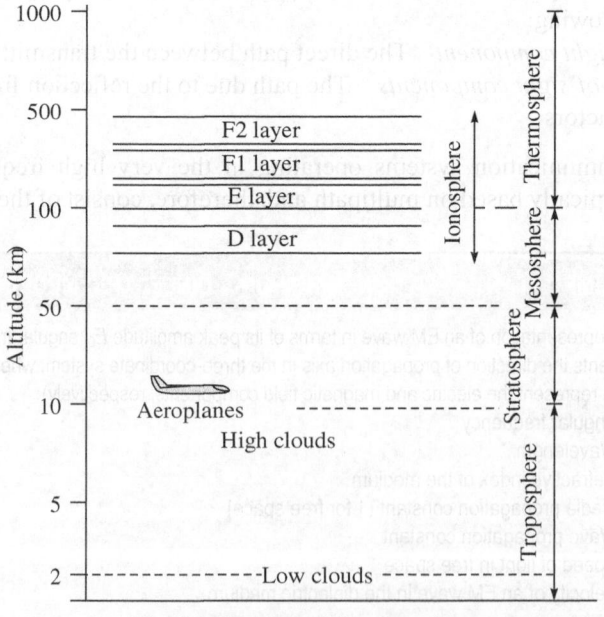

Fig. 3.3   Regions of the atmosphere

Troposphere   The conditions that govern our weather occur here. It is found that low clouds occur at altitudes of up to 2 km, whereas medium-level clouds extend to about 4 km and the highest clouds are found at altitudes up to 10 km. Modern jet airliners fly above this at altitudes of up to 15 km. Within the troposphere, there is generally a steady fall in temperature with height, and this has a distinct bearing on some propagation modes that occur in this region. The fall in the temperature continues in the troposphere until the *tropopause* is reached. This is the area where the temperature gradient levels out and then the temperature starts to rise. At this point, the temperature is approximately −50°C. The refractive index of the air in the troposphere plays a dominant role in radio signal propagation. The refractive index depends on the temperature, pressure, and humidity. When radio signals are affected in the troposphere, it usually occurs at altitudes up to 2 km.

Ionosphere   This is a region where there is a very high level of free electrons and ions. It is found that free electrons affect radio waves. Although ions and electrons are present in low levels even at lower altitudes, the number of ions starts to rise noticeably at an altitude of about 30 km. However, only at about the 60 km mark, it rises to a degree sufficient to have any major effect on radio signals. This layer is mainly responsible for the phase errors in satellite transmissions, especially in global positioning system (GPS).

In the ionosphere, radiation from the ultraviolet (UV) rays of the sun strikes the molecules in the upper atmosphere at very high altitudes, where the gases are very thin. This radiation is so intense that when it strikes the gas molecules, some electrons obtain sufficient energy to leave the molecular structure. This leaves a molecule with a deficit of one electron, which is called an ion, and a free electron (*ionization*). The density of gases and hence ionization vary with altitude. At low altitudes, the effect of the UV rays reduces but the influence of X-rays and cosmic rays increases. As might be expected, the most common molecules to be ionized are nitrogen and oxygen.

Hence, the ionosphere is considered to consist of a number of distinct layers. Each layer overlaps the others with the whole of the ionosphere having some level of ionization. The layers are best thought of as peaks in the level of ionization. These layers are given the designations D, E, F1, and F2.

D layer   The D layer is the lowest layer of the ionosphere. It exists at altitudes from 60 km to 90 km. It is present during the day when radiation is received from the sun. However, because of the high density of the air at this altitude, the ions and electrons recombine relatively quickly. Hence, after sunset, electron levels fall and the layer effectively disappears. This layer is typically produced as a result of X-ray and cosmic ray ionization. It is found that this layer tends to attenuate signals that pass through it.

E layer   The layer beyond the D layer is the E layer. This exists at altitudes between 100 km and 125 km. This layer chiefly reflects the radio signals, although they still undergo some attenuation. In view of the layer's altitude and the density of the air, the electrons and positive ions recombine relatively quickly. This occurs at a rate of about four times that in the F layers, which are higher up, where the air is less dense. Hence, after nightfall, the layer virtually disappears, although there is still some residual ionization. Ionization in this layer takes place through various methods and depends on factors such as the altitude within the layer, the state of the sun, and the latitude. However, X-rays and UV rays produce a large amount of ionization, especially having very short wavelengths.

F layer   The F layer is the most important region for long-distance HF communications. During the day, it splits into two separate layers. These are called the F_1 and F_2 layers, the

The D layer disappears at night; hence, low frequencies like amplitude modulation (AM) can be received better at night compared to day.

F_1 layer being the lower of the two. At night, these two layers merge into a single layer, called the F layer. The altitudes of the layers vary considerably with the time of the day, season, and state of the sun. Typically, in summer, the F_1 layer may be at about 300 km and the F_2 layer at about 400 km or even higher. In winter, these levels may be reduced to about 200 km and 300 km, respectively. At night, the F layer is generally from 250 km to 300 km. Like the D and E layers, the level of ionization falls at night, but in view of the much lower air density, the ions and electrons combine much more slowly and the decay of the F layer is much less. Accordingly, it is able to support communications, although changes are experienced because of the lowering of the ionization levels. The altitudes of the F layers are far more variable than those of the lower layers. As a result, the values that are given here must be taken only as an approximate guide. The UV light causes most of the ionization in this region of the ionosphere.

3.2 RADIO COMMUNICATION CASES

Radio signals are affected in many ways by the objects in their path and by the media through which they travel. Atmospheric variations, discussed in Section 3.1, do not come into the picture for communication near the ground level. The properties of the path over which the radio signals propagate govern the level and quality of the received signal. It is, therefore, very important to know the distance and the radio propagation characteristics that are likely to prevail.

The distances over which radio signals may propagate vary considerably. For some applications, only a short range may be needed. For example, a Wi-fi link may only need to be established over a distance of a few metres. On the other hand, a short-wave broadcast station or a satellite link would need the signals to travel over much greater distances. The radio propagation characteristics would be completely different in these two cases, with the signals reaching their final destinations having been affected in very different ways by the media through which the signals have travelled.

3.2.1 Long-distance Communication

Long-distance radio propagation mechanisms can be categorized into various types depending upon the effects of the media through which the signals propagate.

Free space propagation In this propagation mechanism, the radio signals travel in free space without any obstacles and interferences. If the wavelength is fixed, the distance from the source is the only factor that affects the way in which the field strength varies. This type of radio propagation is mainly encountered by signals travelling to and from satellites where the signal penetrates the atmospheric layers (with atmospheric effects practically) or in LOS propagation with ideal conditions. This is a rather hypothetical case (refer Section 3.3).

Ground wave propagation When signals travel close to the ground, they get modified by the terrain over which they travel. They also tend to follow the earth's curvature. Signals heard on the medium-wave band during daytime use this form of propagation. It operates in the frequency range of below 2 MHz (refer Section 3.4).

Ionospheric propagation It is also known as *sky wave propagation*. It operates in the range of 2–30 MHz. Here, the radio signals are modified and influenced by the action of the free electrons in the upper reaches of the earth's atmosphere in the ionosphere. This form of radio propagation is used by stations on the short-wave bands for their signals to be heard around the globe (refer Section 3.5).

Tropospheric propagation   It is also known as *space wave propagation*. It is suitable for the frequencies in the VHF and UHF ranges. In this type of propagation, the signal is influenced by the variations of the refractive index in the troposphere, which is just above the earth's surface. Tropospheric radio propagation is often the means by which the signals at VHF and above (maybe television signals) are heard over extended distances (refer Section 3.6).

Optical Horizon and Radio Horizon

> Due to radio horizon, effective distance covered between transmitter and receiver increases, which decides the antenna height requirement for LOS communication for long distances

A special case of LOS propagation above 30 MHz, as shown in Fig. 3.4(a), gives rise to two different terminologies, optical horizon and radio horizon. *Optical horizon* is just like the visual limit of the transmitter to see the receiver. *Radio horizon* is the locus of points at which the direct rays from an antenna are tangential to the surface of the earth. Typically, the refractive index in the troposphere falls slowly with height, and the resulting refraction causes the radio horizon to appear to be 1.33 times farther than the optical horizon. In space wave communication, the wave undergoes the earth curvature limitations as shown in the Fig. 3.4(b) (optical limit). It can be observed that slightly more coverage is achieved due to radio horizon.

Most of the long-distance communication applications are broadcast applications through satellite, terrestrial television transmissions, and mobile satellite communication. The multipath component is very low in these cases.

In addition to these categories, many short-range radio or wireless systems have radio propagation scenarios that do not fit neatly into these categories because they may also have indoor models. However, for outdoor models, ground wave concepts can be correlated. Wi-fi and cellular systems, for example, need to have their radio propagation models generated for office or for urban situations such as buildings, vegetation, and vehicles. Under these circumstances, the approach of the analysis is modified because the multipath (NLOS) component is important in these cases.

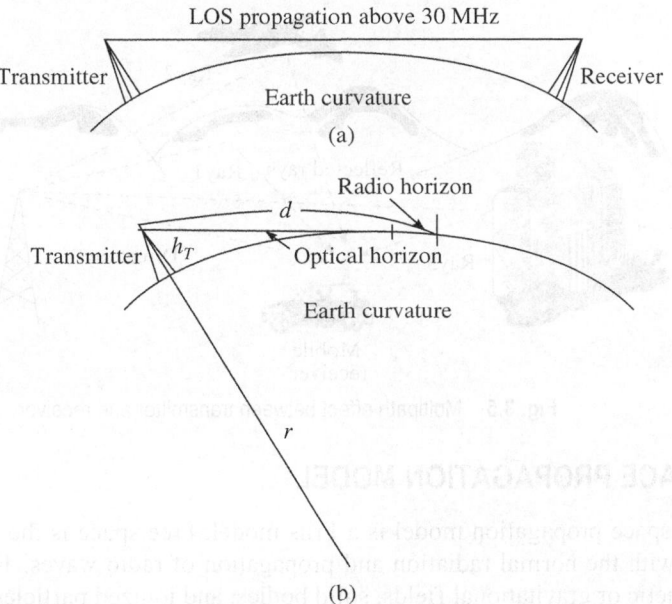

Fig. 3.4   Effect of earth's curvature (a) LOS propagation mode (b) Limitations on antenna height due to optical horizon and radio horizon

3.2.2 Short- and Medium-distance Communications

Both long- and short-distance communications are affected by additive white Gaussian noise (AWGN), which adds spurious frequency components in the signal.

In ideal radio channel conditions, the received signal would consist of only a single direct path signal, which would be a perfect reconstruction of the transmitted signal. However, in a real channel, the signal is modified during transmission through the channel. For NLOS multipath, the received signal consists of a combination of attenuated, reflected, refracted, and diffracted replicas of the transmitted signal.

Consider the urban area situation depicted in Fig. 3.2, where a receiver in a moving automobile receives a signal from a single transmitter, which has propagated along two paths. One propagation path is the direct path (which may be slightly diffracted) from the transmitter to the mobile. The second path is due to the reflection from the building. A majority of such systems work on above 800 MHz up to 2–11 GHz. It is a short-distance scenario. The scenario of receiving a signal within a room or a building is also that of short-distance communication. Figure 3.5 shows the typical scenario of medium-distance multipath or NLOS communication. Rural area communication also represents a medium-distance scenario.

Effects of Multipath

Due to multipath effect, multiple versions of the transmitted signal reach the receiver, which cause *intersymbol interference* (ISI) or *delay spread*. It becomes very hard to extract the original information without complicated *equalizers* (explained in Chapter 8) or some modulation schemes designed to combat ISI (refer Chapter 7). The common representation of a multipath channel is the CIR, which is the response at the receiver if a single impulse is transmitted. It can be calculated by *channel estimation* procedures, which are elaborated in Chapter 8. Mobility of users, and hence mobile receivers, or mobility of surrounding objects can cause a shift in the received carrier frequency, which is called *Doppler effect*.

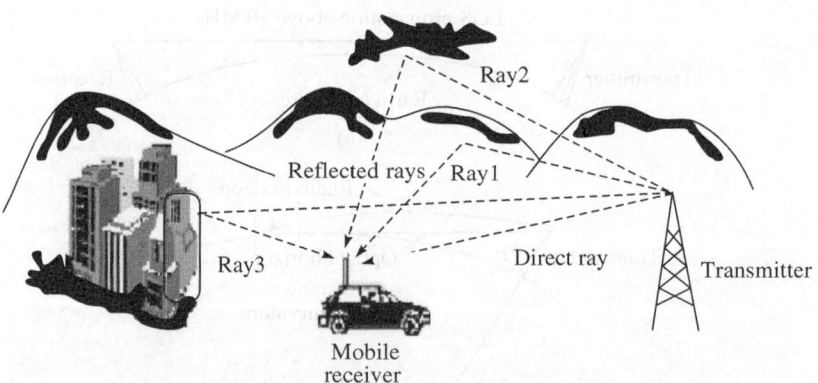

Fig. 3.5   Multipath effect between transmitter and receiver

3.3 FREE SPACE PROPAGATION MODEL

Free space propagation model is a Friis model. Free space is the space that does not interfere with the normal radiation and propagation of radio waves. Thus, it does not have any magnetic or gravitational fields, solid bodies, and ionized particles. Apart from the fact that free space is unlikely to exist anywhere (due to ideal conditions), it certainly does not exist near the earth. The concept of free space is used to study wave propagation in a simplified manner, and the same conditions are then applied to different categories of long-distance communication with the actual scenario.

Any power escaping into free space is governed by the characteristics of the free space. If such power is released purposely, it is said to have been radiated and it then propagates in space in the form of an EM wave. This action can also be related to the term *power density*, which is radiated power per unit area. Power density reduces with distance.

The free space propagation model assumes the transmitting and receiving antennas to be located in an empty environment. Neither absorbing obstacles nor reflecting surfaces are there. In particular, the influence of the earth's surface is assumed to be entirely absent. For propagation distances d much larger than the antenna size, the far field of the EM wave dominates all other components; that is, we are allowed to model the radiating antenna as a point source with negligible physical dimensions. In such a case, the energy radiated by an omnidirectional antenna is spread over the surface of a sphere. This allows us to analyse the effect of the distance on the received signal power (Fig. 3.6a). In the figure, the transmit power is spread over the surface area of a hypothetical sphere at a distance d. The receiver antenna has an aperture A_r, illustrated over the area $4\pi d^2$. Spherical wavefronts are observed in far field, as shown in Figure 3.6(b).

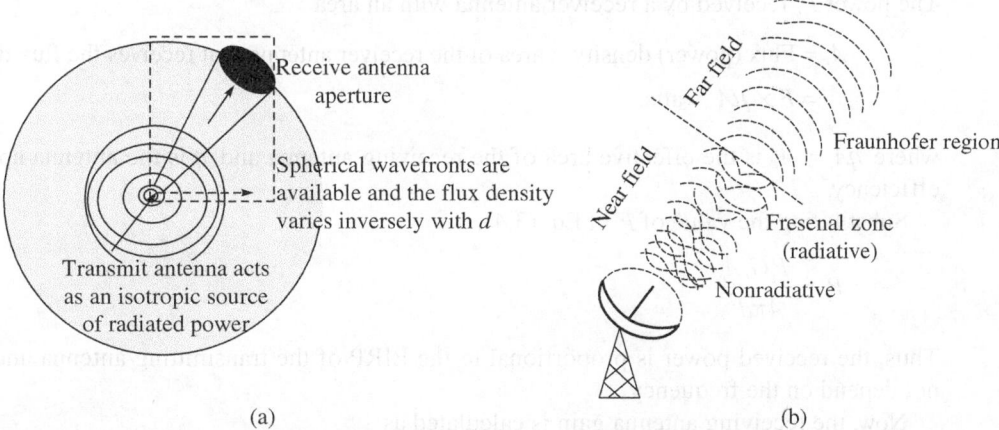

(a)   (b)

Fig. 3.6   Free space propagation (a) Transmit antenna modelled as a point source (b) Near- and far-field concepts

---

FRAUNHOFER DISTANCE

Shorter antennas with dimensions less than half the wavelength are treated as point antennas. Typical wavefronts are generated due to their radiations near the antenna and at far distance. The definition of near and far fields is different for them. In the free space propagation model, though long transmit antennas have physical dimensions larger than half the wavelength, they are considered too small compared to the distance. Hence, simplified assumption as a point antenna. Thus, it adds an isotropic source of radiated power [Fig. 3.6(a)]. Hence, near and far fields are distinguished in terms of Fraunhofer distance.

The far field is called *Fraunhofer region*, from the name of its inventor. It is defined as the region beyond the threshold value of the far-field distance d_f, which relates the largest linear dimension of the transmitter antenna aperture A_t and the carrier wavelength λ as given in Eq. (3.1). The Fraunhofer distance is given by

$$d_f = \frac{2A_t^2}{\lambda} \tag{3.1}$$

In addition, to be in the far-field region, $d_f \gg A_t$ and $d_f \gg \lambda$.

This Friis free space model is valid only for the values of d that are in the far field of the transmitting antenna (Fig. 3.6b) and less than the radio horizon.

As with most large-scale radio wave propagation models, the free space model predicts that the received power decays as a function of the distance between the transmitter and the receiver raised to some power (a power law function). It is seen that the power density is inversely proportional to the square of the distance from the source. This is the inverse square law, which applies universally to all forms of radiation in free space. Considering the isotropic source of the transmitted power (which may be a half-wave dipole antenna), the flux density at a distance d can be calculated as

$$F = \frac{P_t}{4\pi d^2} \text{ W/m}^2 \tag{3.2}$$

Considering the gain of the transmitting antenna, we get

Effective isotropic radiated power (EIRP) = $P_t \times G_t$

or
$$F = \frac{P_t G_t}{4\pi d^2} \text{ W/m}^2 \tag{3.3}$$

The power P_r received by a receiver antenna with an area

A_r = Flux (power) density × area of the receiver antenna that receives the flux density

$$= F \times \mu A_r \text{ watts} \tag{3.4}$$

where $\eta A_r = A_e$ is the effective area of the receiving antenna and η is the antenna aperture efficiency.

Substituting the value of F in Eq. (3.4),

$$P_r = \frac{P_t G_t A_e}{4\pi d^2} \tag{3.5}$$

Thus, the received power is proportional to the EIRP of the transmitting antenna and does not depend on the frequency.

Now, the receiving antenna gain is calculated as

$$G_r = \frac{4\pi A_e}{\lambda^2} \tag{3.6}$$

Using the relationship of Eq. (3.6) and substituting for A_e in Eq. (3.5), we get

$$P_r = \frac{\lambda^2}{(4\pi d)^2} G_t P_t G_r \tag{3.7}$$

The wavelength λ (in metres) is given by

$$\frac{c}{f_c} = \frac{2\pi c}{\omega_c} \tag{3.8}$$

where c = velocity of light
f_c = carrier frequency in Hz
ω_c = angular frequency

Friis equation does not hold for $d = 0$; hence, it is possible to represent receiver power in terms of a reference distance called the close-in distance.

Now, the path loss in general is defined as

$$P_{LdB} = 10 \log \frac{P_t}{P_r} \tag{3.9}$$

$$= 10 \log \left[\frac{(4\pi d)^2}{\pi^2} \right] \text{ (for unity gain conditions)} \tag{3.10}$$

$$= -10 \log \frac{\lambda^2}{(4\pi d)^2} \tag{3.11}$$

If the wavelength is constant, the loss simply depends upon the distance d between the isotropic transmitter and the receiver. Considering the real systems, according to Friis free space equation,

$$P_r = \frac{P_t G_t G_r \lambda^2}{(4\pi d)^2 L} \tag{3.12}$$

where L is the system loss factor, which is not related to propagation. L may be greater than or equal to one. These miscellaneous losses are usually due to transmission line attenuation, filter losses, antenna losses, and so on. $L = 1$ implies that there are no losses in the system hardware.

In practice, *effective radiated power* (ERP) is used instead of EIRP to denote the maximum radiated power [ERP (watt) = EIRP (watt)/1.64]. Moreover, antenna gains are given in units of dBi (dB gain with respect to an isotropic antenna) or dBd (dB gain with respect to a half-wave dipole).

Example 3.1 A satellite link is established between an earth station and a satellite transponder at an RF frequency of 4 GHz. For the earth station transmitter, the transmitted power is 1 kW, and the transmitter and receiver antenna gains are 0 dB. The free space distance is 30,000 m. Find the received power at the transponder. Assume $L = 1$.

Solution $f_c = 4$ GHZ

$\Rightarrow \lambda = c/f = (3 \times 10^8 \text{ m/s})/4 \times 10^9 \text{ Hz} = 3/40 \text{ m}$

Now, $P_r = \dfrac{P_t G_t G_r \lambda^2}{(4\pi d)^2 L}$

$P_r = \dfrac{1000.1.1.(3/40)^2}{(4\pi 30,000)^2}$

$= 5.625/142,22,303,375$

$= 3.96 \times 10^{-11} \text{ W} = -104 \text{ dB}$

20logd Path Loss Law

As the propagation distance increases, the radiated energy is spread over the surface of a sphere of radius d. Hence, the power received decreases proportional to d^{-2}. For the unity gain antennas, the received power p_{rdB} expressed in dB is

$$P_{rdB} = p_{tdB} - 20 \log \frac{d}{\lambda/4\pi} \tag{3.13}$$

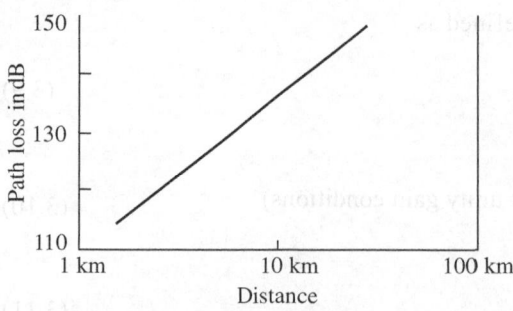

Fig. 3.7 Typical average path loss (in dB) versus distance

This is called the *path loss law*, where p_{tdB} is the transmitted power in dB. The characteristic plot is given in Fig. 3.7.

Furthermore, it will be useful to understand large-scale propagation models using a close-in distance d_o as a known received power reference point. The received power at any distance $d > d_o$ may be related to P_r at d_o as follows:

$$P_r(d) = P_r(d_o)\left(\frac{d_o}{d}\right)^2 \quad \text{for } d \geq d_o \geq d_f \quad (3.14)$$

> Ground wave is the resultant of space waves and surface waves.

In mobile radio systems, it is common to find that P_r has large variations of amplitude values over a typical coverage area of several square kilometres. As the dynamic range of the received power levels is large, often dBm or dBW units are used to express the received power levels. Equation (3.14) can be expressed in dBm or dBW units by taking the logarithm of both sides and multiplying by 10. If P_r is represented in units of dBm, the received power is given by

$$P_r(d)\text{dBm} = 10\log\left[\frac{P_r(d_o)}{0.001W}\right] + 20\log\left(\frac{d_o}{d}\right) \quad \text{for } d \geq d_o \geq d_f \quad (3.15)$$

Here, $p_r(d_o)$ is expressed in watts.

The reference distance (d_o) for practical systems using low-gain antennas in the 1–1.5 GHz region is typically chosen to be 1 m for indoor environments and 1 km to 100 km for outdoor environments.

Path loss causes attenuation between the transmitter power amplifier and the receiver front end. The following are some other effects to be considered during the link design:
- Losses in the antenna feeder (0–4 dB)
- Losses in transmit filters, particularly if the antenna radiates a signal of multiple transmitters (0–3 dB) and antenna directionality gain (0–12 dB)
- Losses in duplex filter
- Fade margins to anticipate for multipath (9–19 dB) and shadow losses (~5 dB)
- Penetration losses if the receiver is indoors, typically about 10 dB for 900 MHz signals

Field Strength

While cellular telephone operators mostly calculate the received power, broadcasters plan the coverage area of their transmitters using the (CCIR-recommended) *electric field strength* E (e-field) at the location of the receiver. The conversion formula is

$$E = \sqrt{\frac{120\pi P_r}{A_e}} \quad (3.16)$$

3.4 GROUND WAVE PROPAGATION

Ground wave propagation is particularly important on the LF and medium-frequency (MF) portions of the radio spectrum. It is used to provide relatively local coverage, especially by radio broadcast stations that require covering a particular locality. It can also be considered for land mobile telecommunication.

Ground wave radio signal propagation is ideal for relatively short-distance propagation on these frequencies during the daytime. Sky wave ionospheric propagation is not possible during the day because of the attenuation of the signals on these frequencies caused by the D region in the ionosphere. In view of this, stations need to rely on ground wave propagation to achieve their coverage.

A ground wave signal is composed of numerous constituents (Fig. 3.8a). If the antennas are in the LOS, then there will be a *direct wave* as well as *reflected waves* (together known as *space waves*). As mentioned already, a direct wave is one that travels directly between two antennas and is not affected by the locality. Reflected waves are generated as the transmission is reflected by a number of objects including the earth's surface and any hills or large buildings. In addition to this, there is a *surface wave*, which is a type of seismic wave that travels across the surface of the earth. Surface waves can be of mechanical or EM nature, travelling along the interface between the differing media or refractive index gradient. They tend to follow the curvature of the earth and enable coverage to be achieved beyond the optical horizon.

Beyond the horizon, the direct and reflected waves are blocked by the curvature of the earth, and the signal is purely made up from the diffracted surface wave. Then, the diffractive action is more pronounced. It is for this reason that surface wave propagation is commonly called ground wave propagation.

Signals spread out from the transmitter along the surface of the earth. Instead of travelling in a straight line, the signals tend to follow the curvature of the earth as shown in Fig. 3.8(b). This is because currents are induced in the surface of the earth, which slows down the wavefront in this region, causing the wavefront to tilt downwards towards the earth. With the wavefront tilted in this direction, the signals are able to curve around the earth and are received well beyond the horizon.

Effect of Frequency

As the wavefront of the ground wave travels along the earth's surface, it is attenuated. The degree of attenuation is dependent upon various factors. Frequency is one of the major

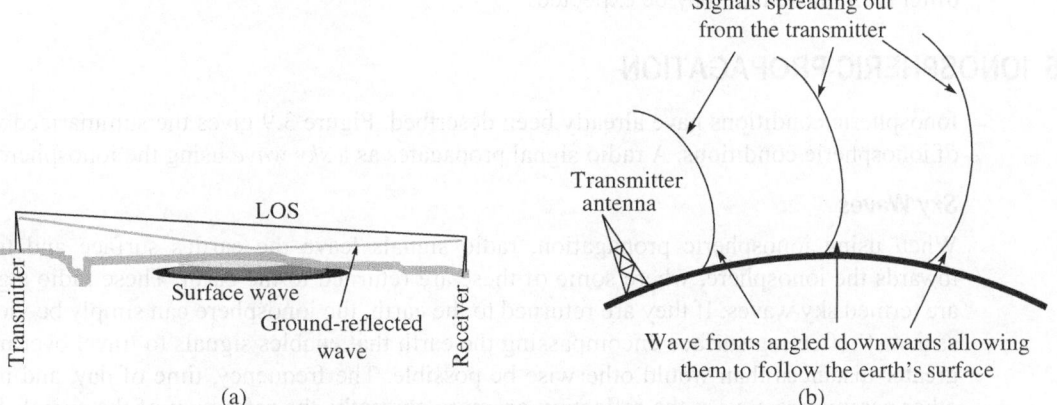

Fig. 3.8 Ground wave propagation (a) Surface wave, LOS, and ground reflected wave
(b) Space wave propagation following earth's curvature

determining factors, as losses rise with increase in the frequency. As a result, it makes this form of propagation impracticable above the bottom end of the HF portion of the spectrum (3 MHz). Typically, a signal at 3.0 MHz will suffer an attenuation that may be in the range of 20–60 dB more than one at 0.5 MHz. This is why even high-power HF broadcast stations may be audible for only a few miles from the transmitting site via the ground wave.

Effect of Ground

The surface wave is also highly dependent upon the nature of the ground over which the signal travels. Ground conductivity, terrain roughness, and the dielectric constant all affect signal attenuation. In addition to this, ground penetration varies, becoming greater at lower frequencies, which means that it is not just the surface conductivity that is of significance. Ground penetration is not important at higher frequencies, but at lower frequencies, the penetration means that ground strata down to 100 m may have an effect.

Despite all these variables, it is found that the terrain with good conductivity gives the best result. Thus, the soil type and moisture content are of importance. Salty sea water is the best, and rich agricultural or marshy land is also good, whereas dry sandy terrain and city centres are by far the worst. This means that sea paths are optimum, although even these are subject to variations due to the roughness of the sea, resulting in path losses being slightly dependent upon the weather. It should also be noted that in view of the fact that signal penetration has an effect, the water table may also have an effect, dependent upon the frequency in use.

Effect of Polarization

The type of antenna used has a major effect on propagation. Vertical polarization is subject to considerably less attenuation than horizontally polarized signals. In some cases, the difference can be of several tens of decibels. It is for this reason that medium-wave broadcast stations use vertical antennas, even if they have to be made physically short by adding inductive loading.

At distances that are typically towards the edge of the ground wave coverage area, some sky wave signal may also be present, especially at night when the D layer attenuation is reduced. This may serve to reinforce or cancel the overall signal, resulting in figures that will differ from those that may be expected.

3.5 IONOSPHERIC PROPAGATION

Ionospheric conditions have already been described. Figure 3.9 gives the summarized view of ionospheric conditions. A radio signal propagates as a *sky wave* using the ionosphere.

Sky Waves

When using ionospheric propagation, radio signals leave the earth's surface and travel towards the ionosphere, where some of these are returned to the earth. These radio signals are termed sky waves. If they are returned to the earth, the ionosphere can simply be viewed as a vast reflecting surface encompassing the earth that enables signals to travel over much greater distances than would otherwise be possible. The frequency, time of day, and many other parameters govern the reflection or, more correctly, the refraction of the signals back to the earth through its layers.

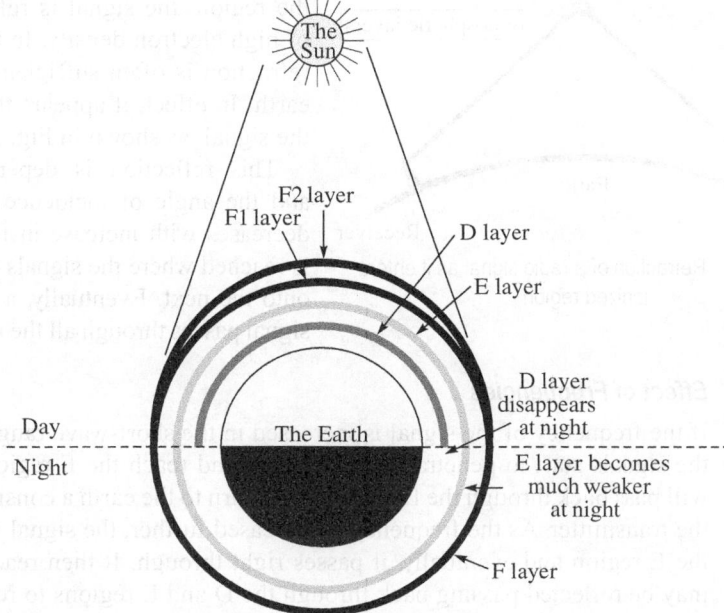

Fig. 3.9 Simplified view of the layers in the ionosphere over periods of day

Effect of D Region

> In the D region, attenuation of the sky wave varies as the inverse square of the frequency; that is, doubling the frequency reduces the level of attenuation by a factor of four.

When a sky wave leaves the earth's surface and travels upwards, the first region that it reaches in the ionosphere is the D region. This region attenuates the signal as it passes through it. The level of attenuation depends on the frequency; low frequencies are attenuated more. This means that LF signals are often prevented from reaching the higher regions except at night, when the region disappears.

The D region attenuates the radio signals because they cause the free electrons in the region to vibrate. During the vibration, millions of electrons collide with the molecules, and at each collision, there is a small loss of energy. The amount of signal loss depends on the following factors:

- *Number of gas molecules present*: As the number of gas molecules increases, the number of collisions and hence the attenuation increase.
- *Level of ionization*: As the level of ionization increases, the number of electrons that vibrate and collide with the molecules increases.
- *Frequency of signal*: As the frequency increases, the wavelength of the vibration shortens and the number of collisions between the free electrons and gas molecules decreases. As a result, LF signals are attenuated far more than HF signals. Even so, HF signals still suffer some reduction in signal strength.

Effect of E and F Regions

The air density in the E and F regions is much lesser than in the D region, which means that when the free electrons are excited by the radio signals and vibrate, fewer collisions occur. As a result, the way in which these regions act is somewhat different. The electrons are again set in motion by the radio signal, but they tend to re-radiate it. The signal is travelling in an area where the density of electrons is increasing; hence, as it progresses further into

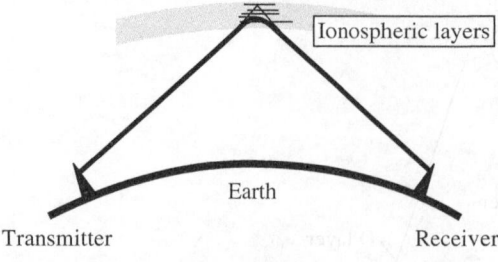

Fig. 3.10 Refraction of a radio signal as it enters ionized region

the region, the signal is refracted away from the area of high electron density. In the case of HF signals, this refraction is often sufficient to bend them back to the earth. In effect, it appears that the region has *reflected* the signal, as shown in Fig. 3.10.

This reflection is dependent upon the frequency and the angle of incidence. The amount of refraction decreases with increase in frequency, until a frequency is reached where the signals pass through the region and onto the next. Eventually, a point is reached where the signal passes through all the regions and into outer space.

Effect of Frequencies

If the frequency of the signal is increased in the short-wave range, a point is reached where the signals start to penetrate the D region and reach the E region. Here, it is reflected and will pass back through the D region and return to the earth a considerable distance away from the transmitter. As the frequency is increased further, the signal is refracted less and less by the E region and eventually, it passes right through. It then reaches the F1 region; here, it may be reflected passing back through the D and E regions to reach the earth again. As the F1 region is higher than the E region, the distance reached will be greater than that for an E region reflection. Finally, as the frequency increases still further, the signal will eventually pass through the F1 region onto the F2 region. This is the highest of the regions in the ionosphere, and the distances reached using this are the greatest. The whole scenario is shown in Fig. 3.11. Sky wave communication suffers from a blackout of signals to some parts of the earth, which are known as *skip zones*. As a rough guide, the maximum skip distance on the earth's curvature is approximately 2500 km for the E region and 5000 km for the F2 region.

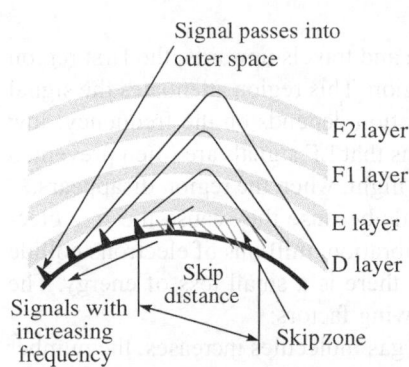

Fig. 3.11 Signals reflected by E and F regions

For a given atmospheric layer, the highest frequency that is reflected back for vertical incidence is given by

$$f_c = \sqrt{81N_m} \qquad (3.17)$$

where N_m is the maximum ionization density in the layer.

Multiple Hops

In the ionospheric propagation, certain frequency signals get reflected towards the ground. Once the signals are returned to the earth from the ionosphere, they are reflected back upwards by the earth's surface, and again they are able to undergo another reflection by the ionosphere. Naturally, the signal strength is reduced at each reflection. It is also found that different areas of the earth reflect radio signals differently. As shown in Fig. 3.12, multiple hops of a signal will be formed due to these multiple reflections. One of the reasons for this is that the signal has to pass through the D region twice for every reflection by the ionosphere.

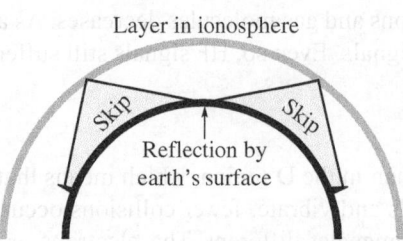

Fig. 3.12 Formation of multiple hops due to reflection by earth's surface

Critical, Maximum Usable, and Lowest Usable Frequencies

In sky waves, multihop propagation losses are introduced by both the earth's surface and the D region even at high frequencies.

When looking at ionospheric or short-wave radio signal propagation for planning a radio network or system, or when predicting the propagation conditions, several frequencies are important and are often mentioned in radio signal propagation predictions and in other literature associated with signal propagation. These include the critical frequency, the lowest usable frequency (LUF), and the maximum usable frequency (MUF).

Critical frequency This is important to give an indication of the state of the ionosphere. It is obtained by sending a signal pulse directly upwards, which may be reflected back and can be received by a receiver on the same site as the transmitter. If the pulse is reflected back to earth, the time between transmission and reception is measured to give an indication of the height of the layer. As the frequency of the signal pulse is increased, a point is reached where it will pass right through the layer and on to the next one or onto outer space. The frequency at which this occurs is called the *critical frequency*. The equipment used to measure the critical frequency is called an *ionosonde*.

Lowest usable frequency As the frequency of transmission is reduced, further reflections from the ionosphere and the losses from the D layer increase. The *LUF* is defined as the frequency below which the signal falls below the minimum strength required, that is, the threshold level, for satisfactory reception.

Thus, it can be seen that the LUF is dependent upon the earth stations at either end of the path. Their antennas, receivers, and transmitter powers, the level of noise in the vicinity, and so on, all affect the LUF. The type of modulation used also has an effect, because some types of modulations can be detected more at lower strengths than at others. If it is necessary to use a frequency below the LUF, then as a rough guide, a gain of 10 dB must be made to decrease the LUF by 2 MHz. This can be achieved by methods such as increasing the transmitter powers and improving the antennas.

Maximum usable frequency As the signal frequency increases, the signal passes through more layers and eventually, travels into outer space. As it passes through a layer, the communication *may* be lost because the signal then propagates over a greater distance than is required. However, when the signal passes through all the layers, the communication *will* be lost. The frequency at which the communication just starts to fail is known as the *MUF*. It is generally three to five times the critical frequency, dependent upon the layer being used and the angle of incidence. It is given by $f_c \sec \theta$, where f_c is the critical frequency and θ is the angle of incidence at the ionosphere measured with respect to the normal.

IONOSPHERIC COMMUNICATION FREQUENCY SELECTION

In general, selecting the higher frequency is the better option. When using higher frequencies, it is necessary to ensure that communications are still reliable. In view of the ever-changing state of the ionosphere, a general rule of thumb is to use a frequency that is about 20 per cent below the MUF. This should ensure that the signal remains below the MUF despite the short-term changes. However, it should be remembered that the MUF will change significantly according to the time of day, and therefore, it will be necessary to alter the frequency periodically to take account of this.

3.6 TROPOSPHERIC PROPAGATION

Elevated duct propaga-
tion increases the
coverage distance up to
2000–3000 km, whereas
troposcatter covers
approximately 800 km.

For frequencies at VHF and above, different modes of propagation prevail. At frequencies above 30 MHz, it is found that the troposphere has an increasing effect on radio signals, which are able to travel over greater distances than would be suggested by LOS calculations. Radio signals may be detected over distances of 500 km or even 1000 km and more. This is normally done by a form of tropo-spheric enhancement, often called tropo for short. At times, signals may even be trapped in an elevated duct in the form of radio signal propagation known as *tropospheric ducting* or *duct propagation*.

Tropospheric ducting is of great importance for those looking at radio coverage of systems such as cellular telecommunications, mobile radio, and other wireless systems as well as other users, including radio hams. Most radio communications at VHF and above are assumed to follow an LOS path. This is not strictly true, and it is found that even under normal conditions, radio signals are able to travel or propagate over distances that are greater than the LOS. From Fig. 3.4, it can be seen that the reason for the increase in the distance travelled by the radio signals is that they are refracted by small changes that exist in the earth's atmosphere close to the ground. It is found that the refractive index of the air near the ground is very slightly higher than that of the air higher up. As a result, the radio signals are bent towards the area of higher refractive index, that is, near the ground. It thereby extends the range of the radio signals.

The refractive index of the atmosphere varies according to the temperature, atmospheric pressure, and water vapour pressure.

Enhanced Conditions for Greater Distances

Under certain conditions, elevated ducts are created (Fig. 3.13a), which results in extended coverage. This form of *lift* in conditions is less pronounced on the lower portions of the VHF spectrum but is more apparent on some of the higher frequencies. These extended distances result from much greater changes in the values of the refractive index over the signal path. These elevated tropospheric ducts occur when a mass of air with a high refractive index has a mass of air with a lower refractive index underneath and above it because of the movement

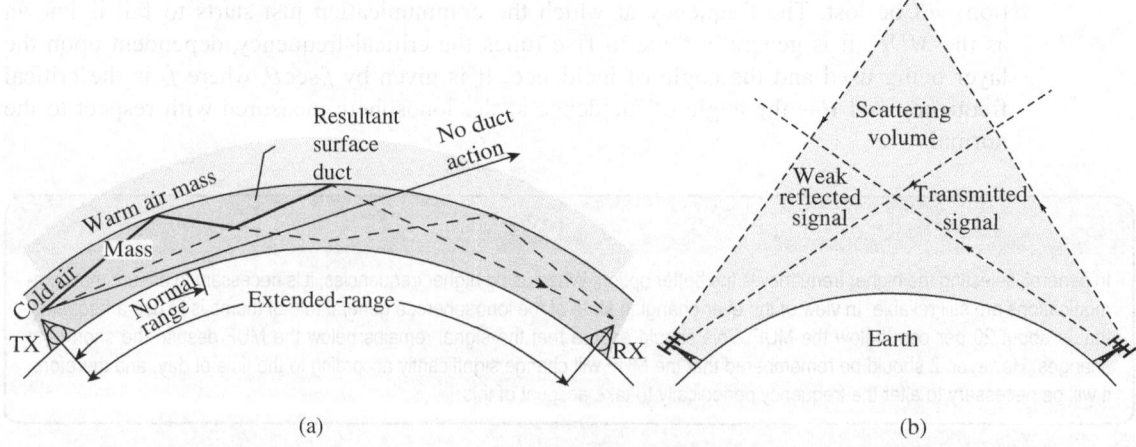

(a)   (b)

Fig. 3.13 Principles of tropospheric propagation (a) Duct and temperature inversion (b) Troposcatter propagation

of air that can occur under some conditions. When these conditions occur, the signals may be confined within the elevated area of the air with the high refractive index and they cannot escape and return to earth. They may also not be audible to the stations underneath the duct and, in this way, create a skip or dead zone similar to that experienced with HF ionospheric propagation.

There is a strong link between weather conditions and radio propagation conditions and coverage. Normally, the temperature of the air closest to the earth's surface is higher than that at a greater altitude. This effect tends to reduce the air density gradient (and hence the refractive index gradient) as air with a higher temperature is less dense. However, under some circumstances, a *temperature inversion* occurs. This happens when the hot air near the earth rises, allowing the colder and denser air to come closer to the earth. This gives rise to a greater change in the refractive index with height.

When signals are propagated over extended distances, they are normally subject to slow deep fading due to multipath. As the winds in the atmosphere move the air around, the different paths will change over a period of time, hence the multiple paths.

Troposcatter Propagation

A useful form of communication for applications where path lengths of about 800 km are needed is *tropospheric scatter* or *troposcatter*. Though reliable, troposcatter links require high powers, high antenna gains, and sensitive receivers. It is often used for commercial applications, normally on frequencies above 500 MHz for over the horizon links. It is ideal for remote telemetry or other links where low- to medium-rate data needs to be carried. Where viable, troposcatter provides a means of communication that is much cheaper than using satellites. The whole scenario is shown in Fig. 3.13(b). In the figure, the resultant duct is shown with a thick black line, no duct action is shown with a thin black line, and other lines are dotted. The densities of warm air mass and cold air mass are shown in faint grey and dark grey colours.

Troposcatter uses the troposphere to return the radio signals to the earth so that they can be received by the distant receiver. It relies on the fact that there are areas of slightly different dielectric constants in the atmosphere at an altitude between 2 km and 5 km. At these heights, even dust in the atmosphere adds to the reflection of the signal. When a transmitter launches a high-power signal, most of it passes through the atmosphere into outer space and only a small amount is scattered back to the earth. Additionally, the angles through which the signals can be reflected are normally small.

The area within which scattering takes place is called the *scatter volume*, and its size is dependent upon the gain of the antennas used at either end. Scatter results in multipath that limits the data rate. It is also found that there are large short-term variations in the signal because of turbulence and changes in the scatter volume. As a result, commercial troposcatter propagation systems use multiple diversity systems. This is achieved by using vertically and horizontally polarized antennas as well as different scatter volumes (angle diversity) and different frequencies (frequency diversity) (refer Chapter 8).

3.7 CHANNEL NOISES AND LOSSES

Both channel noises and losses are unavoidable parameters of any communication system. We have already discussed free space losses in Section 3.3; a few others are discussed in this section. For calculation of losses, a specific case of macrocellular environment or ground wave propagation is considered here. A similar approach can be developed for the calculation of losses for the other types of propagation. We will first look at the different types of noises. It should be remembered that AWGN is always present in any type of channel model.

3.7.1 Different Types of Noises

Noises in communication systems can be categorized into the following types:

External noise   This is caused by the surroundings of the transmitter and the receiver.

Atmospheric noise   This noise is the result of spurious radio waves that include voltages in the antenna as it picks them up. The majority of these radio waves come from natural sources of disturbance. The atmospheric noise is generally called *static* and it is caused by the natural disturbances occurring in the atmosphere. It originates in the form of amplitude-modulated impulses, and because such processes are random in nature, it is spread over most of the RF spectrum normally used for broadcasting. It is additive in nature. As atmospheric noise has an infinite spectrum, just like white light, it is also called *white noise*. Its characteristics are described in Section 3.7.2. It cannot be eliminated, but its effect can be reduced by various methods. Atmospheric noise becomes less severe at frequencies above 30 MHz.

Extraterrestrial noise   This is of two types: solar and cosmic.

Solar noise   Under normal conditions, there is constant radiation from the sun, which is a large body at a very high temperature (greater than 6000°C). The solar noise, therefore, radiates over a very broad frequency spectrum, which includes the frequencies we use for communication.

Cosmic noise   Since distant stars are also at high temperatures, they radiate RF noise in the same manner as the sun. Though they are far from the earth, they are many in number and hence in combination become significant. Galaxies are also responsible for cosmic noise.

Industrial noise   In urban, suburban, and other industrial areas, where machines are used significantly, the noise affecting the signal may be in the range of 1–600 MHz. Automobiles and aircraft ignition, electric motors and switching equipment, leakage from high-voltage lines, a multitude of other heavy electric machines, and many such factors cause this type of noise.

Internal noise   This is caused by the components and connections used in the hardware of transmitters and receivers.

Thermal agitation noise   The noise generated in the resistance or the resistive component is random and is referred to as the thermal, agitation, white, or Johnson noise. It is due to the random and rapid motion of the molecules, atoms, and electrons inside the component. The amount of thermal noise to be found in a bandwidth of 1 Hz in any device or conductor is $N_o = KT$, where N_o is the noise power density in watts/hertz, K is the Boltzmann's constant (1.3803×10^{-23} J/K), and T is the temperature in kelvin (absolute temperature).

> The AWGN channel exhibits thermal noise with Gaussian PDF and the frequency components are spread over the whole frequency spectrum; hence, it is called white.

Shot noise   The most important of all the other sources is the *shot effect*. It leads to shot noise in all amplifying devices and virtually all active devices such as transistors. It is caused by the random variations in the arrival of electrons (or holes) at the output electrode of an amplifying device and appears as a randomly varying noise current superimposed on the output. When amplified, it sounds as though a shower of lead shots were falling on a metal sheet, hence the name shot noise.

Transit time noise   If the time taken by an electron to travel from the emitter to the collector of a transistor becomes significant to the period of the signal being amplified, it means that at frequencies in the upper VHF range and above, transit time effect takes place, and the noise input admittance of the transistor increases. The minute currents induced in the input of the device by random fluctuations in the output current become important at such frequencies, and create random noise and hence frequency distortion.

Intermodulation noise   Apart from this, when signals at different frequencies share the same transmission medium, it may result in *intermodulation noise*. For example, the mixing of signals at frequencies f_1 and f_2 might produce energy at the frequency $f_1 + f_2$. This derived signal can interfere with an intended signal at the frequency $f_1 + f_2$. Crosstalk occurs when a person conversing with someone over the telephone hears another unrelated conversation. This is an unwanted coupling between signal paths.

3.7.2 Noise Parameters

There are mainly three noise-related terms: noise factor, noise figure, and noise temperature. Though the three terms are different, they can be defined in related ways as follows:

Noise factor   For components such as resistors, it is the ratio of the noise produced by the real resistor to the simple thermal noise of an ideal resistor. For a system, it is the ratio of output noise power to input noise power. To make comparisons easier, the noise factor F_{noise} is always measured at the standard temperature (T_o) of 290 K (standardized room temperature). Input noise power is defined as the product of the source noise at the standard temperature (T_o) and amplifier gain (G).

$$P_{ninput} = GKT_oB \tag{3.18}$$

or $\quad F_{noise} = P_{noutput}/P_{ninput}$ 　　　　　　　　　　　　　　　　　　　　　　　　(3.19)

It is also possible to define noise factor in terms of output and input signal-to-noise (SNR) ratios.

$$F_{noise} = SNR_{input}/SNR_{output} \tag{3.20}$$

assuming absence of signal power. Therefore, $F_{noise} = (GKT_oB + \Delta N)/GKT_oB$

Noise figure   It is a figure of merit. Noise figure (NF) is the noise factor converted to decimal notation.

$$NF = 10\log_{10}(F_{noise}) \tag{3.21}$$

> Noise is also caused by flicker and resistance. Flicker noise is very significant at low frequencies.

Noise temperature   It is a means for specifying noise in terms of an equivalent temperature. It was shown earlier that noise power is directly proportional to temperature in kelvin, and it collapses to zero at absolute zero (0 K). The equivalent noise temperature T_e is not a physical temperature of the amplifier or antenna, but rather a theoretical or hypothetical construct that represents the proportional amount of the noise power.

$$T_e = (F_{noise} - 1)T_o \qquad (3.22)$$

Moreover

$$T_e = [\text{antilog}(NF/10) - 1]T_o \qquad (3.23)$$

3.7.3 Ground Reflection Loss

The following model for loss calculation is developed using Fig. 3.8(a).

TWO-RAY MODEL

Two-ray model has been found to be reasonably accurate for calculating large-scale signal strength over distances of several kilometres, especially for mobile radio systems using tall antenna towers (heights > 50 m) as well as for LOS microcellular channels for urban areas. It is the simplest analytical model for analysing multipath reception.

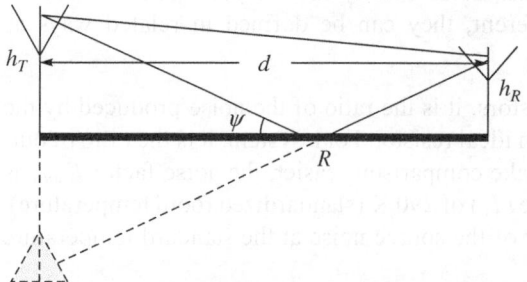

Fig. 3.14 Two-ray model for UHF propagation over plane reflecting earth

The three components of a ground wave are *LOS wave*, *ground-reflected wave*, and *surface wave* (Fig. 3.8a). The surface wave has negligible amplitudes. The relative amplitude of a surface wave is very small for most cases of mobile UHF communication (<< 1). Its contribution is relevant only at a few wavelengths above the ground. So, mostly a *two-ray model* is used, as shown in Fig. 3.14. If we consider the effect of the earth's surface, the expressions for the received signal become more complicated than in the case of free space propagation. The main effect is that signals reflected off the earth's surface may (partially) cancel the LOS wave. The following can be observed from Fig. 3.14:

- Transmit antenna height = h_T
- Receive antenna height = h_R
- Path length of LOS wave = d
- Reflection component = R

The length can be found by taking the mirror image of the transmitting antenna and extending the line to it from the point of reflection.

For isotropic antennas above a plane earth, the received electric field strength E is

$$E = E_0(1 + R_c e^{j\delta} + (1 - R_c)s(\cdot)e^{j\delta} + \ldots \qquad (3.24)$$

where

R_c = ground reflection coefficient
E_0 = field strength for propagation in free space
δ = phase difference between direct and reflected rays
$s(\cdot)$ = surface attenuation factor << 1

The expression in Eq. (3.24) can be interpreted as the complex sum of a direct LOS wave, a ground-reflected wave, and a surface wave. The phasor sum of the first and second terms is known as the *space wave*, as discussed earlier. Referring to Fig. 3.14, for a horizontally polarized wave incident on the surface of a perfectly smooth earth,

$$R_c = \frac{\sin\phi - \sqrt{(\varepsilon_r - jx) - \cos^2\phi}}{\sin\phi + \sqrt{(\varepsilon_r - jx) - \cos^2\phi}} \tag{3.25}$$

where ε_r is the relative dielectric constant of the earth, ϕ is the angle of incidence between the radio ray and the earth's surface, and

$$x = \sigma_g/(\omega_c\varepsilon_0) \approx 18 \times 10^9\,\sigma_g/f_c$$

where σ_g is the conductivity of the ground, ω_c the angular carrier frequency, and ε_0 the dielectric constant of vacuum.

For vertical polarization,

$$R_c = \frac{(\varepsilon_r - jx)\sin\phi - \sqrt{(\varepsilon_r - jx) - \cos^2\phi}}{(\varepsilon_r - jx)\sin\phi + \sqrt{(\varepsilon_r - jx) - \cos^2\phi}} \tag{3.26}$$

The value of δ in Eq. (3.24) can be determined in the following manner:

An LOS and a ground-reflected wave arrive at the receiving antenna. These two rays are out of phase due to differences in the path length and due to phase shifts at the reflection area. The phase difference between the direct and ground-reflected waves can be found using two-ray approximation by considering only an LOS and a ground-reflected wave. Denoting the transmit and receive antenna heights as h_T and h_R, respectively, the phase difference can be expressed as

$$\delta = \frac{2\pi}{\lambda}\left[\sqrt{d^2 + (h_T + h_R)^2} - \sqrt{d^2 + (h_T - h_R)^2}\right] \tag{3.27}$$

The term within the brackets in Eq. (3.27) represents the path difference $\Delta \cong \dfrac{2h_T h_R}{d}$. The exact expression for the phase difference is not very convenient for further analysis. For large propagation ranges, that is, for large d, one can use an approximation based on

$$\sqrt{1 + m} \cong 1 + \frac{m}{2} \tag{3.28}$$

Though this is an approximation, it mostly results in a quite accurate phase difference.

$$\delta \cong \frac{4\pi}{\lambda}\frac{h_T h_R}{d} \tag{3.29}$$

As *d* becomes large, the path difference Δ becomes very small and the amplitudes of the LOS as well as ground-reflected signals become virtually nearly identical and differ only in phase.

This equation and an approximation for the reflection coefficient give an often-used two-ray model for plane earth propagation. For large d, $(d \gg 5h_T h_R)$, the reflection coefficient tends to -1; hence, the received signal power becomes

$$P_r = \frac{\lambda^2}{(4\pi d)^2}\left[2\sin\frac{2\pi}{\lambda}\frac{h_T h_R}{d}\right]^2 P_t G_t G_r \tag{3.30}$$

If

$$d = \frac{4}{\lambda} h_T h_R \qquad (3.31)$$

then the distance d acts as a turnover point d_t, and for this distance, the argument of the sine becomes equal to $\pi/4$.

Example 3.2 For a carrier frequency of 1 GHz, the height of a base station is 30 m and of a mobile antenna is 2 m. Find the turnover distance when the signal is transmitted from the base station to the mobile.

Solution $h_T = 30$ m

$h_R = 2$ m

$\lambda = c/f = (3 \times 10^8 \text{ m/s})/(1 \times 10^9 \text{ Hz})$

$d = (4/3) \times 10 \times 30 \times 2 = 800$ m

For 900 MHz, the turnover distance is about 720 metres. Hence, for macrocellular systems, the distances beyond the turnover distance are more relevant.

The full path loss expression shows an interference pattern of the LOS and ground-reflected waves for relatively short ranges, and a rapid decay of the signal power beyond the turnover distance. For propagation distances substantially beyond the turnover point, the path loss tends to the fourth power distance law as per Egli's model:

$$P_r \to \frac{(h_T h_R)^2}{d^4} P_t G_t G_r \qquad (3.32)$$

Experiments have confirmed that in macrocellular links over a smooth, plane terrain, the received signal power (expressed in dB) decreases with 40log$d$. In addition, a 6 dB/octave height gain is experienced—doubling the antenna height increases the received power by a factor of four. The two-ray model correctly predicts these two effects.

However, in contrast to the theoretical plane earth loss, Egli measured a significant increase in the path loss with increase in the carrier frequency f_c for ranges 1 km $< d <$ 50 km. He proposed the following semi-empirical model:

$$P_r = \frac{(40\text{MHz})^2}{f_c} \frac{(h_T h_R)^2}{d^4} P_t G_t G_r \qquad (3.33)$$

That is, a frequency dependent empirical correction (40 MHz/f_c)^2 was introduced for carrier frequencies 30 MHz $< f_c <$ 1 GHz.

For communication at short range, this formula loses its accuracy because the reflection coefficient is not necessarily close to -1. For $d \ll 4h_T h_R/\lambda$, free space propagation is more appropriate, but a number of significant reflections against the earth's surface must be anticipated. However, these are often hard to distinguish from the reflections against obstacles in the vicinity of the antenna site. Moreover, in streets with high buildings, guided propagation may occur.

Note: Egli's model is derived from data measured to a distance of approximately 50 miles and over gently rolling terrain with average hill heights of approximately 50 feet. It relies on the frequency, distance, and heights of the transmitting and receiving antennas.

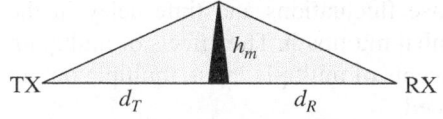

Fig. 3.15 Path profile model for (single) knife-edge diffraction

3.7.4 Diffraction Loss

If direct LOS is obstructed by a single knife-edge type of obstacle, with height h_m (Fig. 3.15), then the diffraction parameter v can be defined as follows:

$$v = h_m \left(\sqrt{\frac{2}{\lambda} \left(\frac{1}{d_T} + \frac{1}{d_R} \right)} \right) \qquad (3.34)$$

where d_T and d_R are the terminal distances from the knife-edge. The diffraction loss, additional to the free space loss and expressed in dB, can be closely approximated by the following:

$$P_{difrdB} = 0 \qquad \text{if } v < 0$$
$$P_{difrdB} = 6 + 9v + 1.27v^2 \qquad \text{if } 0 < v < 2.4$$
$$P_{difrdB} = 13 + 20\log v \qquad \text{if } v > 2.4$$

The attenuation over rounded obstacles is usually higher than the value of P_{difrdB} given by this formula.

3.7.5 Total Path Loss

The previously presented methods for ground reflection loss and diffraction loss suggest an interpretation of the path profile. Obstacles occur as straight vertical lines whereas horizontal planes cause reflections. That is, the propagation path is seen as a collection of horizontal and vertical elements. However, accurate computation of the path loss over NLOS paths with ground reflections is a complicated task and does not allow such simplifications.

Many measurements of propagation losses for paths with combined diffraction and ground reflection losses indicate that the knife-edge type of obstacles significantly reduces ground wave losses. Blomquist suggested two methods to find the total loss:

$$P_{totaldB} = P_{fsdB} + \sqrt{P_{grdB}^2 + P_{difrdB}^2} \qquad (3.35)$$

and the empirical formula

$$P_{totaldB} = P_{fsdB} + P_{grdB} + P_{difrdB} \qquad (3.36)$$

where P_{fsdB} is the free space loss, P_{grdB} is the ground reflection loss, and P_{difrdB} is the multiple knife-edge diffraction loss in dB values.

3.8 FADING IN LAND MOBILE SYSTEMS

Since there are obstacles and reflectors in the wireless propagation channel, the transmitted signal arrives at the receiver from various directions over a multiplicity of paths. As mentioned earlier, such a phenomenon is called *multipath*. The multipath effect is generally present in all wireless communications and it depends upon the distance and surroundings. However, because of the dynamic environment, multipath is a serious problem, specifically for very short distance communication systems such as indoor/outdoor communication and urban area systems. It is an unpredictable set of reflections and/or direct waves, each with its own degree of attenuation and delay as shown in Fig. 3.2.

The multipath effect will cause amplitude and phase fluctuations and time delay in the received signals. We can use various schemes to combat multipath. The effects of multipath are also discussed in Chapter 8, where the basic concepts of multiple input, multiple output (MIMO) and its exploitation of multipath are explained.

> *Note:* The major effect of multipath, noise, and losses together is fading, also called multipath fading.

The quality of communications between a base station transmitter and a mobile (or stationary) receiver depends on several factors, including the general quality of the propagation channel through which the signal passes. In wireless applications, especially cellular communications, the propagation channel is the terrestrial air, with a significant number of man-made and natural objects that get in the way. As the transmitted signal is absorbed by the atmosphere, penetrated through objects of various materials, and reflected off buildings and trees, it experiences fluctuations in its amplitude as well as phase. Wireless channels, in short, exhibit highly irregular amplitude or signal strength behaviour. The phenomenon of time variation of the received signal power caused by changes in the transmission medium or path is generally referred to as *fading*. It is essentially caused by the reception of multiple reflections of the transmitted signal (illustrated in Fig. 3.16). Fading is a key inherent problem of wireless channels.

As shown in Fig. 3.17, the mechanism of fading is normally broken down into the following two categories based on the position of the receiver relative to the transmitter:

(a) Large-scale fading for channel propagation over long distances (of the order of a few kilometres)

(b) Small-scale fading due to time-varying reflections from the surroundings near the receive antenna (of the order of a few hundred metres)

However, both the effects are overlapped and their distinct identification at the receiver is difficult. From the plot of signal strength versus distance, one can approximate them. Figures 3.18(a) and (b) differentiate between the small- and large-scale fading effects and the reasons for these effects.

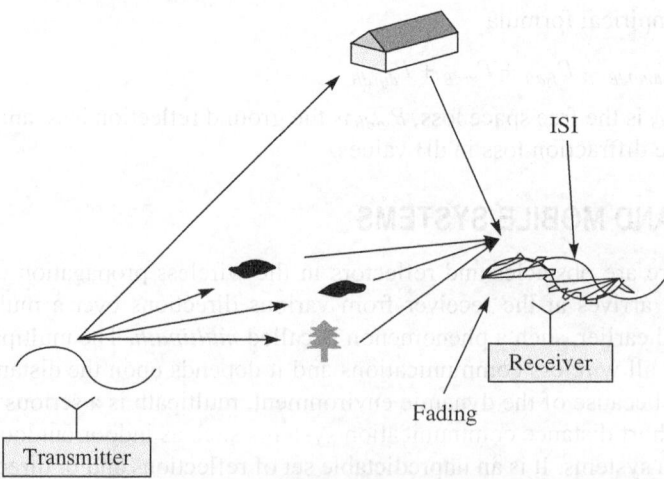

Fig. 3.16 Multipath fading effect

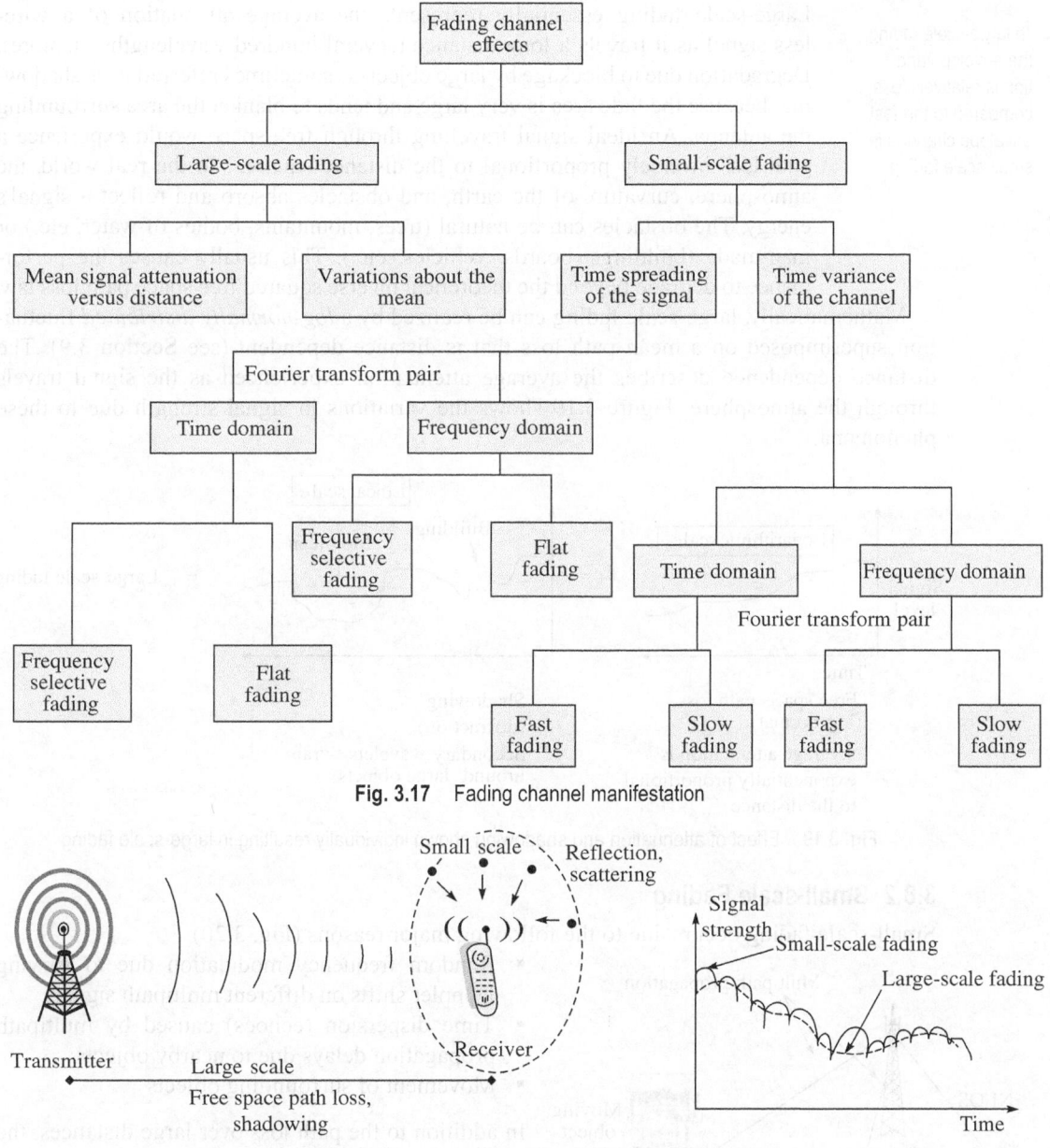

Fig. 3.17 Fading channel manifestation

Fig. 3.18 Fading (a) Difference between large- and small-scale fading environments (b) Large- and small-scale fading shown by variations in signal strength with respect to time

3.8.1 Large-scale Fading

Large-scale fading occurs due to the following major reasons:

- *Attenuation in free space*: Power degrades with the increase in distance.
- *Shadows*: Signals are blocked by obstructing structures.

> In large-scale fading, the envelop variation is relatively less compared to the fast envelope changes in small-scale fading.

Large-scale fading essentially represents the average attenuation of a wireless signal as it travels a long distance (several hundred wavelengths or more). Degradation due to blockage by large objects is sometimes referred to as shadowing, because the fade area is very large and tends to blanket the area surrounding the antenna. An ideal signal travelling through free space would experience a path loss inversely proportional to the distance squared. In the real world, the atmosphere, curvature of the earth, and obstacles absorb and reflect a signal's energy. The obstacles can be natural (trees, mountains, bodies of water, etc.) or man-made (buildings, boards, vehicles, etc.). This usually causes the performance to degrade beyond the theoretical inverse squared free space path loss law.

Mathematically, large-scale fading can be realized by a *log-normally distributed* fluctuation superimposed on a mean path loss that is distance dependent (see Section 3.9). The distance dependence describes the average attenuation experienced as the signal travels through the atmosphere. Figure 3.19 shows the variations in signal strength due to these phenomena.

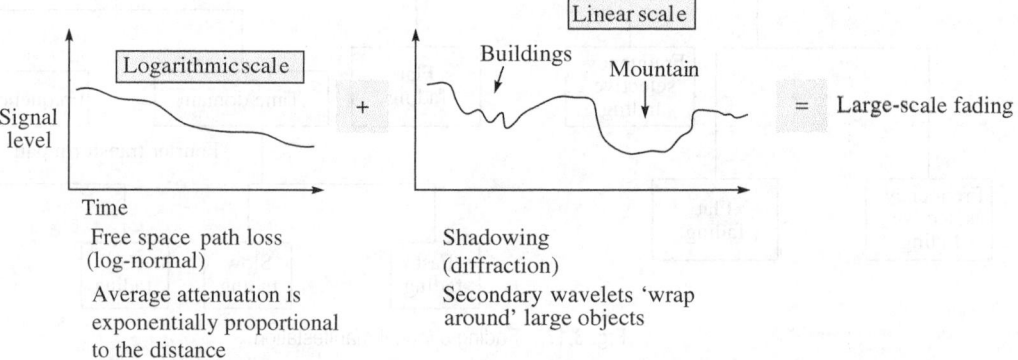

Fig. 3.19 Effect of attenuation and shadowing shown individually resulting in large-scale fading

3.8.2 Small-scale Fading

Small-scale fading occurs due to the following major reasons (Fig. 3.20):

- Random frequency modulation due to varying Doppler shifts on different multipath signals
- Time dispersion (echoes) caused by multipath propagation delays due to nearby objects
- Movement of surrounding objects

In addition to the path loss over large distances, the receive antenna will also experience fluctuations in signal levels that vary significantly over small distances (in the order of one to tens of wavelengths). The fluctuation is a result of two distinct processes: multipath propagation and Doppler shift. A signal transmitted from the base station can take different paths to the receiver due to reflection, diffraction, and local scattering. Different paths have different lengths associated with them, which causes the

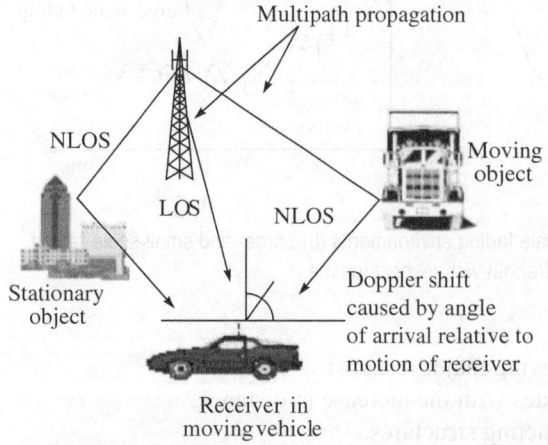

Fig. 3.20 Various reasons for small-scale fading

> Small-scale fading is less predictable and more destructive than large-scale fading.

receiver to see multiple copies of the signal at different times of arrival. In addition, the signal can shift in phase as it is reflected and scattered off local objects. All these signals at different power levels and phase converge on the receiver antenna with constructive or destructive interference. As the mobile device moves, its antenna will experience peaks and valleys of signal strength as these interfering wavelets add and subtract at the receiver. Another adverse effect of motion is Doppler shift. As the receiver antenna moves in relation to a fixed transmitter, the incoming signal will modulate in frequency according to the direction of movement. The copies of the signal that arrive via paths directly in front of the moving receiver will seem to have a higher frequency, whereas the copies of the signal arriving via paths behind the moving receiver will seem to have a lower frequency. This will also cause small-scale fading.

Thus, time spread or delay spread and time variance or Doppler effect are the two main effects observed on the signal due to small-scale fading and leads to the concept of coherence bandwidth and coherence time. The contribution of large-scale fading is also unavoidable in the resultant signal.

Now that we have a general idea of what causes fading, let us move on to the effects observed on the signal. Since small-scale fading is less predictable and potentially more destructive than large-scale fading, most of the discussion here will focus on small-scale fading. The following sections discuss how small-scale fading is observed on the signal and how delay spread and Doppler shift alter the transmitted signal and make it more difficult for the receiver to accurately detect the altered signal.

3.8.3 Delay Spread and Intersymbol Interference

Delay spread effect is mainly due to small-scale fading. As multiple reflections of the transmitted signal may arrive at the receiver at different times and all get added constructively or destructively, this can result in ISI or bits *crashing* or *smearing* into one another, as shown in Fig. 3.21, which the receiver cannot sort out.

This time dispersion of the channel is called *multipath delay spread,* which is an important parameter to asses the performance capabilities of wireless systems. A common measure of multipath delay spread is the *root mean square* (RMS) *delay spread*. There is some finite delay between the time at which the antenna receives the first copy of the signal on the shortest path and that at which it receives the last copy of the same signal on the longest path.

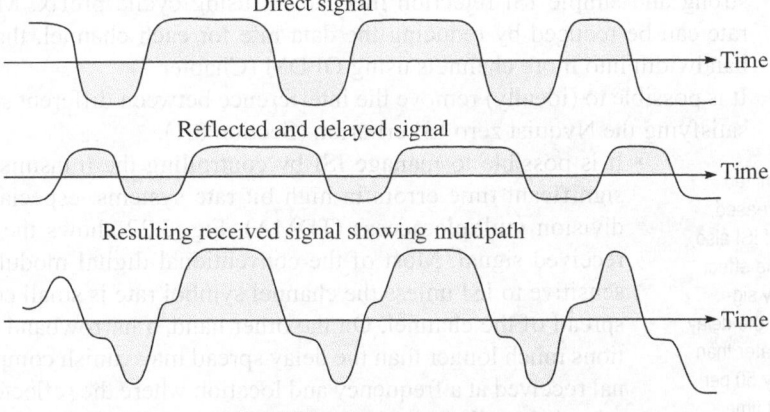

Direct signal

→ Time

Reflected and delayed signal

→ Time

Resulting received signal showing multipath

→ Time

Fig. 3.21 Effect of delay spread

Spreading of pulse is observed due to channel delay spread.

The maximum excess delay time is represented by T_m. (The RMS value of delay spread, T_{rms}, is more commonly used in practice than the maximum excess delay time but is mathematically a little more complex.) Both these time dispersion parameters can be represented in the power delay profile, which is a plot of the power strengths of the received impulses (from the CIR) for the maximum excess delay time (explained in Chapter 4). These parameters can be defined as follows:

Maximum delay time spread   It is the total time interval during which reflections with significant energy arrive.

Root mean square delay spread   It is the standard deviation (or root mean square) value of the delay of reflections weighted proportional to the energy in the reflected waves.

For a digital signal with a high bit rate, this dispersion is experienced as frequency-selective fading (which will be considered in Section 3.9) and ISI. No serious ISI is likely to occur if the symbol duration is longer than, say, ten times the RMS delay spread.

Table 3.1 shows the typical delay spread for various environments. The maximum delay spread in an outdoor environment is approximately 20 µs. Thus, significant ISI can occur at bit rates as low as 25 kbps.

Table 3.1   Typical delay spreads

Environment or cause	Delay spread	Maximum path length difference
Indoor (room)	40–200 ns	12–60 m
Outdoor	1–20 µs	300 m–6 km

As such, delay spread and ISI have similar effects, though there is a difference. ISI is caused by the reception of a limited number of reflections from remote objects, whereas delay spread is caused by the reception of a large number of reflections from nearby objects, often resulting in fading. ISI causes the receiver to receive the original signal, overlapped by some delayed versions of the signal. It may be measured by eye patterns (Chapter 7).

Ways to Reduce Intersymbol Interference

The following are some measures to mitigate ISI:

- Various types of equalizers are used to reject ISI.
- The orthogonal frequency division multiplexing (OFDM) modulation provides a fairly strong and simple ISI rejection mechanism using cyclic prefix. Moreover, the symbol rate can be reduced by reducing the data rate for each channel, that is, by splitting the bandwidth into more channels using OFDM (Chapter 7).
- It is possible to (ideally) remove the interference between different symbols using a filter satisfying the Nyquist zero ISI criterion (Section 8.1).
  - It is possible to manage ISI by controlling the transmission rate. ISI causes significant time errors in high bit rate systems, especially while using time division multiple access (TDMA). Fig. 3.22 shows the effect of ISI on the received signal. Most of the conventional digital modulation techniques are sensitive to ISI unless the channel symbol rate is small compared to the delay spread of the channel. On the other hand, a narrowband signal with bit durations much longer than the delay spread may vanish completely in fade. A signal received at a frequency and location where the reflected waves cancel each other is heavily attenuated, and may thus suffer large bit error rates (BERs).

As the transmitted bit rate is increased, the amount of ISI also increases. The effect becomes very significant when the delay spread is greater than approximately 50 per cent of the bit time.

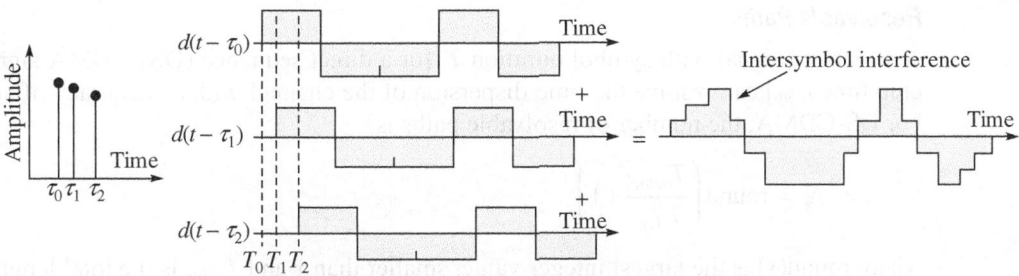

Fig. 3.22 ISI due to delay spread effect

Table 3.2 Measures to eliminate delay spread

System	Measure to eliminate delay spread
Analog	Use narrowband transmission
Global system for mobile communication (GSM)	Use adaptive channel equalization and channel estimation training sequence
DECT (Digitally enhanced cordless telephone)	Use the handset only in small cells with small delay spreads. Diversity and channel selection can help a little (pick a channel where late reflections are in a fade)
IS95 CDMA	Rake receiver separately recovers signals over paths with excessive delays. CDMA array processing can further improve performance, because it also exploits angle spreads
Digital audio or video broadcasting	*OFDM multicarrier modulation*: The radio channel is split into many narrowband subchannels with orthogonality. There is no ISI
MIMO	It exploits multipath diversity and uses multiple transmitting and receiving antennas

- Using a coding scheme that is tolerant to ISI, such as that used in code division multiple access (CDMA), can also help reduce interference.

Table 3.2 lists out some measures to eliminate delay spread in a system.

NARROWBAND TRANSMISSIONS

Delay spread restricts the maximum allowable symbol transmission rate. One can define *narrowband* transmission in the time domain, considering the inter-arrival times of multipath reflections and the time scale of variations in the signal caused by modulation: A signal sees a narrowband channel if the bit duration is sufficiently larger than the inter-arrival time of the reflected waves. In such cases, ISI is small. This outcome leads to few more definitions, which are given in Section 3.9.

Coherence Bandwidth

For a reliable communication without using adaptive equalization or other anti-multipath techniques, the transmitted data rate may be much smaller than the inverse of the RMS delay spread, called the *coherence bandwidth*. The channel transfer function remains virtually constant over the coherence bandwidth.

We shall see in this chapter that narrowband transmission uses radio signals that see flat fading. The channel may be considered relatively constant over the transmit bandwidth. This criterion is found to be satisfied if the transmission bandwidth does not substantially exceed the coherence bandwidth B_c of the channel.

Resolvable Paths

A wideband signal with symbol duration T_s [or a direct sequence (DS)-CDMA signal with chip time t_{chip}] can *resolve* the time dispersion of the channel with an accuracy of about T_s. For DS-CDMA, the number of resolvable paths is

$$N_p = \text{round}\left(\frac{T_{delay}}{t_{chip}} + 1\right)$$ (3.37)

where round(x) is the largest integer value, smaller than x and T_{delay} is the total length of the delay profile. A DS-CDMA rake receiver can exploit N_p-fold path diversity.

Doppler Shift or Spread

Figures 3.23 and 3.24 show the scenarios of Doppler effect. Along with multiple reflected signals, the receiver undergoes the effect of Doppler spread due to its own mobility, especially vehicular mobility. Doppler spread is the width of the received spectrum when a single tone is transmitted and is related to the rate at which fading occurs. It is important in determining the minimum adaptation rate for an adaptive receiver. The motion of a receiving antenna produces Doppler shifts of incoming received waves. Let us consider a signal received over a multipath channel, with many incoming waves. Let the nth reflected wave with amplitude c_n and phase ϕ_n arrive from an angle α_n relative to the direction of motion of the antenna.

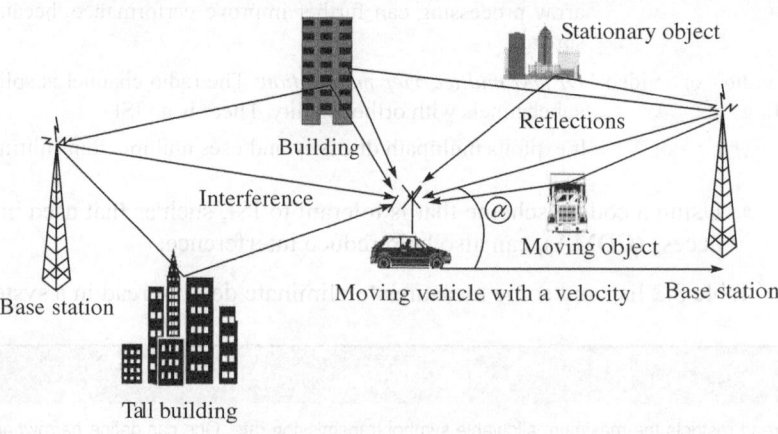

Fig. 3.23 Scenario for Doppler shift

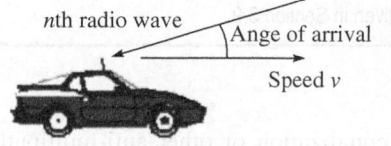

Fig. 3.24 Jake's model for Doppler effect

The Doppler shift of each wave is given by

$$\delta f_n = \frac{v}{\lambda}\cos\alpha_n$$ (3.38)

where v is the speed of the receiving antenna.

The maximum Doppler shift f_d occurs for a wave coming from the direction opposite to that in which the antenna is moving. It has the following frequency shift, with f_c being the carrier frequency and c the velocity of light:

$$f_d = \frac{v}{c} f_c$$ (3.39)

Such motion of the antenna leads to (time-varying) phase shifts of individual reflected waves. This minor shift does not bother radio system designers, as a receiver oscillator can easily compensate for it. However, the problem is that many waves arrive with different shifts. Thus, their relative phases change all the time, and this affects the amplitude of the resulting composite signal.

> Doppler effects determine the rate at which the amplitude of the resulting composite signal changes.

If the same effect is applied to a multicarrier transmission (Chapter 7), all the orthogonal tones placed nearer to each other will be received with a spread. This means that all the carriers will be received with offsets. This will be the worst-case scenario because the carriers will not remain orthogonal, and as they are placed near each other, limiting conditions will be generated. Hence, in a multicarrier environment, the allowable Doppler spread and subcarrier spacing requires more attention.

Example 3.3 A vehicle receives a 910 MHz transmission while travelling at a constant velocity for 15 s. The average fade duration for a signal level 10 dB below the RMS level is 1 ms. How far does the vehicle travel during the 15 s time duration? Assume that the local mean remains constant during travel.

Solution The average fade duration is defined as the average period of time for which the received signal is below the specified (here, RMS) level and can be expressed as

$$\tau_f = \frac{e^{l^2} - 1}{l f_d} \frac{1}{\sqrt{2\pi}} \tag{3.40}$$

where l is the minimum signal strength below the specified (RMS) level due to fade.

we have $l = -10 \text{ dB} = 0.316$

or $\tau_f = \frac{1}{0.316 \times f_d} \frac{e^{0.316^2} - 1}{\times \sqrt{2\pi}} = 1 \text{ ms}$

or $f_d = 132.8 \text{ Hz}$

Now, $v = f_d \times \frac{c}{f_c} = 132.8 \times \frac{3 \times 10^8}{910 \times 10^6} = 43.78 \text{ m/s}$

Total distance $= 15 \times 43.78 = 656.7 \text{ m}$

Doppler Power Spectrum

The models behind Rayleigh or Rician fading (Chapter 4) assume that many waves arrive independently of other waves. Each of these waves has its own random angle of arrival (and thus its own Doppler shift), which is uniformly distributed within [0, 2]. This allows us to compute a probability density function of the frequency of incoming waves. Moreover, we can obtain the Doppler spectrum of the received signal.

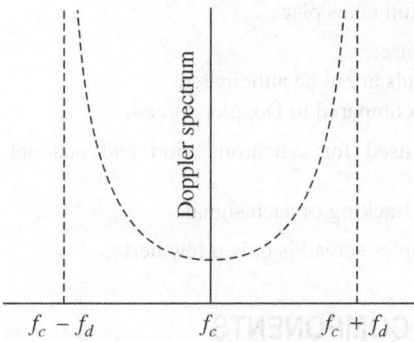

Fig. 3.25 Power density spectrum of a sine wave suffering from a Doppler spread (ideally)

The Doppler effect leads to a U-shaped power spectrum for isotropic scattering as shown in Figures 3.25 and 3.26.

$$S(f) = \frac{1}{4\pi f_d} \frac{1}{\sqrt{1 - \frac{(f - f_c)^2}{f_d^2}}} \tag{3.41}$$

where we assume a unity local mean power.

If a sinusoidal signal is transmitted (represented by the central line at f_c in the frequency domain in Fig. 3.25), then after transmission over a fading channel, we will receive a power spectrum that is spread around the single-frequency

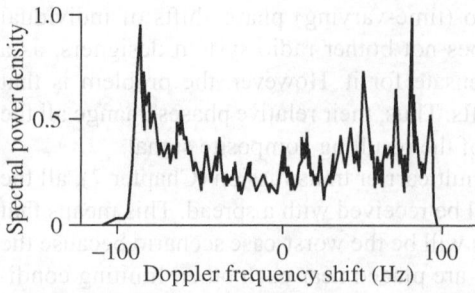

Fig. 3.26   Doppler spread at 1800 MHz = 60.3 Hz (practically achieved)

tone. The frequency range where the power spectrum is non-zero defines the Doppler spread. The Doppler spread is useful, for instance, for computing threshold crossing rates and average fade durations.

Table 3.3 lists out some measures to tackle Doppler spread.

Excess Delay

There are two different forms of multipath scattering depending on the excess time delay of the given channel tap: small excess and large excess. The corresponding effects of each type of excess delay on the Doppler spectrum are discussed here.

Small excess time delays   The channel taps may be modelled as the successive of multipath components that are received from the scatterers close to the mobile. This gives rise to the classical Doppler power spectrum of the received multipath components.

> Delay spread is better studied in time domain, whereas Doppler spread is studied in frequency domain.

Large excess time delays   The classical Doppler model does not provide a satisfactory geometric model for this type of scattering. Instead, multipath energy is more likely to have a narrow Doppler spread, having risen from the reflections of isolated obstacles, such as buildings or hills. The instantaneous variation of signal power in space for a channel depends on the angles of arrival of the multipath components.

Table 3.3   Measures to handle Doppler spread

System	Countermeasure
Analog	Doppler causes random frequency modulation that may be audible; carrier frequency is low enough to avoid problems.
GSM	• Channel bit rate is well above Doppler spread. • TDMA occurs during each bit/burst transmission; the channel is fairly constant. • Receiver training or updating occurs during each transmission burst. • Feedback frequency correction takes place.

DECT	• It is intended for pedestrian use:
	    □ Only small Doppler spreads are to be anticipated.
	    □ The bit rate is very large compared to Doppler spread.

IS95 Cellular CDMA	• *Downlink*: Pilot signal is used for synchronization and channel estimation • *Uplink*: There is continuous tracking of each signal.
Wireless local area networks	Mobility is slow and thus, Doppler spread is only a few hertz.

3.9 FADING EFFECTS ON SIGNAL AND FREQUENCY COMPONENTS

The fading effects of the frequency components are related to bit or symbol transmission rate and time spreading of those pulses. When the transmitted data rate is much smaller than

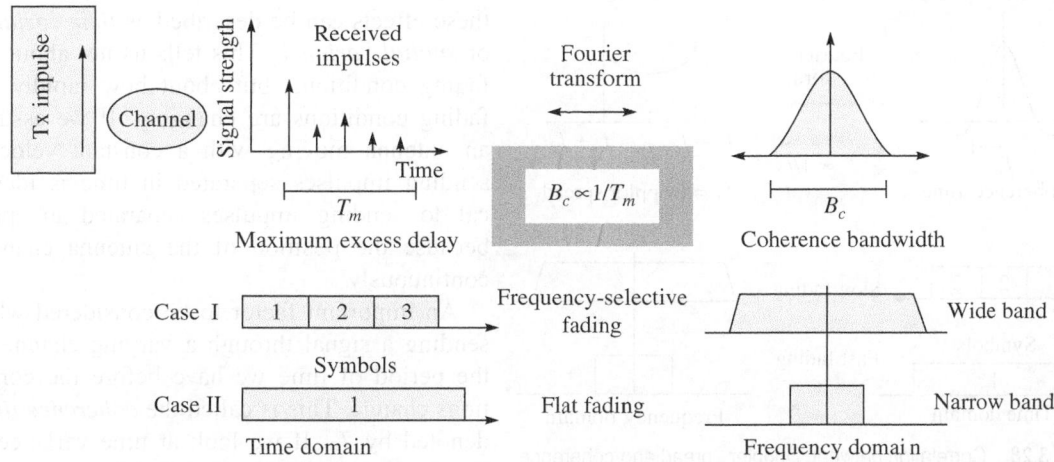

Fig. 3.27 Correlation between delay spread and coherence bandwidth (time spreading) as well as definition of flat and frequency-selective fading

> The coherence bandwidth B_c sets the threshold condition and is the range of frequencies within which the signal impairments of the channel do not vary significantly.

the coherence bandwidth, the wireless channel is referred to as a *flat channel* or *narrowband channel*, and the effect that the signal undergoes is called *flat fading*. In flat fading, all the frequency components of the transmitted signal will be equally affected and will be received with almost equal power level. When the transmitted data rate is closely equal to or larger than the coherence bandwidth, such a channel is called a *frequency-selective channel* or *wideband channel*, and the fading effect in this situation is called *frequency-selective fading*. In this case, one can find the different power levels of the different frequency components. The whole scenario is shown in Fig. 3.27.

When observed in the frequency domain, the delay spread manifests itself as a sort of frequency notch filter. In other words, if two signals that are more than B_c apart from each other in frequency are sent, they will experience different channel conditions. The coherence bandwidth B_c is inversely proportional to T_m.

If the time taken to transmit one symbol T_s is longer than the maximum delay spread T_m ($T_s > T_m$), the path is said to exhibit flat fading, which means that all the multipath components of the transmitted symbol are received within one symbol period. Since signalling rate is inversely proportional to the symbol period, this is equivalent to the case where the symbol bandwidth W is less than the coherence bandwidth ($W < B_c$). On the other hand, frequency-selective fading occurs when $T_s < T_m$ or when $W > B_c$. In this case, the channel will alter the different spectral components of the signal in different ways. As a result, a wideband signal can experience a large variation in the received power over its bandwidth.

We have discussed the effect of time spreading on the signal. Now let us examine time variance. The relationship between coherence time and Doppler spread is shown in Fig. 3.28.

The multipath effects seen previously can affect a stationary receiver, since a transmitted signal will always experience reflection when encountering objects. For receivers that change their location relative to the transmitter, other factors in addition to multipath reflection affect the signal's amplitude and phase. To distinguish them from time spreading,

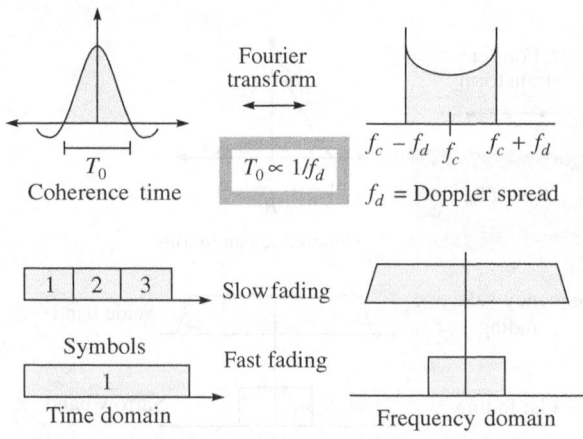

Fig. 3.28 Correlation between Doppler spread and coherence time (time variance) as well as definition of slow and fast fading

> Coherence time sets the threshold condition and is the time period over which the CIR is highly correlated.

these effects can be described as *time variance* or *spatial variance*. This tells us not about the fading conditions, but about how rapidly the fading conditions are changing. If we assume an antenna moving with a constant velocity, sending impulses separated in time is identical to sending impulses separated in space because the position of the antenna changes continuously.

An important factor to be considered while sending a signal through a varying channel is the period of time we have before the conditions change. This is called the *coherence time*, denoted by T_0. If we look at time variance in the frequency domain, we see that a receiver antenna in constant motion will experience shifts in frequency that are dependent on the angle of arrival of the incoming signal and on the speed of motion. Instead of the signal being spread in time, we shall see it being spread in frequency. This Doppler spread, which is represented by f_d, is reciprocally related to the coherence time T_0.

If the symbol period is longer than the coherence time ($T_s > T_0$), the channel exhibits *fast fading*, because the rate at which fading conditions are changing is faster than that at which the symbols are being transmitted. This is analogous to the user bandwidth being smaller than the Doppler spread, ($W < f_d$). If $T_s < T_0$ (or $W > f_d$), the channel exhibits *slow fading*, meaning that the channel conditions are stable and predictable during the time that the symbol is transmitted. Note that for most cellular applications, the order of magnitude of the user bandwidth is generally greater than that of the frequency spread. Therefore, we usually see slow fading effects. However, an important point needs to be remembered. Since the symbol frequency is inversely related to the symbol period, changing the signalling rate to compensate for frequency-selective fading can also alter its performance.

Example 3.4 Measurements for a typical wireless system found that the delay in receiving the first component of a signal over the channel between a steady transmitter and a receiver is 50 ms. The maximum excess delay observed is 65 µs. The transmission bandwidth is 10 MHz with RF carrier frequency 2.4 GHz. Check whether frequency-selective or flat fading will occur over the channel. If the mobile receiver is in vehicular mobility condition at a speed of 50 km/h, check whether fast fading or slow fading will occur.

Solution Bandwidth: $W = 10$ MHz
Maximum excess delay: $T_m = 65$ µs

Coherence bandwidth: $B_c = 1/T_m = 1/(65 \times 10^{-6}) = 15.4$ kHz

Now, as $W > B_c$, frequency-selective fading will occur.

The vehicle moves at a speed of 50 km/h, that is, 13.88 m/s.

Doppler spread: $f_d = \dfrac{v}{c} f_c$

$$= (13.88/3 \times 10^{-8})(2.4 \times 10^9)$$

$$= 111 \text{ Hz}$$

Now, as $W > f_d$, slow fading will occur.

Comparison of Flat and Frequency-selective Fading

In practice, in any radio transmission, the channel spectral response is not flat. There are dips or fades in the response due to reflections causing cancellation of certain frequencies at the receiver. Reflections off nearby objects (e.g., ground, buildings, and trees) can lead to multipath signals of similar signal power as the direct signal. This can result in deep nulls in the received signal power due to destructive interference.

For narrow bandwidth transmissions, if the null in the frequency response occurs at the transmission frequency, then the entire signal can be lost. This can be partly overcome in two ways. One method is to transmit a wide bandwidth signal or spread spectrum as CDMA. In this way, any dip in the spectrum will result in a small loss of signal power rather than a complete loss. Another method is to split the transmission into many small bandwidth carriers, as is done in a coded OFDM (COFDM) or OFDM transmission. The original signal is spread over a wide bandwidth, and so, nulls in the spectrum are likely to affect only a small number of carriers rather than the entire signal. Thus, frequency-selective fading sometimes becomes advantageous.

> *Note:* Flat fading and slow fading degrade the SNR because the reflections cause the phasor components to cancel each other out. On the other hand, fast fading distorts the phase of the transmitted baseband pulse, which can cause synchronization problems in phase-locked loops.

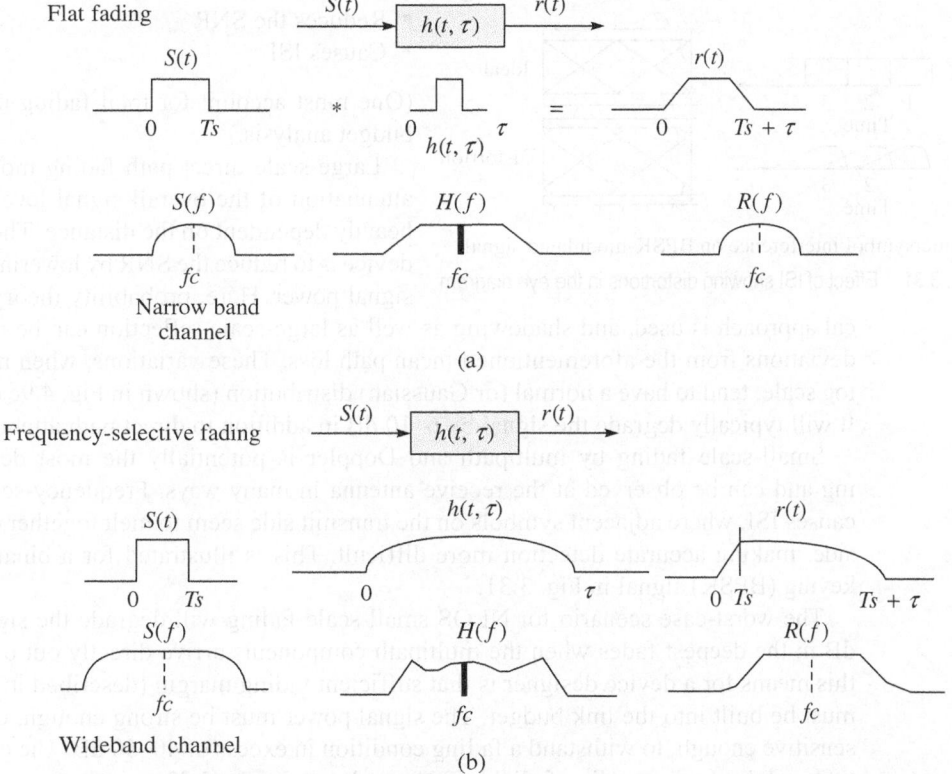

Fig. 3.29 Channel responses (a) Flat fading response (b) Frequency-selective fading response

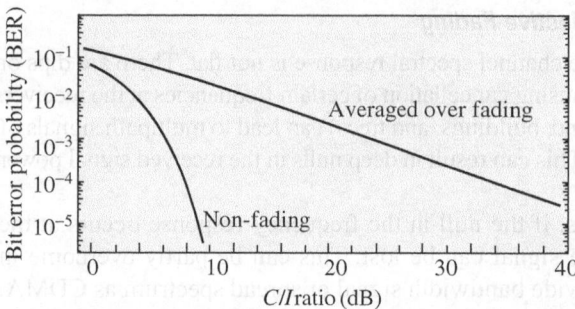

Fig. 3.30 Channel response over fading and non-fading environments

The information lost due to delay and Doppler spread can be recovered up to a certain extent by using forward error correction techniques.

Fading and Bit Errors

A typical characteristic is observed on the wireless channel within fading and non-fading environment. Two typical responses on a pulse under flat and frequency selective fading channels are shown in Figure 3.29. These effects cause bit errors. In Fig. 3.30, the bit error probability for fading and non-fading channels is shown. In a non-fading channel, a slight increase in the carrier-to-interference (C/I) ratio will lead to a considerable drop in bit errors.

In a non-fading radio channel, the BER decreases rapidly with an increase in the C/I ratio. In a fading channel, the received signal is very weak and many bit errors occur. This phenomenon remains even if the (average) SNR is large. Therefore, the BER improves very slowly, and with a fixed slope, if plotted on a log-log scale. (Diversity or error correction can help to make the slope steeper and hence improve performance.)

Removal of Fading Effect

Fading has the following effects on a device:

- Reduces the SNR
- Causes ISI

(One must account for total fading margin in link budget analysis.)

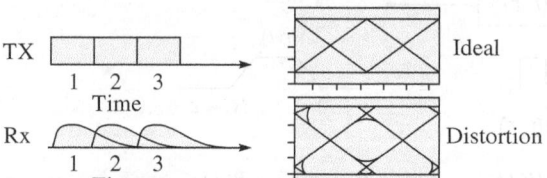

Intersymbol interference on BPSK-modulated signal

Fig. 3.31 Effect of ISI showing distortions in the eye diagram

Large-scale direct path fading mostly results in attenuation of the overall signal level. Path loss is heavily dependent on the distance. The effect on the device is to reduce the SNR by lowering the received signal power. Here, probability theory or a statistical approach is used, and shadowing as well as large-scale reflection can be represented as deviations from the aforementioned mean path loss. These variations, when measured on a log scale, tend to have a normal (or Gaussian) distribution (shown in Fig. 4.9c of Chapter 4). It will typically degrade the signal by 6–10 dB in addition to direct path attenuation.

Small-scale fading by multipath and Doppler is potentially the most destructive fading and can be observed at the receive antenna in many ways. Frequency-selective fading causes ISI, where adjacent symbols on the transmit side seem to melt together on the receive side, making accurate detection more difficult. This is illustrated for a binary phase shift keying (BPSK) signal in Fig. 3.31.

The worst-case scenario for NLOS small-scale fading will degrade the signal by 20–30 dB in the deepest fades when the multipath components arrive directly out of phase. What this means for a device designer is that sufficient fading margin (described in Section 3.11) must be built into the link budget. The signal power must be strong enough, or the receiver sensitive enough, to withstand a fading condition in excess of 40–50 dB. The channel condition and the corresponding fading effect are shown in Fig. 3.32.

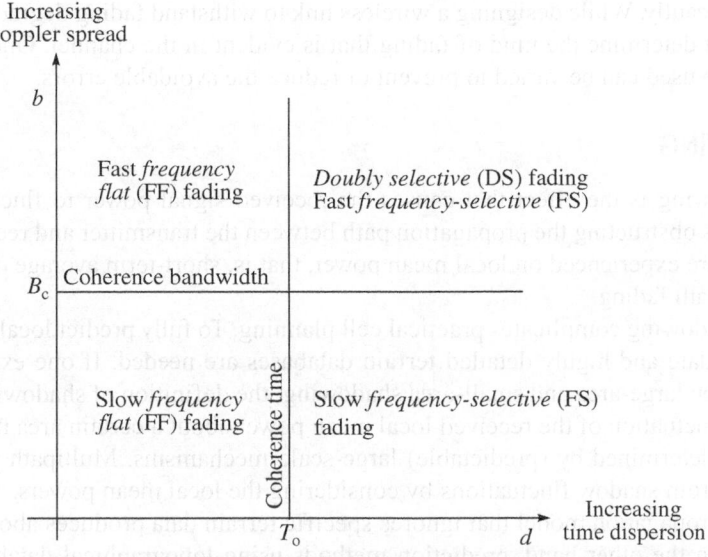

Fig. 3.32 Types of channels and their effects

WAYS TO REDUCE FADING EFFECTS*

- *Signal conditioning*: Limits should be set on bandwidth and bit rate, which is mainly controlled by information theory.
- *Error correction—interleaving and coding*: This is mainly controlled by channel coding. Coding provides redundancy by sending systematically added bits for error detection. Interleaving provides robustness to the link by spreading out errors in time, avoiding large amounts of contiguous data loss.
- There are transmission techniques (line coding) whose signalling properties avoid the most common effects of fading.
- *Device design*:
  - □ *Adaptive equalizer*: Equalization is a commonly used technique to fight ISI caused by frequency-selective fading. The process invokes a filter that has an impulse response opposite to that of the propagating channel. Thus, the combination of the transmission path and the receive filter yields a flat linear response. GSM uses adaptive equalization to mitigate distortion.
  - □ *Rake receiver*: CDMA technologies use a rake receiver to mitigate ISI. It involves using special filters (called *fingers*) that sniff out the spread-out signal components, collect them, and add them coherently by delaying the earlier arriving paths longer than the later arriving paths. Interleaving and coding can be used to increase the E_b/N_o ratio required for accurate detection.
- *Modulation format*: An important example is OFDM, which combats frequency-selective fading by breaking the carrier signal into subcarriers with lower bit rates and thereby longer symbol durations.

* *The techniques will be clear after reading Chapters 5–8.*

> The standard deviation of shadowing will depend on the geographical resolution and the estimate of the area mean power.

The best performance of a wireless link will be realized only in the absence of any type of channel impairment. The presence of Gaussian noise prevents a wireless channel from being completely clear, but many techniques can be employed in the design of wireless devices to reduce the effects of fading. These techniques reduce the error probability from the worst-case Rayleigh curve and bring it closer to the best-case AWGN curve. As we have seen earlier, fading can take a variety of forms. Frequency-selective and fast fading significantly affect the BER probability, whereas flat fading and slow fading affect it less significantly. While designing a wireless link to withstand fading degradation, it is important to first determine the kind of fading that is evident in the channel. Once this is known, the bit rate used can be varied to prevent or reduce the avoidable errors.

3.10 SHADOWING

Shadowing is the effect that causes the received signal power to fluctuate because of the objects obstructing the propagation path between the transmitter and receiver. These fluctuations are experienced on local mean power, that is, short-term average of fluctuations due to multipath fading.

Shadowing complicates practical cell planning. To fully predict local shadow attenuation, up-to-date and highly detailed terrain databases are needed. If one extends the distinction between large-area and small-area shadowing, the definition of shadowing covers any statistical fluctuation of the received local mean power about a certain area mean power, with the latter determined by (predictable) large-scale mechanisms. Multipath propagation is separated from shadow fluctuations by considering the local mean powers.

A propagation model that ignores specific terrain data produces about 12 dB of shadowing. On the other hand, prediction methods using topographical databases with unlimited resolution can, at least in theory, achieve a standard deviation of 0 dB. Thus, standard deviation is a measure of the impreciseness of the terrain description. If, for generic system studies, the (large-scale) path loss is taken to be of a simple form depending only on the distance but not on the details of the path profile, the standard deviation will necessarily be large. On the other hand, while planning a practical network in a certain (known) environment, the accuracy of the large-scale propagation model may be refined. This may allow spectrally more efficient planning if the cellular layout is optimized for the propagation environment.

With shadowing, the interference power accumulates more rapidly than proportional to the number of signals. The accumulation of multiple signals with shadowing is a relevant factor in the planning of cellular networks.

Table 3.4 lists the measures taken by a system to handle shadowing.

Table 3.4 Measures to handle shadowing

DECT	• By proper base station location
IS95	• By power control, which is also needed to achieve sufficient performance of the CDMA receiver
	• By proper base station locations
Digital audio broadcasting	• By single frequency networks as shadow fades are filled in by signals from other co-channel transmitters

3.11 SIGNAL OUTAGES AND FADING MARGIN

Signal outage probability is fairly simple to compute if one knows the probability distribution of the fading.

A mobile radio channel is characterized by rapidly changing channel characteristics. As typically a certain minimum (threshold) signal level is needed for acceptable communication performance, the received signal will experience the following:

• Periods of sufficient signal strength or *non-fade intervals*
• Periods of insufficient signal strength or *fades*

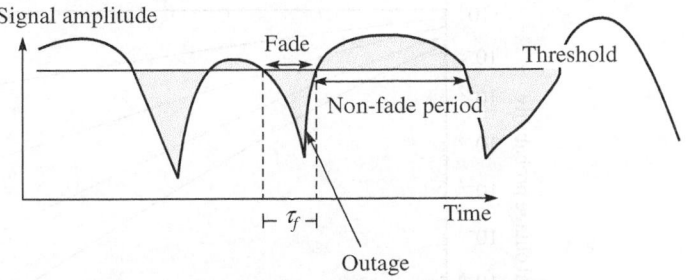

Fig. 3.33 Signal outages occurring during channel fades

During fades, the user experiences a *signal outage*. The term *outage* can be understood using Fig. 3.33.

Fade margin is the ratio of the average received power and some threshold power needed for reliable communication (Fig. 3.34). Outage occurs if the signal drops below the noise power level. If noise is also significant, an outage occurs when the (carrier) signal to noise-plus-interference ratio, $C/(I + N)$, drops below the threshold z. In mobile telephony, the quality of service is often expressed in terms of the probability of outage experienced by subscribers near the boundary of the base station service area. Due to limited spectrum availability, radio networks become increasingly limited by mutual interference between the users. Therefore, outage probability is usually determined in terms of the probability that the signal-to-joint-interference ratio drops below the minimum required level z.

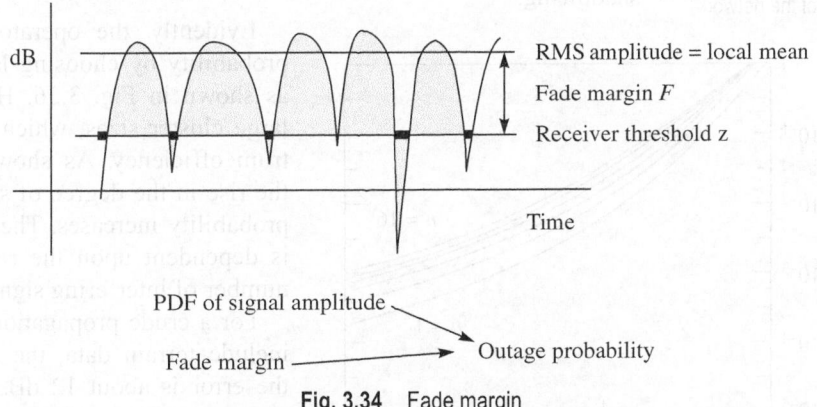

Fig. 3.34 Fade margin

The choice of the value of z, in general, depends on the required quality of service and may, therefore, be somewhat arbitrary. For instance, the outage criterion can be a certain figure of merit subjectively determined by a representative panel of listeners or, in a digital system, an instantaneous BER.

Link budget calculation provides an estimate of the margin for how deep a signal fade can be before the receiver loses the signal. The calculation becomes much more complicated if the (multiple) interfering signals also exhibit fading. Moreover, shadowing also affects outage probability because it depends upon the fade margin, as shown in Fig. 3.35.

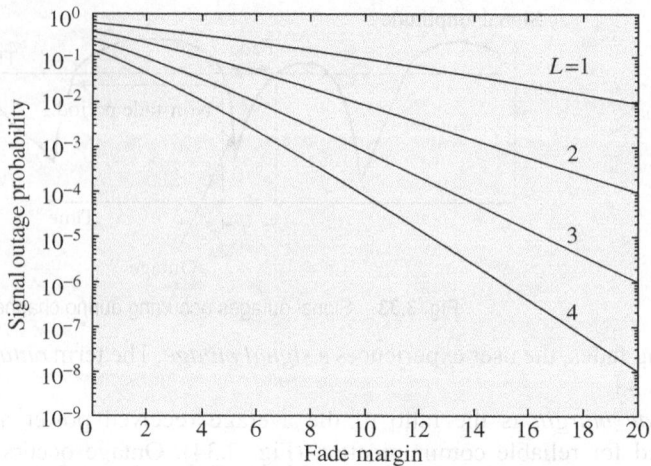

Fig. 3.35 Signal outage probability versus fade margin

Cellular Planning under Fading and Outages

The amount of interference that can be tolerated determines the required separation distance between co-channel cells and, therefore, also the efficiency of the network.

A crucial aspect in the evaluation and planning of radio networks is the computation of the effect of co-channel interference in radio links.

A cellular operator will use topographical databases to estimate outage probabilities in the area covered by the telephone service, but the results for idealized hexagonal cell layouts nonetheless illustrate the effects of reuse distance and shadowing.

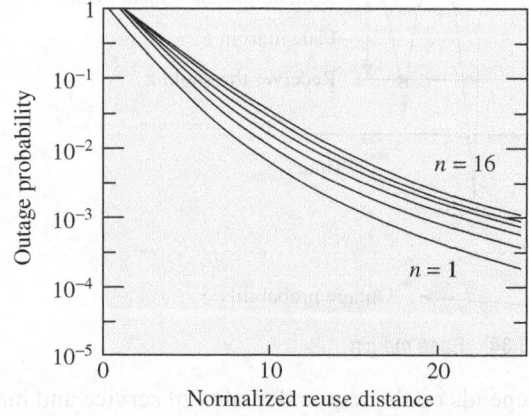

Fig. 3.36 Signal outage probability versus normalized reuse distance for $n = 1, ..., 6$ interfering signals [Shadowing 6 dB, receiver threshold 10 dB, and plane earth loss (40log$d$)]

Evidently, the operator can reduce outage probability by choosing larger reuse distances, as shown in Fig. 3.36. However, this requires large cluster sizes, which results in poor spectrum efficiency. As shown in Fig. 3.37, with the rise in the degree of shadow fading, outage probability increases. The degree of shadowing is dependent upon the reuse distance and the number of interfering signals.

For a crude propagation model that does not include terrain data, the standard deviation of the error is about 12 dB. In such cases, either unacceptably large outage probabilities would be experienced or the cluster size would have to be unacceptably large. Hence, improved cell planning is required, and virtually no operator plans its services considering idealized hexagonal cells. Topographical databases are mostly used to estimate the area mean signal power and interference powers as accurately as possible.

Computation of outage probability requires that we find the probability that the signal-to-interference ratio drops below a certain threshold. As all (fading) signals have fluctuating signal powers, we must integrate over the probability density functions of all the signals involved:

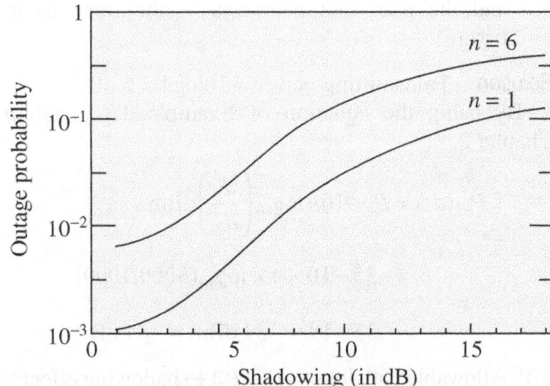

Fig. 3.37 Outage probability versus degree of shadow fading for $n = 1$ and $n = 6$ interfering signals [Shadowing 6 dB, receiver threshold 10 dB, reuse distance equal to 10 times the cell radius, and plane earth loss (40log$d$)]

- Fading and shadowing of the wanted signal
- Fading and shadowing of each interfering signal

Mostly, the following assumptions are made:
- Interfering signals fade independently and independent of the wanted signal.
- Interfering signal powers accumulate incoherently, that is, we may add powers.

In such a case, the PDF of the total interference power is the convolution of the PDF powers of the individual interfering signals.

The number of interfering signals in a typical hexagonal reuse pattern is six, but interfering signals may use discontinuous voice transmission. In such a case, the terminal switches of the carrier pause during speech to reduce the amount of interference caused to other users.

IMPORTANT UNIT CONVERSIONS

dBW (decimal-watt) If the value of 1 W is selected as the reference, then the absolute level of power is defined as

$$\text{Power in dBm} = 10\log_{10} (\text{power in W/1mW})$$

dBm (decimal-milliwatt) If the value of 1 mW is selected as the reference, then the absolute level of power is defined as

$$\text{Power in dBm} = 10\log_{10} (\text{power in W/1mW})$$

These units are extensively used in microwave applications.

MORE SOLVED EXAMPLES

Example 3.5 With the help of the geometry of Fig. 3.3(a), derive the equation to calculate the radio horizon distance. For an antenna of height 80 m, find the effective coverage distance.

Solution If the curved path is approximated by a straight line, Fig. 3.3(a) can be redrawn as shown in Fig. 3.38.

The dotted line is the virtual line connecting the top of the transmitting antenna tower and the cut-off point of the radio horizon that coincides with the earth's surface. This approximately forms a right angle with respect to the earth's radius r at that point, and though fictitious, this approximation is acceptable. So, we can write

$$d^2 + r^2 = (r + h_T)^2$$

$$= r^2 + h_{T^2} + 2rh_T$$

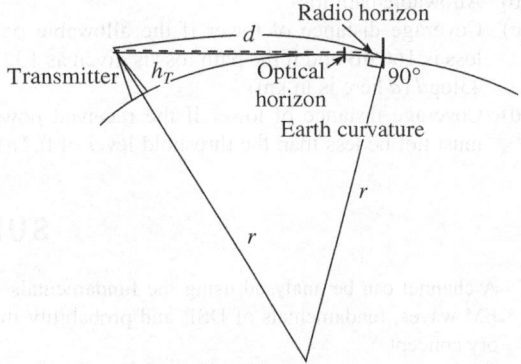

Fig. 3.38 Figure for Example 3.5—Earth's curvature effect

or $\quad d^2 = h_{T^2} + 2rh_T$

Because $2rh_T \gg h_{T^2}, d = \sqrt{2rh_T}$

Now, for $h_T = 80$ m and the earth's radius from its centre is $r = 6371$ km. Therefore,

$$d = \sqrt{2 \times 80 \times 6371 \times 10^3} = 31927.4177 \text{ m}$$

$$= 31.927 \text{ km}$$

Example 3.6 Consider a scenario of multipath with a direct ray and a reflected ray as shown in Fig. 3.39. In addition, consider the following data:

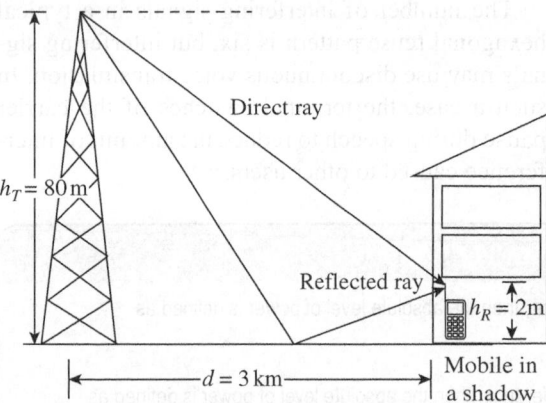

Fig. 3.39 Scenario and data for Example 3.6

The mobile is 3 km away from the tower, reference distance $d_0 = 1000$ m, received power at that distance = −35 dBm, path loss exponent $n = 4$, transmitting power at 1800 MHz = 4 W, and shadow effect = 10 dB.

Find the following:
(a) Received signal power at mobile receiver antenna
(b) Allowable path loss
(c) Coverage distance of tower if the allowable path loss is 160 dB and if the path loss is given as 132 + 43log$d$ (d here is in km)
(d) Coverage distance of tower if the received power must not be less than the threshold level of 0.2 nW

and the path loss behaviour is the same as in part (c)

Solution Transmitting power = 10log4 = 6 dB
By using the equation of Example 2.3 given in Chapter 2,

$$P_{r,i}(d_i) = P_o - 10n\log_{10}\left(\frac{d_i}{d_o}\right) \text{dBm}$$

$$= -35 - 10 \times 4 \times \log_{10}(3000/1000)$$

$$= -35 - 19 = -54 \text{ dBm} = -84 \text{ dB}$$

(b) Allowable path loss = $(P_t - P_r)$ + shadowing effect = [6 − (−84)] + 10 = 100 dB
(c) Allowable path loss = 132 + 43log$d$ + 10 = 160
   or $\quad$ log$d$ = 18/43 = 0.419
   or $\quad\quad$ d = 2.62 km
(c) If received power is to be 0.02 nW = −107 dB
   or $\quad\quad$ $P_t - P_r$ = 132 + 43log$d$
   or $\quad$ 6 − (−107) = 132 + 43log$d$
   or $\quad\quad\quad$ d = 0.361 km

Example 3.7 A vehicle travelling at a speed 40 km/h receives a carrier frequency of 910 MHz. The RMS delay spread observed at the receiving end is 100 ns. What will be the coherence time and coherence bandwidth?

Solution $\lambda = c/f = 3 \times 10^8/910 \times 10^6 = 0.329$ m
Maximum Doppler shift = $f_d = v/\lambda = (40 \times 1000)/(3600 \times 0.329) = 33.77$ Hz
Now, coherence time = $T_0 = 1/2\pi f_d = 1/(2 \times 3.14 \times 33.77) = 0.0047$ s
Coherence bandwidth $B_c = 1/2\pi T_{rms} = 1/(2 \times 3.14 \times 100 \times 10^{-9}) \approx 1.59$ MHz

Note: The relationship between coherence time and Doppler spread is provided in Fig. 3.28 and that between coherence bandwidth and delay spread is given in Fig. 3.27. The equation for coherence bandwidth is given in Chapter 4 while discussing channel modelling (Eq. 4.3a).

SUMMARY

- A channel can be analysed using the fundamentals of EM waves, fundamentals of DSP, and probability theory concept.
- Radio frequency ranges from 0.003 MHz to 3,00,000 MHz.
- Radio waves propagate with reflection, diffraction, scattering, and refraction.

- Polarization of an EM wave is important in radio wave propagation.
- There are many differences between long-distance and short-distance wireless link considerations.
- The earth's atmosphere has various layers with different characteristics.

- The four important types of propagation are free space, ground wave, ionospheric, and tropospheric.
- AWGN is significant over any type of channel. Channel losses can be considered according to the situation.
- For short-distance communication, a direct path and a number of reflected rays are considered to be received, the resultant of which causes ISI effect.
- Multipath causes fading, which is observed in terms of delay spread and Doppler shift.
- Large-scale fading is mainly due to attenuation and shadowing.

- Power delay profile gives the CIR and maximum excess delay, which decides the coherence bandwidth.
- Coherence bandwidth decides the occurrence of frequency-selective or flat fading.
- Doppler spectrum decides the coherence time.
- Coherence time decides the occurrence of fast or slow fading.
- Shadowing effect can be considered over a small or large area.
- Reduction in signal strength due to fading gives signal outage, and it is reduced by increasing the fade margin.

EXERCISES

Multiple-choice Questions

3.1 The electric field of an EM wave at a point in free space is in the positive Y direction and the magnetic field is in the positive X direction. The direction of power flow will be in the
   (a) positive X direction   (c) positive Z direction
   (b) positive Y direction   (d) negative Z direction

3.2 In isotropic propagation, if f_c is the critical frequency and θ is the angle of incidence at the ionosphere measured with respect to the normal, the MUF equals
   (a) $f_c \cos\theta$ (c) $f_c \sec\theta$
   (b) $\sqrt{f_c \cos\theta}$ (d) $\sqrt{f_c \sec\theta}$

3.3 The skip distance is
   (a) independent of the frequency
   (b) independent of the state of ionization
   (c) independent of the transmitting power
   (d) dependent on the transmitting power

3.4 In sky wave propagation, the skip distance is used
   (a) to not exceed the critical frequency
   (b) to avoid the Faraday effect
   (c) to prevent sky wave and upper ray interference
   (d) to obey tilting

3.5 The critical frequency of an ionospheric layer depends upon
   (a) only the height
   (b) only the electron density
   (c) only the height and electron density and nothing else
   (d) the height, electron density, and angle of incidence

3.6 Polarization of EM waves is due to
   (a) reflection
   (b) transverse nature of EM waves
   (c) longitudinal nature of EM waves
   (d) spherical wavefronts of EM waves

3.7 The virtual height of an ionospheric layer is
   (a) more than the height a wave actually reaches
   (b) less than the height a wave actually reaches
   (c) the same as the height a wave actually reaches
   (d) none of these

3.8 The ionosphere roughly extends from
   (a) 50 km to several earth radii
   (b) 50 km to 80 km
   (c) 50 km to 400 km
   (d) 50 km to 150 km

3.9 The total noise of a satellite earth station receiving system consists of
   (a) sky noise
   (b) parametric amplifier noise
   (c) antenna and feeder noise
   (d) all of these

3.10 The ground wave eventually disappears as one moves from the transmitter because of
   (a) surface attenuation   (c) diffraction
   (b) loss of LOS   (d) tilting

3.11 If the reflected wave has to travel one-half wavelength more than the direct wave, the two waves will
   (a) arrive at the receiving antenna in phase
   (b) arrive at the receiving antenna out of phase
   (c) arrive at the receiving antenna in exactly the same time
   (d) not arrive at the receiving antenna

3.12 In the atmosphere, the absorption of the radio waves depends on
   (a) their distance from the transmitter
   (b) their frequency
   (c) their polarization
   (d) the electron density of the ionosphere

3.13 Sky wave propagation is used for
   (a) long-distance transmission
   (b) medium-distance transmission

(c) short-distance transmission

(d) mobile transmission

3.14 Which of the following effects occur in tropospheric scatter propagation?

(a) Faraday effect   (c) Super-refraction

(b) Fading   (d) Atmospheric storm

3.15 The spectral density of white noise

(a) varies with frequency

(b) is constant for all frequency ranges

(c) varies with bandwidth

(d) is constant for a limited range of frequencies

3.16 If η is the positive frequency power density, the power spectrum density of white noise $\delta(\omega)$ is equal to

(a) $1/\eta$ (b) $\eta/2$ (c) η (d) 2η

3.17 Which of the following happens when the transmission bandwidth is greater than coherence bandwidth?

(a) Frequency-selective fading

(b) Flat fading

(c) Large-scale fading

(d) Small-scale fading

3.18 Signal outages may occur due to

(a) destructive addition of the reflected rays

(b) shadowing

(c) fading

(d) all of these

3.19 What happens when the coherence time is less than the symbol duration?

(a) Fast fading

(b) Slow fading

(c) Slow and flat fading

(d) Fast and frequency-selective fading

3.20 Doppler effect is

(a) a tone received with delay

(b) a tone received with reduced amplitude

(c) a tone received with a spectrum spread

(d) motion of the mobile

3.21 Power delay profile can be used for

(a) receiving the resultant phase of the signal

(b) the measurement of the CIR

(c) velocity estimation

(d) coherence time

3.22 Which of the following theories is/are applied to land mobile systems?

(a) Troposcatter propagation

(b) Free space propagation path loss law

(c) Ground wave propagation

(d) All of these

3.23 What makes the wireless channel unpredictable?

(a) Shadowing   (c) Multipath

(b) Doppler spread   (d) All of these

Review Questions

3.1 What is the importance of polarization of EM waves?

3.2 What is the mathematical relationship between the wavelength and the frequency of a sinusoidal EM wave?

3.3 What are the channel problems to be taken into account for long- and short-distance propagation? What are the common problems faced by both types of propagation?

3.4 Show that for a reference transmitter with an EIRP of 1 kW in free space, the usable field strength

$$E_u = \sqrt{\frac{30P_tG_t}{d}}.$$

3.5 For ground wave propagation, show that the reflection coefficient tends to ™1 for angles close to 0. For horizontal polarization, verify that $R_c > 0.9$ for $\Phi < 10°$. Also, for vertical polarization, verify that $R_c > 0.5$ for $\Phi < 5°$ and $R_c > 0.9$ for $\Phi < 1°$. Here, R_c is the absolute value.

3.6 Why is it required to have distinct uplink and downlink frequencies in satellite communication? (Refer Appendix E for the satellite system.)

3.7 What is attenuation? Does it degrade the signal? How?

3.8 List the applications that use microwave communication. Find the advantages and disadvantages of microwave communication.

3.9 What is the difference between ionospheric propagation and tropospheric propagation?

3.10 In which applications will the multihop transmission be useful? How?

3.11 What is the difference between the mean excess delay and maximum excess delay for the multipath power delay profile?

3.12 Define the following terms:

(a) Delay spread

(b) Doppler shift

(c) Coherence bandwidth

(d) Power delay profile

3.13 Differentiate between the following terms:

(a) Large-scale fading and small-scale fading

(b) Fast fading and slow fading

(c) Frequency-selective fading and flat fading

(d) Time spreading and time variance

3.14 How does time variance become significant in deciding fast or slow fading?

3.15 Define the terms *fading margin* and *outage probability*. Mention their significance in cellular planning.

3.16 Write a note on the CIRs for flat and frequency-selective fading.

3.17 Comment on the observations of Doppler shift when a moving vehicle travels towards a transmitter and away from the transmitter.

3.18 Discuss the effect of path loss on the performance of a cellular radio network.

3.19 How does shadowing affect reception? Discuss various situations of shadowing. How does shadowing in a hilly area differ from that in a crowded urban area?

Numerical Problems

3.1 Show that the path loss L between two isotropic antennas ($G_R = 1$ and $G_t = 1$) can be expressed as $L = -32.44 - 20\log f_c/1\text{MHz} - 20\log d/1\text{ km}$, where the loss is found in dB.

3.2 Assume a receiver located 10 km away from a 50 W transmitter. The carrier frequency is 6 GHz. Free space propagation is assumed.
   (a) Find the power received at the receiver.
   (b) Find the magnitude of the e-field at the receiver antenna.
   (c) Find the RMS voltage applied to the receiver input, assuming that the receiving antenna has a real impedance of 50 Ω and is matched to the receiver. (Hint: $v_{rms} = \sqrt{4P_r R}\ V$)

3.3 A transmitter produces 55 W of power. If this power is fed to a unity gain antenna with 890 MHz carrier frequency, determine the received power in dBm and dBW. The free space distance between the transmitting and receiving antennas is 200 m. Also, find the received powers at distances of 500 m, 1 km, and 10 km. Will there be a linear relationship between the received power and distance? Comment on your results.

3.4 Consider a base station antenna with $h_T = 30$ m and a mobile antenna with $h_R = 2$ m. The mobile is separated from the base station by a distance $d = 1000$ m. The carrier frequency is 1 GHz.
   (a) Compute the length of the LOS.
   (b) Compute the length of the ground-reflected wave.
   (c) Compute the phase difference due to path length differences.
   (d) Can we apply Egli's formula here? Explain.

3.5 In a cellular system, a mobile receiver is 2 km away from the base station using $\lambda/4$ monopole antenna with a gain of 2 dB. The e-field, at a distance of 1 km from the transmitter, is measured to be 10^{-3} V/m and the carrier frequency used is 910 MHz. Calculate the length L and the effective aperture of the receiving antenna.

3.6 In Problem 5, assuming a two-ray ground reflection model, calculate the phase difference between the direct and reflected rays as well as the received power at the mobile if the height of the transmitting antenna is 72 m and the receiving mobile antenna height is 1.6 m above the ground. Gain of base station antenna can be assumed as unity.

3.7 Find the far-field distance for an antenna with a maximum dimension of 1 m and an operating frequency of 900 MHz.

3.8 Find whether flat or frequency-selective fading will occur in the following conditions: The RMS delay spread is 1.3 ms and the transmission bandwidth is 1.5 MHz. What is the allowable symbol rate to permit only the flat fading condition for BPSK?

3.9 A transmitter carrier frequency is 890 MHz. For a vehicle moving at 30 km/h, calculate the received carrier frequency if the mobile is moving in the following directions:
   (a) Towards the transmitter
   (b) Away from the transmitter
   (c) In a direction that is perpendicular to the direction of arrival of the transmitted signal

3.10 Determine the isotropic free space loss at 4 GHz for the shortest path to a synchronous satellite from the earth at 35,864 km.

3.11 A spacecraft uses a transmitter power of 15 W at 3 GHz frequency. The transmitting and receiving antenna gains are 25 dB and 35 dB, respectively. The distance from the spacecraft to the earth station is 20,000 km. The data bit rate is 100 kbps for non-return-to-zero transmissions. The effective antenna noise temperature is 14 K. Find the SNR of the received signal.

3.12 Base station A communicates with base station B through LOS communication and through feeder cables. The feeder cable losses are 10 dB approximately at both the sites. The gains of both the transmitting and receiving antennas are approximately the same and equal to 12 dB. Base station A transmits a power of 10 W. Determine the EIRP and allowable path loss.

4 Wireless Channel Modelling

Theme of the Chapter

Accurate characterization of the radio channel and its parameters through a mathematical or simulator channel model is important for predicting reception of the signal, signal coverage, achievable data rates, bit error rate performance, and specific performance attributes of alternative signalling and reception schemes. Channel modelling is the process of correlating mathematics with channel statistics. It allows the analysis of interference from different systems and determination of the optimum location for installing base station antennas. In applications such as mobile satellite communication, channel modelling helps to determine the ideal location to install earth stations.

Chapter 3 discussed various methods to model the channel and analysed their effects on the transmitted signal. This chapter discusses some popular channel modelling techniques including the Rayleigh, Rician, and Nakagami models. In addition, this chapter also throws light on various channel representations such as additive white Gaussian noise (AWGN)-based channel, channel as a filter, and flat fading channel.

Key Topics

- Considerations for modelling a wireless channel
- Power delay profiles
- Additive white Gaussian noise

- Realization of channels by digital filter
- Wideband channel model
- Phase error calculation by mathematical model

- Stochastic channel modelling
- Rayleigh fading channel model
- Rician channel model
- Nakagami channel model
- Okumura–Hata model

4.1 CHANNEL MODELLING

Multipath effect makes the channel time varying, and depending upon the constructive or destructive interference, the quality of the received signal will vary.

A channel can be modelled by two approaches: physical and statistical. Reflection, multipath, attenuation, and so on are considered in the physical approach, whereas input–output elements and transaction probabilities are considered in the statistical approach. The appropriate approach is selected based on whether mathematical calculations are required or simulation of the channel is expected. Sometimes, a combined modelling approach is preferred.

Perfect channel modelling is required to study or analyse a wireless system. This involves the identification of various channel properties, of which one or more must be considered at a time for modelling the channel. Based on the assumptions for the channel, the performance of the system will vary. Some important properties of a channel are summarized as follows:

- Channels may be *time varying* or *static*. The effect of mobility is that the channel varies with the user's location and time, which results in rapid fluctuations of the received power. The slower one moves, the lesser variations will be observed.
- Channels may be *time dispersive* or *non-dispersive*. Due to dispersion, pulse spreading will be observed, which will result in the intersymbol interference (ISI) effect.
- Channels may be *linear* or *non-linear*.
- All channels act as *low-pass filters* under certain conditions as they show the pulse spreading effect.
- Channels may be *fast fading* or *slow fading*, and may be *frequency selective* or *flat fading*.

Depending upon the scenario or system requirements, mobile systems must follow an appropriate model for the analysis. We have two major scenarios: large area, non-congested rural or suburban scenario and small area, congested urban scenario. Hence, channel models are primarily divided into two main categories:

- Outdoor channel model (e.g. Longley–Rice model, Okumura–Hata model, Nakagami model, and Rician model)
- Indoor channel model (e.g. Rayleigh model)

Table 4.1 gives an idea about channel modelling constraints in cellular systems.

Cellular transmissions mostly utilize path loss models. They are discussed in Section 4.11.

Table 4.1 Specific environment for modelling wireless channels for cellular systems

Cell type	Typical cell radius	Location	Typical base station antenna installation height
Large macrocell	1 km–30 km	Outdoor	Above medium rooftop level: all surrounding buildings are below antenna height
Small macrocell	0.5 km–3 km	Outdoor	Above medium rooftop level: heights of some surrounding buildings are above antenna height
Microcell	Up to 1 km	Outdoor	Below medium rooftop level
Picocell	Up to 500 m	Indoor/outdoor	Below rooftop level

There are three categories of path loss models:
(a) *Macrocell path loss models*: Empirical model, semi-empirical model, deterministic model
(b) *Microcell path loss models*: Empirical model, deterministic model
(c) *Picocell path loss models*: Empirical model, semi-empirical model

Hypothetically, if a transmitted impulse is received under the fading and multipath effects with delayed and attenuated versions, it is the channel impulse response and can be plotted on the time axis.

Empirical models are based on measurement data. They are simple and use few parameters; sometimes, they use statistical properties. These models are not very accurate and excessive computational efforts are required.

Deterministic models are site-specific and require a large amount of geometrical information about the site. Such models show very less computational effort and are very accurate.

Semi-empirical or *semi-deterministic models* are based on empirical models and deterministic aspects.

4.1.1 Channel Impulse Response

> The impulse response of the channel indirectly represents the shape of the power delay profile.

It is very convenient to represent channel behaviour with its impulse response, considering the channel as a system. The channel for mobile system applications is characterized by *multipath reception*, as discussed in Chapter 3. The signal reaching the receiver contains not only a direct line-of-sight (LOS) radio wave but also a large number of reflected radio waves. Even worse, in urban centres, the LOS is often blocked by obstacles, and a collection of differently delayed waves is received by a mobile antenna. The reflected waves interfere with the direct wave, which causes significant degradation in the link performance. If the antenna moves, the channel varies with location and time, because the relative phases of the reflected waves change. This leads to fading—time variations of the received amplitude and phase.

In this situation, channel impulse response is the best option for modelling the channel. The impulse response of a wireless channel looks like a series of impulses received from a stationary channel with decreasing amplitudes, as shown in Fig. 4.1. In practice, the number of impulses that can be distinguished is very large and depends on the time resolution of the communication or measurement system.

In system evaluations, we typically prefer to address a class of channels with properties that are likely to be encountered rather than one specific impulse response. Therefore, we define the average (local mean) power that is received with an excess delay that falls within the interval $(T, T + d\tau)$. Such characterization for all T gives the delay profile of the channel. The delay profile determines the frequency dispersion, which is the extent to which the channel fading at two different frequencies f_1 and f_2 are correlated.

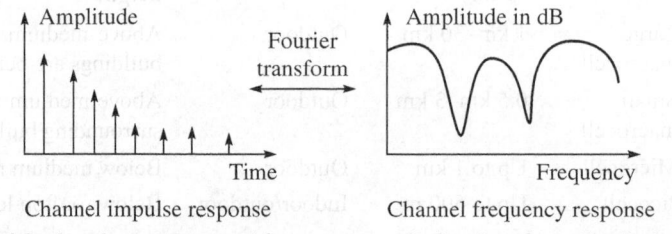

Fig. 4.1   Impulse response and frequency transfer function of a multipath channel

4.1.2 Power Delay Profiles

The *delay profile* is the expected power variation per unit of time received with a certain excess delay. It is obtained by averaging a large set of impulse responses. Thus, it is also called the set of mean power profile of each path. Figures 4.2, 4.3, 4.4, and 4.5 show the delay profiles with various environmental or surrounding conditions.

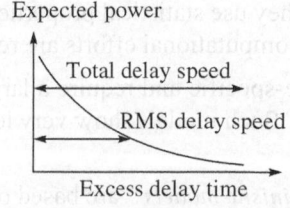

Fig. 4.2   Typical exponential delay profile

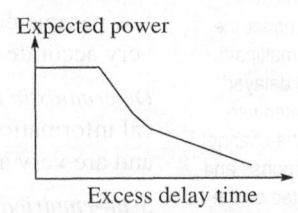

Fig. 4.3   Typical indoor delay profile

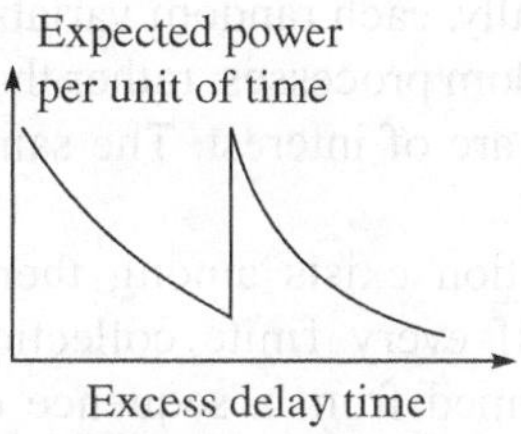

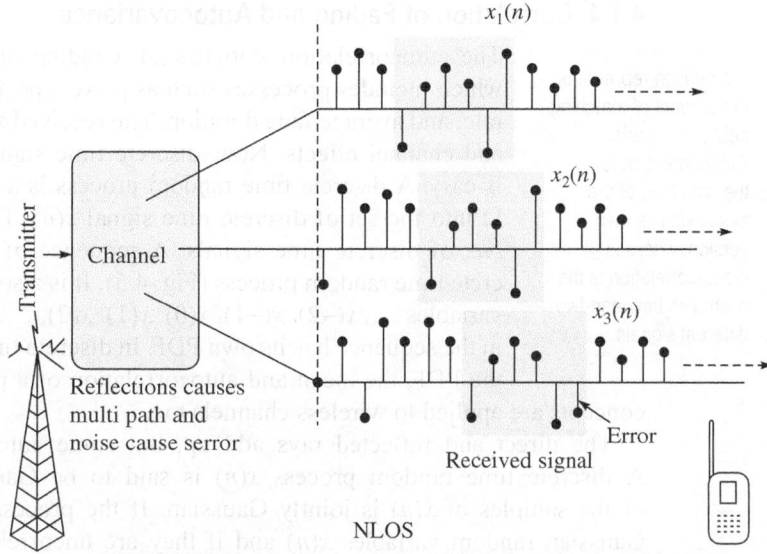

Fig. 4.4 Typical *bad urban* delay profile with tall buildings

Fig. 4.5 Autocorrelation among multiple reflected rays at the receiver

> Channel models such as Rayleigh and Rician are based on typical probability distribution functions.

In an indoor environment (Fig. 4.3), early reflections often arrive with almost identical power. This gives a fairly flat profile up to some point and a tail of weaker reflections with larger excess delay. The correlation of the fading at different carrier frequencies can be computed from the power delay profile.

4.1.3 Channel Modelling and Probability Theory

There are many constraints in modelling a channel. There may be different environments for different situations and different weather conditions. There may be different objects of different materials in a room or different surrounding conditions. Hence, it is very difficult to predict the number of reflected rays and whether constructive (add—in phase) or destructive interference will occur. Many researchers have measured and analysed the first-order statistics of these processes. As a result, channel modelling is based on probability theory and statistics, and its behaviour is represented or characterized by the probability density function (PDF) and cumulative distribution function (CDF). The following are a few definitions that are useful for understanding the statistical way of channel modelling:

Probability density function It describes the probability that a real-valued random variable m will be found at the value m. In other words, it describes the relative likelihood for this random variable to take on a given value. PDF is non-negative everywhere. Its integral over the entire space is equal to one.

Cumulative distribution function It describes the probability that a real-valued random variable m with a given probability distribution will be found at the value less than or equal to m. Thus, it is the area under the probability distribution. CDF is the integration of PDF.

Gaussian process It is a stochastic process, and its realization consists of random values associated with every sample point of space, such that the random variable has a normal distribution (Fig. 4.9c). The channel is mostly treated as a Gaussian channel.

4.1.4 Correlation of Fading and Autocovariance

> Correlation represents the amount of matching between signals. Autocorrelation is the matching of the signal with its own version, whereas cross-correlation is the mismatch between two different signals.

The autocorrelation statistics of a fading process is second-order statistics, which includes processes such as power spectral density (PSD), level crossing rate, and average fade duration. The received signal is random because of noise and channel effects. Now, discrete time signal analysis of random processes is easy. A discrete time random process is a mapping from the sample space Ω into the set of discrete time signal $x(n)$. Thus, it is a collection, or *ensemble*, of discrete time signals. A sequence of coefficients $x(n)$ forms the discrete time random process (Fig. 4.5). It is also the indexed sequence of random variables $..., x(-2), x(-1), x(0), x(1), x(2), ...$. Normally, each random variable in the sequence has its own PDF. In discrete time random processes, rather than the PDF, the mean and autocorrelation of a process are of interest. The same concepts are applied to wireless channels too.

The direct and reflected rays add up, but some autocorrelation exists among them. A discrete time random process $x(n)$ is said to be Gaussian if every finite collection of the samples of $x(n)$ is jointly Gaussian. If the process is formed from a sequence of Gaussian random variables $x(n)$ and if they are uncorrelated, then the process is known as *white Gaussian noise*. For each $x(n)$, we can find the mean and variance. Two other ensemble averages are *autocovariance* and *autocorrelation*. These are calculated as follows:

- Mean, $m_x(n) = E\{x(n)\}$
- Variance, $\sigma_x^2 = E\{|x(n) - m_x(n)|^2\}$
- Autocovariance, $c_x(k,l) = E\{[x(k) - m_x(k)][x(l) - m_x(l)]*\}$
- Autocorrelation, $r_x(k,l) = E\{x(k)x*(l)\}$

For the sake of simplicity, random processes are always assumed to have zero mean, so that the autocovariance and autocorrelation sequences may be used interchangeably. This assumption is acceptable, because if $x(n)$ has a non-zero mean, then a zero-mean process can always be formed.

As in the case of random variables, the autocorrelation and autocovariance functions provide information about the degree of linear dependence between two random variables $x(k)$ and $x(l)$. For example, if $c_x(k,l) = 0$ for $k \neq 0$, then the random variables $x(k)$ and $x(l)$ are uncorrelated, and the knowledge of one does not help in estimating the other using a linear

---

Notes: A random process $x(n)$ due to channel impulse response $h(n)$ is said to be in wide-sense *stationary* if the following three conditions are satisfied.
1. The mean of the process is a constant $m_x(n) = m_x$
2. The autocorrelation $r_x(k, l)$ depends only on the difference $k - l$
3. The variance of the process is finite

---

estimator.

Now, the *correlation of fading* at different carrier frequencies can be computed from the delay profile. The transform of the delay profile gives the autocorrelation of the complex amplitudes of sinusoidal signals at frequencies f_1 and f_2, as given in Eq. (4.1).

$$\text{Delay profile} \leftrightarrow E[x(f_1)x*(f_2)] \tag{4.1}$$

The received signal is a random process. As mentioned already, random processes (due to channel) are always assumed to have zero mean, and hence, the terms autocovariance and autocorrelation are used interchangeably. After some algebraic manipulations, a relationship can be derived to express the autocorrelation or autocovariance of the amplitude versus frequency separation, $f_1 - f_2$.

In Eq. (4.1), if $x(f_1) = R_1$ and $x(f_2) = R_2$, then the normalized autocovariance of the amplitudes R_1 and R_2 of two carriers, one at f_1 and another at f_2, is

$$C = \frac{E\{R_1 R_2\} - E\{R_1\}E\{R_2\}}{s(R_1)s(R_2)} \tag{4.2}$$

> For a reliable communication without using adaptive equalization or other anti-multipath techniques, the transmitted data rate may be much smaller than the inverse of the RMS delay spread. It is called coherence bandwidth.

where $s(R_1)$ is the standard deviation of R_1 and $s(R_2)$ is the standard deviation of R_2.

From this discussion, coherence bandwidth (defined in Chapter 3) is the bandwidth for which the autocovariance of the signal amplitudes at two extreme frequencies reduces from 1 to 0.5. For a Rayleigh fading channel with an exponential delay profile, one finds

$$B_c = 1/(2\pi T_{rms}) \tag{4.3a}$$

where T_{rms} is the root mean square (RMS) delay spread. Sometimes,

$$B_c \cong 1/(5T_{rms}) \tag{4.3b}$$

A typical power delay profile is shown in Figure 4.6 for the given delay spread and coherence bandwidth.

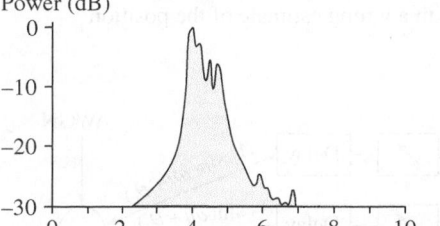

Power (dB)

Delay time (μs)

Fig. 4.6 Power delay profile for delay spread = 1.2 ms and coherence bandwidth $B_c = 1.3$ MHz

---

Example 4.1 In a measurement of power delay profile, the maximum excess delay is 50 ns. Assuming exponentially decaying profile and Rayleigh fading channel, find the maximum transmission bandwidth for which the data can be transferred with minimum ISI.

Solution RMS delay spread is approximately maximum excess delay/$\sqrt{2}$

or RMS delay spread = $0.707 \times 50 = 35.35$ ns

The maximum transmission bandwidth for minimum ISI is the coherence bandwidth.

$$B_c = 1/(2\pi T_{rms})$$
$$= 1/2\pi \times 35.35 \times 10^{-9}$$
$$= 0.00450225 \times 10^9$$
$$\approx 4.5 \text{ MHz}$$

4.1.5 Model of Multipath Effect

A few effects due to multipath reception are listed in Table 4.2 in a simplified manner.

As shown in Fig. 4.7, the model of many randomly phased sinusoids with variations in amplitudes appears to describe the wireless radio channel appropriately and to allow the calculation of outage probabilities, fade durations, and many other critical parameters of wireless links. It greatly facilitated the development systems that can reliably communicate despite the anomalies and unpredictability of the mobile communication channel. However, the number of sinusoids should be kept limited due to inconvenience in the addition of all.

Table 4.2 Effects due to multipath reception

Application	Effect
Fast-moving user	Rapid fluctuations of the signal amplitude and phase
Wideband (digital) signal	Dispersion and ISI
Analog television signal	Ghost images (shifted slightly to the right)
Multicarrier signal	Different attenuation at different (sub-)carriers and at different locations
Stationary user of a narrowband system	Good reception at some locations and frequencies and poor reception at other locations and frequencies
Satellite positioning system	Strong delayed reflections that may cause a severe miscalculation of the distance between the user and the satellite and can result in a wrong estimate of the position

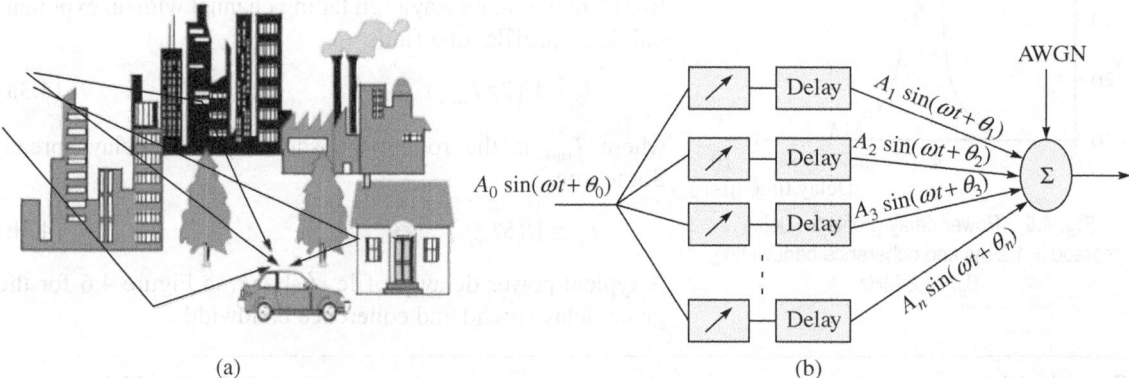

(a)   (b)

Fig. 4.7 Multipath reception modelling (a) Scenario of multipath reception (b) Summation of phase delayed versions of a signal

> Multipath shape factors are influenced by second-order fading statistics and use angular distribution of multipath power for spatial channel modelling.

Although channel fading is experienced as an unpredictable and stochastic phenomenon to the user's device or the system planner, powerful models that can accurately predict the average system performance have been developed. Countermeasures can be used to avoid system failure, even if the channel exhibits fades at particular frequencies of particular locations. Chapter 8 describes such counter measures and multipath mitigation techniques, in order to reduce the bit errors and improved the system performance.

4.1.6 Multipath Shape Factors

Many researchers now utilize a new concept of spatial channel modelling. It relates small-scale fading characteristics of a wireless channel to the multipath angle of arrival. The concept of multipath shape factors allows the quantitative analysis of any distribution of non-omnidirectional multipath waves in a local area, where the local average signal strength is assumed to be wide-sense stationary. The following three shape factors have geometrical interpretations of multipath channel with arbitrary spatial complexity:

(a) Angular spread
(b) Angular constriction
(c) Azimuthal angle of maximum fading

Angular spread is a measure of how multipath concentrates about a single azimuthal direction of arrival. *Angular constriction* is a measure of how multipath concentrates about two azimuthal directions. *Azimuthal angle of maximum fading* is the directional or orientation parameter. Shape factors are derived from the angular distribution of multipath power $P(\Phi)$. This representation includes antenna gains and polarization mismatch effects. Shape factors are based on the complex Fourier coefficients of $P(\Phi)$. If the shape factor approach is used for the analysis of mobile receivers with a velocity v, quick, comprehensive, and accurate solutions can be found. A few examples are given here.

Case 1 This involves a narrowband receiver operating in a local area with multipath arriving from all directions such that the angular distribution of power $P(\Phi)$ is constant. Here, the receiver antenna is an omnidirectional whip, oriented perpendicular to the ground. Due to vertical electric field polarization of the whip antenna, this propagation scenario is referred to as the E_z case (Fig. 4.8a).

Case 2 This involves the same type of narrowband receiver as in case 1 with the same omnidirectional antenna but with a small loop antenna mounted at the top of the receiver such that the plane of the loop is perpendicular to the ground. The antenna pattern of the small loop antenna attenuates the arriving multipath, and the angular distribution of power becomes $P(\Phi) = A \sin^2 \Phi$, where A is the arbitrary gain constant. Unlike the E_z case, the statistics of the propagation scenario will depend upon the direction of travel by the receiver. The H_x and H_y cases will be generated now (Figs 4.8b and c, respectively). In these cases, the receiver will travel in a direction perpendicular to the main lobes of the loop antenna pattern ($\Phi = 0$) and parallel to the main lobes of the loop antenna pattern ($\Phi = \pi/2$), respectively.

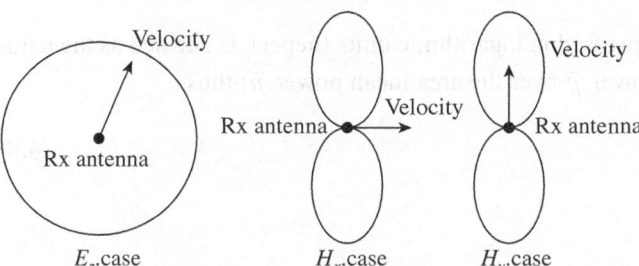

Fig. 4.8 Three cases of angular distribution of power
(a) E_z (b) H_x (c) H_y

The spatial parameters can be derived with the help of the following equations. The nth complex Fourier coefficients of $P(\Phi)$ is given by

$$F_n = \int_0^{2\pi} P(\Phi)\exp(-jn\Phi)d\Phi \tag{4.4}$$

Angular spread Λ is calculated by

$$\Lambda = \sqrt{1 - \frac{|F_1|^2}{F_0^2}} \tag{4.5}$$

Angular constriction γ is calculated by

$$\gamma = \frac{|F_0 F_2 - F_1^2|}{F_0^2 - |F_1|^2} \tag{4.6}$$

> Due to shadowing, fluctuations are experienced on the local mean power, which is log-normal in nature.

Azimuthal direction of maximum fading $\Phi_{\max}$ is given by

$$\Phi_{\max} = \frac{1}{2}\arg\{F_0 F_2 - F_1^2\} \qquad (4.7)$$

The spatial parameters for the E_z case are $\Phi_{\max} = 0$, and since this is the case of omnidirectional antenna, the angular spread at a maximum is $\Lambda = 1$, and the angular constriction at a minimum is $\gamma = 0$. For the H_x and H_y cases, the spatial parameters are $\Lambda = 1$, $\gamma = \frac{1}{2}$, and $\Phi_{\max} = \pi/2$. Since the impinging multipaths have no clear bias in one direction, the angular spread is at a maximum just like the E_z case. However, there is clearly a bias in two directions, resulting in an angular constriction of $\gamma = \frac{1}{2}$.

4.1.7 Considerations for Shadowing Effect

Experiments reported by Egli in 1957 showed that for paths longer than a few hundred metres, the received (local mean) power fluctuates with a *log-normal* distribution about the area mean power. By log-normal is meant that the local mean power expressed in logarithmic values, such as dB or neper, has a normal (i.e., Gaussian) distribution.

We distinguish between local and average means as follows:

- *Local means* Average over about 40 λ, to remove multipath fading
- *Area means* Average over tens or hundreds of metres, to remove multipath fading and shadowing

The mean received power $P_{\log}$, expressed in logarithmic units (neper), is defined as the natural logarithm of the local mean power $\bar{p}$ over the area mean power $\bar{\bar{p}}$; thus,

$$\bar{P}_{\log} = ln\frac{\bar{p}}{\bar{\bar{p}}} \qquad (4.8)$$

It has the normal PDF

$$f_{\bar{P}_{\log}}(\bar{P}_{\log}) = \frac{1}{\sqrt{2\pi}\sigma}e^{-\left\{\frac{1}{2\sigma^2}\bar{P}_{\log}^2\right\}} \qquad (4.9)$$

where σ is the *logarithmic standard deviation* in natural units. If we convert nepers to watts, the log-normal distribution of the received (local mean) power is found as

$$f_{\bar{p}}(\bar{p}) = \frac{1}{\sqrt{2\pi}\sigma\bar{p}}e^{-\left\{\frac{1}{2\sigma^2}\left(ln\frac{\bar{p}}{\bar{\bar{p}}}\right)^2\right\}} \qquad (4.10)$$

Here, the factor '1/local mean power' occurs due to the conversion of the PDF of $P_{\log}$ to the local mean power. There may be large as well as small area shadowing, which can be modelled as two independent Markovian processes.

4.2 ADDITIVE WHITE GAUSSIAN NOISE

Transmission is always corrupted by noise regardless of the type of channel assumed. The simplest mathematical model of the radio channel is the additive white Gaussian noise (AWGN) channel. It is a very good model for physical reality as long as the thermal noise at the receiver (due to atmosphere or thermal agitation or its resultant) is the only source of disturbance. Nevertheless, because of its simplicity, it is often used to model man-made noise or multi-user interference. The AWGN channel model can be characterized as follows:

- The noise is *additive*. The received signal equals the transmit signal plus some noise, where the noise is statistically independent of the signal. The noise $w(t)$ is an additive random disturbance of the useful signal $s(t)$; that is, the received signal is given by

$$r(t) = s(t) + w(t) \tag{4.11}$$

- The noise is *white*. It has constant PSD. Therefore, the autocorrelation of the noise in time domain is zero for any non-zero time offset. The one-sided PSD is usually denoted by N_0. Thus, $N_0/2$ is the two-sided PSD and WN_0 is the noise inside the noise bandwidth W (see Fig. 4.9a). For thermal resistor noise, $N_0 = kT_0$, where k is the Boltzmann constant ($\approx 1.38 \times 10^{-23}$ J/K) and T_0 is the absolute temperature. The unit of N_0 is W/Hz, which is the same as the unit J for the energy. Usually, N_0 is written as dBm/Hz.

$$\text{For } T_0 = 290\,K \quad N_0 \approx -174\,\text{dBm/Hz}$$

However, this is only the ideal physical limit for an ideal receiver. In practice, some decibels have to be added according to the *noise figure*. Typically, N_0 will be a value slightly above −170 dBm/Hz.

- The noise samples have a *Gaussian distribution*. The Gaussian PDF with variance σ^2 is given by

$$p(x) = \frac{1}{\sqrt{2\pi\sigma^2}} e^{-(x-m)^2/2\sigma^2} \tag{4.12}$$

The output of every (linear) noise measurement is a zero-mean Gaussian random variable that does not depend on the time instant when the measurement is done.

One must keep in mind that the AWGN model is a mathematical function, because it implies that the total power (i.e., the PSD integrated over all frequencies) is infinite. Thus, a time sample of white noise has infinite average power, which is certainly not a physically reasonable property. It is known from statistical physics that the thermal noise density decreases exponentially at very high frequencies. However, to understand the physical situation in communications engineering, we should keep in mind that every receiver limits the bandwidth as well as every physical noise measurement. Therefore, it makes sense to think of the noise process to be white, but it cannot be sampled directly without an input device. Each input device filters the noise and leads to a finite power. Instead of the two-sided PSD, if noise over W is represented with an equivalent single band, it will look like Fig. 4.9(b).

The mean of the multiplication of two noise samples (autocorrelation) is

> A channel can be modelled as an FIR filter, and this forms the basis for equalizer design using a transversal filter.

$$E\{\omega(t_1)\omega(t_2)\} = \frac{N_0}{2}\delta(t_1 - t_2). \tag{4.13}$$

where $\delta(t)$ represents the unit impulse at time t.

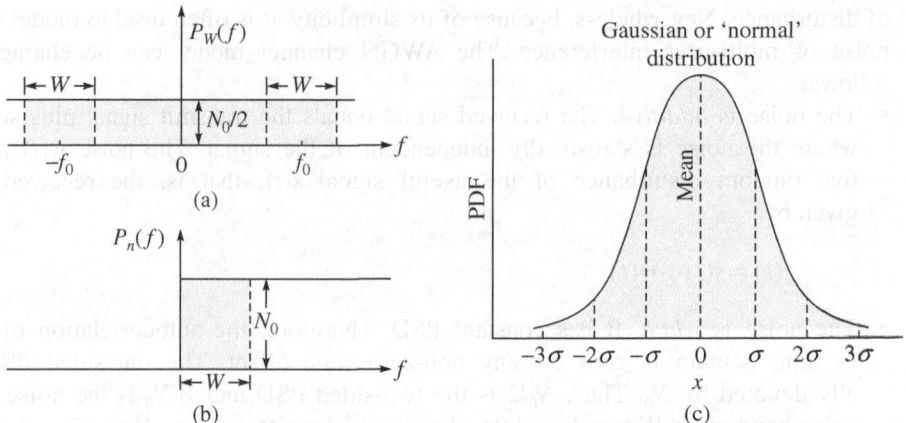

Fig. 4.9 Noise representation (a) Two-sided PSD (b) Equivalent one-sided PSD
(c) Typical Gaussian distribution with zero mean and σ standard deviation

4.3 REPRESENTATION OF DISCRETE CHANNEL BY FILTER

Wireless channels for mobile radio communications are challenging media that require careful system design for reliable transmission. Conceptually, if an impulse is transmitted on a stationary channel, then at the receiver, multiple delayed versions of impulses will be received at different instants of time due to multipath. These impulses are non-correlated and with reducing amplitudes with time. They are just like delayed samples, and hence, a difference equation with coefficient values can be used to represent this concept, which is the equation for a finite impulse response (FIR) filter:

$$r(n) = w_0 s(n) + w_1 s(n-1) + w_2 s(n-2) + \ldots \tag{4.14}$$

In Eq. (4.14), the coefficients w's are the complex channel impulse response or fading coefficients. For time-varying channels, this whole scenario becomes dynamic. The equation forms the basis for an equalizer design. The channel model for a time-varying channel can be approximated by the autoregressive (AR) model. See Fig. 4.10.

$$h(n+1) = w_1 h(n) + w_2 h(n-1) + \ldots w_n h(n-N) + w(n) \tag{4.15}$$

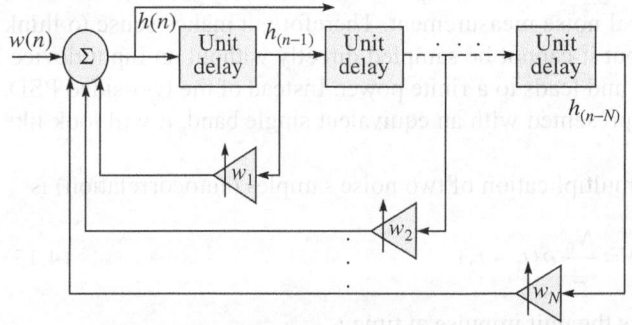

Fig. 4.10 Autoregressive model for a channel

where w's are the weights or coefficient values to estimate the channel.

The concept can be understood better while discussing adaptive channel equalization and channel estimation in Chapter 8 where weights are to be adjusted. In this case, the received time domain signal $r(n)$ is a function of the transmitted signal $s(n)$, the estimated channel transfer function, and AWGN $w(n)$. It can be expressed as

$$r(n) = s(n) * h_e(n) + w(n) \tag{4.16}$$

where * denotes convolution and $h_e(n)$ is the estimated channel impulse response. The whole time domain scenario can be mapped into frequency domain representation.

4.4 STOCHASTIC RADIO CHANNEL MODELLING

In principle, the following three different domains determine radio signal transmission:
(a) Physical conditions selected or operational scenario
(b) Dispersion phenomena of wave propagation
(c) Transceiver characteristics

The *operation scenario* implies some fundamental data such as frequency range, system bandwidth, and environments, (urban, rural, indoor etc.), which even impresses the general character of a wireless transmission link due to the relationship between the topographical features and the wavelength or the time resolution.

Dispersion in frequency, time, direction, and polarization is a crucial aspect of radio communication. We have to distinguish between multipath propagation and temporal fluctuations. In multipath propagation, each path suffers from a multiplicity of well-known effects, Temporal fluctuations can be split into short-term or small-scale fluctuations (fast fading) and long-term or large-scale fluctuations, comprising gradual and sudden changes of path parameters mainly because of movements of the mobile terminals (MT) and the reflectors and scatters, respectively.

Transceiver characteristics contributing to the stochastic radio channel modelling (SRCM) are the parameters that describe the MT movement and the antenna configuration with its radiation pattern and diversity properties.

Table 4.3 shows a generic scheme of all the effects and parameters that are to be considered with regard to the set-up of SRCM.

The deterioration of the transmission quality is also strongly dependent on the signal processing, that is, modulation, coding, detection, and so on.

From Table 4.3, it can be concluded that the consideration of all aspects leads to a rather high computational complexity of the SRCM. The dedication of the SRCM to a certain class of operation scenario, that is, the delivery of indoor services with a given maximum bit rate (or bandwidth) at a prescribed frequency range, is the first step towards a simplified model approach. Moreover, the identification of a few types of environments allows a remarkable reduction

> The allocated frequency range plays a key role in exhibiting path loss, whereas room dimensions influence maximum delay spread in direct proportion.

Table 4.3 Parameters to be considered for stochastic radio channel model

Operation scenario	Dispersion			Transceiver characteristics
	Multipath propagation	**Short-term fluctuations**	**Long-term fluctuations**	
–Frequency range	–Number of paths	For each path:	For each path:	–Trajectory
–Bandwidth	For each path:	–Fast fading	–Path loss	–Velocity
–Type of environment	–Mean power		–Shadowing	–Antenna configuration
	–Delay		–Transactions	
	–Doppler shift		–Delay drift	
	–Incidence direction		–Direction drift	
	–Scattering function			
	–Polarization			

Note: The various parameters extracted for SRCM are complex amplitude, delay, incidence direction, and Doppler shift of the impinging wave components.

Attenuation variation due to fading is represented mathematically by PDF rather than time function due to its random nature.

in complexity. Thus, four important indoor situations can be chosen: small rooms, large rooms, factory halls, and corridors.

Referring to the uncertainty relation, the system bandwidth is inversely proportional to the time resolution and, therefore, determines the necessary sampling rate of the model. The evaluation of future generations of indoor systems equipped with smart antennas and intended for unrestricted mobile operation will require full complexity channel modelling.

A stochastic radio channel model has been developed in order to simulate realistic channel impulse responses according to a wide range of possible physical situations within a given category of environments. If the channel is estimated on the basis of channel statistics, it is called the *blind method* of channel estimation. The complexity of this method is very high.

The transceiver characteristics as well as nearly all dispersion effects, that is, the different phenomena of multipath propagation and short- and long-term fluctuations, are implemented in the SRCM. Under certain conditions, some effects that are expected to have only little impact on the system performance can be neglected. This may result in a considerable reduction of the model complexity. Further improvements of the SRCM are envisaged, especially with regard to the consideration of terminal movements along arbitrary trajectories with varying velocity.

SAGE algorithm   Among the different algorithms for the calculation of the SRCM parameters from the measured data, the space-alternating generalized expectation-maximization (SAGE) algorithm is found to be a powerful tool. Due to advanced semiconductor signal processing devices, even an online multiple parameter extraction seems to be possible in the near future. Thus, the SAGE algorithm will also be a promising signal processing scheme for the next generation of wireless communication systems using smart antennas. Figure 4.11

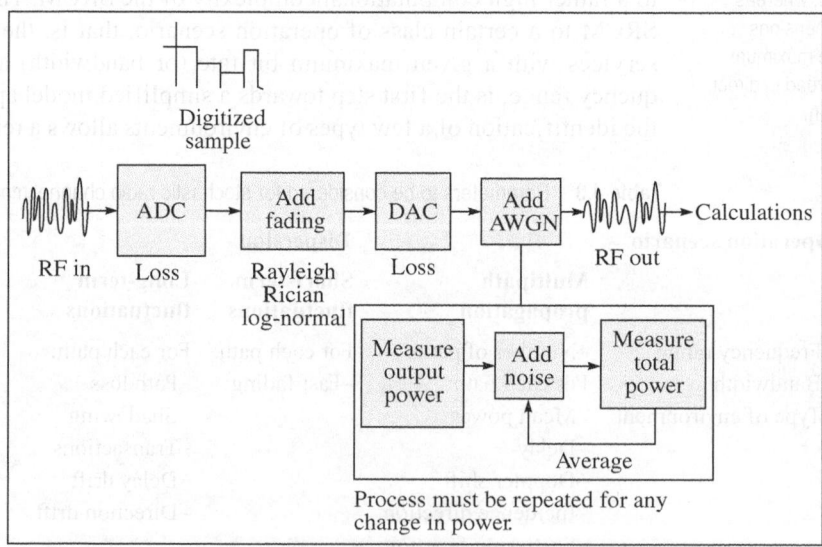

Fig. 4.11   Processing stages for SAGE algorithm to calculate channel parameters

shows some processing stages of the algorithm in which fading and AWGN can be added artificially to measure the channel effects.

4.5 FLAT FADING CHANNEL MODELLING

It is a fundamental requirement to know the properties of flat fading channels before going into details of complicated channel models. Rayleigh, Rician, and Nakagami models, which are based on signal fading statistics, are described in detail later in this chapter. However, a few equations help understand how flat fading can be incorporated in a signal while modelling it mathematically. In a flat fading channel, the received signal $r(t)$ is obtained by the addition of the transmitted signal $s(t)$ multiplied by a time-varying attenuation $\alpha(t)$ and the noise contribution $w(t)$:

$$r(t) = \alpha(t)s(t) + w(t) \tag{4.17}$$

As studied earlier, the time variation of the attenuation is known as fading. The time function of attenuation variation is difficult to represent mathematically, as it is a matter of probability and depends upon the channel environment. For such cases, it is better to represent the probability distribution function for α. It has been shown experimentally as well as argued theoretically that it usually follows a Rayleigh distribution (the Rayleigh model is explained in detail in Section 4.7).

$$\text{pdf}(\alpha) = (\alpha/\sigma^2) \times \exp(-\alpha^2/2\sigma^2) \quad \text{for } 0 < \alpha < \infty \tag{4.18}$$

where σ^2 is the variance of the underlying Gaussian process. The condition for this to be valid is that there are many statistically independent scatterers and no single scatterer makes a dominant contribution. If there is one dominant contribution (usually a LOS component), the distribution of α is a Rice-distributed variable (the Rician model is explained in detail in Section 4.8) characterized by

$$\text{pdf}(\alpha) = (\alpha/\sigma^2) \times \exp[-(\alpha^2 + A^2)/2\sigma^2] \times I_0(\alpha A/\sigma^2) \quad 0 < \alpha < \infty \tag{4.19}$$

where $I_0(x)$ is the modified Bessel function of the first kind and zero order. The parameter A is the amplitude of the dominant component. The Rice parameter K is defined as $A^2/2\sigma^2$.

The multipath propagation also introduces a phase shift. If the amplitude is Rayleigh fading, the phase shift is statistically independent of the amplitude distribution and is uniformly distributed between 0 and 2π. If the amplitude fading is Rician, the joint PDF of amplitude and phase ψ is

$$\text{pdf}_{\alpha,\psi} = (\alpha/2\pi\sigma^2) \times \exp[-(\alpha^2 + A^2 - 2\alpha A \cos\psi)/2\sigma^2] \tag{4.20}$$

An alternative amplitude distribution, which has gained popularity, especially for the evaluation of measurements, is Nakagami-M distribution (the Nakagami model is explained in detail in Section 4.9) given by

Fading is the resultant effect of the time domain delay spread and frequency domain Doppler spread.

$$\text{pdf}_\alpha(\alpha) = \frac{2}{\Gamma(m)}\left(\frac{m}{\Omega}\right)^m \alpha^{2m-1} \exp\left(-\frac{m}{\Omega}\alpha^2\right) \quad \text{for } \alpha \geq 0 \text{ and } m \geq 1/2 \tag{4.21}$$

Here, m is the shape factor. Actually, statistical analysis of Rician and Nakagami-m fading is done with multipath shape factors. For $m = 1$, this distribution reduces to the Rayleigh distribution. The parameter Ω is the mean square value $\Omega = \alpha^2$ and the parameter m is given by

$$m = \frac{\Omega^2}{(\alpha^2 - \Omega)^2} \tag{4.22}$$

Nakagami and Rice distributions are quite similar and each can be approximately converted to the other for $m \geq 1$:

$$m = \frac{(K+1)^2}{(2K+1)} \tag{4.23}$$

and $\quad K = \dfrac{\sqrt{m^2 - m}}{m - \sqrt{m^2 - m}} \tag{4.24}$

The movement of the mobile station leads to a frequency shift of the arriving waves (Doppler effect). If the sinusoidal wave of frequency f_c is transmitted, the spectrum of the received signal is

$$Y(f) \propto \left[\mathbf{pdf}_\gamma(\gamma)G(\gamma) + \mathbf{pdf}_\gamma(-\gamma)G(-\gamma)\frac{1}{\sqrt{\left(\dfrac{v}{c}f_c\right)^2 - (f - f_c)^2}} \right] \tag{4.25}$$

Here, f is the variable for the frequency, $\dfrac{v}{c}f_c$ is the extreme new frequency due to Doppler effect as explained previously, other than f_c and it varies in the range $-\dfrac{v}{c}f_c < f < \dfrac{v}{c}f_c$ and 0 elsewhere, where v is the velocity or speed of movement, γ is the angle between the direction of incidence of the move and the direction of movement, and $G(\gamma)$ is the antenna pattern. For the case when the waves are all incident horizontally and are uniformly distributed in azimuth and the antenna has a uniform pattern in azimuth, we get

$$v(f) \propto \frac{1}{\sqrt{\left(\dfrac{v}{c}f_c\right)^2 - (f - f_c)^2}} \tag{4.26}$$

The Rayleigh or Rice fading is small-scale fading since it describes the variation of the amplitude within an area of about ten wavelengths. Over a larger scale, it has been shown experimentally that the small-scale averaged amplitude obeys a log-normal distribution. Free space propagation model and the related equations can be applied to calculate the path losses.

4.6 WIDEBAND TIME-DISPERSIVE CHANNEL MODELLING

Most of the latest communication systems are digital and wideband, such as code division multiple access (CDMA) and orthogonal frequency division multiplexing (OFDM) explained in Chapter 7. At the same time, wireless channel is a dielectric medium, and hence, the refractive index of the channel, phase velocity, and wave and media propagation constants are the key parameters. Therefore, over a wide range of frequencies in the transmission bandwidth, frequency-dependent performance is obtained.

> Rayleigh fading model does not have an LOS component.

> *Note*: All models might not be suitable for all systems; for example, a delay-based model is well suited for unequalized systems, but it is usually not suitable for CDMA systems.

Research in this area has been going on for more than 25 years and has concentrated mainly on the following two areas (modelling may be done mostly by using measurements):

(a) Results are collected from extensive measurement campaigns. Wideband measurements are much more complicated than simple field strength (i.e., narrowband) measurements.

(b) From the measurements, the channel models are derived, which should fulfil the following two criteria:

   (i) They must be simple enough to allow an analytical computation of basic system performance.

   (ii) They must be very close to the physical reality; in other words, the performance computed by these models must be close to the performance measured in actually existing mobile radio channels.

These requirements are contradictory. Therefore, models of different complexity and accuracy have been developed. Models based on multipath delay and hence phase addition are not applied directly to CDMA systems, because in CDMA, the rake receiver handles multipath.

A further field of research, which is an intermediate step between measurement and modelling, is *information condensation*. As a wideband measurement campaign gives a large amount of data, for further interpretation and processing, a few characteristic parameters must be extracted. This will reduce the amount of data while keeping the loss of information as less as possible. Stanford University Interim (SUI) 1 to 6 channels and international telecommunication union (ITU) pedestrian A, B as well as ITU vehicular A, B channels are now used for wideband system simulations.

4.7 RAYLEIGH FADING MODEL

When the waves of multipath signals are out of phase, there can be a reduction of the signal strength at the receiver. This may result in *deep fading*. The basic model of Rayleigh fading assumes a received multipath signal to consist of a (theoretically infinite) large number of reflected waves with independent and identically distributed (IID) in-phase and quadrature amplitudes. This model has played a major role in our understanding of mobile propagation. The model was first proposed in a comment paper written by Lord Rayleigh in 1889. A signal amplitude (in dB) versus time graph for an antenna moving at constant velocity exhibiting Rayleigh fading is shown in Fig. 4.12(a). Note that the deep fades occur occasionally. Although fading is a random process, deep fades have a tendency to occur approximately every half a wavelength of motion.

Rayleigh distribution is a good model for channel propagation where there is no strong LOS path from the transmitter to the receiver and where the base station is hidden behind a building several blocks away and the arriving signal is bouncing off many scattering objects in the local area.

In the time domain, Rayleigh fading looks like periodic peaks of 10 dB or less interspersed between deep troughs of 40 dB or more. These deep fades (nulls in signal power) will typically occur at separations of half a wavelength. The dense scatterer model, which explains cellular communications propagation, states that the amplitude of multipath rays follows a Rayleigh distribution

Rayleigh distribution is ideal for representing channel conditions seen on a busy street in a city, an indoor environment, an urban area with crowded tall buildings, or a shadowed region.

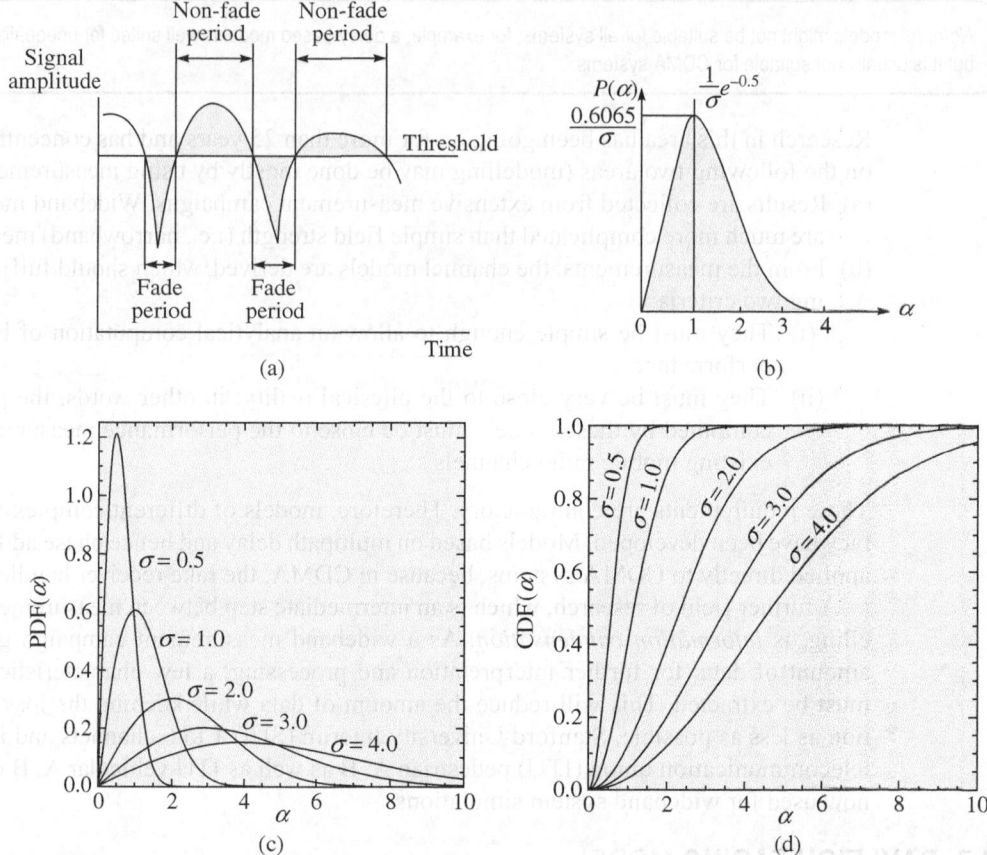

Fig. 4.12 Rayleigh fading (a) Sample of a Rayleigh fading signal (b) Rayleigh PDF (c) Variations in Rayleigh PDF with change in standard deviation (d) Variations in Rayleigh CDF with change in standard deviation

whereas the angle of arrival (multipath phase) follows a uniform distribution. The PDF of Rayleigh distribution shown in Fig. 4.12(b) is given in Section 4.5, which explains flat fading. The variations of PDF and CDF with respect to standard deviation is shown in the plots given in Figure 4.12(c) and (d) respectively, for the various values of σ, which is the square root of variance.

The Doppler shift of each incident ray is given by Eq. (3.38) in Chapter 3. The maximum spread f_d is determined by the speed of the vehicle and is experienced only by the spectral components arriving on paths directly in front of (maximum positive) or directly behind (maximum negative) the motion of the receiver. No shift is seen at 90° and 270° relative to the motion. This is why most of the spectral density $s(f)$ is observed at the edges of the Doppler spread, as given by Eq. (4.27), which gives the proportionate relationship correlating with the rays that are coming directly in front of and directly behind the motion of the antenna. (Equation 3.41 in Chapter 3 gives the exact equation.)

$$S(f) \infty \left[\frac{1}{f_d \sqrt{1 - \left(\dfrac{f - f_c}{f_d}\right)^2}} \right], \quad \text{where} \quad |f - f_c| < f_d \tag{4.27}$$

Phasor cumulation is achieved by placing each multipath phasor (with amplitude and phase) over each previous phasor end, starting from the origin with the first phasor and finally joining the origin with the last phasor tip.

In theory, these edges are infinitely high, but for wireless design verification purposes, the cut-off is typically 6 dB between power at f_c and power at $f_c \pm f_d$, as shown in Fig. 3.25. Here, the proportionality factor is $1/\pi$. The Doppler spread is used to calculate the level crossing rate with respect to a certain threshold level. Figure 4.12(a) shows the fade duration and crossing of the threshold level. In addition, one can calculate the average fade duration, which is the average time over which the signal below the threshold level can be used for deciding diversity schemes or signalling methods. The calculations are given in the examples at the end of the chapter.

4.7.1 Multiple Rayleigh Fading Signals

In a wireless system, typically, interference from multiple transmitters is experienced. Each signal may experience multipath fading and shadowing. Cumulating multiple Rayleigh fading signals requires investigation of the nature of the signals contributing to the interference. We consider the signal behaviour during an observation interval of duration T, which is shorter compared to the rate of channel fading. This implies that fading does not affect the amplitudes and phases of the signals. However, modulation can affect the amplitudes and phases during T. Two extreme cases are distinguished—*coherent cumulation* and *incoherent addition*.

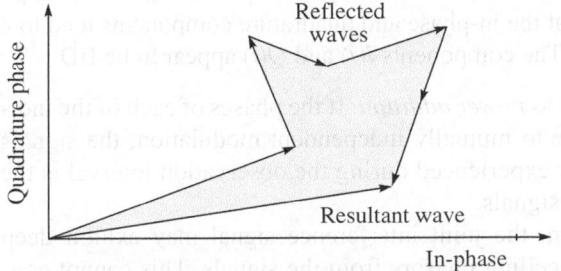

Fig. 4.13 Phasor diagram of a set of scattered waves resulting in a Rayleigh fading envelope

Coherent cumulation This is also known as *phasor cumulation*. It occurs if, during the observation interval, the fluctuations of the phases caused by the modulating signals are sufficiently small and the carrier frequencies of the signals are exactly equal. The joint signal behaves as a Rayleigh phasor, with Gaussian in-phase and quadrature phase components. The instantaneous power is exponentially distributed. The local mean power is equal to the sum of the local mean powers of the individual signals, as shown in Fig. 4.13.

To understand Fig. 4.13, consider a case of an unmodulated carrier. The transmitted signal has the form

$$s(t) = \cos(\omega_c t + \Psi) \tag{4.28}$$

Considering the Doppler effect in Fig. 3.23 given in Chapter 3, the received unmodulated signal $r(t)$ can be expressed as

$$r(t) = \sum_{n=1}^{N} \alpha_n \cos(2\pi f_c t + \Psi + \phi_n + 2\pi \Delta f_n t) \tag{4.29}$$

An in-phase quadrature representation of the form

$$r(t) = I(t)\cos \omega_c t - Q(t)\cos \omega_c t \tag{4.30}$$

can be found with the in-phase component including the Doppler effect as

$$I(t) = \sum_{n=1}^{N} \alpha_n \cos\left(\frac{2\pi v f_c t}{c}\cos \alpha_n + \Psi + \phi_n\right) \tag{4.31a}$$

and the quadrature phase component can be found as

> **Coherent cumulation can occur only if phase modulation with a very small deviation is applied.**

$$Q(t) = \sum_{n=1}^{N} \alpha_n \sin\left(\frac{2\pi v f_c t}{c} \cos \alpha_n + \Psi + \phi_n \right) \tag{4.31b}$$

Coherent cumulation can occur only if the observation interval taken is short with respect to the rate of modulation. This occurs, for instance, in digital systems if the joint interference signal is studied during one-bit interval or during the lock-in of a carrier-recovery loop in a synchronous detector.

Let us consider a stationary user, that is, $v = 0$. An in-phase and quadrature representation reduces to

$$I(t) = \sum_{n=1}^{N} \alpha_n \cos(\Psi + \phi_n) \tag{4.32a}$$

$$Q(t) = \sum_{n=1}^{N} \alpha_n \sin(\Psi + \phi_n) \tag{4.32b}$$

Thus, both the in-phase and quadrature components, $I(t)$ and $Q(t)$, can be interpreted as the sum of many (independent) small contributions. Each contribution is due to a particular reflection, with its own amplitude α_n and phase. For sufficiently large number of reflections (large N), the central limit theorem now says that the in-phase and quadrature components tend to a Gaussian distribution of their amplitude. The components $I(t)$ and $Q(t)$ appear to be IID.

Incoherent addition This is also known as *power addition*. If the phases of each of the individual signals substantially fluctuate due to mutually independent modulation, the signals add incoherently. The interference power experienced during the observation interval is the power sum of the individual signals.

> **With incoherent cumulation, the joint interference signal behaves as a band-limited Gaussian noise source if the number of components is sufficiently large.**

With coherent cumulation, the joint interference signal may exhibit deep fades, caused by mutual cancelling phasors from the signals. This cannot continue for a sustained period due to the phase variations caused by the angle modulation of each signal or by slightly different carrier frequencies due to Doppler shifts and free-running oscillators. Moreover, any fade of one of the signals is likely to be hidden by other interfering signals. Hence, the joint interference signal tends to exhibit less multipath fluctuations per unit of time than the signal from one individual interferer.

MULTIPLE INCOHERENT RAYLEIGH FADING SIGNALS WITH EQUAL AND DIFFERENT MEAN POWER

If the interference is caused by the power sum of n Rayleigh fading signals with *identical* local mean power, the PDF of the joint interference power is the nth convolution of the exponential distribution of the power of an individual interfering signal. The PDF of the joint interference power caused by interfering signals with *different* local mean powers can be approximated by a gamma distribution, which is a two-parameter family of continuous probability distributions, as shown in Fig. 4.14(a).

It may not be fully appropriate to speak of the envelope of such a joint interference signal, but if one defines the amplitude to be proportional to the square root of the power, then one finds that the amplitude has a Nakagami distribution (see Section 4.9). Nakagami-M distribution is a probability distribution related to the gamma distribution that has two parameters—shape parameter m and controlling spread Ω, as shown in Fig. 4.14(b) (defined in Section 4.5).

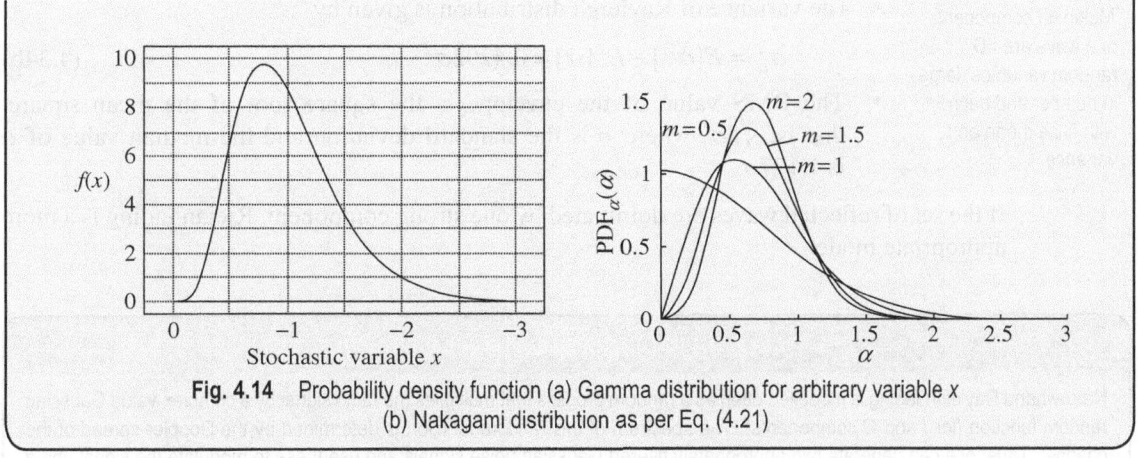

Fig. 4.14 Probability density function (a) Gamma distribution for arbitrary variable x (b) Nakagami distribution as per Eq. (4.21)

4.7.2 Probability Density Function of Rayleigh Signal Amplitude

> An important application of the PDF is the calculation of outage probabilities, that is, the probability that the signal strength drops below a certain threshold level.

In order to obtain the PDF of the signal amplitude r of a Rayleigh fading signal, we observe the random processes of the in-phase and quadrature components, $I(t)$ and $Q(t)$, respectively, at one particular instant t_0.

As per central limit theorem, given certain conditions, the mean of a sufficiently large number of independent random variables, each with finite mean and variance, will nearly be the normal (or Gaussian) distribution. If the number of received waves N becomes very large and all are IID, the central limit theorem says that $I(t_0)$ and $Q(t_0)$ at a particular instant t_0 are (zero-mean) Gaussian random variables, each with variance s^2. Then Lord Rayleigh argued that the received signal is of the form mentioned in Eq. (4.17) and has a Rayleigh amplitude $a(t)$, which is found from $\sqrt{I^2(t) + Q^2(t)}$, and a uniform phase $q(t)$ between 0 and $2p$. The probability density of the amplitude is described by the Rayleigh PDF represented in Eq. (4.18). A received signal consisting of $N = 6$ (Fig. 4.15) reflected waves (the resulting signal amplitude r consists of an in-phase component I and a quadrature component Q. If the antenna moves, the relative phases of the reflected waves change over time. So, r, I, and Q become functions of time t).

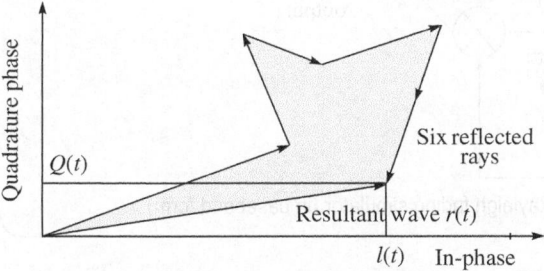

Fig. 4.15 Phasors creating time varying received signal

A few more relationships for Rayleigh fading distribution are as follows:

- The probability that the envelope of the received signal does not exceed a specified (threshold) value J is given by the corresponding CDF.

$$P(J) = \int_0^J p(\alpha)d\alpha = 1 - \exp\left(-\frac{J^2}{2\alpha^2}\right) \tag{4.33}$$

- The mean value of Rayleigh distribution is given by

$$E[\alpha] = \alpha\sqrt{\frac{\pi}{2}} = 1.2533\sigma \tag{4.34a}$$

> Multipath components of a wave are IID random variables, large in number and each with finite mean and variance.

- The variance of Rayleigh distribution is given by

$$\sigma_\alpha^2 = E[\alpha^2] - E^2[\alpha] = 0.4292\sigma^2 \qquad (4.34b)$$

- The RMS value of the envelope is the square root of the mean square, that is, $\sqrt{2}\sigma$, where σ is the standard deviation and the median value of α is 1.177σ.

If the set of reflected waves are dominated by one strong component, Rician fading is a more appropriate model.

RAYLEIGH FADING SIMULATOR: AN EXAMPLE

Narrowband Rayleigh fading is modelled often as a random process that multiplies the radio signal by a complex-value Gaussian random function (for *I* and *Q* components). The spectrum of this random function is determined by the Doppler spread of the channel. Thus, one can generate two appropriately filtered Gaussian noise signals and use these to modulate the signal and a 90° phase shifted version of the signal as shown in Fig. 4.16.

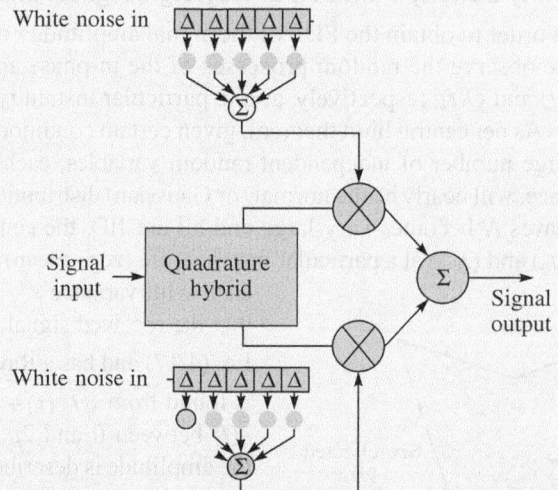

Fig. 4.16 Block diagram of a narrowband Rayleigh fading simulator (in baseband form)

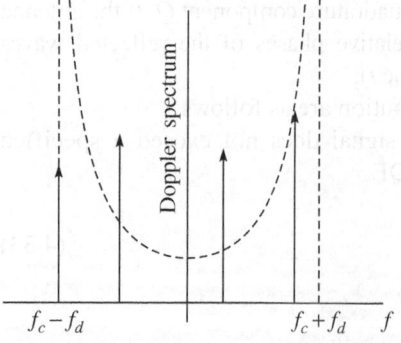

Fig. 4.17 *N* frequency components chosen from the Doppler spectrum for simulation

It is a common practice to generate two filtered noise components by adding a set of six or more sinusoidal signals. Their frequencies are chosen to approximate the typical U-shaped Doppler spectrum, as shown in Fig. 4.17.

The *N* frequency components are taken at

$$f_i = f_d \cos[2\pi i/2(2N+1)] \quad \text{where } i = 1, 2, .., N \qquad (4.35)$$

This specific set of frequencies is chosen to approximate the U-shaped Doppler spectrum. All amplitudes are taken equal to unity. One component at the maximum Doppler shift is also added but at an amplitude of $1/\sqrt{2}$, that is, at about 0.707.

As the demand for mobile communication increases, systems have to be more efficient and cell sizes smaller and smaller. However, the Rayleigh model lacked the effect of a dominant LOS component to describe microcellular propagation, and the Rician model appeared to be more appropriate for this purpose.

4.8 RICIAN FADING MODEL

The Rician fading model is similar to the Rayleigh fading model, except that in Rician fading, a strong dominant component is present. This dominant component can, for instance, be the LOS wave. The refined Rician models also consider the following points:

- The dominant wave can be a phasor sum of two or more dominant signals, such as the LOS and a ground reflection. This combined signal is then mostly treated as a deterministic (fully predictable) process.
- The dominant wave can also be subject to shadow attenuation. This is a popular assumption in the modelling of satellite channels.

Besides the dominant component, the mobile antenna receives a large number of reflected and scattered waves, as shown in Fig. 4.18.

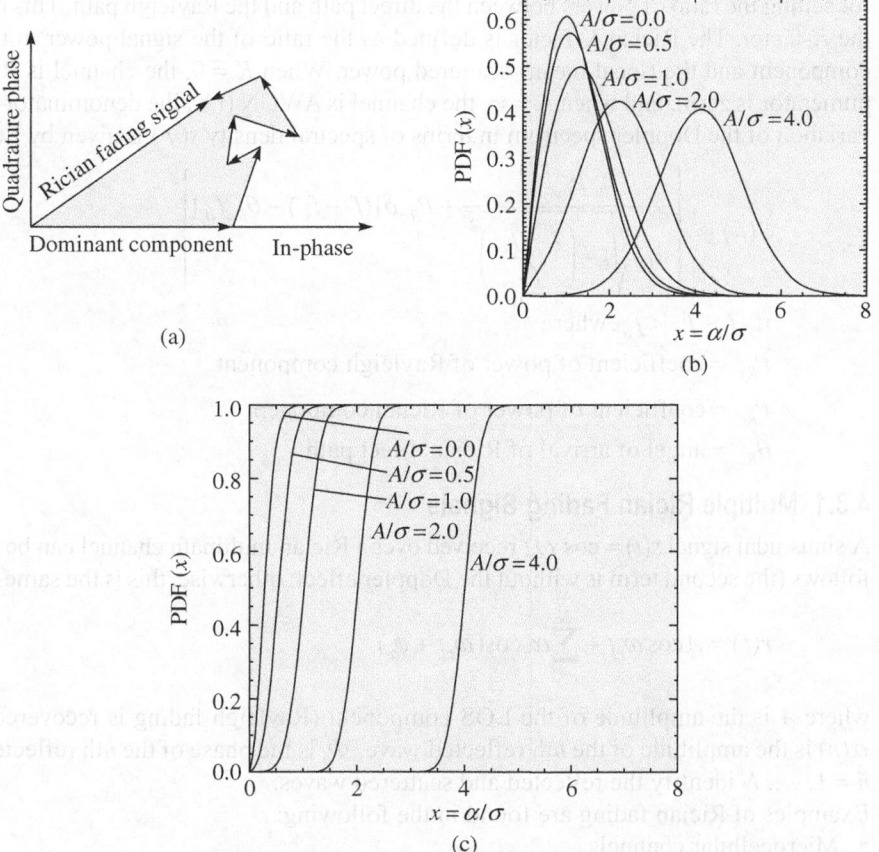

(a)

(b)

(c)

Fig. 4.18 Rician fading model (a) Phasor diagram of a Rician fading signal (b) Rician PDF (c) Rician CDF with change in standard deviation

In rural environments, where the multipath profile includes a few reflected paths combined with a strong LOS path, the spectral power follows a Rician distribution.

Consider again the Doppler shift theory explained in the Rayleigh fading model. The maximum spread f_d is determined by the speed of the vehicle and is only experienced by the spectral components arriving on paths directly in front of (maximum positive) or directly behind (maximum negative) the motion of the receiver, because this is where the cosine of the angle of arrival is one. At any other angle, the cosine is less than one. No shift is seen at 90° and 270° relative to the motion. This is why most of the spectral density is observed at the edges of the Doppler spread, which correlates with the rays that are coming in directly in front of and directly behind the motion of the antenna.

Rayleigh fading is considered as a worst-case scenario of the Rician model with zero LOS component. The angle of arrival of the direct ray and the ratio of the powers between the direct ray and the multipath rays determine how much effect the energy from the direct path has on the normal Rayleigh model. Observing this effect in the frequency domain, what we see is a spike in the power corresponding to the frequency shift attributed to the direct ray. As an example, in the global system for mobile communication (GSM) specifications, the angle of arrival of the direct path is set to 45°. We take the cosine of the angle to see how much statistical weight we give to that particular path. In this case, it is 0.7. Thus, we have a spike in the received signal power at 0.7 times the maximum frequency shift. The Rician model also allows for setting the ratio of powers between the direct path and the Rayleigh path. This ratio is called the K-factor. The Rician K-factor is defined as the ratio of the signal power in the dominant component and the (local mean) scattered power. When $K = 0$, the channel is Rayleigh (i.e., numerator is zero), and when $K = \infty$, the channel is AWGN (i.e., the denominator is zero). The variation of the Doppler spectrum in terms of spectral density $s(f)$ is given by Eq. (4.36).

$$S(f) \propto \left\{ \frac{P_{Ry}}{\pi f_d \sqrt{1 - \left(\frac{f - f_c}{f_d}\right)^2}} + P_{Rc}\delta[(f - f_c) - \theta_{Rc}f_d] \right\}$$

if $|f - f_c| < f_d$, where   (4.36)

P_{Ry} = coefficient of power of Rayleigh component

P_{Rc} = coefficient of power of Rician component

θ_{Rc} = angel of arrival of Rician direct path

4.8.1 Multiple Rician Fading Signals

A sinusoidal signal $s(t) = \cos \omega_c t$ received over a Rician multipath channel can be expressed as follows (the second term is without the Doppler effect; otherwise, this is the same as Eq. 4.29):

$$r(t) = A\cos \omega_c t + \sum_{n=1}^{N} \alpha_n \cos(\omega_c t + \phi_n)$$ (4.37)

where A is the amplitude of the LOS component (Rayleigh fading is recovered for $A = 0$), $\alpha(n)$ is the amplitude of the nth reflected wave, Φ_n is the phase of the nth reflected wave, and $n = 1, \ldots, N$ identify the reflected and scattered waves.
Examples of Rician fading are found in the following:

- Microcellular channels
- Vehicle-to-vehicle communication, for example, for aerial vehicle communications system (AVCS)

In Rician fading multipath model, at least one component is with a non-zero mean whereas others may have a zero mean due to a deterministic strong wave.

- Indoor propagation (special cases)
- Satellite channels

Like in Rayleigh fading, the in-phase and quadrature phase components of the received signal are IID Gaussian random variables. However, in Rician fading, the mean value of (at least) one component is non-zero.

For a large fade margin $F \gg 1$, the probability that the instantaneous power drops below the noise threshold z tends to

$$\text{Outage probability} = \frac{(1 + K)\exp(-K)}{F} \tag{4.38}$$

where the fade margin F is defined as the local mean power minus the threshold z.

4.8.2 Probability Density Function of Rician Signal Amplitude

The derivation of the PDF of the amplitude is more involved in Rician fading (Fig. 4.18b) than in Rayleigh fading, and a Bessel function occurs in the mathematical expression. It has been proposed to approximate this expression by the model for Nakagami fading. However, approximations that focus on the behaviour near the mean value will divert by orders of magnitude in predicting the probability of deep fades.

The derivation is similar to that for Rayleigh fading. In order to obtain the probability density of the signal amplitude α, we observe the random processes $I(t)$ and $Q(t)$ at one particular instant t_0. If the number of scattered waves is sufficiently large and the waves are IID, the central limit theorem says that $I(t_0)$ and $Q(t_0)$ are Gaussian but are no longer zero mean because of the deterministic dominant term. The Rician PDF is given by Eq. (4.19). The transformation of variables shows that the amplitude and phase have a joint PDF represented by Eq. (4.20).

In the expression for the received signal, the power in the LOS equals $A^2/2$. In indoor channels with an unobstructed LOS between the transmitter and receiver antennas, the K-factor is between, say, 4 and 12 dB. Rayleigh fading is recovered for $K = 0$ ($-\infty$ dB). The total local mean power $\overline{p}$ is the sum of the power in the LOS and the local mean scattered power. The local mean scattered power equals $\overline{p} / (K + 1)$. The amplitude of the LOS is $A = \text{sqrt}[2K\overline{p}/(K + 1)]$.

4.9 NAKAGAMI FADING MODEL

Besides Rayleigh and Rician fading, a refined model suggested for the PDF of a signal amplitude exposed to mobile fading is the Nakagami fading model. The distribution of the amplitude and signal power can be used to find the probabilities on signal outages. The following points should be noted:

- If the envelope is Nakagami distributed, the corresponding instantaneous power is gamma distributed.
- The parameter m (as per Eqs 4.22, 4.23, and 4.24) is called the *shape factor* of the Nakagami or the gamma distribution.
- In the special case $m = 1$, Rayleigh fading is recovered, with an exponentially distributed instantaneous power.
- For $m > 1$, the fluctuations of the signal strength reduce compared to Rayleigh fading.

The Nakagami fading model was initially proposed because it matched the empirical results for short-wave ionospheric propagation. In current wireless communication, the main role of the Nakagami model can be summarized as follows:

Though being almost similar models, the behaviours of Nakagami and Rician fading in deep fades are essentially different.

- It describes the amplitude of the received signal after maximum ratio diversity combining (explained in Chapter 8). After k-branch maximum ratio combining (MRC) with Rayleigh fading signals, the resulting signal is Nakagami with $m = k$. The MRC combining of Nakagami-M fading signals in k branches gives a Nakagami signal with a shape factor mk.

- The sum of multiple IID Rayleigh fading signals has a Nakagami-distributed signal amplitude. This is particularly relevant to model interference from multiple sources in a cellular system.

- Nakagami distribution matches some empirical data better than the other models.

- Nakagami fading occurs for multipath scattering with relatively large delay time spreads and with different clusters of reflected waves. Within any one cluster, the phases of individual reflected waves are random, but the delay times are approximately equal for all waves. As a result, the envelope of each cumulated cluster signal is Rayleigh distributed. The average time delay is assumed to differ significantly between clusters. If the delay times too significantly exceed the bit time of a digital link, the different clusters produce serious ISI; therefore, the multipath self-interference then approximates the case of co-channel interference by multiple incoherent Rayleigh fading signals.

- The Rician and Nakagami models behave approximately equivalently near their mean value. This observation has been used to advocate the Nakagami model as an approximation for situations where a Rician model would be more appropriate. While this may be accurate for the main body of the probability density, it becomes highly inaccurate for the tails. As bit errors or outages mostly occur during deep fades, the tail of the PDF mainly determines these performance measures (for probability to receive a low power).

4.10 COMPARISON OF RAYLEIGH, RICIAN, AND NAKAGAMI MODELS

Rican and Nakagami fadings are two generalizations of the Rayleigh fading model. In literature, a Nakagami model is often used for analytical simplicity in cases where Rician fading would be a more appropriate model. However, in contrast to common belief, the Nakagami model is not an appropriate approximation for Rician fading. It has an essentially different behaviour for deep fades such that results on outage probabilities or error rates can differ in orders of magnitude.

To describe microcellular propagation, the Rayleigh model lacked the effect of a dominant LOS component and the Rician model appeared to be more appropriate. For analytical and numerical evaluation of system performance, the expressions for Rician fading are less convenient, mainly due to the occurrence of a Bessel function in the Rician PDF of the received signal amplitude. Hence, approximations by a Nakagami distribution, with simpler mathematical expressions, have become popular.

Both Rician and Nakagami models behave approximately equivalently near their mean value.

In the analysis of outage probabilities or error rates, the behaviour of the model for signals in deep fades has the determining effect. As the behaviour of the PDFs for amplitudes near zero differs significantly, approximations based on the behaviour near the mean are inappropriate.

The Rician and Nakagami models have fundamentally different densities for deep fades. Modelling a Rician fading signal by a Nakagami distribution of the amplitude leads to overly optimistic results, and the discrepancies can be in many orders of magnitude. However, the Nakagami model is sometimes used to approximate the PDF of the power of a Rician fading signal. Matching the first and second moments of the Rician and Nakagami PDFs, we get m as per Eq. (4.23), which tends to $m = K/2$ for large K. In the special case that the dominant component is zero ($K = 0$) or $m = 1$, Rayleigh fading occurs with an exponentially distributed power.

The results are strikingly different for m larger than one. As the relation between K and m was based merely on the first and second moments, it is likely to be most accurate for the values near the mean. Outage probabilities, however, highly depend on the tail of the PDF for small power of the wanted signal. The probability of deep fades differs for these two models, so an approximation of the PDF of a Rician fading wanted signal by a Nakagami PDF can be highly inaccurate. Results differ even in the first-order behaviour. For Rician fading, the slope of the outage probability versus carrier-to-interference ratio C/I is the same as that for Rayleigh fading. For Nakagami fading, the slope is steeper, similar to that of the m-branch diversity reception of a Rayleigh fading signal.

4.11 OKUMURA–HATA PATH LOSS MODEL

The Okumura model for urban areas was developed based on measurements made in and around Tokyo, Japan, in 1968 between 150 MHz and 1920 MHz. There were a number of predictions from a series of graphs, which were approximated in a set of formulae by Hata later on. Thus, the Okumura model served as a base for the Hata model.

Three types of prediction areas were considered in the Okumura model and hence the model was built in the following three modes:

(a) *Open area*: Open spaces with no tall trees or buildings in path
(b) *Suburban area*: Village highways scattered with trees and houses and some obstacles near the mobile but not very congested
(c) *Urban area*: Built-up cities or large towns with large buildings and houses and villages with close houses and tall trees

The Okumara model considers several parameters. The output parameter is the mean path loss (median path loss) L_{dB} for which the calculations are made as follows:

- Frequency f_c—between 150 MHz and 1920 MHz (typically extrapolated up to 3000 MHz
- TX height h_t—between 30 m and 1000 m
- RX height h_r—between 1 m and 10 m
- TX – RX distance d—between 1 m and 100 km

Okumura developed a set of curves giving median attenuation relative to free space. These curves were developed from extensive measurements using vertical omnidirectional antennas at both the base station and mobile. They are plotted as a function of frequency in the afore-mentioned range and as a function of distance from the base station. To determine path loss using Okumura model, the free space path loss between the points of interest is first determined, and then the value of f and d, as read from the curves, is added to it along with correction factors to account for the type of terrain. The model is based on measured data and does not provide any analytical explanation. For many situations, extrapolations of the derived curves can be made to obtain values outside the measurement range, although the validity of the extrapolations depends upon the circumstances and the smoothness of the curve in question.

> Okumura and Hata channel models are based on path loss measurements that are dependent on the frequency and path distance.

The Okumura model is formally expressed as

$$L_{dB} = L_{fsl\_dB} + A_{m\_dB} + H_{mg} + H_{bg} - \sum K_{correction} \quad (4.39)$$

where

L_{fsl_dB} = free space loss in dB
A_{m_dB} = median attenuation
H_{mg} = mobile station antenna height gain factor
H_{gb} = base station antenna height gain factor
$K_{correction}$ = correction factor gain (such as type of environment, water surfaces, and isolated obstacle)

The following scenario is assumed for defining the Okumura–Hata model and the related terminology (Fig. 4.19):

h_m = mobile station antenna height above local terrain height in metres
d_m = distance between the mobile and building in metres

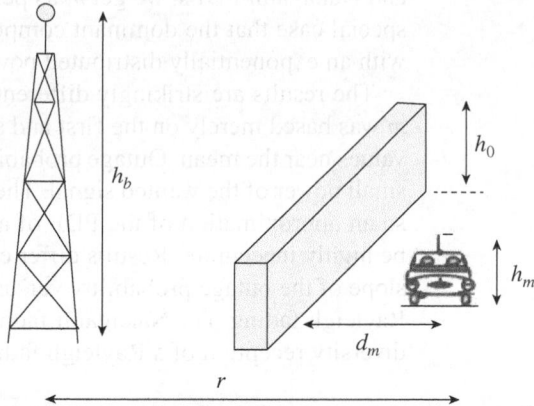

Fig. 4.19 Scenario for developing the Okumura–Hata model

h_0 = height of a building or object above local terrain height in metres
h_b = base station antenna height above local terrain height in metres
r = great circle distance between the base station and mobile in metres
$d = r \times 10^{-3}$ great circle distance between the base station and mobile in kilometres
f = carrier frequency in Hz
$f_c = f \times 10^{-6}$ carrier frequency in MHz
λ = free space wavelength in metres

The Hata model for urban areas is also known as the Okumura–Hata model as it is a developed version of the Okumura model. It is widely used for predicting the behaviour of cellular transmissions in built-up areas. This model collects graphical information from the Okumura model and develops it further to realize the effects of diffraction, reflection, and scattering caused by city structures. Though based on the Okumura model, the Hata model does not cover the whole range of frequencies and covers only up to 1500 MHz, whereas Okumura supports up to 1920 MHz. The Walfisch and Bertoni model is a further advancement of the Hata model. It considers the impact of rooftops and building height by using diffraction to predict average signal strength at the street level.

The following are the parameter ranges for the Okumura–Hata model:

- Frequency f_c—between 150 MHz and 1500 MHz
- TX (base station) height h_t—between 30 m and 200 m
- RX (mobile) height h_r—between 1 m and 10 m
- TX − RX distance d—between 1 m and 20 km

Hata model is suited for both point-to-point and broadcast transmissions.

The Okumura–Hata model takes urban areas as a reference and applies correction factors.

For urban areas:

$$L_{dB} = A + B \log_{10} d - E \quad (4.40)$$

For suburban areas:

$$L_{dB} = A + B\log_{10} d - C \qquad (4.41)$$

For open areas:

$$L_{dB} = A + B\log_{10} d - D \qquad (4.42)$$

where

$A = 69.55 + 26.16\log_{10}f_c - 13.82\log_{10}h_b$
$B = 44.9 - 6.55 \log_{10}h_b$
$C = 2[\log_{10}(f_c/28)]^2 + 5.4$
$D = 4.78(\log_{10}f_c)^2 + 18.33\log_{10}f_c + 40.94$
$E = 3.2[\log_{10}(11.7554h_m)]^2 - 4.97$ for large cities, $f_c \geq 300$ MHz
$E = 8.29[\log_{10}(1.54h_m)]^2 - 1.1$ for large cities, $f_c < 300$ MHz
$E = (1.1\log_{10}f_c - 0.7)h_m - (1.56\log_{10}f_c - 0.8)$ for medium and small cities

WALFISCH–IKEGAMI AND LONGLEY–RICE MODELS

Walfisch–Ikegami (WI) model This model is an extension of the Hata model and takes the following characteristics of the city structure into account:

- Heights of buildings, h_{Roof}
- Widths of roads, w
- Building separation, b
- Road orientation with respect to the direct radio path, α

The WI model increases the accuracy of the propagation estimation even though the model is more complex. The *TX – RX* distance allows estimation from 20 m instead of 1 km for the Okumura–Hata model. The output parameter for calculation is the mean path loss as in the earlier cases. In this model, there are two cases for calculation: LOS and non-line of sight (NLOS). The equations are different for the two cases. The scenario of the model is shown in Fig. 4.20.

Frequency f_c—between 800 MHz and 2000 MHz
TX (base station) height h_{Base}—between 4 m and 50 m
RX (mobile) height h_{Mobile}—between 1 m and 3 m
TX – RX distance d—between 0.02 m and 5 km

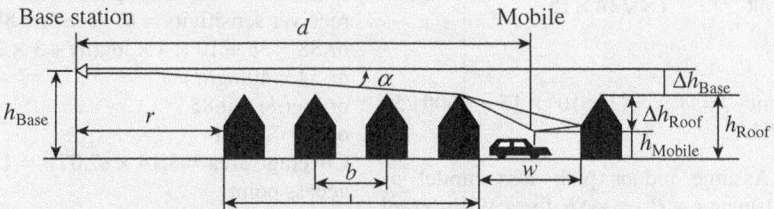

Fig. 4.20   Scenario for Walfisch–Ikegami model

Longley–Rice (LR) model This model predicts the attenuation of radio signals for a telecommunication link in the frequency range of 20 MHz to 20 GHz for different scenarios and different heights of *TX* and *RX* antennas. LR model is also known as the irregular terrain model (ITM). It was created for frequency planning in television broadcasting. The LR model has two parts: a model for predictions over an area and a model for point-to-point link predictions.

The LR model presents a generalization of the received signal power without a detailed characterization of the channel. It also uses statistical resources to characterize the channel, which depends upon the variables. The statistical estimates or attenuation variables are situation variability, time variability, and location variability. Due to these variabilities, there could be deviations more or less significant to the attenuation of the transmitted signal.

The Okumura–Hata model for medium and small cities has been extended to cover 1500–2000 MHz in 1999.

$$L_{dB} = F + B\log_{10}d - E + G \tag{4.43}$$

where

$F = 46.3 + 33.9\log_{10}f_c - 13.82\log_{10}h_b$

E = designed for medium to small cities

$G = 0$ dB for medium-sized cities and suburban areas

$G = 3$ dB for metropolitan areas

MORE SOLVED EXAMPLES

Example 4.2 Consider a Rayleigh flat fading channel and find the level crossing rate (number of fades/s) if the Doppler spread is 15 Hz. The threshold level r_{thres} is the same as the RMS value of the fading envelope r_{rms}. The level crossing rate is proportional to the Doppler spread with proportionality factor $\sqrt{2\pi}\rho e^{-\rho^2}$, where $\rho = r_{thres}/r_{rms}$. In addition, calculate the average fade duration, which is inversely proportional to the Doppler spread with proportionality factor $e^{-\rho^2}/\rho\sqrt{2\pi}$. At what carrier frequency is the Rayleigh behaviour observed if the vehicle moves at a speed of 50 km/h.

Solution Here, $\rho = 1$, because $r_{thres} = r_{rms}$.

Level crossing rate $= \sqrt{2\pi}\rho e^{-\rho^2}f_d$

$\sqrt{2\pi} \times 1 \times e^{-1} \times 15 = 13.83$ fades/s

Average fade duration

$= \dfrac{e^{-\rho^2}}{\rho\sqrt{2\pi}} \times \dfrac{1}{f_d} = \dfrac{e^{-1}}{1 \times \sqrt{2\pi} \times 15} = 0.0098 s$

$f_d = f v/c$

or carrier frequency $f = cf_d/v = (3 \times 10^8 \times 15 \times 3600)/50 \times 1000 = 324$ MHz

Example 4.3 Assume indoor path loss model as $P_{LdB} = P_{L(do)} + 10n\log d + P_{L(f)} + SE$ for a WiFi-based home environment with two floors. Assume path loss exponent $n = 3$. In addition,

$P_{L(do)}$ = reference path loss at 1 m distance from access point = 38 dB

$P_{L(f)} = 4N_f$ = signal attenuation due to N_f floors

SE = Shadow effect = 10 dB

If the mobile is 10 m away from the transmitter of 0 dB, what is the received power?

Solution Received power $P_r = P_t - P_{LdB}$

$P_{LdB} = (P_{L(do)} + 10n\log d + P_{L(f)} + SE) = 38 + 10 \times 4 \times \log(10) + 3 \times 2 + 10 = 94$ dB

or $P_r = 0 - 94 = -94$ dB

Example 4.4 For Example 4.3, if the minimum signal-to-noise ratio (SNR) requirement is 12 dB and the background noise at 2.4 GHz frequency is −100 dBm, what is the coverage area of the access point as per the SNR requirement?

Solution Receiver sensitivity = Background noise + SNR requirement = −100 dBm + 12 = −88 dB

Maximum allowable path loss = Transmit power − receiver sensitivity = 0 − (−88) = 88 dB

or $88 = 38 + 10 \times 4 \times \log(d) + 3 \times 2 + 10$

or $34 = 40\log(d)$

or $\log(d) = 0.85$

or $d = 7.07$ m

Coverage area $= 3.14 \times (7.07)^2 = 156.95$ m^2 around the access point

SUMMARY

- Channel modelling is required to simulate a channel during performance analysis of wireless systems.
- Assumptions for the channel in the modelling will decide the performance of the system. With change in assumptions, the performance will change.

- A channel is of time-varying, dispersive, non-linear, and fading type when it undergoes multipath.
- For long- and short-distance communication channels, assumptions must be different.
- Channel path loss models are empirical, semi-empirical,

or deterministic, and they are applied depending upon the area.

- Channel behaviour is random in nature and hence is represented by its PDF and CDF.
- Power delay profile represents the channel response with multipath.
- Channel modelling utilizes the concept of autocorrelation and second-order statistics.
- Shadowing can be calculated by local mean and area mean.
- In the absence of multipath, the most appropriate channel model is AWGN.
- A multipath channel can be represented by a digital

filter equation due to direct and/or multiple reflected delayed components received at the receiver.
- A time-varying channel can be approximated by an AR model.
- Wideband channels exhibit frequency selective fading.
- The Rayleigh fading model has multiple reflected rays added at the receiver, whereas the Rician model has a strong direct ray with multiple reflected rays. Nakagami fading occurs when there are large delay spreads. Many times, the Nakagami model is used for cellular systems.
- The Okumura–Hata model is an empirical model for path loss calculation in urban or semi-urban areas.

EXERCISES

Multiple-choice Questions

4.1 Gaussian probability density is defined as

(a) $p(x) = \dfrac{1}{\sqrt{2\pi\sigma^2}} e^{-(x-m)^2/2\sigma^2}$

(b) $p(x) = \dfrac{1}{2\pi\sigma^2} e^{-(x-m)^2/2\sigma^2}$

(c) $p(x) = \sqrt{2\pi\sigma^2}\, e^{-(x-m)^2/2\sigma^2}$

(d) $p(x) = 2\pi\sigma^2 e^{-(x-m)^2/2\sigma^2}$

4.2 If x and y are two independent Gaussian random variables, each with the average value zero and variance σ^2, the joint density function is

(a) $p(x, y) = p(x) + p(y)$ (c) $p(x, y) = p(x) - p(y)$
(b) $p(x, y) = p(x) \cdot p(y)$ (d) $p(x, y) = p(x)/p(y)$

4.3 Direct (LOS) component is not considered in

(a) Rician fading   (c) Rayleigh fading
(b) Nakagami fading   (d) all of these

4.4 Which of the following model(s) is/are a good approximation for cellular mobile communication with vehicular mobility?

(a) Rician model   (c) Rayleigh model
(b) Nakagami model   (d) All of these

4.5 *K*-factor is the ratio of powers between

(a) the direct path and the reflected path
(b) the direct paths of two base stations
(c) the direct path and the Rayleigh path
(d) the scattered path and the direct path

4.6 When *K*-factor is zero, the model becomes a

(a) Rician model   (c) Rayleigh model
(b) Nakagami model   (d) Hata model

4.7 For indoor propagation, the most suitable model(s) is/are

(a) Rician model   (c) Rayleigh model
(b) Nakagami model   (d) all of these

4.8 Matching the first and second moments of Rician and Nakagami PDFs, the shape factor of the Nakagami model and the Rician factor can be related as

(a) $m = \dfrac{K^2 + K + 1}{2K + 1}$

(b) $m = \dfrac{K^2 + 2K + 1}{2K + 1}$

(c) $K = \dfrac{m^2 + m + 1}{2m + 1}$

(d) $m(K + 1) = K^2 + 2K + 1$

Review Questions

4.1 Why do we require channel modelling? What are the difficulties one might face while modelling a channel?

4.2 What are the different characteristics of a channel taken into consideration while modelling mathematically? Correlate this with the mathematical models developed for the various types of systems in the field of digital signal processing (DSP) (e.g., linear time-invariant system).

4.3 Can a power delay profile help in the modelling of a channel? If so, then explain how this profile becomes useful.

4.4 What is the concept of multipath shape factors? Define the three multipath shape factors and their significance.

4.5 What are the fundamental concepts of DSP you observed in this chapter?

4.6 What do you mean by first-order and second-order statistics?

4.7 What will be the channel considerations for long- and short-distance channel modelling?

4.8 Find the assumptions for the channel in the case of an indoor propagation and an outdoor propagation model. Identify the models suitable for indoor and outdoor mobile applications.

4.9 'Wideband signal transmission is a serious problem in wireless communication.' Justify this statement.

4.10 What do you mean by wide-sense stationary processes?

4.11 Collect the Poisson and Markovian processes and distributions. How are they concerned with the wireless channel?

4.12 Describe the nature of AWGN.

4.13 Compare the Rayleigh, Rician, and Nakagami fading models. Why do we use Rayleigh fading model in case of short-distance communication?

4.14 Correlate the Rician K-factor and Nakagami-M shape factor. Give its significance.

4.15 Which PDFs are applicable for describing the wireless fading channel? Explain.

4.16 How can we differentiate between the Okumura and Hata models?

Numerical Problems

4.1 The RMS delay spread and mean excess delay are defined from a power delay profile, which is the spatial average of consecutive impulse response measurements collected and averaged over a local area. The power delay profile of a typical outdoor channel is given in Fig. 4.21.

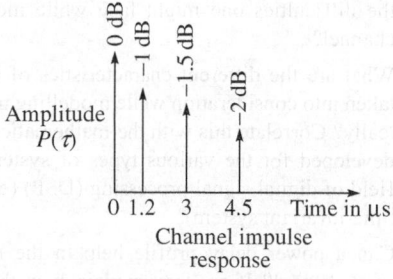

Fig. 4.21 Figure for Problem 1

Calculate the RMS delay spread over the channel. If binary phase shift keying (BPSK) modulation is used, what is the maximum bit rate that can be sent over the channel without equalization? Estimate the coherence bandwidth of the channel.

Hint: use the following equations:

(a) Mean excess delay $\bar{\tau} = \dfrac{\sum\limits_k P(\tau_k)\tau_k}{\sum\limits_k P(\tau_k)}$

(b) RMS delay spread $T_{rms} = \sqrt{\overline{\tau^2} - (\bar{\tau})^2}$

(c) $\overline{\tau^2} = \dfrac{\sum\limits_k P(\tau_k)\tau_k^2}{\sum\limits_k P(\tau_k)}$

Further, note that the RMS delay spread or symbol time should be less than or equal to 0.1. This condition will be helpful to decide allowable bit rate. (Note: RMS delay spread is of the order of microseconds in outdoor channels and nanoseconds in indoor radio channels.)

4.2 A pure sinusoidal wave of frequency 100 MHz with zero reference phase is transmitted through a channel. If it undergoes different delays over three different paths and if the delay over each path creates a phase difference of $\pi/8$, $\pi/5$, and $\pi/6$, respectively, what will be the phase of the resultant signal at the receiving end? Will it be constructive addition?

4.3 For the Rayleigh channel model, if the PDF has the standard deviation for a particular amplitude as 0.125, find the mean and variance of the PDF.

4.4 If the typical Rayleigh fading envelope at 900 MHz, with receiver speed 120 km/h, is between +7 dB and −25 dB about the RMS value of the envelope, get the approximate PDF.

4.5 For an indoor mobile communication system, find the outage probability in case of Rician fading if the local mean power is 15 dB and the receiver noise threshold level is 6 dB. The Rician factor is 5. Prove the relationship that you use.

4.6 If Example 4.3 is given for an office environment with reference distance path loss 38 dB and the number of floors equal to four, find the received signal power at a distance of 10 m. Compare and comment the results of both the cases.

4.7 For video transfer between two mobile devices using Bluetooth with NLOS, the SNR requirement is 10 dB. The background noise is −95 dBm. The transmit power of sender device is 25 dBm. What is the range within which the video transfer is possible? Assume the indoor path loss model of Example 4.3 with the number of floors equal to one. Neglect the shadow effect.

4.8 For a Doppler spread of 20 Hz, velocity of the mobile vehicle 45 km/h, and carrier frequency 960 MHz, find the Doppler spectrum. If this Doppler spread causes fading, determine the coherence time and the average fade duration. Are they both similar? Comment.

4.9 If the Nakagami shape factor $m = 3$, what is the corresponding Rician factor K? Give the Rician distribution with this value of K-factor.

4.10 For a large city urban area, calculate the path losses using Hata model for cellular radio signals of 910 MHz. The tower height is 120 m, and the mobile user is standing 5 km away from the transmitter and is holding the receiver at a height of around 2.3 m. Can these parameters be used to find the Okumura losses? If yes, compare the results with justification.

4.11 If the multipath components are received with the phasors $2\angle30$, $1.5\angle50$, $1.2\angle35$, $4\angle45$, and $3.2\angle75$, find the resultant received signal with the phase and amplitude conditions using phasor diagrams. Verify the resultant amplitude mathematically.

PART 3
Wireless Communication Techniques

The performance of the wireless link totally depends upon how a signal is processed, transmitted, and received. Processing of a signal is done as per the nature of the channel, available spectrum, and bit rate requirements.

There are techniques by which the spectrum can be conserved, phase errors and bit errors can be controlled, and signal-to-noise ratio can be improved. Also, spectrum is always a scarce resource for a multiuser scenario when the available spectrum and channel are to be shared; hence, management techniques are required.

This part, comprising of five chapters, is critically important as it explains the physical layer and multiple access sub-layer, because the overall efficiency of the wireless link and multiuser system depends upon the various techniques adopted. Thus, knowledge about such techniques at different levels of the link along with their pros and cons are important.

5 Source Coding Techniques

Theme of the Chapter

The information source in wireless communication may be a real-time audio or video signal; hence, digitization is always required. Digitization is a lossy process but can be achieved successfully with some constraints. However, appropriate sampling, quantization, and digitization methods should be followed. At the same time, the generated database is tremendously large with absolute analog-to-digital conversion, which cannot always be transmitted with the desired high speed. Instead, if source coding and compression is applied to the database, speedy communication can be achieved.

Source coding is also entropy coding. While sending information to the other end of the system, minimum energy should be consumed for useful information alone. This chapter discusses most of the source coding methods utilized in present wireless systems. Coding may be done by removing redundancy from the data so that only useful data is optimally preserved and minimum database is transmitted. However, the channel coding stage again adds redundant bits for error-handling purpose at the cost of reduction in information transmission efficiency.

Key Topics

- Analog-to-digital conversion and its constraints
- Wireless real-time communication: voice, audio, image, and video
- Basic properties of digital speech or audio signal, image signal, and video signal

- Quantization techniques in general: uniform, non-uniform, adaptive, and vector
- Pulse code modulation (PCM), differential PCM, and adaptive differential PCM
- Information sources and entropy
- Information source coding fundamentals

- Huffman coding
- Voice coders
- Linear predictive coding
- Methods for coding in frequency domain: sub-band coding and transform coding (discrete cosine transform and wavelet transform)
- Encrypters

5.1 ANALOG-TO-DIGITAL CONVERSION

Most wireless applications deal with real-time analog signals, which should be converted into digital form. This is known as analog-to-digital (ADC) conversion. The analog signal, a continuous variable defined with infinite precision, is converted into a discrete sequence of measured values that are represented digitally but with loss of precision. The converted signal is shown in Fig. 5.1.

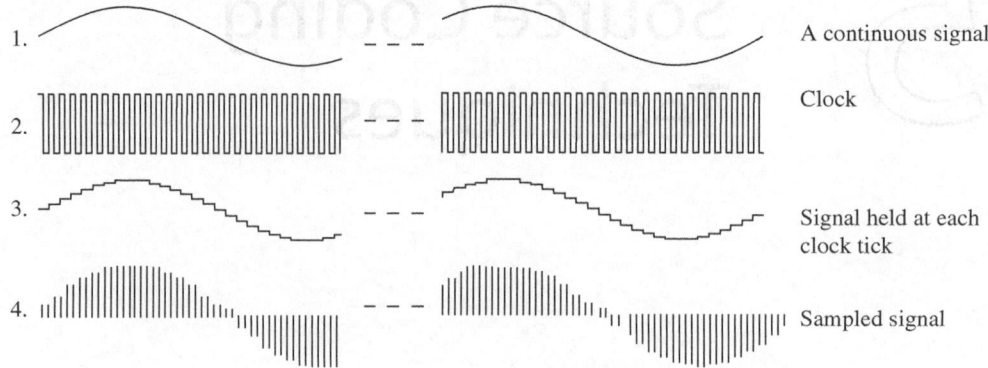

Fig. 5.1 Sampling procedure for ADC

The completion of the sampling process (for sample-based data) gives a discrete time signal with natural or quantized amplitudes. Information is lost in converting the signal from analog to digital, due to the following factors:

- Inaccuracies in the measurement of the input signal
- Timing jitter in samples that introduces extra spectrum as well
- Discretization in the time and in the levels
- Limits on the duration of the measurements

Discretization in levels results in quantization errors. *Quantization* is an intermediate process while performing ADC to assign values near predefined values. This is discussed in detail in Section 5.3.

There are timing constraints in ADC. The continuous analog signal has to be held before it can be sampled. Only after it is held can the signal be measured and the measurement converted to a digital value. Samples are usually taken at equal intervals of time. Initially, we do not know what we measure or what we do not. In the process of measuring the signal, some information may be lost. Sometimes, we may have some prior knowledge of the signal or may be able to make some assumptions that will let us reconstruct the loss. Figure 5.2 gives the constraints for measuring the analog signal, and converting it into the sampled signal with timing constraints, which results in *sampling error*.

Note that sampling takes place after the hold. This means that we can sometimes use a slower ADC than might seem required at first sight. The hold circuit must act fast enough to

Quantization is an intermediate process while performing ADC to assign values near predefined values.

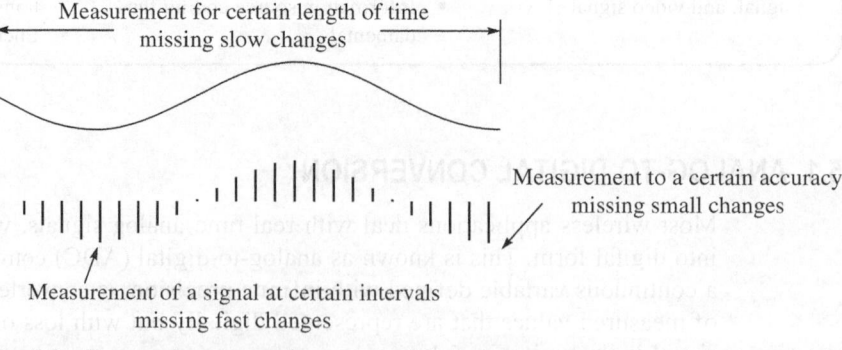

Fig. 5.2 Reasons for errors during sampling (clock errors may have some timing uncertainty)

prevent the signal from changing during the time the circuit is acquiring the signal value, but the ADC has all the time that the signal is held to make its conversion.

5.1.1 Aliasing

Now let us see what happens to the input analog voice signal that is undersampled. Undersampling can be understood by the Nyquist criterion or sampling theorem, which states the following:

$$f_s \geq 2W_{signal} \tag{5.1}$$

It means that the sampling frequency must be equal to or greater than twice the input analog signal bandwidth. Any sampling frequency that is less than twice the bandwidth of the input analog signal creates an overlap between the frequency spectrum of the samples and the associated input analog signal bandwidth. The low-pass output filter, used to reconstruct the original input signal, is not good enough to detect this overlap. Hence, it creates a new signal that does not originate from the source. This creation of a false signal during undersampling is called *aliasing*.

We might question the need for such criteria given by Nyquist. We sample the signal only at intervals and do not know what happens between the samples. The sampling rate must be decided based on how many minor details of the time waveform are to be maintained. This situation is illustrated in Fig. 5.3.

Figure 5.3 shows a *fast change* or *spike* that happened to fall between adjacent samples. Again, we cannot track the rapid intersample variations. Hence, we must sample fast enough to see most of the rapid changes in the signal; thus, sampling rate depends on the highest frequency (found from the spectrum). This is what is described in the Nyquist criteria or sampling theorem. The mathematical proof for the criteria is not provided here. However, it can be referred in any book on signals and systems or basic communication. Nyquist criteria helps avoid aliasing. The highest and lowest frequencies of the signal decides the bandwidth. If the signal is of band-pass type, it can be mapped to 0 Hz as the starting frequency; the maximum frequency of the bandwidth deciding sample rate is called the *Nyquist frequency*.

> To prevent aliasing effect, the sampling rate must be faster than twice the maximum frequency content in the signal.

Aliasing effect can be observed in frequency domain as overlapping of consecutive spectral components. This is better understood with discrete time Fourier transform (DTFT). The frequency domain representation of the sampled signal and aliasing effect are shown in Fig. 5.4. While converting a sampled signal into frequency domain, periodic energy bundles are created, as shown in Fig. 5.4. The gap between the bundles depends upon the sampling rate. If the frequency domain response is observed carefully, it can be seen that

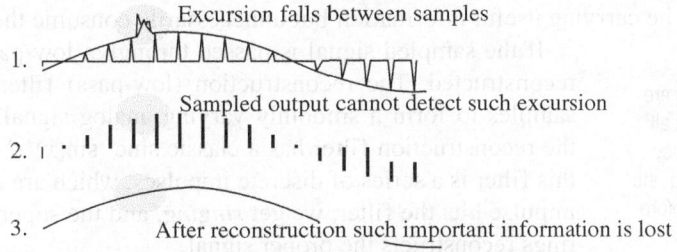

1. Excursion falls between samples

2. Sampled output cannot detect such excursion

3. After reconstruction such important information is lost

Fig. 5.3 Importance of sampling rate selection

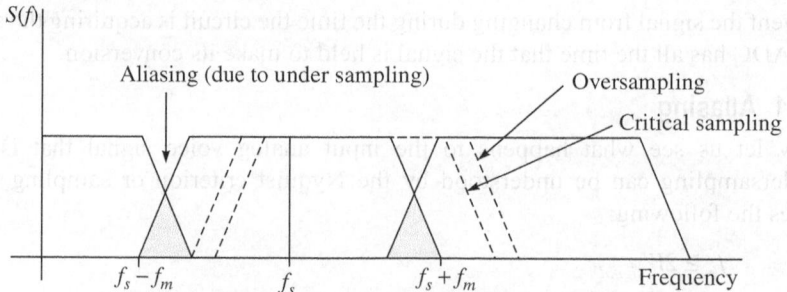

Fig. 5.4 Aliasing effect in the frequency domain

multiple harmonic blocks are generated, centred at the multiples of sampling frequency f_s due to DTFT.

Notes: The following points summarize the foregoing discussion:
1. If the sampling rate of a time domain signal is lower than the Nyquist rate, aliasing is observed in the frequency domain, which causes overlapping of two consecutive blocks. Aliasing shows interference between corner frequencies, which results in loss.
2. If critical sampling is done with the Nyquist rate, the two consecutive spectra critically touch each other.
3. With oversampling, a gap is observed between two consecutive blocks. Critical sampling or oversampling reduces the losses due to gaps between two consecutive spectral blocks.

5.1.2 Antialiasing

Aliasing can be avoided by filtering out before sampling all the high-frequency components from the source signal; that is, the antialiasing filter is a low-pass filter. For example, to prevent aliasing, a band-limiting filter is used to limit the frequency spectrum of the input analog voice signal before it is sampled. This can be seen in the block diagram of pulse code modulation (PCM) (see Section 5.5).

Most of the energy of the spoken language is somewhere in between 200–300 Hz and about 2700–2800 Hz. Roughly, a 3000 Hz bandwidth is required for standard speech and standard voice communication. A bandwidth of 4000 Hz is made from the equipment point of view. If the highest input frequency were 4 kHz, then the sampling frequency would be set to twice this amount, or 8 kHz. This sampling frequency is the standard used by voice-digitizing systems.

The band-limiting filter also makes the signal spectral efficient, as only useful frequency components are considered, removing unwanted ones. Figure 5.5 shows the effect of an antialiasing filter, which is a low-pass filter. It removes the glitches in the signal, which may not be carrying useful information but unnecessarily consume the spectrum.

Low-pass filters are used to filter out all the high-frequency components from the source signal before sampling.

If the sampled signal is passed through a low-pass filter, the signal can be reconstructed. The reconstruction (low-pass) filter interpolates between the samples to form a smoothly varying analog signal. The impulse response of the reconstruction filter has a classic sinc 'sin(x)/x' shape. The stimulus fed to this filter is a series of discrete impulses, which are the samples. Every time an impulse hits the filter, we get *ringing*, and the superposition of all these peaky rings reconstructs the proper signal.

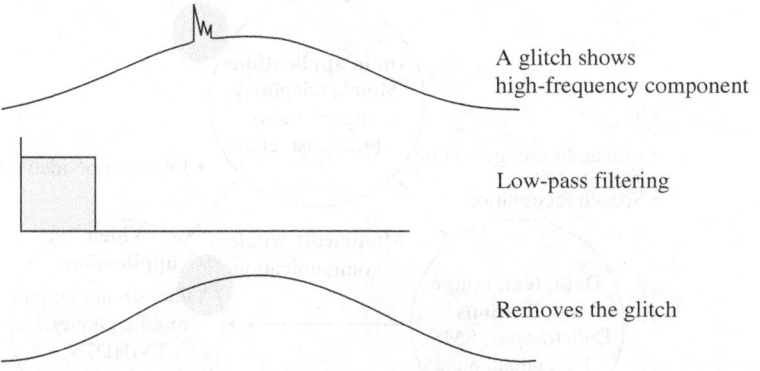

A glitch shows
high-frequency component

Low-pass filtering

Removes the glitch

Fig. 5.5 Effect of an antialiasing filter

FREQUENCY RESOLUTION DURING SAMPLING

The source signal is sampled only for a certain time. Any slow changes in the signal will not be noticed if the sampling time is not long enough. In fact, the sampling time should be long enough to detect not only the low frequencies in the signal but also the small differences between frequencies. The length of time for which the signal is to be sampled determines the ability to resolve adjacent frequencies, which is called *frequency resolution*. The source signal must be sampled for at least one complete cycle of the lowest frequency that is to be resolved.

It is evident that there is a case of forced compromise. The sampling time should be fast to avoid aliasing and for a long time to achieve a good frequency resolution. However, sampling fast for a long time implies many samples and many samples mean numerous computations. Therefore, there has to be a compromise between *resolving frequency components* of the signal and *being able to see high frequencies*.

5.2 WIRELESS MULTIMEDIA COMMUNICATION

> If the signal contains frequency components close to Nyquist, the reconstruction filter has to be very sharp.

Most of the wireless applications are voice, audio, text, image, or video communication, which in general is called multimedia communication. Multimedia information may be real-time, probabilistic (non-deterministic), continuous analog, and energy signals in time domain. As far as frequency domain is concerned, audio bandwidth (20 Hz to 20 kHz) is comparatively less than image and video bandwidth (which is in MHz and depends upon the resolution aspects). Integration and interaction among different multimedia types create new challenging opportunities. The concept is represented in Fig. 5.6.

The first task for a multimedia transmitter is *source coding*, which moderates and optimizes the database before modulation. Source coding helps in reducing the energy of total transmissions and improves the speed of useful data transfer. Source coders may be of three types: *waveform coders*, *parametric coders*, and *hybrid coders*. The data compression technique is an inherent part of a source coder most of the times; if not, sometimes an additional compression scheme is utilized to remove the redundancy present in the data. This increases source efficiency. Information theory helps in optimizing the size of the final data to be transmitted. To apply source coding and compression, it is imperative to know the signal to be communicated and its properties. Speech signal is the simplest signal and so let us begin with the study of the properties of a speech signal.

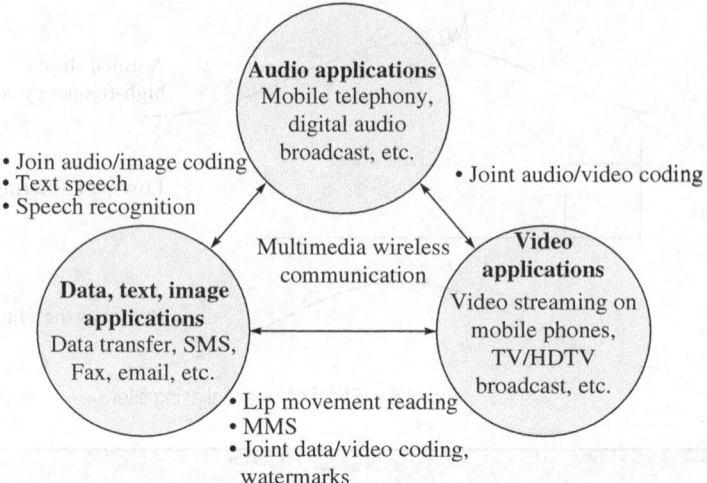

Fig. 5.6 Multimedia communication and few media interactions in existing wireless communication systems (applied on wired lines as well)

5.2.1 Basic Properties of Speech Signal

> Parametric source coders exploit the human capabilities of speaking and hearing.

Speech waveforms have many useful properties that can be exploited while designing quantizers and coders. The most basic property of speech waveforms is that they can be band-limited, and this property is used by all speech coders. Finite bandwidth means that it can be time discretized at a finite rate and reconstructed completely from its samples, provided that the sampling rate follows the Nyquist criteria. Some properties of speech waveform are discussed here.

Non-uniform Probability Distribution of Speech Amplitude

The PDF of a speech signal in general is characterized by the following factors:

- Very high probability of non-zero amplitudes
- Significant probability of very high amplitudes
- Monotonically decreasing function of amplitudes between these extremes

The exact distribution is, however, dependent upon the recording conditions and input bandwidth. Equation (5.2) gives the Laplacian function, which is a good approximation to the long-term PDF of telephone-quality speech signals.

$$\text{PDF}(x) = \frac{1}{\sqrt{2}\sigma_x} \exp(-\sqrt{2}\,|x|/\sigma_x) \tag{5.2}$$

This PDF shows a peak at zero, which is due to the existence of frequent pauses and low-level speech segments, and then shows a decrease within a 4 kHz spectrum. Short-time PDFs of speech segments are also single-peaked functions and are usually approximated as a Gaussian distribution. Speech signals can have separate PDFs for voiced and unvoiced portions of the speech. A typical short-term speech PDF is shown in Fig. 5.7(a) and the power spectral density (PSD) of a speech signal is given in Fig. 5.7(b).

In non-uniform quantizers (Section 5.4.2), an attempt is made to match the distribution of quantization levels to that of the PDF of the input speech signal by allocating more quantization levels in regions of high probability and fewer levels in the regions where the probability is low.

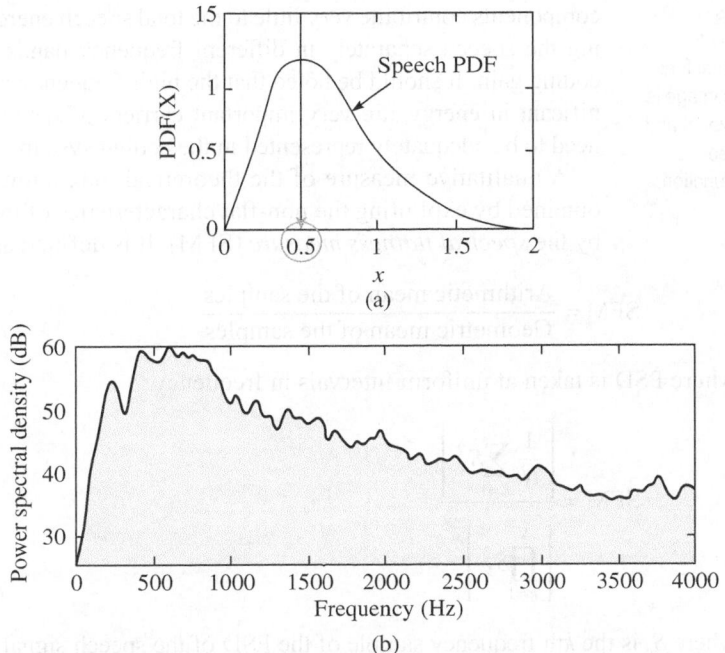

Fig. 5.7 Speech signal (a) Typical PDF of a short-term speech signal
(b) PSD of a speech signal

Non-zero Autocorrelation between Successive Speech Samples

> Speech sample amplitudes do not change abruptly, and hence, there is good autocorrelation with the nearest samples, which gradually reduces for successive ones.

There is much correlation between the adjacent samples of a speech segment; hence, the correlation must be non-zero. This implies that in every sample of speech, there is a large component that is easily predicted from the value of the previous samples with a small random error. All differential and predictive coding schemes are based on exploiting this property. Following the autocorrelation function gives the quantitative measure of the closeness or similarity between successive samples of speech signal as a function of their time separation index.

$$C_{auto}(k) = \frac{1}{N} \sum_{n=0}^{N-k-1} s(n)s(n+k) \qquad (5.3)$$

Here, $s(k)$ represents the kth speech sample. Normally, the autocorrelation function is normalized to the variance of the speech signal and hence is restricted to have values in the range $[-1, 1]$ with $C_{auto}(0)$, which is due to the correlation of a sample with itself. Typical signals have an adjacent sample correlation $C_{auto}(1) = 0.8$ to 0.9, at the most.

Non-flat Nature of Speech Spectra

The non-flat nature of the PSD of speech makes it possible to obtain significant compression by coding the speech in the frequency domain (sub-band coding). It is achieved by dividing the total band into sub-bands and finding the significant frequencies.

The non-flat nature of the PSD is due to the frequency domain manifestation of the non-zero autocorrelation property. Typical long-term averaged PSDs of speech show that high-frequency

> Audio signal is a one-dimensional function, whereas image is a two-dimensional and video is a three-dimensional function.

components contribute very little to the total speech energy. This indicates that coding the speech separately in different frequency bands can lead to a significant coding gain. It should be noted that the high-frequency components, though insignificant in energy, are very important carriers of speech information, and hence need to be adequately represented in the coding system.

A qualitative measure of the theoretical maximum coding gain that can be obtained by exploiting the non-flat characteristics of the speech spectra is given by the *spectral flatness measure* (SFM). It is defined as follows:

$$\text{SFM} = \frac{\text{Arithmetic mean of the samples}}{\text{Geometric mean of the samples}} \tag{5.4}$$

where PSD is taken at uniform intervals in frequency.

$$\text{SFM} = \frac{\left[\dfrac{1}{N}\displaystyle\sum_{k=1}^{N} S_k^2\right]}{\left[\displaystyle\prod_{k=1}^{N} S_k^2\right]^{\frac{1}{N}}} \tag{5.5}$$

where S_k is the kth frequency sample of the PSD of the speech signal. Typically, speech signals have a long-term SFM value of 8 and a short-term SFM value varying widely between 2 and 500.

Existence of Voiced and Unvoiced Speech Segments

A speech signal contains both voiced and unvoiced segments. While producing a voiced sound, the vocal chords vibrate, which can be felt on the throat. With unvoiced sound, the vocal chords do not vibrate. This is discussed further in Section 5.10.1, where the process of human speech generation is explained.

Quasiperiodicity of Voiced Speech Signals

A sample speech signal is shown in Fig. 5.8.

It can be observed from the figure that the signal is quasiperiodic (periodic in an irregular fashion). The spectrum for speech (combined voiced and unvoiced sounds) has a total bandwidth of approximately 7000 Hz with an average energy at about 3100 Hz. The auditory canal optimizes speech detection by acting as a resonant cavity at this average frequency. Note that the power of speech spectra and the periodic nature of formants drastically diminish above 3500 Hz,

Fig. 5.8 Sample speech signal in analog form representing voiced and unvoiced intervals and quasiperiodicity

Note: Periodic behaviour is defined as recurring at regular intervals; quasiperiodicity property displays irregular periodicity and its Fourier transform exhibits peaks at certain points.

PROPERTIES OF IMAGE AND VIDEO

The following are the important properties of images and videos:
- Non-uniform probability distribution
- Non-flat nature of image spectra
- A good correlation between successive samples (or pixels) in images and between successive image shots in videos

as shown in Fig. 5.7(b). Speech encoding algorithms can be less complex than general encoding by concentrating (through filters) on this region. Furthermore, since line quality telecommunications employ filters that pass frequencies up to only 4000 Hz, high frequencies produced by fricatives are removed. This forms one basic bandwidth for speech, which is a compromised selection.

5.2.2 Digital Baseband

The audio or video signals are generated naturally, and are of analog nature. To transmit them by digital techniques, they are converted into the digital baseband, that acts as a source of the database. All the digital techniques are then applied to this database.

Digital Speech or Audio

Digital speech or audio signal is a one-dimensional signal, similar to analog speech or audio but discrete in nature, which is then coded. It is achieved by sampling the analog input and then performing ADC. The fundamentals are discussed in Sections 5.4 and 5.5 while explaining quantization and PCM. Nyquist criterion is the fundamental that supports selecting the sampling frequency for discretization of the analog speech or audio, as described in Section 5.1.

Digital Image

Digital images are recorded or captured directly from digital cameras or indirectly by scanning a photograph or document with a scanner. The image is divided into a matrix or array of small picture elements called *pixels* (Fig. 5.9a). Pixels are samples of the image and decide the resolution of the image. Each pixel is represented by value of the example and its unique position in a two-dimensional array as shown in Fig. 5.9(a).

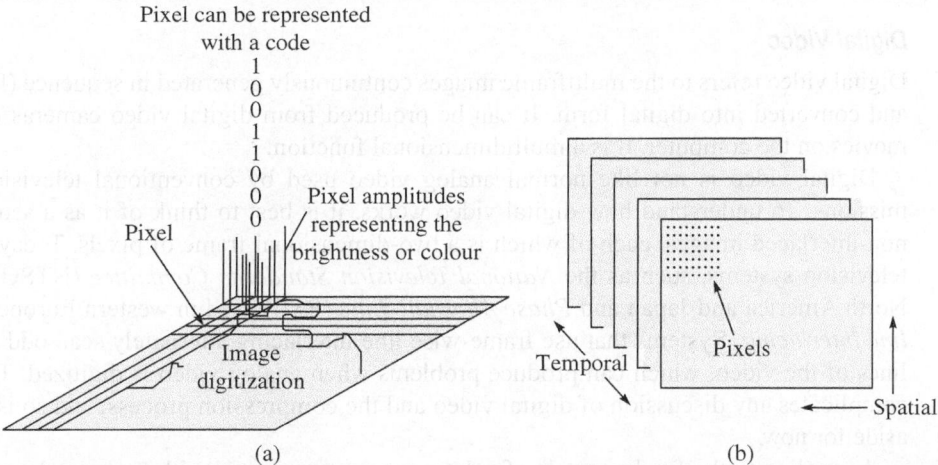

Fig. 5.9 Visualizing digital image and video (a) Image converted into matrix of pixels (b) Multiple images transmitted in video

> Audio, image, and video have redundant information, which is removed for efficient source coding and compression and to increase the speed of communication.

The analog value can be converted into a code. The advantage of digital images is that they can be processed in many ways by computer systems.

Mathematically, a *digital image* can be defined as a two-dimensional function $f(x, y)$, where x and y are the spatial (image plane) coordinates representing a pixel; the amplitude of $f(x, y)$ at any pair of coordinates (x, y) is called the *intensity* or *grey level* of the image at that point. In colour images, each pixel is associated with a colour. The collection of pixels is arranged in a two-dimensional matrix (spatial representation) form, the size of which decides the image resolution. The image can also be represented in frequency domain. In general, the pixel value is related to the intensity or colour that can be understood when the digital image is converted into an analog image for display and viewing.

The numerical size (number of bits) of an image is the product of the following two factors:

(a) *Pixels in an image*: The number of pixels in an image is found by multiplying the number of pixels in a row (pixel length) and the total number of rows created in the image due to resolution of the image.

(b) *Pixel depth*: It is actually the number of bits per pixel. This is usually in the range of 8–16 bits or 1–2 bytes/pixel.

Dots per inch (dpi) is a measure of printing resolution. Dots can vary in size to accommodate the number of pixels or pixels per inch (ppi), which is used to scale existing digital images to a specified print size. This will affect the print size of the photo and quality of the print image.

Example 5.1 When an image is digitized, it contains 256 different kinds of pixels. Find the pixel depth. If 256×256 resolution is used, what will be the size of the storage required for one digital image?

Solution There are 256 different kinds of pixels, which means that there are 256 different quantized levels while converting the image into digital form.

Now, $2^8 = 256$. This means that the pixel depth will be 8 bits/pixel. (It is similar to 8 bits/sample.)

Thus, $256 \times 256 \times 8 = 5,24,288$ bits/image will be required for the storage.

In other words, 65,536 bytes will be occupied. ∎

Digital Video

Digital video refers to the multiframe images continuously generated in sequence (Fig. 5.9b) and converted into digital form. It can be produced from digital video cameras or stored movies on the computer. It is a multidimensional function.

Digital video is not like normal analog video used by conventional television transmissions. To understand how digital video works, it is best to think of it as a sequence of non-interlaced images, each of which is a two-dimensional frame of pixels. Today's analog television systems, such as the *National Television Standards Committee* (NTSC) used in North America and Japan and *Phase Alternate Line* (PAL) used in western Europe, employ *line interlacing*. Systems that use frame-wise line interlacing alternately scan odd and even lines of the video, which can produce problems when analog video is digitized. This issue complicates any discussion of digital video and the compression process, and so is best left aside for now.

According to the fundamentals of colour images, associated with each pixel are two values: *luminance* and *chrominance*. Luminance is a value proportional to the pixel's intensity,

---

POINTS TO REMEMBER

1. Speech, audio, image, and video signals are of stochastic nature.
2. Different methods of source coding will use the appropriate form of the following two options:
   - In the form of bits—1's and 0's, which can be compressed directly (entropy coding) using algorithms
   - In the form of real waveforms, which must be digitized and then compressed (waveform + entropy coding) using signal property
3. Source coding may be done in time or frequency domain.

After sampling and ADC, signal reconstruction is always a lossy process because of quantization errors.

whereas chrominance is a value that represents the colour of the pixel and there are a number of representations to choose from. There are different colour models. Any colour can be synthesized by an appropriate mixture of three properly chosen primary colours. Red, green, and blue (RGB) are usually chosen as the primary colours.

Digital video can be characterized by the following variables.

- *Frame rate*: It is the number of frames displayed per second. The illusion of motion can be experienced at frame rates as low as 12 frames/s.
- *Frame dimensions*: It is the width and height of the image expressed in number of pixels. Digital video comparable to television requires dimensions of about 640 × 480 pixels.
- *Pixel depth*: It is the number of bits per pixel. In some cases, it might be possible to separate the bits dedicated to luminance from those used for chrominance. In others, all the bits might be used to reference one of a range of colours from a known palette.

Table 5.1 provides the possible values of parameters for typical applications of digital video.

Table 5.1 Possible values of parameters for typical applications of digital video

Application	Frame rate	Frame dimensions	Pixel depth
Computer monitor (liquid crystal display) 800 × 600 to 1280 × 1024	15	1024 × 768	16
Entertainment digital television	25	640 × 480	16
Surveillance	5	640 × 480	12
Video telephony	10	320 × 240	12
High-definition television	25	1920 × 1080	24

5.3 SOURCE CODING STAGES

Figure 5.10 shows the general block diagram of the stages involved in source coding. It must be noted that all stages are incorporated as per the requirement of the application . In general, these stages have functions of signal processing and compression.

5.3.1 Analog Signal

As shown in the figure, information source output is obtained through the measurement and observation stages. Signal processing generally involves two tasks. First, it is a vehicle

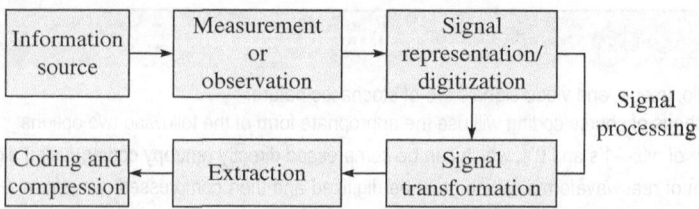

Fig. 5.10 General block diagram indicating source coding steps

for obtaining a general representation of a speech, audio, image, or video signal in either waveform or parametric form. Second, signal processing aids in transforming the signal representation into alternate forms, which are less general in nature but more appropriate to the coding method.

In order to convert a signal into a more convenient form for some new transform-based efficient source coding techniques, such as discrete cosine transform (DCT) and wavelet, appropriate signal processing is applied. The last step in the process is the extraction of useful message information. This step may be performed either by human listeners or automatically by machines. On the basis of useful information, source coding with compression is applied. Finally, the source-coded data may be available as output, which goes to the error-handling or channel coding stage.

It is necessary to know the method of digitization of the signal for waveform coders. In Section 5.1, the sampling process in practical aspects is explained along with its limitations. Before we study digitization methods, such as PCM, we must first study the types of quantization processes, so that the best method can be adopted for the digitization of the information signal and the conversion may be near lossless. There are various waveform coders, which are explained in Section 5.7. Normally, waveform coders generate huge databases, especially for video, which are sometimes difficult to handle, occupying large memory space and reducing spectral efficiency. In order to optimize this, the fundamentals of information theory are applied to represent the data in compressed form.

5.3.2 Digital Signal

Some of the early processing stages are omitted here. Various compression techniques are applied on the digital database as per information theory, as in the analog case.

Compression schemes may be lossy or lossless. An entropy encoding is a lossless data compression scheme.

5.4 QUANTIZATION TECHNIQUES

Quantization is the process of mapping a continuous range of the amplitudes of a signal into a finite set of discrete amplitudes. When a signal is converted into a digital form, its precision is limited by the number of converted bits. The smoothly varying analog signal can be represented only as a *stepped* waveform due to the limited precision, as shown in Fig. 5.11. The errors introduced by digitization are both non-linear and signal dependent, the effects of which cannot be calculated using normal mathematics. We cannot reduce the errors by simple means, and even if we calculate their effects, we will have to do so separately for every type of signal we expect. Figure 5.11 shows the quantization error as if it were a source of random noise. Therefore, quantization error is called *quantization noise*. In simplified form,

$$\text{Quantization error } q_e = \text{Actual value of the sample} \\ - \text{Quantized value of that sample} \tag{5.6}$$

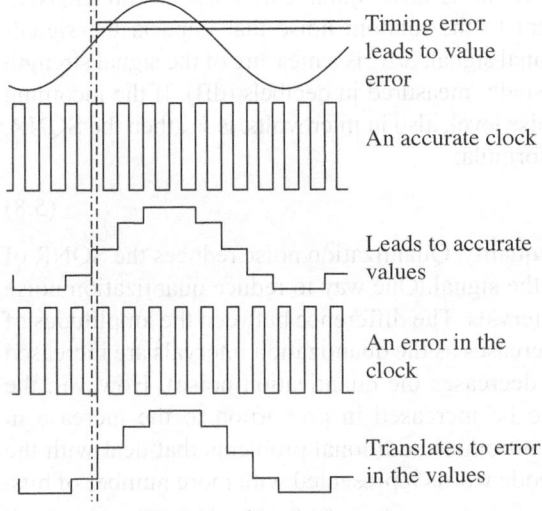

Timing error leads to value error

An accurate clock

Leads to accurate values

An error in the clock

Translates to error in the values

Fig. 5.11 Quantization noise

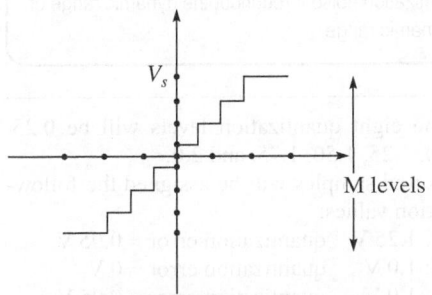

Fig. 5.12 Uniform quantization

In practice, ADC by quantization is always a lossy process. Its quality of output is always dependent upon the maximum frequency content of the signal and the sampling rate applied to it. To convert natural samples into binary codes, discrete levels are to be allocated to samples. This results in the noiselike effect in the signal artificially. Quantization can be thought of as a process that removes irrelevancies in a signal, and the operation is irreversible.

Uniform quantization, non-uniform quantization, adaptive quantization, and vector (statistical data based) quantization are four approaches for quantization. The first two methods are conventional and simple. Adaptive is the improvement over the conventional one and statistical quantization methods can improve on the adaptive techniques. Sometimes, quantization becomes an integral part of speech and source coders.

5.4.1 Uniform Quantization

As the sampled input signal enters the quantization phase, each input sample is assigned a quantization interval that is nearest to its amplitude height. If all quantization intervals are equally spaced throughout the dynamic range of the input analog signal, it is called *uniform quantization*. Each quantization interval is assigned a discrete value in the form of a binary code word. Figure 5.12 shows an example of uniform quantization.

In uniform quantization, the total range (signal strength $\pm V_s$) is uniformly divided into M discrete levels (equal step size = $2V_s/M$). A quantizer that uses n bits can have $M = 2^n$ discrete amplitude levels. The distortion introduced due to quantization operation is directly proportional to the square of the step size, which in turn is inversely proportional to the number of levels for a given amplitude range. The mean square error is the measure for the distortion introduced by the quantization process.

$$E_{ms} = E[(s(t) - s_q(t))^2] = \frac{1}{T}\int_0^T [s_q(t) - s(t)]^2 \, dt \tag{5.7}$$

where $s(t)$ is the analog signal and $s_q(t)$ is the quantized signal. For calculating the quantization error in discrete form, each sample is considered separately. That is, the difference between

Note: Only a finite number of bits can represent a limited dynamic range of measurement values. Arbitrarily larger positive or smaller negative values are not representable by the quantizer and are typically represented by codes related to upper and lower limits. This leads to quantizer saturation error.

> Quantization noise results from quantization errors, is treated as additive, and cannot be removed fully.

the natural amplitude and quantized amplitude of the sample will give the quantization error (Eq. 5.6). All such sample errors must be squared, added, and averaged for calculating the mean square error. The performance of a quantization process is measured by another term, *signal-to-quantization noise ratio* (SQNR). Quantization noise is equivalent to the random noise that impacts the signal-to-noise ratio (SNR) of the original signal. S/N_q is a measure of the signal strength relative to background noise. The ratio is usually measured in decibels (dB). If the incoming signal strength in microvolts is V_s and the noise level, also in microvolts, is V_n, then the SQNR, S/N_q, in decibels, is given by the following formula:

$$S/N_q = 20\log_{10}(V_s/V_n) \tag{5.8}$$

Higher the SQNR, better will be the signal quality. Quantization noise reduces the SQNR of a signal and hence degrades the quality of the signal. One way to reduce quantization noise is to increase the number of quantization intervals. The difference between the amplitudes of the input signal and the quantized signal decreases as the quantization intervals are increased (increase in the number of discrete levels decreases the quantization noise). However, the amount of code words would also have to be increased in proportion to the increase in quantization intervals. This process would introduce additional problems that deal with the capacity of a PCM system to handle more code words represented with more number of bits.

> *Note*: Due to amplitude compression, logarithmic increase in quantization noise throughout the dynamic range of a sampled signal will keep the SQNR constant throughout this dynamic range.

Example 5.2 The following are the readings for the measurement of quantization error in five consecutive samples. The number of quantization levels in the dynamic range of 2 V is eight.

Sample 1: 1.2 V
Sample 2: 1 V
Sample 3: 0.95 V
Sample 4: 1.41 V
Sample 5: 1.65 V

Find the quantization error in terms of its mean square value.

Solution If the dynamic range is 2 V, then the smallest step size will be 2/8 = 0.25 V.

Hence, the eight quantization levels will be 0.25, 0.5, 0.75, 1.0, 1.25, 1.50, 1.75, and 2.0.

The measured samples will be assigned the following quantization values:

Sample 1: 1.25 V   quantization error = 0.05 V
Sample 2: 1.0 V   quantization error = 0 V
Sample 3: 1.0 V   quantization error = 0.05 V
Sample 4: 1.5 V   quantization error = 0.09 V
Sample 5: 1.75 V   quantization error = 0.1 V

Mean square error
$$= [(0.05)^2 + 0 + (0.05)^2 + (0.09)^2 + (0.1)^2]/5$$
$$= [0.0025 + 0.0025 + 0.0081 + 0.01]/5$$
$$= 0.0231/5 = 0.00462$$

Root mean square error = 0.0678

Most voice signals generated are of the low kind. To improve voice quality at lower signal levels, uniform quantization is replaced by a non-uniform quantization process.

5.4.2 Non-uniform Quantization

When the step size is fixed and uniform, it decides the resolution of that quantizer stage. The method fails if the speech signal is weak or if it is necessary to maintain the smallest amplitude information whose amplitude is less than the uniform step size. Here, beginning from the smallest possible step size at the lowest amplitude level, the step size should be increased gradually; thus, there is non-uniform distribution of levels. This is a more efficient method than the uniform quantization method (see Fig. 5.13).

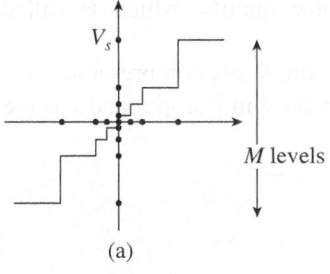

(a)

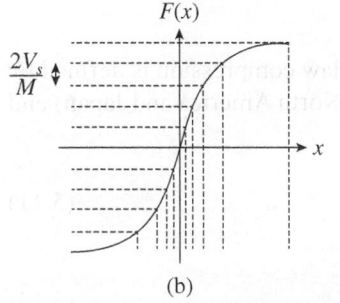

(b)

Fig. 5.13 Non-uniform quantization
(a) Gradually increasing steps sizes
(b) Companding process example

Non-uniform quantizers distribute the levels in accordance with the PDF of the input waveform. For an input with the PDF $pdf(x)$, the mean square error is given by

$$E_{ms} = E[(s(t) - s_q(t))^2]$$
$$= \int_{-\infty}^{\infty} [s_q(t) - s(t)]^2 pdf(x) dt \qquad (5.9)$$

This is the same relationship as mentioned in Eq. (5.7) with necessary addition of the PDF function. From this equation, it is clear that the total distortion can be reduced by decreasing the quantization noise $[s(t) - s_q(t)]^2$. This means that quantization levels need to be concentrated in amplitude regions of high probability and that are at lower amplitudelevels.

Companding   There are many ways to design non-uniform quantizers by companding. *Companding* refers to the process of first *compressing* an analog signal at the source and then *expanding* this signal back to its original size when it reaches its destination. The concept is shown in Fig. 5.13(b). Each envelope segment between samples can be treated as piecewise approximation of the original envelope. There are many algorithms for compression and expansion, but a simple and robust method used in commercial telephony is the *logarithmic quantizer*. During the companding process, the input analog signal samples are compressed into logarithmic segments, and then, each segment is quantized and coded using uniform quantization. In other words, larger amplitude signals are compressed more than the smaller amplitude signals, causing the quantization noise to increase as the signal amplitude increases.

The International Telecommunication Union (ITU-T) standards for companding are called A-law and μ-law. The A-law standard is used in European countries and μ-law is used in North America and Japan. These are audio compression schemes (codecs) defined by the Consultative Committee for International Telephony and Telegraphy (CCITT).

> Companding is the joint process of compressing an analog signal amplitude at the transmitter and then expanding it at the receiver.

Example 5.3   Give the procedure for A-law and μ-law standard companding processes for 256 quantization levels.

Solution

(a)   Calculate the linear approximation of the logarithmic curve of a sample input signal. The logarithmic curve is created during the compression part of companding. A-law and μ-law calculate this linear approximation differently.

(b)   Divide the logarithmic curve into 16 segments (8 positive and 8 negative), with 16 quantization intervals per segment (total of 256 quantization intervals). Each successive segment is twice the length of the previous segment (logarithmic increase). A-law and μ-law have different segment lengths because of the different calculations of linear approximation.

(c)   Use eight-bit code words for each quantization interval (maximum of 256 code words, one for each quantization interval). The first bit of the code word represents the polarity of the quantization interval; the second, third, and fourth bits represent the segment number, and the last four bits indicate the quantization interval within the segment. The eight-bit code words allow for a bit rate of 64 kbps, calculated by multiplying the sampling rate (twice the maximum input frequency) by the size of the code word (2×4 kHz $\times 8$ bits = 64 kbps).

The *A*-law and μ-law PCM produce acceptable voice quality, which is called *toll quality*.

Limiting the linear sample values to 12 magnitude bits, the *A*-law compression is defined by Eq. (5.10), where *A* is the compression parameter ($A = 87.7$ in Europe) and x is the normalized (normalized to one) integer to be compressed.

$$F(x) = \begin{bmatrix} \dfrac{A \times |x|}{1 + \ln(A)} & 0 \leq |x| < \dfrac{1}{A} \\ \dfrac{\mathrm{sgn}(x) \times (1 + \ln(A \,|x|))}{1 + \ln(A)} & \dfrac{1}{A} \leq |x| \leq 1 \end{bmatrix} \qquad (5.10)$$

Limiting the linear sample values to 13 magnitude bits, the μ-law compression is defined by Eq. (5.11), where μ is the compression parameter ($\mu = 255$ in North America and Japan) and x is the normalized integer to be compressed.

$$F(x) = \frac{\mathrm{sgn}(x) * \ln(1 + \mu\,|x|)}{\ln(1 + \mu)} \qquad 0 \leq |x| \leq 1 \qquad (5.11)$$

Here, $\quad \mathrm{sgn}(x) = \begin{cases} +1 & \text{for } x \geq 0 \\ -1 & \text{for } x < 0 \end{cases}$ \hfill (5.12)

Similarities between A-law and μ-law

The following are some similarities between *A*-law and μ-law:
- Both provide linear approximations of logarithmic input–output relationship.
- Both are implemented using eight-bit code words (256 levels, one for each quantization interval), which allow for a bit rate of 64 kbps.
- Both break a dynamic range into a total of 16 segments (piecewise approximated companding curve):
  - There are eight positive and eight negative segments.
  - Each segment is twice the length of the preceding one.
  - Uniform quantization is used within each segment.
- Both use similar approaches for coding the eight-bit word.

Differences between A-law and μ-law

The following are the differences between *A*-law and μ-law:
- Different linear approximations lead to different lengths and slopes.
- The numerical assignment of the bit positions in the eight-bit code word to the segments and the quantization levels within the segments are different.
- The *A*-law provides a greater dynamic range than the μ-law.
- The μ-law provides better signal or distortion performance for low-level signals than the *A*-law. It also has a better SQNR at the lower signal levels.
- The *A*-law requires 12 bits for a uniform PCM equivalent. The μ-law requires 13 bits for a uniform PCM equivalent.
- An international connection needs to use the *A*-law, and μ to *A* conversion is the responsibility of the country using the μ-law.

5.4.3 Adaptive Quantization

Adaptive means that when the input signal is of low peak-to-peak amplitude, keeping the same number of intervals with a small step size, as may be in the case of large peak-to-peak amplitude signal with larger step size.

Adaptive quantization is an upgraded version of non-uniform quantization. There is a difference between the long-term and the short-term PDFs of speech waveform because speech is a non-stationary stochastic process. The time-varying and random nature of the speech signal results in a dynamic range of approximately 35–40 dB or more. Similarly, image and video signals are also of random and time-varying nature. An efficient way to accommodate a large dynamic range is to adopt a time-varying quantization technique. Adaptive quantization scheme is an important scheme that is suitable for audio and image as well as for very low bit rate video coders. In recent years, many adaptive quantization schemes have been proposed. Broadly, these schemes can be divided into two categories: *backward* and *forward*.

Backward Adaptation

The basis to adjust the quantization step of backward adaptive quantization schemes is the fullness of the buffer and the bit rate of the channel. This kind of method can make the output bit rate of the encoder fit the channel effectively and the decoded audio quality is also good, but it causes much fluctuation of the decoded image or video quality when the bit rate is low. Backward adaptation uses past decoded speech to estimate the *linear prediction* (LP) *model*, but the following issues are also important due to which the method is not popular nowadays:

- The LP parameters are not transmitted directly, because of the following reasons:
  - The LP model is poorly trained on delayed data.
  - The LP parameters are hard to quantize and consume considerable bandwidth.
  - The LP model has a simpler bitstream format.
- Longer model order is possible and is not limited by the quantization error.
- The decoder needs more power to generate the LP model.
- The backward adaptation technique works best for low-delay systems.

The method can be used in linear predictive coders, which are economical with 2.4 kbps bit rate and are lossy.

Forward Adaptation

Forward adaptive quantization schemes adjust the quantization step by predicting the output bit amount or iterative trial. Here, statistical concepts or blind methods can be followed. Although the computation cost of forward adaptive quantization schemes is higher than that of backward schemes, the decoded signal quality is better because they take considerations of signal contents. Figure 5.14 shows a typical example of a forward adaptive quantizer.

Many new techniques have been derived for adaptive quantization. Compared with traditional backward schemes, the new adaptive quantization schemes take considerations of not only the status of the buffer but also the information contents and human characteristics for perceptions of multimedia. Moreover, the new schemes take advantage of fuzzy theory to avoid complex computation and iterative trial. With these adaptive quantization schemes, we can keep the quality of the decoded information consistent with good audio-visual effects as well.

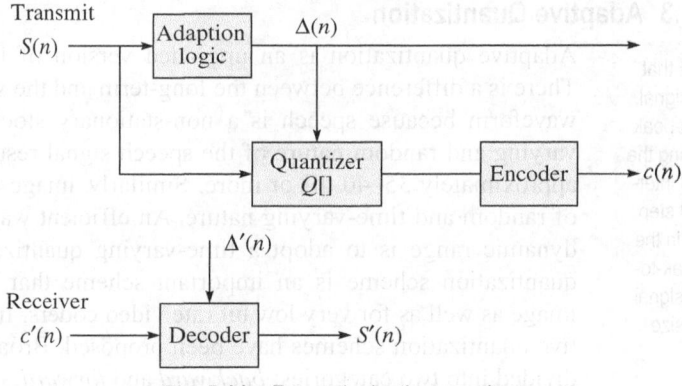

Fig. 5.14 Forward adaptive quantizer

5.4.4 Vector Quantization

The methods studied so far are scalar quantization, in which the scalar value is selected from a finite list of possible values to represent an input sample, which is close (in some sense) to the sample it is representing. In vector quantization (VQ), a vector is selected from a finite list of possible vectors to represent an input vector of samples. The selected vector is chosen to be close to the vector it is representing. VQ is a lossy data compression method based on the principle of block coding. It is a *fixed-to-fixed length algorithm*. In the earlier days, the design of a vector quantizer was considered to be a challenging problem due to the need for multidimensional integration. In 1980, Linde, Buzo, and Gray (LBG) proposed a VQ design algorithm based on a training sequence. The use of a training sequence bypasses the need for multidimensional integration. A VQ that is designed using this algorithm is referred to in the literature as an LBG-VQ and is only an approximator. The idea is similar to that of *rounding off* (say, to the nearest integer).

Vector quantization is an example of *competitive learning*. Normally, a basic competitive learning network has one layer of input nodes and one layer of output nodes. Binary valued outputs are often (but not always) used. There are as many output nodes as there are classes. The goal here is to have the network discover the structure in the data by finding how the data is clustered. The results can be used for data encoding and compression.

> Vector quantization is a lossy data compression method based on approximating all the samples in a region with a code vector, treated as a quantized sample.

Each input vector (sample) can be visualized as a point in an N-dimensional space. The quantizer is defined by a partition of this space into a set of non-overlapping volumes. These volumes are called *intervals*, *polygons*, and *polytops*,

---

WHY VECTOR QUANTIZATION

If we talk in terms of waveforms, Shannon's rate distortion theorem states that there exists a mapping from a source waveform to output code words such that for a given distortion D, $R(D)$ bits per sample are sufficient to reconstruct the waveform with an average distortion arbitrarily close to D. Therefore, the actual rate R has to be greater than $R(D)$. This function, called the *rate-distortion function*, represents a fundamental limit on the achievable rate for a given distortion. Scalar quantizers do not achieve performance close to this information theoretical limit, and thus VQ becomes important. Shannon predicted that better performance could be achieved by coding many samples simultaneously instead of one sample at a time because of better approximation or averaging.

respectively, for one-, two-, and N-dimensional vector spaces. The task of the vector quantizer is to determine the interval, polygon, or volume in which the input vector is located. The output of the optimal quantizer is the vector identifying the centroid of that volume for an N-dimensional case. As in scalar quantizers, the mean square error is a function of the boundary locations for the partition and the multidimensional PDF of the input vector. The mentioned terms will be clarified in the following example. Quantization in one- and two-dimensional cases is given in Example 5.4.

Example 5.4 With the help of a numerical example, explain the difference between one-dimensional VQ and two-dimensional VQ.

Solution Typical numerical values are selected for the explanation here.

For one-dimensional VQ:
Every number less than −2 is approximated by −3.

Every number between −2 and 0 is approximated by −1.

Every number between 0 and 2 is approximated by +1.

Every number greater than 2 is approximated by +3.

Thus, four quantization levels are chosen. Note that the approximate values are uniquely represented by two

bits. This is a one-dimensional, two-bit VQ. It has a rate of 2 bits/dimension (Fig. 5.15a).
For two-dimensional VQ:
In contrast to one-dimensional VQ, here, every pair of numbers falling in a particular region (polygon) are approximated by a centroid (•)(shown in Fig. 5.15b) associated with that region. For example, if there are 16 regions, each of them can be uniquely represented by four bits. Thus, this is a two-dimensional, four-bit VQ. Its rate is also 2 bits/dimension (Fig 5.15b). Figure 5.15 shows the concept of encoding regions, input vectors, and code words; the input vectors are marked with crosses (×), code words are marked with dots, and the Voronoi regions are separated by boundary lines.

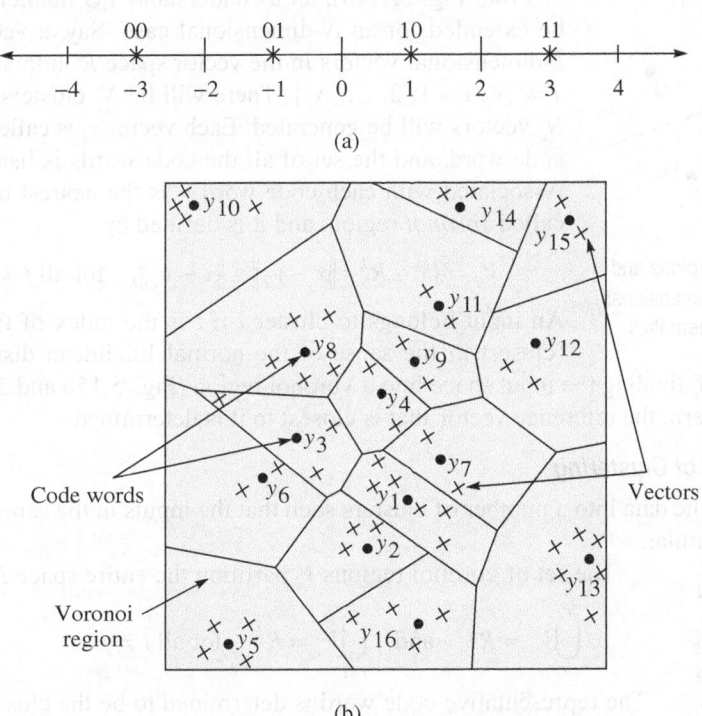

Fig. 5.15 Concept of VQ (a) One-dimensional (b) Two-dimensional

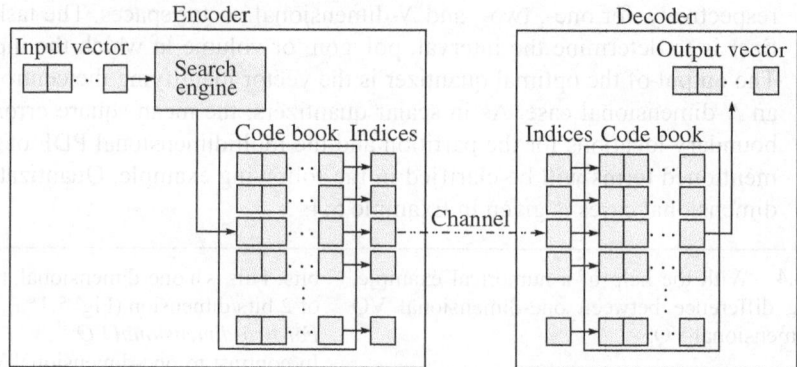

Fig. 5.16 Vector-quantized source encoding and decoding processes

In Example 5.4, the stars are called *vectors* and the regions are called *encoding regions*, which are polygons, as mentioned previously. The regions are also called the *clusters*. The dots represent the code vectors. The set of all code vectors is called the *codebook* and the set of all encoding regions is called the *partition* of the space. The codebook has a set of addresses too. Each code word has an address. We transmit only the addresses and not the code. The source encoder and the decoder using VQ are shown in Fig. 5.16. In this figure, such addresses are represented by indices.

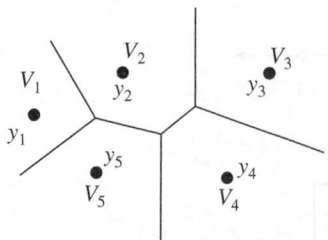

Fig. 5.17 Voronoi regions and code vectors for two-dimensional VQ correlating mathematics

From Fig. 5.15(b), let us understand VQ mathematically, which can be extended for an N-dimensional case. Say, a vector quantizer maps k-dimensional vectors in the vector space R^k into a finite set of vectors $Y = \{y_i : i = 1, 2, ..., N_c\}$. There will be N_c clusters, and corresponding N_c vectors will be generated. Each vector y_i is called a code vector or a code word, and the set of all the code words is listed in a codebook Y. Associated with each code word y_i is the nearest neighbouring region, called *Voronoi* region, and it is defined by

$$V_i = \{x \in R^k : \|x - y_i\| \le \|x - y_j\|, \quad \text{for all } j \ne i\} \qquad (5.13)$$

An input belongs to cluster i if i is the index of the closest prototype (closest in the sense of the normal Euclidean distance). This has the effect of dividing the input space into a Voronoi region (Figs 5.15b and 5.17). For each training pattern, the reference vector that is closest to it is determined.

Method of Clustering

Divide the data into a number of clusters such that the inputs in the same cluster are in some sense similar.

The set of Voronoi regions V_i partition the entire space R^k such that

$$\bigcup_{i=1}^{N_c} V_i = R^k \quad \text{and} \quad \bigcup_{i=1}^{N_c} V_i = R^k \quad \text{for all } i \ne j \qquad (5.14)$$

> Vector quantizers yield better performance even when the samples are independent of one another; however, performance can improve further if there is a strong correlation between the samples in the group.

The representative code word is determined to be the closest in Euclidean distance from the input vector x. The Euclidean distance is defined by

$$d(x, y_i) = \sqrt{\sum_{j=1}^{k} (x_j - y_{ij})^2} \qquad (5.15)$$

> Codebooks are look-up table algorithms that store a list of code words corresponding to the code vectors.

where x_j is the jth component of the input vector and y_{ij} is the jth is component of the code word y_i.

Vector quantization is a *delayed decision coding technique* that maps a group of input samples (typically a speech frame), called a vector, to a codebook index. In each quantizing interval, the codebook is searched and the index of the entry that gives the best match to the input signal frame is selected.

The number of samples in a block (vector) is called the dimension L of the vector quantizer. The rate R of the vector quantizer is defined as

$$R = \frac{\log_2 n}{L} \text{ bits/dimension} \tag{5.16}$$

where n is the size of the VQ codebook and R may also take fractional values. All the quantization principles used in scalar quantization apply to VQ as a straightforward extension. Instead of quantization levels, we have quantization vectors, and distortion is measured as a squared Euclidean distance between the quantization vector and the input vector. VQ is known to be most efficient at a very low bit rate ($R = 0.5$ bits/sample or less).

Example 5.5   A speech frame with 24 samples is to be quantized using the VQ method and 8 bits/sample will be used for representing the quantized samples. Find the rate of quantization for one-dimensional VQ and two-dimensional VQ.

Solution   For one-dimensional VQ, the rate of quanti-zation is

$$R = \log_2 8/1 = 3 \text{ bits/dimension}$$

For two-dimensional VQ, it is

$$R = \log_2 8/2 = 3/2 = 1.5 \text{ bits/dimension}$$

Codebook Preparation and its Use

There are two main tasks in VQ. The first is the code design and the second is the code use. The form of the algorithm selected to control the complexity of encoding and decoding may couple both the tasks—*partition* and *search*.

Codebooks are nothing but look-up table algorithms. A list of code words is stored in the codebook memory. Each pattern is identified by an address or a pointer index (Fig. 5.16). The algorithm tries to find the best match from the codebook and transmits the pointer index. The code vectors stored in the codebook, tree, or trellis are the likely or typical vectors. The task of identification of the likely code vectors is called *populating* the code. The methods of determining the code population are classically deterministic, stochastic, and iterative.

Given an input vector and a populated codebook, tree, or trellis, the coder algorithm must conduct a search to determine the best-matching contender vector. The coder performance improves for larger dimensional spaces but its complexity increases at the same rate. An exhaustive search over a large dimension improves the coder performance but is also pro-hibitively time consuming. VQ can be used for (lossy) data compression.

5.5 PULSE CODE MODULATION

Digital signals, having only *one-bit* and *zero-bit* states are more easily separated from noise and can be amplified without corruption; hence, communication systems have converted to

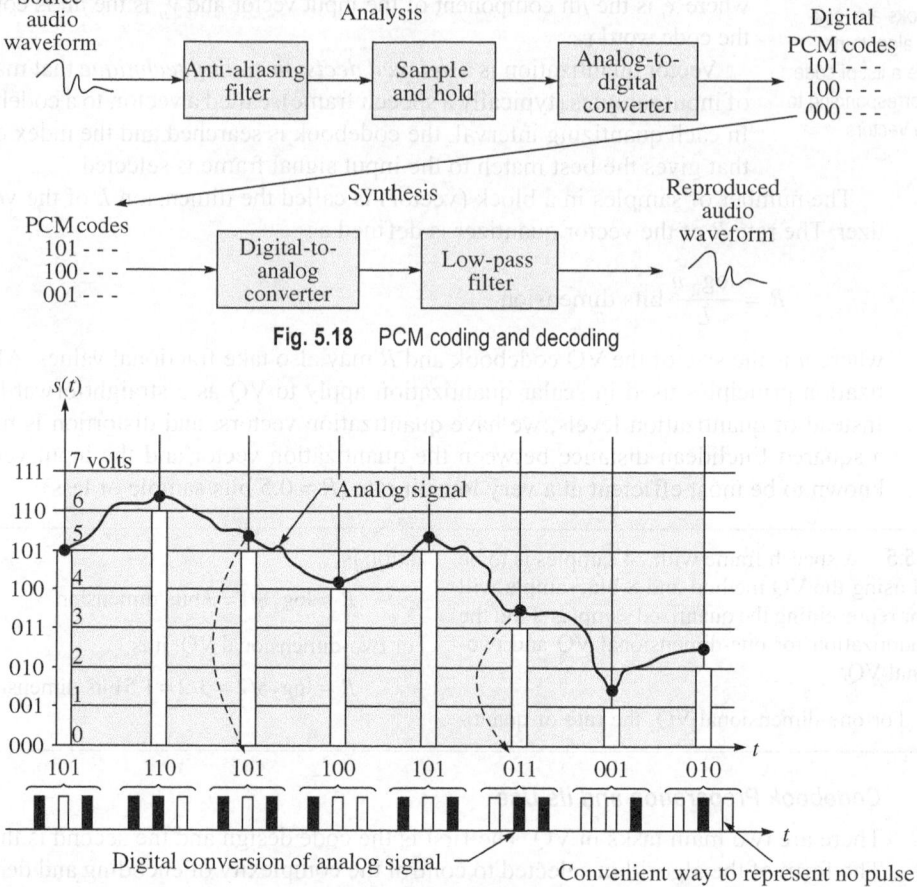

Fig. 5.18 PCM coding and decoding

Fig. 5.19 Analog signal converted into PCM signal by discrete level assigned to the sample and ADC

a digital transmission format called PCM. Figure 5.18 shows the process of PCM as a block diagram.

Scrambling prevents long sequences of identical bits by randomizing the sampled data or bitstream and combating synchronization issues up to a certain extent.

Pulse code modulation is formed from an analog signal by operations of antialiasing filtering, sampling, quantization, and ADC or encoding. The input analog signal, denoted as $s(t)$ in Fig. 5.19, is continuous in both time and amplitude. PCM is a process that begins by antialiasing low-pass filtering the analog signal to ensure that no frequencies above f_{max} are present. The next step is to sample the filtered analog input. A clock signal generates pulses at the Nyquist sampling rate of at least $2f_{max}$, which are used by a sampler to produce $2f_{max}$ samples of the analog signal per second. The sampling operation will keep the signal continuous in amplitude but discrete in time (refer Section 5.1).

Sampling is followed by quantizing and rounding off each sampled value. The result of sampling and quantizing—called *digitization*—is a series of varying-amplitude pulses at the sampling rate. Such a pulse amplitude modulated (PAM) signal, discrete in time and amplitude, is shown in Fig. 5.19. This is the stage where errors are introduced at each step of the process.

Antialiasing filter is a low-pass filter. It ensures that no frequencies above f_{max} are present in the input analog signal.

This signal is basically a step-envelope amplitude modulated signal and subject to degradation by noise as in any AM signal. To convert these steps or pulses into a digital signal, the pulses are grouped according to the clusters of bits in binary numbers and are then transmitted in rapid sequence. It can be verified that the values at the sampling moments are 5, 6, 5, 4, 5, 3, 1, 2, and so on. The integers are translated into a binary representation; the PCM data encoding would look like this in binary: 101, 110, 101, 100, 101, 011, 001, 010, and so on. After PCM, the digital signal can be either return-to-zero (RZ), non-return-to-zero (NRZ), or any other line coding format. For an NRZ system to be synchronized using in-band information, there must not be long sequences of identical symbols, such as 1's or 0's, so data *scrambling* is applied. Scrambling is nothing but data randomization.

Pulse code modulation is a lossy conversion. It may be noted that there are two sources of impairment:

(a) Rounding the analog signal to the nearest integer value results in *quantization error* or noise.

(b) The frequency range of the analog signal is higher than half the sampling rate, which is known as *aliasing error*.

As the transmitted on–off pulses are of equal amplitude, they are immune to additive noise in the sense that only the presence or absence of a pulse needs to be determined at the receiver. This process is reversed at the receiver, which converts the binary data into analog levels, decoding the received binary number into a staircase analog signal, which is then smoothed by passing it through a low-pass filter with bandwidth f_{max}.

> *Note*: For 4 kHz voice digitization, the standard word size used is 8 bits. If an input analog signal is sampled 8000 times/s and each sample is given a code word that is 8 bits long, then the maximum transmission bit rate for telephony systems using PCM will be 64,000 bits/s.

Pulse code modulation is used in baseband transmissions of digital telephone systems (trunk lines) and is also the standard form for digital audio in computers and various compact disc formats, digital video, and so on. The transmission bandwidth occupied by the rectangular pulses of the PCM signal is equal to the PCM word size in bits per sample multiplied by the sampling rate in samples per second. PCM can also be applied to companded signals.

Several PCM streams may be multiplexed into a larger-aggregate data stream. This technique is called time division multiplexing, which will be discussed in Chapter 9. For wireless communication, the PCM signal can be produced from the band-limited audio, image, or video signals. If necessary, the PCM signal may be line coded and further processed. It can be passed through raised cosine filter for pulse shaping.

Example 5.6 A sinusoidal signal with 8 bits/sample and 24 ksamples/s is converted into the PCM form. What will be the SNR for the full range of sinusoid? If companding is applied before PCM conversion, what will be the new SNR value with $\mu = 100$?

Solution For uniformly quantized sinusoid, for fine quantization, the quantized signal power can be approximated to the second moment of the clean signal $s(t)$. So,

Quantized signal power $= \overline{s^2(t)}$

The quantization noise power is given by $\dfrac{S_{pq}^2}{3M^2}$, where M is the number of quantization levels and S_{pq} is the peak quantization level of the uniform quantizer (all values of s_p are replaced by s_{pq}).

Moreover, $\left[\dfrac{S_{pq}^2}{S^2(t)}\right]^{1/2}$ is known as the crest factor.

Now,

$$SNR = 3M^2 \frac{\overline{S^2(t)}}{S_{pq}^2} = 3(2^n)^2 \frac{\overline{S^2(t)}}{S_{pq}^2}$$

or $\frac{\overline{S^2(t)}}{S_{pq}^2} = \frac{1}{2}$ for sinusoid

or $SNR = 3(2^8)^2 \times \frac{1}{2}$

or $= 3(256)^2 \times \frac{1}{2}$

or $= 98,304 = 49.93$ dB

For companded sinusoid,

$$SNR = \frac{3M^2}{[\ln(1+\mu)]^2} = \frac{3(2^8)^2}{[\ln(1+100)]^2}$$

$$= \frac{196608}{21.3} = 9230.423 = 39.65 \text{ dB}$$

Example 5.7 Find the bandwidth expansion factor for a PCM signal with sync pulses and with rectangular pulses, whose basic bandwidth is 4 kHz and the number of bits per sample is 8.

Solution According to Nyquist criteria, the sampling rate is 8 ksamples/s.

For sync pulses:
Transmission bandwidth = $8 \times 8/2 = 64/2 = 32$ kHz
Bandwidth expansion factor = $32/4 = 8$

For rectangular pulses:
Transmission bandwidth = $8 \times 8 = 64$ kHz
Bandwidth expansion factor = $64/4 = 16$

The SNR can be improved by trading the bandwidth because SNR increases with the transmission bandwidth of the PCM signal. Here, SNR includes SQNR.

5.6 DELTA MODULATION

> Delta modulation is a method to reduce the data rate used in PCM voice reproduction and requires only one-bit buffer.

Pulse code modulation is an absolute coding method and uses no data compression. On the other hand, voice waveforms contain much redundant data. Long periods of silence are interspersed with sounds that slowly vary in pitch. If some time is taken to analyse the analog-to-digital samples, it will be noticed that the changes are, for the most part, gradual. Moreover, the variations in the signal between adjacent samples are a limited portion of the full dynamic range. One method of reducing the data rate used in PCM voice reproduction is called delta modulation.

Delta modulation assumes that the input signal's waveform has a fairly uniform and predictable slope (rate of rising and falling). Rather than storing an 8-bit or a 12-bit quantity for each sample, a delta modulator stores only a single bit. When the computer samples the input signal from the ADC, it compares the current reading to the preceding sample. Each sample of the source waveform is tested to see if its amplitude is higher or lower (within the resolution of a fixed quantization value Δr) than that of the previous sample. If the amplitude is higher, the single-bit delta-modulated encoding value is set to 1; if it is lower, the encoding value is set to 0. This is shown in Fig. 5.20.

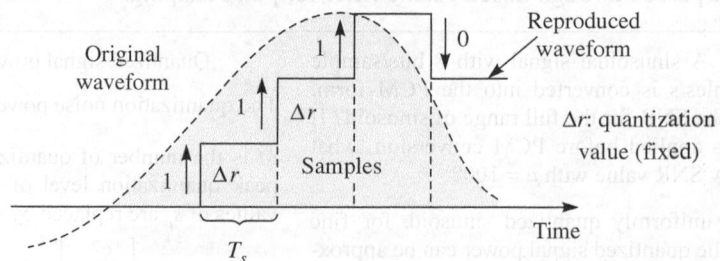

Fig. 5.20 Waveform sampling, coding, and reproduction concept for delta modulation

Reproduction of the waveform is accomplished by sending the stored bits in sequence to the output, where their values are integrated. However, like other techniques, delta modulation has limitations, one of which is sampling-rate restriction. As only a single bit changes between samples, the rate at which samples are taken must be sufficiently fast such that no significant information is lost from the input signal. When the source waveform changes too rapidly, the fixed quantization value may be too small to express the full change in the input; this slope overload causes a *compliance error*. On the other hand, when the input waveform does not change much (at the extreme, a DC signal), the vertical deflection in the quantization value results in *granular noise* in the output due to the small signal amplitude. Slope overload and granular noise are shown in Fig. 5.21.

Furthermore, if the slope of the input waveform varies a lot, the reproduced waveform may be audibly distorted. Hence, using delta modulation may not reduce the data rate much, although there are many different variant schemes, and it is difficult to predict which is optimal in a given situation.

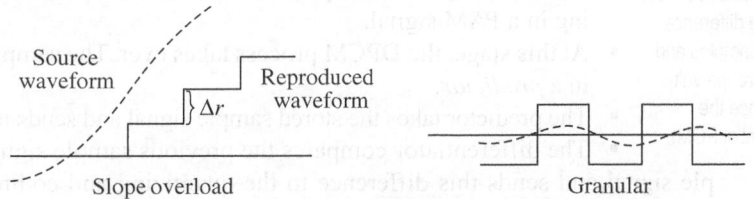

Fig. 5.21 Two potential problems occurring in delta modulation

5.7 MODIFICATIONS TO PULSE CODE MODULATION

In conventional PCM, the analog signal is processed before being digitized. Once the signal is digitized, the PCM signal is usually subjected to further processing (e.g., data compression). Some modified forms of PCM combine signal processing with coding. These simple techniques have been largely rendered obsolete by modern transform-based signal compression techniques. There are two modifications to the conventional PCM.

(a) *Differential* (or *delta*) *PCM* (DPCM) encodes the PCM values as differences between the current and the previous values. For audio, this type of encoding reduces the number of bits required per sample by about 25 per cent compared to PCM.

(b) *Adaptive DPCM* (ADPCM) is a variant of DPCM that varies the size of the quantization step to allow further reduction of the required bandwidth for a given SQNR.

> *Note*: Delta modulation is popular not for commercial use but because it forms the basis for DPCM and ADPCM when it is combined with PCM, which gives appreciable performance.

5.7.1 Differential Pulse Code Modulation

In telephony, a standard audio signal for a single phone call is a 64 kbps digital signal known as DS0. The default encoding on a DS0 is either μ-law PCM or A-law PCM. DPCM was designed to calculate the difference between consecutive samples and then transmit this small difference signal instead of the entire input sample signal. Since the difference between the input samples is less than the entire input sample, the number of bits required for transmission is reduced. DPCM can reduce the bit rate of voice transmission down to 48 kbps. Figure 5.22 shows the block diagram of the DPCM.

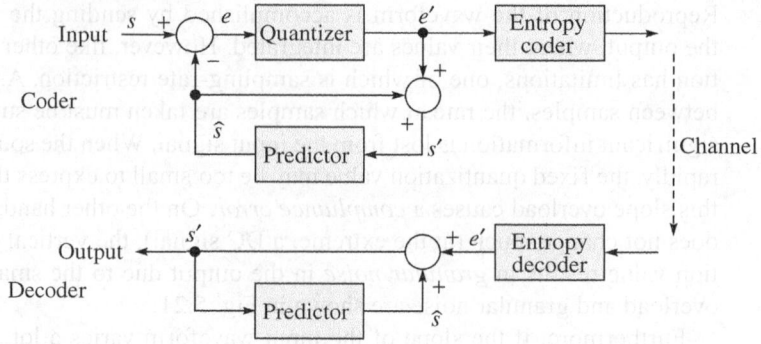

Fig. 5.22 Block diagram of DPCM

DPCM works exactly like PCM; quantization is applied to the difference value of the samples and then the difference value is coded (hence the name differential PCM).

The following points explain the process step by step:

- The input signal is sampled at a constant Nyquist sampling frequency, resulting in a PAM signal.
- At this stage, the DPCM process takes over. The sampled input signal is stored in a *predictor*.
- The predictor takes the stored sample signal and sends it through a *differentiator*.
- The differentiator compares the previous sample signal with the current sample signal and sends this difference to the quantizing and coding phase of PCM. (This phase could be uniform quantizing or companding with A-law or μ-law.)
- After quantizing and coding, the difference signal is transmitted to its final destination.
- At the receiving end of the network, the entire process is reversed: First, the difference signal is decoded and de-quantized. Then, this difference signal is added to a sample signal stored in a predictor and sent to a low-pass filter, which reconstructs the original input signal.

Figure 5.23 shows the calculation of DPCM values for the final transmission. Three consecutive samples are shown ($k - 1$, k, and $k + 1$); these samples have amplitudes nearer to each other, and due to correlation between consecutive samples, the next one can be predicted, which gives the difference signal to be coded. It is clear that three different PCM codes are required for representing the absolute levels, whereas only two differential values are coded for DPCM. Thus, reduction in data is achieved.

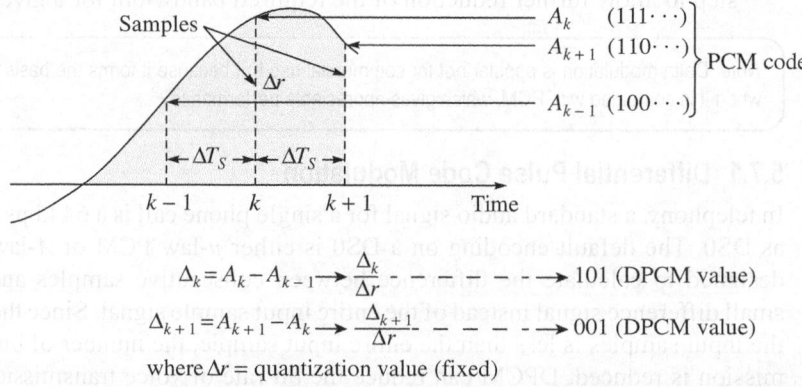

Fig. 5.23 DPCM generation shown with the help of sampled waveform

In ADPCM, prediction is based on the knowledge of the autocorrelation properties of speech or fuzzy logic may be applied.

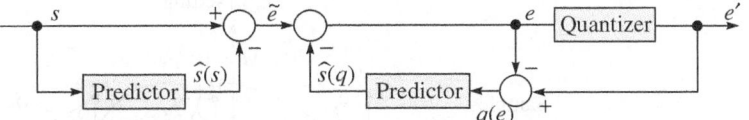

Fig. 5.24 Quantization error feedback in DPCM coder and prediction error

Thus, DPCM is a good way of reducing the bit rate for voice transmission by exploiting the redundancies present in the speech signal, but it causes some other problems related to voice quality. As stated earlier, DPCM quantizes and encodes the difference between the previous sample input signal and the current sample input signal. If it quantizes the difference signal using uniform quantization, the possible sources of error are quantization error and prediction error. This is shown in Fig. 5.24. Some sort of adaptiveness is required to overcome this problem and is achieved through adaptive DPCM, discussed in Section 5.7.2.

Differential PCM exhibits some of the same limitations as simple delta modulation but to a lesser degree. Only when the difference between the samples is greater than the maximum DPCM encoding value does distortion (compliance error) occur. The solution to this problem is to reduce the input bandwidth or raise the sampling frequency.

5.7.2 Adaptive Differential Pulse Code Modulation

Adaptive DPCM codec is a waveform codec. It quantizes the difference between consecutive speech signal samples, or a prediction is made of the speech signal as in DPCM but quantization is adaptive. In practice, ADPCM encoders are implemented using signal prediction techniques accommodating forward adaptive quantizers, instead of simply encoding the difference between adjacent samples. A linear predictor is used to predict the current sample using previous ones. The difference between the predicted and actual samples, called the prediction error, is then encoded for transmission.

If the prediction is accurate, then the difference between the real and predicted speech samples will have a lower variance than the real speech samples and will be accurately quantized with fewer bits than would be needed to quantize the original speech samples. At the decoder, the quantized difference signal is added to the predicted signal to produce the reconstructed speech signal or recovers an approximation to the original speech signal by integrating the quantized adjacent sample differences. Since the *quantization error variance* for a given number of bits per sample is directly proportional to the *input signal variance*, the reduction obtained in the quantizer input variance reduces the *reconstruction error variance* for a given value of bits per sample. Figure 5.25 shows the ADPCM generation. The difference in the method can be compared with Fig. 5.23 showing DPCM generation.

The performance of the DPCM codec is aided by the use of adaptive prediction and quantization. In ADPCM, each sample's encoding is derived by a procedure that includes the following steps (refer to Fig. 5.25):

- A differential-k (δk) is obtained by subtracting the previous PCM code value from the current value.
- The quantization value (Δk) is obtained by multiplying the previous quantization value with coefficient times the absolute value of the previous ADPCM code value.
- The PCM-valued differential-k is then expressed in terms of the quantization value (Δk) and encoded in four bits.

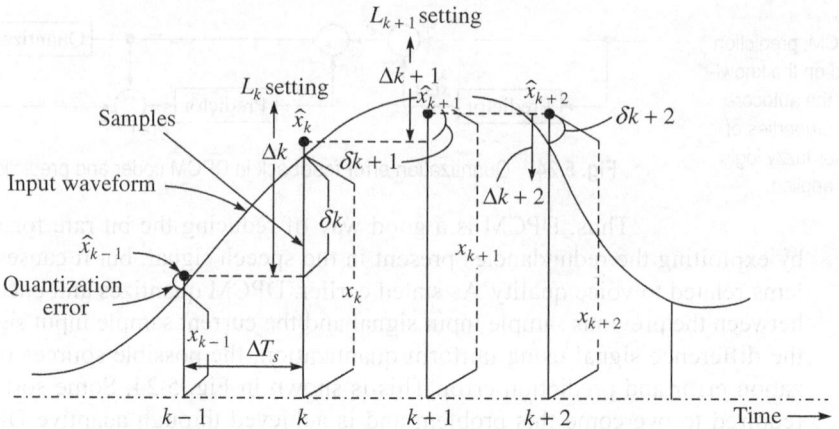

$$\delta k = x_k - \hat{x}_{k-1}$$
$$\Delta k = \Delta k - 1 \times M(|L_{k-1}|)$$

where x_k is the PCM code value $\hat{x}_k$ is the reproduced PCM code value
δk is the differential (PCM code value) Δk is the quantization value
L_k is the ADPCM code value M_k is a coefficient

Fig. 5.25   ADPCM generation concept

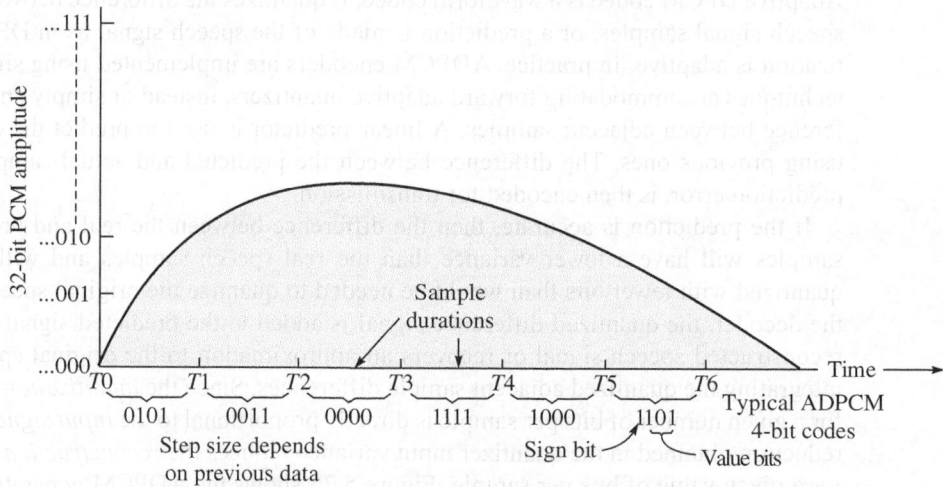

Fig. 5.26   Typical ADPCM four-bit codes from PCM

Using this procedure, typical values of ADPCM are derived from the PCM values and are illustrated in Fig. 5.26 and Table 5.2.

Adaptive DPCM condenses 12-bit PCM samples into only 3 or 4 bits. It can reduce the bit rate of voice transmission down to 32 kbps, half the bit rate of *A*-law or *μ*-law PCM. ADPCM produces *toll quality* voice just like *A*-law or *μ*-law PCM.

An ADPCM encoder makes best use of the available dynamic range of four bits by varying its step size in an adaptive manner. The step size of the quantizer depends on the dynamic range of the input, which is speaker dependent and varies with time. The adaption is in practice achieved by normalizing the input signals via a scaling factor derived from a prediction of the

Table 5.2 ADPCM code structure

Sign value	Value bits	Comments
1	000	Minimum excursion (−ve direction)
1	100	Medium excursion (−ve direction)
1	111	Maximum excursion (−ve direction)
0	000	Minimum excursion (+ve direction)
0	100	Medium excursion (+ve direction)
0	111	Maximum excursion (+ve direction)

dynamic range of the current input. This prediction is obtained from two components: a fast component for signals with rapid amplitude fluctuations and a slow component for signals that vary more slowly. The two components are weighted to give a single quantization scaling factor.

In the same manner, ADPCM can also be applied to image and video signals. Where circuit costs are high and loss of voice quality is acceptable, it sometimes makes sense to compress the voice signal even further. An ADPCM algorithm is used to map a series of 8- to 12-bit PCM samples into a series of 4-bit ADPCM samples. In this way, the capacity of the line is almost doubled.

5.8 INFORMATION SOURCES AND ENTROPY

As the digital baseband signal is obtained from analog sources, the database may not be completely useful and some redundant information may be present (here, we are discussing discrete memoryless sources that generate sequence of messages without any influence of past data generation). Hence, the actual outcome may be different from the useful source data. In source coding, the amount of useful information must be measured and data compression should be applied by removing redundant information. There is a channel coding stage in the communication link (explained in Chapter 6) where the redundant bits are added systematically with the purpose of error detection and correction activity. Removal of redundant information at the source coding stage and addition of redundant information at the channel coding stage must be balanced. At the same time, channel capacity is also limited, according to Shannon, who discovered long back that information transmission, noise, and communication channel capacity have a relationship.

The first point of consideration is that an information source has a *probability distribution*, that is, a set of probabilities $\{P(\omega_1), P(\omega_2), \ldots, P(\omega_n)\}$ assigned to a set of outcomes, say $X = \{\omega_1, \omega_2, \ldots, \omega_n\}$, containing the useful information (the outcomes may be message states or symbols). This reflects the fact that the information contained in an outcome is determined not only by the outcome but also by its uncertainty. An almost certain outcome contains little information. In 1927, Hartley introduced a measure of the information contained in an outcome (sometimes called *self-information*) ω_i, which he defined as

$$I(\omega_i) = \log_2\left(\frac{1}{P\{\omega_i\}}\right) = -\log_2(P\{\omega_i\}) \qquad (5.17)$$

Some adaptive DPCM techniques are used in Voice over Internet Protocol (VoIP) communications.

This measure satisfies our requirement that the information contained in an outcome is proportional to its uncertainty. If $P\{\omega_i\} = 1$, then $I\{\omega_i\} = 0$, indicating that a certain event contains no information. The definition also satisfies the requirement that the total information in independent events should be added,

More the surety of a state, more will be the probability of that state and hence less will be the entropy contained in that message state. This means that the amount of useful information is also less.

which is shown in Eq. (5.18). Clearly, a rain forecast for two days contains twice as much information as that for one day. From Eq. (5.17), for two independent outcomes of a source, ω_i and ω_j,

$$I(\omega_i \text{ and } \omega_j) = \log_2\left(\frac{1}{P\{\omega_i \text{ and } \omega_j\}}\right) = \log_2\left(\frac{1}{P\{\omega_i\}P\{\omega_j\}}\right)$$

$$= \log_2\left(\frac{1}{P\{\omega_i\}}\right) + \log_2\left(\frac{1}{P\{\omega_j\}}\right) = I(\omega_i) + I(\omega_j) \qquad (5.18)$$

Traditionally, engineers have used the term *bit* to describe the information content. Hartley's measure defines the information in a single outcome. The measure *entropy* $H(X)$, sometimes called *absolute entropy*, defines the average information content of the source X as a whole. It is the mean information provided by the source per source output or symbol and is measured in bits per symbol. If multiplied with the symbol per sample rate, entropy can be given in bits per second. The entropy equation is given by

$$H(X) = \sum P\{\omega_i\}I\{\omega_i\} = \sum -P\{\omega_i\}\log_2(P\{\omega_i\}), \text{ for all messages, states or}$$
$$\text{symbols} \qquad (5.19)$$

> *Note:* If a discrete channel has a capacity C and a discrete source has an entropy H (average information content) and if $H \le C$, then there exists a coding system such that the output of the source can be transmitted over the channel with an arbitrarily small possibility of errors.

For the sake of our discussion, we shall consider entropy per second to be bits per second of information transmitted. The word *discrete* refers to the information transmitted in symbols, that is, digital information. Shannon also showed that if our information source is sending the data at the rate the communication channel can handle, then we can add some extra bits to the data stream to push the error rate down to an arbitrarily low level. This is the concept of channel coding and is discussed in detail in Chapter 6. Thus, source coding must be an efficient technique with good compression. Reduced source data makes channel coding possible, albeit with a trade-off. Communication delay is increased due to channel coding. The cost of using channel coding to protect the information is a reduction in data rate or an expansion in bandwidth. However, for every situation, there are enough choices of channel coding such that there exists some satisfactory compromise between delay and error performance.

Binary symmetric source A *binary symmetric source* (BSS) is a source with two outputs whose probabilities are p and $(1 - p)$, that is, $P = \{p, (1 - p)\}$. The rain forecast mentioned earlier is a BSS. A binary signal with output zero or one is a BSS. The entropy of the source is given by

$$H(X) = -p\log_2(p) - (1 - p)\log_2(1 - p) \qquad (5.20)$$

The function of Fig. 5.27 takes the value zero when $p = 0$. When one outcome is certain, so is the other, and the entropy is zero. As p increases, so does the entropy, until it reaches a

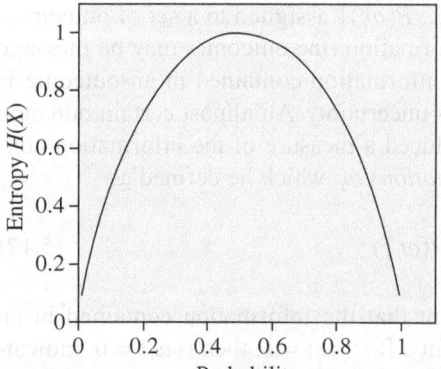

Fig. 5.27 Entropy of a binary symmetric source

> *Note:* The term *entropy* is borrowed from thermodynamics, where too it is a measure of the uncertainty or disorder in a system.

maximum when $p = 1 - p = 1/2$. When p is greater than $1/2$, the curve declines symmetrically and reaches zero when $p = 1$. When $p = 1/2$, then $H(X) = 1$. We conclude that the average information in the BSS is maximized when both outcomes are equally likely.

An equally probable BSS has an entropy, or average information content per symbol, of 1 bit/symbol. Sometimes, there may be more than 1 bit/symbol. Hence, more information can be read at a time, which is a necessity of high-speed communications.

If an information source X has J symbols, its maximum entropy is $\log_2(J)$, and this is obtained when all J outcomes are equally likely. Thus, for a J symbol source,

$$0 \leq H(X) \leq \log_2(J) \tag{5.21}$$

It seems intuitively reasonable that an information source of entropy $H(X)$ needs on average only H binary bits to represent each symbol.

5.9 INFORMATION SOURCE CODING FUNDAMENTALS

More the entropy, less will be the probability of occurrence and hence more will be the number of bits and vice versa. This fundamental is used in variable length coding.

Some measures are applied to the database before the modulation stage in order to optimize it, that is, to reduce the number of bits to be transmitted in an effective manner. One can try to reduce the redundant information, and thus, increase the entropy of the source. This as a result, increases the information transmission efficiency.

5.9.1 Entropy Coding

The replacement of discrete message states (e.g., rain and no rain) with a binary representation is termed *source coding*, as per the discussion in Section 5.3. In any coding operation, we replace the message state or symbol with a code word. The purpose of source coding is to reduce the number of bits required to convey the useful information provided by the information source by correlating it with entropy.

Table 5.3 Variable length coding

Probability of occurrence of a code	No. of bits per code
0.729	0
0.081	1
0.081	01
0.081	10
0.009	11
0.009	00
0.009	000
0.001	111

Variable length coding   Table 5.3 shows the probability of occurrence of each code word in a variable length coding technique. It can be seen that this code will on average use 1.2 bits/sequence instead of 3 bits/sequence. This example shows how to decrease the average number of bits per symbol on the basis of probability. Moreover, without difficulty, we have found a code that has an average bit usage less than the source entropy. However, there is a difficulty with the code in Table 5.2. Before a code word can be decoded, it must be parsed. Parsing describes the activity of breaking the message string into its component code words. After parsing, each code word can be decoded into its symbol sequence.

The code in Table 5.4, however, is an instantaneously parseable code. It satisfies the prefix condition.

The code in Table 5.4 uses on average 1.568 ($\approx$1.6) bits/sequence. This is a 47 per cent improvement on the code identifying each symbol with bits.

Huffman coding   The variable length coding concept forms the basis for *Huffman code* for the sequence set. The code for each sequence is found by generating the *Huffman code tree* for

Table 5.4 Instantly parseable variable length coding

Sequence	Probability of occurrence of a code	Code word	Letter for identification in the tree
000	0.729	1	A
001	0.081	011	B
010	0.081	010	C
011	0.081	001	D
100	0.009	00011	E
101	0.009	00010	F
110	0.009	00001	G
111	0.001	00000	H

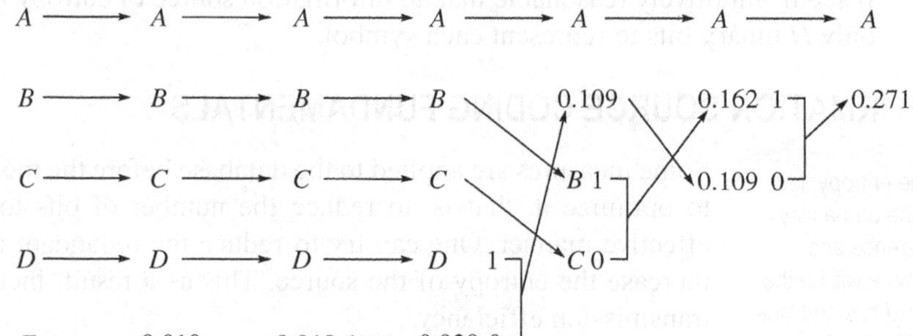

Fig. 5.28 Derivation of Huffman code tree

the sequence. A Huffman code tree is an unbalanced binary tree. The derivation of the Huffman code tree is shown in Fig. 5.28 and the tree itself is shown in Fig. 5.29. In both these figures, the letters A to H have been used in place of the sequences in Table 5.4 to make them easier to read.

In Fig. 5.29, the sequences are ordered with respect to the probability of occurrence of the sequence, with the highest probability at the top of the list. (The probabilities are effectively used as *weights* in the process to be explained.) The tree is derived bottom up, in terms of *branch nodes* and *leaf nodes* by combining weights and removing leaf nodes in progressive stages. As shown in Fig. 5.29, the two lowest leaf nodes G and H have their weights added, and the top node is labelled 1 and the lower one 0. The next stages are represented by that weight, and the list is rewritten in order of the weights. Now, the two lowest leaf nodes are E and F, and they are labelled 1 and 0, respectively, and their weights are added to be taken on to the next stage. This continues until only two nodes remain. The Huffman tree shown in Fig. 5.29 is then produced by following backwards along the arrows in the figure. To derive the code words from the tree, descending from the top node (the root node), the 1's and 0's should be listed in order until the leaf node for one of the letters is reached.

> Before performing Huffman coding, it is necessary to arrange state probabilities in descending order and then create the tree.

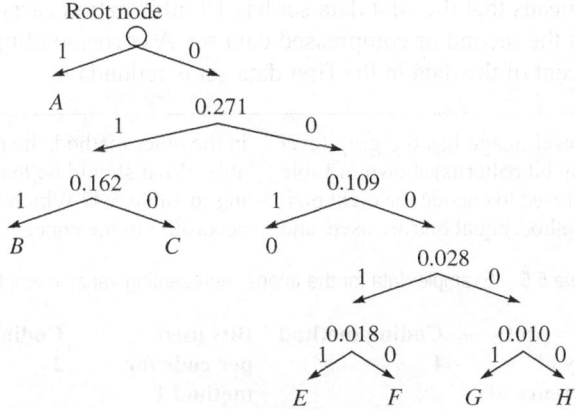

Fig. 5.29 Huffman code tree

Huffman coding relies on the fact that both the transmitter and the receiver know the sequence set (or dataset) before communicating and can build the code table. Where this is not possible, *dynamic Huffman coding* can be used to build the code table as the data is transmitted or received.

The *noiseless source coding theorem* (also called Shannon's first theorem) states that an instantaneous code can be found that encodes a source of entropy $H(X)$ with an average number of bits per symbol B_s such that

$$B_s \geq H(X) \tag{5.22}$$

Ordinarily, the longer the sequences of symbols, the closer B_s will be to $H(X)$. Like many theorems of information theory, the theorem tells us nothing but how to find the code.

5.9.2 Data Compression

The term *data compression* refers to the process of reducing the amount of data required to represent a given quantity of information. A clear distinction needs to be made between *data* and *information* to help understand that they are not synonymous. A story (information) can be prepared by using different words (data). There may be some redundant data that may not have relevance to or significance in that story and can be removed from the total database. Data redundancy is a critical issue in data compression. It is not an abstract concept but is a mathematically quantifiable entity.

Let n_1 and n_2 denote the number of information-carrying units in two data sets that represent the same information. The relative data redundancy R_d of the first data set n_1 can be defined as

$$R_d = 1 - (1/CR) \tag{5.23}$$

where CR is called the *compression ratio* and is given by

$$CR = n_1/n_2 \tag{5.24}$$

> Information is conveyed by means of data. However, using minimum data to convey the same information will increase the speed of communication.

For the case $n_2 = n_1$, $CR = 1$ and $R_d = 0$, indicating that relative to the second data set the first representation of the information contains no redundant data. When $n_2 \ll n_1$, $CR \to \infty$ and $R_d \to 1$, implying significant compression and highly redundant data. Finally, when $n_2 \gg n_1$, $CR \to 0$ and $R_d \to -\infty$, indicating that the second data sets contains more data than the original information. This is the case of data expansion instead of compression. A practical compression ratio such as 10 means that the first data set has 10 information carrying units or bits for every 1 unit or bit in the second or compressed data set. A corresponding redundancy of 0.9 implies that 90 per cent of the data in the first data set is redundant.

Example 5.8 An eight-level image has the grey-level distribution (corresponding bit patterns) shown in Table 5.5. Two methods are employed to encode the eight possible grey levels. In one method, equal bits are used, and in the other method, the number of bits or levels is variable. What should be the average length of bits according to Table 5.4? Which is the more efficient technique according to the concept of information theory?

Table 5.5 Example data for the image representing variable length coding

L_k = no. of possible levels	$P(L_k)$ = probability of occurrence of kth level	Coding method 1	Bits used per code for method 1	Coding method 2	Bits used per code for method 2
$L0 = 0$	0.18	000	3	11	2
$L1 = 1/7$	0.25	001	3	01	2
$L2 = 2/7$	0.21	010	3	10	2
$L3 = 3/7$	0.17	011	3	001	3
$L4 = 4/7$	0.08	100	3	0001	4
$L5 = 5/7$	0.06	101	3	00001	5
$L6 = 6/7$	0.03	110	3	000001	6
$L7 = 1$	0.02	111	3	000000	6

Solution Average no. of bits for coding method 1 = 3 bits (as they are equal)

Average no. of bits for coding method 2
= 2(0.18) + 2(0.25) + 2(0.21) + 3(0.17) + 4(0.08)
+ 5(0.06) + 6(0.03) + 6(0.02)
= 2.71 bits

The resulting compression ratio, $CR = 3/2.71 = 1.107$.

Thus, approximately 10% of the data resulting from the use of code 1 is redundant. Hence, the second method of variable length coding is the more efficient technique. The exact level of redundancy can be determined as follows:
$$R_d = 1 - (1/1.107) = 1 - 0.903 = 0.097$$

5.9.3 Lossy and Lossless Compression

There are two types of compression techniques—lossy and lossless. Both the techniques use various methods to achieve data compression. In the lossless technique, the decompression fully recovers the data. The techniques discussed until now in this chapter—DCT, VQ, and Huffman coding—are all of the lossy type. Lossless compression techniques include run-length encoding and string table compression. Compression methods are otherwise known as algorithms. Compression techniques are used to improve the effective use of the spectrum by sending more data within the specified duration.

5.10 VOCODERS

Wireless communication deals with voice or speech, especially for mobile telephony. Transmission of compressed speech requires special and efficient speech coding rather than conventional waveform coding. The coder is a hardware circuit (chip) or software routine that converts the spoken word into digital code and vice versa. There are two types of speech coding techniques:

(a) We have previously seen waveform coders using PCM/ADPCM techniques. These are relatively simple and have better adaptive capability and speech quality. They are widely used in the 16–64 kbit/s range.

(b) Other types of coders are based on *parametric coding*. These encoders take advantage of the predictable elements in human speech; that is, they are specialized for human voice characteristics. They analyse vocal tract sounds and send a recipe for rebuilding the sound at the other end rather than sending the actual sound waves. As a result, the speech codec is able to achieve a much higher compression ratio, which yields a smaller amount of digital data for transmission, compared to other techniques. However, if music is encoded with such a speech codec, it will not sound as good as the original when decoded at the other end. Speech or voice coders of this type are called *vocoders* in short form.

Several low data rate encoders are discussed here with an assessment of their subjective quality.

A speech signal may be processed before coding; for example, noise may be removed by digital signal processing (DSP) techniques. Similarly, a few frequency domain operations such as upsampling, downsampling, and removal of certain harmonics can be performed over the speech signal, depending on the requirements before speech coding.

Vocoders may be classified on the basis of bit rate as follows:

- *Medium rate*: 8–16 kbps
- *Low rate*: between 8 kbps and 2.4 kbps
- *Very low rate*: below 2.4 kbps

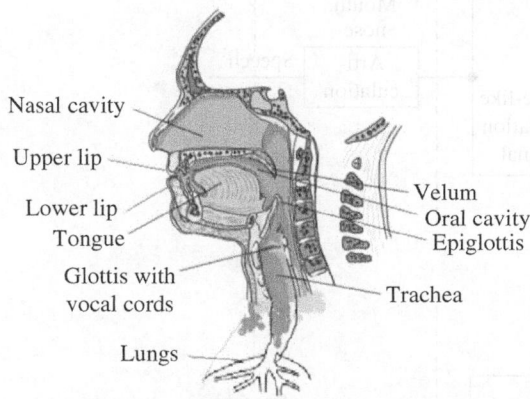

Fig. 5.30 Cross section of the human speech organ

Nasal cavity
Upper lip
Lower lip
Tongue
Glottis with vocal cords
Lungs
Velum
Oral cavity
Epiglottis
Trachea

Vocoders exploit speech properties, derive the important speech parameters, and perform parametric coding so that redundancies will be removed automatically.

5.10.1 Theory of Vocoders

A vocoder in general is a speech analyser and synthesizer. A cross-sectional view of the human speech organ is shown in Fig. 5.30.

Speech is produced by the cooperation between the lungs, glottis (with vocal cords), and articulation tract (mouth and nose cavities). The human voice consists of the sounds generated by the opening and closing of the glottis by the vocal cords. For the production of voiced sounds, the lungs press air through the epiglottis, and the vocal cords vibrate and interrupt the air stream and produce a quasiperiodic pressure waveform with many harmonics. This basic sound is then filtered by the nose and throat (a complicated resonant piping system) to produce differences in harmonic content (formants) in a controlled way, creating the wide variety of sounds used in speech. The pressure impulses are commonly called *pitch impulses* and the frequency of the pressure signal is the *pitch frequency* or *fundamental frequency*.

In Fig. 5.31, a typical impulse sequence (sound pressure function) produced by the vocal cords for a voiced sound is shown. It is the part of the voice signal that defines the speech melody. Figure 5.32 depicts the variation of the pitch frequency with time, also called the pitch frequency contour.

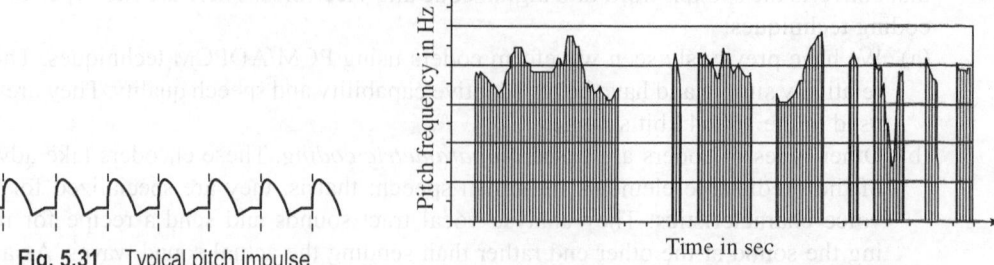

Fig. 5.31 Typical pitch impulse sequence

Pitch frequency in Hz

Time in sec

Fig. 5.32 Variation of pitch frequency

In parametric coding, the parameters are formants and formant frequency, pitch impulses and pitch frequency, and so on.

The pitch impulses stimulate the air in the mouth cavity and for certain sounds also the nasal cavity. When the cavities resonate, they radiate a sound wave, which is the speech signal. Both cavities act as resonators with characteristic resonance frequencies, called *formant frequencies*. Since the mouth cavity can be greatly changed, we are able to pronounce many different sounds. There is another set of sounds, known as the unvoiced and plosive sounds, which are not modified by the mouth in the same fashion. In the case of unvoiced sounds, the excitation of the vocal tract is more noiselike.

The aforementioned speech generation process is modelled in the form of a block diagram in Fig. 5.33(a), and the equivalent synthetic speech production blocks are shown in Fig. 5.33(b).

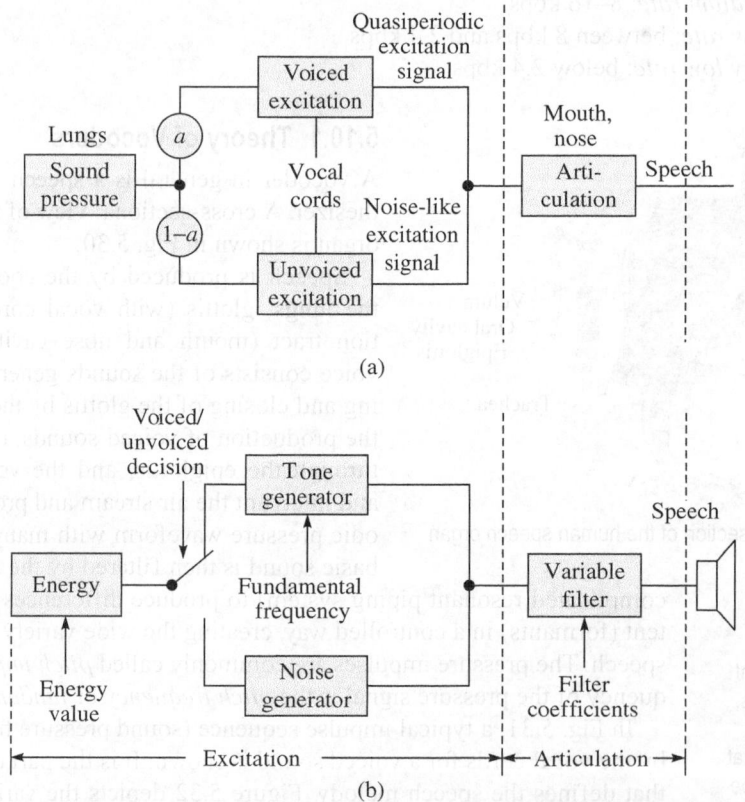

Fig. 5.33 Speech production (a) Human speech production modelling (b) Equivalent synthetic speech production blocks

Compared to AaS, the AbS approach is capable of producing better speech quality at low data rates, but the encoder in the transmitter is more complex.

The synthetic sound is analysed and generated as follows. The components of the human speech organ, namely the excitation and the vocal tract parameters, are computed. The components are then fed into the synthesis part of a vocoder, which generates a synthesized speech signal. The user can replay the signal and compare it with the reference speech signal. The pitch frequency contour is graphically presented for the reconstructed signal as well, and the user can directly manipulate this contour.

Vocoders examine speech by finding this basic carrier wave, which is at the *fundamental frequency*, and measuring how its spectral characteristics are changed over time by recording someone speaking. In doing so, the vocoder dramatically reduces the amount of information needed to store speech, from a complete recording to a series of numbers. The user can manipulate the fundamental frequency contour, the number of prediction coefficients, the signal energy, and so on and he or she can then hear the result of these manipulations.

Vocoders may be used in musical instruments, television and films, robots, or talking computers.

Quality and bit rate variations can be plotted for waveform as well as parametric vocoders, as shown in Fig. 5.34. Parametric vocoders use different approaches for coding a voice signal or they exploit the various parameters for coding purpose; therefore, they can be further classified as explained in Section 5.10.2. In fact, speech coding system development continues to this day to be a vigorous area of research and development.

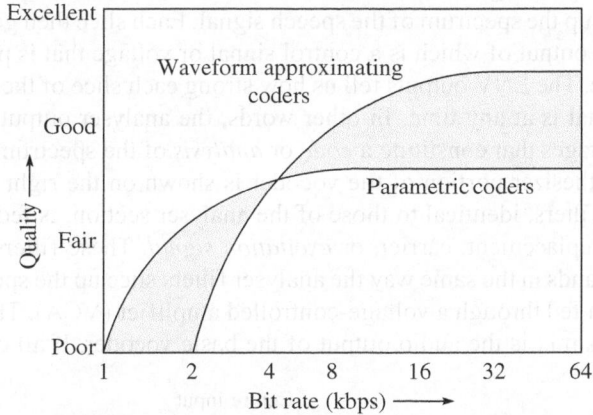

Fig. 5.34 Quality versus bit rate for speech coders

5.10.2 Types of Vocoders

The two basic speech coding methods for data rates between 4.8 kbps and 16 kbps are *analysis and synthesis* (AaS) and *analysis by synthesis* (AbS).

In the AaS approach, an analyser in the transmitter analyses the original speech and extracts a set of parameters that represent some kind of source filter model. These parameters are then transmitted to the receiver, where a synthesizer reconstructs the speech based on the received parameters. In this system, the distortion throughout the whole coding process is difficult to check and control, because the analyser and the synthesizer are located separately in the transmitter and the receiver.

In the AbS approach, an analyser and a local synthesizer are introduced in the transmitter. The synthesized speech is now available in the transmitter for analysis. A trial-and-error procedure, similar to a closed loop, determines the optimum parameters in the transmitter. In the receiver, these parameters reconstruct the synthetic speech, which match the real signal with the minimum perceptual error.

Different types of speech coders are discussed in this section.

Channel Vocoder

Channel vocoders are parametric, frequency domain vocoders. It was the first among the analysis–synthesis systems of speech demonstrated practically. Channel vocoders encode and multiplex the sampled envelope of the speech signal for a number of frequency bands. Here, just like frequency band allocation for different channels, the range of frequency is divided, hence the name channel

> Channel vocoders typically operate between 1 kbps and 2 kbps. Even though these coders are efficient, they produce a synthetic quality sound and, therefore, are not generally used in commercial mobile telephone systems.

vocoder. The channel vocoder uses a bank of filters or digital signal processors to divide the signal into several sub-bands, as shown in Fig. 5.35. It is a source filter model representing the statistical characteristics of the speech signal. The left side of the figure shows the analyser portion of the device. A speech signal is fed through a series of band-pass filters. (The power levels are transmitted together with a signal that represents a model of the vocal tract.)

Sampling is done synchronously every 10–30 ms. Along with the energy information about each band, the pitch frequency for the voiced speech and the voiced or unvoiced decision are also transmitted.

The centre frequencies $F1$, $F2$, $F3$, …, Fn of the filters are spaced one-quarter to one-half octave apart; together, the filter bands cover most of the audio spectrum. Thus, the filter band slices up the spectrum of the speech signal. Each slice then goes to an envelope follower (ENV), the output of which is a control signal or voltage that is proportional to the strength of that slice. The ENV outputs tell us how strong each slice of the frequency spectrum of the speech signal is at any time. In other words, the analyser output is a set of slowly varying control voltages that constitute a *code* or *analysis* of the spectrum of the speech signal.

The synthesizer portion of the vocoder is shown on the right side of Fig. 5.35. A set of band-pass filters, identical to those of the analyser section, is fed by a second audio signal, called the replacement, carrier, or *excitation signal*. These filters slice up the carrier spectrum into bands in the same way the analyser filters slice up the speech signal spectrum. Each slice is then fed through a voltage-controlled amplifier (VCA). The outputs of the VCAs are mixed. This mix is the audio output of the basic vocoder. If all of the VCA control signals

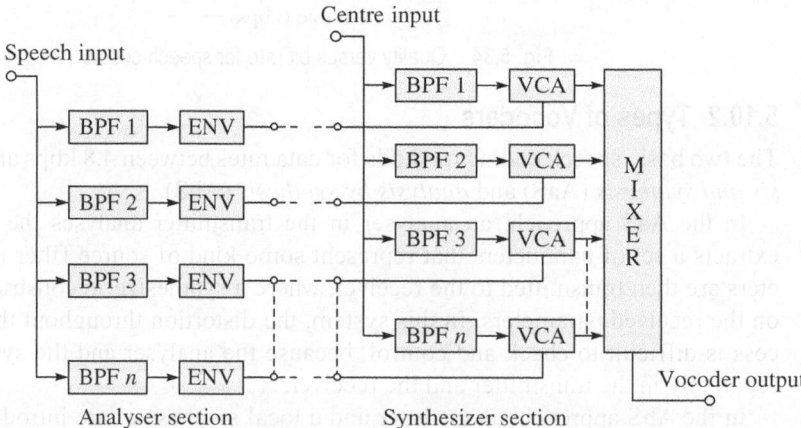

Fig. 5.35   Basic analysis synthesis function in a channel vocoder

are of the same voltage, the vocoder output will in principle be the same as the carrier input. If, on the other hand, the VCA control inputs are connected to the analyser envelope follower outputs, the spectral variations of the speech signal are impressed on the carrier signal.

Formant Vocoder

The formant vocoder is also a parametric vocoder. As mentioned earlier, the two types of speech sounds, voiced and unvoiced, produce different sounds and spectra due to their differences in sound formation. With voiced speech, air pressure from the lungs forces the normally closed vocal cords to open and vibrate. The vibrational frequencies (pitch) vary from about 50 Hz to 400 Hz (depending on the person's age and sex) and form resonance in the vocal track at odd harmonics. These resonance peaks are called formants and can be seen in the voiced speech in Figs 5.36(a) and (b).

Unvoiced sounds, called *fricatives* (e.g., *s*, *f*, *sh*), are formed by forcing air through an opening (hence the term *fricative*, derived from the word *friction*). Fricatives do not vibrate the vocal cords and therefore do not produce as much periodicity as seen in the formant structure in voiced speech; unvoiced sounds appear more noiselike (see Figs 5.37a and b).

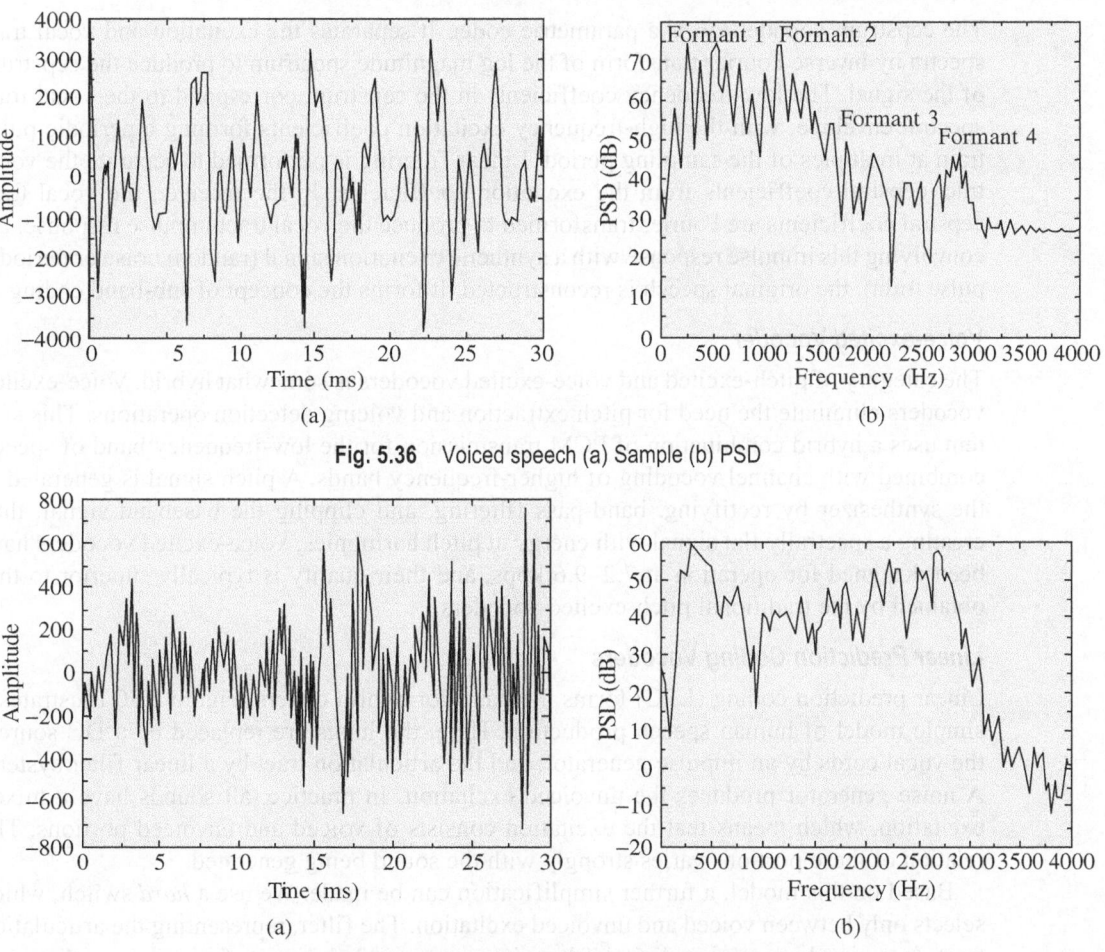

Fig. 5.36   Voiced speech (a) Sample (b) PSD

Fig. 5.37   Unvoiced speech (a) Sample (b) PSD

> LPC vocoders require very low bit rate for speech transmission (about 3 kbps compared to PCM (64 kbps).

Time domain samples lose periodicity, and the PSD does not display the clear resonant peaks that are found in voiced sounds.

Unlike channel vocoders, formant vocoders transmit formant information, that is, the positions of the peaks (formants) of the spectral envelope (Fig. 5.36b), instead of sending samples of the entire power spectrum envelope. Since speech signal information is primarily contained in the formants, a vocoder that can predict the position and bandwidths of the formants could achieve high quality at very low bit rates.

Formant vocoders typically operate in the range of 1 kbps. Theoretically, the bit rate of formant vocoders is lower than that of channel vocoders because they use fewer control signals. Typically, a formant vocoder must be able to identify at least three formants for representing a speech sound. It must control the intensities of the formants.

The two main drawbacks of formant vocoders are (a) lower information rate and (b) greater distortion. At the moment of release of the stop constriction, the resonances of the vocal tract change rapidly, these changes are traditionally called format transitions. Due to these reasons, formant vocoders have not been very successful.

Cepstrum Vocoder

The cepstrum vocoder too is a parametric coder. It separates the excitation and vocal tract spectra by inverse Fourier transform of the log magnitude spectrum to produce the cepstrum of the signal. The low-frequency coefficients in the cepstrum correspond to the vocal tract spectral envelope, with the high-frequency excitation coefficients forming a periodic pulse train at multiples of the sampling period. Linear filtering is performed to separate the vocal tract cepstral coefficients from the excitation coefficients. In the receiver, the vocal tract cepstral coefficients are Fourier transformed to produce the vocal tract impulse response. By convolving this impulse response with a synthetic excitation signal (random noise or periodic pulse train), the original speech is reconstructed. It forms the concept of sub-band coding.

Voice-excited Vocoder

The category of pitch-excited and voice-excited vocoders is somewhat hybrid. Voice-excited vocoders eliminate the need for pitch extraction and voicing detection operations. This system uses a hybrid combination of PCM transmission for the low-frequency band of speech combined with channel vocoding of higher-frequency bands. A pitch signal is generated at the synthesizer by rectifying, band-pass filtering, and clipping the baseband signal, thus creating a spectrally flat signal with energy at pitch harmonics. Voice-excited vocoders have been designed for operation at 7.2–9.6 kbps, and their quality is typically superior to that obtained by the traditional pitch-excited vocoders.

Linear Prediction Coding Vocoders

Linear prediction coding (LPC) forms the basis for hybrid coders. Figure 5.30 illustrates a simple model of human speech production. Here, the lungs are replaced by a DC source, the vocal cords by an impulse generator, and the articulation tract by a linear filter system. A noise generator produces the unvoiced excitation. In practice, all sounds have a mixed excitation, which means that the excitation consists of voiced and unvoiced portions. The relation of these portions varies strongly with the sound being generated.

Based on this model, a further simplification can be made. We use a *hard* switch, which selects only between voiced and unvoiced excitation. The filter, representing the articulation tract, is a simple recursive digital filter; its resonance behaviour (frequency response) is

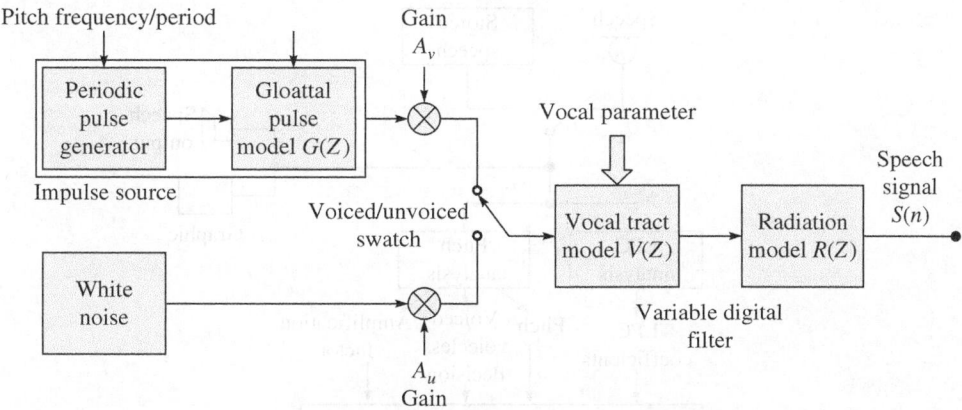

Fig. 5.38 Synthetic speech generation model

Channel, formant, cepstrum, and voice-excited vocoders do not have many applications in mobiles but they are good examples of how speech properties are utilized.

defined by a set of filter coefficients. Since the computation of the coefficients is based on the mathematical optimization procedure of LPC, they are called *LPC coefficients* and the complete model is called the *LPC vocoder*. In practice, the LPC vocoder is used for speech telephony.

Important advantages of the LPC vocoder are the manipulation facilities and the narrow analogy to human speech production. Since the main parameters of speech production, namely the pitch and the articulation characteristics, expressed by the LPC coefficients are directly accessible, the audible voice characteristics can be widely influenced. For example, the transformation of a male voice into that of a female or a child is very easy. In addition, the number of filter coefficients can be varied to influence the sound characteristics and the formant characteristics.

Today, LPC vocoders are the most popular vocoders and are used mainly in digital personal communication services. The LPC algorithm assumes that each speech sample is a linear combination of previous samples. Speech is sampled, stored, and analysed. Coefficients calculated from the sample are transmitted and processed in the receiver. With long-term correlation from samples, the receiver accurately processes and categorizes voiced and unvoiced sounds. The LPC family uses pulses from an excitation pulse generator to drive filters whose coefficients are set to match the speech sample. The excitation pulse generator differentiates the various types of LPC vocoders discussed here. LP filters are simple to implement and simulate filtering and acoustic pulses produced in the mouth and throat. LPC coder development is shown in Figs 5.38–5.41.

Regular pulse excited LPC The excitation source signal is modelled as either a periodic impulse train for voiced speech such as vowel sounds or a random noise for unvoiced speech such as consonants. A regular pulse excited (RPE) coder analyses the signal to determine if it is voiced or unvoiced. After determining the period for voiced sounds, the periodicity is encoded and the coefficient is transmitted. When the signal changes from voiced to unvoiced, a code is transmitted that stops the receiver from generating periodic pulses and makes it start generating random pulses to correspond to the noiselike nature of the fricatives.

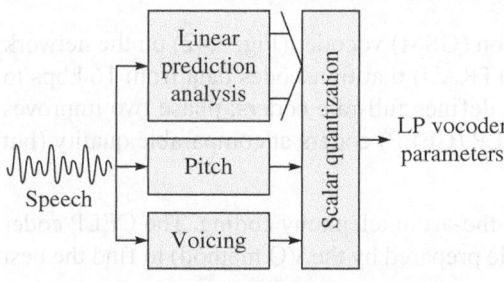

Fig. 5.39 Vocoder parameter extraction

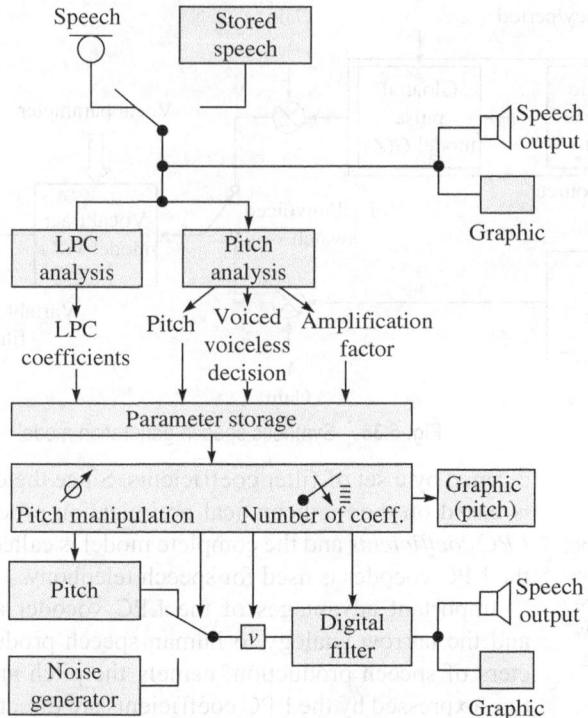

Fig. 5.40 Combining the concept of Figs 5.38 and 5.39 for an LPC

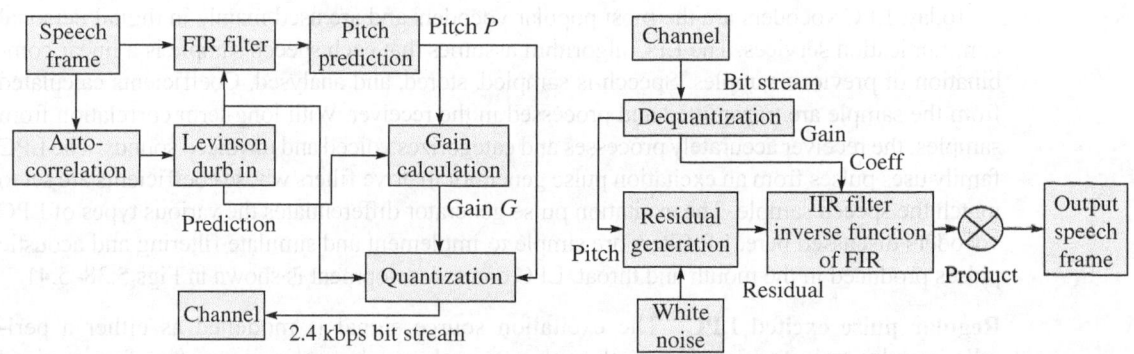

Fig. 5.41 LPC encoder and decoder using Levinson–Durbin recursion

The global system for mobile communication (GSM) vocoder (Fig. 5.42) on the network side is in the transcoder and rate adapter unit (TRAU) that transcodes data from 16 kbps to 64 kbps. Phase one of the GSM specification defines full-rate coders; phase two improves capacity by supporting half-rate code excited LP (CELP) coders at comparable quality (but requires more processing capability).

Code excited LPC These coders are state-of-the-art in telephony coding. The CELP coder is optimized by using a codebook (look-up table prepared by the VQ method) to find the best

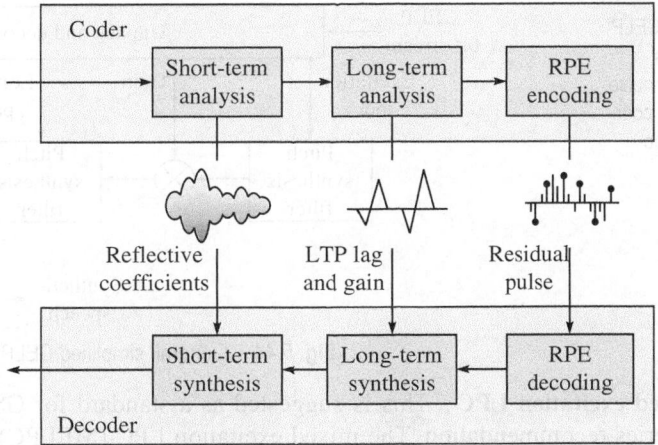

Fig. 5.42 GSM RPE coder and decoder

match for the signal. The LP residual is vector quantized, typically 40 samples (5 ms), using 1024 codebook entries. Coding requires more computations than decoding (need for codebook searches). This method reduces processing complexity, and the required data transmission rate is achieved. The resultant bit rate is about 4 kbps.

The theory says that after all the linear correlations have been removed, the prediction residuals are uncorrelated values with a Gaussian distribution, which therefore can generate the excitation codebook as a set of random numbers, where all parameters of the system are obtained by minimizing the final mean squared error. Sequential parameter estimation is non-optimal. Figures 5.43(a) and (b) give the simplified diagram of CELP coders with two different approaches, Fig. 5.44 shows the diagram of a general CELP decoder.

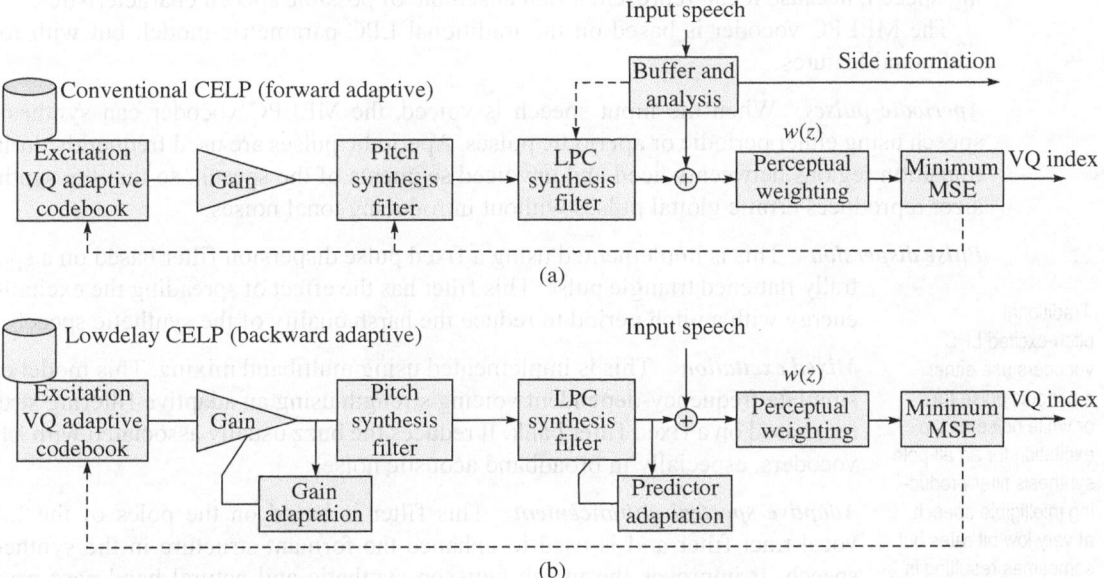

Fig. 5.43 Adaptive systems for CELP (a) Forward (b) Backward

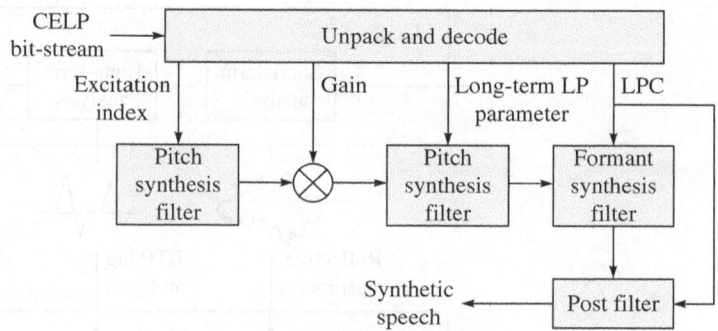

Fig. 5.44 *General, simplified CELP decoder*

RPE, MPE, and CELP are hybrid coders based on regular pulse, mixed pulse, and code excitations, respectively.

Mixed excitation LPC This is suggested as a standard for GSM and is explained in the G series recommendation. The mixed excitation LPC (MELPC) algorithm is the new 2400 bps Federal Standard hybrid speech coder. The United States Department of Defense (DoD) Digital Voice Processing Consortium (DDVPC) selected it after an extensive testing program for many years.

The selection test concentrated on four areas: intelligibility, voice quality, talker recognizability, and communicability. The selection criteria also included hardware parameters such as processing power, memory usage, and delay. MELPC was selected as the best of seven options. MELPC is robust in difficult background noise environments, such as those frequently encountered in commercial and military communication systems. It is very efficient in its computational requirements. This translates into relatively low power consumption, an important consideration for portable systems.

In LPC, there is the inability of a simple pulse train to reproduce all kinds of voiced speech. The MELPC vocoder uses a mixed excitation model that can produce natural sounding speech, because it can represent a rich ensemble of possible speech characteristics.

The MELPC vocoder is based on the traditional LPC parametric model, but with four additional features.

Aperiodic pulses When the input speech is voiced, the MELPC vocoder can synthesize speech using either periodic or aperiodic pulses. Aperiodic pulses are used frequently during transition regions between voiced and unvoiced segments of the speech, so that the synthesizer reproduces erratic glottal pulses without introducing tonal noises.

Pulse dispersion This is implemented using a fixed pulse dispersion filter based on a spectrally flattened triangle pulse. This filter has the effect of spreading the excitation energy with a pitch period to reduce the harsh quality of the synthetic speech.

Traditional pitch-excited LPC vocoders use either a periodic pulse train or white noise as the excitation for an all-pole synthesis filter producing intelligible speech at very low bit rates but sometimes resulting in noisy outcomes.

Mixed excitation This is implemented using multiband mixing. This model can simulate frequency-dependent voicing strength using an adaptive filtering structure based on a fixed filter bank. It reduces the buzz usually associated with LPC vocoders, especially in broadband acoustic noise.

Adaptive spectral enhancement This filter is based on the poles of the LPC vocal tract filter and is used to enhance the formant structure in the synthetic speech. It improves the match between synthetic and natural band-pass waveforms and introduces a more natural quality to the speech output.

QCELP vocoders   The basic CELP algorithm is one of the AbS methods widely used in the low bit rate of speech coding. Qualcomm code excited linear prediction (QCELP) is also a CELP algorithm, but it differs from the traditional CELP in that it dynamically adjusts the encoded data rate based on speech signal energy, background noise, and other speech characteristics. Therefore, the average data rate of the compressed speech is significantly reduced, while the voice quality is not affected. The QCELP vocoder consists of an encoder and decoder. When using a general DSP chip to implement a QCELP vocoder, approximately 20–25 MIPS are needed, of which 90 per cent are for the encoder and the remaining 10 per cent are for the decoder.

> QCELP vocoders are mainly used in code division multiple access (CDMA)-based systems.

5.11 SOURCE CODING IN FREQUENCY DOMAIN

Frequency domain speech or image coders take advantage of speech perception and generation models. However, the algorithms will not be totally dependent on the models used. Sub-band coding and transform coding are the two main methods used in this domain.

5.11.1 Sub-band Coding

The basic principle of sub-band coding is that most of the energy of a speech signal is contained in the lower frequencies, and therefore, one should encode the low-frequency band with more number of bits than the high-frequency band. Sub-band coding is a method in which the speech signal is subdivided into several frequency bands and each band is digitally encoded separately according to some criteria.

In other words, sub-band coding is a method of controlling and distributing quantization noise across the signal spectrum. Due to the non-linear quantization process, distortion products are typically broad in spectrum. The human ear does not detect the quantization noise equally well at all frequencies. It is, therefore, possible to achieve considerable improvement in quality by coding the signal into narrower bands.

Each sub-band is sampled at the Nyquist rate. Decimation by a factor of two is performed after frequency subdivision. By allocating a different number of bits per sample to the signal in the four sub-bands, reduction in the bit rate of digitized speech signal can be achieved; hence, effective coding with compression is achieved.

Band splitting can be done in many ways, of which two approaches are suggested here:

Approach 1   Divide the entire speech band into unequal bands using perceptual criteria such as critical bands that contribute equally to the articulation index. The method is suggested by Crochiere et al. (1976). An example is shown in Fig. 5.45.

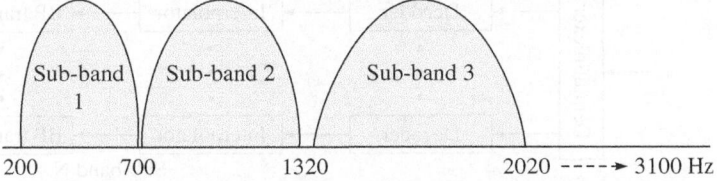

Fig. 5.45   Split of spectrum into unequal bands using perceptual criteria

> Sub-band coding of signals is an effective method for achieving bandwidth compression in a digital representation of the signal when the signal energy is concentrated in a particular region of the frequency band.

Approach 2 Split the speech such that it is divided into equal width sub-bands, and assign to each sub-band a number of bits proportional to perceptual significance while encoding them.

Instead of partitioning into equal width bands, octave band splitting, matching the first approach, is often employed. As the human ear has an exponentially decreasing sensitivity to frequency, this kind of splitting is more in tune with the perception process.

Octave Band Splitting

An example of a frequency subdivision is given here. Let us assume that the speech signal is sampled at a rate of f_s samples per second. The first frequency subdivision splits the signal spectrum into the following two equal width bands:

(a) A low-pass signal $(0 \leq f \leq f_s/4)$
(b) A high-pass signal $(f_s/4 \leq f \leq f_s/2)$

The second frequency subdivision splits the low-pass signal from the first stage into two equal bands again as follows:

(a) A low-pass signal $(0 \leq f \leq f_s/8)$
(b) A high-pass signal $(f_s/8 \leq f \leq f_s/4)$

As the third stage is accommodated, the third frequency subdivision further splits the low-pass signal from the second stage into two equal bandwidth signals. Thus, the signal is subdivided into four frequency bands, covering three octaves.

There are various methods for generating and processing sub-band signals. One way is to make a low-pass translation of the sub-band signal to zero frequency by a modulation process equivalent to single-sideband (SSB) modulation. This type of translation makes it possible to reduce the sampling rate and possesses other benefits from coding low-pass signals. An LP translator-based sub-band coder–decoder is shown in Fig. 5.46. Decimation and interpolation processes make the coding process possible.

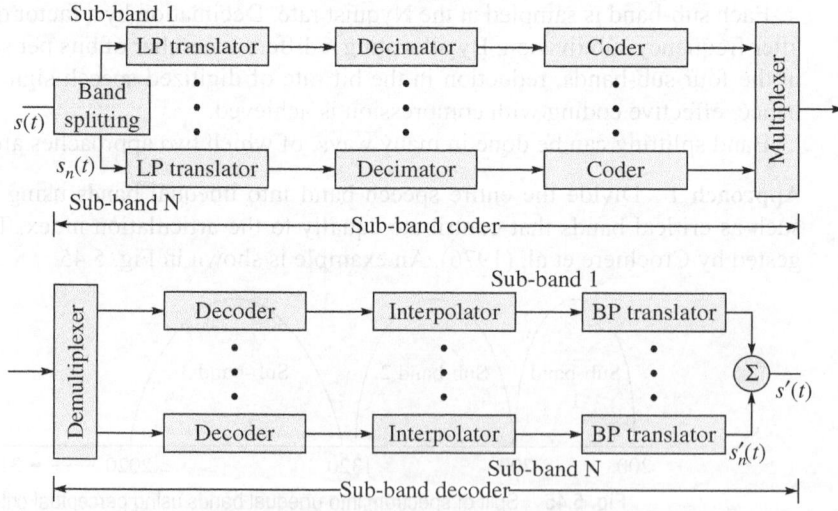

Fig. 5.46 Block diagram of sub-band coder–decoder

The input signal is filtered and translated to sub-bands. Each band has a different width. Hence, the sampling rate is decided accordingly. The resulting signal $s_n(t)$ (n stands for the nth channel of sub-bands) is modulated by a cosine wave and filtered using a *low-pass filter*. This LP-translated signal undergoes the decimation process and is then digitally encoded. Finally, all channels are multiplexed. At the receiver, the data is demultiplexed into separate channels, decoded, and band-pass translated to give the final estimate of $s'_n(t)$.

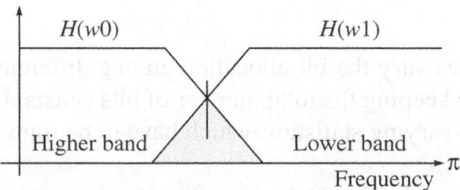

Fig. 5.47 Response characteristics of a QMF

Filter design is particularly important in achieving good performance in sub-band coding. Aliasing resulting from the decimation of the sub-band signals must be negligible. We cannot use sharp cut-off filters, as they are practically unrealizable. Therefore, the practical solution is to use a quadrature mirror filter (QMF). This filter has the frequency response characteristics as shown in the Fig. 5.47.

The synthesis method for the sub-band encoded speech signal is basically the reverse of the encoding process. The signals in adjacent low-pass and high-pass frequency bands are interpolated, filtered, and combined, as shown in Fig. 5.46. A pair of QMFs are used in the signal synthesis for each octave of the signal.

Sub-band coding is also an effective method to achieve data compression in image signal processing. It can be used for coding the speech at bit rates of about 9.6–32 kbps.

5.11.2 Transform Coding

A transform coder codes the short-time transform of a windowed sequence of samples and encodes them with the number of bits proportional to its perceptual significance. Discrete cosine transform (DCT) and wavelet transform are the two main techniques of transform coding.

Discrete Cosine Transform

The DCT is closely related to the discrete or fast Fourier transform. It is an energy compression technique and plays a role in coding audio signals and images (e.g., in the widely used standard JPEG compression). All transform operations are mathematically related. One-dimensional DCT can be represented as follows:

$$t(k) = c(k) \sum_{n=0}^{N-1} \delta(n) \cos \frac{\pi(2n+1)k}{2N} \tag{5.25}$$

where s is the array of N original values, t is the array of N transformed values, and the coefficients c are given by

$$c(0) = \sqrt{1/N}, \ c(k) = \sqrt{2/N} \quad \text{for} \quad 1 \le k \le N-1 \tag{5.26}$$

The DCT in two dimensions (for images), for a square matrix, can be written as

$$t(i,j) = c(i,j) \sum_{n=0}^{N-1} \sum_{m=0}^{N=1} \delta(m,n) \cos \frac{\pi(2m+1)i}{2N} \cos \frac{\pi(2n+1)i}{2N} \tag{5.27}$$

with an analogous notation for N, s, and t, and the $c(i,j)$ given by $c(0,j) = 1/N$, $c(i,0) = 1/N$, and $c(i,j) = 2/N$ for both i and $j \ne 0$.

The DCT has an inverse, defined by

$$\delta(n)) = \sum_{k=0}^{N=1} c(k)t(k)\cos\frac{\pi(2n+1)k}{2N} \tag{5.28}$$

for the one-dimensional case and

$$\delta(m,n) = \sum_{i=0}^{N-1}\sum_{i=0}^{N-1} c(i,j)t(i,j)\cos\frac{\pi(2m+1)i}{2N}\cos\frac{\pi(2n+1)i}{2N} \tag{5.29}$$

for two dimensions.

Most of the practical transform coding schemes vary the bit allocation among different coefficients adaptively from frame to frame while keeping the total number of bits constant. This dynamic bit allocation is controlled by time-varying statistics, which have to be transmitted as side information.

Entropy coding achieves additional compression losslessly by encoding the quantized DCT coefficients more compactly based on their statistical characteristics. The JPEG proposal specifies both Huffman coding and arithmetic coding. The baseline sequential codec uses Huffman coding, but codecs with both methods are specified for all modes of operation. Arithmetic coding, though more complex, normally achieves five to ten per cent better compression than Huffman coding.

Wavelet Transform

Along with speech compression, image and video compression are equally important nowadays. Use of wavelet transform in speech applications is less. Certain ideas of the wavelet theory appeared quite a long time ago. In recent years, there even appeared a separate scientific area to deal with wavelet analysis and the wavelet transformation theory. Unlike the Fourier transform whose basis functions are sinusoids (composition of some combination of sine and cosine signals), wavelet transforms are based on small waves, called wavelets of varying frequency and limited duration. Fourier analysis, using the Fourier transform, is a powerful tool for analysing the components of a stationary signal (a stationary signal is a signal that repeats). The Fourier transform is less useful in analysing non-stationary data, where there is no repetition within the region sampled. Wavelet transforms (of which there are, at least formally, an infinite number) allow the components of a non-stationary signal to be analysed. Wavelets also allow filters to be constructed for stationary and non-stationary signals.

Fig. 5.48 Pyramidal image processing

Wavelets are extensively used for the purposes of filtration and preprocessing data, analysis and predictions, and image recognition as well as for processing and synthesizing various signals such as speech or medical signals, for compressing and processing images, for training neural networks, and so on.

Just like quadrature mirror filtering from digital speech recognition, sub-band coding from signal processing, and pyramidal image processing (Fig. 5.48), wavelet transform is based on multiresolution theory. Many different types of wavelets are defined and all can exhibit different features (Fig. 5.49).

The mathematics of wavelets is much larger than that of the Fourier transform. In fact, the mathematics of

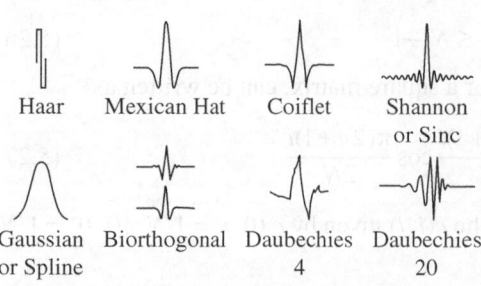

Fig. 5.49 Wavelets

wavelets encompasses the Fourier transform. Wavelets can be symmetrical, asymmetrical, or non-symmetrical. They are also grouped into those having compact domain and those not having it. Some functions have analytical form; others have the fast algorithm for calculating the wavelet transform associated with them.

Of late, the wavelet transform theory has been developing at an incredible pace. It has led to the development of fields such as orthogonal and biorthogonal wavelets, discrete wavelets, multiwavelets, and wavelet packages.

Example 5.9 Give the scheme of finding the discrete wavelet transform (DWT) of a signal $x(n)$ by passing it through a three-level filter bank of QMF.

Solution Let us first draw three levels of the filter bank.

At each level, as shown in Fig. 5.50, the signal is

decomposed into low and high frequencies. Due to the decomposition process, the input signal must be a multiple of 2^n where n is the number of levels. For example, for a signal with 16 samples, frequency range 0–f_n, and three levels of decomposition (Table 5.6), four output scales are produced (Fig. 5.51):

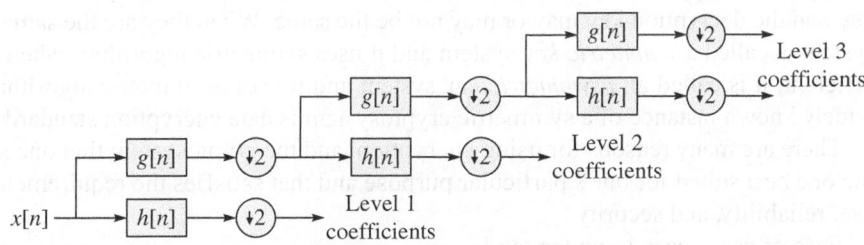

Fig. 5.50   A three-level filter bank

Table 5.6   Decomposition levels

Level	Frequencies	Samples
3	0 to $f_n/8$	4
	$f_n/8$ to $f_n/4$	4
2	$f_n/4$ to $f_n/2$	8
1	$f_n/2$ to f_n	16

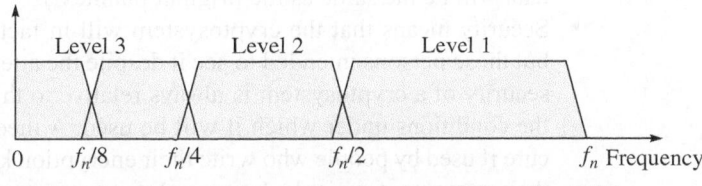

Fig. 5.51   Frequency domain representation of the DWT

5.12 ENCRYPTION AND DECRYPTION

Source-coded information may be given to the encrypter stage before passing to the channel coding part, or the encrypted data may come from the application.

Cryptography is the art and science of secret writing by encryption. The original data is to be revealed by decryption only to those who are authorized to see it and is hidden from all others.

> The most widely known instance of an asymmetric key cryptosystem is PGP (pretty good privacy).

Cryptosystem is a method to accomplish cryptography.
Cryptanalysis is the practice of defeating such attempts to hide information.
Cryptology includes both cryptography and cryptanalysis.
- The source-coded information to be hidden is called *plaintext*.
- The hidden information is called *ciphertext*.

- Encryption is any procedure to convert plaintext into ciphertext by means of an encryption engine (generally, a computer program) whose operation is determinate (the encryption method) but which functions in practice in a way dependent on a piece of information (the encryption key) that has a major effect on the output of the encryption process.
- Encryption process may or may not increase the size of the data but it simply converts the data into a secret text.
- Decryption is any procedure to convert ciphertext into plaintext.
- A cryptosystem is designed such that decryption can be accomplished only under certain conditions, which generally means only by persons in possession of both a decryption engine (these days, generally a computer program) and a particular piece of information, called the decryption key, that is supplied to the decryption engine in the process of decryption.

Encryption–decryption methods should be exactly opposite to regain the original data, and encryption–decryption keys must be as per the scheme adopted. In this process, the encryption key and the decryption key may or may not be the same. When they are the same, the cryptosystem is called a *symmetric key* system and it uses symmetric algorithm; when the keys are different, it is called an *asymmetric key* system and it uses asymmetric algorithm. The most widely known instance of a symmetric cryptosystem is data encryption standard (DES).

There are many reasons for using encryption, and the cryptosystem that one should use is the one best suited for one's particular purpose and that satisfies the requirements of ease of use, reliability, and security.

- Ease of use is easy to understand.
- Reliability means that the cryptosystem will reveal exactly the information hidden when it is needed (in other words, the ciphertext will always be recoverable and the recovered data will be the same as the original plaintext).
- Security means that the cryptosystem will in fact keep the information hidden from all but those persons intended to see it despite the attempts of others to crack the system. The security of a cryptosystem is always relative to the task it is intended to accomplish and the conditions under which it will be used. A theoretically secure system becomes insecure if used by people who write their encryption keys on pieces of paper that they stick to their computer terminals. In general, a cryptosystem can never be shown to be completely secure in practice.

MORE SOLVED EXAMPLES

Example 5.10 If a 1 V peak-to-peak sinusoidal signal is to be modulated by the PCM technique and uniform quantization is used, then for $M = 16, 32, 64, 128$, and 256, find the SQNR and comment on the result.

Solution Signal power $= (V_{pp}/2)^2 = (M \times s_{pq}/2)^2$, where s_{pq} = resolution between quantization levels

Average quantization noise power $= s_{pq}^2/12$ (derived from probability of error $= 1/s_{pq}$)

$SQNR = 3M^2$

For $M = 16$, $SNR = 28.95$ dB

$M = 32$, $SNR = 34.87$ dB

$M = 64$, $SNR = 40.89$ dB

$M = 128$ $SNR = 46.92$ dB

$M = 256$ $SNR = 52.94$ dB

Comment: As the number of levels increases, the value of the quantized sample becomes closer to the exact value: hence, the quantization noise reduces and the SNR increases.

SNR is also calculated as follows:

$SQNR = 6n + 1.76$ dB

Now, for $M = 16$, $n = 4$ or $6 \times 4 + 1.76 = 25.76$ dB

$M = 32$, $n = 5$ or $6 \times 5 + 1.76 = 31.76$ dB and so on

Example 5.11 An information of maximum frequency 3.1 kHz is transmitted over a PCM system, with certain quantization levels. The quantization distortion should not exceed ±1% of the peak value of the analog signal.

Find the following:

(a) Minimum number of bits per sample
(b) Minimum sampling rate
(c) Minimum PCM pulse transmission rate

Solution

(a) d_q = quantization distortion
V_p = peak voltage
$d_q(\text{max}) \le 0.01 V_p$
or $\quad (2V_p/M) \le 0.01 V_p$
or $\quad 2/0.01 \le 2^n$
or $\quad 200 \le 2^n$
The nearest value of n is 8.

(b) Sampling rate should be at least $2 \times 3.1 = 6.2$ ksamples/s

(c) PCM pulse transmission rate = bits/sample × samples/s = $8 \times 6200 = 49.6$ kbps

Example 5.12 The total spectrum is split into three bands with the frequency ranges as shown in Fig. 5.52.

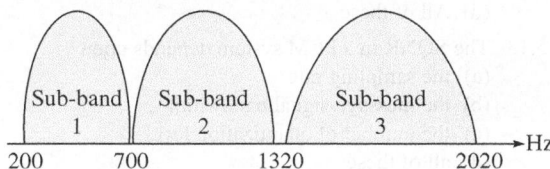

Fig. 5.52 Figure for Example 5.12

Find the final bit rate if the total band of 200–2020 Hz is converted into PCM with 7 bits/sample. If sub-band coding is used with band 1 coded with 5 bits/sample, band 2 with 6 bits/sample, and band 3 with 7 bits/sample as per the importance of the spectral components. prove that this coding technique provides compression for the same information if 10 s frame is to be transmitted.

Solution For simple PCM case, sampling frequency should be $1820 \times 2 = 3640$ samples/s

Transmission rate in this case = $7 \times 3640 = 25.48$ kbps

Bit duration is 0.039 ms within 10 s or ($10 \times 1000/0.039$) = 2,56,410 bits are required.

For sub-band coding case, sampling rates are split as per bandwidth (the total is 3640 sample/s)

Band 1 = $700 - 200 = 500$ Hz
Sampling frequency = $2 \times 500 = 1000$ sample/s
Bit rate = $5 \times 1000 = 5$ kbps
Band 2 = $1320 - 700 = 620$ Hz
Sampling frequency = $2 \times 620 = 1240$ sample/s
Bit rate = $6 \times 1240 = 7.44$ kbps
Band 3 = $2020 - 1320 = 700$ Hz
Sampling frequency = $2 \times 700 = 1400$ sample/s
Bit rate = $7 \times 1400 = 9.8$ kbps

The total bit rate is 22.24 kbps, hypothetically assuming the same bit duration for all three bands.

Bit duration = 0.044 ms or total bits within 10 s frame = 2,27,273 bits

Thus, compression is achieved to send the same information.

SUMMARY

- Analog-to-digital and digital-to-analog conversions of signals are frequently required in the wireless systems and are always lossy.
- Sampling is unavoidable for digitization process.
- A signal should be sampled at the minimum Nyquist frequency or at a higher frequency to reconstruct the signal back at the receiver; otherwise, aliasing can be observed.
- Antialiasing filters must be analog and should be used before sampling stage.
- In audio, image, and video, real-time signals will follow the envelope of stochastic nature.
- Speech or audio, image, and video signals have unique properties, which can be exploited in source coding.
- The information signal should be source coded so that the final minimum digital base is created with maximum information.

- For digitization, the analog signal must be converted into a digital form by sampling and quantization, which is a lossy process.
- The quantization process also adds quantization noise. A good quantization technique can make it possible to recover data at the receiver. It may be uniform, non-uniform, adaptive, or vector.
- Vector quantization is a statistical base approach and used in modern coder designs.
- The various methods for digitizing analog waveforms are PCM, DM, and DPCM-ADPCM.
- Information theory correlates the entropy of the information and the capacity of the channel for reliable and fruitful transmission.
- Coding methods help compress the data size. Coding redundancy and compression ratio are the measures for the amount of compression.

- Low bit rate parametric speech coders are known as vocoders. These methods take advantage of the speech signal properties.
- Sub-band coding is a frequency domain coding method. It uses speech properties and divides the bands accordingly to achieve the compression effect.

- Discrete cosine transform and wavelet transform are transform coding methods. The wavelet transform method is much suitable for coding of images for which the multiresolution analysis approach is used. (It is still a research topic and is less suitable for real-time communication due to heavy processing.)

EXERCISES

Multiple-choice Questions

5.1 A narrowband signal occupying a frequency range of nf_0 to $(n + 1)f_0$ is to be sampled. Consider n as an integer. The minimum sampling rate is
(a) $2nf_0$
(c) $2f_0$
(b) $2(n + 1)f_0$
(d) none of these

5.2 The signal that shows the aliasing effect must have a sampling rate that is
(a) less than the signal bandwidth
(b) greater than the Nyquist rate
(c) equal to the Nyquist rate
(d) all of these

5.3 In pulse modulation systems, the number of samples required to ensure no loss of information is given by
(a) Fourier transform
(c) Parseval's theorem
(b) Nyquist theorem
(d) Shannon's theorem

5.4 If p_k is the probability of a message being received or transmitted, the amount of information I_k associated in bits is given by
(a) $I_k = \log_2 p_k$
(c) $I_k = 2\log_2 p_k$
(b) $I_k = \log_2(1/p_k)$
(d) $I_k = \log_{10}(p_k)$

5.5 Entropy is basically a measure of
(a) rate of information
(b) average information
(c) disorder of information
(d) probability of information

5.6 A PCM receiver receives
(a) quantization noise
(c) interference noise
(b) channel noise
(d) all of these

5.7 In PCM for q quantizing levels, the number of pulses p in a code group is given by
(a) $\log_{10} q$ (b) $\log_2 q$ (c) $\ln q$ (d) $2\log_2 q$

5.8 The principal merit of a PCM system is its
(a) lower bandwidth
(b) lower noise
(c) lower power requirement
(d) lower cost

5.9 A compander is used in communication systems to
(a) compress the bandwidth
(b) improve the frequency response

(c) reduce the channel noise
(d) improve SNR

5.10 The channel capacity is exactly equal to the
(a) noise rate in channel
(b) bandwidth of demand
(c) amount of information per second
(d) available bandwidth

5.11 Which of the following systems is digital?
(a) PPM   (b) PWM   (c) PCM   (d) PFM

5.12 Which of the following is unlikely to happen when quantizing noise is decreased in PCM?
(a) Increase in the bandwidth
(b) Increase in the number of standard levels
(c) Increased channel noise
(d) All of these

5.13 The SQNR in a PCM system depends upon
(a) the sampling rate
(b) the message signal bandwidth
(c) the number of quantization level
(d) all of these

5.14 Which of the following systems is not digital?
(a) Differential PCM
(c) ADPCM
(b) DM
(d) PAM

5.15 Which of the following methods of source coding achieves maximum compression?
(a) A-law PCM
(c) DM
(b) μ-law PCM
(d) ADPCM

5.16 Which of the following error(s) is/are likely to occur in the differential PCM schemes?
(a) Prediction error
(c) Quantization error
(b) Compliance error
(d) All of these

5.17 For successful information transmission over a channel with capacity C, the source entropy must be
(a) less than or equal to C
(b) equal to C
(c) greater than C
(d) less than C

5.18 If a source outcome has a probability of one, the information content in that is
(a) one
(c) one-half
(b) zero
(d) none

5.19 The average information in the binary symmetric source is maximized when both outcomes with probability p and $1 - p$ are
(a) 1 and 0, respectively
(b) zero
(c) equally likely to occur
(d) maximum

5.20 A Huffman tree algorithm is used for
(a) redundancy removal
(b) coding gain
(c) variable length coding
(d) none of these

5.21 Which of the following vocoders is based on filter banks?
(a) Formant
(b) Cepstrum
(c) Channel
(d) LP

5.22 Which of the following vocoders requires past record of speech samples?
(a) Formant
(b) Cepstrum
(c) Channel
(d) LP

5.23 Which of the following is a time domain source coding method?
(a) DCT
(b) Sub-band
(c) Wavelet
(d) LPC

5.24 Which of the following methods uses the multiresolution analysis method?
(a) Sub-band coding
(b) Wavelet transform
(c) QMF filtering
(d) All of these

5.25 Which of the following vocoders is based on speech properties?
(a) Formant
(b) LPC
(c) CELP
(d) QCELP

5.26 Which of the following vocoders is based on LP vocoder?
(a) Channel
(b) Formant
(c) Voice excited
(d) MELP

5.27 Fourier transform is suitable for _____ signals, whereas wavelet transform is suitable for _____ signals.
(a) stationary, non-stationary
(b) stationary, stationary as well as non-stationary
(c) stationary as well as non-stationary, stationary
(d) non-stationary, stationary

Review Questions

5.1 Discuss the effects of undersampling and oversampling.

5.2 Why should the sampled signal be quantized?

5.3 Observe real-time signal waveforms and list their general characteristics. What is the nature of a speech signal?

5.4 List the different quantization methods and compare them. If we have to choose any one method for high bit rate communication, which will be the most suitable one?

5.5 How does the VQ method differ from conventional methods?

5.6 Explain how the concept of VQ is equally applied in case of intervals, polygons, or polytops. What will be the code vector in each case?

5.7 How can a VQ exhibit compression as well?

5.8 What are the drawbacks of PCM that can be removed by differential methods? What are the drawbacks of differential PCM methods?

5.9 What are the waveform coding methods by which information transmission efficiency can be improved?

5.10 A binary source $S = \{s1, s2\}$ has $P = \{p, (1 - p)\}$. Plot the entropy of the source versus p as p varies from 0 to 1 and comment on the result.

5.11 What is the difference between waveform coding and source coding? Give one example of each.

5.12 Why are vocoders designed separately and the conventional PCM is not used?

5.13 List the various vocoders and the speech property exploited by them.

5.14 How do DCT and DFT differ?

5.15 How can you achieve compression by sub-band coding?

5.16 What do you mean by multiresolution analysis? How is this method used in wavelet transform?

Numerical Problems

5.1 What is the least sampling rate required to sample the signal $f(t) = \sin^3(\omega_0 t)$? Show graphically the effect caused by a reduction of the sampling rate below the Nyquist rate.

5.2 The spectral range of a modulated signal extends from 1.0 MHz to 1.2 MHz. Find the minimum sampling rate and maximum sampling time.

5.3 Find the number of quantization levels for sending 8 bits/sample. If more bits are used for encoding purpose, what will be the effect on the bandwidth? If fewer bits are used, comment on the reconstruction of the signal.

5.4 Consider an audio signal with spectral components in the range 300 Hz–3 kHz. Assume that a sampling rate of 8 kHz will be used to generate a PCM signal.

(a) For $SNR = 30$ dB, what is number of uniform quantization levels needed?

(b) What is the required data rate?

5.5 The bandwidth of a television video plus audio signal is 5 MHz. If the signal is quantized at 512 levels, determine the data rate of the resulting PCM signal. Assume that the signal is sampled at a rate of 20% above the Nyquist rate.

5.6 In a PCM conversion, an analog signal is sampled at the Nyquist sampling rate $1/T_s$, quantized using L quantization levels, and converted into binary digital form for sending over the channel. Prove that the bit duration T of the transmitted signal must satisfy the following condition:

$$T \leq T_s / \log_2 L$$

Also comment on this.

5.7 A voice signal bandwidth is between 300 Hz and 3.1 kHz, considering significant frequency components. Assume 8 ksamples/s is used to generate a PCM signal and if peak signal power to average quantization noise power requirement is 32 dB.

(a) What are the minimum number of bits per sample and minimum number of uniform quantization levels required?

(b) What will be the final bit rate?

(c) What will be the transmission bandwidth to make the PCM signal detectable?

5.8 In an eight-bit uniform quantizer with a dynamic range of 2 V and a span of -1 V to 1 V, what may be the step size? Find the SQNR for the sinusoidal signal that spans the full dynamic range. [Hint:

$$SQNR = 3M^2 \frac{\overline{S^2(t)}}{S_p^2}, \text{ where } \overline{s^2(t)} \text{ is the signal power,}$$

M is the number of quantization levels, and S_p^2 is the peak quantization level of the uniform quantizer]

5.9 A signal band-limited to 100 kHz is sampled at a rate of 50% higher than the Nyquist rate and quantized, so that 8 bits/sample can be assigned. A μ-law quantizer with $\mu = 255$ is used. Determine SQNR. If SQNR found is to be increased at least by 10 dB, how will you achieve the same?

5.10 Consider a source $S = \{s_1, s_2, s_3, s_4\}$ with $P = \{1/2, 1/4, 1/8, 1/8\}$. Find the self-information of each message and entropy of source S.

5.11 Given the messages x_1, x_2, x_3, x_4, x_5, and x_6 with respective probabilities of 0.4, 0.2, 0.2, 0.1, 0.06, and 0.04, construct the Huffman code. Determine the efficiency and redundancy of the code.

5.12 A source emits eight messages with probabilities 0.5, 0.25, 0.1, 0.07, 0.05, 0.02, 0.005, and 0.005. Find the entropy of the source. Obtain the Huffman code and find the average length of the code. Determine the efficiency and redundancy of the code.

5.13 Following is the data given:
$S = \{s_1, s_2, s_3, s_4, s_5, s_6, s_7, s_8, s_9, s_{10}\}$
$P(S) = \{0.18, 0.17, 0.16, 0.15, 0.10, 0.08, 0.05, 0.05, 0.04, 0.02\}$
The encoding alphabet is $\{0, 1, 2, 3\}$.
Apply Huffman coding and determine coding gain and redundancy.

5.14 In a delta modulation system, the input is a 8 kHz sine wave with 1 V peak-to-peak amplitude. The sampling rate is 10 times the Nyquist rate. Find the step size required to prevent the slope overload condition and to minimize granular noise. What is the PSD for granular noise? Find the SNR of this DM system.

5.15 Over a period of 15 samples, the sample amplitudes are 1.5, 2, 2.5, 2, 1.5, 1, 0.5, 1, ... Find the spectral flatness measure.

6 Channel Coding Techniques

Theme of the Chapter

Source coding represents the source information with the minimum number of symbols by systematically removing redundant bits. When a code is transmitted over a channel in the presence of noise, errors are likely to occur. Hence, channel coding adds redundant bits to the source information to minimize the probability of errors. It is very important over wireless channels. It reduces the bit error rate in the final reception, thereby improving the quality of reception. With this aim, the various channel coding schemes are highlighted in this chapter. Other related terms important for performance comparison for different channel coding schemes are also discussed. Channel coding in general can be done by error-detecting or error-correcting codes. The usefulness of error-correcting codes for a wireless digital communication system is also dealt with in the chapter. Basically, coding methods are based on mathematical or logical operations and their corresponding decoding.

Key Topics

- Channel coding and decoding
- Shannon limit
- Trade-offs in error detection and error correction
- Channel capacity and coding gain
- Error-detecting and error-correcting capabilities
- Various block codes, such as Hamming, Bose–Chaudhuri–Hocquenghem, and Reed–Solomon codes
- Hard and soft decoding methods
- Convolutional coding and Viterbi decoding
- Interleaver, puncture coding, and turbo coding

6.1 CHANNEL CODING AND DECODING

Coding theory, in general, deals with the transmission of data over noisy channels in the best-optimized way by adopting various source and channel coding–decoding schemes. Coding theory was formulated in 1948 by Claude Shannon, which appeared in his paper *A Mathematical Theory of Communication*. He studied entropy and source coding, capacity of the channel, and so on. Later, Huffman and Hamming also worked on this theory. In the last 50 years, due to the growth of digital communication, coding theory has grown into a discipline with applications in almost every area of communication, such as satellite, cellular telephony, internet, digital recording, and data storage.

Note: Channel coding refers to the class of signal transformations designed to improve communication performance by enabling the transmitted signals to better withstand the effects of various channel impairments, such as noise, interference, and fading.

The existence of noise as well as the variation in propagation characteristics of different communication channels makes the correct handling of errors vital. Some of the ways in which errors are handled include error detection, error correction, data acknowledgment, and data resends. Error coding (for detection and correction purpose) can be classified into two areas (as in source coding):

(a) Waveform (or signal) coding

(b) Structured sequences (or structured redundancy) coding

Data acknowledgement and resends are parts of error control over bidirectional links. *Waveform coding* deals with the transformation of waveforms into *better* waveforms to make the detection process less subject to errors. This approach is mainly suitable for signalling purposes, for example, *m*-ary signalling, antipodal signalling, orthogonal signalling, and trellis coded modulation (TCM). *Line coding* is a kind of waveform coding method for handling channel impairments, synchronization between transmitter and receiver, maintaining power level, and so on. Frequently, line coding is treated as a modulation technique, and hence, it is discussed in Chapter 7.

Structured sequences transform data sequences into *better* sequences having structured redundancy (addition of redundant bits). The redundant bits can be used for detecting and/or correcting errors. Examples include cyclic redundancy check (CRC), block coding, convolutional coding, and turbo coding. The focus in this chapter is on this type of coding.

There are two basic approaches towards controlling errors:

(a) Automatic repeat request (ARQ)

(b) Forward error correction (FEC)

Frame error rate (FER) is the important parameter in ARQ, while bit error rate (BER) is the important parameter in FEC.

Automatic repeat request The ARQ, or error detection, error correction (sometimes), and retransmission, utilizes parity bits or redundant bits for checksum added to the data to detect an error. The errors may be corrected sometimes when single-bit errors or error bit positions can be identified. The receiving terminal does not always attempt to correct the error; rather, it requests the transmitter to retransmit the data. It is clearly noticeable that a two-way link is required for such a dialogue between the transmitter and the receiver. Thus, it controls the error by allowing retransmissions rather than performing error correction every time. This is possible only in packet-based systems with non-real-time data. For real-time transmissions such as multimedia, retransmissions are difficult; hence, structured redundancy coding is a better option in such cases.

ARQ is also known as backward error correction (BEC) and requires at least a half duplex and preferably a full duplex link between the sender and the receiver.

The techniques for error detection, such as performing parity check or CRC on the data, can guarantee detection of errors with a greater certainty on a wired link than wireless system with normal BER (bit error rate). This detection process can then be used to send an acknowledgement (ACK) or a failed/not acknowledgement (NAK) back to the transmitting node to request retransmission. Typically, a system based on a combination of error detection and acknowledgements can perform well in a stationary environment where data latency is not a problem.

Modern wireless systems use a combination of error-detecting codes and error-correcting codes.

The following are the three popular types of ARQ schemes:

(a) Stop and wait

(b) Go-back-n

(c) Selective repeat

The *stop and wait* scheme is the simplest scheme. The transmitter transmits a frame and waits until the receiver sends the acknowledgement. Then, it sends the next frame and the process continues. This achieves error and flow control simultaneously.

In the *go-back-n* strategy, a bidirectional link is required. The transmitter transmits the frames on the forward link while receiving the ACKs on the reverse link. If an error is found in a frame, the receiver sends an NAK frame with that frame number. Though the transmitter has continued the transmissions, it will go back to the faulty frame and will repeat transmissions from that frame onwards (Fig. 6.1). In selective repeat, if NAK is received for a particular frame, that frame alone will be resent by the transmitter, though it may be out of turn (Fig. 6.2). In Figs 6.1 and 6.2, frame 2 is shown to be received with error.

The principle advantages of this type of system are the simplicity and the high data throughput relative to the data rate for good signal-to-noise ratio (SNR) conditions. This type of system does not tend to work well in poor SNR conditions due to the large number of re-sends necessary. In order to improve the typical BER of a mobile communication system, error detection and correction are usually employed. This involves the transmission of some additional data to allow the detection, identification, and correction of errors. This is known as FEC.

Forward error correction   This method uses redundant bits added systematically to detect and correct errors independently. FEC schemes do not send ACK/NAK signals and are more useful in the wireless environment. In this case, a one-way link will suffice for proper communication. Error handling is done by the receiver if the appropriate method according to the application is selected.

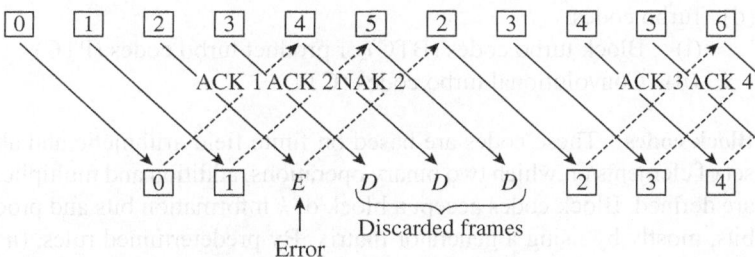

Fig. 6.1   Go-back-n scheme of ARQ

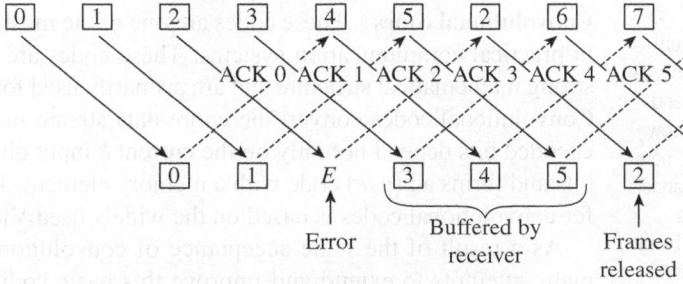

Fig. 6.2   Selective repeat scheme of ARQ

In short, there are two types of codes in general:

(a) *Error-detecting codes*: Examples include parity check, longitudinal redundancy check (LRC), vertical redundancy check (VRC), and CRC.

(b) *Error-correcting codes*: These are generally used over a wireless medium or time-dispersive channel. Examples are block codes [such as Hamming code, Hadamard code, Golay code, cyclic code, Bose–Chaudhuri–Hocquenghem (BCH code), and Reed–Solomon (RS) code], convolutional codes, and turbo codes.

Both schemes commonly use the following basic concept. At the transmitter side, an encoder adds redundancy to the data. Then, at the receiver, a decoder exploits the redundancy in such a way that a reasonable number of channel errors can be detected or corrected.

A binary encoder takes in k bits at a time and produces an output (or code word) of n bits, where $n > k$. While there are 2^n possible sequences of n bits, only a small subset of them, 2^k to be exact, will be valid code words. The ratio k/n is called the *code rate*.

Lower code rates can generally correct more channel errors than higher code rates and thus, are more *energy efficient*. However, higher code rates are more *bandwidth efficient* than lower code rates because the amount of overhead (in the form of parity bits) is less. Thus, selection of the code rate involves a trade-off between energy efficiency and bandwidth efficiency.

Error-correcting codes used in wireless communication are broadly categorized into the following three types:

(a) Block codes:
   (i)   Hamming code
   (ii) BCH codes
   (iii) RS codes
(b) Convolutional codes
(c) Trellis coded modulation
(d) Turbo codes:
   (i) Block turbo codes (BTC) or product turbo codes (PTC)
   (ii) Convolutional turbo codes (CTC)

Block codes   These codes are based on finite field arithmetic and abstract algebra. A finite set of elements on which two binary operations, addition and multiplication, can be performed are defined. Block codes accept a block of k information bits and produce a block of n coded bits, mostly by using a generator matrix. By predetermined rules, $(n - k)$ redundant bits are added to the k information bits to form the n coded bits. Commonly, these codes are referred to as (n, k) block codes. The same basic concept is applied to different block coding schemes.

Most communication engineers try to minimize the transmission bandwidth, because narrow bandwidths permit more communication channels to be packed into a defined frequency band and have the advantages of a good SNR.

Convolutional codes   These codes are one of the most widely used channel codes in practical communication systems. These codes are developed with a separate strong mathematical structure and are primarily used for real-time error correction. Convolutional codes convert the entire data stream into a single code word. The encoded bits depend not only on the current k input bits but also on the past input bits and forms an (n, k) code with a memory element. The main decoding strategy for convolutional codes is based on the widely used Viterbi algorithm.

As a result of the wide acceptance of convolutional codes, there have been many attempts to extend and improve this basic coding scheme. This advancement resulted in two new coding schemes, namely TCM and *turbo codes*.

Random coding of long block lengths may also perform close to channel capacity, and this forms the basis of the turbo code, which however is very hard to decode.

Trellis coded modulation   This scheme adds redundancy by combining coding and modulation into a single operation (as the name implies), and it is categorized as a *waveform coder*. The unique advantage of TCM is that there is no reduction in the data rate or expansion in the bandwidth, as required by most of the other coding schemes.

Turbo codes   These codes are near channel capacity error-correcting codes. These codes are able to transmit information across the channel with an arbitrary low (approaching zero) BER. It has been shown that a turbo code can achieve performance within 1 dB of channel capacity. Without doubt, the performance of a turbo code is partly due to the random interleaver used to give the turbo code a *random* appearance. However, one main advantage of a turbo code is that there is enough code structure (from the convolutional codes) to decode it efficiently.

All these channel coding techniques are discussed in this chapter; however, an introduction is first provided to the general concepts of channel coding.

6.2 CHANNEL CAPACITY

After the channel coding stage, the modulated signal with the final transmission bandwidth has to be exposed to the channel, and hence, channel capacity is an important concept in this chapter. In Chapters 1 and 5, channel capacity was introduced as per requirement. Here, it is explained in a better way to understand its practical significance. Shannon showed that the system capacity C (maximum data rate in bits per second) of a channel perturbed by additive white Gaussian noise (AWGN) is a function of the average received signal power S, the average AWGN power N, and the bandwidth W. The only option available to increase a channel's capacity is to increase either the bandwidth (W) or the SNR. The capacity relationship (popularly known as *Shannon–Hartley theorem*) can be stated as

$$C = W \log_2 \left(1 + \frac{S}{N} \right) \tag{6.1}$$

where W is in hertz and the capacity C is given in bits per second. It is theoretically possible to transmit information over such a channel at any rate R, where $R \leq C$, with an arbitrarily small error probability by using a sufficiently complicated coding scheme. For an information rate $R > C$, it is not possible to find a code that can achieve an arbitrarily small error probability. An increase in the SNR requires an increase in transmitted power, as the noise within the channel is beyond our control. Thus, we can trade either power or bandwidth to achieve a specified channel data rate.

Like binary symmetric source, the *binary symmetric channel* is a channel with binary input and output. Associated with each output is a probability p that the output is correct and

Note: One of the well-known results of information theory is the Shannon's channel coding theorem, which states that for a given channel with capacity C, there exists a code that will permit error-free transmission across the channel at a rate R, provided $R \leq C$. Equality is achieved only when the SNR is infinite.

a probability $(1 - p)$ that it is not. For such a channel, the channel capacity turns out to be

$$C = 1 + p \log_2(p) + (1 - p) \log_2(1 - p) \tag{6.2}$$

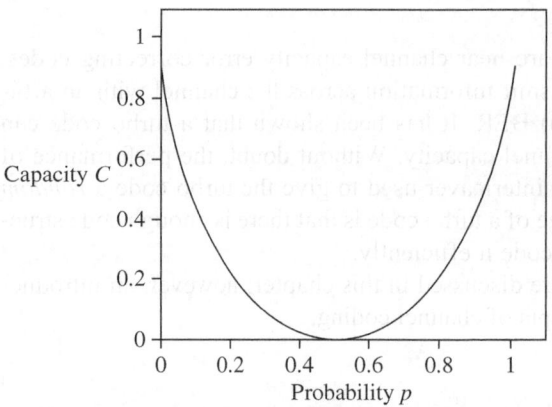

Fig. 6.3 Capacity of a binary symmetric channel

Here, p is the bit error probability. If $p = 0$, then $C = 1$. If $p = 0.5$, then $C = 0$. Figure 6.3 shows the diagrammatic representation of the capacity of a binary symmetric channel. Thus, if there is equal probability of receiving a 1 or 0, irrespective of the signal sent, the channel is completely unreliable and no message can be sent across it.

The graphs in Fig. 5.27 of Chapter 5 and Fig. 6.3 are exactly opposite. For a channel of bandwidth W, we can transmit at most $2W$ changes per second. Thus, the maximum capacity in bits per second is $2W$. For binary channels, we have

$$C = 2W[1 + p \log_2(p) + (1 - p) \log_2(1 - p)] \tag{6.3}$$

Thus, for a binary channel, the maximum bit rate is $2W$. We note that the capacity is always less than the maximum bit rate. The final data rate R, or the information rate, describes the rate of transfer of data bits across the channel. Theoretically, we can write

$$W \geq C \geq R \tag{6.4}$$

[Instead of $H(X)$, R is used here purposely (we leave it to the readers to find the reason).]
As a matter of practical fact,

$$W > C > R \tag{6.5}$$

If the source is optimally coded, we can rephrase the channel coding theorem as follows: *A source of information with entropy $H(X)$ can be transmitted error free over a channel provided $H(X) \leq C$.*

Note: The capacity of a binary channel is much less than that calculated using the Hartley–Shannon Law (Eq. 6.1). The reason is that the equation applies to the systems whose outputs may take any values. We use the systems obeying the equation because they are technically convenient and not because they are desirable.

6.3 SHANNON LIMIT

Claude Shannon worked extensively on the fundamental information transmission capacity of a communication channel. He showed that 'there exists a limiting value of E_b/N_o below which there can be no error-free communication at any information rate'. This limiting value of E_b/N_o is called the *Shannon limit*. This theory is applicable to both wired and wireless channels. Here, E_b is the bit energy and N_o is the noise power per hertz or noise density.

Shannon's work showed that the values of S, N, and W set a limit on the transmission rate but not on error probability; for the case where the transmission bit rate is equal to the channel capacity, $R = C$ (limiting value), we can use the identity

$$\frac{E_b}{N_o} = \left(\frac{S \cdot T_b}{N/W}\right) = \frac{\left(S/R\right)}{\left(N/W\right)} \tag{6.6}$$

Simplifying,

$$\frac{E_b}{N_o} = \frac{S}{N}\left(W/R\right) \tag{6.7}$$

Substituting $R = C$, it can be derived

$$\frac{E_b}{N_o} = \frac{S \cdot W}{(N \cdot C)} \tag{6.8}$$

Now, using Eqs (6.1) and (6.8),

$$\frac{C}{W} = \log_2\left[1 + \frac{E_b}{N_o}\left(\frac{C}{W}\right)\right] \tag{6.9}$$

where C/W denotes the *bandwidth efficiency*. If C/W tends to 0, using identity

$$\lim(1+x)^{1/x} = e, x \to 0$$

we get

$$E_b/N_o = -1.6\,\text{dB} \tag{6.10}$$

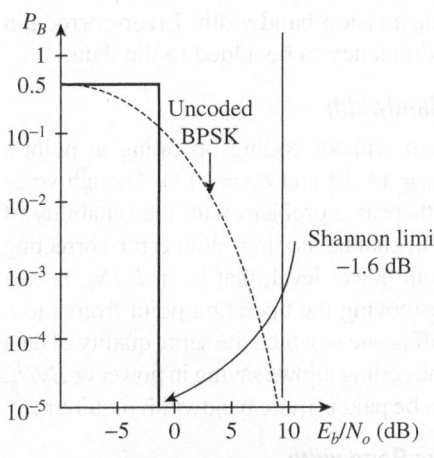

Fig. 6.4 Shannon limit

Equation (6.10) is popularly known as the *Shannon limit*, below which there can be no error-free communication at any information rate.

In Fig. 6.4, the Shannon limit is shown in the undashed bit error probability P_B versus E_b/N_o curve. The curve is discontinuous, going from a value of $P_B = 0.5$ to $P_B = 0$ at $E_b/N_o = -1.6$ dB. For a bit error probability of 10^{-5}, the optimum uncoded binary phase shift keying (BPSK) modulation requires an E_b/N_o of typically 9.6 dB. Therefore, for this case, Shannon's work promised the existence of a theoretical

---

CODING PARADOX

Shannon showed that capacity could be achieved with bit error probability approaching zero by a *completely random code*, (i.e., a randomly chosen mapping set of code words) only when the block length n approaches infinity. However, random codes are not practically feasible.

In order to be able to encode and decode with reasonable complexity, codes must possess some sort of structure. Unfortunately, *structured codes perform considerably worse than random codes*. This is the basis of *coding paradox*. Surprisingly, for practical purposes, the capacity limit applied was a few decibels lower than the Shannon limit, called the *cut-off rate bound*. Finally, turbo codes have become popular for achieving the Shannon limit.

performance improvement of 11.2 dB [9.6 − (−1.6) = 11.2 dB] over the performance of an optimum uncoded binary modulation, through the use of coding techniques. Today, most of that promised improvement (as much as 10 dB) is realizable with turbo codes.

6.4 CHANNEL CODING AND TRADE-OFFS

Error-correcting codes can be regarded as a vehicle for effecting various system trade-offs. Figure 6.5 compares two curves depicting *bit error performance* (P_B) versus *bit energy-to-noise density ratio* (E_b/N_o) (which is the same as the SNR; E_b/N_o is normally a parameter for a digital signal representing the signal strength). One curve represents a typical modulation scheme without coding; the other represents the same modulation with coding.

Though channel coding is incorporated in wireless links to improve the BER, there exists a few trade-offs, as described here, especially in terms of transmission bandwidth.

Error Performance vs Bandwidth

Consider a communication system operating at point *p* in Fig. 6.5 ($P_B = 10^{-2}$ and $E_b/N_o =$ 8 dB) where voice quality is not good; that is, the bit error probability (P_B) has to be reduced, say, from 10^{-2} to 10^{-4}. Without coding, the path is to be followed through *p* to *q* with increased E_b/N_o. However, with coding, keeping the same E_b/N_o, the BER can be reduced to the desired level moving from *p* to *r*. Channel coding requires new components (encoder and decoder) to be added, and the price to be paid is more transmission bandwidth. Error-correction coding needs redundancy to be added to the data.

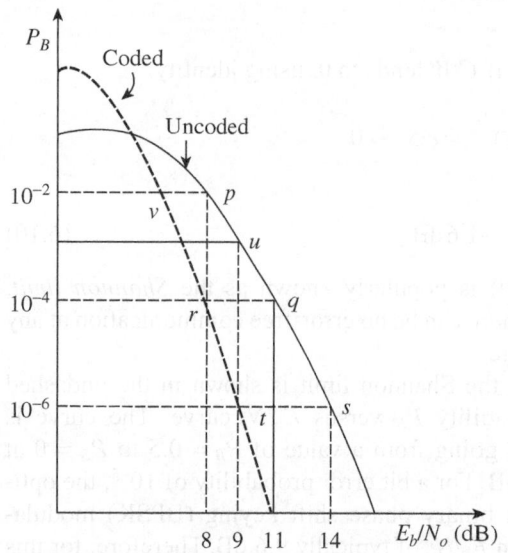

Fig. 6.5 Comparison of typical coded versus uncoded error performance

Power versus Bandwidth

Consider a system without coding operating at point *s* in Fig. 6.5 ($E_b/N_o = 14$ dB and $P_B = 10^{-6}$). Though voice quality is good, there is a problem with the reliability of the system without channel coding. With error-correcting codes, reduction in power level, that is, in E_b/N_o, is easily achievable by moving the operating point from *s* to *t*. Thus, the trade-off is one in which the same quality of data is achieved but the coding allows saving in power or E_b/N_o. Again, the cost to be paid is more bandwidth requirement.

Data Rate versus Bandwidth

Now, consider the system is operating at the same point *s*. Assume that there is no problem with the data quality and no particular need for reducing power. However, the requirement is increased data rate. Now, there is a well-known relation of data rate *R* with E_b/N_o:

$$\frac{E_b}{N_o} = \frac{P_r}{N_o}\left(\frac{1}{R}\right) \tag{6.11}$$

To detect, and possibly correct errors, the channel code sequence must be longer than the source sequence due to added redundancy; hence, the code rate is always less than one.

where P_r is the received power. If we do nothing to the system except increasing the data rate *R*, this expression shows that the received E_b/N_o would decrease, and in Fig. 6.5, the operating point would move upwards from point *s* to, say, some point u. Now, envision walking down the vertical line to point t on the curve that represents coded modulation. Increasing the data rate degrades the quality of the data. However, the use of error-correction coding brings back the same quality at the same power level (P_r/N_o). The value of E_b/N_o is reduced, but the code facilitates getting the same error probability with a lower E_b/N_o. The price we pay for getting this higher data rate is an increased bandwidth. Similar is the case with the capacity and bandwidth trade-off.

It should be noted that in each of the aforementioned trade-off examples, a *traditional* code involving redundant bits and faster signalling for a real-time communication system has been assumed; hence, in each case, the cost is *expanded bandwidth*. However, TCM does not require faster signalling or expanded bandwidth for real-time systems.

> *Note*: It is clear that for non-real-time communication systems, error-correction coding can be used with a somewhat different trade-off. In this case, it is possible to obtain an improved bit error probability or reduced power (similar to the first two trade-offs mentioned) but only by paying the price of *delay* instead of *bandwidth*.

6.5 PERFORMANCE TERMINOLOGIES

We will discuss various performance parameters that are related to the channel coded technique or required to evaluate the performance of the link due to the channel coding stage.

Coding Gain

For a given *bit error probability*, coding gain is defined as the reduction in E_b/N_o that can be realized through the use of the code. Coding gain, generally expressed in decibels, is calculated as

$$G(\text{dB}) = \left(\frac{E_b}{N_o}\right)_{uncoded} (\text{dB}) - \left(\frac{E_b}{N_o}\right)_{coded} (\text{dB}) \qquad (6.12)$$

Code Rate

The rate of a channel code is the average ratio of the source sequence length to the channel code length, as mentioned previously.

Hamming Weight

The Hamming weight $W(U)$ of a code word U is defined to be the number of non-zero elements in U. For a binary vector, this is equivalent to the number of 1's in the vector. For example, if $U = 10010101$, then $W(U) = 4$.

Hamming Distance

A good channel code is designed such that even if a few errors occur in transmission, the output can still be identified with the correct input. This is possible because the output, although incorrect, is sufficiently similar to the input to be recognizable. The idea of similarity is made more firm by the definition of a *Hamming distance*. Let x and y be two binary sequences of the same length. The Hamming distance between these two codes is the number of symbols or bits that disagree. Suppose the code x is transmitted over the channel. Due to errors, y is received. The decoder will assign to y the code x that minimizes the Hamming distance between x and y.

In the CDMA scheme, as all the users share the same spectrum, each user interferes with each of the other users in the same or nearby cells. Hence, the capacity is inversely proportional to E_b/N_o. It means lower E_b/N_o and, therefore, more capacity; the codes achieve a reduction in each user's power, which in turn allows for more number of users.

If the transmitter sends 10000 but there is a single-bit error and the receiver gets 10001, it can be seen that the *nearest* code word is in fact 10000 and so the correct code word is found. It can be shown that to detect *n*-bit errors, a coding scheme requires the use of code words with a Hamming distance of at least $n + 1$. It can also be shown that correcting *n*-bit errors requires a coding scheme with at least a Hamming distance of $2n + 1$ between the code words. By designing a good code, we try to ensure that the Hamming distance between possible code words *x* is larger than the Hamming distance arising from the errors. The Hamming distance between two code words *U* and *V*, denoted by $d(U, V)$, is defined to be the number of elements in which they differ.

Example 6.1 If $U = 1001000101$ and $V = 1011001101$, find the Hamming distance between two code words.

Solution Comparing the sequences, the bit positions that differ in terms of bits are shown in bold.

$U = 1001000101$

$V = 1011001101$

Thus, the Hamming distance is

$d(U, V) = 2$

The Hamming distance between two code words is equal to the Hamming weight of their sum. In mathematical notation,

$$d(U,V) = W(U + V) \tag{6.13}$$

Example 6.2 Prove the following equation for the values of *U* and *V* given in Example 6.1:

$$d(U,V) = W(U + V)$$

Solution Adding *U* and *V* means an exclusive OR operation.

$U = 1001000101$

$V = 1011001101$

———

0010001000

Now, the Hamming weight of this sum is equal to the number of 1's, which is two.

Thus, $d(U, V) = 2$

Minimum Distance of Linear Code

The minimum distance of a linear code will indirectly give the code word selection criteria. Using the property of linear codes, if *U* and *V* are code words, then $Z = U + V$ must also be a code word. Hence, the distance between two code words is equal to the weight of a third code word; that is,

$$d(U,V) = W(U + V) = W(Z) \tag{6.14}$$

> All error patterns can never be correctly decoded and corrected. There are some restrictions and conditions that decide the error-detecting and error-correcting capabilities.

There are a number of possible values of *U* and *V*. Thus, the minimum distance of a linear code can be ascertained without examining the distance between all combinations of code word pairs. We only need to examine the weight of each code word (excluding the all-zeros code word) in the subspace; the minimum weight of a code word in a subspace corresponds to the minimum distance d_{min} between two code words in the subspace.

Equivalently, the minimum distance, d_{min}, corresponds to the smallest of the set of distances between the all-zeros code word and all other code words.

If the code used is binary, the distance is known as the Hamming distance. Now, the *error-correcting capability* t_c of a code is defined as the maximum number of guaranteed correctable errors per code word and is given as

$$t_c = \left\lfloor \frac{d_{\min} - 1}{2} \right\rfloor \qquad (6.15)$$

where $\lfloor x \rfloor$ means the largest integer not to exceed x. The *error-detecting capability* of a code t_d is defined as the maximum number of guaranteed detectable errors per code word prior to correction and is given as

$$t_d = d_{\min} - 1 \qquad (6.16)$$

Therefore, the number of errors that can be detected and corrected depends on the minimum Hamming distance (or *free distance*).

6.6 STATISTICAL CONCEPTS FOR DECODING

The statistical concepts are necessary to properly understand Viterbi and turbo decoding.

Maximum Likelihood Decoding

If all message sequences are equally likely (equiprobable), *maximum likelihood decoding* achieves the minimum probability of error by choosing the maximum out of all possible likelihood functions $P(Y|S_m)$, for all S_m, where Y is the received sequence and S_m is one of the possible transmitted sequences.

The maximum likelihood decoder is an optimal decoder that chooses S_m if

$$P(Y|S_m) = \max P(Y|S_m) \text{ over all } S_m \qquad (6.17)$$

For example, let us say that there are only two equally likely possible sequences, $s_1(t)$ and $s_2(t)$, that might have been transmitted. Therefore, from the received signal, we need to decide whether $s_1(t)$ was transmitted [if $p(Y|s_1) > p(Y|s_2)$] or whether $s_2(t)$ was transmitted (maximum likelihood decision).

Hard Decision Decoding versus Soft Decision Decoding

Hard decision decoding Consider a two-level signal for a digital system. One bit can be transmitted per symbol, with, say, a 0 being sent as −1 V and a 1 as +1 V. At the receiver, assuming that the gain is correct, we should expect to receive a signal always in the vicinity of either −1 V or +1 V, depending on whether a 0 or a 1 was transmitted, the departure from the exact values ±1 V being caused by the inevitable noise added in transmission.

A simple receiver might operate according to the rule that negative signals should be decoded as 0 and positive ones as 1. This is an example of a *hard decision*, with 0 V as the decision boundary. However, noise usually has a continuous distribution such as Gaussian and may occasionally have a large amplitude, although with a lower probability than for smaller values. Thus, if, say, +0.5V is received, it most probably means that a 1 was transmitted, but there is a smaller yet still finite probability that actually 0 was sent.

For multilevel transmission, hard decision decoding or soft decision Viterbi decoding uses a quantization process.

Soft decision decoding Soft decision Viterbi decoders use a quantization process. These maintain a history of many possible transmitted sequences, building up a view of their relative likelihoods and finally selecting the value 0 or 1 for

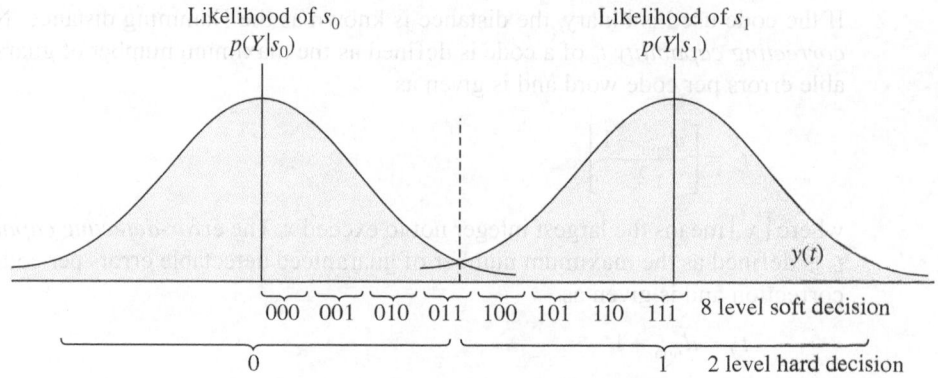

Fig. 6.6 Hard and soft decoding decisions

each bit according to which the *maximum likelihood* is decided. For convenience, a Viterbi decoder adds *log-likelihoods* (rather than multiplying probabilities) to accumulate the likelihood of each possible sequence. It can be shown that in the case of BPSK or quadrature phase shift keying (QPSK) the appropriate log likelihood measure or *metric* of the certainty of each decision is simply proportional to the distance from the decision boundary on constellation. (The slope of this linear relationship also depends directly on the SNR.) Thus, the Viterbi decoder is fed with a soft decision comprising both the hard decision (the sign of the signal) and a measure of the amplitude of the received signal.

With other rectangular-constellation modulation systems, such as 16 quadrature amplitude modulation (QAM) or 64 QAM, each axis carries more than one bit. At the receiver, a soft decision can be made separately for each received bit. The metric functions are more complicated than for QPSK, being different for each bit, but the principle of the decoder exploiting the knowledge of the expected reliability of each bit remains the same.

Suppose a binary signal transmitted over a symbol interval $(0, T)$ is represented by $s_1(t)$ for binary 1 and $s_0(t)$ for binary 0. The received signal is $y(t) = s_m(t) + n(t)$, where $n(t)$ is zero-mean Gaussian noise process. Hence, at the receiver side, decision is to be made on the basis of a comparison between $y(t)$ and a threshold. The conditional probabilities of $y(t)$ as well as $p(Y|s_1)$ and $p(Y|s_0)$, are shown in Fig. 6.6, labelled likelihood of s_1 and likelihood of s_0, respectively. The demodulator of the wireless link converts the set of time random variables $\{y(t)\}$ into a code sequence Y and passes it on to the decoder. In short, the demodulator output can be configured in a variety of ways.

Demodulator Output Quantized to Two Levels—Hard Decision

In this case, the output of the demodulator is quantized to two levels, zero and one, and fed into the decoder (same as threshold detection using a comparator or a one-bit analog-to-digital converter). As the decoder operates on the firm or hard decisions made by the demodulator, the decoding is called *hard decision decoding*.

Demodulator Output Quantized to More than Two Levels—Soft Decision

The price paid for soft decision decoding is an increase in the required memory size at the decoder and possibly a speed penalty.

The demodulator can also be configured to feed the decoder with a quantized value of $y(t)$ greater than two levels. Such an implementation furnishes the decoder with more information than is provided in the hard decision case. When the quantization level of the demodulator output is greater than two (same as an n-bit analog-to-digital converter), the decoding is called *soft decision decoding*.

> The simplest error-detecting code is the parity code for which $n - k = 1$. It can be in one- and two-dimensional scenarios for horizontal and vertical parity check.

Eight levels (3 bits) of quantization are illustrated on the abscissa of Fig. 6.4:

- When the demodulator sends a hard binary decision, it sends it as a single binary symbol.
- When the demodulator sends a soft binary decision, quantized to three bits, that is, eight levels, it sends the decoder a three-bit word describing an interval along $y(t)$. In effect, sending such a three-bit word in place of a single binary symbol is equivalent to sending the decoder a measure of confidence along with the code symbol decision.

Referring to Fig. 6.6, if the demodulator sends 111 to the decoder, this is tantamount to declaring the code symbol to be a 1 with very high confidence, whereas sending a 100 is tantamount to declaring the code symbol to be a 1 with very low confidence. It should be clear that ultimately every message decision out of the decoder must be a hard decision. The idea behind the demodulator not making hard decisions and sending more data (soft decisions) to the decoder can be thought of as an interim step to provide the decoder with more information, which it then uses for recovering the message sequence (with better error performance than it could experience in the case of hard decision decoding).

In Fig. 6.6, the eight-level soft decision metric often shows $-7, -5, -3, -1, 1, 3, 5, 7$. Such a designation lends itself to a simple interpretation of the soft decision; the sign of the metric represents a decision (0 or 1) and the magnitude of the metric represents the confidence level of that decision.

Hence, the following can be inferred:

$$\text{Soft decision} = \text{Hard decision} + \text{Confidence level}$$

A Posteriori Probability and A Priori Probability

When there is prior knowledge about the data, the possibility or probability of receiving that data is called a priori probability. For an AWGN channel, *a posteriori probability* (APP) of decision of data bit can be thought of as a *refinement* of the prior knowledge about the data, brought about by examining the received signal *s* (which is a *continuous valued random variable* or a *test statistic* obtained at the output of a demodulator or some other signal processor).

6.7 CHANNEL CODING SCHEMES

This section discusses various channel coding techniques. In Section 6.7.1, error-detection codes are introduced in a summarized way. They are described in detail in Section 6.8 and in the various sections that follow. Parity check and polynomial codes are sometimes sufficient in wired channels.

6.7.1 Error-detection Codes

The simplest error-detection mechanism is the repetition code. Repetition of a sequence means that instead of sending 0, 1, ..., we shall send 00,11, ... This is an inefficient way. However, the results of information theory impose theoretical limitations. These results offer little clue as to how the coding should be performed. Error-detection coding is designed to permit the detection of errors. Once detected, the receiver may ask for a retransmission of the erroneous bits or it may simply inform the recipient that the transmission was corrupted.

Practical codes are normally (n, k) binary block codes. For these codes, the rate of the code is the ratio k/n, and the redundancy of the code is $1 - (k/n)$. Here, data will be in the same sequence order and additional bit or bits will be added to it. Our ability to detect errors depends on the rate. A low rate has a high detection probability as well as a high redundancy. The receiver will assign to the received code word the pre-assigned code word that minimizes the Hamming distance between the two words. If we wish to identify any pattern of m or less errors, the Hamming distance between the pre-assigned code words must be $m + 1$ or greater.

A very common block code is the *single parity check code*. This code appends to each k data bits an additional bit, whose value is taken to make the $n = k + 1$ word even (or odd). Such a choice is said to have even (odd) parity. With even (odd) parity, a single-bit error will make the received word odd (even). The pre-assigned code words are always even (odd) and hence are separated by a Hamming distance of two or more. A single error in a binary code can be detected without much loss of efficiency. For the given data bits, we count the number of times the digit 1 appears in the string and then append to the string an extra check bit (digit) that would make this number even. If the number of 1's is odd, the check digit should be 1, changing the data 0010 to the code word 00101, which is a (5, 4) parity check code. For the data 0011, the check digit will be 0, making it 00110.

Now, a single error in any position will make the parity odd. When receiving a code word, we do a parity check. We once again count the numbers of 1's. If this number is odd, we know that an error has occurred. If the number is even, we simply remove the last digit and decode the original code word. Here, it is not possible to detect more than one bit error. More generally, no even number of errors will be detected and every odd number of errors will be detected. Such a code has $n - 1$ message bits and one check bit. It is a $(n, n - 1)$ block code. The redundancy of this code is

$$\frac{n}{n-1} = 1 + \frac{1}{n-1} \tag{6.18}$$

The information rate of a (5, 4) parity check code is $4/5 = 0.80$, whereas that of a (8, 4) block repetition code is $4/8 = 0.50$. Since both are used to detect errors in codes with four message bits and both detect only a single error, parity check is clearly more efficient.

Example 6.3 illustrates how addition of a parity bit can improve error performance.

Example 6.3 Show that addition of parity bits in the data bits $k = 8$ improves the bit error performance, where the BER is $p = 10^{-4}$ without addition of parity its.

Solution For no addition of parity bits:

Probability of a single-bit error $= p$

Probability of no error in 1 bit $= (1 - p)$

Probability of no error in 8 bits $= (1 - p)^8$

Probability of an unseen error in 8 bits $= 1 - (1 - p)^8$

$$= 7.9 \times 10^{-4}$$

Hence, the probability of a transmission with an error is 7.9×10^{-4}.

With the addition of a parity error bit, we can detect any single-bit error.

Therefore,

Probability of no error in 1 bit $= (1 - p)$

Probability of no error in 9 bits $= (1 - p)^9$

Probability of a single error in 9 bits $= 9(p)(1 - p)^8$

Probability of an unseen error in 9 bits

$$= 1 - (1 - p)^9 - 9(p)(1 - p)^8 = 3.6 \times 10^{-7}$$

As can be seen, the addition of a parity bit has reduced the uncorrected error rate by three orders of magnitude. ∎

Single parity bits are common in asynchronous, character-oriented transmission. Where synchronous transmission is used, additional parity symbols are added, which checks the parity

> A polynomial code can detect any error burst of length less than or equal to the length of the generator polynomial.

of not only each *row* but also of each *column* (preferably each of eight bits). The column is formed by listing each successive eight-bit word one beneath the other. This type of parity checking is called *block sum checking*, or LRC and VRC, and it can correct any single two-bit error in the transmitted block of rows and columns. However, some combinations of errors will go undetected in such a scheme.

Parity checking in this way provides good protection against single- and multiple-bit errors when the probability of errors is independent. However, in many circumstances, errors occur in groups or bursts. Parity checking of the kind just described then provides little protection. In these circumstances, a polynomial code is used.

Polynomial Codes

The mechanism of *polynomial codes* is not covered in detail in this book; however, any book related to data link control can be referred for information on this topic. Polynomial codes are also block codes. They work on each frame (for frame-based data). Additional digits are added to the end of each frame. These digits depend on the contents of the frame. The number of added digits depends on the length of the expected error burst. Typically, 16 or 32 digits are added. The computed digits are called the *frame check sequence* (FCS) or CRC. Before transmission, each frame is divided by a generator polynomial. The remainder of this division is added to the frame. On reception, the division is repeated. Since the remainder has been added, the result should be zero. A non-zero result indicates that an error has occurred. These codes are sometimes categorized under error-correction coding, as they can correct single-bit errors.

The technique requires the addition of hardware to perform division. However, with modern integrated circuitry, this hardware is now available inexpensively. CRC error checking is quite common for both wired and wireless systems. Most of these codes are suitable for non-real-time data and are used to decide the retransmissions.

6.7.2 Error-correction Codes

Real-time transmission often precludes retransmission. It is necessary to get it right the first time. In these special circumstances, the additional bandwidth required for the redundant check bits is an acceptable price. We will discuss the important channel codes (error-correcting codes) in Sections 6.8–6.11 under three main categories: block codes, convolutional codes, and turbo codes.

6.8 BLOCK CODES

The properties used in block codes are linear, systematic, and cyclic. Let $C1$ and $C2$ be any two n-bit code words belonging to a set of (n, k) block code. If $C1 \oplus C2$ is also an n-bit code word belonging to the same set of (n, k) block code, then such a block code is called a *linear block code*. This linear block code is said to be *systematic* if k message bits appear either at the beginning or at the end of the code word.

An (n, k) linear block code C is said to be a *cyclic code* if every cyclic shift of the code is also a code vector of C. For example, if $C1 = 0111001$ is a code vector of C and $C2 = 1011100$ (the last 1 of $C1$ is shifted into the first position) is also a code vector of C, then it is called a cyclic code. Cyclic codes have two advantages over linear codes. First, encoding circuits can be easily implemented with a simple shift register with feedback connections and some basic gates. Second, cyclic codes have a fair amount of mathematical structure (algebraic structure) that makes it possible to design codes with useful error-correcting properties.

Table 6.1 Truth table for exclusive OR

A	B	$A \oplus B$
0	0	0
0	1	1
1	0	1
1	1	0

6.8.1 Hamming Codes

A Hamming code is a block code capable of identifying and correcting any single-bit error occurring within the block. It is identified as an (n, k) Hamming code. Hamming codes employ modulo-2 arithmetic (exclusive OR). Table 6.1 shows the truth table for this operation.

The Hamming code is generated by inserting systematically calculated bits at 2^m bit positions where $m = 0, 1, 2, 3, \ldots$ (this will be clear after going through Example 6.4) and has the following parameters with conditions to be satisfied:

Code length: $\quad n \le 2^{n-k} - 1$ (6.19)

Number of message bits: $\quad k \le n - \log_2(n + 1)$ (6.20)

Number of parity check bits: $\quad n - k$ (6.21)

Error-correcting capability: $\quad t_c = \left\lceil \dfrac{d_{min} - 1}{2} \right\rceil$ as in Eq. (6.15)

Example 6.4 For the code word bit pattern 10 01 10 1, find the Hamming code to detect and correct single-bit errors, assuming each code word contains an ASCII character, that is, a seven-bit data field.

Solution This type of a coding scheme requires four check bits, because in this scheme, the check bits occupy all bit positions that are powers of two. Such a code is thus known as an (11, 7) block code with a rate of 7/11 and a redundancy of (1 − 7/11). For example, the bit positions of the value 1001101 are as follows:

Bit position 11 10 9 8 7 6 5 4 3 2 1
Bit value 1 0 0 x 1 1 0 x 1 x x

The four bit positions marked with *x* are used for the check bits, which are derived as follows. The four-bit binary numbers corresponding to those bit positions having a binary 1 are added together using modulo-2 arithmetic, and the four check bits are then the four-bit sum:

$$11 = 1011$$
$$7 = 0111$$
$$6 = 0110$$
$$3 = 0011$$
$$\overline{1001}$$

The transmitted code word is now as follows:
Bit position 11 10 9 8 7 6 5 4 3 2 1
Bit value 1 0 0 1 1 1 0 0 1 0 1
Similarly, at the receiver, the four-bit binary numbers corresponding to those bit positions having a binary 1,

including the check bits, are again added together, and if no error has occurred, the modulo-2 sum should be zero:

$$11 = 1101$$
$$8 = 1100$$
$$7 = 0111$$
$$6 = 0110$$
$$3 = 0011$$
$$1 = 0001$$
$$\overline{0000}$$

Now, consider a single-bit error, say bit 11 is corrupted from 1 to 0. The new modulo-2 sum would now be

$$8 = 1100$$
$$7 = 0111$$
$$6 = 0110$$
$$3 = 0011$$
$$1 = 0001$$
$$\overline{1011}$$

Firstly, the sum is non-zero, which indicates an error, and secondly, the modulo-2 sum, equivalent to decimal 11, indicates that bit 11 is the erroneous bit. The latter would, therefore, be inverted to obtain the correct code word and hence data bits.

The Hamming codes suffer from the same difficulty as parity codes. They offer good protection against single-bit errors but little protection against burst errors. Convolutional codes are designed to deal with these circumstances. Convolutional codes are different from the previous codes we have examined in that they work in a statistical sense. By this, we mean that it is not possible to say, for example, that every single-bit error will be corrected. We can only say that on average the use of the convolutional code will improve the error rate.

6.8.2 Bose–Chaudhuri–Hocquenghem Codes

One of the major considerations in the design of optimum codes is to design the smallest block size (n) for a given message block of k so as to obtain a desired value of d_{min}, or for a given n and k, one may wish to design codes with the largest d_{min}. One of the most important and powerful classes of linear block codes are the BCH codes, whose properties are discussed in this section.

For any integer $m \geq 3$ and $t_c < (2^m - 1)/2$, there exists a BCH code with the following parameters:

$$\text{Block length} \qquad n = 2^m - 1 \tag{6.22}$$

$$\text{Number of message bits: } k \geq n - mt_c \tag{6.23}$$

$$\text{Minimum distance: } \qquad d_{min} \geq 2t_c + 1 \tag{6.24}$$

Clearly, BCH codes are t_c random error-correcting codes. The major advantage of these codes is the flexibility in choosing the code parameters such as block length and code rate.

Let us analyse the single error-correcting BCH codes. It can be proven that a single error-correcting BCH code is isomorphic to a Hamming code, which is a single error-correcting linear code. We will assume that the code length is $n = 2^m - 1$. We want to construct codes of length n over a finite field Φ. Let us start by factoring $(x^n - 1)$ into the irreducible form over Φ using the following rule:

$$f(x) = f_1(x)f_2(x)\cdots f_k(x) \tag{6.25}$$

If $(x^n - 1)$ is not separable in Φ, extend Φ to Φ' and write

$$x^n - 1 = (x - 1)(x - \alpha^1)(x - \alpha^2)(x - \alpha^3)\cdots(x - \alpha^{n-1}) \tag{6.26}$$

where α^n are the n distinct roots of unity in the extension of Φ. Each of the powers of α is also a root of some $f_i(x)$. Then, for each αq^i, define $q_i(x)$ to be the polynomial $f_k(x)$ such that $f_k(\alpha) = 0$. Note that $q_i(x)$ are the minimal polynomials for α^i and that they need not be distinct. This follows since α^i and α^j may have the same minimal polynomial.

A BCH code of length n with a designed distance d is a code with the generating polynomial

$$g(x) = \text{LCM}\ (q_{k+1}(x), q_{k+2}(x), \ldots, q_{k+d-1}(x)) \tag{6.27}$$

for some integer k.

Example 6.5 Find the generating polynomial for a BCH code if $n = 4$.

Solution We start by factoring $x^4 - 1$ to get

$$x^4 - 1 = (x - 1)(x - 2)(x - 3)(x - 4)$$

Then, of all the roots $\{1, 2, 3, 4\}$, we must determine which is the primitive fourth root of unity. This means that there is some element such that $\alpha^4 = 1$ and $\alpha^k \neq 1$ for $1 \leq k < 4$. Root 2 satisfies this condition. Let $\alpha = 2$. Thus, $q_i(x)$ are labelled as follows:

$q_0(x) = x - 1, q_1(x) = x - 2,$ or

$q_2(x) = x - 4, q_3(x) = x - 3$

since

$|q_0(\alpha^0) = 0 \quad q_1(\alpha^1) = 0$

$q_2(\alpha^2) = 0 \quad q_3(\alpha^3) = 0$

Now, suppose we were looking for a code of distance 3. We can choose k arbitrarily, and so assuming $k = 0$, the generating polynomial is

$g(x) = \text{LCM}(q_{0+1}(x), q_{0+2}(x))$

$= (x - 2)(x - 4)$

$g(x) = x^2 + 4x + 3$

Then, the first row of the generating matrix is the row vector (3, 4, 1, 0).

As we have defined a BCH code to be cyclic, we get the complete generating matrix from the cyclic shifts of the generating code vector.

$$\begin{pmatrix} 3 & 4 & 1 & 0 \\ 0 & 3 & 4 & 1 \end{pmatrix}$$

which is a cyclic [4, 2] code.

In Example 6.5, the weight of the first row vector is three, and since it is a cyclic code, we know that the minimum weight is exactly three. Hence, the result tells us that the exact minimum weight is one.

One of the advantages of this type of code is that there is a relatively simple way to decode.

6.8.3 Reed–Solomon Codes

Reed–Solomon codes are a specific type of BCH codes—linear block-based error-correcting codes just described. The encoder for an RS code differs from binary encoders in that it operates on multiple bits rather than on individual bits. The t_c random error-correcting RS codes have the following parameters:

Block length: $n = 2^m - 1$

Message size: k symbols   (6.28)

Parity check size: $r = (n - k) = 2t_c$ symbols   (6.29)

Minimum distance: $d_{\min} = 2t_c + 1$ symbols   (6.30)

Number of correctable symbol in error $t_c = (n - k)/2$ (6.31)

The encoder for an RS (n, k) code on m-bit symbols divides the incoming binary data stream into blocks, each of $k \times m$ bits long. Each block is treated as k symbols with each symbol having eight bits. The encoding algorithm expands a block of k symbols by adding $(n - k)$ redundant symbols, when m is an integer power of two. For bytes, the popular value of m is eight. Eight-bit RS codes are extremely powerful. For any (n, k) linear block code, $d_{\min}$ $(n - k + 1)$, but for an RS code, $d_{\min} = n - k + 1$. It means that the minimum distance is always equal to the design distance of the code. Therefore, the code is also called the *maximum distance separable* code. Figure 6.7(a) shows a typical systematic RS code in that the data is left unchanged and the parity symbols are appended. RS code generation is based on Galois field arithmetic. An RS code word is generated using a special polynomial. All valid code words are exactly divisible by the generator polynomial. The general form of the generator polynomial is the same as Eqs (6.25) and (6.26). The schematic diagram of an RS (255, 249) code is given in Fig. 6.7(b).

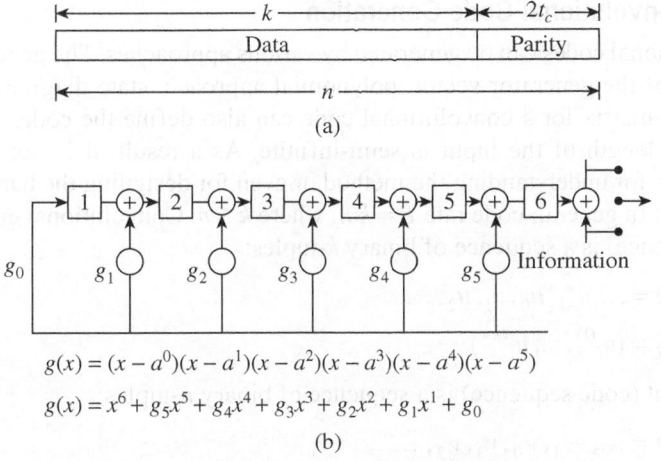

$$g(x) = (x - a^0)(x - a^1)(x - a^2)(x - a^3)(x - a^4)(x - a^5)$$

$$g(x) = x^6 + g_5 x^5 + g_4 x^4 + g_3 x^3 + g_2 x^2 + g_1 x^1 + g_0$$

(b)

Fig. 6.7 RS code (a) Typical code word (b) Encoder for an RS (255, 249) code

The code words received with errors are inserted at the decoder input. A syndrome calculator calculates $2t_c$ syndromes with calculations similar to parity calculations. These can be found by submitting $2t_c$ roots of the generator polynomial into the received code words with error. Next, the symbol error locations are identified by finding an error locator polynomial and then finding the roots of this polynomial. Once the symbol error values are determined, the error corrector uses these findings to correct the code as the recovered one. Various algorithms are used for these calculations.

Example 6.6 For a (31, 15) RS code, answer the following questions:
(a) How many bits per symbol of the code are there?
(b) What is the block length in terms of bits?
(c) What is the minimum distance of the code?
(d) How many symbols in error can the code correct?
(e) What is the length of an in-phase burst that the code can correct?

Solution
(a) We have

$$n = 2^m - 1 = 31 \quad \text{or} \quad 2^m = 32$$

So, the number of bits per symbol $m = 5$ bits
(b) Block length $= 31 \times 5 = 155$ bits
(c) The minimum distance of the code

$$d_{\min} = n - k + 1 = 31 - 15 + 1 = 17$$

(d) The number of correctable symbols in error

$$t_c = (n - k)/2 = (31 - 15)/2 = 8$$

(e) The length of an in-phase burst that the code can correct is $m = 5$.

6.9 CONVOLUTIONAL CODES

Convolutional codes are often used to improve the performance of wireless links. A convolutional encoder is called so because it performs a *convolution* of the input stream with the encoder's *impulse responses*: Every output of the encoder can be described by its own transfer function, which is closely related to the generator polynomial. Convolutional codes are of importance as they are used in most of the latest mobile networks. These codes are more appropriate for error correction than block codes. In this section, convolutional codes are explained in detail.

> A convolutional encoder is a discrete linear time-invariant system generating non-linear codes.

6.9.1 Convolutional Code Generation

Convolutional codes can be generated by various approaches. The generation is possible with the help of the generator vector, polynomial approach, state diagram, tree, and trellis. The generator matrix for a convolutional code can also define the code, and it is semi-infinite, since the length of the input is semi-infinite. As a result, it is not a convenient method. Moreover, for understanding the method or even for designing the hardware, mathematics is necessary. In general, code rate $R = k/n$, where $k \leq n$. Convolutional encoder input (information sequence) is a sequence of binary k-tuples:

$$u = \dots, u_{-1}, u_0, u_1, u_2, \dots \tag{6.32}$$

where $\quad u_i = (u_i^{(1)}, \dots, u_i^{(k)}) \tag{6.33}$

The output (code sequence) is a sequence of binary n-tuples:

$$v = \dots, v_{-1}, v_0, v_1, v_2, \dots \tag{6.34}$$

where $\quad v_i = (v_i^{(1)}, \dots, v_i^{(n)}) \tag{6.35}$

Generator sequences specify the convolutional code completely by the associated generator matrix. An encoded convolutional code is produced by matrix multiplication of the input and the generator matrix. The sequences must start at a finite (positive or negative) time and may or may not end. The relation between the information sequences and the code sequences is determined by the equation

$$v = u * G \tag{6.36}$$

where

$$G = \begin{pmatrix} G_0 & G_1 & \cdots & G_m & & & \\ & G_0 & G_1 & \cdots & G_m & & \\ & & G_0 & G_1 & \cdots & G_m & \\ & & & \ddots & \ddots & & \ddots \end{pmatrix} \tag{6.37}$$

is the semi-infinite generator matrix and the submatrices G_i, $0 \leq i \leq m$, are binary $k \times n$ matrices. The arithmetic in it is carried out over the binary field, and the parts left blank in the generator matrix G are assumed to be filled with zeros.

The right-hand side of Eq. (6.42) defines a discrete time convolution between u and $G = (G_0 \ G_1 \dots G_m)$, hence the name convolutional code. As in many other situations where convolutions appear, it is convenient to express the sequences in some sort of transform. In information theory and coding theory, it is common to use the delay operator D, or the D-transform. The information and code sequences become

$$u(D) = \cdots + u_{-1}D^{-1} + u_0 + u_1D + u_2D^2 + \cdots \tag{6.38}$$

and $\quad v(D) = \cdots + v_{-1}D^{-1} + v_0 + v_1D + v_2D^2 + \cdots \tag{6.39}$

They are related through the equation

$$v(D) = u(D)G(D) \tag{6.40}$$

where $\quad G(D) = G_0 + G_1D + \cdots + G_mD^m \tag{6.41}$

is the generator matrix. According to the formula for convolution, encoder outputs are formed by *modulo-2 discrete convolutions*.

$$v^{(1)} = u*g^{(1)}, v^{(2)} = u*g^{(2)}, \ldots, v^{(j)} = u*g^{(j)} \tag{6.42}$$

where u is the information sequence.

Therefore, the lth bit of the jth output branch is

$$v_l^{(j)} = \sum_{i=0}^{m} u_{l-i} g_i^{(j)} = u_l g_0^{(j)} + u_{l-1} g_1^{(j)} + \cdots + u_{l-m} g_m^{(j)} \tag{6.43}$$

Here, L is number of registers used to generate the code sequence, $m = L + 1$, $u_{l-i} \neq 0$, $l < i$.

Example 6.7 An information source is $u = [1, 0, 1, 1, 1]$. Assume the generator matrix as follows:

$$\begin{pmatrix} g^{(1)} = [1 \quad 0 \quad 1 \quad 1] \\ g^{(2)} = [1 \quad 1 \quad 1 \quad 1] \end{pmatrix}$$

Find the convolutional codes.

Solution We have $v = u * G$

Therefore, the following equations result:

$$v_l^{(1)} = u_l + 0 + u_{l-2} + u_{l-3}$$
$$v_l^{(2)} = u_l + u_{l-1} + u_{l-2} + u_{l-3}$$

$$G = \begin{bmatrix} g_0^{(1)} g_0^{(2)} & g_1^{(1)} g_1^{(2)} & g_2^{(1)} g_2^{(2)} & \cdots & g_m^{(1)} g_m^{(2)} \\ & g_0^{(1)} g_0^{(2)} & g_1^{(1)} g_1^{(2)} & \cdots & g_{m-1}^{(1)} g_{m-1}^{(2)} & g_m^{(1)} g_m^{(2)} \\ & & g_0^{(1)} g_0^{(2)} & \cdots & g_{m-2}^{(1)} g_{m-2}^{(2)} & g_{m-1}^{(1)} g_{m-1}^{(2)} & g_m^{(1)} g_m^{(2)} \\ & & & \ddots & & & & \ddots \end{bmatrix}$$

where the blank areas are all zeros, the encoding equations can be rewritten in matrix form as

$$v = uG,$$

where all operations are module-2. G is called the *generator matrix* of the code. Note that each row of G is

$$v = uG$$

$$= (1 \quad 0 \quad 1 \quad 1 \quad 1) \begin{bmatrix} 1 & 1 & 0 & 1 & 1 & 1 & 1 & 1 \\ & 1 & 1 & 0 & 1 & 1 & 1 & 1 & 1 \\ & & 1 & 1 & 0 & 1 & 1 & 1 & 1 & 1 \\ & & & 1 & 1 & 0 & 1 & 1 & 1 & 1 & 1 \\ & & & & 1 & 1 & 0 & 1 & 1 & 1 & 1 & 1 \end{bmatrix}$$

$$= (1 \quad 1, \quad 0 \quad 1, \quad 0 \quad 0, \quad 0 \quad 1, \quad 0 \quad 1, \quad 0 \quad 1, \quad 0 \quad 0, \quad 1 \quad 1)$$

Hence, the encoder output becomes

$$v = [v_0^{(1)} v_0^{(2)} v_1^{(1)} v_1^{(2)} v_2^{(1)} v_2^{(2)} \ldots]$$

Let the information sequence $u = (1\ 0\ 1\ 1\ 1)$. Then the output sequences are

$$v^{(1)} = (1\ 0\ 1\ 1\ 1)*(1\ 0\ 1\ 1) = (1\ 0\ 0\ 0\ 0\ 0\ 0\ 1)$$
$$v^{(2)} = (1\ 0\ 1\ 1\ 1)*(1\ 1\ 1\ 1) = (1\ 1\ 0\ 1\ 1\ 1\ 0\ 1)$$

and the code word is

$$v = (1\ 1,\ 0\ 1,\ 0\ 0,\ 0\ 1,\ 0\ 1,\ 0\ 1,\ 0\ 0,\ 1\ 1).$$

If the generator sequences $g^{(1)}$ and $g^{(2)}$ are interlaced and then arranged in the matrix

identical to the preceding row but shifted $n = 2$ places to the right and that G is a semi-infinite matrix, corresponding to the fact that the information sequence u is of arbitrary length. If u has finite length L, then G has L rows and $2(m + L)$ columns, and v has length $2(m + L)$.

If $u = (1\ 0\ 1\ 1\ 1)$, then

The codes described so far are time invariant with a fixed generator matrix. Another type is the time-varying convolutional codes, which are beyond the scope of this book. Convolutional code generation by hardware is explained in Section 6.9.2.

6.9.2 Convolutional Encoder

Convolutional codes are used in applications that require good performance with low implementation cost. They operate on code streams (not in blocks). A continuous sequence of information bits is mapped into a continuous sequence of encoder output bits. This mapping is highly structured and systematic so that decoding is possible. The decoding methods for these codes are different from those for the block codes. Convolutional codes have a memory that utilizes previous bits to encode or decode the following bits (block codes are memoryless, as mentioned earlier). They achieve good performance by expanding their memory depth.

Convolutional codes are denoted by (n, k, L), where n is the number of output bits (coded), k is the number of input bits (uncoded), and L is the code (or encoder) memory depth, which represents the number of register stages (Fig. 6.8a).

$$\text{Constraint length, } C = (L + 1) \tag{6.44}$$

Constraint length is defined as the number of encoded bits a message bit can influence. The constraint length concept is shown in Fig. 6.8(b).

To convolutionally encode data, start with L memory elements shift register, each element holding one input bit. Pass the information sequence through a finite state shift register. Unless otherwise specified, all memory registers start with the value of 0. The encoder has modulo-2 adders and n generator polynomials—one for each adder. Figure 6.9 shows a typical convolutional coder (here, $x_{out} = x_1' x_1'' x_2' x_2'' x_3' x_3'' \ldots$ is the convolutional-coded output ready for modulation). An input bit m_j is fed into the rightmost register. Using the generator polynomials and the existing values in the remaining registers, the encoder outputs n bits. Now, bit shift all register values to the left (m_j moves to m_{j-1}, m_{j-1} moves to m_{j-2}) and wait for the next input bit. If there are no remaining input bits, the encoder continues output until all registers have returned to the zero state.

> It is found that convolutional codes can have a larger coding gain than what can be achieved through block coding with the same complexity.

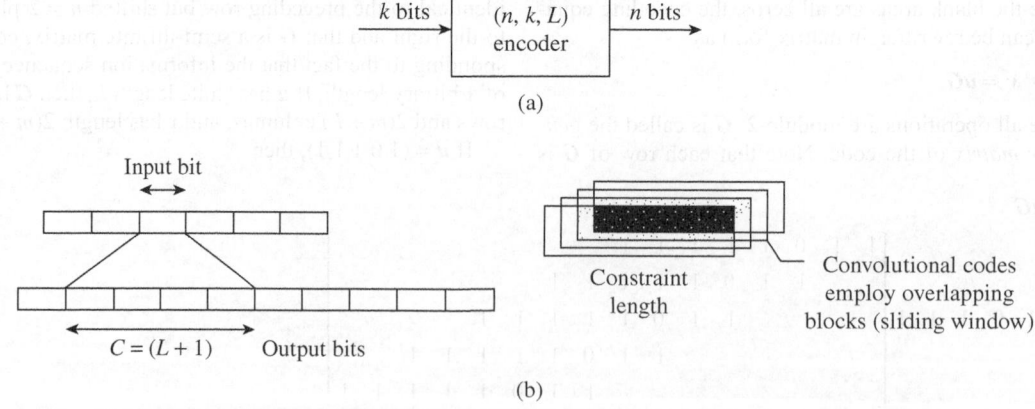

Fig. 6.8   Convolutional encoder (a) Simple conceptual diagram (b) Constraint length concept

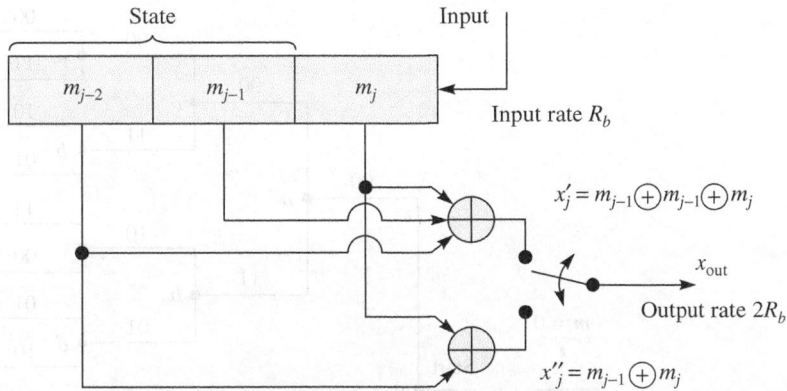

Fig. 6.9 Typical convolutional encoder

> *Note*: To provide the extra bits needed for error control, an output rate greater than the message bit rate could be achieved by connecting two or more modulus-2 summers to the register and interleaving the encoded bits via a switch. In this situation, each message bit influences a span of $n(L+1)$ successive output bits, which is the constraint length measured in terms of encoded output bits.

> *Note*: Convolutional encoder is a *finite state machine* (FSM), processing information bits in a serial manner. Thus, the generated code is a function of the input and the state of the FSM. Hence, convolutional code generation can also be represented using a state diagram.

Example 6.8 For $(n, k, L) = (2, 1, 2)$ convolutional encoder, find the constraint length.

Solution Each message bit influences a span of $C =$ $(L+1) = 3$ successive output bits = constraint length C. Thus, for the generation of a two-bit output, we require two shift registers and one bit input.

6.9.3 Trellis Diagram

The trellis diagram can be used for the decoding of convolutional codes. A typical tree is given in Fig. 6.10.

The tree is useful to create a trellis. The example discussed in Section 6.9.4 will explain how to use the trellis diagram.

6.9.4 Decoding Methods for Convolutional Codes

> The goal of selecting the optimum path can be expressed, equivalently, as choosing the code word with the maximum likelihood metric or as choosing the code word with the minimum distance metric.

As compared to coding, the decoding part is difficult to design. The stream received is error-laden, and the decoder tries to resolve as many errors as possible. Link performance depends on decoding techniques. The following section describes the basic Viterbi decoder along with some simple examples.

Viterbi Algorithm

Viterbi algorithm is mainly used for convolutional decoding. The Viterbi decoding algorithm was developed and analysed by Viterbi in 1967.

The Viterbi algorithm removes from consideration those paths that could not possibly be candidates for the maximum likelihood choice. When two paths

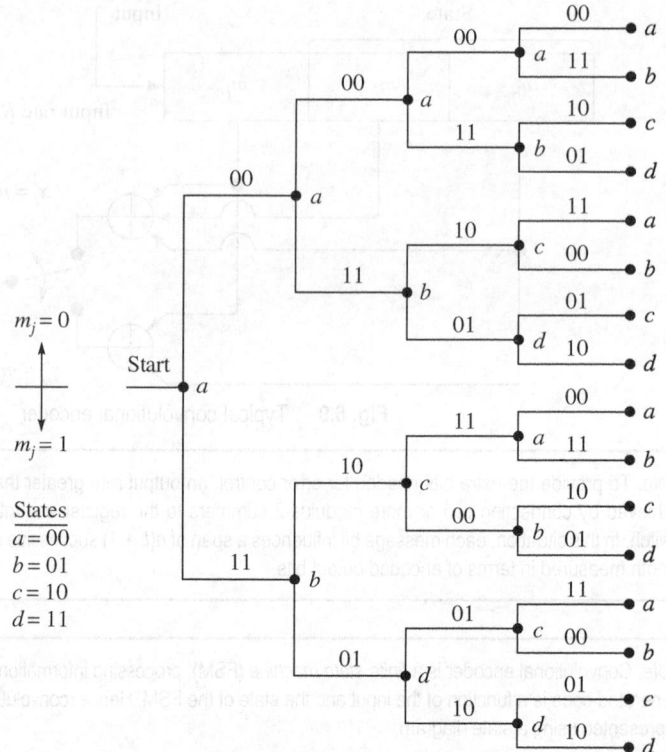

Fig. 6.10 Typical tree diagram

enter the same state, the one having the best metric is chosen; this path is called the *surviving path*.

This selection of surviving paths is performed for all the states. The decoder continues in this way to advance deeper into the trellis, making decisions by eliminating the least likely paths. The early rejection of the unlikely paths reduces the decoding complexity.

Hard Decision Viterbi Algorithm for (2, 1, 3) Coder: Example

For simplicity, Hamming distance is a proper measure to consider the difference between the received and expected patterns. The encoder for this example is given in Fig. 6.11. The generator vectors are 111 (7) and 101 (5). The basic idea behind the decoding procedure can best be understood by examining Fig. 6.12 encoder trellis in concert with Fig. 6.13 decoder trellis

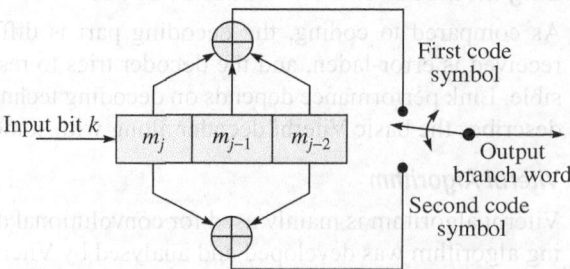

Fig. 6.11 Convolutional encoder (rate ½, C = 4, generator polynomials = [7, 5])

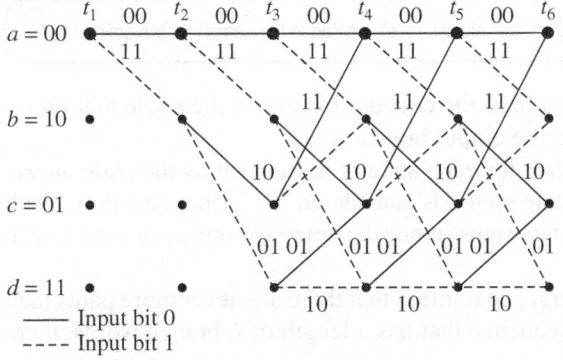

— Input bit 0
---- Input bit 1

Fig. 6.12   Encoder trellis diagram (rate ½, C = 4)

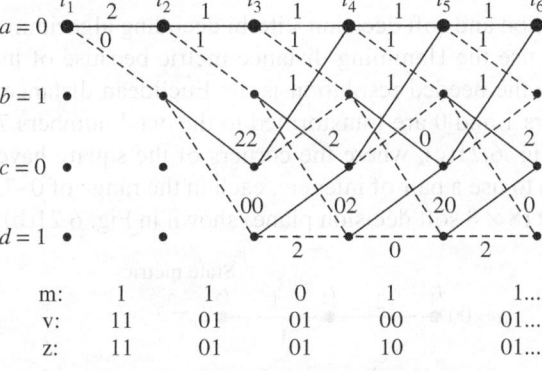

Fig. 6.13   Decoder trellis diagram (rate ½, C = 4)

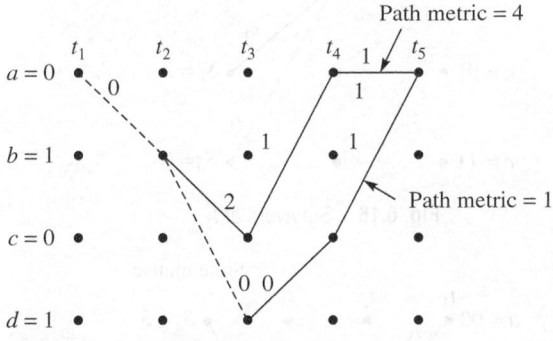

Fig. 6.14   A case of two merging paths

> The metric entered on a decoder trellis branch represents the difference (distance) between what was received and what should have been received.

for the Hamming distance between the received code symbols and the branch word corresponding to the same branch from the encoder trellis.

The branch words seen on the encoder trellis branches characterize the encoder in Fig. 6.11 and are known a priori to both the encoder and the decoder. These encoder branch words are the code symbols that would be expected to come from the encoder output as a result of each of the state transitions. As the code symbols are received, each branch of the decoder trellis is labelled with the Hamming distance between the received code symbols and each of the branch words for that time interval.

Survivor path calculation   The trellis branches with solid lines for the input 0's and dashed lines for the input 1's. Note that the bit cannot be decoded until the path metric computation has proceeded to a much greater depth into the trellis. The decoder trellis in Fig. 6.13 shows the following:

A message sequence, $m = 11011$

The corresponding code word sequence,
$$V = 1101010001$$
The received sequence, $Z = 1101011001 \ldots$

Labelling procedure   From the received sequence Z, it is seen that the code symbol received at time t_1 is 11. On the decoder trellis, we label the state 00 $\to$ 00 transition with the Hamming distance between them, namely 2. Looking at the encoder trellis again, we see that a state $a = 00 \to b = 10$ transition yields an output branch word of 11, which corresponds exactly with the code symbols we received at time t_1. Therefore, on that decoder trellis, the Hamming distance labelled is 0.

We continue labelling the decoder trellis branches in this way as the symbols are received at each time t_i. Thus, the most likely (minimum distance) path through the trellis can be found, which gives the decoded output.

Optimal path selection   A typical example of Fig. 6.14 shows two paths merging at time t_5 to state 00. Let us define the *cumulative Hamming path metric* of a given path at time t_i as *the sum of the branch Hamming distance metrics along the path up to time t_i.* In Fig. 6.14, the upper path has metric 4, and the lower has metric 1. The upper path cannot be a portion of the *optimum path* because the lower path, which enters the same state, has a *lower metric*. This observation holds true because of the *Markov (chain process) nature* of

Note: If any two paths in the trellis merge to a single state, one of them can always be eliminated in the search for the optimal path.

the encoder state: The present state summarizes the encoder history in the sense that previous states cannot affect future states or future output branches.

At a given time, the *winning path metric* for each state is designated as the *state metric* for that state at that time. The minimum state metric is considered for further decoding. With reference to Figs 6.12 and 6.13, the first few steps in our decoding example are shown with the help of Figs 6.15–6.20.

Pruning the trellis, because paths remerge, guarantees that there are never more paths than there are states. For a binary code word sequence that has a length of L branch words, there are 2^L possible sequences.

Soft Decision Viterbi Algorithm

The primary difference between hard decision and soft decision Viterbi decoding algorithms is that the soft decision algorithm cannot use the Hamming distance metric because of its limited resolution. A distance metric with the needed resolution is the Euclidean distance, and to facilitate its use, the binary numbers 1 and 0 are transformed to the octal numbers 7 and 0, respectively. This can be seen in Fig. 6.21(c), where the corners of the square have been relabelled accordingly; this allows us to use a pair of integers, each in the range of 0–7, for describing any point in the 64-point set (8×8 soft decision plane) shown in Fig. 6.21(b).

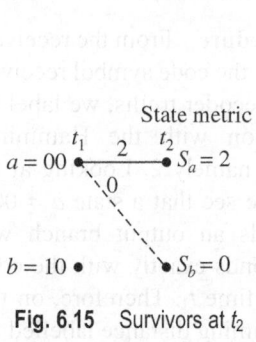

Fig. 6.15 Survivors at t_2

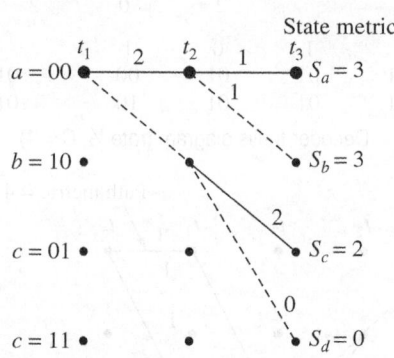

Fig. 6.16 Survivors at t_3

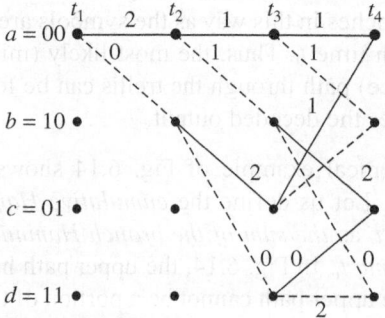

Fig. 6.17 Metric comparisons at t_4

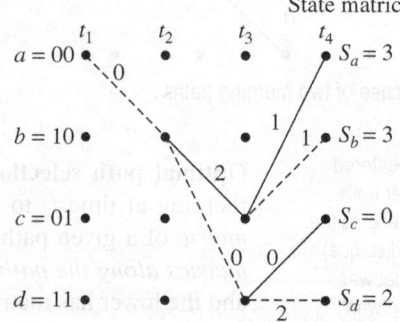

Fig. 6.18 Survivors at t_4

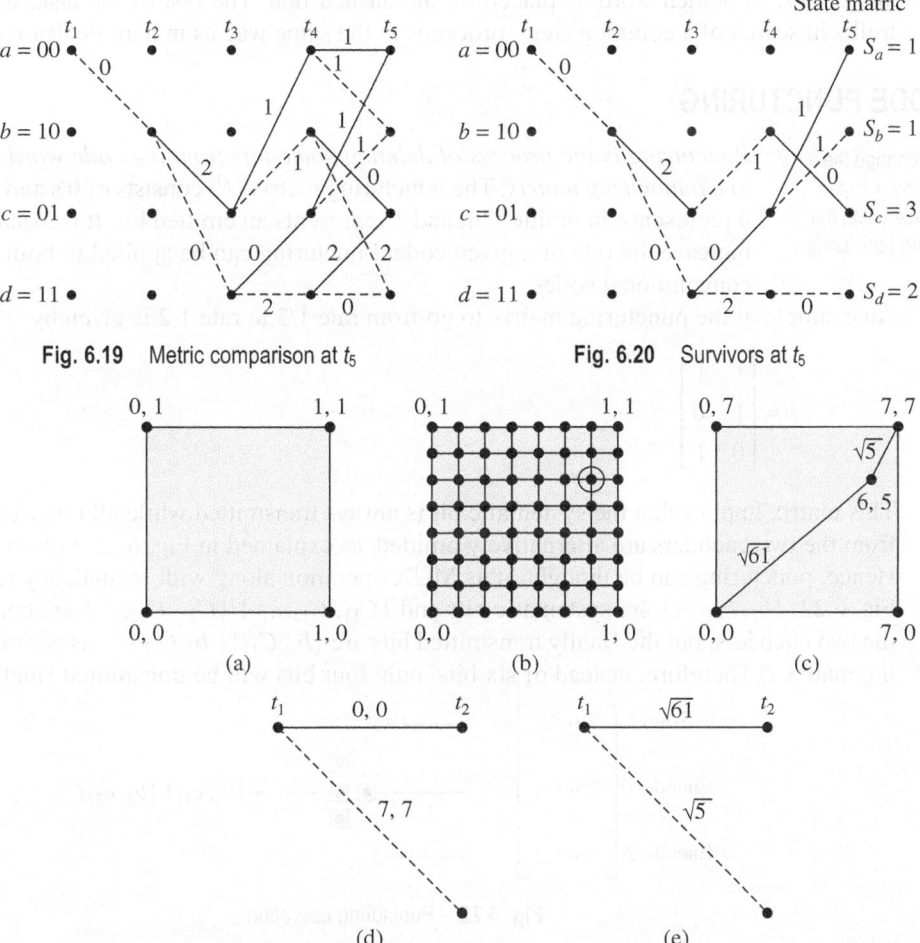

Fig. 6.19 Metric comparison at t_5 **Fig. 6.20** Survivors at t_5

Fig. 6.21 Viterbi algorithm (a) Hard decision plane (b) Eight-level by eight-level soft decision plane (c) Example of soft code symbols (d) Encoding the trellis section (e) Decoding the trellis section

Figure 6.21(c) also shows the point (6, 5) representing an example of a pair of noisy code symbol values that might stem from a demodulator. Imagine that the square in Fig. 6.21(c) has the coordinates x and y. Then, the Euclidean distance between the noisy point (6, 5) and the noiseless point (0, 0) is $\sqrt{(6-0)^2 + (5-0)^2} = \sqrt{61}$. Similarly, the Euclidean distance between the noisy point (6, 5) and the noiselewss point (7, 7) is $\sqrt{(6-7)^2 + (5-7)^2} = \sqrt{5}$.

Soft decision Viterbi decoding, for the most part, proceeds in the same way as hard decision decoding. The only difference is that Hamming distances are not used. Consider how soft decision decoding is performed with the use of Euclidean distances. Figure 6.21(d) shows the first section of an encoding trellis, originally presented in Fig. 6.10, with the branch words transformed from binary to octal. Suppose a pair of soft decision code symbols with values (6, 5) arrive at the decoder during the first transition interval. Figure 6.21(e) shows the first section of a decoding trellis. The metric ($\sqrt{61}$), representing the Euclidean distance between the arriving (6, 5) and the (0, 0) branch word, is placed on the solid line. Similarly, the metric ($\sqrt{5}$), representing the Euclidean distance between the arriving (6, 5) and the (7, 7) branch word, is placed on the dashed line. The rest of the task, pruning the trellis in search of a common stem, proceeds in the same way as in hard decision decoding.

6.10 CODE PUNCTURING

One can change the code rate by puncturing the code using the appropriate puncturing matrix.

Puncturing is the process of deleting some bits from the code word according to a puncturing matrix. The puncturing matrix (P) consists of 0's and 1's, where 0 represents an omitted bit and 1 represents an emitted bit. It is usually used to increase the rate of a given code. Puncturing can be applied to both block and convolutional codes.

An example of the puncturing matrix to go from rate 1/3 to rate 1/2 is given by

$$p = \begin{bmatrix} 1 & 1 \\ 1 & 0 \\ 0 & 1 \end{bmatrix}$$

This matrix implies that the systematic bit is always transmitted while all other bits coming from the two encoders are alternatively omitted, as explained in Fig. 6.22 with more clarity. Hence, puncturing can be thought of as MUX operation along with redundancy removal. In Fig. 6.22, $\{b_1, b_2, ...\}$ are systematic bits and $\{C_{11}, C_{12}, ...\}$ $\{C_{21}, C_{22}, ...\}$ are coming from the two encoders, but the finally transmitted bits are $\{b_1, C_{11}\}\{b_2, C_{22}\}...$ as per the puncturing matrix P. Therefore, instead of six bits, only four bits will be transmitted finally.

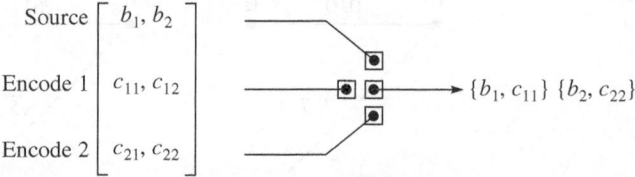

Fig. 6.22   Puncturing operation

6.11 TURBO CODES

Turbo codes were first introduced in 1993 by Berrou, Glavieux and Thitimajshima, when the need arose for a scheme that achieves a bit error probability of 10^{-5} using a rate ½ code over an AWGN channel and BPSK modulation at an E_b/N_o of 0.7 dB.

Turbo codes are constructed by using two or more component codes on different interleaved versions of the same information sequence. For conventional codes, the final step at the decoder yields hard decision decoded bits, whereas for a *concatenated scheme*, such as a turbo code, to work properly, the decoding algorithm should not limit itself to passing hard decisions among the decoders. To best exploit the information learned from each decoder, the decoding algorithm must be based on soft decisions rather than hard decisions. For a system with two component codes, the concept behind turbo decoding is to pass soft decisions from the output of one decoder to the input of the other decoder and to *iterate* this process several times through *feedback* so as to produce more reliable decisions.

Proper design of a turbo code may approach the Shannon limit after an infinite number of iterations of decoding (with unlimited latency) or (with large interleaver).

According to Fig. 6.23, the principle of concatenated codes is to feed the output of one encoder (outer encoder) to another encoder and so on. The final

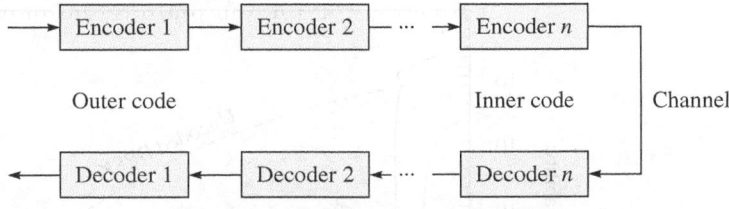

Fig. 6.23 Principle of concatenated codes

For turbo codes, the same decoder may serve for various coding rates by means of puncturing, provided that the decoder knows the puncturing tables.

encoder before the channel is known as the inner encoder. The resulting composite code is clearly much more complex than any individual code. However, it can be decoded step by step as shown. We simply apply each of the component decoders in turn, from the inner to the outer. This scheme suffers from many drawbacks, the most significant of which is error propagation. If a decoding error occurs in a code word, it usually results in a number of data errors. When these are passed on the next decoder, they may overwhelm the ability of that code to correct the errors. The performance of the outer decoder might be improved if these errors were distributed among a number of separate code words. This can be achieved using the interleaver–de-interleaver (explained in Section 6.12).

The interleaver may be placed between the outer and inner encoders of a concatenated code that uses two component codes and the de-interleaver between the inner and outer decoders. Provided that the rows of the interleaver are at least as long as the outer code words and the columns at least as long as the inner data blocks, each data bit of an inner code word falls into a different outer code word. Hence, the outer code is able to correct at least one error, and it can always cope with a single decoding error in the inner code.

When the redundant information of a given encoder is not transmitted, the corresponding decoder input is set to zero. However, the decoder needs to know the current puncturing table. This function is performed by the DEMUX/INSERTION block in the turbo decoder. The DEMUX will demultiplex the stream between the decoders and the INSERTION will insert an analog zero if the corresponding bit is omitted. When the code is punctured, the branch metric corresponding to the punctured bits need not be computed. However, determining the best puncturing pattern for turbo codes is still an open problem.

Although turbo codes have the potential to offer unprecedented power efficiencies, they have some peculiarities that should be taken into consideration. The typical performance of turbo coding is shown in Fig. 6.24.

The performance of turbo coding systems is characterized by two distinct regions—the *turbo cliff* region, where the BER drops within a fraction of a decibel of the SNR to a very low value, and the *error floor/flare* region, which is characterized by a slow decrease of the error rate with increasing SNR.

Error floor is a phenomenon of flattening out the BER curve at higher SNRs (Fig. 6.24), which hinders the ability of a turbo code to achieve extremely small BERs. The error floor is mainly due to the presence of a few low-weight code words. At a low SNR, these code words are insignificant, but as the SNR increases, they begin to dominate the performance of the code. Hence, sometimes it is preferred to use a convolutional code for high SNR values.

Though there can be several arguments in favour of turbo codes, its decoding algorithm is no doubt more complex than any other FECs. One easy way to reduce the complexity is to halt the decoder iterations once the entire frame has been completely corrected. This will

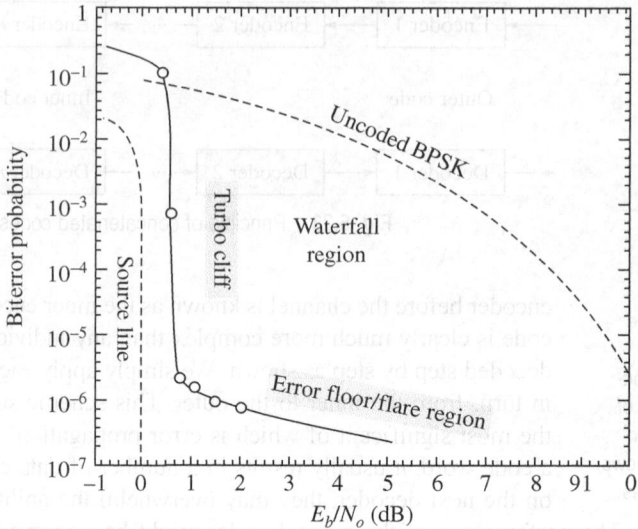

Fig. 6.24 Typical performance of turbo coding

prevent over-iteration, which corresponds to wasted hardware clock cycles. Other methods such as implementing the entire decoder in analog circuitry rather than in digital hardware and sliding window algorithms for reducing memory requirement are also used. Turbo-coded systems typically experience significant latency due to interleaving and iterative decoding. For example, an interleaver used in a turbo code uses 65,536 bits. Since this number of bits must inevitably be stored in the interleaver in the encoder and/or decoder at any given time, there is latency. Thus, for an information rate of (say) 8 kbps (appropriate for speech transmission), there is a delay of 65,536/8 = 8192 ms or more than 8 s. This delay is quite unacceptable in a telephone system, because the conversation would be highly disruptive. In fact, the delay limitation for speech services on wireless links usually is 40 ms, which corresponds to a latency of 320. Thus, if a turbo code is to be used for speech services at this data rate, then this interleaver can no longer be used. However, to reduce latency, one can think of a low interleaver size, that is, low latency, but there will be some performance penalty such as departing from the Shannon limit and error floor discussed earlier.

6.12 INTERLEAVER

The errors in a memoryless channel can be characterized as single randomly distributed bit errors whose occurrences are independent from bit to bit, whereas a channel with memory exhibits mutually dependent signal transmission impairments. A channel that exhibits mul-
tipath fading (Chapter 3), where signals arrive at the receiver over two or more paths of different lengths, is an example of a channel with memory. The effect is that the signals can arrive out of phase with each other, and the cumulative received signal is distorted. Wireless mobile communication channels, as well as ionospheric and tropospheric propagation channels, suffer from such phenomenon. All such time-correlated impairments result in statistical dependence among successive symbol transmissions. That is, the disturbance tends to cause errors that occur in *bursts* instead of as *isolated events*.

> An interleaver tries to randomize and distribute or spread out the burst errors; hence, after de-interleaving, there will be an improved BER.

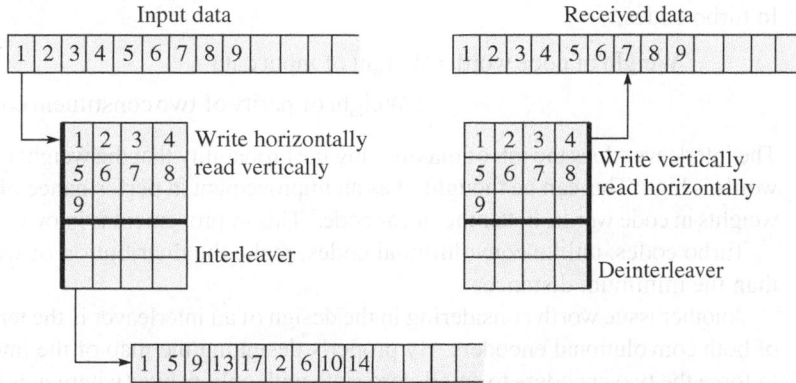

Fig. 6.25 Interleaving and de-interleaving operation

As the channel is assumed to have memory, the errors no longer can be characterized as single randomly distributed bit errors whose occurrences are independent from bit to bit. Most channel coding techniques are designed to combat random independent errors. The effect of channel memory on such coded signals is to cause degradation in error performance. Some coding techniques have been proposed for channels with memory, but the greatest problem with such coding is the difficulty in obtaining accurate models of the often time-varying statistics of such channels. The use of *time diversity* or *interleaving* is a *technique* that requires knowledge of only the duration or span of the channel memory and not its exact statistical characterization.

As shown in Fig. 6.25, an interleaver is a device that arranges the ordering of sequence of symbols in a deterministic manner. For a typical case, it can be thought of as a storage memory where the data is read in row-wise and read out column-wise. There may be other algorithms as well. Associated with the interleaver is a de-interleaver, which applies inverse permutation to restore the original sequence. The most critical part in the design of a turbo code is the interleaver design, in which two main issues are the *interleaver size* and the *interleaver map*. The size of the interleaver plays an important role in the *trade-off between performance and time* (*delay*) since both are directly proportional to size. On the other hand, the map of the interleaver plays an important role in setting the code performance.

Separating the symbols in time effectively transforms a channel with memory to a memoryless one and thereby enables the random error-correcting codes to be useful in a burst noise channel.

Notes: In turbo codes, the interleaver has more functions (this must be clear after the discussion on turbo codes):

1. The interleaver is used to feed the encoders with permutations so that the generated redundancy sequences can be assumed to be independent. The validity of the assumption that the generated redundant sequences are independent is a function of the particular interleaver used. This will exclude a number of interleavers that generate regular sequences such as a cyclic shift.

2. Another key role of the interleaver is to shape the weight distribution of the code, which, ultimately, controls its performance. This is so because the interleaver will decide which word of the second encoder will be concatenated with the current word of the first encoder and hence what weight the complete code word will have. Therefore, the aim of the designer is to produce (by manipulating the weights of the second redundancy part through interleaver mapping) whole code words with the overall weights as large as possible.

In turbo codes,

$$\text{Weight of code word} = \text{Weight of input data}$$
$$+ \text{Weight of parity of two constituent codes} \tag{6.45}$$

The interleaver does the job of maximizing the probability that the weight of two constituent code words is high. This can be thought of as an improvement in performance of the code, as more the weights in code words, better the linear code.[†] This improvement is known as the *interleaver gain*.

Turbo codes, unlike convolutional codes, make the distribution of weight more important than the minimum distance.

Another issue worth considering in the design of an interleaver is the termination of the trellis of both convolutional encoders. By properly designing the map of the interleaver, it is possible to force the two encoders to an all-zero state with only *m* bits (where *m* is the memory length of the convolutional encoder, assuming that the same convolutional code is used in both encoders).

The interleaver shuffles the code symbols over a span of several block lengths (for block codes) or several constraint lengths (for convolutional codes). Several types of interleavers are commonly used, including *block*, *convolutional*, and *pseudo-random* interleavers. A typical interleaver used in an orthogonal frequency division multiplexing (OFDM) system (Chapter 7) is shown in Fig. 6.26.

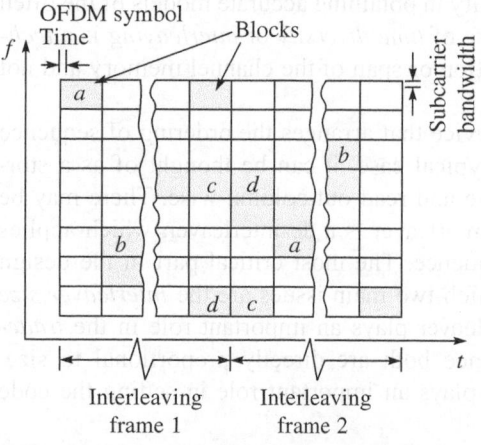

Fig. 6.26 A typical example of burst error handling by block interleaving in OFDM

Data transmission in this OFDM system concept is performed by blocks of several subcarriers in several consecutive OFDM symbols. The blocks of subcarriers and the OFDM symbols used for transmission are determined by a frequency hopping pattern (Chapter 7). This pattern is repetitive with some period. We refer to this period as the interleaving frame. Coding and interleaving are performed over all blocks of data symbols transmitted during such an interleaving frame. Thus, one way of interpreting the OFDM concept is by the time and frequency grids shown in Fig. 6.26, where each transmission block (rectangle) is orthogonal to all other transmission blocks in the grid. The size of a transmission block depends on the choice of parameters for the OFDM system. This may eliminate the probability of the whole block of frame in error. The concept is just like writing vertically and reading it horizontally. This incorporates randomness in the blocks.

6.13 PERFORMANCE OF TURBO CODES

Performance of turbo codes differs based on various parameters or various changes made in the input or in the system design. This section discusses the variations in performance. The general performance comparison between convolution and turbo codes is given in Fig. 6.27, in terms of BER vs E_b/N_o.

Number of Iterations

Obviously, the number of iterations increases the performance and at the same time increases the latency period (Fig. 6.27). Therefore, there should be a trade-off between the latency

---

[†] Turbo code is a modified linear block code as both recursive systematic convolutional (RSC) encoders are linear.

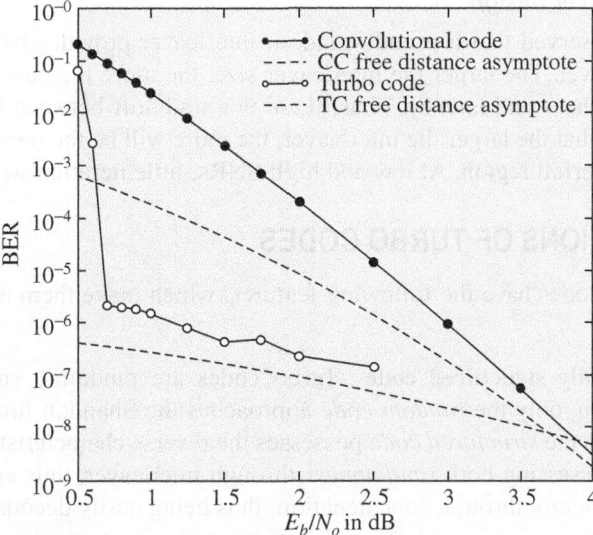

Fig. 6.27 Typical performance comparison between turbo code ($k = 5$, $L = 65,536$) and convolutional code ($C = 15$)

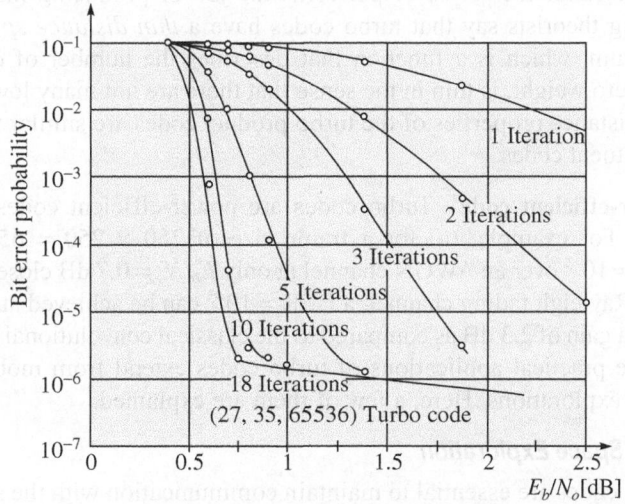

Fig. 6.28 Effect of number of iterations

period and the performance gain. Normally, the number of iterations is 18, but in most of the cases, 6 will suffice.

Block Size

It is observed that an increase in the block size results in a steady increase in the performance gain. If the block size is increased, then the size of the interleaver must be correspondingly increased. Since a non-uniform random interleaver is used, a large interleaver size will improve the randomization between adjacent bits (the more is the randomness, the more it will approach towards the Shannon limit); moreover, low-weight code words are hardly likely to occur.

Interleaver Design

It is observed that a pseudo-random interleaver provides better performance than a block interleaver. The larger the interleaver size, the more the performance gain due to increased randomness and latency. Thus, there is a trade-off between latency and performance. It is shown that the larger the interleaver, the more will be the iterations necessary to converge in the waterfall region. At low and high SNRs, little iteration will suffice.

6.14 APPLICATIONS OF TURBO CODES

Turbo codes have the following features, which make them attractive for the latest applications.

Randomly structured code Turbo codes are randomly structured codes. According to Shannon, only the *random code* approaches the Shannon limit but it is difficult to decode, whereas the *structured code* possesses the reverse characteristics. A turbo code is an optimal one, possessing both *randomness* through interleaver, thus approaching the Shannon limit, and *structure* through concatenation, thus being easily decodable.

Good linear code with thin distance spectrum Turbo codes are good linear codes. A *good* linear code is one that has mostly high-weight (Hamming weight) code words. The pseudo-random interleaver performs the job of producing mostly high Hamming weight. Coding theorists say that turbo codes have a *thin distance spectrum*. That is, the distance spectrum, which is a function that describes the number of code words of each possible non-zero weight, is thin in the sense that there are not many low-weight code words present. The distance properties of the turbo product codes are similar to those of the product of the constituent codes.

Power-efficient code Turbo codes are power-efficient codes with high correction capability. For example, (a) for a frame size of $250 \times 250 = 65,536$ bits, we can achieve a BER $= 10^{-5}$ over an AWGN channel at only $E_b/N_o = 0.7$ dB close to the Shannon limit, and (b) for a Rayleigh fading channel, a BER $= 10^{-5}$ can be achieved at $E_b/N_o = 4.3$ dB, which represents a gain of 2.3 dB as compared to the classical convolutional code with similar complexity.

The practical applications of turbo codes extend from mobile communications to deep space explorations. Here, a few of them are explained.

Deep Space Exploration

Turbo codes are essential to maintain communication with the spacecraft exploring the solar system, because over the vast distances involved, signal power is at a premium. An improvement of a small fraction of a decibel can make a difference of millions of miles to the operational range of a mission. The Jet Propulsion Laboratory (JPL), which carries out research for NASA, was among the first to realize the potential of turbo codes, and as a result, these codes were used in the Mars Pathfinder mission.

Mobile Communications

Turbo codes are one of the options for FEC coding in the universal mobile telecommunication system (UMTS) third-generation mobile radio standard. A great deal of development has been carried out here, especially on the design of interleavers of different lengths, for applications both to speech services, where latency must be minimized, and to data services, which must provide very low BER. Turbo codes are mainly considered for WiMAX (Chapter 12) and long-term evolution (LTE) (Chapter 11) systems described later on. Reed–Solomon Convolution Code (RS-CC) concatenation is used in the channel coding stage.

Terrestrial Digital Video Broadcast Standard

Recently, turbo codes have been incorporated into the standard, which will include a return channel in digital broadcast systems.

MORE SOLVED EXAMPLES

Example 6.9 If a rate ½ channel-coded PCM signal is required to transmit with an SNR of 10 dB and a bandwidth of 200 kHz, what should be the maximum bit rate of the PCM signal for error-free transmissions?

Solution $SNR = 10$ dB or $SNR = 10$

$$C = W \log_2\left(1 + \frac{S}{N}\right)$$

Now, the channel-coded transmission rate must be less than or equal to the channel capacity. Considering the worst-case scenario,

$$\text{Channel} - \text{coded rate} = (200 \times 10^3)\log_2(1 + 10)$$

$$= (200 \times 10^3) \times 3.46 = 692 \text{ kbps}$$

$$\text{Maximum PCM bit rate} = \frac{1}{2} \times 692$$

$$= 346 \text{ kbps (for error} - \text{free transmissions)}$$

Beyond this rate, errors may occur.

Example 6.10 For Fig. 6.29, find the coded output for the given input 10110111001, with the least significant bit (LSB) first. What is the code rate? What is the constraint length?

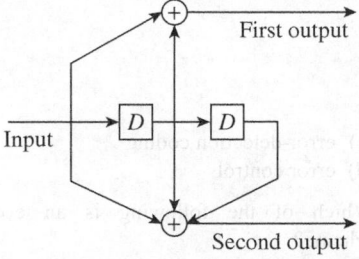

Fig. 6.29 Convolutional encoder for Example 6.10

Solution Assuming reset condition zero at the beginning, Tables 6.2 and 6.3 show the first and second output values.

Table 6.2 First output

Present state	State after delay	Output
1 (LSB)	0	1
0	1	1
0	0	0
1	0	1
1	1	0
1	1	0
0	1	1
1	0	1
1	1	0
0	1	1
1	0	1

Table 6.3 Second output

Present state	State after delay	State after second delay	Output
1 (LSB)	0	0	1
0	1	0	1
0	0	1	1
1	0	0	1
1	1	0	0
1	1	1	1
0	1	1	0
1	0	1	0
1	1	0	0
0	1	1	0
1	0	1	0

Output of the encoder n is 1–1 1–1 0–1 1–1 0–0 0–1 1–0 1–0 0–0 1–0 1–0
Number of shift registers $L = 2$
Code rate = ½

Constraint length = $L + 1 = 2 + 1 = 3$

Constraint length measured in terms of number of bits = $n(L + 1) = 22(3) = 66$ bits

Example 6.11 A convolutional encoder is given with the following generator polynomial:

$$G = [1 + D^2, 1 + D + D^2]$$

If puncture coding is applied with the puncture matrix $\begin{bmatrix} 1 & 0 \\ 1 & 1 \end{bmatrix}$ for rate 2/3 encoding, find the code for the input 110111.

Solution Table 6.4 provides the output values:

Table 6.4 Output

Input	State after D	State after D^2	$1 + D^2$	$1 + D + D^2$
1	0	0	1	1
1	1	0	1	0
0	1	1	1	0
1	0	1	0	0
1	1	0	1	0
1	1	1	0	1

For applying puncturing, remove the third bit of the four. Therefore, the final pattern will be 1–1 0–1 0–0 1–0 1. (Rate ½ divided by rate ¾ gives the 2/3 rate code.)

SUMMARY

- Through channel coding, we add redundant bits to the data in a systematic way.
- Channel coding improves the BER performance in general.
- There are two types of channel coding—waveform and structured sequence coding.
- There are two basic approaches to adding structured redundancy—ARQ and FEC. Hence, we may have error-detecting codes and error-correcting codes.
- Selection of the code rate involves a trade-off between energy efficiency and bandwidth efficiency.
- Communication engineers always try to reduce the transmission bandwidth but at the same time, $W > C > R$.
- The minimum distance between the codes gives the number of bits in an error.
- Parity check, CRC, and so on are error-detecting codes and are mostly used with retransmissions.

- Block, convolutional, and turbo codes are error-correcting codes and do not allow retransmissions. These are mainly used with real-time systems.
- In the latest systems, error-detecting and error-correcting codes are used together appropriately.
- Convolutional codes are suitable for a channel with memory.
- Viterbi algorithm is used for decoding convolutional codes. Decoding is a complex task performed at the receiver and its capabilities decide the performance of the overall system.
- Turbo codes are random-natured, concatenated, and structured codes and undergo iterative decoding.
- Turbo codes can approach the Shannon limit.
- Interleavers and puncture codes play an important role in turbo coding. The interleaver increases the randomness in the code, and puncturing increases the code rate.

EXERCISES

Multiple-choice Questions

6.1 Through channel coding, the information transmission efficiency
   (a) increases
   (b) reduces
   (c) remains the same
   (d) sometimes increases and sometimes reduces

6.2 An error-correction coding method is also called
   (a) ARQ
   (b) FEC

   (c) error-detection coding
   (d) error control

6.3 Which of the following is an error-detection scheme?
   (a) Turbo coding   (c) Cyclic coding
   (b) Convolutional coding (d) Parity check

6.4 Which of the following is an error-correction scheme?
   (a) Parity check   (c) RS coding
   (b) LRC   (d) CRC

6.5 Code rate is nothing but
   (a) *nk*   (b) *n/k*   (c) *k/n*   (d) *n − k*

6.6 The Hamming weight in the sequence 1110100 is
   (a) 3   (b) 1   (c) 4   (d) 7

6.7 The error-detecting capability of a code t_d is defined as
   (a) $t_d = d_{min} − 1$
   (b) $t_d = (d_{min} − 1)/2$
   (c) $t_d = 1 − d_{min}$
   (d) $t_d = d_{min} − 1/2$

6.8 Which of the following codes can approach the Shannon limit?
   (a) Convolutional code   (c) Turbo code
   (b) Hamming code   (d) Block code

6.9 Quantization process is used in
   (a) Maximum likelihood decoding
   (b) soft decision decoding
   (c) Viterbi decoding
   (d) hard decision decoding

6.10 Which of the following stages adds randomness to the turbo code?
   (a) Interleaver   (c) Inner encoder
   (b) Code puncturing   (d) Convolutional coding

6.11 Which of the following changes the code rate?
   (a) Interleaver   (c) Code puncturing
   (b) Hamming bits   (d) Convolutional coding

6.10 Find the various algorithms for decoding a convolutional code.

6.11 What will be the role of the number of iterations in turbo codes?

6.12 How can we say that the interleaver plays a trade-off between performance and time delay?

6.13 How can the interleaver be incorporated in a concatenated code generation and its decoding? What will be its advantage?

6.14 Find the turbo coding methods or diagrams in which puncture codes are incorporated.

6.15 Compare the BER performance of convolutional-coded and turbo-coded BPSK modulators. In addition, comment on the limitations of turbo coding.

6.16 Why is the turbo coding method considered to be the most efficient method?

6.17 Explain in detail cyclic redundancy with an example. In addition, explain why CRC check is often used with convolutional coding.

6.18 Simulate the algorithm for soft decision making for decoding purpose. List the decoding algorithms in which soft decision making is used.

Review Questions

6.1 Why is information theory applied to both source and channel coding? What do you mean by removal of redundancy in source coding? What do you mean by addition of redundancy in channel coding?

6.2 How do data and information differ?

6.3 Describe the methods for controlling errors. How can error control be achieved with the error-detection schemes?

6.4 What types of errors may occur in the data? What may be the reasons behind them?

6.5 Explain the error-detection and error-correction capabilities of linear block codes.

6.6 State and prove the Shannon–Hartley law. State the significance of the Shannon limit. Discuss the overall work of Shannon, who introduced many laws.

6.7 What do you mean by rate ½ and rate 1/3 convolutional coders?

6.8 How can you represent a trellis diagram in terms of a state diagram?

6.9 For error handling, what is the limit up to which redundancy can be added to the source-coded data to be transmitted?

Numerical Problems

6.1 If a channel coding scheme is incorporated using a block code such that it is to transmit three times the binary information symbols, what may be the code rate and the redundancy added?

6.2 Find a generator polynomial $g(x)$ for a (7, 4) cyclic code and find the code vectors for the following data vectors: 1010, 1111, 0001, and 1000.

6.3 Consider rate 4/7 (7, 4, 3) Hamming codes with the following parity check matrix:

$$H = \begin{bmatrix} 1 & 0 & 1 & 0 & 1 & 0 & 1 \\ 0 & 1 & 1 & 0 & 0 & 1 & 1 \\ 0 & 0 & 0 & 1 & 1 & 1 & 1 \end{bmatrix}$$

Find the set of code words for all the combinations of message bits starting from 0000 to 1111.

6.4 Consider the (3, 1, 3) convolutional code with impulse responses $g^{(1)} = (110)$, $g^{(2)} = (101)$, and $g^{(3)} = (111)$.
   (a) Draw the encoder block diagram.
   (b) Find the generator matrix.
   (c) Find a code word corresponding to the information sequence (11101).
   (d) Draw the code tree.

6.5 In order to construct a (15, 7) BCH code for the coefficients given in the following table, what will be the generator polynomial? Show all the steps.

n	k	t_c	Generator polynomial coefficients
7	4	1	1011
15	7	2	111010001
15	5	3	10100110111

6.6 In the example of hard decision Viterbi decoding discussed in Section 6.9.4, if the message bit is 110110, the coded bits are 111001111001, and the received bits are 111001110001, find the trellis and the decoding path on it.

6.7 With an RS code RS (255, 223) with eight-bit symbols, how many errors can be corrected? What will be the maximum block length? If the code is shortened to (200, 168), how will the transmission be?

7

Modulation Techniques

Theme of the Chapter

Modulation is required to make a baseband signal suitable for transmission over a channel. In this chapter, we will concentrate on single-carrier and multicarrier digital modulation schemes. Among these schemes, a few are power efficient and others are spectrally efficient. The general feature of these schemes is phase mapping in the in-phase and quadrature (IQ) vector plane representing the typical constellations. The system performance is different for different modulation schemes. Hence, depending upon the application, environment or channel, and signal-to-noise ratio condition, the suitable scheme must be selected. Different diagrams, such as eye, polar or vector, constellation, and trellis, are explained in the chapter, as they are important in the performance analysis and comparison of the modulation schemes. Spectrum efficiency and power considerations are also detailed. For all the schemes discussed here, the input signal will be in digital form and the output signal will be in analog form. Spread spectrum modulation (SSM) and orthogonal frequency division multiplexing (OFDM) are the latest modulation techniques and hence are given more weightage than the conventional modulation techniques in this chapter. SSM is achieved with the help of a unique code. The code repeatedly becomes pseudo-noise sequences, which must be known at the receiver end to despread the signal again. The advantage of the system is highly secure communication. OFDM is a digital modulation scheme that can support high-speed video communication, with the elimination of intersymbol interference and interchannel interference. At the same time, it can accommodate more number of users, exhibiting spectral efficiency. It is a multiplexing as well as a multiple access scheme.

Key Topics

- Digital modulation performance parameters
- Coherent and non-coherent systems
- Polar and constellation diagrams
- Eye diagram
- Trellis diagram
- Line coding or signalling
- Constant envelope modulation schemes: BPSK, QPSK, M-PSK, FSK, MSK, and GMSK

- Variable envelope modulation schemes: ASK, QAM, and M-QAM
- Differential modulation schemes: DQPSK and DQAM
- IQ offset modulation scheme: OQPSK
- Bandwidth and spectrum efficiency
- Issues related to transmission power
- Spread spectrum modulation

- Pseudo-noise code and properties
- DSSS, FHSS, THSS, and hybrid techniques
- SSM advantages
- Multicarrier modulation techniques
- OFDM subcarrier allocations and spectrum setting
- OFDM block diagram
- OFDM advantages

7.1 DIGITAL MODULATION AND PERFORMANCE PARAMETERS

Let us begin with a revision of the fundamentals. A digital information signal in the time domain can be viewed as a periodic sequence of pulses depending upon the bit pattern of 1 and 0. The bit rate is simply the inverse of the bit period. Digital signals have equivalent representations in the frequency domain, where the energy of the signal is spread across a set of frequencies. This representation is called the *power spectrum* or *spectrum*. The signal bandwidth is a measure of the width of the spectrum. The bandwidth and bit rate of a digital signal are related but are not exactly the same. The relation between them depends on the type of modulation used. The ratio of bit rate to available bandwidth is called the *spectral efficiency* and is measured in bits per second per hertz (bits/s/Hz).

Typically, the objective of a digital communication system is to transport digital data between two or more nodes. In wireless communications, this is usually performed by adjusting the physical characteristic of a sinusoidal carrier—frequency, phase, amplitude, or a combination thereof. The carrier characteristic in real systems is adjusted with a modulator at the transmitting end to impose the physical change to the carrier and with a demodulator at the receiving end to detect such changes and thereafter to get the data back.

Developers of communication systems face the following constraints: (a) available bandwidth, (b) permissible power, and (c) inherent noise level of the system.

7.1.1 Coherent and Non-coherent Systems

The terms *coherent* and *non-coherent* a frequently used when discussing the generation and reception of digital modulation. When linked to the process of modulation, the term *coherence* relates to the ability of the modulator to control the phase of the signal and not just the frequency. The example of frequency shift keying (FSK) given in Fig. 7.1 shows that an FSK-modulated

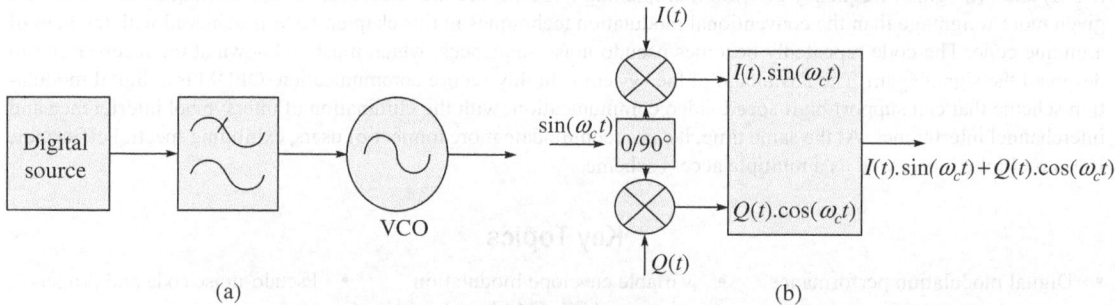

Fig. 7.1 FSK-modulated signal (a) Non-coherent generation (b) Coherent generation

---

SELECTION CRITERIA OF DIGITAL MODULATION SCHEMES

- Compared to analog modulation, digital modulation provides more information capacity, compatibility with digital data services, higher data security, better quality communication, and quicker system availability.
- The choice of the digital modulation scheme will significantly affect the characteristics and performance of the system. There are several digital modulation schemes, but there is no *universally* best choice of scheme. Depending on the link design, some will prove to be a better fit than others.
- While selecting the modulation scheme, consideration must be given to channel quality, required data rate, acceptable level of latency, available bandwidth, link budget, hardware cost, size, and current consumption.
- Rather than bits, symbols are considered for digital modulation schemes. It is because symbols can be easily converted into modulo values for implementation purpose. In addition, symbols are actually mapped into constellation diagram.

signal can be generated both coherently with an in-phase and quadrature (IQ) modulator and incoherently with a voltage-controlled oscillator (VCO) and a digital voltage source.

In a system as shown in Fig. 7.1(a), the instantaneous frequency of the output waveform is determined by the modulator (within a tolerance set by the VCO, data amplitude, etc.), but the instantaneous phase of the signal is not controlled and can have any value. Alternatively, coherent generation of the modulation is achieved as shown in Fig 7.1(b). Here, the phase of the signal is controlled rather than the frequency.

The modulator shown in Fig. 7.1 offers the possibility to shape the resultant carrier phase trajectory at baseband with either analog filtering or digital signal processing and a digital-to-analog converter (DAC). This can be used to generate both constant amplitude signals and amplitude-modulated signals. The use of the term *coherent* with respect to the act of demodulation refers to a system that makes a demodulation decision based on the received signal phase and not the frequency. The additional *information* due to forward error correction (FEC) results in an improved bit error rate (BER) performance.

To understand a coherent system, it is necessary to study first the polar representation of the signal and then its IQ components. For phase control based systems, on the basis of the received phase, hard or soft decision-making is to be incorporated at the receiver end to form the bit sequence again before channel decoding is performed.

7.1.2 Polar Representation and In-phase–Quadrature Diagrams

Polar representation and rectangular representation of a polar diagram (*IQ*) plays an important role in finding the BER.

A signal can be represented by its amplitude, frequency, and phase. A carrier signal is an analog signal, and in digital modulation schemes, the amplitude and phase of the carrier are very important once its frequency is chosen. Amplitude is essential for the identification of logic levels and phase for synchronization. These can be viewed in a simple way with a polar diagram.

Generally, the carrier becomes a frequency and phase reference, and the received signal is interpreted relative to the carrier. The signal can be expressed in polar form as a magnitude and a phase. The phase is relative to the reference carrier signal, and the magnitude is either an absolute or a relative value. Polar diagrams are the basis of many displays used in digital communications, although it is common to describe the signal vector by its rectangular coordinates of *I* (in-phase) and *Q* (quadrature).

Figure 7.2 shows different forms of modulation in polar form. There may be magnitude changes (7.2 (b)), phase (7.2 (c)) or frequency changes (7.2 (d)), or both magnitude and phase changes (7.2 (e)). Zero degree is the reference set due to carrier with zero phase.

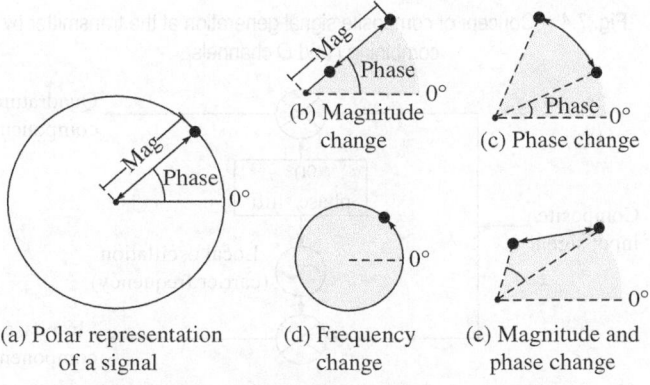

(a) Polar representation of a signal   (b) Magnitude change   (c) Phase change   (d) Frequency change   (e) Magnitude and phase change

Fig. 7.2 Signal changes or modifications in polar form

In digital communications, the modulated signal is often expressed in terms of the *I* and *Q* components. This is a rectangular representation of the polar diagram. On a polar diagram, the *I* axis lies on the 0° phase reference, and the *Q* axis is rotated by 90°. The signal vector's projection onto the *I* axis is its *I* component and the projection onto the *Q* axis is its *Q* component (Fig. 7.3). This representation is very important to understand the shift keying techniques as well as the orthogonal frequency division multiplexing (OFDM) technique. These *I* and *Q* components are orthogonal and do not interfere with each other.

The *IQ* diagrams are particularly useful because they mirror the way most transmission signals are created using an IQ modulator. Signals in the *I* and *Q* channels that are separated by 90° are also known as being orthogonal to each other or in quadrature. They are two independent components of a signal. When recombined, they give a composite output signal or complex signal whose magnitude is found by following the trigonometry rules. Examples for this are provided in Figs 7.4 and 7.5.

The composite signal with information about the magnitude and phase (*I* and *Q*) arrives at the receiver input. The input signal is mixed with the local oscillator signal at the carrier frequency in two forms, as shown in Fig. 7.5. One is at an arbitrary zero phase, while the other has a 90° phase shift. The composite input signal is thus broken into an in-phase, *I*, component and

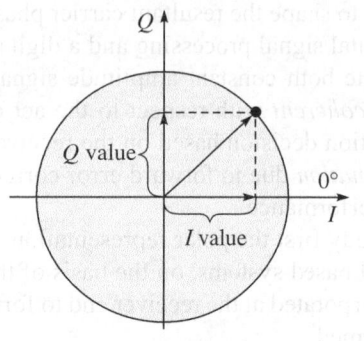

Polar-to-rectangular conversion

Fig. 7.3 IQ vector diagram representation

Note: The main advantage of IQ modulation is the symmetric ease of combining independent signal components into a single composite signal and later splitting such a composite signal into its independent components.

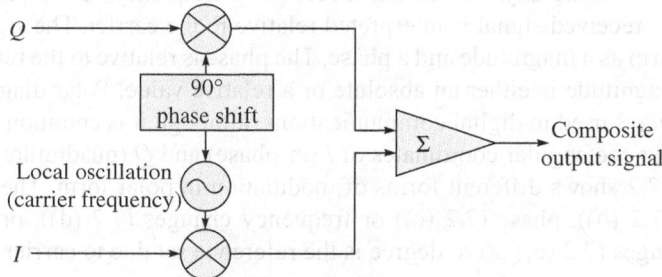

Fig. 7.4 Concept of composite signal generation at the transmitter by combining *I* and *Q* channels

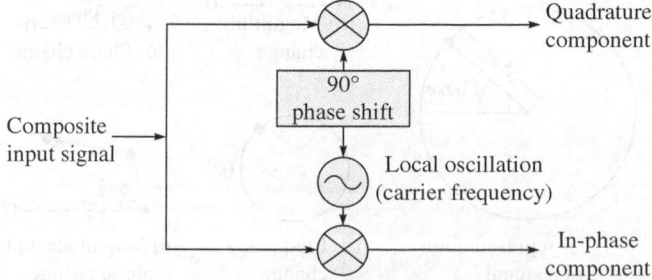

Fig. 7.5 Concept of splitting *I* and *Q* channels at the receiver

The signal may, in principle, circle the origin in one direction forever, necessitating infinite phase shifting capability.

a quadrature, Q, component. One can be changed without affecting the other. Normally, information cannot be plotted in a polar format and reinterpreted as rectangular (Cartesian) values without doing a polar-to-rectangular conversion. This conversion is done by the in-phase and quadrature mixing processes in a digital radio. A local oscillator, a phase shifter, and two mixers can perform the conversion accurately and efficiently.

Most digital modulators map the data to a number of discrete points on the IQ plane. These points are known as *constellation points*. As the signal moves from one point to another, simultaneous amplitude and phase modulation usually results. It is difficult and complex to accomplish this with an amplitude modulator and a phase modulator. Alternatively, simultaneous amplitude modulation (AM) and phase modulation is easy with an IQ modulator.

For IQ modulation purposes, symbols are inserted as an input. Symbols are the multiple bits to be processed together as per the requirement of the constellation points. A symbol clock represents the frequency and exact timing of the transmission of the individual symbols. At the symbol clock transitions, the transmitted carrier is at the correct IQ value to represent a specific symbol (a specific point in the constellation).

7.1.3 Constellation Diagrams

A polar diagram shows several symbols at a time. It shows the instantaneous value of the carrier at any point on the continuous line of the circle, represented as IQ or magnitude–phase values (Fig. 7.2).

A constellation diagram shows a repetitive *snapshot* of the same burst, with the values shown only at the discrete decision points on the IQ plane.

The relationship between the constellation points and bits per symbol is given by

$$M = 2^n$$

where M is the number of constellation points and n is the number of bits per symbol, that is,

$$n = \log_2(M) \tag{7.1}$$

Constellation diagrams can also be considered as signal space diagrams represented in terms of the basis functions, which are one dimensional for amplitude shift keying (ASK) and two dimensional for M-ary phase shift keying (M-PSK) schemes. In general, an M-PSK constellation is circular except the special case of $M = 2$. Constellation diagrams are shown along with different modulation schemes in later sections of this chapter.

> Notes:
> 1. Constellation diagrams display phase errors, as well as amplitude errors, at the decision points. Transitions from decision points degrade the quality of reception.
> 2. Constellation diagrams indirectly provide insight into the varying power levels and the effects of filtering and intersymbol interference (ISI).

7.1.4 Eye Diagrams

Another way to view a digitally modulated signal is with an eye diagram. Separate eye diagrams can be generated for the I channel and Q channel data. Eye diagrams display I and Q magnitude versus time in an infinite persistence mode, with retraces.

In Fig. 7.6, the I and Q transitions are shown separately and an *eye* (or eyes) is formed at the symbol decision times. Quadrature phase shift keying (QPSK) has four distinct IQ states, one

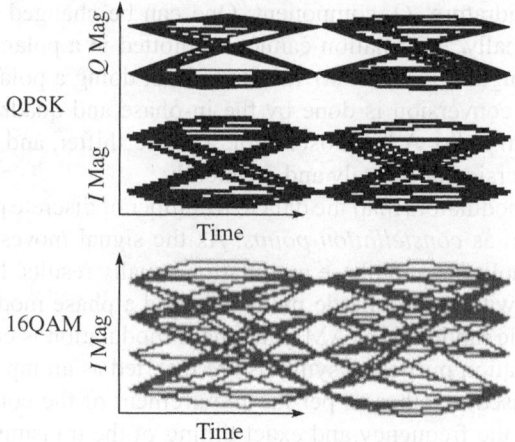

Fig. 7.6 Example of an eye diagram

in each quadrant. There are only two levels for I and two levels for Q. This forms a single eye for each I and Q. Other schemes use more levels and create more nodes in time through which the traces pass. The second example is a 16 quadrature amplitude modulation (QAM) signal, which has four levels forming three distinct eyes. The eye is open at each symbol. A *good* signal has wide open eyes with compact crossover points.

7.1.5 Trellis Diagrams

A *trellis* diagram resembles a garden trellis, hence the name. We have already discussed trellis diagrams for convolution decoding in Chapter 6. Figure 7.7 shows the possible paths for decoding, indirectly representing different phases versus time, of which the best path is to be recovered following an algorithm so that minimum phase errors can be observed at the receiver end.

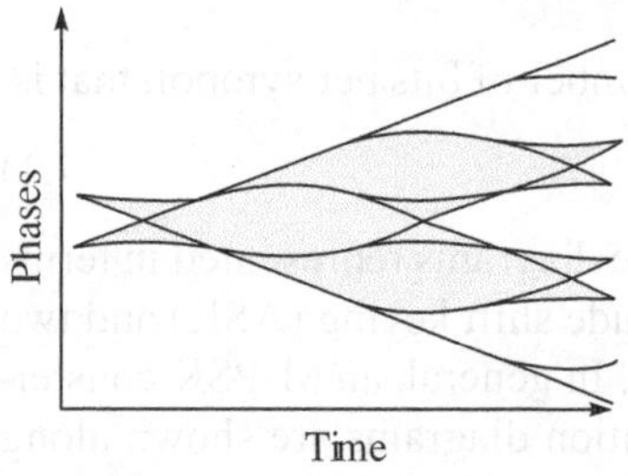

In general, a trellis diagram shows the time on the x-axis and the phase on the y-axis. This allows the examination of the phase transitions with different symbols. If a long series of binary 1's is sent, the result is a series of positive phase transitions of certain degrees per symbol. If a long series of binary 0's is sent, there is a constant declining phase of 90° per symbol. Typically, there are intermediate transmissions with random data. When troubleshooting, trellis diagrams are useful in isolating missing transitions, missing codes, or blind spots in the IQ modulator or mapping algorithm.

Fig. 7.7 Trellis diagram

There are many ways of looking at a digitally modulated signal, some of which are as follows:

Time and frequency domain view Signals can be analysed either in the time domain or in the frequency domain. Signals can be converted from one domain into another to get the relation between the data rate and the corresponding bandwidth occupied.

Power versus frequency view Signals can also be analysed in terms of the amount of power assigned to each frequency component (power spectral density).

Crest factor is an important parameter and is defined as peak-to-average power ratio.

Frequency versus time view This view approximates the data rate or bandwidth variations with time.

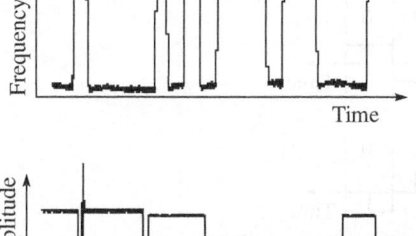

Fig. 7.8 Examples of frequency and power amplitude variations with time

Power versus time view This view is very useful for examining the power level changes involved in pulsed or burst carriers and for understanding how transmitters are turned on and off.

Figure 7.8 shows the variation of frequency and power magnitude with time.

> *Notes:*
> 1. Very fast power changes will result in frequency spreading or spectral growth.
> 2. Very slow power changes waste valuable transmit time, as the transmitter cannot send data when it is not fully on.
> 3. Turning on too slowly can also cause high BERs at the beginning of the burst.
> 4. Asking for excessive power from an amplifier can lead to compression or peak power clipping; hence, peak and average power levels must be well understood to avoid signal distortion.

7.2 LINE CODING OR SIGNALLING

We have seen the digital input–digital transmission type of communication system. In digital baseband, the binary form of signals is not always sufficient to match the channel requirements. A digital baseband signal is often converted from the binary form to another form. Sometimes, extra pulses or transitions may be added in this conversion. Hence, it is considered as a line coding technique.

Line coding provides particular spectral characteristics of a pulse train. The most common codes for mobile communications are return-to-zero (RZ), non-return-to-zero (NRZ), and Manchester encoding. All of these may be either unipolar (with voltage levels 0 or V) or bipolar (with voltage levels $+V$ or $-V$).

- In the RZ code, the pulse returns to zero within every bit period. This widens the spectrum but improves timing synchronization.
- The NRZ codes do not return to zero during a bit period; the signal stays at a constant level throughout a bit interval. These codes are more spectrally efficient than RZ codes but are poor in synchronization capabilities. Unipolar signals may contain a DC component, which may be blocked in some part of the hardware; hence, normally, bipolar NRZ is preferred.
- The Manchester code is an NRZ code with special features, may be without a DC level and ideally suitable for signalling purpose. It has synchronization capabilities and offers at least one transition per bit interval; hence, these codes provide two pulses to represent each binary symbol. Easy clock recovery is possible with Manchester coding.

> Line coding is regarded as a modulation technique in baseband systems.

Figure 7.9(a) given in Example 7.1 shows examples of waveforms of these line codes. There are also other forms of line codes, such as differential Manchester encoding and multilevel coding.

Example 7.1 For the binary pattern 01001110, draw the waveforms of NRZ, RZ, and Manchester coding.

Solution Figure 7.9(a) shows the NRZ, RZ, and Manchester coded binary signals and Fig. 7.9(b)

shows the bandwidth comparison for these signalling schemes.

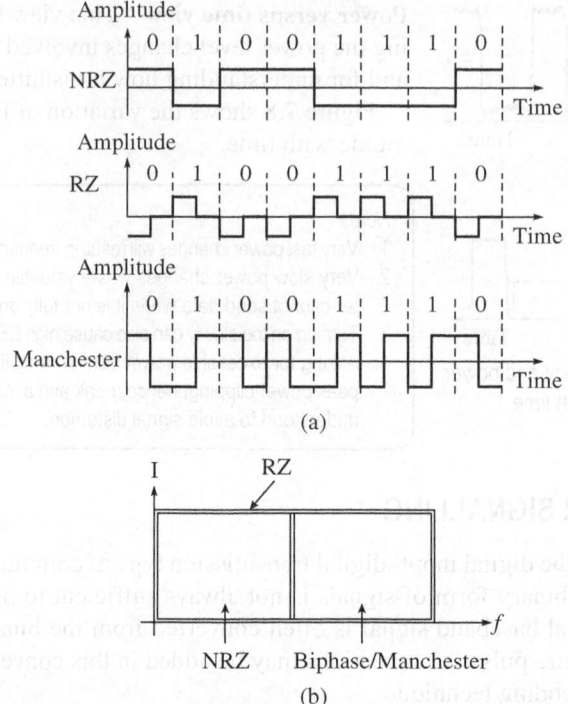

Fig. 7.9 Line coding schemes (a) NRZ, RZ, and Manchester coded binary signals (b) Bandwidth comparison

Let us classify the digital modulation schemes into the following four main categories in general:

(a) Constant envelope modulation, such as M-PSK, FSK, minimum shift keying (MSK), and Gaussian MSK (GMSK)
(b) Variable envelope modulation, such as ASK and M-ary QAM (M-QAM)
(c) Differential modulation, such as differential phase shift keying (DPSK) and differential QAM (DQAM)
(d) IQ offset modulation, such as offset QPSK (OQPSK) and offset keyed QPSK (OKQPSK)

BANDWIDTH COMPARISON FOR SIGNALLING SCHEMES

NRZ Maximum frequency occurs when the data is 10101010101.... This sequence produces a square wave. From the Fourier series, the waveform contains the first, third, fifth, and so on (odd) harmonics. If T is the period of one cycle and if the signal is passed through a low-pass filter, then the maximum bandwidth would be $1/T$ Hz; that is, if the fundamental harmonic is allowed to pass, the data sequence 101010... can then be completely recovered. Hence, the minimum channel bandwidth $W_m = 1/T$. The lower frequency limit is DC in this case, because the continuous 1's transmitted decide this.

RZ Maximum frequency occurs when continuous 1's are transmitted. In this case, one cycle of the square wave occurs within one-bit duration. If the sequence is of continuous 0's, then the $-V$ signal level decides the DC. Thus, for RZ, the bandwidth is twice the bandwidth of NRZ.

Manchester Biphase (Manchester) signalling also has the maximum frequency case similar to RZ when continuous 1's or 0's are transmitted. However, minimum frequency occurs when the sequence is 101010.... This relative comparison is shown in Fig. 7.9(b).

7.3 CONSTANT ENVELOPE MODULATION

> Constant envelope modulation is not suitable in systems where bandwidth efficiency is more important than power efficiency.

In constant envelope modulation, the amplitude of the carrier is constant, regardless of the variation in the modulating signal. It is a power-efficient scheme that allows efficient class-C amplifiers to be used without introducing degradation in the spectral occupancy of the transmitted signal. However, constant envelope modulation techniques occupy a larger bandwidth than the schemes in which the amplitude of the transmitted signal varies with the modulating digital signal.

The M-PSK and FSK modulation schemes are of constant envelope type. MSK is a special type of FSK where the peak-to-peak frequency deviation is equal to half the bit rate. GMSK is a derivative of MSK, where the bandwidth required is further reduced by passing the modulating waveform through a Gaussian filter. The Gaussian filter minimizes the instantaneous frequency variations over time. All these schemes are discussed in detail in Sections 7.3.1–7.3.6.

7.3.1 Binary Phase Shift Keying

One of the simplest forms of digital modulation is *binary* or *biphase shift keying* (BPSK). One application that uses this scheme is deep space telemetry. BPSK is achieved by multiplying the data by a carrier $A\sin\omega_c t$, where the data has two possible levels, +1 and −1 and where A is the peak amplitude of the carrier and ω_c is the frequency of the carrier. The phase of a constant amplitude carrier signal moves between 0 and 180° or $\pm\pi/2$.

On an IQ diagram, the *I* state has two different values. There are two possible locations in the state diagram, so a binary 1 or 0 can be sent. The symbol rate is one bit per symbol. BPSK demonstrates better performance than ASK and FSK. Figure 7.10 shows the waveform and constellation diagram of BPSK.

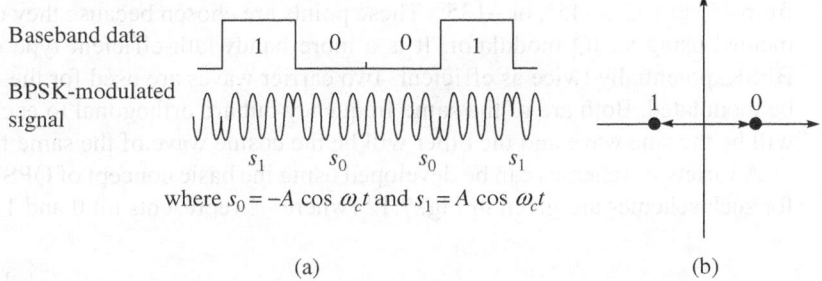

(a)   **(b)**

Fig. 7.10 BPSK (a) Waveform (b) Constellation diagram

7.3.2 Quadrature Phase Shift Keying

Another common type of phase modulation is QPSK. The transmitter and receiver diagrams for QPSK are given in Figs 7.4 and 7.5, along with the explanation of *I* and *Q* vectors; the transmitter diagram is repeated with a better representation in Fig. 7.11(a). Actually, BPSK can be expanded to an M-ary scheme, employing multiple phases and amplitudes as different states, and QPSK is a special case with $M = 4$. In any phase shift keying (PSK) scheme, the symbols are mapped into phases. In QPSK, two bits are used per symbol to be mapped into four phases.

The QPSK scheme is used extensively in applications such as code division multiple access (CDMA), cellular services, wireless local loops, Iridium (a voice or data satellite

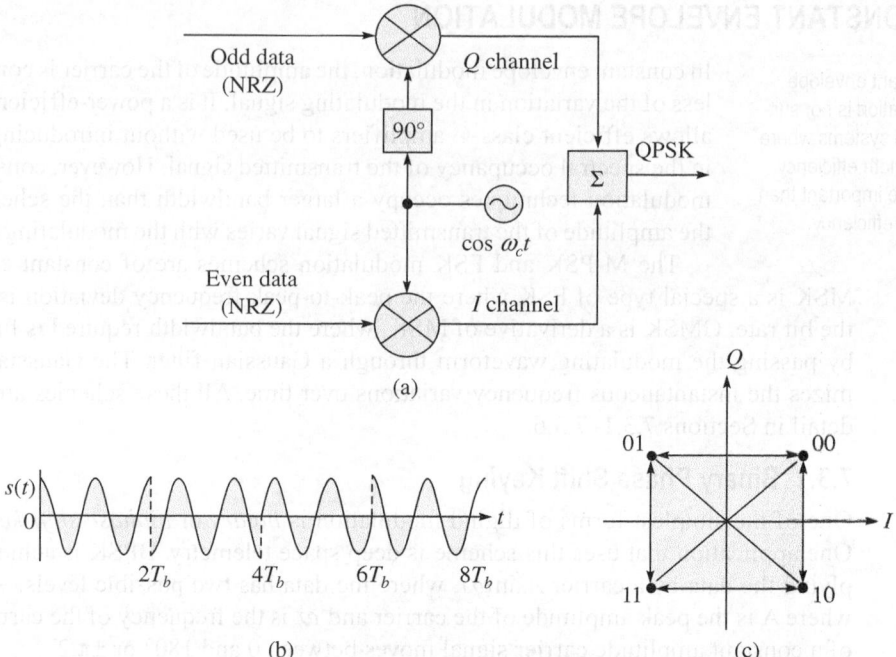

Fig. 7.11 QPSK (a) Transmitter (b) Waveforms on time axis (c) Two bits per symbol representing one phase on the constellation

system), and digital video broadcasting-satellite (DVB-S). Quadrature refers to the signal shifts between phase states that are separated by 90°. The signal shifts in increments of 90° from 45° to 135°, −45°, or −135°. These points are chosen because they can be easily implemented using an IQ modulator. It is a more bandwidth-efficient type of modulation than BPSK, potentially twice as efficient. Two carrier waves are used for the *I* and *Q* channels to be modulated. Both are of the same frequency but are orthogonal to each other; that is, one will be the sine wave and the other will be the cosine wave of the same frequency.

A variety of schemes can be developed using the basic concept of QPSK. The IQ diagrams for such schemes are given in Fig. 7.12 (where −1 represents bit 0 and 1 represents bit 1).

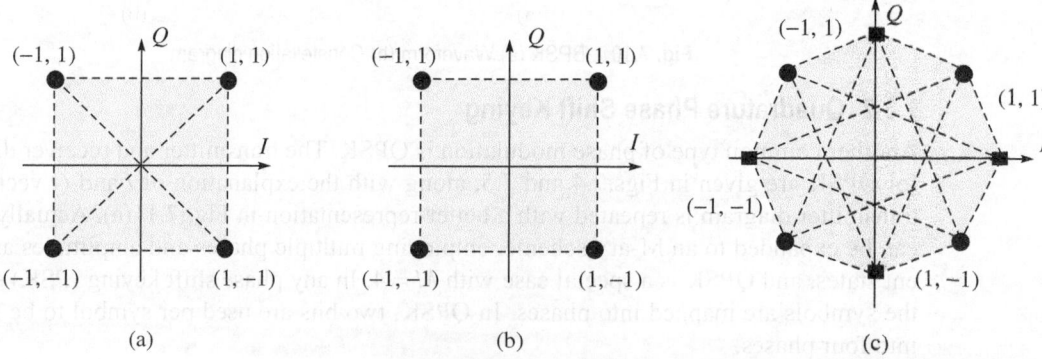

Fig. 7.12 Various modulation schemes based on QPSK but with different constellations (a) Conventional QPSK (b) OQPSK (c) π/4 QPSK

Conventional QPSK has transitions through zero (i.e., $\pm\pi$ or $180°$ phase transition). Sudden phase reversals can throw the amplifiers into saturation. As shown in Fig. 7.11(b), phase reversals cause the envelope to go to zero momentarily. This may cause non-linearities in the amplifier circuitry. This can be prevented using linear amplifiers but they are expensive and power consuming. The solution to these problems is the variations of QPSK—OQPSK and $\pi/4$ differential QPSK (DQPSK) explained in Sections 7.5 and 7.6.

7.3.3 M-ary Phase Shift Keying

In general, an M-PSK waveform is represented by

$$s(t) = f(t)\cos\left[\omega_c t + \frac{2\pi}{M}(m-1)\right] \quad 0 \le t \le T_s \tag{7.2}$$

where $m = 1, 2, ..., M$ denotes the M possible phases of the carrier $f_c = \omega_c/2\pi$ corresponding to the M possible data symbols represented by Eq. (7.1) and $f(t)$ is a real-valued pulse waveform (normally rectangular). All the M different M-PSK waveforms have the same energy:

$$E_s = \frac{1}{2}E_{f(t)} \tag{7.3}$$

where $E_{f(t)}$ denotes the energy of the basic pulse $f(t)$.

This signal set is two dimensional. Hence, PSK waveforms may be represented by a linear combination of two basis functions $\psi_1(t)$ and $\psi_2(t)$.

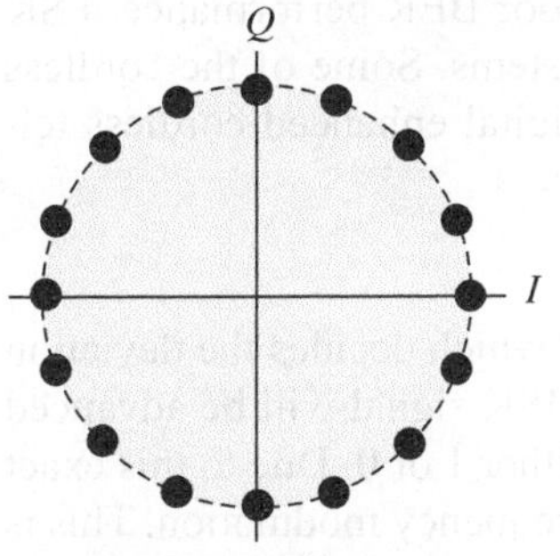

In general, M-PSK schemes are more spectrally efficient. As we increase the value of M, the consumption of the spectrum reduces. M is related to the powers of two—8, 16, 32, and so on. In 8PSK, three bits per symbol are read together to be mapped into phases. As there are eight combinations of three bits, eight points will be created on the constellation and so on. The constellation diagram for 16PSK is given in Fig. 7.13.

7.3.4 Frequency Shift Keying

The generic name for the digital family of frequency modulation is FSK. In this modulation, the frequency of the carrier is changed as a function of the modulating signal (data) being transmitted. The amplitude remains unchanged. In binary FSK (BFSK or 2FSK), a 1 is represented by one frequency and a 0 is represented by another frequency (Fig. 7.14).

Frequency modulation and phase modulation are closely related. A static frequency shift of +1 Hz means that the phase is constantly advancing at the rate of $360°$ per second (2π rad/s) relative to the phase of the unshifted signal. Bandwidth occupancy of FSK is dependent on the spacing of two symbols.

Fig. 7.13 Constellation of 16PSK

MFSK requires a considerably increased bandwidth in comparison to M-PSK.

SELECTION CRITERIA FOR M-PSK SCHEMES

- M-PSK is a power-efficient technique. It can be used wherever there are large amplitude variations.
- M-QAM schemes are utilized if the possibility of phase errors is high.
- Higher M-ary schemes have less spacing between the constellation points. Hence, they are greatly affected by noise.
- M-ary schemes are bandwidth efficient but are also susceptible to noise; hence, a higher value of M can be used with a higher signal-to-noise ratio (SNR) value, that is, when the mobile is near the tower.

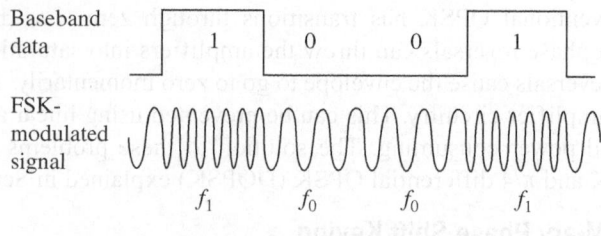

where $f_0 = A \cos(\omega_c - \Delta\omega)t$ and $f_1 = A \cos(\omega_c + \Delta\omega)t$

Fig. 7.14 FSK-modulated time domain waveform

A frequency spacing of inverse of 0.5 times the symbol period is typically used. FSK can be expanded to an M-ary scheme called M-ary FSK (M-FSK), employing multiple frequencies as different states, as given in Eq. (7.4).

$$s_i(t) = A\cos 2\pi[f_c + (2i - 1 - M)f_d]t, 1 \leq i \leq M \tag{7.4}$$

where f_c = carrier frequency
$\quad f_d$ = difference frequency
$\quad M$ = signalling elements = 2^n
$\quad n$ = bits per symbol or signalling element

An advantage of the FSK scheme is that it is very simple to generate and demodulate. In addition, due to the constant amplitude, it can utilize a non-linear power amplifier. Significant disadvantages, however, are the poor spectral efficiency and poor BER performance. FSK is used in many applications including cordless and paging systems. Some of the cordless systems using FSK are CT2 (cordless telephone 2), DECT (digital enhanced cordless telephone) and Bluetooth, which uses Gaussian FSK.

7.3.5 Minimum Shift Keying

Minimum shift keying is FSK with a modulation index of 0.25, which decides the deviation from the carrier frequency. Therefore, the carrier phase of an MSK signal will be advanced or retarded 90° over the course of each bit period to represent either 1 or 0. Due to this exact phase relationship, MSK can be considered as either phase or frequency modulation. This is why it is also called *continuous phase frequency shift keying* (CPFSK).

The result of this exact phase relationship is that MSK cannot be generated with a VCO and a digital waveform. Instead, an IQ modulation technique, as for PSK, is usually implemented. Coherent demodulation is usually employed for MSK due to its superior BER performance.

If we have two orthogonal signals representing bit 0 and bit 1, then

$$s_0(t) = \cos(2\pi f_0 t + \theta) \tag{7.5a}$$

$$s_1(t) = \cos(2\pi f_1 t + \theta) \tag{7.5b}$$

> MSK is widely used in real systems as it has exact phase relationship between each bit.

where θ is some arbitrary phase offset. It can be shown that the minimum frequency separation will be according to the orthogonality constraints—that is, $\Delta f = f_1 - f_0$ will be equal to $(1/2T_b)$, where T_b is the bit duration. This result is not dependent on the phase offset θ, and hence, we may design the system to have continuous phase (no discontinuity as in the case of M-PSK). This will result in

GMSK is used in the global system for mobile communication (GSM) cellular standard.

lower spectral side lobes. In other words, the value of θ in each symbol period is such that the phase continues smoothly from the previous symbol. If f_c is the centre frequency, then f_1 and f_0 can be represented as

$$f_0 = f_c - (1/4T_b) \tag{7.6a}$$

$$f_1 = f_c + (1/4T_b) \tag{7.6b}$$

Since a frequency shift produces an advancing or a retarding phase, frequency shifts can be detected by sampling the phase at each symbol period. Phase shifts of $(2N + 1)\pi/2$ radians $(N = 0, 1, 2, \ldots)$ are easily detected with an IQ demodulator. At even-numbered symbols, the polarity of the *I* channel conveys the transmitted data, whereas at the odd-numbered symbols, the polarity of the *Q* channel conveys the data. This orthogonality between *I* and *Q* simplifies detection algorithms and hence reduces power consumption in a mobile receiver. The minimum frequency shift that yields orthogonality of *I* and *Q* is that which results in a phase shift of $\pm\pi/2$ radians per symbol. FSK with this deviation is called MSK. The deviation must be accurate in order to generate repeatable 90° phase shifts. A phase shift of +90° represents a data bit equal to 1, whereas −90° represents a 0. The polar diagram of MSK is shown in Fig. 7.15. The peak-to-peak frequency shift of an MSK signal is equal to one-half of the bit rate.

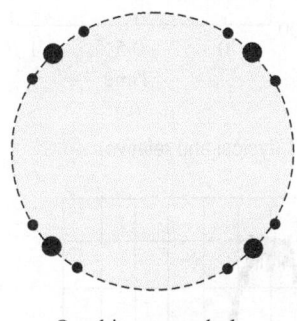

One bit per symbol

Fig. 7.15 Polar diagram of MSK

Both FSK and MSK produce constant envelope carrier signals, which is a desirable characteristic for improving the power efficiency of transmitters. Due to amplitude variations, non-linearities are found in an amplifier's amplitude transfer function, generating spectral growth, a component of adjacent channel power. Therefore, more efficient amplifiers (which tend to be less linear) can be used with constant envelope signals.

In practice, waveforms are filtered with a Gaussian filter. Hence, instead of rectangular pulses, Gaussian pulses enter the modulator, resulting in a narrow spectrum. In addition, the Gaussian filter has no time domain overshoot, which would broaden the spectrum by increasing the peak deviation.

7.3.6 Gaussian Minimum Shift Keying

Gaussian MSK is a derivative of MSK where the bandwidth required is further reduced by passing the modulating waveform through a Gaussian low pass filter (GLPF), as shown in Fig. 7.16. The Gaussian filter minimizes instantaneous frequency variations over time. The major advantages of GMSK are that it has a constant envelope, spectral efficiency, and a good BER performance and is self-synchronizing.

Like MSK, GMSK is a constant amplitude scheme. Moreover, GMSK can be viewed as either frequency or phase modulation. The phase of the carrier is advanced or retarded up to 90° over the course of a bit period depending on the data pattern, although the rate of change of phase is limited with a Gaussian response. The net result of this is that depending on the bandwidth time product (BT), effectively the severity of the shaping, the achieved

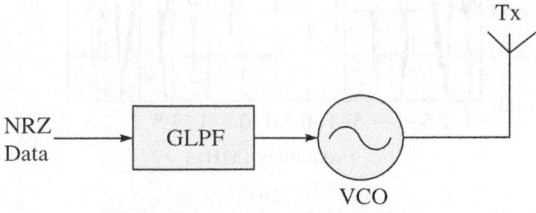

Fig. 7.16 Conceptual diagram of GMSK

phase change over the bit may fall short of 90°. This will obviously have an impact on the BER, although the advantage of this scheme is the improved bandwidth efficiency. The extent of this shaping can be clearly seen from the eye diagrams given in Fig. 7.17.

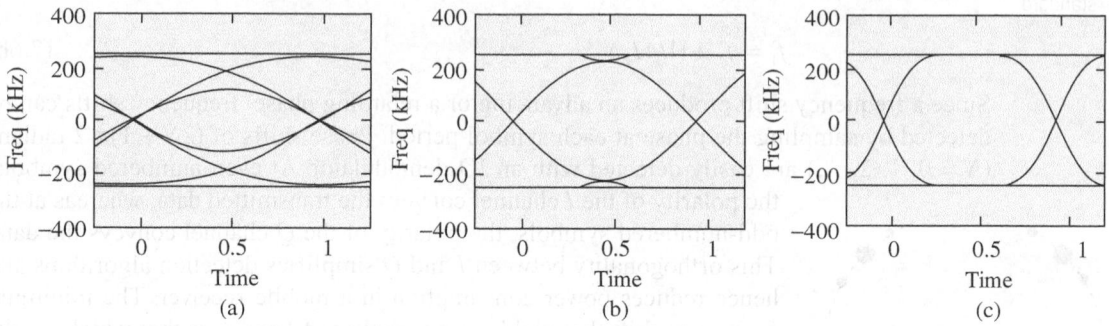

Fig. 7.17 Eye diagrams for GMSK schemes (a) $BT = 0.3$ (b) $BT = 0.5$ (c) $BT = 1$ (typical and relative)

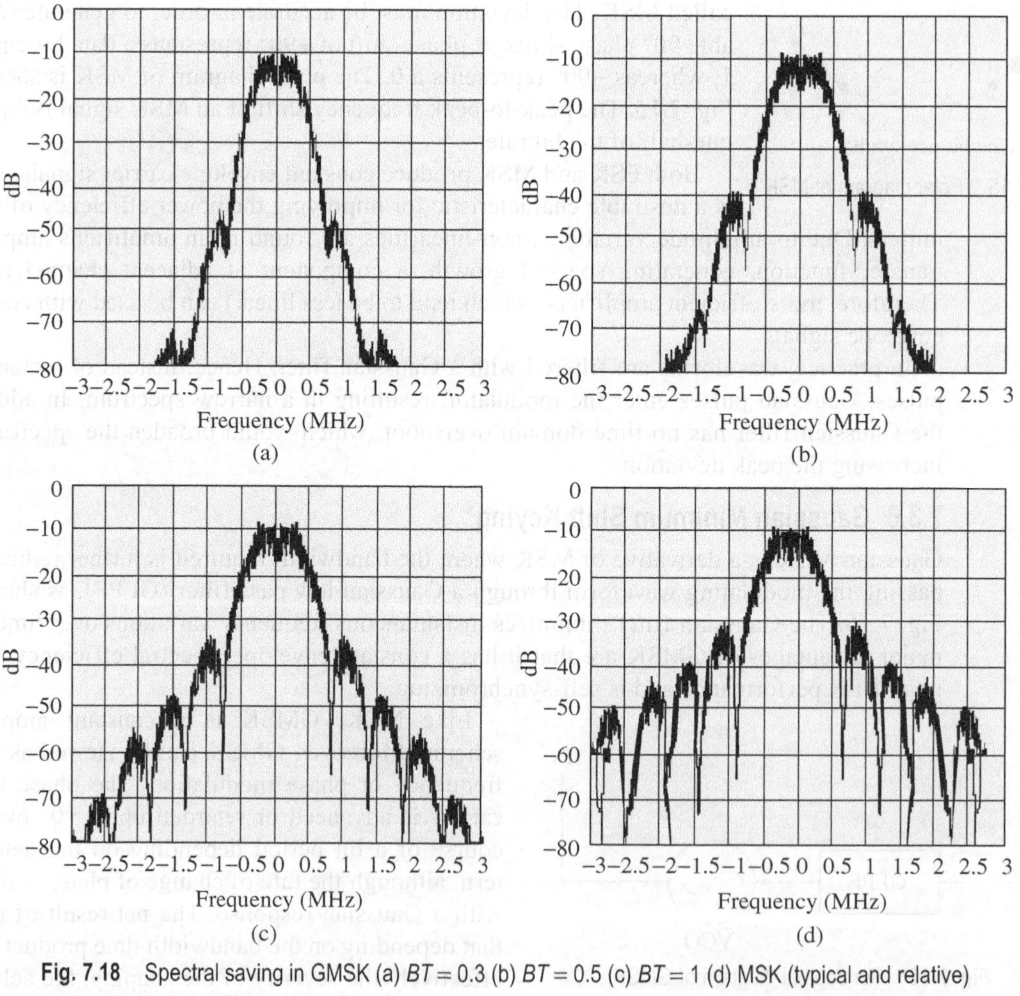

Fig. 7.18 Spectral saving in GMSK (a) $BT = 0.3$ (b) $BT = 0.5$ (c) $BT = 1$ (d) MSK (typical and relative)

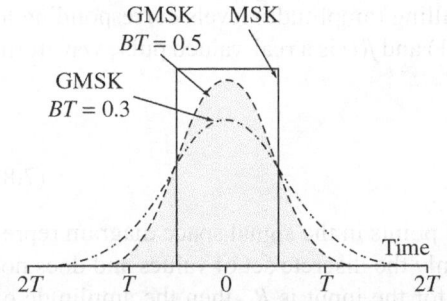

Fig. 7.19 Pulse shaping and ISI possibilities in GMSK

This resultant reduction in the phase change of the carrier for the shaped symbols (i.e., 101 and 010) will ultimately degrade the BER performance, as less phase has been accrued or retarded; therefore, less noise will be required to transform a 0 to a 1 and vice versa. The spectral efficiency of this scheme can be proved with the help of the results given in Figs 7.18 and 7.19, which compare the spectra as well as pulse shaping and ISI capabilities of GMSK with different BT values and MSK.

7.4 VARIABLE ENVELOPE MODULATION SCHEMES

Variable envelope modulation schemes require both amplitude and phase information to detect the signal properly.

7.4.1 Amplitude Shift Keying

The ASK scheme is a modified version of AM in which the amplitude levels are a number of signalling levels (instead of continuous envelope followed in the case of an analog signal), as shown in Fig. 7.20(b). The figure is that of an AM system with a digital modulating signal. ASK demonstrates poor performance, as it is heavily affected by noise and interference. In Fig. 7.20(b), the mapping of the symbols to the information bits is in the form of grey code and the adjacent signal levels differ in one-bit position.

In general, an ASK waveform is represented by the following equation:

$$s(t) = [A_m f(t)]\cos \omega_c t \quad 0 \le t \le T_s \tag{7.7}$$

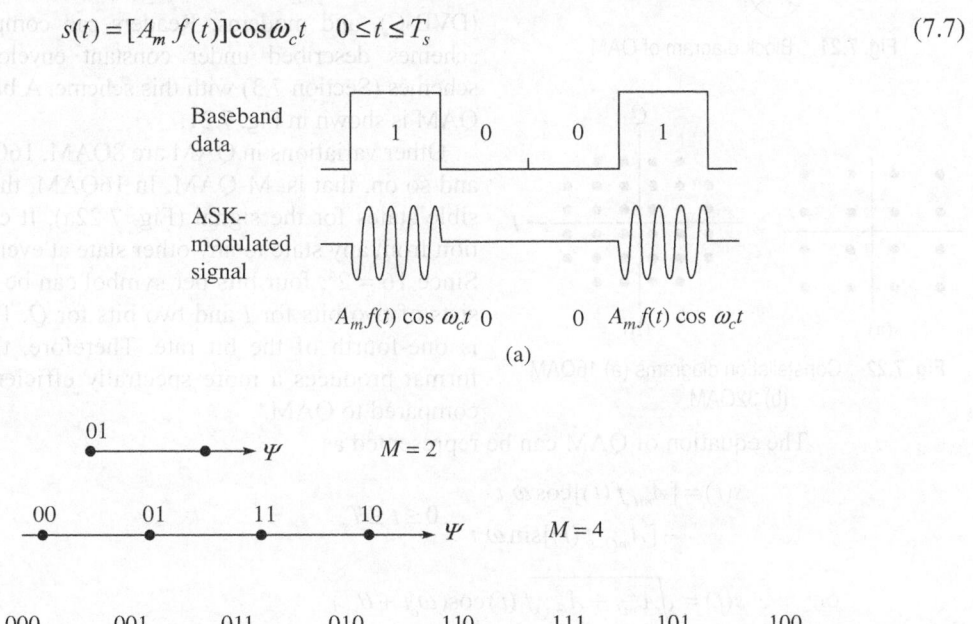

Fig. 7.20 ASK (a) Waveform for a binary case (b) One-dimensional signal space diagram

where $m = 1, 2, ..., M$ denotes the M possible signalling (amplitude) levels corresponding to the M possible data symbols represented by Eq. (7.1) and $f(t)$ is a real-valued pulse waveform (normally rectangular).

Here, A_m takes the following different values:

$$A_m = \frac{(2m - 1 - M)x}{2} \tag{7.8}$$

where x is the distance between the adjacent signal points in the signal space diagram represented by the basis function $\psi(t)$. Here, A_m takes only the discrete set of values and does not change during the bit interval. If the symbol rate for the input is R_s, then the amplitude of the ASK signal changes every $1/R_s$ seconds. The energy of the ASK signal is (from Eq. 1.11)

$$E_s = \frac{1}{2} A_m^2 E_{f(t)} \tag{7.9}$$

where $E_{f(t)}$ denotes the energy of the basic pulse $f(t)$.

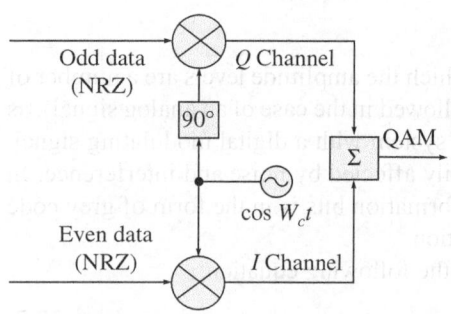

Fig. 7.21 Block diagram of QAM

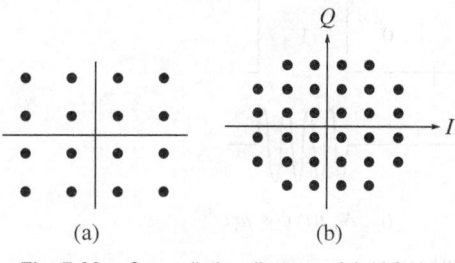

(a)   (b)

Fig. 7.22 Constellation diagrams (a) 16QAM
(b) 32QAM

7.4.2 Quadrature Amplitude Modulation and M-ary Quadrature Amplitude Modulation

Quadrature amplitude modulation is the method of combining two amplitude-modulated signals into one channel. QAM may be analog or digital. Analog QAM combines two amplitude-modulated signals using the same carrier frequency with 90° phase difference. In digital QAM, two ASK signals are combined in the same way. QAM is used in applications such as microwave digital radio, digital video broadcasting-cable (DVB-C), and modems. Readers can compare the QPSK schemes described under constant envelope modulation schemes (Section 7.3) with this scheme. A block diagram of QAM is shown in Fig. 7.21.

Other variations in QAM are 8QAM, 16QAM, 32QAM, and so on, that is, M-QAM. In 16QAM, there are 16 possible states for the signal (Fig. 7.22a). It can have transition from any state to any other state at every symbol time. Since $16 = 2^4$, four bits per symbol can be sent. This consists of two bits for I and two bits for Q. The symbol rate is one-fourth of the bit rate. Therefore, this modulation format produces a more spectrally efficient transmission compared to QAM.

The equation of QAM can be represented as

$$s(t) = [A_{mI} f(t)] \cos \omega_c t \\ - [A_{mQ} f(t)] \sin \omega_c t \qquad 0 \le t \le T_s \tag{7.10a}$$

or $$s(t) = \sqrt{A_{mI}^2 + A_{mQ}^2} f(t) \cos(\omega_c t + \theta_m) \tag{7.10b}$$

where $m = 1, 2, ..., M$, and A_{mI} and A_{mQ} are the information-dependent amplitudes of the I and Q components. W_c is the angular frequency of the carrier and θ_m are the phases associated with QAM signal constellation points.

KEY ADVANTAGE OF 32QAM

In the case of 32QAM, there are six *I* values and six *Q* values, resulting in 36 possible states ($6 \times 6 = 36$)—too many states for a power of two (the closest power of two is 32). Hence, the four corner symbol states, which take the most power to transmit, are omitted (Fig. 7.22b). This reduces the amount of peak power the transmitter has to generate. Since $2^5 = 32$, there are five bits per symbol and the symbol rate is one-fifth of the bit rate.

TRADE-OFF

With higher M-ary schemes, we get a high bit rate communication; however, we should note the following:
- For higher M-ary modulation schemes, the symbols are very close together and are thus more subject to errors due to noise and distortion and need to be transmitted with extra power.
- This reduces power efficiency compared to simpler schemes.
- Higher M-ary scheme makes the radio complex.
- In a noisy environment, M-QAM degrades faster than M-PSK and degradation appears as more BER. Thus, M-QAM needs more power for better quality.
- In low SNR conditions, the case of loss of symbol clock results in further bit errors; thus, clock recovery is frequently needed.

The current practical limits are approximately 256QAM. A 256QAM system uses 16 *I* values and 16 *Q* values, giving 256 possible states. Since $2^8 = 256$, each symbol can represent eight bits. A 256QAM signal that can send eight bits per symbol is spectrally very efficient.

7.5 DIFFERENTIAL MODULATION SCHEMES

Differential modulation can be seen in DQPSK, differential 16QAM (D16QAM), and so on. The term *differential* means that the information is not carried by the absolute state but is carried by the transition between states. In some cases, there are also restrictions on allowable transitions. This occurs in $\pi/4$ DQPSK where the carrier trajectory does not go through the origin. A DQPSK transmission system can have transition from any symbol position to any other symbol position.

> In OQPSK, phase transitions take place every T_b s, whereas in QPSK, the transitions take place every $2T_b$ s; thus, the phase transition above the origin is eliminated.

The $\pi/4$ DQPSK modulation format uses two QPSK constellations offset by $45°$ ($\pi/4$ radians) as shown in Fig. 7.23(a). Fig. 7.23(b) shows the phase shifts associated with possible bit patterns with two bits. Transitions must occur from one constellation to the other. This guarantees that there is always a change

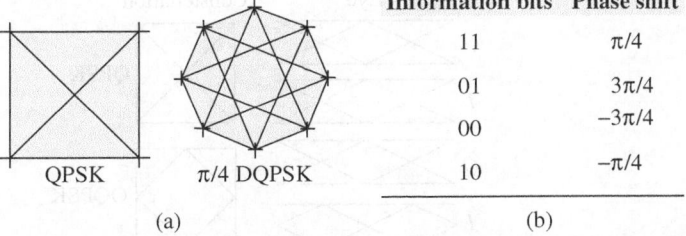

Information bits	Phase shift
11	$\pi/4$
01	$3\pi/4$
00	$-3\pi/4$
10	$-\pi/4$

(a)   (b)

Fig. 7.23 $\pi/4$ DQPSK modulation (a) difference in constellation diagrams between QPSK and $\pi/4$ QPSK (b) bit combinations and corresponding phases in $\pi/4$ DQPSK

in phase at each symbol, making clock recovery easier. The data is encoded in the magnitude and direction of the phase shift, not in the absolute position on the constellation. One advantage of $\pi/4$ DQPSK is that the signal trajectory does not pass through the origin, thus simplifying transmitter design. Another is that $\pi/4$ DQPSK, with root raised cosine filtering, has better spectral efficiency than GMSK, the other common cellular modulation type. Differential modulation schemes establish a correlation between consecutive symbols, and hence, it becomes easier to detect errors.

7.6 OFFSET MODULATION SCHEMES

> Spectral efficiency can be defined at a physical link level, area level or a system level depending upon the allocated spectrum to be considered for calculation.

An example of offset modulation is OQPSK. As previously discussed, the potential for $180°$ phase shift in QPSK results in the requirement of better linearity in the power amplifier and spectral re-growth. OQPSK reduces this tendency by adding a time delay of one bit period (half a symbol) in the Q arm of the modulator. The result is that the phase of the carrier is potentially modulated every bit (depending on the data), not every other bit as for QPSK; hence, the phase trajectory never approaches the origin.

As with the other phase modulation schemes, shaping of the phase trajectory between constellation points is typically implemented with a raised cosine filter to improve the spectral efficiency. Due to the similarities between QPSK and OQPSK, similar signal spectra and probability of error are achieved. OQPSK is utilized in the North American IS-95 CDMA cellular system for the reverse link from the mobile to the base station.

In QPSK, I and Q bit streams are switched at the same time. The symbol clocks, or the I and Q digital signal clocks, are synchronized. In OQPSK, the I and Q bit streams are offset in their relative alignment by one bit period (one-half of a symbol period). This is shown in Fig. 7.24.

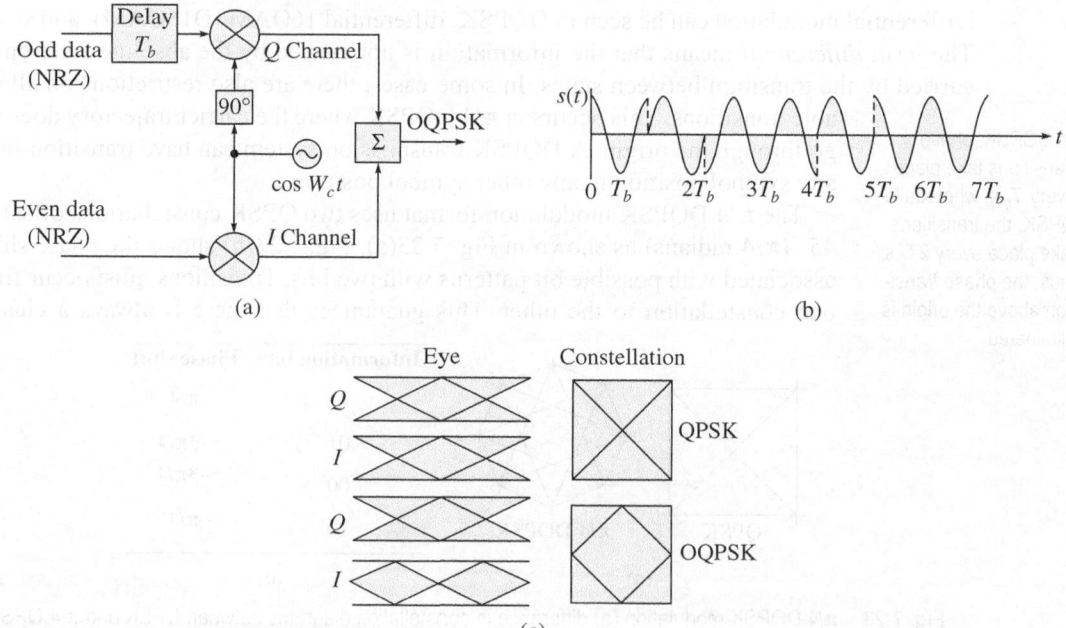

Fig. 7.24 OQPSK modulation (a) OQPSK diagram (only delay part being additional) (b) OQPSK waveforms on time axis (c) Difference between QPSK and OQPSK eye and constellation diagrams

Since the transitions of I and Q are offset, at any given time, only one of the two bit streams can change values. This creates a dramatically different constellation, even though there are still just two I and Q values.

In OQPSK, the signal trajectories are modified by the symbol clock offset so that the carrier amplitude does not go through or near zero (the centre of the constellation). The spectral efficiency is the same with two I states and two Q states. The reduced amplitude variations (perhaps 3 dB for OQPSK versus 30–40 dB for QPSK) allow a more power-efficient and less linear radio frequency (RF) power amplifier to be used.

7.7 MODULATION SCHEMES AND SPECTRUM EFFICIENCY

Spectrum efficiency describes how efficiently the allocated bandwidth is utilized in terms of achievable data rate or the ability of a modulation scheme to accommodate the data rate within a limited bandwidth. Thus, the ratio of bit rate to available bandwidth is called the *spectral efficiency*.

Table 7.1 shows the theoretical spectrum efficiency limits for the main modulation types. Note that these figures cannot actually be achieved in practical radios since they require perfect modulators, demodulators, filters, and transmission paths.

Transmission Bandwidths

If R_b is the bit rate and r is a factor related to the technique by which the signal is filtered to establish a transmission bandwidth, that is, roll-off factor in case of raised cosine filter, $0 \le r \le 1$, then the transmission bandwidths for various modulation schemes are as follows:

ASK: $(1 + r)R_b$

FSK: $2\Delta f + (1 + r)R_b$, $\Delta f = f_2 - f_c$ or $f_c - f_1$

M-PSK: $\left[\dfrac{1+r}{\log_2 M}\right]R_b$, M = signalling level or elements

MFSK: $\left[\dfrac{(1+r)M}{\log_2 M}\right]R_b$

If the radio has a perfect (rectangular in the frequency domain) filter, then the occupied bandwidth could be made equal to the symbol rate.

Table 7.1 Spectrum efficiency limits for different modulation schemes

Modulation format	Theoretical spectrum efficiency limits
MSK	1 bit/s/Hz
BPSK	1 bit/s/Hz
QPSK	2 bits/s/Hz
8PSK	3 bits/s/Hz
16 QAM	4 bits/s/Hz
32 QAM	5 bits/s/Hz
64 QAM	6 bits/s/Hz
256 QAM	8 bits/s/Hz

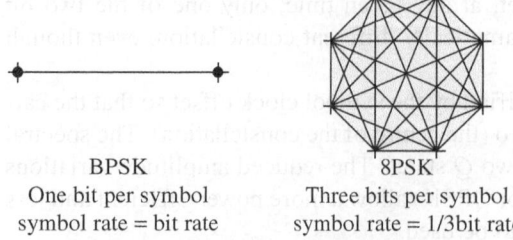

BPSK

One bit per symbol

symbol rate = bit rate

8PSK

Three bits per symbol

symbol rate = 1/3bit rate

Fig. 7.25 Constellation showing the number of bits per symbol and symbol rate of BPSK and 8PSK

An example of how symbol rate influences spectrum requirements in eight-state PSK (8PSK) is shown in Fig. 7.25. It is a variation of PSK. There are eight possible states that the signal can transit to at any time. The phase of the signal can take any of the eight values at any symbol time. Since $2^3 = 8$, there are three bits per symbol. This means that the symbol rate is one-third of the bit rate. This is relatively easy to decode. Figure 7.25 shows that the number of bits per symbol is more in the case of 8PSK and hence the symbol rate reduces compared to BPSK.

Practically, the bandwidth of the transmitting (modulated) signal can be controlled by applying external measures such as pre-modulation or post-modulation filtering. Pulse shaping is very helpful in spectrum saving. This will be discussed in Section 8.1.2 of Chapter 8. Filtering allows the transmitted bandwidth to be significantly reduced without losing the content of the digital data. It improves the spectral efficiency of the signal.

Instead of post-modulation filtering, *windowing* may also be used (refer to Section 8.1.3 of Chapter 8). This technique also saves the spectrum by eliminating out-of-band components.

Other techniques for maximizing spectral efficiency include the following:

- Relate the data rate to the frequency shift (as in global system for mobile communication, or GSM).
- Restrict the types of transitions.

7.8 TRANSMISSION POWER

> Transmission power is indirectly dependent upon transmission bandwidth.

As with any natural resource, it makes no sense to waste the RF spectrum by using channel bands that are too wide. Therefore, narrow filters are used to reduce the occupied bandwidth of the transmission. Narrow filters with sufficient accuracy and repeatability are very difficult to build. Wide filters result in ISI and tighten the requirements on clock accuracy.

Narrow filters also result in more overshoot than wide filters and, therefore, higher peak carrier power. The power amplifier must then accommodate the higher peak power without distortion. The bigger amplifier causes more heat and electrical interference to be produced, as the RF current in the power amplifier will interfere with other circuits. Larger and heavier batteries will be required. The alternative is to have shorter talk time and smaller batteries. Constant envelope modulation, as used in GMSK, can use class-C amplifiers, which are the most efficient. In summary, spectral efficiency is highly desirable, but there are penalties in power, cost, size, weight, complexity, talk time, and reliability.

Now, radios require a significant excess power beyond the power needed to transmit the symbol values themselves. A typical value of the excess power needed at an α of 0.2 for QPSK with Nyquist filtering would be approximately 5 dB. This is more than three times the peak power because of the filter used to limit the occupied bandwidth. These principles apply to QPSK, OQPSK, DQPSK, and the varieties of QAM such as 16QAM, 32QAM, 64QAM, and 256QAM. Not all signals will behave in exactly the same way, and exceptions include FSK, MSK, and others with constant envelope modulation. The power of these signals is not affected by the filter shape. Table 7.2 shows the various modulation formats and their applications.

Table 7.2 Modulation formats and applications

Modulation format	Application
MSK, GMSK	GSM, CDPD
BPSK	Deep space telemetry, cable modems
QPSK, $\pi/4$ DQPSK	Satellite, CDMA, NADC, TETRA, PHS, PDC, LMDS, DVB-S, cable(return path), cable modems, TFTS
OQPSK	CDMA, satellite
FSK, GFSK	DECT, paging, RAM mobile data, AMPS, CT2, ERMES, land mobile, public safety
8PSK	Satellite, aircraft, telemetry pilots for monitoring broadband video systems
16 QAM	Microwave digital radio, modems, DVB-C, DVB-T
32 QAM	Terrestrial microwave, DVB-T
64 QAM	DVB-C, modems, broadband set top boxes, MMDS
256 QAM	Modems, DVB-C (Europe), digital video(US)

AN EXAMPLE IMPLYING MORE SPECTRAL EFFICIENCY WITH HIGHER M-ARY SCHEMES

A radio has an eight-bit sampler, with the sampling rate at 10 kHz for voice.

Basic bit stream rate in the radio = 8 bits × 10 ksamples/s = 80 kbps

(For the moment, we will ignore the extra bits required for synchronization, error correction, etc.)

Symbol rate = Bit rate/Number of bits that can be transmitted with each symbol

If one bit is transmitted per symbol, as in BPSK, the symbol rate will be the same as the bit rate of 80 kbps.

If two bits are transmitted per symbol, as in QPSK, then the symbol rate will be half of the bit rate or 40 kbps.

If more bits can be sent with each symbol, then the same amount of data can be sent in a narrower spectrum, because the modulation spectrum is decided by the symbol rate and not the bit rate. This is why modulation formats that are more complex and use a higher number of states can send the same information over a narrower piece of the RF spectrum.

7.9 SPREAD SPECTRUM MODULATION

Spread spectrum techniques were developed to meet the needs of military communications. They were based on signalling schemes (code or sequence) that greatly expand the transmitted spectrum relative to the data rate. Now, the techniques are also used for commercial applications. There is a growing interest in these techniques for use in mobile radio networks and for both communication and positioning applications in satellites.

Spread spectrum transmission offers the following three main advantages over fixed frequency transmission:

(a) Spread spectrum signals are highly resistant to noise and interference. The process of re-collecting a spread signal spreads out noise and interference, causing them to recede into the background.

(b) Spread spectrum signals are difficult to intercept.

(c) Spread spectrum transmissions can share a frequency band with many types of conventional transmissions with minimal interference. These signals add minimal noise to the narrow-frequency communications, and vice versa. As a result, bandwidth can be utilized more efficiently.

A frequency hopping spread spectrum (FHSS) signal simply sounds like an increase in the background noise to a narrowband receiver.

Spread spectrum techniques also meet the following objectives: (a) operation with a low-energy spectral density, (b) multiple access capability without external control, (c) security (difficult for unauthorized receivers to observe the message), (d) anti-jamming capability, (e) multipath protection, and (f) ranging.

One way of classifying spread spectrum systems is as follows:

Averaging system In this system, interference reduction takes place because the interference can be averaged over a large time interval.

Avoidance system In this system, reduction of interference occurs because the signal is made to avoid the interference for a large fraction of time.

The spread spectrum technique is a wideband modulation technique. Here, the bandwidth expansion factor is very large. In fact, bandwidth expansion does not combat white noise as it does in frequency modulation (FM), pulse code modulation (PCM), and other wideband modulation methods, because bandwidth expansion is achieved by something that is independent of the message rather than being uniquely related to the message.

As far as classification by modulation technique is concerned, the two major techniques used in spread spectrum systems are direct sequence (DS) and frequency hopping (FH). Out of the two, perhaps, FH is easier to visualize.

The following are the other spread spectrum modulation (SSM) techniques:
- Time hopping spread spectrum (THSS)
- Hybrid methods
- Chirped spread spectrum (CSS)

Note: A DS system is an averaging system, whereas FH, time hopping (TH), and chirping systems are avoidance systems.

7.10 PSEUDO-NOISE CODES, PROPERTIES, AND CODE GENERATION

In a DS spread spectrum (DSSS) system, a unique code is used to spread and despread the signal using the logic shown in Fig. 7.26(a). This unique code is the DS, known as the pseudo-noise (PN) code. The PN code, while concatenated, appears as a random sequence to an unauthentic user.

In Fig. 7.26(a), for logic 1, the inverted PN sequence appears in the output, whereas for logic 0, the same sequence appears because of the XOR operation. As shown in Fig. 7.26(b), for the data voltage level, +1 is considered for logic 1 and −1 is considered for logic 0. Here, the multiplication operation rather than XOR is considered. Thus, wherever logic 0 occurs, the inverted PN sequence appears in the output. In the case of FH or TH, the PN sequence is used to generate the hopping frequency or the hopping time slots.

Assume that the clock rate is provided for generating each bit of a PN code (with one clock cycle, one bit of PN code comes out). The bit rate of a PN code is called the *chip rate* ($1/t_{chip}$), which is 10 times or more than the data bit rate. The smallest time increment in the sequences of certain period or duration is t_{chip} and is known as a *time chip*. The total period consists of N_c time chips. Thus, in this technique, the PN code directly does the PSK of the data, increasing its bandwidth. In a typical DS system, a double balanced mixer is driven by the PN code to switch a carrier's phase between 0° and 180°. This is known as BPSK. The DS receiver despreads this wideband signal by using a synchronized pseudo-random sequence identical to that in the transmitter.

The chip rate decides the final transmission spectrum of the DSSS system.

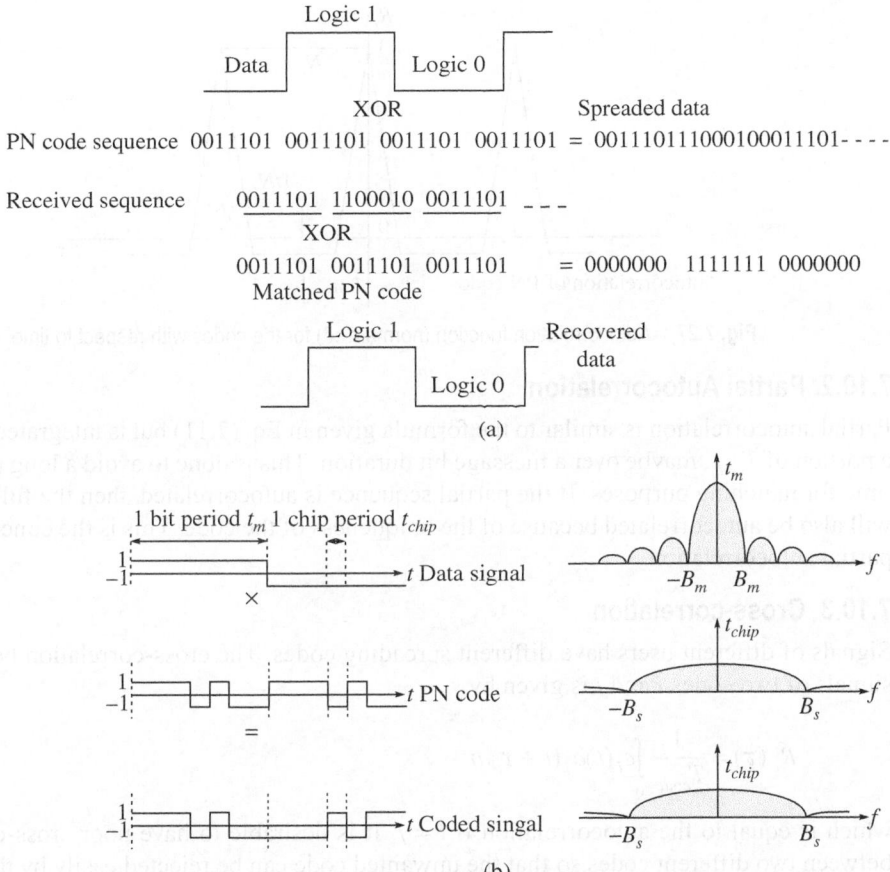

Fig. 7.26 Spread spectrum (a) XOR logic (b) Spreading by multiplication (also shows relation between bit and chip durations)

In a spread spectrum based multiuser system, unique codes are to be assigned to different users to differentiate them. For the identification of a unique user, the code must have some desirable operations at the receiving end, so that the data may not be divulged to a wrong user.

7.10.1 Autocorrelation

The autocorrelation should be maximal for the DS or PN codes so that correct PN signal can be identified at the receiver from the numerous coexisting signals. The autocorrelation function of a typical PN sequence is shown in Fig. 7.27. Note that on a normalized basis, it has a maximum value of 1, which repeats itself every period, but in between these peaks, the level is at a constant value of $1/N_c$. If N_c is a very large number, the autocorrelation function is very small in this region.

The autocorrelation of the spreading waveform (PN signal) $c(t)$ is represented mathematically as

> Autocorrelation is the procedure by which the matching of a signal is done with itself during the detection process.

$$R_a(\tau) = \frac{1}{T_{code}} \int_0^\infty c(t)c(t+\tau)dt \qquad (7.11)$$

where $T_{code} = N_c t_{chip}$ is the code period and τ represents a time shift variable.

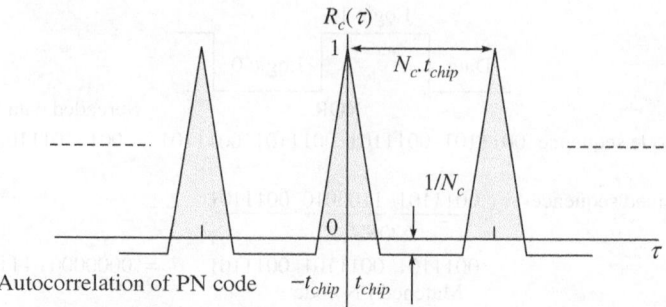

Fig. 7.27 Autocorrelation function (normalized) for the codes with respect to time

7.10.2 Partial Autocorrelation

Partial autocorrelation is similar to the formula given in Eq. (7.11) but is integrated only over a portion of T_{code}, maybe over a message bit duration. This is done to avoid a long processing time for matching purposes. If the partial sequence is autocorrelated, then the full sequence will also be autocorrelated because of the uniqueness of the code. This is the concept behind partial autocorrelation.

7.10.3 Cross-correlation

Signals of different users have different spreading codes. The cross-correlation between the signals of two codes i and j is given by

$$R_c(\tau) = \frac{1}{T_{code}} \int_0^\infty c_i(t)c_j(t + \tau)dt \qquad (7.12)$$

which is equal to the autocorrelation if $i = j$. It is desirable to have poor cross-correlation between two different codes so that the unwanted code can be rejected easily by the receiver of the CDMA system.

7.10.4 Properties of Pseudo-noise Codes

The randomness of a PN sequence is tested by the following code properties observed over one full period.

Balance property In each period of the sequence, the number of binary 1's differs from that of binary 0's by at most one digit. Consider a typical PN code:

0001 0011 0101 111 (seven 0's and eight 1's meet the balance condition)

Run length property Among the runs of 1's and 0's in each period, it is desirable that about one-half of the runs of each type are of length one, one-fourth are of length two, one-eighth are of length three, and so on. Consider the same code again:

Number of runs = 8

$$\underset{3}{\underline{000}}\ \underset{1}{\underline{1}}\ \underset{2}{\underline{00}}\ \underset{2}{\underline{11}}\ \underset{1}{\underline{0}}\ \underset{1}{\underline{1}}\ \underset{1}{\underline{0}}\ \underset{4}{\underline{1111}}$$

Autocorrelation property The autocorrelation function of a maximal length sequence (explained in Section 7.11.3) is periodic and binary valued. We can state the autocorrelation function as

$$R_a(\tau) = \frac{1}{N_c}\Big[\text{Number of agreements (a)} - \text{Number of disagreements (d)}$$
$$\text{in one full period}\Big]$$

0 0 0 1 0 0 1 **1** 0 1 0 1 **1 1 1**

1 0 0 0 1 **0 0 1 1** 0 1 0 **1 1 1**

d a a d d a d a d d d d a a a

$$R_c(\tau) = -\frac{1}{15}$$

Note: In short, a periodic sequence will repeat a PN code of N_c chip durations (one period) infinite times. Such a sequence is said to be pseudo-random if it satisfies the following conditions:
1. In every period, the number of +1's differs from that of −1's by exactly one (balance property). Hence, N_c is an odd number.
2. In every period, half of the runs of the same sign have length one, one-fourth have length two, one-eighth have length three, and so forth. In addition, the number of positive runs equals that of negative runs (run property).
3. The autocorrelation of a periodic sequence is two valued, that is, N_c for shifts 0, N_c, $2N_c$, $3N_c$, and so on and −1 otherwise (without normalization).

7.10.5 Aperiodic and Periodic Sequences

A PN sequence consists of a sequence of plus or minus ones (+1 or −1) that possesses certain specified autocorrelation properties. There are two general classes of PN sequences—aperiodic and periodic. An aperiodic sequence is one that does not repeat itself in a periodic fashion. It is usually assumed that the sequence has a value of zero outside its stated interval. A periodic sequence, however, is a sequence of plus or minus ones that repeats itself exactly with a specified period.

> The reason for using the shift register is that the period of the PN sequence can easily be made very large by increasing the number of register stages, so that apparently it looks random to the users.

An ideal aperiodic sequence of N_c chips has the autocorrelation to be N_c for no shift and 0 or ±1 with a shift of one or more bits. Such sequences (also called Barker sequences) are known to exist for very few values of N_c. Specifically, they are $N_c = 1, 2, 3, 4, 5, 7, 11,$ and 13. No larger sequences have been found. Periodic sequences are much more important in a spread spectrum system than aperiodic sequences because of their ability to communicate continuously for the required time and because of the limitations of aperiodic sequences.

Example 7.2 A seven-digit aperiodic Barkers sequence has the form +1, +1, +1, −1, −1, +1, −1. Determine the autocorrelation for this code. Does a cyclic shift of this sequence (e.g., +1, +1, −1, −1, +1, −1, +1) have the same correlation properties?

Solution Considering a discrete signal for the limited duration, we can establish the autocorrelation as follows:
For the Barker sequence
Theoretically, autocorrelation function is

$$R_a(k) = \sum_{n-1}^{N_c-k} a_n a_{n+k} = \begin{cases} N_c \text{ for } k = 0 \\ 0 \text{ or } \pm 1 \text{ for } k \neq 0 \end{cases}$$

where k is the number of shifts in the bit. For this particular sequence, for $k = 0$:

+1 +1 +1 −1 −1 +1 −1

+1 +1 +1 −1 −1 +1 −1

+1 +1 +1 +1 +1 +1 +1 = N_c (adding up all the values)

Barker sequences are too short for a spread spectrum system but sometimes may be used for synchronization purposes under some conditions.

For one shift, that is, $k = 1$:

$$+1 +1 +1-1 -1 +1-1$$
$$+1 +1 +1-1 -1 +1-1$$

$$\overline{}$$

$0 +1 +1-1 +1-1 -1-0 = 0$
(adding up all the values)

For two shifts, that is, $k = 2$:

$$+1 +1 +1-1 -1 +1-1$$
$$+1 +1 +1-1 -1 +1-1$$

$$\overline{}$$

$0-0 +1-1 -1-1 +1-0\ 0 = -1$ (adding up all the values)

Similarly, for more shifts, we can check the results.
For the cyclic shift
For one cyclic shift, that is, $k = 1$:

$$+1 +1 +1 -1 -1 +1 -1$$
$$-1 +1 +1 +1 -1 -1 +1$$

$$\overline{}$$

$-1 +1 +1 -1 +1 -1 -1 = -1$ (adding up all the values)

With the cyclic shift, this sequence will have the same autocorrelation properties; that is, it will be a two-valued function. With no shift, it is N_c, and for other cases, it is -1 (it is left to the readers to find the reason).

The popular periodic code sequences used in spread spectrum transmission are (a) maximum length (ML) sequences, (b) Walsh–Hadamard (W–H) sequences, (c) Gold codes or sequences, and (d) Kasami sequences.

7.10.6 Maximum Length Sequences

The PN code is generated in a ML shift register, which is a linear feedback shift register (LFSR), and hence, it is called an ML sequence. These PN code generators are periodic in that the sequence that is produced repeats itself after a period of time. Such a typical periodic sequence generator is shown in Fig. 7.28(a). The value of N_c is $2p - 1$, where p is the number of stages in the code generator. The short length sequences are $N_c = 7, 15, 31, 63, 127, 255, ...$

The LFSR can be implemented in two ways:

(a) *Fibonacci implementation*: It consists of a shift register in which a binary-weighted modulo-2 sum of the taps is fed back to the input (Fig. 7.28b).
(b) *Galois implementation*: It consists of a shift register the contents of which are modified at every step by a binary-weighted value of the output stage (Fig. 7.28c).

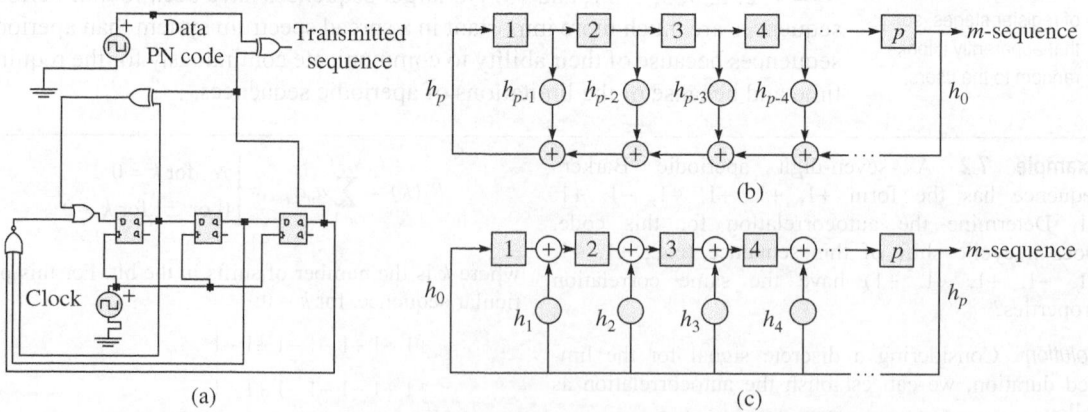

Fig. 7.28 ML sequence operators (a) Typical PN code generator with three shift register stages and its application for data spreading (b) Fibonacci implementation of LFSR (c) Galois implementation of LFSR

FIBONACCI VERSUS GALOIS IMPLEMENTATIONS

Thorough inspection reveals that the order of the Galois weights is opposite to that of the Fibonacci weights. Given the identical feedback weights, the two LFSR implementations will produce the same sequence. However, the initial states of the two implementations must necessarily be different for the two sequences to have identical phases. The initial state of the Fibonacci form is called the *initial fill*, which comprises the first *p* bits output from the generator, whereas the initial state of the Galois generator must be adjusted appropriately to attain the equivalent initial fill. The Galois form is generally faster due to the reduced number of gates in the feedback loop.

In general, a shift register implements a primitive polynomial $h(x)$:

$$h(x) = x^p + h_{(p-1)}x^{(p-1)} + \ldots + h_1 x + 1 \tag{7.13}$$

which is the generator polynomial for a code. In the polynomial $h(x)$, the coefficient h_i takes on the binary value 0 (no connection) or 1 (feedback or connection). Two exceptions are h_0 and h_p, which are always 1 and thus are always connected. Note that h_p is not really a feedback connection but rather is the input of the shift register. In short, the coefficients h_i determine the feedback connections in the code generator.

For the designing code generators, the feedback connection table exists. A few entries of feedback connection are given in Table 7.3. The feedback connections, or the corresponding realized polynomial, are represented by the notation

$$[p, h_{(p-1)}(p-1), \ldots, h_1] \tag{7.14}$$

where zero entries are not written explicitly. The table entries can be extended with an increase in the number of stages, say, 12, 24, 32, and more.

It can be summarized that the code generator produces a binary-valued sequence. This sequence repeats itself after every N_c elements. These elements are called *chips*. One period of the sequence, that is, N_c chips, is called the *PN* or *spreading code*. In DS-CDMA practice, the sequence is transmitted mostly as a bipolar waveform, called the *spreading waveform*.

Thus, with a fairly modest amount of hardware due to the growth of very large scale integration (VLSI), it becomes possible to make arbitrarily longer sequences. The modulation of the spread spectrum carrier can be either biphase or quadriphase. The features of the ML or PN code can be summarized as follows:

- The ML linear feedback shift register sequences have desirable autocorrelation properties for spread spectrum applications.
- The ML sequences can be used in systems that need to operate in channels with large delay spreads in multipath channels.

Table 7.3 Feedback connection table

Number of stages	Taps	Set	Connection notation
3	2	1	[3, 2]
4	2	1	[4, 3]
5	2	1	[5, 3]
5	4	2	[5, 4, 3, 2], [5, 4, 3, 1]
6	2	1	[6, 5]
6	4	2	[6, 5, 4, 1], [6, 5, 3, 2]
11	2	1	[11, 9]
11	4	22	[11, 10, 9, 7], [11,10, 9, 5], and so on

Example 7.3 A clock frequency is 10 MHz; hence, $t_{chip} = 10^{-7}$ s. If a shift register with 32 stages is used for generating the PN sequence, then compute the time to complete one cycle of the sequence.

Solution Total number of chips in a period,

$$N_c = 2^{32} - 1 = 4.29 \times 10^9$$

So, $N_c \times t_{chip} = 4.29 \times 10^9 \times 2 \times 10^{-7}$

$$= 4.29 \times 10^2 \text{ s} = 429 \text{ s}$$

(If the stages are increased to 41, the total time to generate one sequence will be 2.545 days. The exercise is left to the readers.)

Example 7.4 Draw a [3, 1] PN code generator that realizes the polynomial and generates [3, 1] codes. Moreover, draw the waveforms for the m-sequence for three periods.

Solution [3, 1] for ML code generation can be interpreted as follows:

The [3, 1] code has register length $n = 3$.

So, code length $N_c = 2^3 - 1 = 7$

The primitive polynomial $h(x)$ for this case is $h(x) = x^3 + 0 + x + 1$. (The polynomial suggests the stages from which the output is fed back.)

As can be seen, $h_1 = 1$, $h_2 = 0$, and $h_3 = 1$. The generator is shown in Fig. 7.29 and the waveform is given in Fig. 7.30.

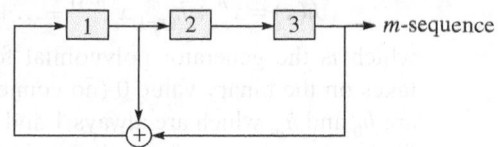

Fig. 7.29 [3, 1] ML code generator

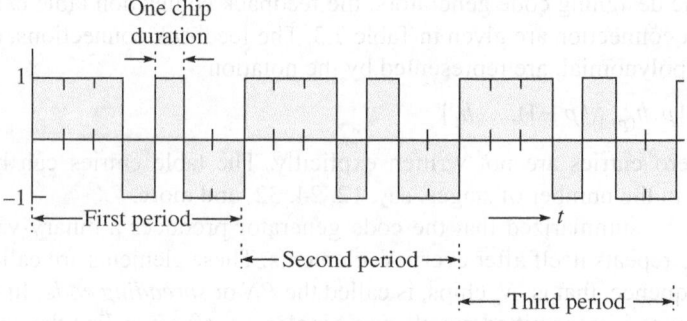

Fig. 7.30 Waveform of [3, 1] code sequence

7.10.7 Walsh–Hadamard Sequences

If perfectly synchronized with respect to each other, W–H codes are perfectly orthogonal. That is, these are optimal codes to avoid interference among users in the link from the base station to the terminals.

The simplest matrix of two orthogonal W–H codes is

$$C(2^1) = \begin{bmatrix} 1 & 1 \\ 1 & -1 \end{bmatrix} \tag{7.15}$$

The code of user 1 is the first column, that is, (1, 1), and the code of user 2 is the second column, that is, (1, −1). Clearly, (1, 1) is orthogonal to (1, −1). This matrix can be extended using a recursive technique. For $K = 2^n$ users, the matrix is found from the code matrix for $2^{(n-1)}$ users, according to

Walsh codes have the best orthogonality features and hence are widely used in CDMA systems.

$$C(2^n) = \begin{bmatrix} C_{n-1} & C_{n-1} \\ C_{n-1} & -C_{n-1} \end{bmatrix} \tag{7.16}$$

Gold sequences are constructed by XOR-ing two selected *m*-sequences of the same length with each other.

7.10.8 Gold Sequences

For a Gold sequence of length $N_c = 2^p - 1$, where p = number of LFSR shift elements we need to use two LFSRs, each of length $2^p - 1$. If the LFSRs are chosen appropriately, Gold sequences have better cross-correlation properties than the ML LSFR sequences, as shown in Table 7.4.

Gold (and Kasami) showed that for certain well-chosen *m*-sequences, the cross-correlation takes on only three possible values, namely -1, $-R_c$, and $R_c - 2$. Two such sequences are called *preferred sequences*. Here, t depends solely on the length of the LFSR used. In fact, for a LFSR with p memory elements,

$$\text{if } p \text{ is odd,} \quad R_c = 2^{(p+1)/2} + 1 \tag{7.17}$$

$$\text{if } p \text{ is even,} \quad R_c = 2^{(p+2)/2} + 1 \tag{7.18}$$

Thus, a Gold sequence formally is an arbitrary phase of a sequence in the set $G(x, y)$ defined by

$$G(x,y) = \{x, y, x \oplus y, x \oplus Ty, x \oplus T^2 y, x \oplus T^{(Nc-1)}y\} \tag{7.19}$$

where T^k denotes the operator that shifts vectors cyclically to the left by k places, $k = 1, 2, \ldots (N_C - 1) \oplus$ is the XOR operator, and x, y are the *m*-sequences of period generated by different primitive binary polynomials.

It is well known that the *partial cross-correlation* values can be altered by changing the phases of the code sequences. In theory, then, it is possible to find the optimal phases that minimize the interference in the desired data signal. However, for K users, each employing a sequence of period N_c, there are a total of $N_c K$ different sets of sequence phases possible.

Table 7.4 Calculation of cross-correlation of Gold sequences

Number of LFSR elements (p)	Sequence length ($N_c = 2^p - 1$)	Number of *m*-sequences	Maximum cross-correlation of *m*-sequence, normalized	Cross-correlation of Gold sequence (R_c)	Cross-Correlation of Gold-sequence, normalized $[R_c/(2^p - 1)]$
3	7	2	0.71	5	0.71
4	15	2	0.60	9	0.60
5	31	6	0.35	9	0.29
6	63	6	0.36	17	0.27
7	127	18	0.32	17	0.13
8	255	16	0.37	33	0.13
10	1023	60	0.37	65	0.06
12	4095	144	0.34	129	0.03

7.11 DIRECT SEQUENCE SPREAD SPECTRUM SYSTEM

In a DSSS system, the user signal is multiplied by a PN code sequence of high bandwidth. The resulting coded signal is transmitted over the radio channel. DS is responsible for the spreading of the bandwidth. This section describes the important aspects of designing various stages of a DSSS system, including the transmitter and receiver block diagrams, rake

The PN sequence is also called the *chip sequence* or DS, which is the concatenated PN code of one period.

receiver aspects, properties of PN signals, and characteristics of a DSSS system over performance parameters.

7.11.1 Transmitter and Receiver

In the transmitter (Fig. 7.31) of a DSSS system, a MOD 2 adder is used for biphase modulation. While using quadriphase modulation, two MOD 2 adders are used with two alternate chips available from the PN code generator. Two balanced modulators are fed with 90° phase shifted carriers (similar to QPSK generation). Adding both the signals, the SSM RF output is obtained. It may be noted that message modulation is performed using the binary data and code.

The diagram of a receiver is shown in Fig. 7.32. The receiver for a spread signal must perform three distinct functions: detection of the presence of a signal, carrier removal, and despreading or demodulation using a PN sequence.

Detection of signal and despreading operations can be either active or passive:

Active method It involves searching for the signal's presence in both the time and frequency domains and tracking the sequence after it has been acquired, despreading the signal with the correlator, and demodulating the signal in the usual way, that is, using a PN sequence again.

Passive method It is required to search the signal in terms of its carrier frequency to be detected, because the passive system will respond whenever the signal occurs. Despreading is accomplished in a matched filter rather than a correlator. Demodulation is performed in the usual manner, as mentioned previously.

The complete process of spread spectrum modulation and demodulation can be understood with the following example.

Note: The choice of the method—active or passive—depends upon the conditions. Active methods are preferred when the sequence is too long and the processing gain is very large. On the other hand, passive methods are preferred when the sequence is short or when used as an aid to acquisition. It is possible to combine both the approaches in a single receiver.

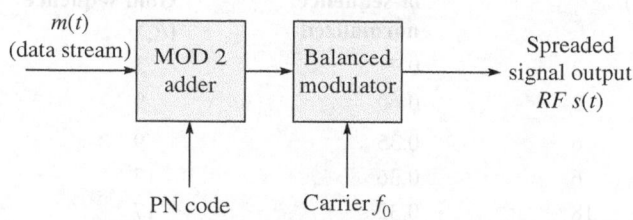

Fig. 7.31 Simplified diagram for biphase modulation

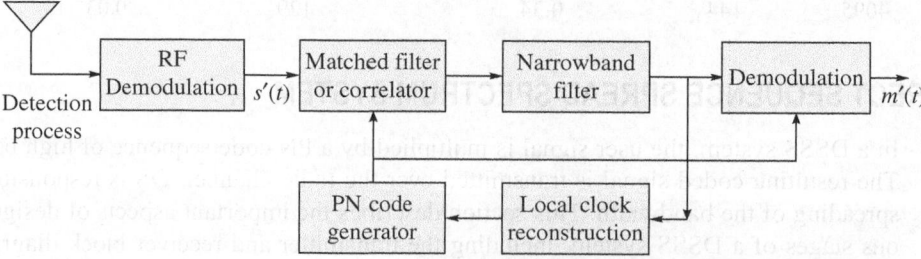

Fig. 7.32 Carrier demodulation and despreading of SSM signal to get original data

Example 7.5 Draw the waveforms of the different stages of a DSSS modulator and demodulator for input data bits 1 followed by 0. The PN sequence during these two bits is (01010001101011100110000 1).

Solution Considering the XOR operation, we get the waveforms as shown in Fig. 7.33

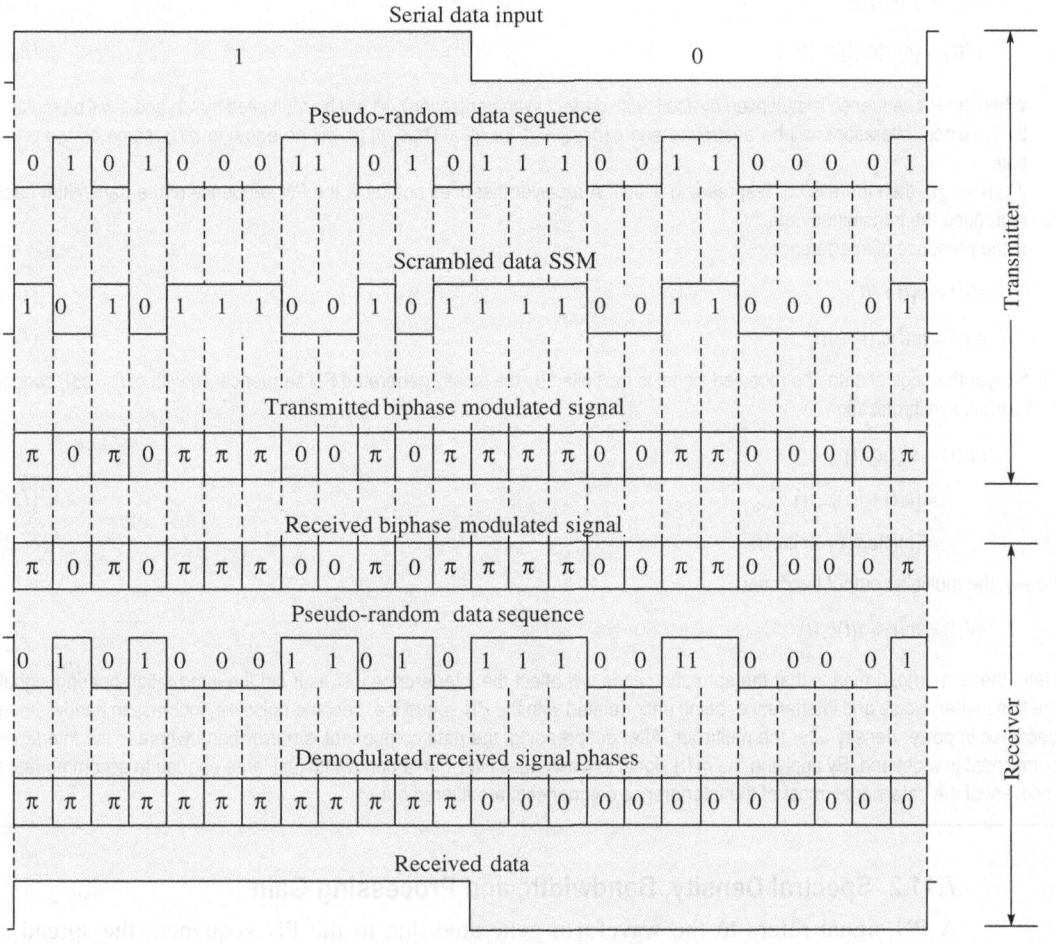

Fig. 7.33 Waveforms for Example 7.5

MATHEMATICAL ASPECTS OF SPREAD SPECTRUM MODULATION

Spreading If the binary data is $m(t)$, PN signal (at the transmitter end) is $c_t(t)$, and transmitted baseband is $s(t)$, then the binary data is directly multiplied by the PN sequence, which is independent of the binary data, to produce the transmitted baseband signal.

$$s(t) = m(t)c_t(t) \tag{7.20}$$

The effect of multiplication of $m(t)$ and the PN sequence is to spread the baseband bandwidth B_m of $m(t)$ to a baseband bandwidth of B_s.

Despreading In the receiver, the received baseband signal $s'(t)$ is multiplied by the PN sequence (at the receiver end) $c_r(t)$.

- If $c_r(t) = c_t(t)$, and is synchronized to the PN sequence in the received data, then the recovered binary data $m'(t)$ is produced. The bandwidth is also despread. We have

$$m'(t) = s'(t)c_r(t) \tag{7.21a}$$

$$m'(t) = (m(t)c_t(t))c_r(t) \tag{7.21b}$$

When the PN sequence is multiplied by itself (with perfect synchronization), −1 will be multiplied by −1, and 1 will be multiplied by 1. Hence, alterations will be destroyed and $c_t(t)c_r(t) = 1$ for all t. Thus, $m'(t)$ will be equal to $m(t)$, same as the original data.

- If $c_t(t) \neq c_r(t)$, then there is no despreading action. A receiver that does not know the PN sequence of the transmitter cannot reproduce the transmitted data.

In the presence of interference,

$$s'(t) = s(t) + i(t) \tag{7.22a}$$

$$s'(t) = m(t)c_t(t) + i(t) \tag{7.22b}$$

To recover the original data, the received signal is multiplied by the locally generated PN sequencer (t) with $c_t(t) = c_r(t)$ condition and perfect synchronization.

$$m'(t) = s'(t)c_r(t)$$

$$= [s(t) + i(t)]c_r(t) \tag{7.23a}$$

$$= [m(t)c_t(t) + i(t)]c_r(t) \tag{7.23b}$$

Finally, the multiplier output becomes

$$m'(t) = m(t) + i(t)c_r(t) \tag{7.24}$$

Here, the term $i(t)c_r(t)$ means that the spreading code will affect the interference just as it did the information-bearing signal at the transmitter. Noise and interference, being uncorrelated with the PN sequence, become noiselike, increase in bandwidth, and decrease in power density after the multiplier. After despreading, the data component is narrowband whereas the interference component is wideband. By applying the $m'(t)$ signal to a low-pass filter with a bandwidth just large enough to accommodate the recovery of the data signal, most of the interference components are filtered out.

7.11.2 Spectral Density, Bandwidth, and Processing Gain

A PN signal refers to the waveform generated due to the PN sequence (the spread spectrum output with carrier allocation will also be of the same nature and can be considered in the same manner). The frequency domain representation of a time domain pulse is a sinc form of envelope. The PN code and data are both of pulsed or square wave nature. If the binary or quadriphase PN sequence is considered to be purely random (due to the long sequence, though it may be repeated) rather than periodic, its spectral density is given by

$$S(f) = \frac{t_{chip}}{2} \left\{ \left[\frac{\sin \pi (f - f_0) t_{chip}}{\pi (f - f_0) t_{chip}} \right]^2 + \left[\frac{\sin \pi (f + f_0) t_{chip}}{\pi (f + f_0) t_{chip}} \right]^2 \right\} \tag{7.25}$$

The expression is normalized to represent a signal having unit average power. The spectral density for the positive and negative frequencies is shown in Fig. 7.34. f_0 represents the

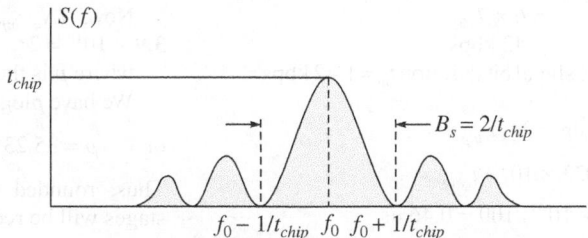

Fig. 7.34 Spectral density of binary PN sequence

centre frequency of the sync-shaped spectrum (which may be the carrier after upconversion as the shape of the bandwidth is retained).

Generally, the bandwidth B_s of a PN signal is defined as the frequency increment between the two zeroes of the spectral density that are closest to the centre frequency. It is clear from Fig. 7.34 that the bandwidth of the signal is $2/t_{chip}$. As the message is binary or a square wave, its spectral density will be similar but centred on zero. Thus, the message signal spectral density is

$$S_m(f) = t_m \left[\frac{\sin \pi f t_m}{\pi f t_m} \right]^2 \tag{7.26}$$

The bandwidth of the message B_m with similar considerations will be $1/t_m$, because it is customary to use only the positive frequency portion of the spectrum in defining the bandwidth. Let B_s be the bandwidth of the SSM or PN signal and B_m be the bandwidth of the message signal. Then, the processing gain is given by

$$PG = B_s/B_m \tag{7.27}$$

This gives the amount of spreading of the signal.

If biphase modulation is used, then the processing gain is given by

$$PG = 2t_m/t_{chip} \tag{7.28}$$

For quadriphase modulation, the processing gain is

$$PG = t_m/t_{chip} \tag{7.29}$$

Example 7.6 If the chip rate of a DSSS transmitter is 20 Mcps and the message bit rate is 10 kbps, find the processing gain achieved finally if biphase modulation is used.

Solution $\quad t_{chip} = \dfrac{1}{20 \times 10^6}$

or SSM bandwidth $B_s = 2/t_{chip} = 40 \times 10^6$ Hz

Message bandwidth $B_m = 1/t_m = 10^4$ Hz

Hence, processing gain $= B_s/B_m = 2/t_m/t_{chip} = 4000$

In decibels, processing gain can be represented as $10\log_{10}(4000) = 36$ dB

Example 7.7 A recorded conversation is to be transmitted by a PN spread spectrum system. Assume that

the spectrum of the speech waveform is band-limited to 3 kHz and uses 128 quantization levels. (a) Find the chip rate required to obtain a processing gain of 20 dB. (b) Given that the sequence length is to be greater than 5 h, find the number of shift register stages required.

Solution
(a) Processing gain $= 20$ dB $= 10\log_{10}PG$

or $\quad PG = 10^2 = 100$

Now, 128 quantization levels means that 7 bits/sample will be utilized.

Moreover, 3 kHz speech signal means that the sampling rate will be 6 kHz (Nyquist criteria).

Message bit rate = Sampling rate × Bits/sample

$$= 6 \times 7$$
$$= 42 \text{ kbps}$$

Now, message signal bit duration $t_m = 1/42 \text{ kbps} = 0.023 \text{ ms}$

Processing gain $= 2t_m/t_{chip}$

or $\quad 100 = 2 \times (0.023 \times 10^{-3})/t_{chip}$

or $\quad t_{chip} = (0.046 \times 10^{-3})/100 = 0.46 \text{ μs}$

(b) Sequence length is $5 \text{ h} = 5 \times 60 \times 60 = 18,000 \text{ s}$

Now, $(N_c t_{chip})/t_{chip} = 18,000/0.46 \times 10^{-6} = 3.9 \times 10^{10} = 2^p$

where p is the number of stages

We have $p\log_{10}2 = \log(3.9 \times 10^{10})$

or $\quad p = 35.23$

Thus, rounded to the nearest higher value, 36 stages will be required.

7.11.3 Rake Receiver

Rake is a long-handled tool with a row of prongs at the end for smoothing the soil. The receiver designed on such a concept for a DSSS system is known as a rake receiver; it exploits multipath and its hardware is mainly based on digital signal processing. A rake receiver does almost the same task of a rake by extracting and smoothing the available multipath signals. The front end looks like a row of prongs (parallel channels as shown in Fig. 7.35), known as the fingers of the rake receiver.

A rake receiver consists of multiple correlators, each synchronized to one of the time offsets of the received signal. It correlates the received signal with different time offsets of the spreading and scrambling codes and performs a channel correction to compensate for the differing channel characteristics of the individual multipaths. These steps are performed for all multipaths in the environment. The results of these operations are combined to drive the decision-making process for the value of the received symbol.

Two configurations with a five-finger rake receiver are shown in Fig. 7.35.

If reflected waves arrive with small relative time delays, self-interference occurs. The rake receiver architecture allows an optimal combining of the energy received over different paths with different delays. It avoids wave cancellation (fades) if the delayed paths arrive with phase differences and appropriately weighs the signals coming in with different SNRs.

In short, the rake receiver is designed to optimally detect a DS-CDMA signal transmitted over a dispersive multipath channel.

> Rake receiver is a spread spectrum signal receiver configuration with the best features for combating channel multipath by exploiting multipath.

The spreading code is chosen to have a very small autocorrelation value for any non-zero time offset. This avoids crosstalk between fingers. In practice, the situation is less ideal. The crosstalk between the signals in different fingers is determined not by the full periodic autocorrelation but rather by two partial correlations, with contributions from two consecutive bits or symbols. There have been attempts to find sequences that have satisfactory partial correlation values,

---

IN CONTRAST TO A RAKE RECEIVER...

In the matched filter receiver, the signal is correlated with a locally generated copy of the signal waveform. If, however, the signal is distorted by the channel, the receiver should correlate the incoming signal by a copy of the expected received signal, rather than by a copy of the transmitted waveform. Thus, the receiver should estimate the delay profile of the channel and adapt to its locally generated copy according to this estimate.

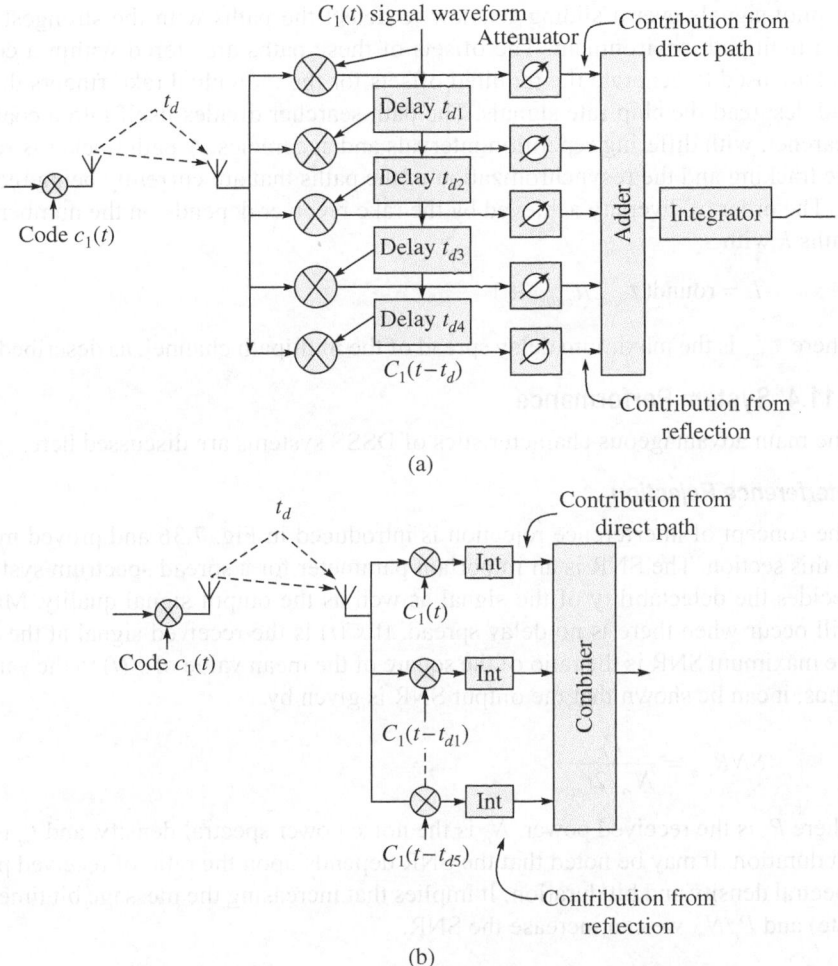

Fig. 7.35 Rake receiver configurations (a) With five fingers, one combiner, and one integrator
(b) Reception with multiple integrators at rake receiver and processing

but the crosstalk due to partial (non-periodic) correlations remains substantially more difficult to reduce than the effects of periodic correlations.

In addition to the actual signal reception tasks, the rake receiver must also perform a set of timing and synchronization tasks. A path searcher performs a correlation of a fixed set

---

RAKE RECEIVER AND SOFT HANDOVER

In soft handover, a mobile terminal is in contact with multiple base stations at the same time. This can occur when a mobile terminal is at a cell border and the signals from all surrounding base stations have near equal strength.

The signals from multiple base stations differentiate themselves in their scrambling code. Thus, to handle the soft handover scenario, the rake receiver in the mobile terminal must be capable of correlating the received signal with the individual scrambling codes of all the base stations involved. Furthermore, the rake receiver must also be capable of collecting and using the energy from the multipath components of a signal.

of pilot signals over a sliding window to detect the paths with the strongest signal values in a multipath environment. The offsets of these paths are stored within a control context and are used to generate the required offsets for the individual rake fingers that descramble and despread the chip rate signals. The path searcher divides itself into a coarse and a fine searcher, with differing repetition intervals and accuracies. A path tracker is responsible for the tracking and the resynchronization of the paths that are currently being received.

The order of diversity achieved by the rake receiver depends on the number of resolvable paths L with

$$L = \text{round}(\tau_{max}/t_{chip}) + 1 \qquad (7.30)$$

where τ_{max} is the maximum delay spread of the multipath channel, as described in Chapter 3.

7.11.4 System Performance

The main advantageous characteristics of DSSS systems are discussed here.

Interference Rejection

The concept of interference rejection is introduced in Fig. 7.36 and proved mathematically in this section. The SNR is an important parameter for a spread spectrum system because it decides the detectability of the signal as well as the output signal quality. Maximum SNR will occur when there is no delay spread. If $s'(t)$ is the received signal at the rake receiver, the maximum SNR is the ratio of the square of the mean value of $s'(t)$ to the variance of $s'(t)$. Thus, it can be shown that the output SNR is given by

$$SNR_{out} = \frac{P_r}{N_o/2t_m} \qquad (7.31)$$

where P_r is the received power, N_o is the noise power spectral density, and t_m is the message bit duration. It may be noted that the SNR depends upon the ratio of received power to noise spectral density and bit duration. It implies that increasing the message bit time (reducing bit rate) and P_r/N_o, we can increase the SNR.

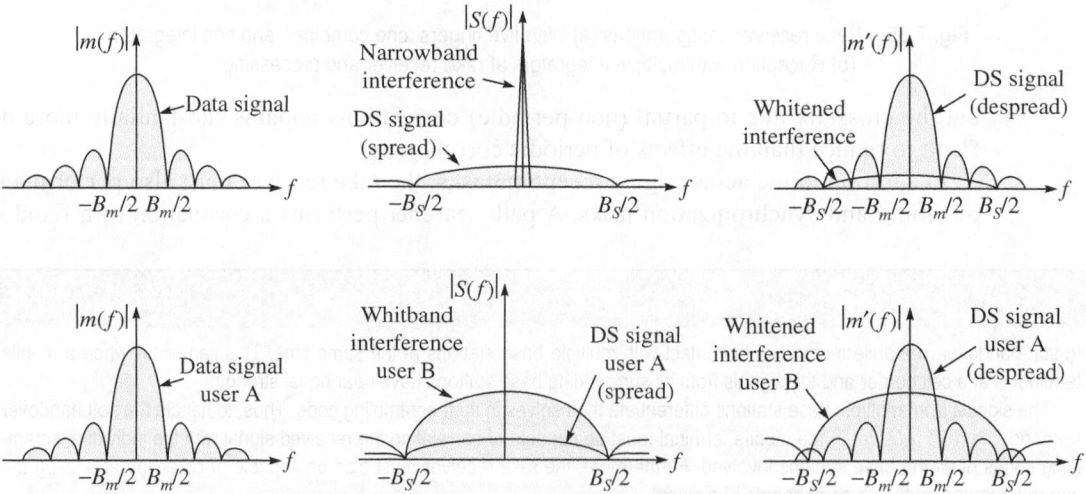

Fig. 7.36 Narrowband and wideband interference rejection

The existence of other users' signal is treated as an interfering signal for the intended one and it reduces the SNR of the intended signal.

Now, the mean square value of the output interference signal is given by

$$\overline{j^2} = \frac{J}{PG} \tag{7.32}$$

where J is the total interference power and PG is the processing gain. This is the case when the interfering signal bandwidth is less than the transmission bandwidth. We can see that the interference power can be reduced by a factor equal to the processing gain. This is an average result; however, the averaging takes place over all phases of interference. For some fixed phase relationships, the mean square value of the output interference may be twice the value given by Eq. (7.32).

It is also interesting to consider the situation in which the interference bandwidth B_j is greater than the spread spectrum bandwidth B_s. In this case,

$$\overline{j^2} = \frac{B_s}{2B_j} \cdot \frac{J}{PG} \tag{7.33}$$

As J is the interference power in its total bandwidth, the factor represents $\frac{B_s}{2B_j} \cdot J$ simply the portion of the interference power that exists in the receiver bandwidth. This power is again reduced by a factor equal to the processing gain. Considering the interference signal as a noise, the SNR is to be calculated as follows:

$$SNR_{out} = \frac{P_r}{N_o/2t_m + J/PG} \tag{7.34}$$

However, $\dfrac{PG}{t_m} = \dfrac{B_s}{B_m} \cdot B_m = B_s$

Hence, the output SNR becomes

$$SNR_{out} = \frac{PG \cdot P_r}{(N_o B_s/2) + J} \tag{7.35}$$

Now, $SNR_{in} = \dfrac{P_r}{(N_o B_s/2) + J} \tag{7.36}$

The new relationship can be written as

$$SNR_{out} = (PG) \cdot SNR_{in} \tag{7.37}$$

Here, the losses in the system are not considered.

Example 7.8 A DS system has a PN code rate of 192×10^6 chips/s and a binary message bit rate at 7500 bps.
(a) If quadriphase modulation is used, find the performance gain.
(b) Assuming the received signal power is 4×10^{-14} W and the one-sided noise spectral density level N_o is 1.6×10^{-20} W/Hz., find the SNR in the input bandwidth of the receiver.

Solution
(a) $t_{chip} = 1/192 \times 10^6 = 0.0052~\mu s$

$t_m = 1/7500 = 0.133~ms$

For quadriphase modulation

$PG = t_m/t_{chip} = 0.133 \times 10^{-3}/0.0052 \times 10^{-6}$
$= 25{,}577$

Processing gain in decibels $= 10\log_{10}(25{,}577) = 44.08~dB$

(b) P_r = Received power = 4×10^{-14} W
One-sided noise spectral density = 1.6×10^{-20} W/Hz

For biphase modulation

Signal bandwidth = $2/t_{chip}$ = $2/0.0052 \times 10^{-6}$ = 384.6 MHz

Noise power = $N_o \times$ Bandwidth = $1.6 \times 10^{-20} \times$ $384.6 \times 10^6 = 615 \times 10^{-14}$ W

$SNR = P_r/N_o = 4 \times 10^{-14}/615 \times 10^{-14} = 0.0065$

In decibel scale, $SNR = 10\log_{10}(P_r/N_o) = -21.86$ dB

For quadriphase modulation

Signal bandwidth = $1/t_{chip}$ = $1/0.0052 \times 10^{-6}$ = 192.3 MHz

Noise power = $N_o \times$ Bandwidth = $1.6 \times 10^{-20} \times$ $192.3 \times 10^6 = 307.7 \times 10^{-14}$ W

$SNR = P_r/N_o = 4 \times 10^{-14}/307.7 \times 10^{-14} = 0.013$

In decibel scale, $SNR = 10\log_{10}(P_r/N_o) = -18.86$ dB

Anti-jam Characteristics

The interference rejection capabilities of a spread spectrum system are of importance primarily because of their ability to evaluate the degree to which such a system can reduce the effects of intentional jamming. This ability can be expressed in terms of the jamming margin, sometimes called the anti-jam (AJ) margin, and it is usually expressed in decibels. If we write the SNRs in terms of decibels, Eq. (7.37) can be rewritten as

$$SNR_{in}(\text{dB}) = SNR_{out}(\text{dB}) - PG(\text{dB}) \tag{7.38}$$

The anti-jam margin can be defined as

$$\text{Margin } (AJ) = -SNR_{in}(\text{dB}) - L(\text{dB}) \tag{7.39}$$

where L represents the losses in the system expressed in decibels. The anti-jam margin can also be written as

$$\text{Margin } (AJ) = PG(\text{dB}) - L(\text{dB}) - SNR_{out}(\text{dB}) \tag{7.40}$$

Example 7.9 Consider a spread spectrum system in which the chip rate is 10^7 chips/s and the message bit rate is 100 bps. If it is desired to obtain an output SNR of 25, which is about 15 dB, and if the losses are determined to be 2 dB, then find the anti-jam margin.

Solution Considering biphase modulation,

Processing gain $PG = 2t_m/t_{chip}= 2 \times 10^7/100 = 2 \times 10^5 = 53$ dB

Margin $(AJ) = 53 - 2 - 15 = 36$ dB

It means that the desired output SNR can be obtained if the jamming signal is less than 36 dB. If we consider some practical circumstances at the receiver, the actual anti-jam margin may be something less than this value.

Energy and Bandwidth Efficiency

The output SNR of the spread spectrum receiver in the presence of noise and interference is given by

$$SNR_{out} = \frac{P_r}{N_o/2t_m + J/PG} \tag{7.41}$$

However, the energy associated with a message bit is

$$E_b = P_r T_m = \text{Energy per bit} \tag{7.42}$$

The output SNR can also be represented by the following expression:

$$SNR_{out} = \frac{P_r \cdot t_m}{N_o/2 + Jt_m/PG} = \frac{E_b}{N_o/2 + J/B_s}$$

$$= \frac{E_b/N_o}{1/2 + J/N_o B_s} \tag{7.43}$$

Rewriting this equation in terms of E_b/N_0,

$$\frac{E_b}{N_o} = \left[\frac{1}{2} + \frac{J}{N_o B_s} \right] SNR_{out} \tag{7.44}$$

Two important observations can be made from this result:
1. When the interference is narrowband, increasing the spreading improves the energy efficiency.
2. When the interference is wideband, such as white noise, increasing the spreading does not improve the energy efficiency because the interference power J increases directly with the signal bandwidth B_s.

The bandwidth utilization efficiency B/R is given by

$$\frac{B}{R} = \frac{B_s}{2R_m} \tag{7.45}$$

where R_m is the message bit rate. The term $B_s/2$ is used since this is the equivalent energy bandwidth of the signal. It is important to note that B/R increases linearly with spreading, and thus, the bandwidth utilization efficiency becomes poorer with the increase in the bandwidth of the spread spectrum. If the interference is narrowband compared to the spread spectrum signal bandwidth, then combining the results of Eqs (7.44) and (7.45),

$$\frac{E_b}{N_o} = \left[\frac{1}{2} + \frac{J}{2N_o R_m (B/R)} \right] SNR_{out} = \frac{1}{2} \left[1 + \frac{J/N_o R_m}{B/R} \right] SNR_{out} \tag{7.46}$$

If the interference is wideband compared to the spread spectrum signal bandwidth, then Eq. (7.46) can be written as

$$\frac{E_b}{N_o} = \frac{1}{2} \left[1 + \frac{J_o}{N_o} \right] SNR_{out} \tag{7.47}$$

The quantity J_o is the spectral density of the interfering signal.

NEAR–FAR PROBLEM

Near–far problem in a spread spectrum system relates to the problem of very strong signals at the receiver swamping out the effects of weaker signals. This is particularly serious in the case of DSSS multi-user systems.

The problem can be considered with the help of a receiver and two transmitters (Fig. 7.37). One transmitter is near the receiver; the other is far away from the receiver. If both the transmitters transmit simultaneously with equal powers, then the receiver will receive more power from the nearer transmitter. This creates a difficulty in detecting the signal from the farther transmitter.

As one transmission's signal is the other's noise, the SNR for the farther transmitter must be much higher. If the nearer transmitter transmits a signal that is orders of magnitude higher than the farther transmitter, then the SNR for the farther transmitter may be below the required value, making the signal undetectable, and the farther transmitter may just as well not transmit. This effectively jams the communication channel. To achieve successful communication, the farther transmitter would have to drastically increase its transmission power, which simply may not be possible. In short, *the near–far problem is one of detecting and receiving a weaker signal among the stronger signals.* Moreover, the received level fluctuates quickly due to fading. In order to maintain the strength of the received signal level at the base station, power control is employed in CDMA systems.

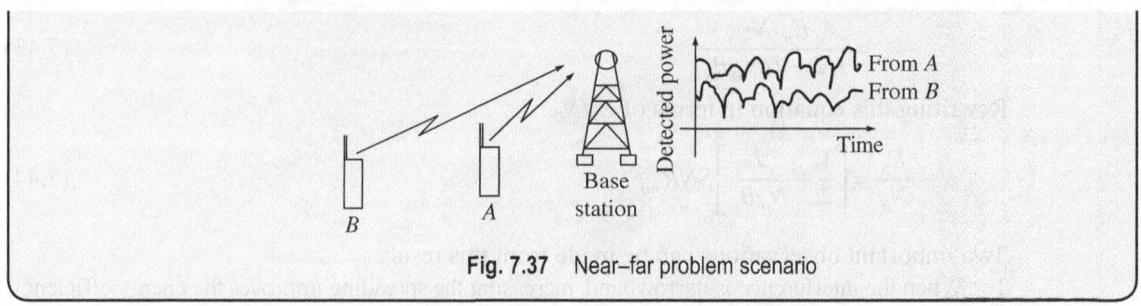

Fig. 7.37 Near–far problem scenario

Power control In CDMA systems or other mobile phone-like networks, the near–far problem is commonly solved by dynamic output power adjustment of the transmitters. That is, the nearer transmitters use less power so that the SNR for all the transmitters at the receiver is roughly the same. As all mobiles transmit at the same carrier frequency, the internal interference generated within the system plays a critical role in determining the system capacity and voice quality.

The objective of power control is to limit transmitted power on the forward and reverse links while maintaining link quality under all conditions (Fig. 7.38).

The following two conditions must be satisfied simultaneously:
(a) The transmit power from each mobile must be controlled to limit interference.
(b) The power level should be adequate for a satisfactory voice quality.

Power control is capable of compensating the fading fluctuation. Received powers from all mobile stations are controlled to be equal. Thus, the near–far problem is mitigated by power control. Now, the detected power strengths of users *A* and *B* shown in Fig. 7.37 will be almost equal.

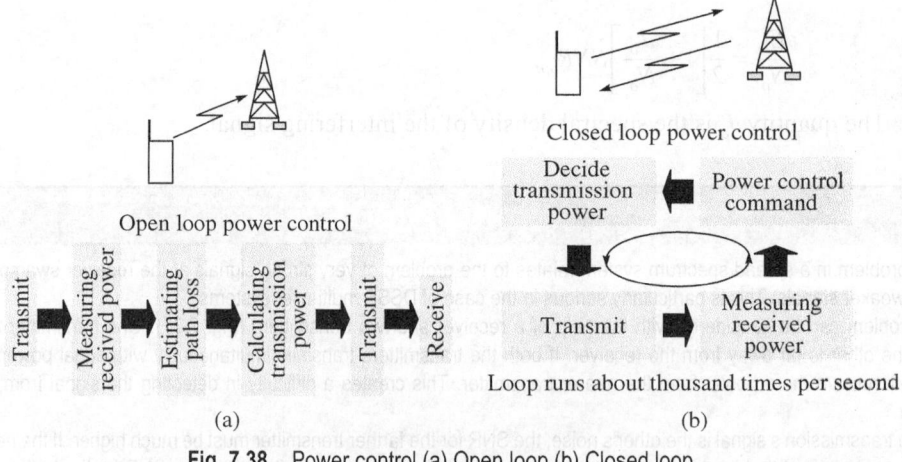

Fig. 7.38 Power control (a) Open loop (b) Closed loop

7.12 FREQUENCY HOPPING SPREAD SPECTRUM—TRANSMITTER AND RECEIVER

In frequency hopping systems, the transmitter changes the carrier frequency according to a certain *hopping* pattern (Fig. 7.39), meaning that the frequency is constant in each time chip but

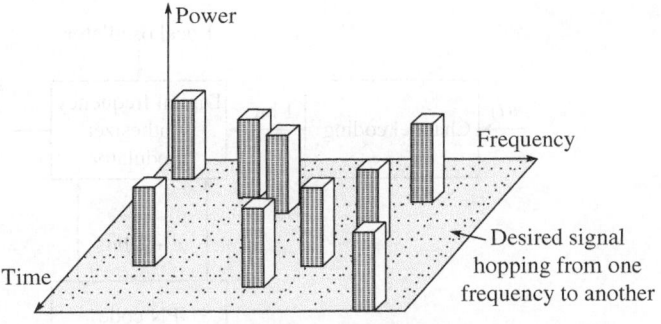

Fig. 7.39   Concept of FH

An FHSS is much less susceptible to the near–far problem because it is an avoidance system rather than an averaging system.

changes from chip to chip. The advantage is that the signal sees a different channel and a different set of interfering signals during each hop. This avoids the problem of failing communication at a particular frequency, because of a fade or a particular interferer.

There are two kinds of FH—slow FH (SFH) and fast FH (FFH)—and there is a considerable difference in performance between the two systems.

• In SFH, one or more data bits are transmitted within one hop; that is, the hopping rate is less than the message bit rate. An advantage is that coherent data detection is possible. Systems using slow hopping often also employ (burst) error-control coding to restore loss of (multiple) bits in one hop.

• In FFH, one data bit is divided over multiple hops; that is, the FH rate is greater than the message bit rate. In fast hopping, coherent signal detection is difficult, and seldom used. Mostly, FSK or MFSK modulation is used.

There is an intermediate situation in which the hop rate and the message bit rate are of the same order. SFH is a popular technique for wireless local area networks (LANs). In GSM telephony, SFH can be used at the discretion of the network control software. It avoids the problem of the stationary terminal that happens to be located in a fade losing its link with the base station. FFH is adopted in Bluetooth.

As nearby hopping interferers are unlikely to continuously transmit in the same frequency slot as the reference user, the near–far problem is less severe in FHSS than in DS-CDMA. This advantage made SFH popular, particularly for wireless LANs, where the terminals can be located anywhere.

For purposes of illustration, consider a fast hop system in which there are k frequency hops in every message bit duration t_m. Thus, the chip duration is

$$t_{chip} = t_m / k \qquad k = 1, 2, 3, \ldots \qquad (7.48)$$

The number of frequencies over which the signal may hop is usually a power of two, although all these frequencies are not necessarily used in a given system. The number of frequencies is in a power of two because it is generated by a PN sequence generator control, and PN sequences are related to powers of two. The ML sequence, that is, c chips, will produce $M = 2^c$ frequencies for each distinct combination of these digits.

As shown in Fig. 7.40, one bit comes from the message and $c - 1$ bits come from the PN code generator. If a bit from the message produces the smallest frequency change, then by

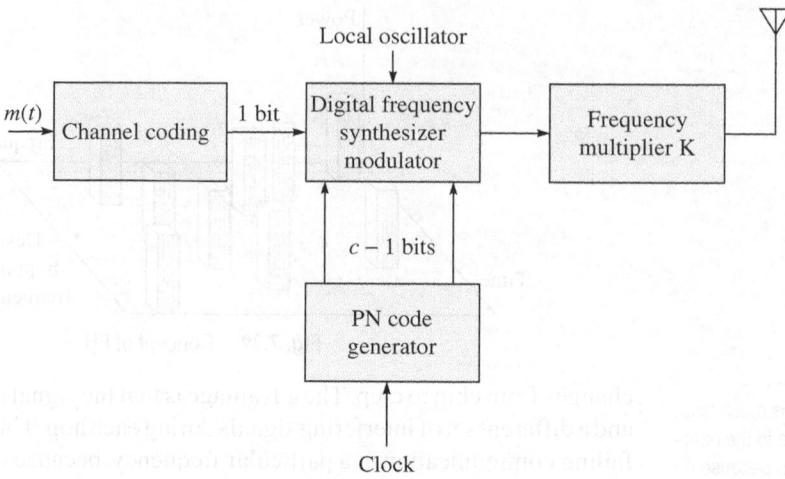

Fig. 7.40 FHSS generator diagram

In FHSS, the FH pattern is accomplished by means of a digital frequency synthesizer, which in turn is driven by a PN code generator.

itself it will produce a BFSK signal. The $c - 1$ bits from the PN code generator then hop this FSK signal over the range of possible frequencies. The data is channel coded to combat bit errors. It may also be noted that there is a frequency multiplier K at the output of the system. It is to increase the bandwidth and thereby increase the processing gain. It also changes the shape of the spectrum.

Considering again the fast hopping case, if M frequencies are separated by $f_1 = 1/t_{chip} = k/t_m$, then the signal bandwidth is given by

$$B_s = KMf_1 = KM/t_{chip} \tag{7.49}$$

Hence, the processing gain is calculated as

$$PG = B_s/B_m = \frac{KM/t_{chip}}{1/t_m} = \frac{kKM/t_m}{1/t_m} = kKM \tag{7.50}$$

A non-coherent FHSS receiver is shown in Fig. 7.41.

Coherent reception is possible in FHSS but it is more difficult to achieve and places some constraints on the nature of the transmitted signal and the transmission medium. Hence, the non-coherent method is normally used. As shown in Fig. 7.41, the locally generated frequency hop signal is multiplied by the incoming signal in a mixer. If the two are in step, the result will be a normal BFSK signal, which is demodulated in the usual way. Error correction is made to improve the BER response. The output of the mixer is also applied to the early and late gates that produce an error signal to control the clock frequency. This keeps the locally generated frequency hop pattern in synchronism with the incoming signal.

Interference results whenever there is simultaneous occupancy of a given frequency slot; in this case, it does not matter much if the interfering signal is much stronger than the desired signal, because within a fraction of time, a new frequency will come up. Another advantage of FH signals is that acquisition is normally much faster than in a DSSS system. This is so because the chip rate is considerably less in the FH system.

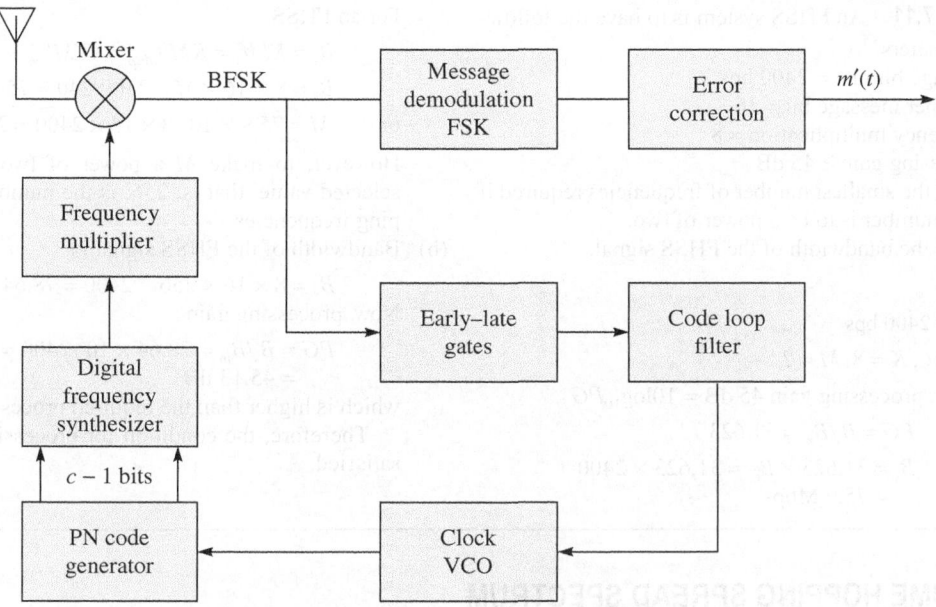

Fig. 7.41   Non-coherent FH receiver

A disadvantage of an FHSS system is that coherent demodulation techniques cannot be used easily. This results in poor performance against thermal noise.

Example 7.10   In an FHSS system for the generation of hopping frequencies, a 3-bit PN code generator is used. Suppose the carrier frequency is 8 kHz and frequency spacing is 0.5 kHz. Corresponding to the code pattern 000, the frequency is 9.75 kHz. Find all the hopping frequencies. Show all these frequencies on the time–frequency plane.

Table 7.5   Data for example 7.10

Control codes	Carrier frequency (kHz)	Derivation of output frequency (kHz)	Output frequency (kHz)
000	8	8 + 1.75	9.75
001	8	8 + 1.25	9.25
010	8	8 + 0.75	8.75
011	8	8 + 0.25	8.25
100	8	8 − 0.25	7.75
101	8	8 − 0.75	7.25
110	8	8 − 1.25	6.75
111	8	8 − 1.75	6.25

Solution   Table 7.5 lists the data, and Fig. 7.42 shows the time–frequency plane.

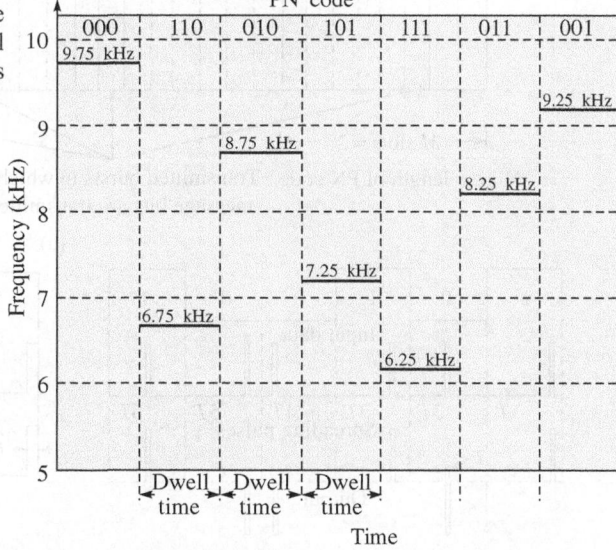

Fig. 7.42   Time–Frequency plane showing FH for Example 7.10

Example 7.11 An FHSS system is to have the following parameters:

Message bit rate = 2400 bps
Hops per message bit = 16
Frequency multiplication = 8
Processing gain ≥ 45 dB

(a) Find the smallest number of frequencies required if this number is to be a power of two.
(b) Find the bandwidth of the FHSS signal.

Solution

(a) $B_m = 2400$ bps

$k = 16, K = 8, M = ?$

Now, processing gain 45 dB = $10\log_{10}PG$

$PG = B_s/B_m = 31,623$

$B_s = 31,623 \times B_m = 31,623 \times 2400$
$= 75.9$ Mbps

For an FHSS,

$$B_s = KMf_1 = KM/t_{chip} = kKM/t_m$$

$$B_s = 8 \times 16 \times M \times 2400/240 = 75.9 \text{ Mbps}$$

or $M = 75.9 \times 10^6/8 \times 16 \times 2400 = 247$

However, to make M a power of two, the next selected value, that is, 256, is the number of hopping frequencies.

(b) Bandwidth of the FHSS signal is

$$B_s = 8 \times 16 \times 256 \times 2400 = 78.64 \text{ Mbps}$$

Now, processing gain

$$PG = B_s/B_m = 78.64 \times 10^6/2400 = 32,768$$
$$= 45.13 \text{ dB}$$

which is higher than the required processing gain.

Therefore, the condition for processing gain is satisfied.

7.13 TIME HOPPING SPREAD SPECTRUM

The TH concepts can be understood with the help of Fig. 7.43.

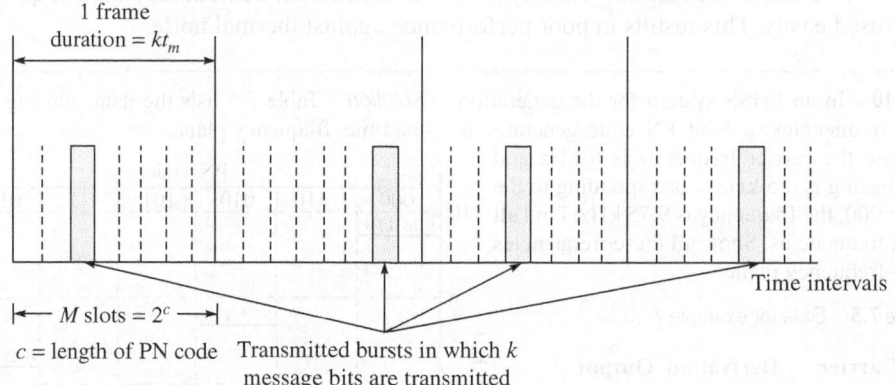

(a)

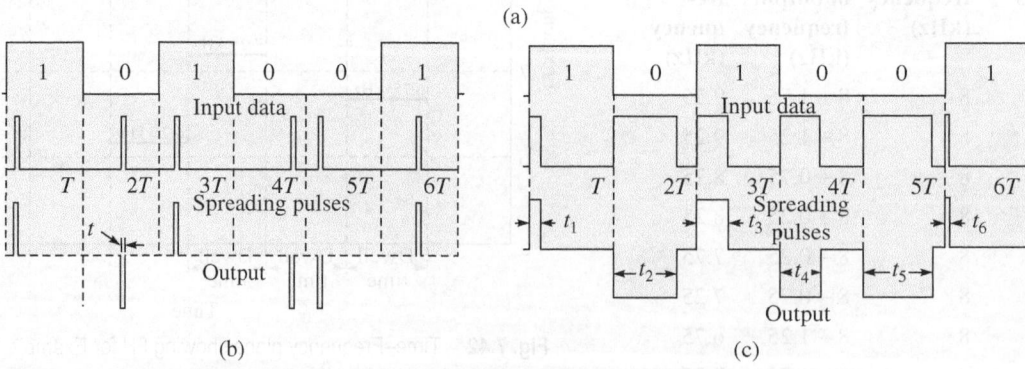

(b)   (c)

Fig. 7.43 TH (a) Concept (b) Waveforms showing THSS signal formation on bit-by-bit basis (c) TH with variable time slots (bit by bit)

In THSS, non-coordination in timing overlaps transmission bursts, which will result in message errors and require the use of error-correction coding to restore the message bits.

The time axis is divided into intervals known as frames, and each frame is subdivided into M time slots (Fig. 7.43a). The total number of slots is decided on the basis of the PN code or the length of the m-sequence. The slots and length of the m-sequence are related to each other by the relation $M = 2^c$. During each frame, one and only one time slot is modulated with a message by any reasonable modulation method. The particular time slot chosen for a given frame is selected by means of a PN code generator. All the message bits accumulated in the previous frame are transmitted in a burst during the selected time slot. The frame duration T_f, the number of message bits k, and the message bit duration t_m are related to each other by

$$T_f = kt_m \tag{7.51}$$

The width of each time slot in a frame is T_f/M and the width of each bit in the time slot is T_f/kM or simply t_m/M.

$$\text{Processing gain } PG = \begin{cases} B_s/B_m = 2t_m/(t_m EM) = 2M & \text{for biphase modulation} \tag{7.52a} \\ t_m/(t_m/M) = M & \text{for quadriphase modulation} \tag{7.52b} \end{cases}$$

This indicates that the transmitted signal bandwidth is $2M$ times the message bandwidth, and hence, the processing gain of the TH system is twice the number of time slots in each frame when biphase modulation is used and half this when quadriphase modulation is used. Figures 7.43(b) and (c) show other possible variations.

Interference among simultaneous users in a TH system can be minimized by coordinating the times at which each user can transmit a signal. This also avoids the near–far problem. The acquisition time is similar to that of DS systems for a given bandwidth. Implementation is simpler than in an FH system.

Example 7.12 A certain THSS system is allocated a maximum signal bandwidth of 8 MHz. Assuming that the message bit rate after channel coding is 3200 bps, find the number of bits necessary from the code generator to control the ON/OFF switching time. If biphase modulation is used, what is the processing gain?

Solution $PG = B_s/B_m = 2M$ for a biphase modulation system

or $PG = 8 \times 10^6/3200 = 2500 = 2M = 33.97$ dB

or $M = 1250 = $ Total number of slots

It is not a power of two; but as the bandwidth must not exceed 8 MHz, we shall choose $M = 1024$ (previous lowest power of two).

Now, $M = 2^c$ implies $c = 10$ bits are required from the code generator to control the ON/OFF switching time.

The processing gain will now be $2M = 2048 = 33$ dB. (This value does not differ much from the previous value of the processing gain.)

Table 7.6 compares the features of all the three SSM methods.

Table 7.6 Comparison of SSM methods

DS method	FH method	TH method
Averaging type	Avoidance type	Avoidance type
Single-carrier system for all times	Different hopping carriers, one at a time	Single carrier but transmitted at different times

(Contd)

Table 7.6 *(Contd)*

DS method	FH method	TH method
Simplest implementation	Complex implementation because synthesizers are required	Simpler than FH
Long acquisition time	Short acquisition time	Long acquisition time
Near–far problem	No near–far problem	Near–far problem is avoided by proper time synchronization

Advantages	*Advantages*	*Advantages*
• Best noise and anti-jam performance • Most difficult to detect • Best discrimination against multipath	• Greatest amount of spreading • Can be programmed to avoid portion of the spectrum	• Highest bandwidth efficiency • Useful when transmitter is average power limited but not peak power limited

Limitations	*Limitations*	*Limitations*
• Requires wideband channel with little phase distortion • Requires fast code generator • Requires long chip sequence	• Not useful for range and range rate measurement • Requires error correction	• Requires error correction • Critically requires synchronization • Can be jammed easily hence not used generally

7.14 HYBRID SPREAD SPECTRUM SYSTEMS

The use of hybrid techniques attempt to capitalize upon the advantages of a particular method while avoiding the disadvantages. For example, evaluation of DS and FH shows that both techniques have their specific advantages and disadvantages. DS, on one hand, suffers heavily from the near–far effect, which makes this technique hard to apply to systems without the ability of power control. On the other hand, its implementation is inexpensive. The *PN code* generators are easy to implement and the spreading operation itself can be simply performed by XOR ports. FH effectively suppresses the near–far effect and reduces the need for power control. However, implementation of the (fast) hopping frequency synthesizer required for a reasonable spreading gain is more problematic in terms of higher silicon cost and increased power consumption.

Applying both techniques allows for combining their advantages while reducing the disadvantages. This results in a reasonable near–far resistance at an acceptable hardware cost.

Many different hybrid combinations are possible, some of which are PN/FH, PN/TH, FH/TH, and PN/FH/TH.

While designing a hybrid system, the designer should decide FFH or SFH is to be applied. FFH increases the cost of the frequency synthesizer but provides more protection against the near–far effect. SFH combines a less expensive synthesizer with a poor near–far rejection and the need for a more powerful error-correction scheme (several symbols are lost during a *hit* jamming). Such considerations can be applied to other types of systems as well.

Example 7.13 A PN/FH hybrid spread spectrum system has the following parameters:

Message bit rate = 9600 bps after error-correction coding

PN code rate = 1,53,600 chips/s

Chips per hop = 16

Hops per message bit = 1

Number of hopping frequencies used = 4096

(a) Find the processing gain.

(b) Find the bandwidth of the spread spectrum signal.

Solution

(a) *PN* code rate = 1,53,600 bps, which is the spacing between the two frequencies.

Number of hopping frequencies, $M = 4096$

Hopping rate = 1,53,600/16 = 9600 hops/s

$$B_s = Mf_1 = 4096 \times 1,53,600 = 629 \text{ MHz}$$

Now, $PG = 629 \times 10^6 / 9600 = 65,536$

$$PG \text{ in decibels} = 10\log_{10} PG = 10\log_{10}(65,536)$$
$$= 48.16 \text{ dB}$$

(b) Bandwidth of the spread spectrum signal is 629 MHz without RF multiplication factor.

7.15 MULTICARRIER MODULATION TECHNIQUES

In all the modulation techniques discussed until now, a single carrier was used. In multicarrier modulation, multiple orthogonal carriers are used to send the information signal. The advantages of using multiple carriers will be clear after going through this section.

7.15.1 Basic Principles of Orthogonality

Orthogonality between two signals means that the two coexisting signals are independent of each other in a specified time interval and do not interact with each other. The concept of orthogonal signals is essential for understanding the OFDM system. Loss of orthogonality results in blurring between these information signals and degradation in communications.

Two periodic signals are orthogonal when the integral of their product over one period is equal to zero and they have an integral number of cycles in the fundamental period; that is, the peak of a carrier must occur at the null of the previous carrier. The mathematical representation of the orthogonal continuous time and discrete time signals is shown in Eqs (7.53) and (7.54); these are the conditions for orthogonality.

Continuous time:

$$\int_0^T \cos(2\pi n f_0 t) \times \cos(2\pi m f_0 t) dt = 0 \quad (n \neq m) \tag{7.53}$$

Discrete time:

$$\sum_{k=0}^{N-1} \cos\left(\frac{2\pi k n}{N}\right) \times \cos\left(\frac{2\pi k m}{N}\right) = 0 \quad (n \neq m) \tag{7.54}$$

where N = period of k samples. Consider the orthogonal signals shown in Fig. 7.44.

In a multicarrier modulation scenario, symbols are assigned to multiple orthogonal carriers and all the carriers are transmitted in a combined manner; hence, the carriers are called *subcarriers*. After symbol assignment, these subcarriers

> Orthogonality is a property that allows multiple information signals to be transmitted perfectly over a common channel and to be detected without interference.

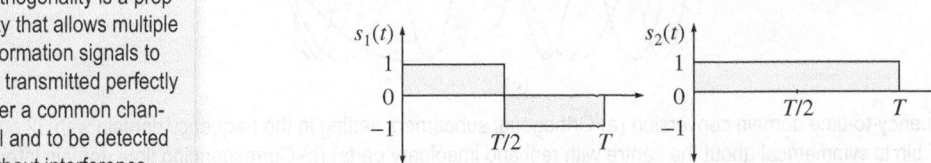

Fig. 7.44 Example of two orthogonal signals

are combined by the inverse fast Fourier transform (IFFT) technique. This combined signal is known as the OFDM baseband. For symbol mapping, conventional M-ary techniques (e.g., QPSK and 16QAM) are used. Equation (7.55) shows a set of orthogonal signals that represent the unmodulated subcarriers in the time domain. From these subcarriers, by deciding the value of Δf, the IFFT bin setting can be achieved.

$$sk^{(t)} = \begin{cases} \cos(2\pi k \Delta f t) & 0 < t < T_s, k = 1, 2, 3 \dots N_c \\ 0 & \text{(otherwise)} \end{cases} \qquad (7.55)$$

where

$$\Delta f = 1/T_s \qquad (7.56)$$

is the subcarrier spacing, T_s is the symbol duration, N_c is the number of subcarriers, and $(N_c + 1)\Delta f$ is the transmission bandwidth of the OFDM baseband signal.

7.15.2 Subcarrier Setting in Spectrum

Figure 7.45 illustrates how the orthogonal subcarriers are considered in the frequency and time domains and how they are added up to get the OFDM baseband signal.

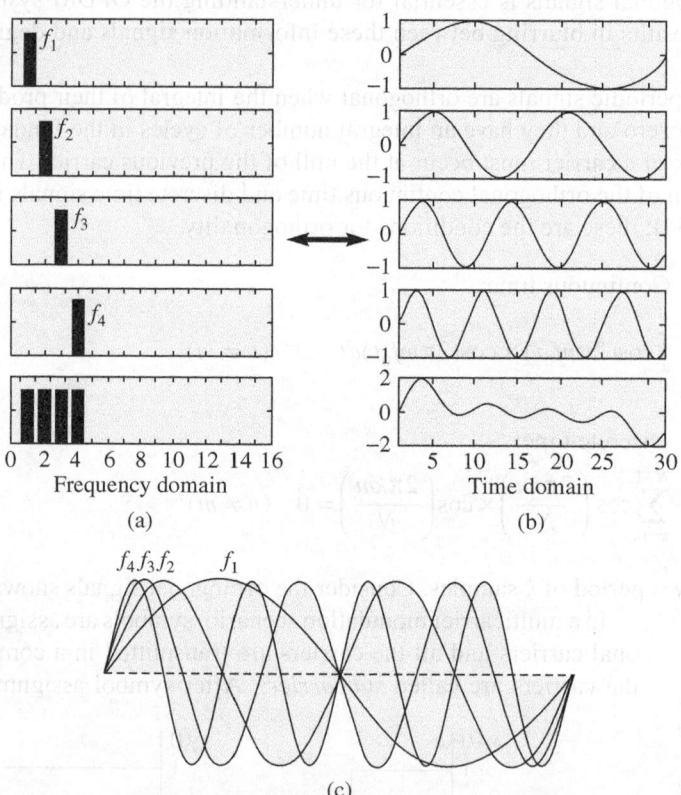

Fig. 7.45 Frequency-to-time domain conversion (a) Orthogonal subcarriers setting in the frequency domain with 32 point IFFT bin (the IFFT bin is symmetrical about the centre with real and imaginary parts) (b) Corresponding time domain interpretations for interval N = 32 samples (c) Plot of four subcarriers on the same time axis for the addition of subcarriers in the time domain to get the OFDM baseband (conceptual representation)

The subcarriers are assigned in two ways. Consider one OFDM frame and a set of subcarriers assigned to it.

One symbol assigned to one subcarrier   As per the symbol mapping scheme adopted, the bits per symbol are read and assigned to one subcarrier. Then, another group of bits per symbol are read and assigned to another subcarrier that is orthogonal to the previous one. This process is continued. T_s will be the duration of the symbol in this case. This is done for all parallel subcarriers that are to be combined together and IFFT is applied.

Multiple symbols assigned to one subcarrier   As per the symbol mapping scheme adopted, the symbols of the whole OFDM frame are read and then a symbol matrix is formed. The whole frame is represented in terms of time versus frequency plot (a two-dimensional lattice) in which each point of lattice represents a symbol. In this matrix, the frequency axis will support the total number of subcarriers to be assigned to the OFDM symbol and the time axis will represent the number of symbols assigned to that subcarrier within the OFDM symbol duration. This is formed by breaking the OFDM frame into blocks and prolonging each block to the OFDM symbol duration T_s.

> *Note*: OFDM blocks contain one or more symbols as per the two possibilities discussed. Thus, it can be said that 'OFDM is a block modulation scheme'.

An Example of Four Subcarriers with Three Symbols per Subcarrier

Figure 7.46 represents a typical case of the second method of carrier assignment. The block duration will be T_s after the serial-to-parallel conversion procedure. Thus, the effective bit duration and hence the duration between two consecutive pulses on all parallel lines are prolonged. (Bits per symbol are as per the symbol mapping scheme used.) The OFDM signals are made up of a sum of sinusoids, each representing a modulated subcarrier.

Figure 7.46 shows a typical case of a block of three symbols that are assigned to one subcarrier, that is, three symbols modulating a subcarrier. Hence, bits per symbol and symbols per carrier are necessary to pre-decide or plan such that the total OFDM frame is broken up appropriately to match the total number of subcarriers to be assigned to it.

Once the bin setting is over, each OFDM block with OFDM symbol duration is assigned to those subcarriers according to the pre-decided pattern; that is, the subcarriers are modulated now. If a pulse (in the time domain) is assigned to a subcarrier, then it will take the sinc shape in the frequency domain while centring that subcarrier. Instead of a narrow pulse, if symbols are assigned to that carrier, its sinc shape (Fig. 7.46b) will shrink because time expansion will consume less spectrum. The same concept is applied to OFDM while modulating each subcarrier by the OFDM block assigned to it, and this is followed for all the subcarriers assigned to the whole OFDM frame. The carriers can be placed as near as possible, maintaining the orthogonality; that is, the peak of one carrier should coincide with the null of its nearest subcarrier. As the width of the sinc shape depends upon T_s, it places a limit on the subcarrier spacing. Each narrow bandwidth corresponding to each subcarrier is decided by the symbol duration T_s. Hence, the subcarrier spacing will be inversely proportional to the symbol duration. Figure 7.47(a) represents four subcarriers modulated by

As the subcarrier bandwidths are very narrow, they act as samples at the points of subcarrier frequencies in the spectrum and hence the bin setting is visualized in the frequency domain.

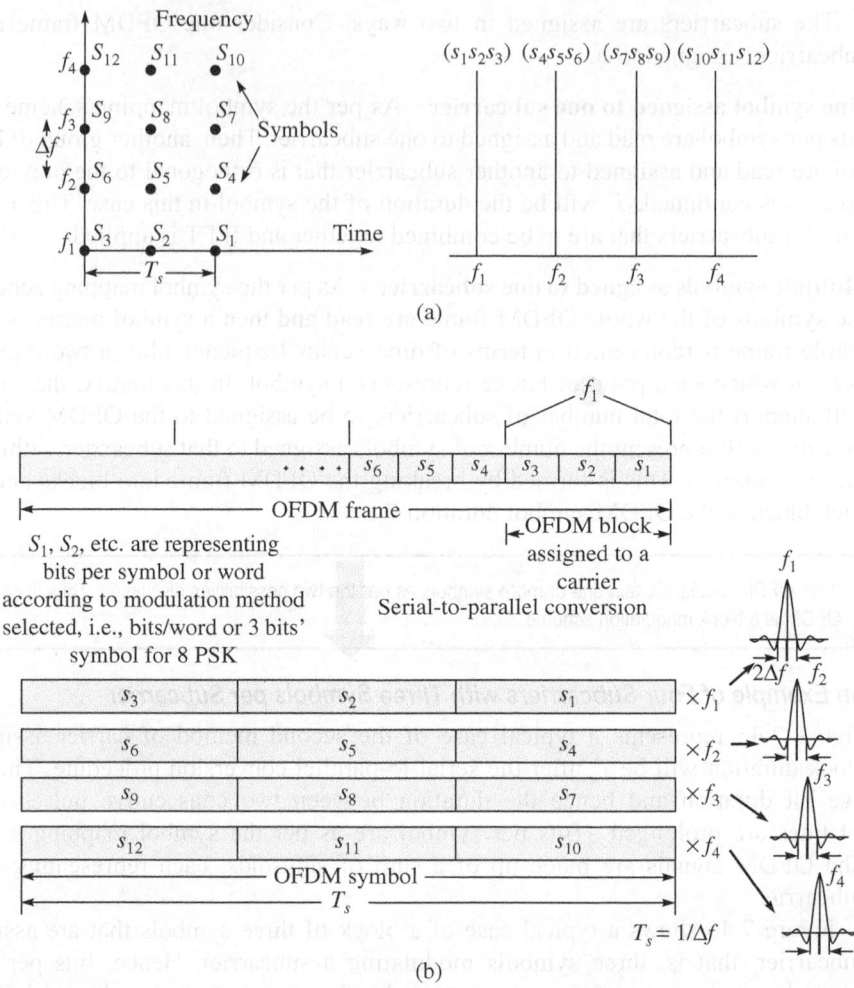

Fig. 7.46 OFDM modulation stage (a) Lattice representation carrier assignment planning (b) Concept of carrier frequency allocation to the symbols (a typical case in which three symbols are assigned to a carrier), serial-to-parallel conversion, and mapping into frequency domain components after modulation

the symbols assigned to them and hence narrow channels exhibit a sinc shape. Subcarriers also maintain the orthogonality. Figure 7.47(b) represents the overall transmission bandwidth.

In general, in OFDM, in the IFFT bin setting at each subcarrier sinc has a peak at the centre frequency and nulls evenly spaced with a frequency gap equal to the carrier spacing, as shown in Fig. 7.48.

The sinc shape has a narrow main lobe, which decays fast, and many side lobes. When the spectrum setting is done for the IFFT stage, the spectrum is not treated as continuous, as shown in the Figs 7.47 and 7.48, but is considered in terms of discrete samples. The sampled spectrum is shown with *o*'s in Fig. 7.47(a). If the FFT is time synchronized, the frequency samples of the IFFT correspond to the peaks of the subcarriers; thus, the overlapping frequency region between subcarriers does not affect the receiver.

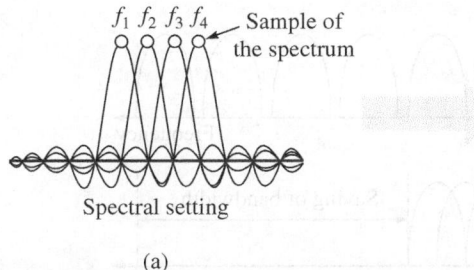
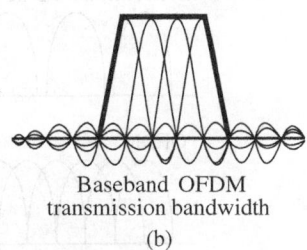

f_1 f_2 f_3 f_4 — Sample of the spectrum

Spectral setting

Baseband OFDM transmission bandwidth

(a)

(b)

Fig. 7.47 Four subcarriers making the total occupied bandwidth for an OFDM baseband signal (a) Modulated subcarriers with spectral setting (b) Overall bandwidth

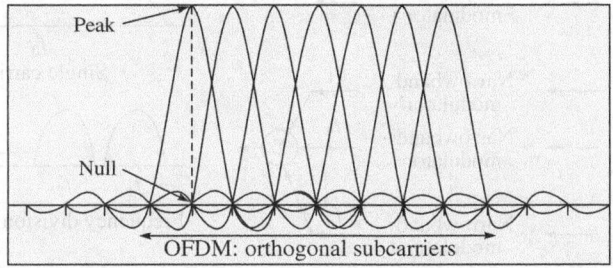

Peak

Null

OFDM: orthogonal subcarriers

Fig. 7.48 Orthogonal signals in the frequency domain: peak of one signal occurs at the null of its nearer subcarrier

In short, in OFDM modulation, the following take place:

- The available channel is divided into several independent subcarriers.
- All the subcarriers in a channel are transmitted at a time by making all the subcarriers orthogonal to each other, preventing interchannel interference (ICI).
- The received signals are retrieved by reverse processing.
- The frequency difference between two consecutive subcarriers is not selected arbitrarily but is related to the data rate or symbol time.

Example 7.14 A 64 kbps voice frame is to be modulated by an OFDM scheme. The duration of the OFDM symbol is 1000 µs. A total of 32 subcarriers is to be assigned to this frame. Find the following:
(a) Null-to-null subchannel bandwidth assuming a square signal
(b) Total bandwidth occupied
(c) Number of bits in the OFDM frame

Solution
(a) T_s = symbol duration = 1000 µs

So, Δf = subcarrier spacing = $1/T_s$ = 1/1000 µs = 0.001 MHz = 100 kHz

Hence, null-to-null subchannel bandwidth = 200 kHz (sinc shape)
(b) N_c = number of subcarriers = 32

Hence, transmission bandwidth = $(N_c + 1)\Delta f$ = 33 × 100 = 3300 kHz
(c) Bit duration = $1/(64 \times 10^3)$ = 15.625 µs

Hence, number of bits per frame = 1000/15.625 = 64 bits

7.15.3 Frequency Division Multiplexing Versus Orthogonal Frequency Division Multiplexing

The following are some differences between FDM and OFDM:

- OFDM signals must be time and frequency synchronized, which is not necessary in case of FDM.

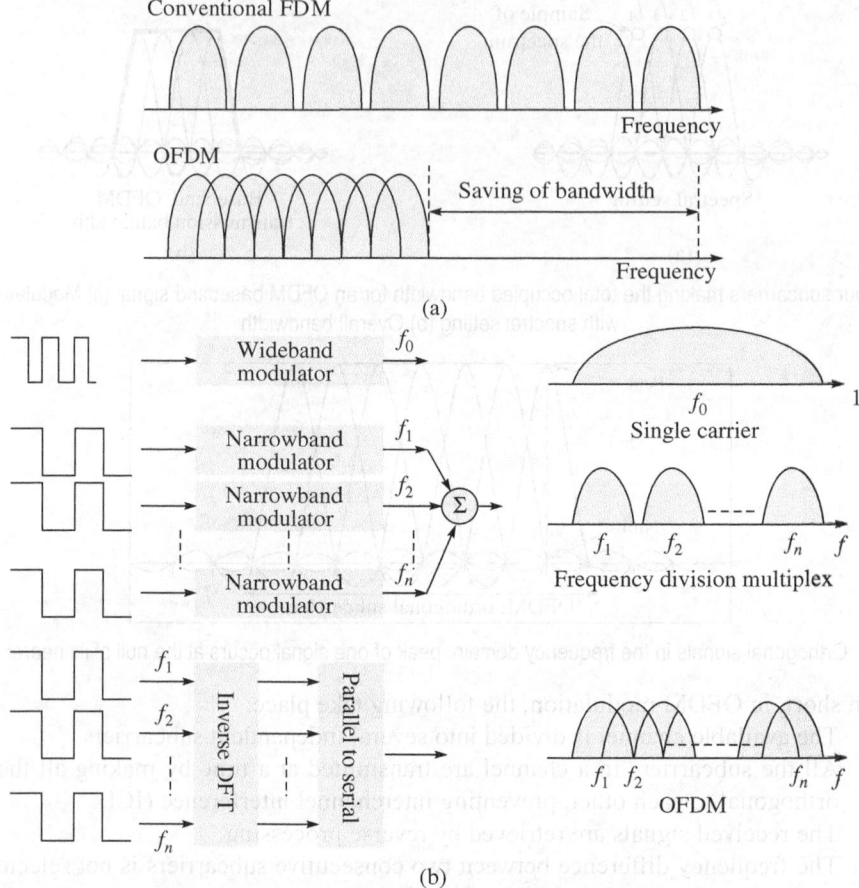

Fig. 7.49 FDM vs OFDM—comparing spectral efficiency (a) Spectrum saving due to multi-carrier modulation (b) Better spectral efficiency of OFDM compared to other techniques

- FDM is a single-carrier oriented system whereas OFDM is a multicarrier system.
- In FDM, transmission signals need to have a frequency guard band between each consecutive channel bandwidth allocated to prevent inter-carrier-interference. In OFDM, the guard band is required only after a set of subchannels comprising transmission bandwidth.
- In FDM, there is no relationship between carriers. Separation between two channels depends upon the basic bandwidth and the modulation scheme adopted.

In OFDM, multiple subcarriers are used for the transmission of the multiplexed multi-user information in a frame. This reduces the overall spectrum requirement compared to individual allocations in FDM. In FDM, windowing is applied on an individual channel basis, while in OFDM, it is applied on the overall multicarrier set, creating the bandwidth. Figure 7.49 shows how OFDM is a spectrally efficient technique.

Example 7.15 Considering the typical case, five 100 kHz channels are placed near each other, modulated by a single-carrier scheme with five different carriers. If the same five carriers are placed orthogonally to each other, compare the occupied bandwidth with respect to the first case and find the saving in bandwidth due to the multicarrier scheme.

Solution   (Note: Here, specific modulation schemes are not considered; otherwise, the results may differ. It is just an arbitrary example with approximate comparison.)

For a single-carrier system:

The total occupied bandwidth will be $100 \times 5 = 500$ kHz without a guard interval.

For orthogonal carriers:

Considering the same bandwidth of 100 kHz as the subcarrier bandwidth, the spacing between the orthogonal carriers will be approximately 50 kHz. So, the total occupied bandwidth will be $5 \times 50 + 50 = 250 + 50 = 300$ kHz.

The saving in bandwidth is 200 KHz.

7.16 ORTHOGONAL FREQUENCY DIVISION MULTIPLEXING TRANSMITTER AND RECEIVER

A simple block diagram of OFDM is shown in Fig. 7.50(a).

> *Note*: The following are a few assumptions about the OFDM system:
> 1. The OFDM symbol time is very large compared to the guard interval (discussed in Section 7.16.3).
> 2. The transmitter and receiver are perfectly synchronized.
> 3. Over each narrow subcarrier band, flat fading is considered, while all the subcarriers are received with different strengths. Therefore, the overall transmission bandwidth is said to undergo frequency-selective fading, as shown in Fig. 7.50(b).

According to the simplified diagram shown in Fig. 7.50(a), the following are the different stages involved.

Channel coding stage   The transmitter of an OFDM link starts from the channel coding stage if the source-coded data is available in the required form. Through the scrambling, convolutional coding, code puncturing, and interleaving operations, redundancy will be added in the data to combat the fading channel. Thereafter, the OFDM modulation stage begins.

Modulation stage   To generate OFDM successfully, the relationship between all the subcarriers must be carefully controlled to maintain the orthogonality of the carriers. For this reason, OFDM is generated by first choosing the size of the spectrum required, based on the input data, and then selecting the modulation scheme used. Each subcarrier to be produced is assigned some data to transmit. The required amplitude and phase of the subcarrier are calculated based on the modulation scheme. Some additional subcarriers called pilot subcarriers are added to create the reference at the receiver end, which carries out the channel estimation procedure to remove channel impairments.

The spectrum setting of the modulated subcarriers acts as the sampled IFFT bin setting. The spectrum setting is then converted into its equivalent time domain signal using IFFT. It is not necessary that the number of subcarriers and the IFFT bin size should be equal.

The major part of an OFDM baseband system is dealt with complex digital signal processing and requires rigorous settings and calculations.

Zero padding can be applied to the IFFT bin to increase the IFFT points and to improve the time domain resolution, or guard carriers can be used. After the IFFT stage, the time domain signal must be received in serial form through P/S. The signal thus generated is called the OFDM baseband.

Upconversion stage   To generate an upconverted RF signal, the signal must be filtered and mixed to the desired transmission frequency.Before RF upconversion, a cyclic prefix is added to remove the ISI effect. In addition, the output of

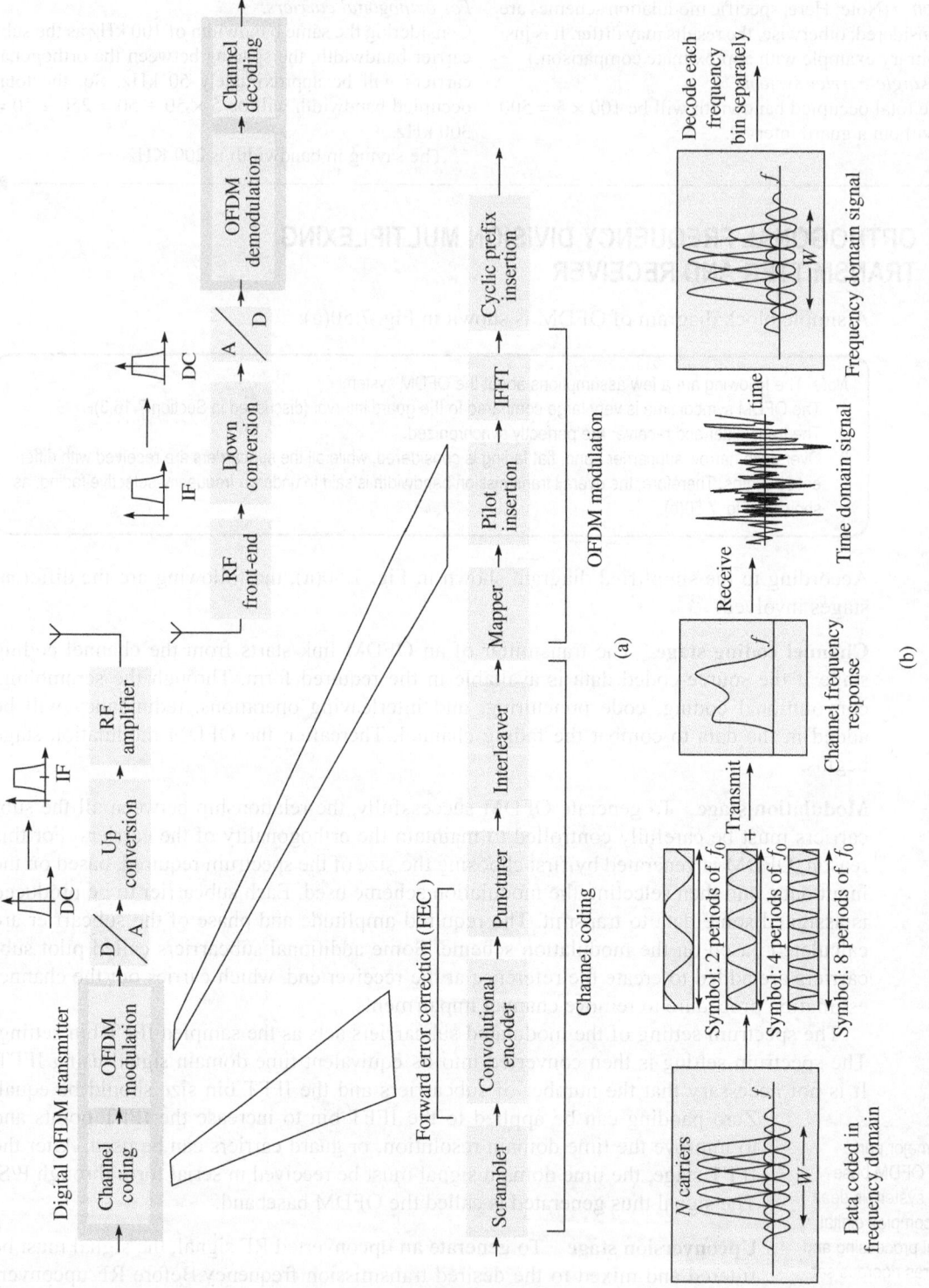

Fig. 7.50 OFDM transmitter and receiver (a) Conceptual block diagram of a discrete OFDM system (channel coding and OFDM modulation blocks are highlighted and an exactly opposite task is followed at the receiver) (b) Effect of frequency-selective fading on the received spectrum

the IFFT stage is in the form of discrete samples. Therefore, digital-to-analog conversion is required before the RF conversion stage.

Downconversion stage   The RF downconversion process is the reverse process performed at the receiver front end. At this stage, the cyclic prefix is removed.

Demodulation stage   At the receiver, after analog-to-digital conversion, the FFT transforms a time domain signal into its equivalent frequency spectrum. The amplitude and phase of the sinusoidal components represent the frequency spectrum of the time domain signal. Since each bin of an IFFT corresponds to amplitude and phase of a set of orthogonal sinusoids, the reverse process guarantees that the subcarriers generated are orthogonal. Of course, Doppler effect may exhibit some shifts in the received carrier frequency, which must be corrected by perfect synchronization.

Note: The OFDM link requires some additional blocks:
1. For subcarrier offset removal, pilot signals are considered, which may have a slightly higher power than the data subcarriers. Some other methods such as automatic frequency control may also be used.
2. Channel estimation is the process by which the channel impulse response is estimated by an adaptive feedback system and hence phase correction is applied. This is done by extracting the pilot subcarriers.
3. Pilot information can be sent in the time or frequency domain, and accordingly, time or frequency domain methods are utilized to estimate the channel.

The important signal processing stages are shown in Fig. 7.51 and are explained in detail in Sections 7.16.1–7.16.5.

7.16.1 Serial-to-parallel Conversion and Symbol Mapping

The input serial data stream (or OFDM block) is read into the symbols—for example, two bits per symbol for QPSK, eight bits per symbol for 256PSK, and so on—and converted into a parallel format just like a conventional M-PSK method. For the lattice style of subcarrier allocation, after reading the symbols, all the symbols are reorganized into N_c parallel lines, which is equal to the number of subcarriers. The data is then transmitted in parallel by assigning different blocks to different subcarriers in the transmission, $f1, f2$, and so on, as

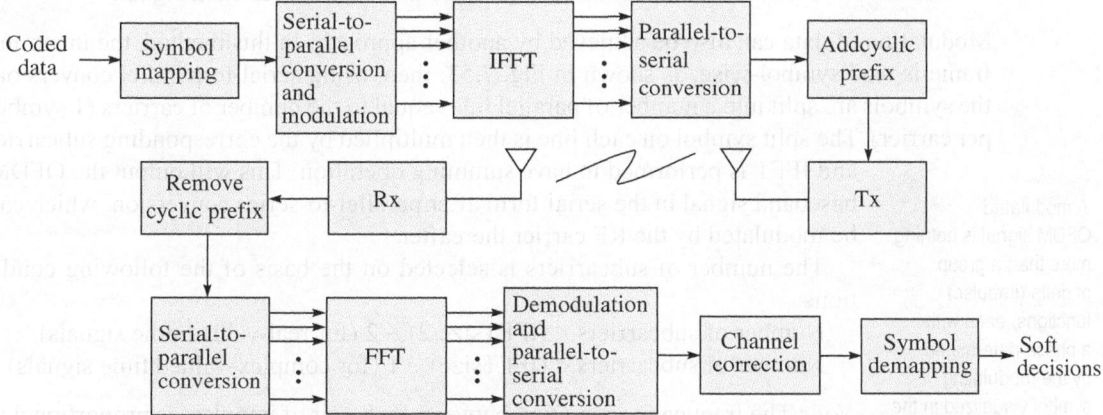

Fig. 7.51   OFDM system diagram showing important signal processing stages

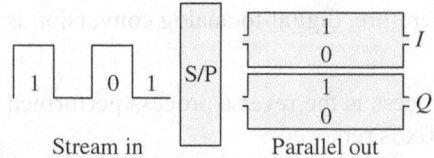

Fig. 7.52 Serial-to-parallel conversion of data

shown in Fig. 7.46.OFDM is a high data rate communication. Therefore, a high probability of ISI is expected as the time duration becomes less than the maximum delay of the RF channel; but here, reading the data in the form of words is advantageous (Fig. 7.52). Indirectly, the time duration between two consecutive pulses is prolonged and hence the ISI can be minimized.

Suppose a symbol transmission takes 4 s. Then, according to Fig. 7.52, each piece of data on the left has a duration of 1 s. On the other hand, OFDM would send the four pieces simultaneously as shown on the right. In this case, each piece of data has a duration of 4 s. The M-ary shift keying schemes or QAM can be used for symbol mapping in OFDM. Symbols are mapped onto the constellation diagram.Thus, this method of parallel transmission can support a very high bit rate.

Sometimes, the data to be transmitted on each subcarrier can be differentially encoded with previous symbols (this is not a compulsory process, but it improves the correlation between consecutive symbols and helps in detecting errors) and then mapped into a PSK format. As differential encoding requires an initial phase reference, an extra (null) symbol is added at the beginning for this purpose. The data on each symbol is then mapped to a phase angle based on the modulation method. For example, for QPSK, the phase angles used are 0°, 90°, 180°, and 270°. The use of PSK produces a constant amplitude signal and is normally chosen for its simplicity and for reducing problems with amplitude fluctuations due to fading.

7.16.2 Modulation of Data

The following steps may be carried out for applying modulation to the carriers in case there are long frames and more symbols per carrier (refer to Fig. 7.46 and the explanation provided with it):

- After reading the data in terms of bits per symbol, convert the serial symbol stream into parallel segments according to the number of carriers and the symbols per carrier.
- Assign an OFDM block to the appropriate subcarrier in the IFFT bin.
- Take the IFFT of the result and this will give the discrete time domain signal.

Modulation of data can also be achieved by another approach. In this method, the incoming frame is read symbol-wise, as shown in Fig. 7.53; then, using serial-to-parallel conversion, the symbols are split into a number of parallel lines equal to the number of carriers (1 symbol per carrier). The split symbol on each line is then multiplied by the corresponding subcarrier and IFFT is performed to have summing operation. This will output the OFDM baseband signal in the serial form after parallel-to-series conversion, which can be modulated by the RF carrier thereafter.

> A modulated OFDM signal is nothing more than a group of delta (impulse) functions, each with a phase determined by the modulating symbol visualized in the frequency domain.

The number of subcarriers is selected on the basis of the following conditions:

Number of subcarriers < (IFFTsize/2) − 2 (for real-valued time signals)
Number of subcarriers < (IFFTsize) − 1 (for complex-valued time signals)

Note: The frequency separation between each pair of impulses is proportional to the inverse of the symbol duration.

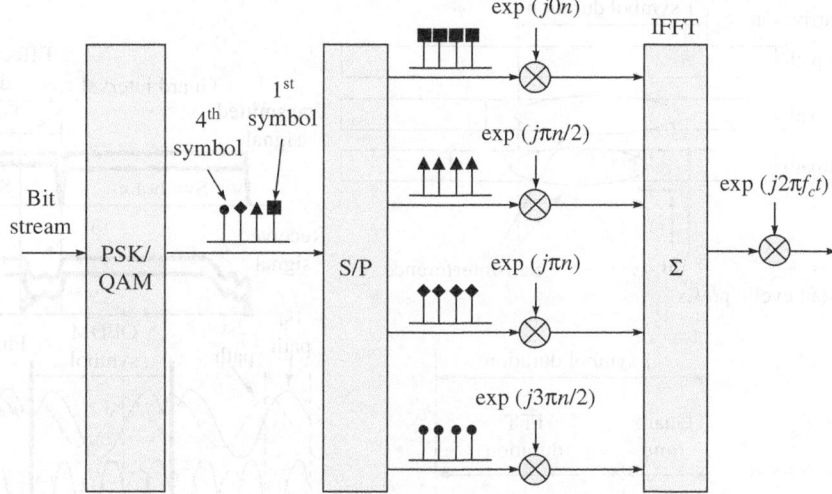

Fig. 7.53 Frequency-to-time domain conversion and IFFT acting as summer

7.16.3 Guard Period

One of the most important properties of OFDM transmissions is its high level of robustness against multipath delay spread. This is a result of the very long symbol period used compared to path delay, which minimizes the ISI. The level of multipath robustness can be further increased by the addition of a guard period between the transmitted symbols.

Intersymbol interference is a common problem found in high data rate communication. It occurs when the transmission interferes with itself and the receiver cannot decode the transmission correctly. This is because an increase in the data rate results in a decrease in the time duration between consecutive pulses. For avoiding the ISI, the pulse time duration should be greater than the maximum delay of the channel. To achieve this, a guard interval is provided along with the data period such that the ISI effect observed in the guard interval can be removed afterwards and the data can be retrieved.

The guard period allows time for the multipath signals from the previous symbol to die away before the information from the current symbol is gathered. The most effective guard period to use is a cyclic extension of the symbol called the *cyclic prefix*. If a mirror in time at the end of the symbol waveform is provided at the start of a symbol as the guard period, this effectively extends the length of the symbol while maintaining the orthogonality of the waveform.

The added guard interval and its effect in reducing the ISI are shown in Figs 7.54 and 7.55.

ADVANTAGES OF CYCLIC PREFIX

Using a cyclic extended symbol, multipath immunity as well as symbol time synchronization tolerance can be achieved.

Transmitting a cyclic prefix of the data during the guard interval transforms the linearly convolutive channel into a circularly convolutive channel. Hence, the channel equalization problem is simplified. Specifically, channel equalization in the frequency domain can be done using one-tap filters. This is because cyclic prefixing makes the channel matrix circulant, which is diagonalized by the IFFT and FFT operations.

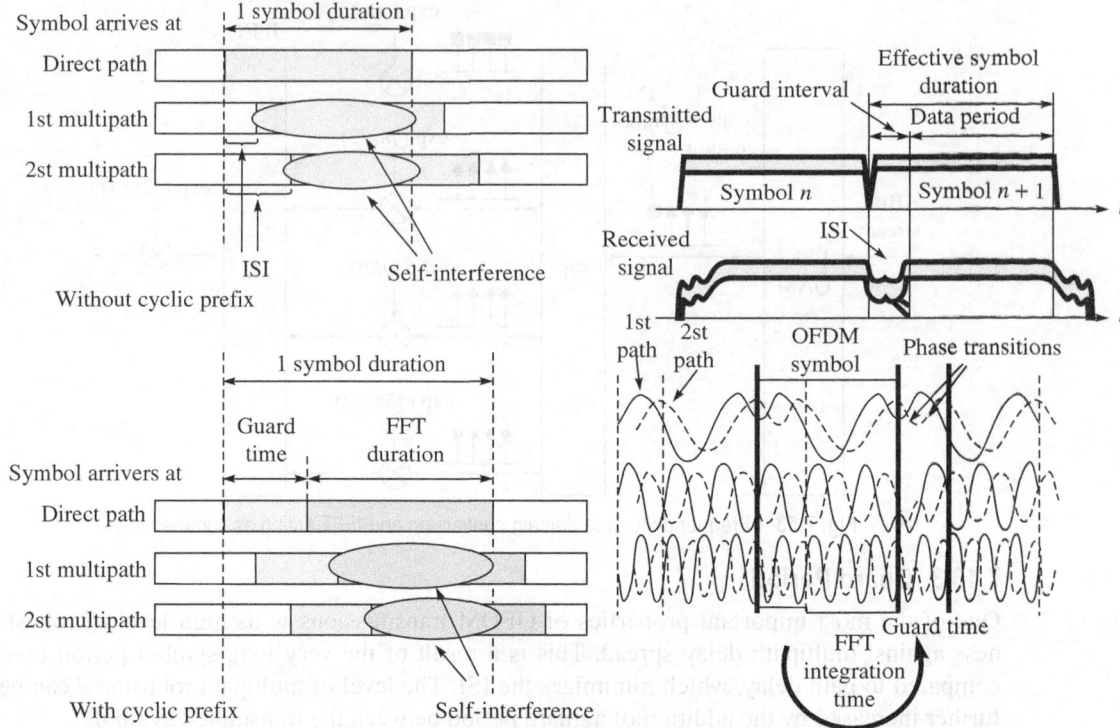

Fig. 7.54 Removal of multipath effect due to addition of guard interval in each symbol block by padding copy of part of the symbol (phase errors are mainly in the guard interval)

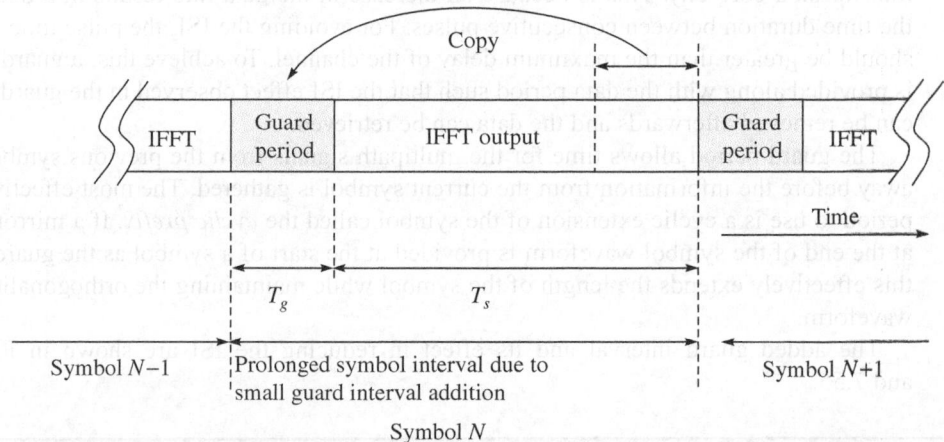

Fig. 7.55 Increase in effective symbol duration due to addition of guard period in terms of cyclic prefix after IFFT, making a circularly convolutive channel

The signal energy from all multipaths adds at the input to the receiver, and because the FFT is energy conservative, the whole available power feeds the decoder.

Although the cyclic prefix introduces a loss in SNR, this may be considered as a small price to pay to mitigate interference. Note that some bandwidth efficiency is lost with the addition of the guard period, but the ensuing advantages are many, and hence a compromising solution is required.

Example 7.16 An OFDM symbol duration is 1280 μs. If such a frame is modulated by an OFDM scheme, transmitted over the channel, and received after a delay of 450 ns at the receiver, what should be the minimum duration of the cyclic prefix? Show that the cyclic prefix does not affect the occupied spectrum much if the number of subcarriers is 64 and 64-point FFT is used. Will ISI occur?

Solution The cyclic prefix minimum duration must be 450 ns.

If the cyclic prefix of 0.45 μs is appended to OFDM symbol duration, 1280.045 is the final frame duration (expected).

Spacing between two consecutive subcarriers without cyclic prefix will be $1/T_s = 781.25$ Hz.

So, the spectrum occupied will be $(64 + 1) \times 781.25 \approx 50.781$ kHz.

The 64-point IFFT reflects in the time domain as 64 samples of the OFDM symbol with the sampling duration of 20 μs ideally. If at least one sample is cyclically prefixed, one point is added in the IFFT bin, and the additional spectrum consumption will be 781.25 Hz; hence, total spectrum ≈ 50.781 kHz + 781.25 Hz ≈ 51.562 kHz. One sample cyclic prefix (CP) is sufficient to combat 450 ns path delay. Therefore, no ISI will occur and the spectrum is also not much affected.

■

7.16.4 Radio Frequency Upconversion

It is not practical to generate the OFDM signal directly at the RF. Therefore, the OFDM baseband signal must be upconverted for transmission. To remain in the discrete form, the OFDM baseband can be upsampled and added to a discrete carrier frequency. This carrier can be an intermediate frequency whose sample rate is handled by current technology. It could then be converted to analog and increased to the final transmit frequency using analog frequency conversion methods. Alternatively, the OFDM modulation can be immediately converted to analog and directly increased to the desired RF transmit frequency. Either way, the selected technique would have to involve some form of linear AM or QAM (possibly implemented with a mixer). For the upconversion process, the OFDM baseband must be in its IQ form as shown in Fig. 7.56.

Radio Frequency Modulation

The output of the OFDM modulator generates a baseband signal, which must be increased to the required transmission frequency. This can be implemented using the following techniques:

- Analog technique (Fig. 7.57)
- Digital technique (Fig. 7.58)

> If the delay spread is longer than the guard interval, it begins to cause ISI. However, if the echoes are sufficiently small within the guard interval, they do not cause significant problems.

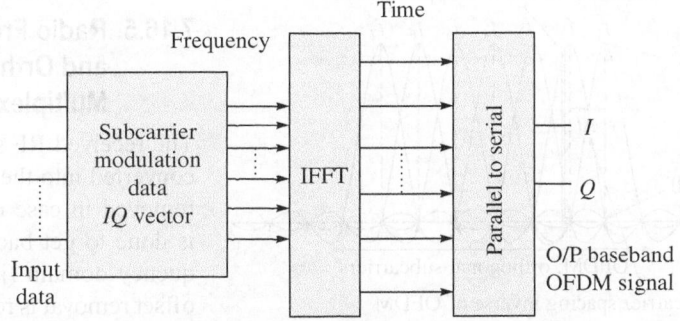

Fig. 7.56 IFFT stage before upconversion stage giving the signal in IQ form

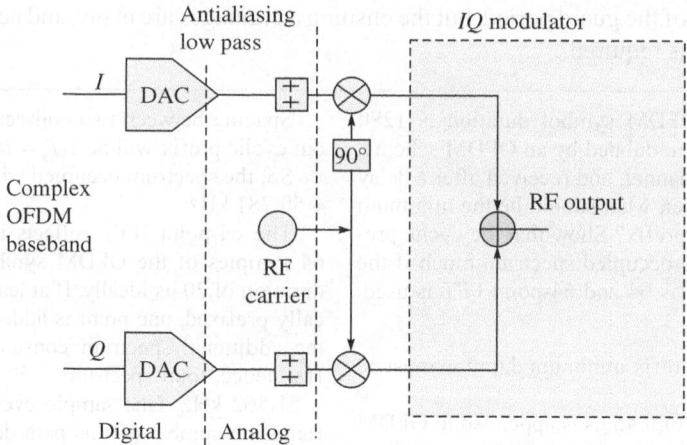

Fig. 7.57 RF modulation of complex baseband OFDM signals using analog techniques such as M-ary schemes

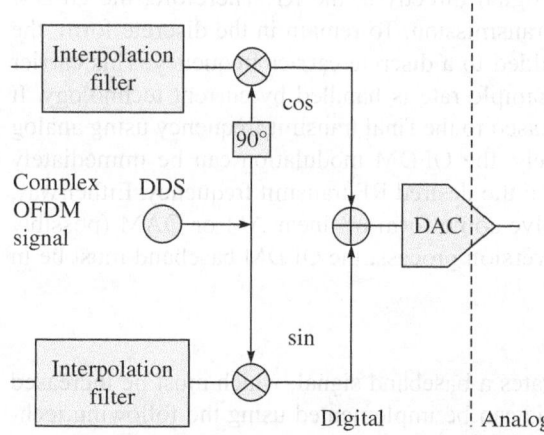

Fig. 7.58 RF modulation of complex OFDM signals using digital techniques

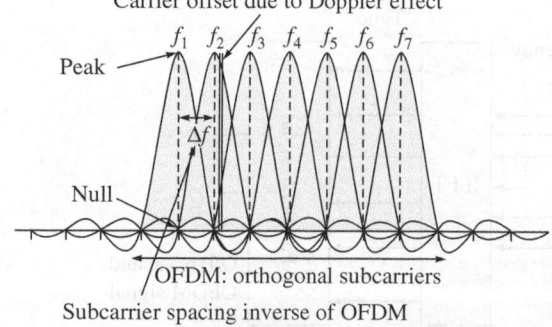

Fig. 7.59 Carrier offset condition at the receiver end

Both techniques perform the same operation. However, the performance of the digital modulation technique will tend to be more accurate due to improved matching between the processing of the I and Q channels and the phase accuracy of the digital IQ modulator. In both the figures, the levels are shown mentioning whether the signal is in analog or digital form.

The role of the interpolation filter is to increase the sampling rate for better analog conversion. Digital techniques use a direct digital synthesizer (DDS), which can generate the required discrete RF carrier by controlling the phase of samples and then converting into analog form. For RF carrier generation, most of the latest application specific integrated circuits (ASIC) designs use a DDS that can generate very high carrier frequencies.

After RF conversion and transmission, the transmitted signal will get corrupted by the various channel effects and then it will be received by the receiver.

7.16.5 Radio Frequency Downconversion and Orthogonal Frequency Division Multiplexing Demodulation

The received RF signal is downconverted and then converted into the digital form. The guard period is removed in case of a cyclic prefix. Demodulation is done to get back the required setting in the frequency domain (just like a transmitter), for which offset removal is required. The subcarrier offset condition is shown in Fig. 7.59. The receiver basically performs the reverse operation of the transmitter.

Due to Doppler effect, carrier offset will take place, resulting in loss of orthogonality.

The FFT of each OFDM symbol is performed to find the original transmitted spectrum. The phase angle of each transmission carrier is then evaluated and converted back to the data word by demodulating the received phase (demapping). The data words are then split back to the same pattern as the original bits to have the serial data again by parallel-to-serial conversion.

The following steps may be taken to demodulate the OFDM:

- Partition the input stream into vectors representing each symbol period.
- Take the FFT of each symbol period vector.
- Extract the carrier FFT bins and calculate the phase of each bin.
- Calculate the phase difference, from one symbol period to the next, for each carrier.
- Decode each phase into binary data.
- Sort the data in the appropriate order.

The efficiency η of the system depends upon the delay spread and symbol duration.

$$\eta = \frac{\text{Delay spread}}{\text{Symbol duration}} \tag{7.57}$$

The effect of the number of subcarriers and guard time duration on the system performance is summarized as follows:

- For a given number of subcarriers, increasing the guard time duration reduces the ISI due to the decrease in the delay spread relative to the symbol time, but it also reduces the power efficiency and bandwidth efficiency.
- For a given signal bandwidth, increasing the number of subcarriers increases the power efficiency, but it also increases the symbol duration and results in a system more sensitive to Doppler spread.

Applications of OFDM are found in the following systems:

- WiFi
- WiMAX
- Ultra-wideband
- Digital audio broadcasting and digital video broadcasting
- Long-term evolution

It is very efficient to use OFDM with the multiple input, multiple output (MIMO) technique.

MORE SOLVED EXAMPLES

Example 7.17 Get the constellation diagram of $\pi/4$ DQPSK. If the bit stream is 1101001110 and the modulated signal is $A\cos(\omega t + \theta_m)$, what are the phase values corresponding to the pattern given?

Solution A $\pi/4$ DQPSK modulation scheme has eight phases (eight constellation points), which are assumed to be created using two QPSK signal constellations, one offset by 45° to the other. During each symbol period, the phase angle from only one constellation is alternately used to transmit every pair of bits. Hence, successive symbols have a relative phase difference $\Delta\theta_m$ of phases, 45° or $\pi/4$ (for symbol 00), 135° or $3\pi/4$ (for symbol 01), −45° or $-\pi/4$ (for symbol 10), and −135° or $-3\pi/4$ (for symbol 11). This is differential encoding of input bits using the previous absolute phase, which is then mapped onto the absolute phase angle.

Hence, the in-phase and quadrature-phase components, I_m and Q_m, respectively, for the mth symbol can be written as follows:

$$I_m = I_{m-1}\cos(\Delta\theta_m) - Q_{m-1}\sin(\Delta\theta_m)$$

and

$$Q_m = I_{m-1}\sin(\Delta\theta_m) + Q_{m-1}\cos(\Delta\theta_m)$$

Now, as the absolute phase of the $(m-1)$th symbol is θ_{m-1}, I_m and Q_m can be rewritten as

$$I_m = \cos\theta_{m-1}\cos(\Delta\theta_m) - \sin\theta_{m-1}\sin(\Delta\theta_m)$$

and $\quad Q_m = \cos\theta_{m-1}\sin(\Delta\theta_m) + \sin\theta_{m-1}\cos(\Delta\theta_m)$

Hence, the constellation will be as shown in Fig. 7.60 with the possible amplitudes of I_m and Q_m as ±1, 0, and ±0.707.

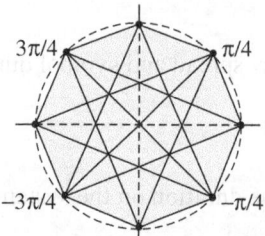

Fig. 7.60 Constellation of π/4 DQPSK

The absolute phases will then be calculated assuming the initial absolute phase to be equal to zero.

	11	01	00	11	10
$\Delta\theta_m$	−3π/4	−3π/4	π/4	−3π/4	−π/4
In degrees	−135	135	45	−135	−45
Absolute angles	−135	0	90	180	−90

Example 7.18 A GSM system with GMSK modulation uses 200 kHz channel bandwidth with a data rate of 270.833 kbps as per standard specifications (refer to Chapter 11). Calculate the minimum frequency shift, bandwidth efficiency, and transmitted upper and lower frequencies if the trunk carrier frequency is 910 MHz. Also, find the 3-dB bandwidth of Gaussian filter that is used for the generation of 0.3, 0.5, and 1 GMSK signals (refer to Figs 7.18a, b, and c).

Solution The minimum frequency shift is between transmitted bits 0 and 1 consecutively, as shown in Fig. 7.61.

For this, the minimum shift $\Delta f = f_1 - f_0 = (1/2T_b) =$ ½ × 270.833 × 10³ = 135.416 kHz

Bandwidth efficiency = Bit rate/Available bandwidth = 270.833/200 = 1.354 bps/Hz

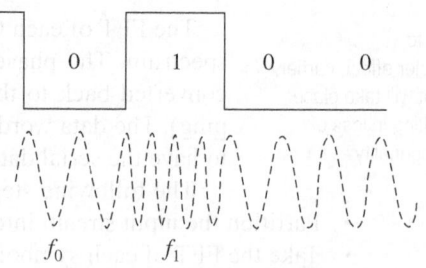

Fig. 7.61 Waveforms of GMSK showing minimum shift

If f_c is the centre frequency, then f_0 (lower) and f_1 (upper) frequencies can be represented as

$$f_0 = f_c - (1/4T_b) = 910 \times 10^6 - (0.25 \times 270.833 \times 10^3)$$
$$= 909.932 \text{ MHz}$$
$$f_1 = f_c + (1/4T_b) = 910 \times 10^6 + (0.25 \times 270.833 \times 10^3)$$
$$= 910.0677 \text{ MHz}$$

Now, for bit rate 270.833 kbps,

$$BT = 0.3, B = 0.3 \times 270.833 = 81.25 \text{ kHz}$$
$$BT = 0.5, B = 0.5 \times 270.833 = 135.42 \text{ kHz}$$
$$BT = 1, B = 1 \times 270.833 = 270.833 \text{ kHz}$$

Example 7.19 For a QPSK signal with SNR 10 dB, find the probability of error. If this error probability is allowed for the transmission of bit rate of 100 kbps over the additive white Gaussian noise (AWGN) channel with one-sided power spectral density of noise 0.7×10^{-11} W/Hz, what should be the amplitude of the signal satisfying these conditions?

Solution $\quad SNR = 10$
In general, for M-PSK systems,

$$SNR = \log_2 M(E_b/N_o) \quad \text{for } M \geq 2$$
$$M = 4 \quad \text{for a QPSK system}$$
or $\quad E_b/N_o = 10/\log_2 M = 10/2 = 5$
General equation for probability of error P_e is

$$P_e = (1/\log_2 M)\text{erfc}\left[\sin\left(\frac{\pi}{M}\right)\sqrt{\log_2 M \times \frac{E_b}{N_o}}\right]$$

For QPSK,

$$P_e = (1/2)\text{erfc}\left[\sin\left(\frac{\pi}{4}\right)\sqrt{\frac{2E_b}{N_o}}\right]$$
$$= (1/2)\text{erfc}\left[(1/\sqrt{2})\times\sqrt{\frac{2E_b}{N_o}}\right]$$
$$= (1/2)\text{erfc}\left[\sqrt{\frac{E_b}{N_o}}\right]$$

(*Note*: $P_e = (1/2)\text{erfc}(\sqrt{x}) = Q(\sqrt{2x})$, erfc = the complimentary error function, derived from the conditional probability density functions of the filter output, showing that bit 0 is detected instead of 1 and vice versa. For a Gaussian channel, $Q(z) = \dfrac{e^{-z^2/2}}{\sqrt{2\pi} \cdot z}$

where, $z \gg 1$.)

So, $P_e = \dfrac{e^{-5^{2/2}}}{\sqrt{2\pi} \cdot 5} = 2.95 \times 10^{-7}$

Now, $E_b/N_o = A^2/(R N_o \log_2 M)$

or $5 = A^2/(100 \times 10^3 \times 0.7 \times 10^{-11} \times 2)$

or $A = 0.00265 \text{ V} = 2.65 \text{ mV}$

SUMMARY

- The choice of a digital modulation scheme is made depending upon the application and the channel environment.
- The expected parameters for digital modulation schemes are spectrum efficiency, power efficiency, and less susceptibility to noise.
- Diagrams such as polar, constellation, eye, and trellis are used for the performance analysis of a system with different modulation schemes.
- Some of the constant envelope schemes are M-PSK, FSK, MSK, and GMSK.
- M-ary PSK schemes are power-efficient techniques and become spectrally efficient for higher values of M.
- ASK and QAM are modulation schemes of variable amplitude type and hence have degraded power efficiency.
- For a fast fading environment, both amplitude and phase changes occur very fast, and so, selection of the scheme is a crucial issue. Adaptive schemes are utilized.
- Other types of modulation are differential and IQ offset modulation.
- Coherent reception provides better performance than differential reception but requires a complex receiver.
- The SSM scheme uses a code of higher rate for the data modulation purpose and therefore it results in spectrum spreading.
- The different types of SSM methods are DSSS, FHSS, THSS, hybrid methods, and chirping.
- The PN code used for the spreading purpose is a random sequence whose smallest element is known as the chip. The chip rate is decided by the clock rate and it is much higher than the data bit rate.
- The receiver designed for DSSS reception is a rake receiver with a number of correlators, and it exploits multipath.
- Near–far problem is a serious problem in DSSS systems and power control is required to eliminate it.

- DSSS exhibits interference rejection and anti-jam characteristics.
- In the FH method, the PN code is used for hopping pattern generation. Complex synthesizers are required for hopping frequency generation, which makes the system complex.
- In the TH method, the slots are decided on the basis of the PN code.
- Hybrid systems can be designed using combinations of PN, FH, and TH systems. This is done to use the advantages of each system and to eliminate the disadvantages.
- OFDM is a wideband scheme based on block modulation.
- OFDM combats ISI by inserting a guard interval or cyclic prefix in the OFDM symbols and combats ICI by maintaining the orthogonality between consecutive carriers.
- The OFDM method can use M-PSK or M-QAM mapping for assigning the amplitude and phase to the set of orthogonal subcarriers that make one transmission bandwidth.
- The OFDM block size, number of subcarriers, and number of IFFT point selection are independent of each other; however, some conditions must be satisfied.
- Each narrow subchannel will have a sinc shape due to Nyquist pulse shaping.
- Narrow subchannels will undergo flat fading while the overall transmission bandwidth will be considered under frequency-selective fading.
- Advantages of OFDM include high spectral efficiency, mitigation to channel problems, simple implementation through FFT, high data rate support, flexibility and adaptation features, and a low-complexity multiple access scheme.
- Disadvantages of OFDM include high sensitivity to time and carrier offset and synchronization and a high peak to average power ratio (PAPR) compared to single-carrier systems.

EXERCISES

Multiple-choice Questions

7.1 A wireless MODEM is
(a) a circuit that carries out the modulation and demodulation of a carrier frequency
(b) an automatic repeat request device for correcting errors
(c) a system for transmitting high-speed burst
(d) an anti-jamming technique invariably installed on all communication devices.

7.2 Which of the following modulation schemes is most spectrally efficient?
(a) 256PSK (c) QPSK
(b) 8PSK (d) 16PSK

7.3 Spectral efficiency can be improved by
(a) increasing the power
(b) using shaping filters
(c) reading more symbols simultaneously
(d) all of these

7.4 In QPSK, phase reversals occur at
(a) T_b interval (c) $T_b/2$ interval
(b) $4T_b$ interval (d) $2T_b$ interval

7.5 Which of the following is a power-efficient technique?
(a) 16PSK (c) ASK
(b) QAM (d) 16QAM

7.6 A linear amplifier is required in
(a) QPSK (c) DQPSK
(b) OQPSK (d) all of these

7.7 MSK is a special case of CPFSK with a modulation index of
(a) 1 (c) 0.25
(b) 0.5 (d) none of these

7.8 On the constellation diagram, the allowable phase margin in an 8PSK scheme is
(a) 45° (b) 90° (c) 22.5° (d) 180°

7.9 In GMSK, of two cases with $BT = 0.3$ and $BT = 0.5$, the ISI effect will
(a) be more in the second case
(b) be equal in both the cases
(c) be more in the first case
(d) not be there

7.10 The amplitude versus time plot with multiple traces is represented on the
(a) polar plot
(b) constellation diagram
(c) trellis diagram
(d) eye diagram

7.11 A rake receiver's front end detects the required signal by
(a) correlation (c) convolution
(b) superheterodyning (d) demodulation

7.12 Which of the following is an averaging type of system?
(a) FHSS (b) PN/FH (c) THSS (d) DSSS

7.13 The chip duration of a PN code
(a) can be any value
(b) must be much less than the message bit duration
(c) must be equal to the message bit duration
(d) must be much greater than the message bit duration

7.14 If two different PN codes are correlated, then
(a) their cross-correlation should be high
(b) their cross-correlation should be low
(c) their autocorrelation should be high
(d) none of these is true

7.15 For good interference rejection, the preferable value of processing gain should be
(a) 0.5 (b) 10 (c) 1(d) 6500

7.16 In every period, the number of +1's differs from the number of −1's by exactly 1. This represents
(a) the autocorrelation of a PN code
(b) the balance property of a PN code
(c) the run property of a PN code
(d) all of these

7.17 N_c must be
(a) a very large number
(b) a very large odd number
(c) an odd number
(d) an even number

7.18 The near–far problem can occur in
(a) FHSS (c) THSS
(b) PN/FH (d) DSSS

7.19 B/R represents the
(a) spectral efficiency
(b) SNR
(c) bandwidth utilization efficiency
(d) processing gain

7.20 The greatest amount of spreading can be obtained in
(a) FHSS (b) PN/FH (c) THSS (d) DSSS

7.21 Which of the following relationships represents interference rejection?
(a) $SNR_{out} = (PG).SNR_{in}$

(b) $\dfrac{E_b}{N_o} = \left[\dfrac{1}{2} + \dfrac{J}{N_o B_s}\right] SNR_{out}$

(c) $\overline{J^2} = \dfrac{J}{PG}$

(d) $SNR_{out} = \dfrac{Pr}{N_o/2t_m + J/PG}$

7.22 OFDM is a method of
(a) multiplying
(b) orthogonal coding
(c) parallel transmissions
(d) domain transformations

7.23 Due to orthogonality,
(a) subcarriers can be demodulated independently
(b) subcarriers cannot be overlapped
(c) the same subcarrier can be used for parallel transmissions
(d) subcarriers interfere

7.24 Orthogonality means
(a) $\int V_1(t)/V_2(t) = 0$ (c) $\int_T v_1(t)v_2(t) = 0$
(b) $v_1(t)v_2(t) = 0$ (d) none of these

7.25 If there are 256 symbols in a frame to be modulated by 16 subcarriers, each subcarrier will modulate
(a) 16 symbols   (b) 256 symbols
(c) 4 symbols   (d) 8 symbols

7.26 For which of the following modulation mappings must the SNR be the highest to maintain the required BER?
(a) QPSK   (b) QAM   (c) 16PSK   (d) 16QAM

7.27 A cyclic prefix is added to
(a) eliminate ICI
(b) eliminate bit errors
(c) increase spectral efficiency
(d) eliminate ISI

7.28 Addition of pilots will be helpful in
(a) frequency offset removal
(b) synchronization
(c) channel estimation
(d) all of these

7.29 In which of the following types of pilot arrangements will all the subcarriers act as pilots?
(a) Comb   (b) Block
(c) Superimposed   (d) Scattered

7.30 The number of constellation points in OQPSK is
(a) 8   (b) 4   (c) 16   (d) 6

Review Questions

7.1 Why does the choice of a modulation scheme depend upon the environment?

7.2 How can we categorize the different modulation schemes?

7.3 Simulate a QPSK scheme using MATLAB and observe the waveforms, polar diagram, constellation diagram, and its spectrum. Also, extend this exercise for all other modulation schemes. (*Hint*: *pskmod* and *pskdemod* as well as *qammod* and *qamdemod* functions are given in MATLAB in which you can set the value of *M*.)

7.4 What is the difference between QAM and QPSK schemes?

7.5 Why is VCO used many times in the diagram of MSK? Is it possible practically?

7.6 Discuss the different types of QPSK schemes with difference in constellations.

7.7 How are the four corner points eliminated from the constellation of 32QAM?

7.8 What is the major difference between the block diagrams of QPSK and OQPSK?

7.9 How can you justify the reduction in spectral consumption for higher M-ary schemes?

7.10 Why does the BER decrease with an increase in the SNR in the case of M-ary PSK based systems while increasing *M*?

7.11 What do you understand by spreading of spectrum?

7.12 How can you say that a spread spectrum system becomes spectrally efficient and in what sense?

7.13 Design any suitable five-stage PN code generator. You can observe that the length of the sequence is an odd number. Why?

7.14 Give the required autocorrelation properties for DSSS in general.

7.15 Discuss the suitability of the ML sequence, Walsh code, and Gold sequence for a DSSS based multi-user system.

7.16 Demonstrate that the codes in an 8×8 Walsh matrix are orthogonal to each other by showing that multiplying any code by any other code produces a result of zero.

7.17 Discuss in detail the quadriphase modulation based DSSS transmitter with all its aspects.

7.18 What is the importance of the multiple fingers with the correlator at the rake receiver?

7.19 How can you say that FSSS is an avoidance type of system?

7.20 What do you mean by partial correlation? Why is partial correlation important for the rake receiver rather than full correlation?

7.21 Comment on the SNR of a DSSS system. Can you say that the anti-jamming characteristics of DSSS depend upon the available SNR? How can the SNR be improved in that case?

7.22 Compare the DSSS and FHSS types of systems with their pros and cons. Find the existing systems and standards in which these schemes are utilized.

7.23 Draw the block diagram of a PN/FH hybrid system and compare it with the block diagram of an FHSS system. Why are hybrid methods suitable in many cases in spite of their complexity?

7.24 In FHSS systems, what are the parameters upon which the processing gain depends?

7.25 What are the reasons for TH not being much used practically?

7.26 Differentiate between single-carrier and multicarrier systems. List out all the single-carrier modulation schemes and compare them with OFDM.

7.27 What is the role of M-PSK mapping in OFDM? Is it the same as the independent M-PSK scheme? When do we use M-PSK mapping and when is M-QAM used?

7.28 What is the role of the IFFT stage in OFDM? Explain the effect of IFFT point selection on the OFDM baseband signal and the implementation of the system.

7.29 List the parameters to be decided in advance before designing an OFDM system. What may be the criteria in deciding these parameters?

7.30 Explain the need to transmit pilot carriers. How will they help the system?

7.31 How are timing and synchronization important in an OFDM system?

7.32 Explain the amplitude limitations in an OFDM system. How can these be eliminated?

7.33 Prove that an OFDM system is a more bandwidth-efficient technique compared to a single-carrier system.

Numerical Problems

7.1 What is the bandwidth efficiency for FSK, ASK, BPSK, and QPSK for a BER of 10^{-5} on a channel with an SNR of 15 dB?

7.2 In a QPSK transmitter, the information bit rate is 100 kbps. If the available bandwidth is 200 kHz, find the spectral efficiency of the scheme. Compare it with the case of 16PSK.

7.3 Calculate the minimum required bandwidth for a QPSK system so that 15% spectral power contributes to out-of-band components. The symbol duration is 0.1 ms.

7.4 Calculate the minimum bandwidth requirement for a non-coherent BFSK system if the symbol duration is 0.2 ms. What is the minimum bandwidth required for an MSK system having the same symbol duration?

7.5 If it is required to send 1.024×10^6 binary digits (bits) per second with $P_e \le 10^{-6}$ with BPSK and 16PSK,

the channel noise power spectral density is $N_o = 10^{-8}$. Determine the transmission bandwidth and the signal power required at the receiver input in both the cases.

7.6 In the case of PSK, prove that bit error probability is

$$P_e = Q\sqrt{\frac{2E_b}{N_o}}.$$

7.7 A QPSK scheme is used to modulate the bit stream 001010011100. Sketch the transmitted waveform. Assume $f_s = f_b/2$. Also, get the waveform for 8PSK for the same stream but assume $f_s = f_b/3$.

7.8 For the sequence in Problem 7, draw the MSK signal waveform.

7.9 Find the equations for determining the distance between the constellation points on the signal space diagram for the ASK, FSK, and M-PSK modulation schemes.

7.10 Prove that the phase in MSK is continuous.

7.11 For a band-pass channel using BPSK signalling, the data rate is 9.6 kbps. Find the transmission bandwidth between (a) the first nulls and (b) the second nulls. Also, find the peak value of the spectrum.

7.12 A telephone line is equalized to allow band-pass data transmission over a frequency range of 300–3100 Hz. Design a 16QAM signalling scheme that will allow a data rate of 9.6 kbps to be transferred over the channel so that the channel is ISI free.

7.13 Show that the periodic PN signal normalized autocorrelation function corresponding to the binary case with equal probability of 1's and 0's (+1 and −1) and with $N = 7$ is of the form shown in Fig. 7.62.

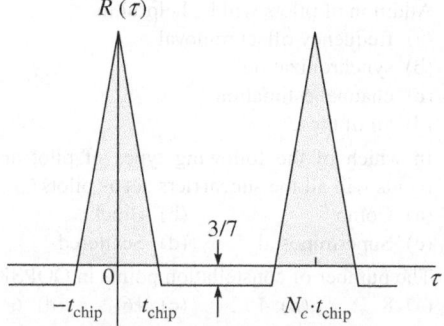

Fig. 7.62 Figure for Problem 13

7.14 An aperiodic PN waveform is defined for the interval $N_c = 7$ with a sequence +1, −1, −1, +1, −1, +1, +1. Find the autocorrelation function of this process. Is this an ideal aperiodic sequence? (*Hint*: Different shifts will give different values of the correlation functions.)

7.15 Evaluate the 15-bit spreading code 100110101111000 for balance, run property, and correlation.

7.16 A DSSS system is operating under the conditions of two independent jammers having very close centre frequency. The chip rate is 125 Mbps and the message bit rate is 2500 bps. If the combined bandwidth of the two jammers is 60 kHz and their respective received signal powers are 0.2×10^{-7} W and 10^{-6} W, what is the approximate interference power at the output of the correlator?

7.17 In a DSSS system, the message bit rate is 4800 bps and the PN clock rate is 220 Mbps.
(a) Find the processing gain.
(b) Find the output SNR if the received signal power is 6×10^{-10} W and one-sided noise spectral density is 10^{-20} W/Hz.
(c) If the coherent reference signal is off by 10° and the synchronization error is 30%, will the code noise term be a determining factor in the output SNR? How?

7.18 A speech signal band-limited to 4 kHz and 256 quantization levels is to be transmitted by a DSSS transmitter. Find the required chip rate to obtain a processing gain of 6400.

7.19 An FHSS systems utilizes a fast hop system that contains 10 hops/message bit and 1024 hopping frequencies. The message bit rate is 2400 bps and the final RF multiplication factor is 10. Find the (a) RF signal bandwidth, (b) processing gain in decibels, (c) PN code generator clock rate, and (d) frequency separation in kHz. For the found value of processing gain, what must be the equivalent PN code rate when $t_m = 1/42,000$.

[*Hint*: PN code generator clock rate = $k(c - 1) \times$ message bit rate]

7.20 In an FH system, the switching speed of the synthesizer is 5 µs and the message bit rate is 5 kbps after error-correction coding. There are 5 hops/message bit and the final frequency multiplication is 8. What is the maximum processing gain obtainable?

7.21 Consider a hybrid system PN/FHSS. The PN code rate is 250 kchips/s. For the FH, 8192 frequencies are used with a spacing of 250 kHz. For error correction, 1/3 rate convolution coding is used. Find the processing gain if the final message rate is (a) 75 bps and (b) 2400 bps.

7.22 For the polynomials $h(x) = x^4 + x^3 + x^2 + 1$ and $h(x) = x^4 + x^2 + x + 1$, assuming initial condition (1111), (a) calculate and plot the autocorrelation functions for each, (b) calculate and plot the cross-correlation functions, and (c) draw the block diagrams using both the Fibonacci and Galois methods.

7.23 For the polynomials in Problem 22, find the Gold sequence.

7.24 In a PN sequence with the length $2^8 - 1$, how many runs of 1111 would be expected?

7.25 In an FHSS system, the total bandwidth $B_s = 400$ MHz and the individual channel bandwidth $B_m = 100$ Hz. What is the minimum number of PN bits required for each frequency hop?

7.26 For an OFDM system, if 52 subcarriers, spaced at 312.5 kHz, are defined, find the total occupied bandwidth excluding the secondary lobes.

7.27 If the OFDM bandwidth is 19.2 MHz and 48 subcarriers are modulated with convolution-coded 2/3 rate data using (a) BPSK and (b) QPSK, in each case what must be the input data rate?

7.28 In an OFDM system, 52 subcarriers are defined with 300 kHz spacing. If, for the channel estimation purpose, 8 pilot carriers are added at equal distance maintaining the same spacing, what will be the percentage rise in the occupied spectrum?

7.29 In an OFDM system, the frequency domain setting is such that the spacing between two subcarriers is 312.5 kHz. If the frequency spacing includes a cyclic prefix of 800 ns added to the OFDM symbol, answer the following:
(a) What will be the final symbol duration with the cyclic prefix?
(b) What will be the OFDM symbol duration without the cyclic prefix?
(c) Will there be any loss of spectrum efficiency due to addition of the cyclic prefix?

7.30 Find the additions and multiplications required for a 256-point FFT and a 1024-point FFT. By selecting the 1024-point IFFT instead of the 256-point FFT, will you be able to get good performance? If yes, at what cost?

7.31 A data rate of 5 Mbps is targeted in a multipath radio environment by using BPSK modulation. The maximum delay spread is 25 µs, and the number of subcarriers is 128 for multicarrier transmission. Compare the ISI effect if the system is (a) single carrier and (b) multicarrier.

7.32 If a modulator transmits 12,800 symbols/s with 32 different combination states, what is the final bit rate? If this stream is to be transmitted over the AWGN channel within a bandwidth of 30 MHz, find the suitable modulation scheme. In this scenario, what is the probability of error if $E_b/N_o = 20$.

7.33 Find the BER probability of an 8PSK system for 114 kbps data with a signal amplitude of 12 mW and noise spectral density of 10^{-9} W/Hz.

8 Zero Intersymbol Interference, Diversity, Estimation, and Equalization

Theme of the Chapter

This chapter discusses a few more applications of digital signal processing at various stages of the wireless link for optimizing the system. A received signal is degraded because of channel effects, which can be reduced up to a certain extent by using some techniques. The earlier part of the chapter explains the methods applied at the transmitting end for an intersymbol interference (ISI)-free reception, and the rest throws light on some important techniques, such as diversity, channel estimation, and channel equalization, that can be applied at the receivers.

The ideal sharp pulse consumes a very large bandwidth that is practically impossible to allocate to the user. Nyquist suggested a sine-shaped pulse by which spectrally efficient transmissions can be achieved with zero ISI reception. Some filters can shape and control the transmission bandwidths to accommodate more users with less ISI effects.

This chapter also covers techniques such as diversity, equalization, and channel estimation that help a receiver combat various channel problems. Ultimately, the goal of these techniques is to improve the bit error rate performance. Multiple input, multiple output is the latest diversity-based technique, which is in high demand due to its capability to provide spectral efficiency and high data rates.

Key Topics

- Nyquist criteria for zero ISI
- Filtering (pulse shaping)
- Windowing techniques
- Detection strategies
- Matched filter

- Diversity techniques and combining
- Introduction to MIMO
- Channel estimation techniques
- Equalization techniques

- LS and LMS algorithms
- Multiuser detection

8.1 ZERO INTERSYMBOL INTERFERENCE COMMUNICATION TECHNIQUES

Intersymbol interference (ISI) is a form of distortion in a signal. Here, a received symbol interferes with the subsequent symbols, which makes the communication less reliable. ISI is caused due to multipath propagation or the inherent non-linear frequency response of a channel. The presence of ISI in a system introduces errors in the decision device at the receiver output. Therefore, in the design of transmitting and receiving filters, the objective is to minimize the effect of ISI. ISI effect is shown in Fig. 8.1.

In ISI, successive symbols interfere with each other due to the pulse broadening effect caused by multipath.

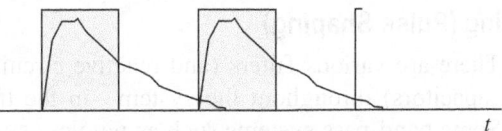

Fig. 8.1 ISI between pulses or symbols (ideally square shape)

8.1.1 Nyquist Criteria for Zero Intersymbol Interference

Nyquist showed that the theoretical minimum bandwidth needed to detect R_s symbols without ISI is $R_s/2$ Hz. The sinc-shaped (t/T) pulse is called the *ideal Nyquist pulse*; its multiple lobes comprise a main lobe and side lobes that are infinitely long. Nyquist established that if each pulse of a received sequence is of the form *sinc* (t/T), the pulses can be detected without ISI.

Figure 8.2 illustrates how ISI is avoided. There are two successive pulses $s(t)$ and $s(t - T)$. Even though $s(t)$ has long tails, the figure shows a tail passing through zero amplitude at the instant $t = T$ when $s(t - T)$ is to be sampled; likewise, all tails pass through zero amplitude when any other pulse of the sequence is to be sampled. They satisfy the orthogonality condition.

The more compact the signalling spectrum, the higher is the allowable data rate or the greater is the number of users that can simultaneously be served. Nyquist has imposed the limitation to such bandwidth reduction. If we operate the system at smaller bandwidths, then according to the Nyquist condition, the pulse would spread in time, which would degrade the system bit error rate (BER) performance due to increased ISI. A prudent goal is to compress the bandwidth of the data impulses to a reasonably small bandwidth greater than the Nyquist minima. This is accomplished with a *Nyquist filter*. Without such a measure, each pulse extends into every other pulse in the entire sequence.

Long-time responses exhibit large-amplitude tails near the main lobe of each pulse. Such tails are undesirable because they contribute zero ISI only when the sampling is performed at exactly the correct sampling time; when the tails are large, small timing errors will result in ISI. Therefore, although a compact spectrum provides optimum bandwidth utilization, it is very susceptible to ISI degradation induced by timing errors.

The terms *Nyquist filter* and *Nyquist pulse* are often used to describe the general class of filtering and pulse shaping that satisfies zero ISI at the sampling points. The basic transversal filter (explained in Section 8.8.1) can be used for pulse shaping for zero ISI. Among the classes of Nyquist filters, the most popular ones are the raised cosine filter and square root raised cosine filter. Others are Gaussian filter and Chebyshev filter.

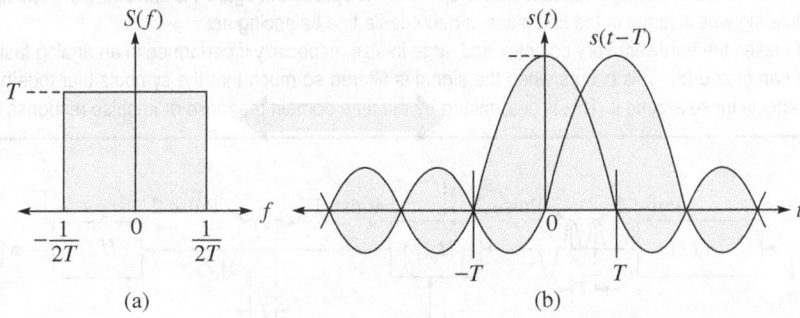

(a)   (b)

Fig. 8.2 Nyquist pulse detection without ISI

8.1.2 Filtering (Pulse Shaping)

> Filtering optimizes the spectral efficiency of the transmission by reducing the transmitted bandwidth of the signal without losing data.

There are various filters (and reactive circuit elements such as inductors and capacitors) throughout the system—in the transmitter, channel, and receiver. Some band-pass systems such as wireless systems are characterized by fading channels that behave like undesirable filters manifesting signal distortion. When the receiving filter is configured to compensate for the distortion caused by both the transmitter and the channel, it is often referred to as a receiving or an equalizing filter. Taking all these filtering effects into one overall equivalent system transfer function,

$$H(f)=H_t(f)H_c(f)H_r(f) \tag{8.1}$$

where $H_t(f)$ characterizes the transmitting filter shaping the pulse, $H_c(f)$ is the filtering within the channel, and $H_r(f)$ is the receiving or equalizing filter.

Due to the effects of system filtering, the received pulses can overlap one another, as shown in Fig. 8.3. The tail of the pulse can *smear* into adjacent symbol intervals, thereby interfering with the detection process and degrading the error performance. Even in the absence of noise, the effects of filtering and channel-induced distortion lead to ISI. Sometimes $H_c(f)$ is specified, and the problem remains to determine $H_t(f)$ and $H_r(f)$ such that the ISI is minimized at the output of $H_r(f)$.

Any fast transition in a signal—it may be amplitude, phase, or frequency—will require a wide occupied bandwidth. Any technique that helps to slow down these transitions will narrow the occupied bandwidth. Filtering serves to smooth these transitions. It reduces interference because it reduces the tendency of one signal or transmitter to interfere with another. At the receiver end, the reduced bandwidth improves sensitivity because more noise and interference are rejected.

Pulse Shaping Using Raised Cosine Filter

Figure 8.4 shows the impulse or time domain response of a raised cosine filter, a class of Nyquist filter. The impulse response of a Nyquist filter rings at the symbol rate. The filter

---

DOWNSIDES OF FILTERING

- Some types of filtering cause the trajectory of the signal to overshoot in many cases. This overshoot path represents the carrier power and phase. Therefore, the carrier requires more power from the transmitter amplifiers to take on these values. It requires more power than would be necessary to transmit the actual symbol itself. Carrier power cannot be clipped or limited (to eliminate the overshoot) without causing the spectrum to spread out again. As narrowing the spectral occupancy was the reason filtering was inserted in the first place, it becomes a fine balancing act.
- Filtering makes the hardware very complex and large in size, especially if performed in an analog fashion.
- Filtering can create ISI. This occurs when the signal is filtered so much that the symbols blur together (Fig. 8.1) and each symbol affects those around it. This is determined by the time domain response or impulse response of the filter.

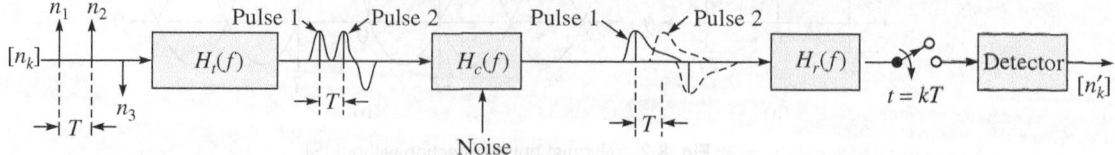

Fig. 8.3 Pulse transmission over a channel

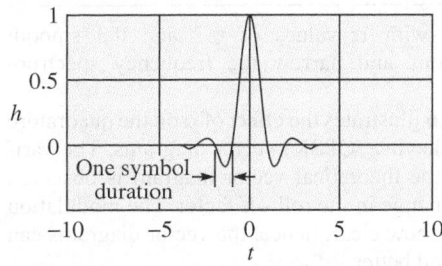

Fig. 8.4 Time domain response of raised cosine filter

The time response of the raised cosine filter goes through zero with a period that exactly corresponds to the symbol spacing; hence, ISI exists at all times during transmission except at symbol decision times.

is chosen to ring, or have the impulse response of the filter cross through zero, at the symbol clock frequency. Adjacent symbols do not interfere with each other at the symbol times because the response equals zero at all symbol times except the centre (desired) one. Nyquist filters heavily filter the signal without blurring the symbols together at the symbol times. This is important for transmitting data without errors caused by ISI.

The sharpness of a raised cosine filter is described by the *roll-off factor α*), which decides the slope of the low-pass filter cut-off region. Figure 8.5 gives the response of the raised cosine filter for different α values. The value of α gives a direct measure of the occupied bandwidth of the system and is calculated as

$$\text{Total bandwidth } W = \text{Symbol rate} \times (1 + \alpha) \qquad (8.2)$$

If the filter had a perfect (ideal) characteristic with sharp transitions or cut off and $\alpha = 0$, the occupied bandwidth for $\alpha = 0$ would be

$$\text{Total bandwidth } W = \text{Symbol rate} \times (1 + 0) = \text{Symbol rate} \qquad (8.3)$$

Figure 8.5 gives the roll-off characteristics of a raised cosine filter. A change in α causes a change in the sharpness of the cut-off region or the bandwidth.

Ideally, with a sharp cut-off region, the occupied bandwidth would be the same as the symbol rate, but this is not practical. An α of zero is impossible to implement. α is sometimes called the *excess bandwidth factor*, as it indicates the amount of occupied bandwidth that will be required in excess of the ideal occupied bandwidth (which would be the same as the symbol rate). At the other extreme, a broader filter with an α of one is easier to implement. The occupied bandwidth for $\alpha = 1$ will be

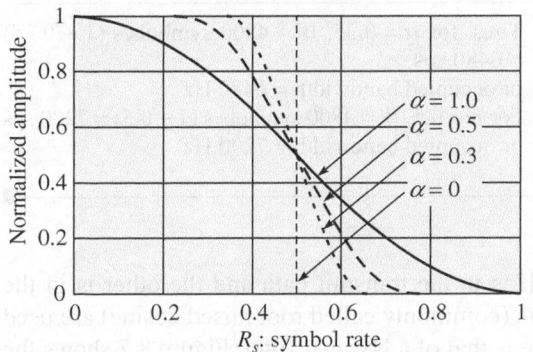

Fig. 8.5 Raised cosine filter response for different α values

$$\text{Total bandwidth } W = \text{Symbol rate} \\ \times (1+1) \qquad (8.4) \\ = 2 \times \text{symbol rate}$$

Here, $\alpha = 1$ uses twice as much bandwidth as used for $\alpha = 0$. In practice, it is possible to implement an α below 0.2 and make good, compact, and practical radios. Typical values range from 0.35 to 0.5, though some video systems use an α as low as 0.11.

Different filter α values also affect the transmitted power. For an unfiltered signal with an α of infinity, the maximum or peak power of the carrier is the same as the nominal power at the symbol states. No extra power is required due to the filtering process. A raised cosine filter is normally used with a pulse code modulated signal for baseband transmissions.

Example 8.1 Show that the different filter bandwidths (change in α) show different effects on the vector diagram of a modulated signal.

Solution If a radio has no transmitter filter, the transitions between states are instantaneous as shown in Fig. 8.6(a). No filtering means $\alpha = \infty$ hypothetically. Transmitting this signal would require infinite bandwidth. Figure 8.6(b) is an example of a signal at $\alpha = 0.7$. Figure 8.6(c) shows the signal at $\alpha = 0.3$.

The filters with α values of 0.7 and 0.3 smooth the transitions and narrow the frequency spectrum required.

Figure 8.6 illustrates the effect of α on the quadrature phase shift keying (QPSK) vector diagrams. The variation from the theoretical vector diagram is observed due to the change in the roll-off factor. The modulation schemes are now clear; hence, the vector diagrams can be understood better.

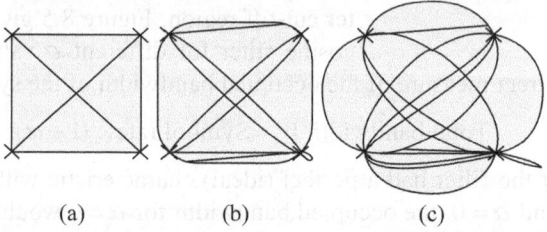

$$(a) \qquad\qquad (b) \qquad\qquad (c)$$

Fig. 8.6 QPSK vector diagrams—effect of α (can be observed by simulation) (a) Without filtering (b) $\alpha = 0.75$ (c) $\alpha = 0.375$

Example 8.2 In a QPSK system, the bit rate is 9600 bps. Check the bandwidth requirement with a roll-off factor of 0.35 and 0.5.

Solution Total bandwidth W = Symbol rate $\times (1 + \alpha)$
    For a QPSK scheme, symbol rate = 4800 bps

Thus, for $\alpha = 0.35$, $W = 4800$ symbols/s $(1 + 0.35)$ = 6480 sps
or occupied bandwidth = 6480 Hz
For $\alpha = 0.5$, $W = 4800$ symbols/s $(1 + 0.5) = 7200$ sps
or occupied bandwidth = 7200 Hz

Square Root Raised Cosine Filter

Usually, the filter is split such that one half is in the transmit path and the other is in the receive path. In this case, root Nyquist filters (commonly called root raised cosine) are used in each part, so that their combined response is that of a Nyquist filter. Figure 8.7 shows the block diagram of a square root raised cosine filter.

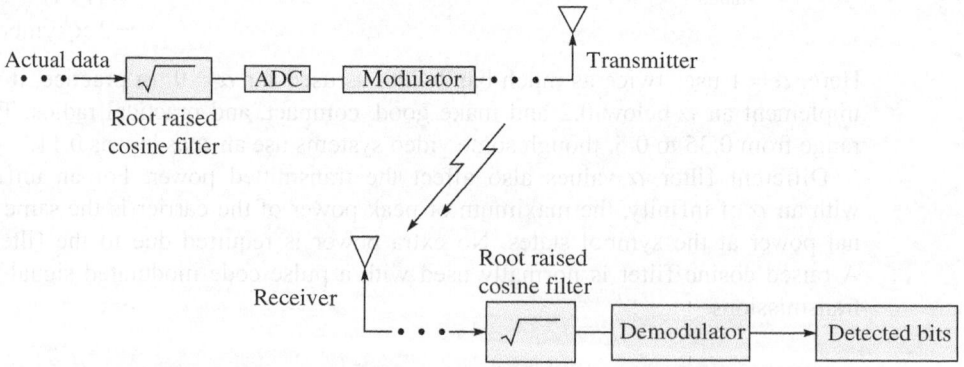

Fig. 8.7 Square root raised cosine filter shown as premodulation filtering and predemodulation filtering

Gaussian filtering is
the result of the convo-
lution of the signal with
the Gaussian function.

Sometimes, filtering is desired at both the transmitter and the receiver. Filtering at the transmitter reduces the adjacent channel power radiation, whereas filtering at the receiver reduces the effects of broadband noise as well as interference from other transmitters in nearby channels. To get zero ISI, both filters are designed such that the combined result of the filters and the rest of the system is a full Nyquist filter. Potential differences can cause problems in manufacturing because the transmitter and receiver are often manufactured by different companies. The receiver may be a small hand-held model and the transmitter may be a large cellular base station. If the design is performed correctly, it results in the best data rate, the most efficient radio, and reduced effects of interference and noise. This is why root Nyquist filters are used in receivers and transmitters as

$$\sqrt{\text{Nyquist filter}} \times \sqrt{\text{Nyquist filter}} = \text{Nyquist filter} \qquad (8.5)$$

Gaussian Pulse Shaping Filter

Gaussian filter is normally used in Gaussian minimum shift keying (GMSK) and Gaussian frequency shift keying (GFSK) modulation schemes. In these schemes, minimum shift keying (MSK) or frequency shift keying (FSK) will be performed after passing the signal through a Gaussian filter.

The GMSK modulation method is used in the global system for mobile communication (GSM). GSM signals have a small blurring of symbols on each of the four states because the Gaussian filter used in GSM does not have zero ISI. The phase states vary to a certain extent, causing a blurring of the symbols. Wireless system architects must decide the amount of ISI that can be tolerated in a system and combine that with noise and interference.

Gaussian filters are used in GSM because of their advantages in carrier power, occupied bandwidth, and symbol clock recovery. The Gaussian filter has a Gaussian shape in both the time and frequency domains, and it does not ring like the raised cosine filters. Its effects in the time domain are relatively short and each symbol interacts significantly (or causes ISI) with only the preceding and succeeding symbols. This reduces the tendency for particular sequences of symbols to interact, which makes amplifiers easier to build and more efficient. Figure 8.8 shows the response of a Gaussian filter. Equation (8.6) represents the time response of the filter with respect to the various values of α, where α can be defined within the filter response equation or as mentioned in Fig. 8.8.

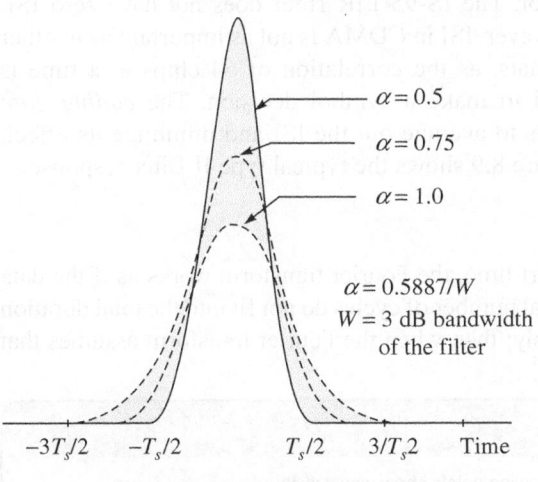

$\alpha = 0.5$
$\alpha = 0.75$
$\alpha = 1.0$

$\alpha = 0.5887/W$
$W = 3$ dB bandwidth of the filter

$-3T_s/2 \quad -T_s/2 \quad T_s/2 \quad 3/T_s2 \quad$ Time

Fig. 8.8 Gaussian filter response

$$h(t) = \frac{\sqrt{\pi}}{\alpha} \exp\left[-\left(\frac{\pi t}{\alpha}\right)^2\right] \qquad (8.6)$$

The *roll-off factor* is very important in a raised cosine filter. The corresponding term for a Gaussian filter is *bandwidth time product* (BT). Occupied bandwidth cannot be stated in terms of BT because a Gaussian filter's frequency response does not go identically to zero, as does a raised cosine. Common values for BT are 0.3–0.5.

Chebyshev Equiripple Finite Impulse Response Filter

> Chebyshev filters have steeper roll-off and more passband ripple (type I) or stopband ripple (type II) compared to raised cosine,

A Chebyshev equiripple finite impulse response (FIR) filter is used for baseband filtering in IS-95 code division multiple access (CDMA). With a channel spacing of 1.25 MHz and a symbol rate of 1.2288 MHz in IS-95 CDMA, it is vital to reduce leakage to adjacent radio frequency (RF) channels. This is accomplished by using a filter with a very sharp *shape factor* using an α value of only 0.113.

An FIR filter's impulse response exists for only a finite number of samples. Equiripple indicates that there is a *rippled* magnitude frequency response envelope of equal maxima and minima in the passband and stopband.

There are two types of Chebyshev low-pass filters, and both are based on Chebyshev polynomials.

(a) The type I filter has an all-pole transfer function, and it has an equiripple passband and a monotonically decreasing stopband.

(b) A type II low-pass filter has both poles and zeros. Its passband is monotonically decreasing and it has an equiripple stopband.

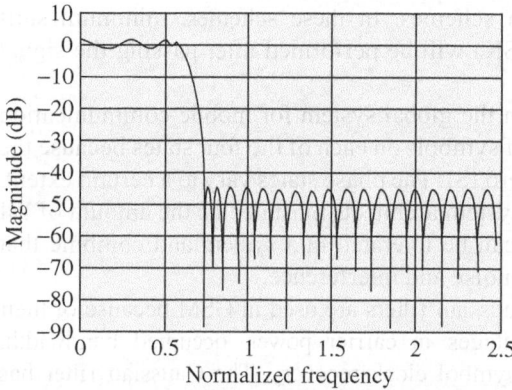

By allowing some ripple in the passband or stopband magnitude response, a Chebyshev filter can achieve a *steeper* pass-to-stop band transition region; that is, the filter roll-off is faster than can be achieved by the same order Butterworth filter.

A Chebyshev FIR filter uses a much lower order than a Nyquist filter to implement the required shape factor. The IS-95 FIR filter does not have zero ISI. However, ISI in CDMA is not as important as in other formats, as the correlation of 64 chips at a time is used to make a symbol decision. The *coding gain* tends to average out the ISI and minimize its effect. Figure 8.9 shows the typical type II filter response.

Fig. 8.9 Chebyshev type II FIR filter response

8.1.3 Windowing Techniques

Though we measure a signal for only a short time, the Fourier transform works as if the data were periodic for the entire time. If an integral number of cycles do not fit into the total duration of the measurement, the cycle ends randomly; then when the Fourier transform assumes that

---

THE COST OF SPECTRAL EFFICIENCY

Spectral efficiency is important, but it is achieved at a cost. The following points show why it is so:

- Channel bands supporting natural resources are too wide and waste a lot of RF spectrum. Therefore, narrow filters are used to reduce the occupied bandwidth.
- However, narrow filters with sufficient accuracy are difficult to construct. Smaller values of α increase the ISI because more symbols can contribute. This tightens the requirements on clock accuracy.
- Narrow filters also result in more overshoot and, therefore, require high peak carrier power. Moreover, we require a large amplifier, which in turn may cause heavy heat and electrical interference. Therefore, we require large and heavy batteries. We can afford small batteries only for portability of devices.

The only compromise in such a case is to have short talktime.

Windowing suppresses discontinuities and avoids the broadening of the frequency spectrum caused by discontinuities.

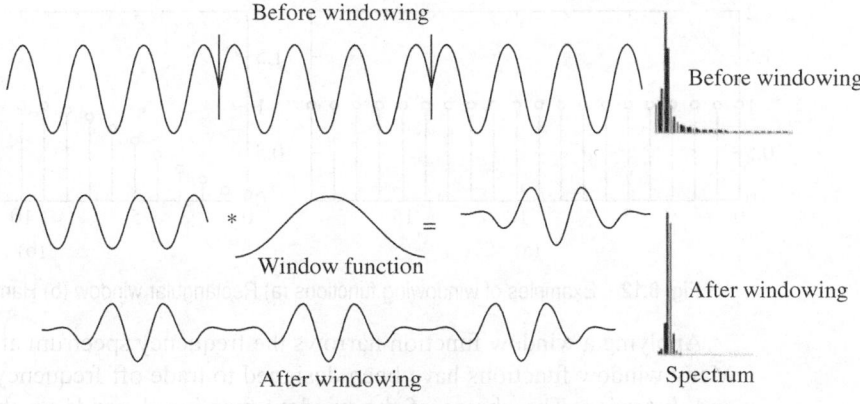

Fig. 8.10   Effect of windowing

the signal repeats, the end of one signal segment does not connect smoothly with the beginning of the next; moreover, the assumed signal is similar to the actual signal but has little discontinuity at regular intervals. The discontinuities can be reduced by shaping the signal such that its ends match more smoothly as shown in Fig. 8.10. As we cannot assume anything about the signal, we need a way to make a signal's ends connect smoothly to each other when repeated.

One way to do this is to multiply the signal by a *window* function, as discussed here. The easiest way to ensure that the ends of a signal match is to force them to be zero; this way, their value is necessarily the same. We should also make sure that the signal is going in the right direction at the ends to match up smoothly. The slope of the signal at its ends should also be zero. Mathematically, the value of a window function and all its derivatives is zero at the ends.

Windowing can narrow the spectrum, but it is important to remember that windowing is really a distortion of the original signal. It adds BER in the performance but improves spectral efficiencies at the same time. Using the windowing function in a system is a compromise.

As per Fig. 8.11, the transient response actually has a broad frequency spectrum, but windowing forces it to look as if it has a narrow frequency spectrum instead. Even worse, the window function has attenuated the signal at the point where it was largest, so it has suppressed a large part of the signal power. This means that the overall signal-to-noise ratio (SNR) has been reduced.

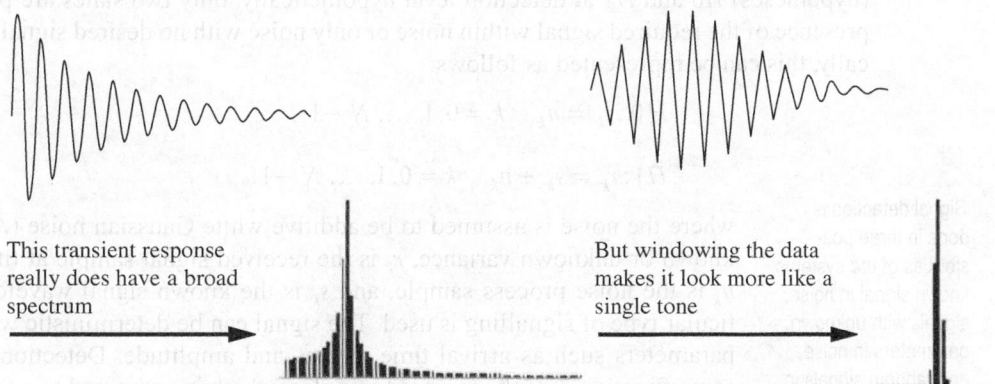

This transient response really does have a broad spectrum

But windowing the data makes it look more like a single tone

Fig. 8.11   The result of applying a window function without proper thought

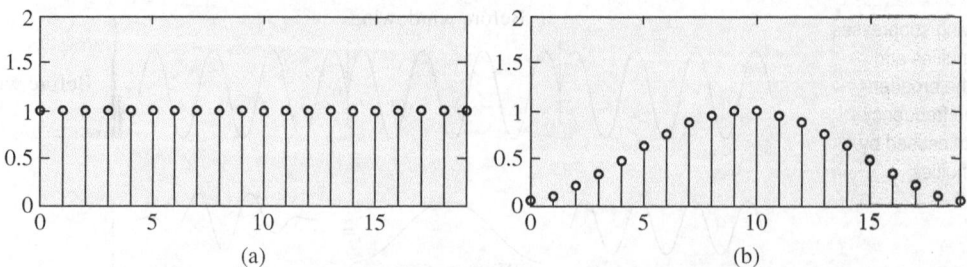

Fig. 8.12 Examples of windowing functions (a) Rectangular window (b) Hanning window

Applying a window function narrows the frequency spectrum at the expense of the SNR. Many window functions have been designed to trade off frequency resolution against SNR and distortion. The choice of the window function depends on the knowledge of the signal and what you want to do with it. Two examples of windowing functions are given in Fig. 8.12. Here, the function is shown in discrete form; the x-axis and y-axis show the index values of the samples and magnitude, respectively.

8.2 DETECTION STRATEGIES

The expected signal must be detected from among the various coexisting signals at the receiving end. Detection theory, or signal detection theory, is well established and is a means to quantify the ability to differentiate between information-bearing energy patterns and random energy patterns (such as noise) that distract from the information.

The basis for the signal detection theory is that nearly all reasoning and decision-making takes place in the presence of some uncertainty. The theory provides a precise language and graphic notation for analysing decision-making in the presence of uncertainty. Graphical interpretations, receiver operating characteristics, and discriminability index are used for decision-making.

Mostly, the threshold-based detection method (simple binary hypothesis test) is used, but according to the theory of detection, there are a number of determiners of how a detecting system will detect a signal and where its threshold level will be. Changing the threshold level will affect the ability to discern between useful and non-useful signals. The aim of the detection problem is to determine whether there exists a signal or not. Suppose there are two states (hypotheses) $H0$ and $H1$ at detection level hypothetically, only two states are possible—the presence of the required signal within noise or only noise with no desired signal; mathematically, this can be represented as follows:

$$H0 : r_k = n_k \quad k = 0, 1, ..., N-1 \tag{8.7a}$$

$$H1 : r_k = s_k + n_k \quad k = 0, 1, ..., N-1 \tag{8.7b}$$

Signal detection is done in three possibilities of the systems: known signal in noise, signals with unknown parameters in noise, and random signals in noise.

where the noise is assumed to be additive white Gaussian noise (AWGN) with known or unknown variance, r_k is the received signal sample at time instant k, n_k is the noise process sample, and s_k is the known signal waveform if a particular type of signalling is used. The signal can be deterministic with unknown parameters such as arrival time, phase, and amplitude. Detection is based on some function T of the received samples, which is compared to a threshold γ. If the threshold is exceeded, it is decided that $H1$ is true. Let r denote the received

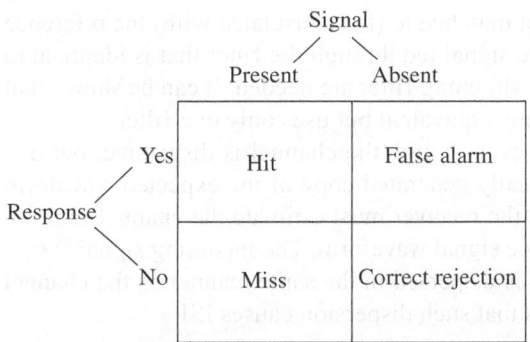

Fig. 8.13 Possible states in signal detection and probabilities that may occur

A matched filter is a filter used in communications to match a particular transit waveform and a correlation detector.

samples as a column vector. The probability of a false alarm P_{FA} is the probability that $H1$ is selected even when $H0$ is actually true; that is, $P_{FA}= P(T(\mathbf{r}) > \gamma, H0)$. The probability of miss P_M is the probability that $H0$ is selected when $H1$ is true. The probability of detection is $P_D = 1 - P_M$ and is the probability that $H1$ is selected when it is actually true; that is, $P_D= P(T(\mathbf{r}) > \gamma, H1)$. The concept is shown in Fig. 8.13.

The two main components of the decision-making process are information acquisition and criterion. The optimal detectors use the maximum likelihood test in which probability density functions (PDFs) are used if the parameters are unknown and random but their PDFs are known.

In some applications, it is possible to assign prior probabilities to the possible hypothesis. In the Bayesian approach to hypothesis testing, costs are assigned to different types of errors and the decision rule is based on minimizing the expected cost (risk). It can be shown that the detector minimizing the Bayes cost also uses the likelihood ratio. If the costs are available, but the priors cannot be realistically assigned, then the min-max criterion can be applied. It minimizes the maximum possible risk.

Energy detectors, which are known as radiometers, are used in several applications. For example, a radiometer can be used for detecting spread spectrum signals, and in cognitive radios, it can be used for finding signal-free bands.

8.3 MATCHED FILTER

Many fundamental theorems in signal detection theory have been developed. However, the theory of the matched filter receiver is of particular interest. An example diagram is given in Fig. 8.14.

The signal is multiplied by a locally stored reference copy and integrated over time (a matched filter correlates the incoming signal with a locally stored reference copy of the transmit waveform). The matched filter maximizes the SNR for a known signal. It can be an optimal detector under the following conditions

• The channel produces AWGN.
• The channel is LTI.
• Exact time reference is available (the signal amplitude as a function of time is precisely known).

The matched filter can be applied even if the noise is non-white. In this case, one can prefilter the incoming signal to make the noise component white. This is called a *whitening filter*. Evidently, this also filters the wanted signal.

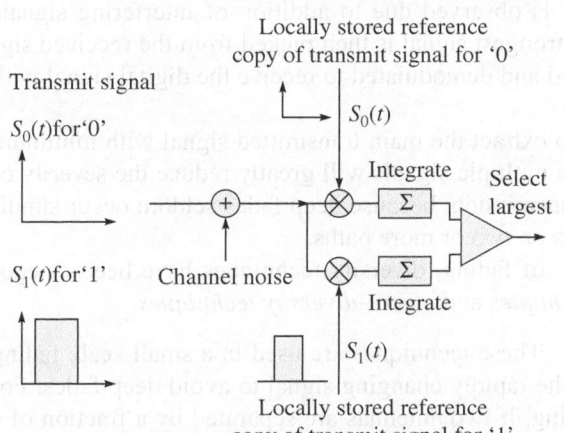

Fig. 8.14 Possible implementation of matched filter receiver

Therefore, the filtered incoming signal is not matched to (i.e., correlated with) the reference transmit signal, but is matched to a reference signal fed through the filter that is identical to the whitening filter. Thus, two copies of the whitening filter are needed. It can be shown that one can build a detector that is mathematically equivalent but uses only one filter.

The matched filter concept can be used even though the channel is dispersive, but one must process the incoming signal with a locally generated copy of the expected waveform after transmission over the channel. That is, the receiver must estimate the channel impulse response (CIR) and apply this to the reference signal waveform. The incoming signal is correlated with the reference waveform, which is dispersed in the same manner as the channel disperses the radio signal. A complication is that such dispersion causes ISI.

8.4 DIVERSITY TECHNIQUES

In telecommunications, diversity techniques are used for improving the reliability of a message signal by utilizing two or more communication channels with different characteristics. Multiple versions of the same signal may be transmitted and/or received and combined in the receiver. Alternatively, a redundant forward error-correction code may be added and different parts of the message may be transmitted over different channels.

A diversity scheme is a method used to develop information from several signals transmitted over independent fading paths. It means that the diversity method requires a number of transmission paths, all carrying the same message but having independent fading statistics. The mean signal strengths of the paths should also be approximately the same. The basic requirement of independent fading is that the received signals should be uncorrelated. Therefore, the success of a diversity scheme depends on the degree to which the signals on the different diversity branches are uncorrelated. The general basic model of the diversity concept is given in Fig. 8.15. Here, different independent fading paths are shown with different CIRs $h_1(t)$ to $h_n(t)$, and the distortion in the signal is observed due to addition of interfering signals

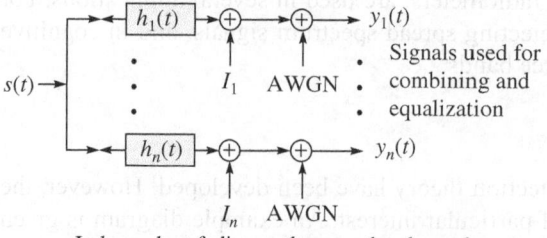

Fig. 8.15 Basic concept of diversity

I_1 to I_n and white Gaussian noise. The strongest signal is then picked from the received signals. This signal can be further equalized and demodulated to receive the digital signal with minimum BER.

Diversity combining is the process to extract the main transmitted signal with minimum channel effects. Properly combining the multiple signals will greatly reduce the severity of fading and improve the reliability of transmission, because deep fades seldom occur simultaneously during the same time intervals on two or more paths.

Depending upon the type of fading, diversity techniques have been categorized as *micro-diversity techniques* and *macro-diversity techniques*.

Micro-diversity techniques These techniques are used in a small-scale fading environment. They exploit the rapidly changing signal to avoid deep fades. For example, in small-scale fading, if two antennas are separated by a fraction of a metre, one may receive a null while the other receives a strong signal. By selecting the best signal at all times, a receiver can mitigate small-scale fading effects.

> Diversity combining process is essential along with diversity reception to get the strongest possible reception.

Macro-diversity techniques These techniques are used in a large-scale fading environment. Here, we generally select the base station that is not shadowed when others are. The mobile can substantially improve the average SNR on the forward link.

> Diversity plays an important role in combating fading and co-channel interference and in avoiding error bursts.

Though a number of diversity schemes are described in the literature, we will briefly consider some basic ones here.

Space diversity This can be a microscopic or macroscopic diversity technique and is used at the transmitter or the receiver. The signal is transferred over several different propagation paths. In wired transmission, this can be achieved by transmitting via multiple wires. In wireless transmission, it can be achieved by antenna diversity using multiple transmitter antennas (transmit diversity) and/or multiple receiving antennas (receive diversity). In the latter case, a diversity combining technique is applied before further signal processing takes place.

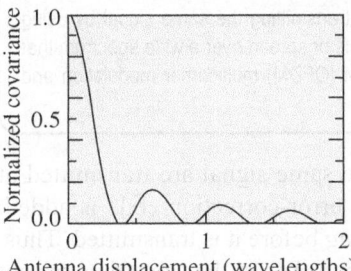

Fig. 8.16 Autocorrelation coefficient versus space separation

If the receiver has multiple antennas, the distance between the receiving antennas is made large enough to ensure independent fading. This arrangement is called *space diversity*. Space separation of *half of the wavelength* is sufficient to obtain two uncorrelated signals. A typical graph of autocovariance of the received signal strength in a Rayleigh fading channel versus the normalized antenna displacement is shown in Fig. 8.16. Based on space diversity, various types of systems are possible. Space diversity and other such systems are explained in Section 8.6.

Polarization diversity This technique exploits the fact that obstacles scatter waves differently depending on their polarization. It is hoped that the received waves do not cancel each other out in one of the branches, resulting in a relatively strong signal. Antennas can transmit either a horizontal or a vertical polarized wave. When both waves are transmitted simultaneously, received signals will exhibit uncorrelated fading statistics. This scheme can be considered as a special case of space diversity because separate antennas are used. However, only two diversity branches are available, as there are only two orthogonal polarizations.

Angle diversity In this technique, directional antennas receive only a fraction of all the scattered energy. As the received signals arrive at the antenna via several paths, each with a different angle of arrival, the signal component can be isolated by using directional antennas. Each directional antenna will isolate a different angular component. Hence, the signals received from different directional antennas pointing at different angles are uncorrelated.

SITE DIVERSITY—A SPECIAL FORM OF SPACE DIVERSITY

In site diversity, the receiving antennas are located at different receiver sites. For instance, consider the following:
- In land mobile radio, where vehicle-mounted and hand-held radios communicate with a base station over a single frequency, space diversity is achieved by having several receivers at different sites.
- As a result, signals from within a cell may be received at different corners of the hexagonal area.
- The advantage is that not only the multipath fading attenuation is independent at each branch but also the shadowing and path losses are uncorrelated to some extent.

For a stationary mobile station, time diversity is useless. This is in contrast to all other diversity types, as they are all independent of the movement of the mobile station.

Frequency diversity In this technique, information is transmitted on more than one carrier frequency, because the frequencies separated by more than the coherence bandwidth of the channel will not experience the same fading. This is often employed in microwave line-of-sight (LOS) links, which carry several channels in a frequency division multiplexing (FDM) mode. In practice a $1:N$ (arbitrary variable) protection switching is provided by a radio licence, wherein one frequency is nominally idle but is available on a standby basis to provide frequency diversity switching for any one of the other N carriers. When diversity is needed, the appropriate traffic is simply switched to the backup frequency.

Note: Frequency diversity is exploited in wideband systems. Instead of transmitting the same signal by using different carriers, the signal is transferred using several frequency channels or spread over a wide spectrum that is affected by frequency-selective fading. Examples are orthogonal FDM (OFDM) multicarrier modulation and frequency hopping spread spectrum techniques.

Time diversity In this technique, multiple versions of the same signal are transmitted at different time instants. Alternatively, a redundant forward error-correction code is added, and the message is spread in time by means of *bit interleaving* before it is transmitted. Thus, error bursts are avoided, which simplifies the error correction. When the same data is sent at different time instants, the received signals can be uncorrelated if the time separations are large enough. The time difference between two transmissions should be large compared to the time the mobile antenna takes to move half a wavelength. The required time separation is at least as great as the reciprocal of the fading bandwidth, which is two times the speed of the mobile station divided by the wavelength. Hence, the time separation is inversely proportional to the speed of the mobile station.

Data can be repeatedly transmitted at time spacing that exceeds the coherence time of the channel; hence, multiple repetitions of the signal will be received with independent fading conditions, thereby providing diversity.

Time diversity will be less effective in systems with stationary antennas, such as indoor wireless communication, as the channel characteristics do not change very much with time. However, this technique may be helpful if uncorrelated interference signals are experienced during successive attempts.

Joint diversity The increased mobility of users often results in fast fading or large Doppler spreads that cause performance degradation.
- Joint time–frequency diversity techniques provide significant performance improvement over existing systems in the single-user receiver design.
- Like time–frequency diversity, we can create diversity dimensions artificially, such as space–time diversity for multi-user detection in CDMA systems.
- Diversity gains can be achieved by using multiple elements for antenna arrays processing and adaptively combining space–time–frequency diversity signals.
- Similarly, space–frequency diversity is exhibited in a multiple input, multiple output (MIMO) OFDM system. However, due to heavy computational complexity, the trade-off between performance and complexity becomes an important issue when applying joint diversity techniques.

8.5 DIVERSITY COMBINING TECHNIQUES

Space diversity reception or combining methods can be classified into four types:
(a) Selection combining
(b) Threshold combining
(c) Maximal ratio combining
(d) Equal gain combining

8.5.1 Selection Combining

> The main drawback of selection combining is that it cannot function on a truly instantaneous basis in dynamic channel conditions.

Selection combining is the simplest combining scheme. It is based on the principle of selecting the best signal (the largest energy or SNR) among all the signals received from different branches.

In this scheme, m modulators are used to provide m diversity branches whose gains are adjusted to provide the same SNR for each branch. The receiver branch having the highest instantaneous SNR is then connected to the demodulator. The conceptual diagram is shown in Fig. 8.17.

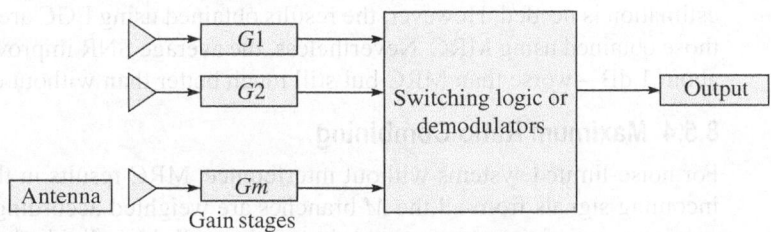

Fig. 8.17 Selection combining

8.5.2 Threshold Combining

Threshold combining is a special form of selection diversity that is less expensive to implement. Here, a limited number of signals are considered for the extraction purpose using a threshold level. Scanning and feedback mechanisms are used here. In this scheme, instead of always using the best of M signals, the M signals are scanned in a fixed sequence until one is found to be above a predetermined threshold. This signal is then received until it falls below the threshold, and the scanning process is again initiated. The receiver switches to another signal when the current signal drops below the predefined threshold. The constraint on this scheme is that the statistics obtained by this method are inferior to those obtained using the other methods. The threshold method is illustrated in Fig. 8.18.

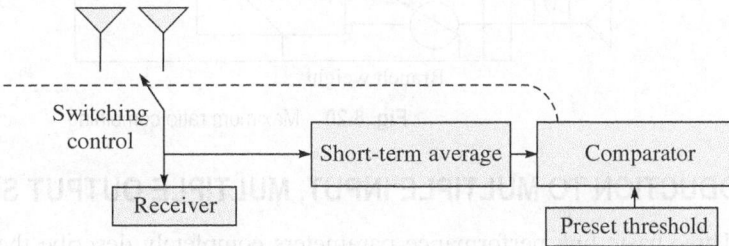

Fig. 8.18 Threshold diversity combining

> *Note*: An equal gain combiner gains all signals with the same value and combines them, whereas a maximum ratio combiner gains more in better branches and less in bad branches.

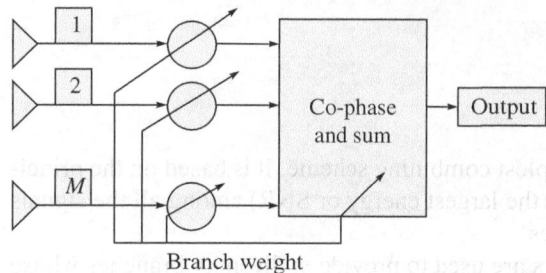

Fig. 8.19 Equal gain combining

8.5.3 Equal Gain Combining

Equal gain combining (EGC) is better than selection diversity and is almost as good as maximum ratio combining (MRC), but is less complex in terms of the signal processing and feedback part. The idea behind this technique will be clear from Fig. 8.19. In EGC, the branch weights are all set to unity (eliminated); that is, the adaptively controlled amplifiers or attenuators are not needed. The signals from each branch are co-phased to provide equal gain combining. This allows the receiver to exploit signals that are simultaneously received on each branch. Thus, EGC is simpler to implement than MRC. Moreover, no channel amplitude estimation is needed. However, the results obtained using EGC are comparatively worse than those obtained using MRC. Nevertheless, the average SNR improvement of EGC is typically about 1 dB—worse than MRC but still much better than without diversity.

8.5.4 Maximum Ratio Combining

For noise-limited systems without interference, MRC results in the best SNR. Here, all the incoming signals from all the M branches are weighted according to their individual signal voltage to noise power ratios and then added. All the individual signals must be co-phased before being added. This requires an individual receiver circuitry and a phasing circuit for each antenna element. It produces an acceptable SNR compared to other techniques. It has extremely good signal processing and complex hardware. Figure 8.20 illustrates this scheme.

The received signals are weighted with respect to their SNR and then added. The resulting SNR yields $\sum_{k=1}^{M} SNR_k$, where SNR_k is the SNR of the kth received signal.

A method called interference rejection combining is used in smart antennas.

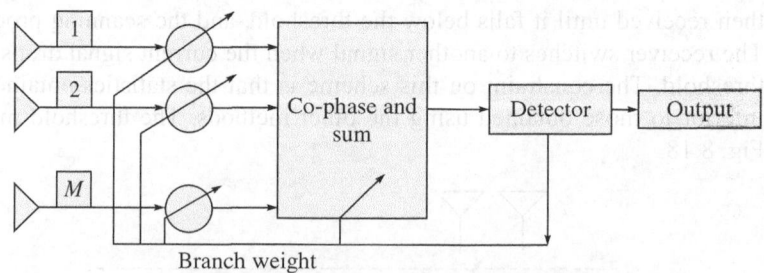

Fig. 8.20 Maximum ratio combining

8.6 INTRODUCTION TO MULTIPLE INPUT, MULTIPLE OUTPUT SYSTEMS

Three basic link performance parameters completely describe the quality of any wireless link: speed (or spectrum), range (or coverage), and reliability (or security). The use of

VARIOUS MIMO SYSTEMS

In a simple smart antenna system (MIMO based on spatial diversity), multiple transmitting antennas may carry a single stream of data and the receiving antennas receive multiple signals of the same information without change in the transmission capacity. The strongest signal is considered for the final reception.

An MIMO system in the true sense differs from a smart antenna system, as it is an efficient spatial multiplexing technique with extra capabilities. In this case, the MIMO system has two or more transmitting antennas and two or more receiving antennas. MIMO transmits and receives two or more streams of radio signals over a matrix channel in parallel where each signal carries unique information and the system delivers two or more times the data rate per channel. This is called spatial multiplexing-based MIMO.

> MIMO offers greater spectral efficiency as compared to SISO, SIMO, and MISO systems. Higher data rates, greater range, increased number of users, and enhanced reliability.

multiple waveform transmission in parallel constitutes a new type of radio communication—communication using multidimensional signals—which is the way to improve all the three performance parameters using a multiple antenna system. MIMO is a multiple antenna system that shows how to achieve higher data rates, wider coverage, and increased reliability, all without using additional frequency spectrum. It exploits the use of multiple signals (space diversity) into the wireless medium and multiple signals received from the wireless medium to improve the wireless channel performance. It can provide a combination of a multi-antenna system with a multicarrier system.

There are two types of MIMO systems—open loop (without channel state information or CSI) and closed loop (with CSI in the feedback). The closed loop system is possible due to packet-based transmissions with a training sequence preamble at the beginning of the packet. This helps in deriving CSI.

In general, MIMO exploits multipath by spatial diversity as well as spatial multiplexing.

8.6.1 Spatial Diversity in MIMO

In space diversity, a signal is transferred over several different propagation paths. As mentioned earlier, in wireless transmission, it can be achieved by antenna diversity using multiple transmitter antennas (transmit diversity) and/or multiple receiving antennas (receive diversity). In the latter case, a diversity combining technique is applied before further signal processing takes place.

Presently, four different types of multi-antenna systems can be categorized based on diversity (input and output refer to the number of antennas):

(a) Single input, single output (SISO)—no diversity
(b) Single input, multiple outputs (SIMO)—receive diversity
(c) Multiple inputs, single output (MISO)—transmit diversity
(d) Multiple inputs, multiple outputs (MIMO)—transmit–receive diversity

The SISO system is very simple and deals with the communication between a transmitter and a receiver. In SISO, error probability is critically damaged by fading. The SIMO, MISO, and MIMO systems are represented conceptually in Fig. 8.21.

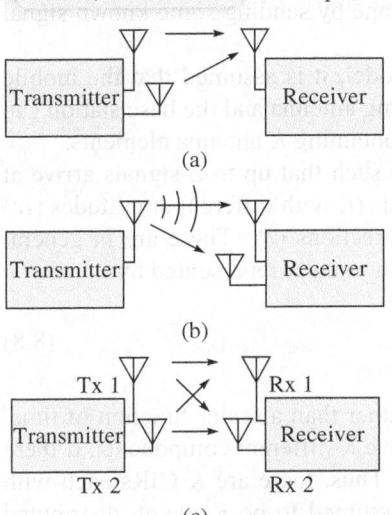

Fig. 8.21 Different types of systems and spatial diversity shown with these systems (both transmitter and receiver must have digital signal processing units) (a) Beamforming (beam steering) employs two transmit antennas to deliver the best multipath signal (b) Diversity (receive combining) uses two receive antennas to capture the best multipath signal (c) There may be physical resemblance between radio systems using a combination of beam steering and diversity and MIMO systems

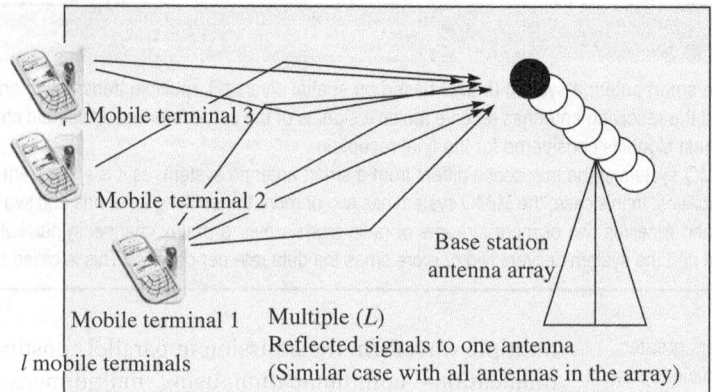

Full CSI means the knowledge of complete channel transfer function, whereas partial CSI provides limited channel information.

Fig. 8.22 Example of SIMO system (the mobile terminal has a single antenna whereas the base station is complex with multiple antennas)

In a SIMO channel, the concept of MRC is offered as a way to exploit the receive diversity. The error probability achieved by MRC is to be much smaller than that corresponding to a SISO channel. To perform MRC, the receiver has to know the fading, or, in other words, the receiver has to have access to the CSI. This is usually done by sending some known signal through the channel.

Figure 8.22 represents the SIMO scenario. In the model, it is assumed that the mobile transmitters are relatively simple with a single transmitting antenna and the base station can have a complex receiver with adaptive smart antennas containing K antenna elements.

As shown in Fig. 8.22, the multipath environment is such that up to L signals arrive at each base station antenna from different mobile terminals (l) with different amplitudes (α_l) and phases (ϕ_l) at different delays (τ_l) from different directions (θ_l). These are in general time-invariant, and as a result, the CIR for each antenna is usually represented by

$$h(t) = \sum_{l=1}^{L} [\alpha_l(t) e^{j\phi(t)}] \delta(t - \tau_l(t)) a(\theta_l(t)) \tag{8.8}$$

Here, it should be noted that the CIR is now a vector rather than a scalar function of time. Further, $a[\theta_l(t)]$ is an array response vector and will have K different components, if there are K antenna elements of the receiving antenna array. Thus, there are K CIRs each with L multipath components. The amplitudes are usually assumed to be Rayleigh distributed although they are now dependent on the array response vector $a[\theta_l(t)]$ as well.

When there are l antenna elements in a mobile terminal and one base station antenna element, it makes a MISO channel. In this case, the CIR is an $l \times 1$ matrix. When there are l mobile terminals transmitting at a time and K base station antenna elements to receive them all, it makes a MIMO channel. In this case, the CIR is an $l \times K$ matrix that associates a transmission coefficient between each pair of antennas for each multipath component.

The beamforming technique in MISO is analysed here. Beamforming is the directional reception of waves. It can be used to increase the average SNR through focusing energy into desired directions. Transmit beamforming achieves a diversity order of K and an antenna gain of K, the same as MRC with K receive antennas. However, for transmit beamforming, the transmitter must have the CSI. This presents us with a bit of a problem, because in order for the transmitter to have the CSI, the receiver must send it to the transmitter, unavoidably reducing the throughput.

A spatial diversity MIMO system does not improve the data rate. It deals only with improving the SNR condition.

Using Alamouti's scheme, we can achieve transmit diversity without having to provide the transmitter with the CSI. In 1998, Alamouti presented the simplest form of space–time block codes (STBC) for two transmitter antennas and one receiver antenna. The cods uses a complex orthogonal design in which the transmission matrix is square and satisfies the condition of complex orthogonality in both space and time dimensions. This is the only code that achieves full diversity gain with code rate $r = k/T = 1$, without sacrificing the data rate, if a block of length T encodes k symbols. The code rate of STBC measures the number of symbols per time slot it transmits on average over the course of one block.

The MIMO channel brings together and takes full advantage of the transmit and receive diversity. The CSI may not be required in this channel. Given multiple antennas, the spatial dimension can be exploited to improve the BER performance of the wireless link and the data rate depending on the application.

Comparison of Channel Capacities of Various Multi-antenna Systems

According to Shannon, the limit on the channel capacity is given by

$$C = W \log_2 (1 + SNR)$$

This is for the SISO system. Here, W represents the channel bandwidth.

For the SIMO system, we have M antennas at the receiver end. Suppose the signals received by these antennas have the same amplitude on average, then they can be added coherently to produce an M^2 times increase in the signal power. Hence, the increase in SNR is equivalent to $\dfrac{M^2 \cdot (\text{signal power})}{M \cdot \text{Noise}} = M \cdot SNR$.

So, the channel capacity becomes

$$C = W \log_2 (1 + M \cdot SNR) \tag{8.9}$$

For the MISO system, we have N transmitting antennas. The total transmitted power is divided into N branches. There is only one receiving antenna and the noise level is the same as in the SISO case.

Thus, the overall increase in SNR is approximately $\dfrac{N^2 \cdot (\text{signal power}/N)}{\text{Noise}} = N \cdot SNR$.

Thus, the channel capacity for this case is

$$C = W \log_2 (1 + N \cdot SNR) \tag{8.10}$$

SMART ANTENNA

Smart antenna systems used with spatial diversity exploit the concept of MIMO intelligently. A smart antenna can automatically change the directionality of its radiation patterns in response to its signal environment. Terms that are commonly associated with various aspects of smart antenna include phased array, space division multiple access (SDMA), spatial processing, digital beamforming, coherent combining, and adaptive antenna systems.

Smart antenna systems fall into two main categories: (a) switched-beam systems and (b) adaptive array systems. Both the systems direct a main lobe (or radio beam) towards individual users and attempt to reject interference or noise from the outside of that main lobe. In smart antennas, the data is transmitted over a vector channel.

Normally, a smart antenna system performs better in LOS or close-to-LOS systems. This is especially true when the optimization criterion depends explicitly on the angle of arrival (or departure) parameter. Alternatively, MIMO technology with spatial multiplexing can perform well in non-LOS (NLOS) but they really try to mitigate multipath rather than exploiting it.

The MIMO system can be viewed, in effect, as a combination of MISO and SIMO systems. In this case, it is possible to get approximately an MN-fold increase in the SNR, yielding a channel capacity equal to

$$C = W \log_2(1 + MN \cdot SNR) \tag{8.11}$$

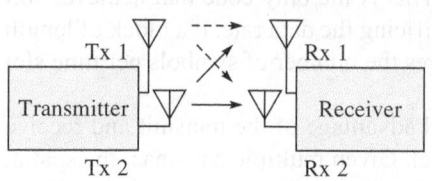

Fig. 8.23 MIMO using multiple transmitting and receiving antennas to send multiple signals over the same channel, multiplying spectral efficiency

MIMO beamforming is the signal processing to control the directionality of transmissions and receptions and depends upon factors such as the distance between Tx and Rx, number of antenna elements and spacing between each, signal strength, and orientation.

By analysing this equation, it can be concluded that the channel capacity is higher for the MIMO system than for the other systems.

8.6.2 Spatial Multiplexing in MIMO

While beamforming and receive combining are valuable enhancements to conventional radio systems, MIMO with spatial multiplexing (in comparison with Fig. 8.21c) is a paradigm shift, dramatically changing the perceptions of and responses to multipath propagation (Fig. 8.23). The information transmitted by both the antennas is different (shown with the help of separate dotted lines in the figure).

The underlying mathematical nature of spatial multiplexed MIMO, where data is transmitted over a matrix rather than a vector channel, creates new and enormous opportunities beyond just the added diversity or array gain benefits—the spectrum efficiency. This is explained in detail with the help of Fig. 8.24.

Here, Ai, Bi, and Ci represent the symbol constellations for the three inputs at the various stages of transmission and reception. A high-rate bit stream (the leftmost part of the block diagram) is decomposed into three independent rate bit sequences, which are then transmitted simultaneously using multiple antennas, thus consuming one-third of the nominal spectrum.

As such, this type of MIMO systems can be viewed as an extension of the smart antennas. A strong analogy can be made with CDMA transmission in which multiple users share the same time or frequency channel, which are mixed upon transmission and recovered through their unique codes. However, the advantage of MIMO is that the unique signatures of the input streams (virtual users) are provided by nature in a close-to-orthogonal manner (depending, however, on the fading correlation) without frequency spreading and hence at no cost of spectrum efficiency.

Another advantage of MIMO is the ability to jointly code and decode multiple streams, as they are intended for the same use. The MIMO channel relies on the presence of a rich multipath, which is needed to make the channel spatially selective.

8.6.3 Channel Modelling

A model for deriving the channel matrix for the system shown in Fig. 8.24 is given in Fig. 8.25.

It is common to model a wireless channel as a sum of two components—an LOS component and an NLOS component.

LOS component model The Rician factor is the ratio between the power of the LOS component and the mean

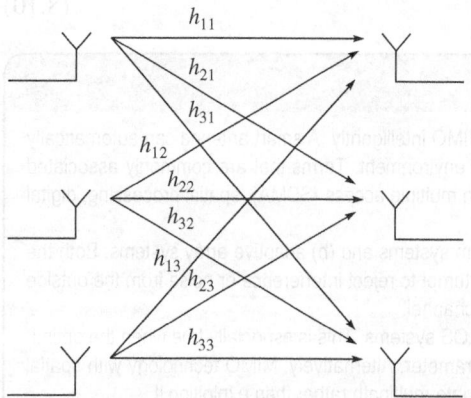

Fig. 8.25 Model for deriving a 3×3 channel matrix

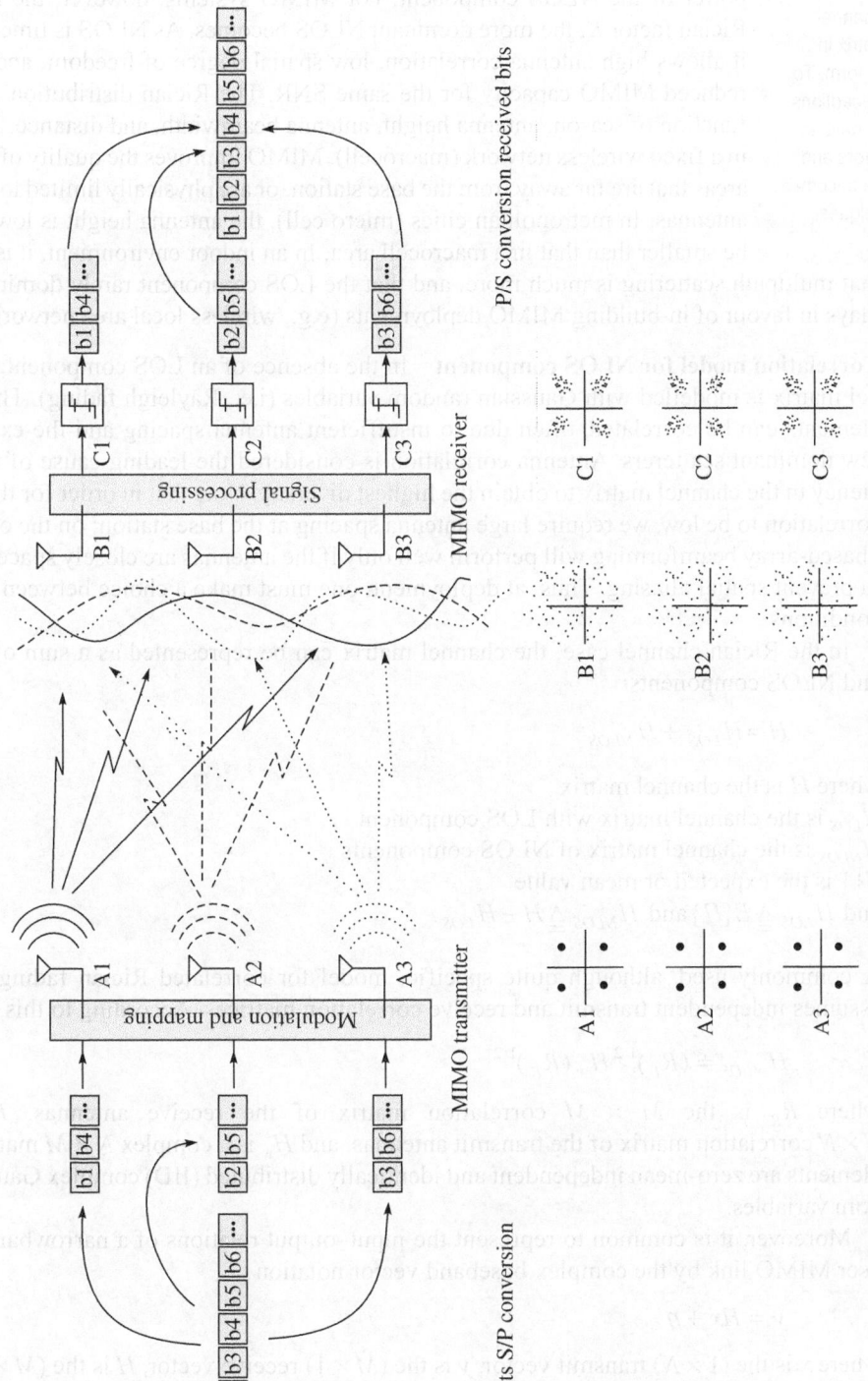

Fig. 8.24 Basic spatial multiplexing scheme with three Tx and three Rx antennas yielding threefold improvement in spectral efficiency

> The MIMO channel model is a matrix in mathematical form. To analyse the receptions through it, we have to use eigenvectors and singular-value decomposition techniques.

power of the NLOS component. For MIMO systems, however, the higher the Rician factor K, the more dominant NLOS becomes. As NLOS is time-invariant, it allows high antenna correlation, low spatial degree of freedom, and, hence, a reduced MIMO capacity for the same SNR. The Rician distribution factor is a function of season, antenna height, antenna beamwidth, and distance. Therefore, in a fixed wireless network (macrocell), MIMO improves the quality of service in areas that are far away from the base station, or are physically limited to using low antennas. In metropolitan cities (micro cell), the antenna height is low, factor to be smaller than that in a macrocell area. In an indoor environment, it is observed that multipath scattering is much more, and that the LOS component rarely dominates. This plays in favour of in-building MIMO deployments (e.g., wireless local area network).

Correlation model for NLOS component   In the absence of an LOS component, the channel matrix is modelled with Gaussian random variables (i.e., Rayleigh fading). The antenna elements can be correlated, often due to insufficient antenna spacing and the existence of few dominant scatterers. Antenna correlation is considered the leading cause of rank deficiency in the channel matrix to obtain the highest diversity. Note that in order for the antenna correlation to be low, we require large antenna spacing at the base station; on the other hand, phased-array beamforming will perform well only if the antennas are closely spaced in order to prevent spatial aliasing. Thus, at deployment, one must make a choice between these two conditions.

In the Rician channel case, the channel matrix can be represented as a sum of the LOS and NLOS components:

$$H = H_{LOS} + H_{NLOS} \tag{8.12}$$

where H is the channel matrix
H_{LOS} is the channel matrix with LOS component
H_{NLOS} is the channel matrix of NLOS components
$E\{\}$ is the expected or mean value
and $H_{LOS} \underline{\triangleq} E\{H\}$ and $H_{NLOS} \underline{\triangleq} H - H_{LOS}$.

A commonly used, although quite specific, model for correlated Rician fading channels assumes independent transmit and receive correlation matrices. According to this model,

$$H_{NLOS} = (R_T)^{1/2} H_w (R_R)^{1/2} \tag{8.13}$$

where R_R is the $M \times M$ correlation matrix of the receive antennas, R_T is the $N \times N$ correlation matrix of the transmit antennas, and H_w is a complex $N \times M$ matrix whose elements are zero-mean independent and identically distributed (IID) complex Gaussian random variables.

Moreover, it is common to represent the input–output relations of a narrowband, single-user MIMO link by the complex baseband vector notation

$$y = Hx + \eta \tag{8.14}$$

where x is the $(1 \times N)$ transmit vector, y is the $(M \times 1)$ receive vector, H is the $(M \times N)$ channel matrix, and η is the $(M \times 1)$ AWGN vector at a given instant of time. It is assumed that the channel matrix is random, the receiver has perfect channel knowledge, and the channel is memoryless. General entropy of the channel matrix is denoted by $\{h_{ij}\}$. This represents the complex gain of the channel between the jth transmitter and the ith receiver or indirectly the channel response. For MIMO system, the channel matrix is written as

$$\begin{pmatrix} h_{11} & \cdots & h_{1N} \\ h_{12} & & h_{2N} \\ \vdots & \ddots & \vdots \\ h_{1M} & \cdots & h_{MN} \end{pmatrix} \tag{8.15}$$

where $h_{ij} = \alpha + j\beta$ (complex representation).

In a rich scattering environment with no LOS, the channel gains $|h_{ij}|$ are usually Rayleigh distributed, where α and β are independent and normal distributed random variables.

8.7 CHANNEL ESTIMATION TECHNIQUES

> Channel estimation is an autoregressive process that may be performed with a number of iterations. An autoregressive model specifies that the output variable depends linearly on its own previous values.

In the detection part, we have two options to inverse the distortion due to channel. We can do channel estimation followed by channel inversion (or equalization, discussed in Section 8.8)—two separate tasks—or we can do channel equalization directly based on certain criterion such as minimum mean square error (MMSE). After one of the aforementioned processes is performed, decision-making is done for the bits based on the constellation of the transmitted signal.

For receivers based on channel estimation, to send the transmitter signal to the receiver with features to combat the channel problems, some characteristics of the channel must be estimated, either in terms of delay between the transmitter and the receiver or in terms of channel response. Channel estimation is the estimation of the CIR at the receiver. Channel estimators help in equalization at the receiver end. They usually need some kind of information as a reference, in terms of known training sequences, pilots, or some behavioural models or natural constraints. Channel estimation techniques may be divided into three categories.

CSI-based channel estimation In this technique, pilots are sent with data symbols. The effects on the pilot symbols represent the CSI indirectly. A pilot is basically a reference carrier/tone or a reference signal/symbol that is known at the receiver end in terms of position or sequence/pattern. It has undergone the most recent channel behaviour along with the other carriers or symbols. Hence, it provides the CSI and is therefore purposely and systematically transmitted and used for channel estimation.

Blind channel estimation Natural constraints are used in this technique. The likelihood function should be the model to pursue a completely blind approach. In most of these cases, Gaussian assumptions are used for the transmitted data, channel, and received data. Channel estimation is achieved by maximizing the likelihood function.

Notes: The following few points should be noted regarding channel estimation:
1. A channel estimate is only a mathematical estimation of what is truly happening in the natural environment. Many times, it utilizes channel statistics.
2. The estimation is performed using the received signal and is possible in both the directions in a bidirectional link.
3. Channel estimation allows the receiver to approximate the effect of the channel on the signal to eliminate the actual channel. It is essential for removing ISI and noise. It is used in diversity combining, maximum likelihood detection, angle of arrival estimation, and so on.
4. Efficient channel estimation strategies are required for coherent detection and decoding.

A detector, a channel
estimator, and an
equalizer may be
integrated in a system
to detect the signal and
compensate for phase
errors and to provide a
good SNR signal for the
demodulation process.

Semi-blind channel estimation   It is the combination of pilots and constraints.

Time-variant and frequency-selective fading channels present a severe challenge to the designer of a wireless communication system. A receiver plays a dual role to tackle this problem: phase correction by channel estimation or equalization and demodulation of the signal. Several choices are possible for the implementation of a receiver depending on the modelling of the channel and the complexity invested in each task. The estimated samples may have an estimation error that depends on the measurement noise, transmitted symbols, properties of the channel, estimation algorithm, and deviation from a time-invariant channel in the estimation interval.

Basics of Channel Impulse Response Estimation

In general, channel processes are wide-sense stationary. Conceptually, on a stationary wireless channel, if an impulse is transmitted, then multiple delayed versions of impulses will be received at the receiver at different instants of time. These impulses are non-correlated and with reducing amplitudes with time. They are just like delayed samples, and hence, a differential equation with coefficient values can be correlated with this concept. We have already studied this concept in wireless channel modelling with power delay profiles (Chapter 4).

Here, the radio channel is described by a discrete time transfer function, that is, a discrete time impulse response. The impulse response $h_e(n)$ is time-varying, and the goal of the identification procedure is to estimate the time-dependent parameters in $\{h_e(n)\}$ as accurately as possible. For the linear time-varying discrete systems, the received time domain signal $y(n)$ is a function of the transmitted signal $x(n)$, channel transfer function, and AWGN $w(n)$. It can be expressed as

$$y(n) = h_e(n) * x(n) + w(n) \tag{8.16a}$$

or

$$y(n) = \sum_{k=0}^{\infty} h_e(n)x(n-k) + w(n) \tag{8.16b}$$

where * denotes the convolution process. This equation represents one channel snapshot for the estimation interval. The whole time domain scenario can be mapped into the frequency domain, because convolution in the time domain represents multiplication in the frequency domain. Over short time intervals (a batch of data), the time-varying channel can be approximately described by a time-invariant impulse response. A further simplification is to assume that the transfer function $\{h_e(n)\}$ can be described by a time-invariant FIR model of length M in each time interval. The model is then modified to

$$y(n) = \sum_{k=0}^{M-1} h_e(n)x(n-k) + w(n) \tag{8.16c}$$

where M has to be chosen large enough to encompass all significant contributing paths. By expressing Eq. (8.16c) using the unit delay operator q^{-1} $[q^{-1}x(n) = x(n-1)]$, we obtain

$$y(n) = H(q^{-1})x(n) + w(n) \tag{8.16d}$$

where $H(q^{-1}) = \sum_k h_e(q^{-1})$. This model is a valid approximation for time segments that are short related to the channel variation.

For time-varying channels, this whole scenario will become dynamic, and in each snapshot or state, the CIR will be different. The next state of the channel can be estimated by the previous impulse responses. The channel model for time-varying channel can be approximated by the autoregressive model, also explained in Chapter 4. For next state estimation based on the present state condition, we can write

$$h(n+1) = ah(n) + w(n) \tag{8.17}$$

where a is the coefficient value to estimate the next channel state. Similarly, many delayed components of the channel will form the differential equation with weight coefficient a_i, where $i = 0,1,2$.

A comparison between the two main channel estimation methods is given in Table 8.1. However, a hidden Markov model based on the states or a linear prediction model based on the past record can also be used to estimate the present condition of the channel. Memory consumption is more in these cases.

Channel estimation methods are widely used in CDMA as well as OFDM systems. Wiener filter, Kalman filter, and so on can be used as estimation filters.

Table 8.1 Comparison between channel estimation methods

Training sequence based method	Blind method
Sequences known to the receiver are embedded into the frame and sent over the channel.	No training sequence is required but the method uses certain underlying mathematical properties of the data being sent.
It is easy and very popular. It consumes high bandwidth.	It is excellent for applications where bandwidth is scarce.
Computationally, it is not too intensive.	Computationally, it is extremely intensive.
The major drawback is that it is a waste of information bandwidth.	The major drawback is that it is hard to implement on real-time systems.

8.8 EQUALIZATION TECHNIQUES

Equalization or deconvolution is the process of recovery of a signal that is distorted by a convolution process during its transmission through a channel observed in additive noise. For this, a stage must be designed in the receiver such that its transfer function is exactly opposite to the channel transfer function. Figure 8.26 represents the basic concept of equalization.

The concept can be represented mathematically using the following equation:

$$r(n) = h(n) * s(n) + w(n) \tag{8.18}$$

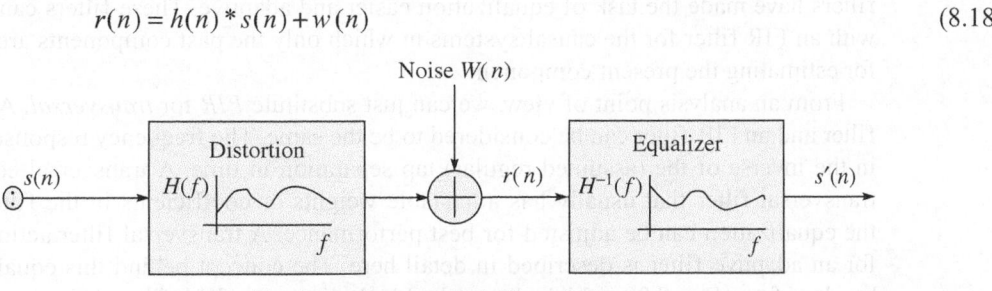

Fig. 8.26 Basic concept of equalization

Fig. 8.27 Channel equalization concept in a digital communication system using matched filter and decision-making

> An equalizer is an inverse channel filter to mitigate the unpredicted channel problems.

If noise is removed by a noise elimination filter and if the following operation is performed, the original signal can be detected with the removal of channel effects and phase ambiguities.

$$s'(n) = h^{-1}(n) * r(n) \tag{8.19}$$

From Eq. (8.19), it is clear that the equalization problem is relatively simple when the channel response is known and invertible and when the channel output is not noisy. However, in most practical cases, the channel response is unknown, time-varying, and non-linear and may also be non-invertible. Moreover, AWGN is observed in the output.

The simplest equalizer is the matched filter correlator. The use of a matched filter in channel equalization is given in Fig. 8.27. The equalized signal can be demodulated later.

The difference between the channel equalizer output and the desired signal is known as the *equalization error*. The equalization error signal or convolution noise is generated due to the white Gaussian noise present in the signal. If the channel distorts the pulse shape, the matched filter will no longer be matched perfectly, ISI may increase, and the system performance will degrade. The least square error (LSE), MMSE, linear mean square (LMS), or some such method may be used for checking the minimum error or maximum matching condition.

Digital communication systems can provide schemes for equalizer training periods during which a training pseudo-noise (PN) sequence, also known at the receiver, is transmitted. A synchronized version of the PN sequence is generated at the receiver, where the channel input and output signals are used for a channel estimation based equalizer by using some adaptive algorithm. The obvious drawback in using training periods for channel equalization is that power, time, and bandwidth are consumed for the equalization process.

8.8.1 Transversal Filters

Before the introduction of truly discrete digital filters, there were continuous time analog filters (such as delay lines) that were *tapped* at discrete points. The signal input appeared in delayed versions; it transversed in time (i.e., went across) as the delay line taps are viewed or accessed at one instant of time. Thus, they were identified as transversal filters. Digital filters have made the task of equalization easier and adaptive. These filters can be realized with an FIR filter for the causal systems in which only the past components are considered for estimating the present component.

From an analysis point of view, we can just substitute *FIR* for *transversal*. A transversal filter and an FIR filter can be considered to be the same. The frequency response is periodic in the inverse of the (assumed regular) tap separation in time. A transversal equalizer is a transversal filter that usually has adjustable weights or coefficients in the filter such that the equalization can be adjusted for best performance. A transversal filter acting as a basis for an adaptive filter is described in detail here. The concept behind this equalization will be clear from Fig. 8.28, which shows the block diagram of the filter. Here, $y_{(n)}$'s (arbitrary constants) represent the delayed versions of a discrete signal and w's represent the weights for coefficient adjustment.

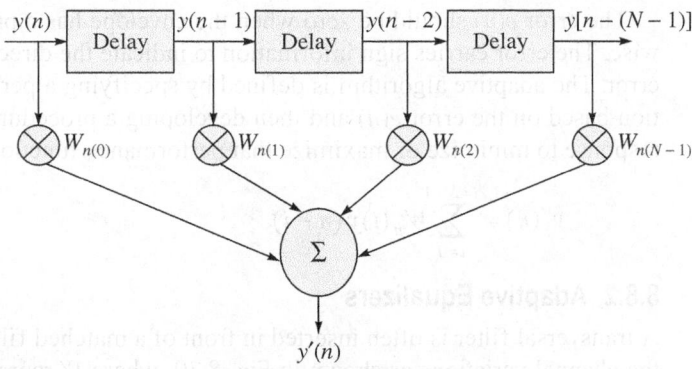

> A transversal filter equalizer is very similar to an autoregressive channel model.

Fig. 8.28 Transversal filter

In an N-tap transversal filter, the value of N is determined by practical considerations. An FIR filter is chosen because of its stability. Equations (4.14) and (4.15) represent the suitability of selection of an FIR filter. The use of the transversal structure allows relatively straightforward construction of the filter. As the input, coefficients, and output of the filter are all assumed to be complex valued, the natural choice for the property measurement is the modulus or instantaneous amplitude. If $y'(n)$ is the complex-valued filter output, then $|y(n)|$ denotes the amplitude. The convergence error $e(n)$ can be defined as follows:

$$e(n) = |y(n)| - A \tag{8.20}$$

where A is the amplitude in the absence of signal degradations.

MULTIPLIER-LESS TRANSVERSAL FILTER

Conventional transversal filters require a multiplying element for each tap. As multiplication is a time-consuming process, a multiplier-less transversal filter can be designed, which can carry out the multiplication in the logarithmic domain to save time. That is, all multiplications are replaced by additions. Such a filter is shown in Fig. 8.29.

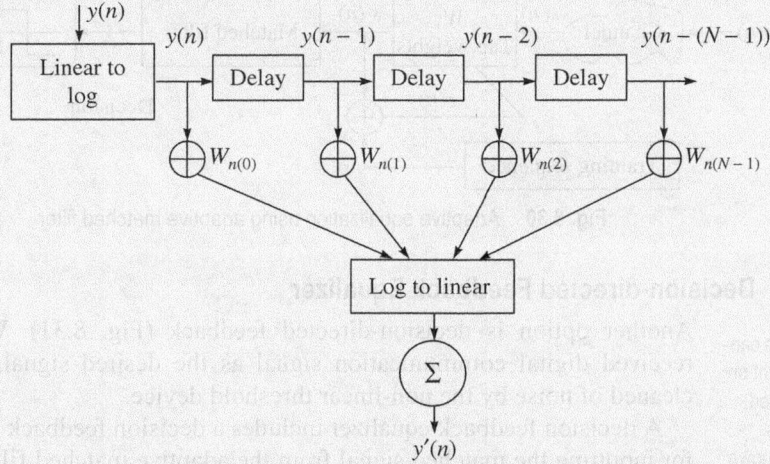

Fig. 8.29 Multiplier-less transversal filter (multiplications are replaced by additions)

The error $e(n)$ should be zero when the envelope has a proper value and non-zero otherwise. The error carries sign information to indicate the direction in which the envelope is in error. The adaptive algorithm is defined by specifying a performance, cost, or fitness function based on the error $e(n)$ and then developing a procedure that adjusts the filter impulse response to minimize or maximize that performance function.

$$y'(n) = \sum_{i=0}^{i=N-1} W_n(i)y(n-i) \tag{8.21}$$

8.8.2 Adaptive Equalizers

A transversal filter is often inserted in front of a matched filter correlator to compensate for the channel variations as shown in Fig. 8.30, where W represents the weights or the coefficients of the filters that can be varied in accordance with the channel conditions so that exact approximation of the channel is possible. In addition, we must note that the filter is usually linear, although modifying the coefficients very rapidly can cause the filter to become non-linear during the change.

This is, of course, unrealizable unless we have access to the original transmitted signal, here the training sequence. The training sequence can be used in this manner. There is a periodical broadcast of a known training signal. The adaptation is switched on only when the training signal is being broadcast and thus $s(n)$ is known.

An adaptive matched filter (which includes a transversal filter for inputting the digital baseband signal and a control signal generating circuit for supplying tap coefficients to the transversal filter) makes symmetric an asymmetric impulse response within the inputted digital baseband signal due to fading in the propagation paths and outputs a matched signal. This is achieved by having control over the transversal filter by the control signal generating circuit using tap coefficients. Tap coefficients are generated through correlation and time-average processing of the polarity signals indicating the polarities of the digital baseband signal and the output signal of the transversal filter.

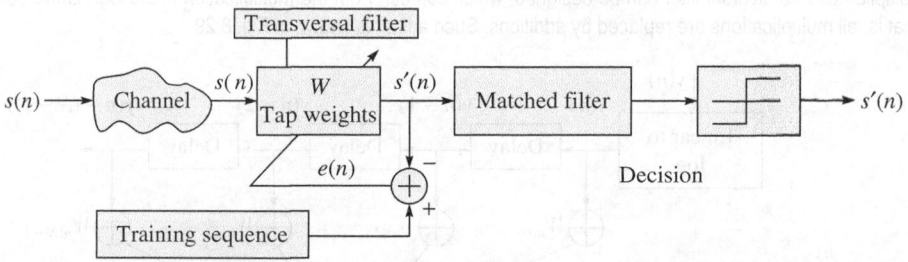

Fig. 8.30 Adaptive equalization using adaptive matched filter

8.8.3 Decision-directed Feedback Equalizer

Adaptive equalizers can control the tap weight on the basis of estimated channel coefficients from the received known training sequences.

Another option is decision-directed feedback (Fig. 8.31). We can use our received digital communication signal as the desired signal, as it has been cleaned of noise by the non-linear threshold device.

A decision feedback equalizer includes a decision feedback transversal filter for inputting the matched signal from the adaptive matched filter and a control signal generating circuit for supplying tap coefficients to the decision feedback

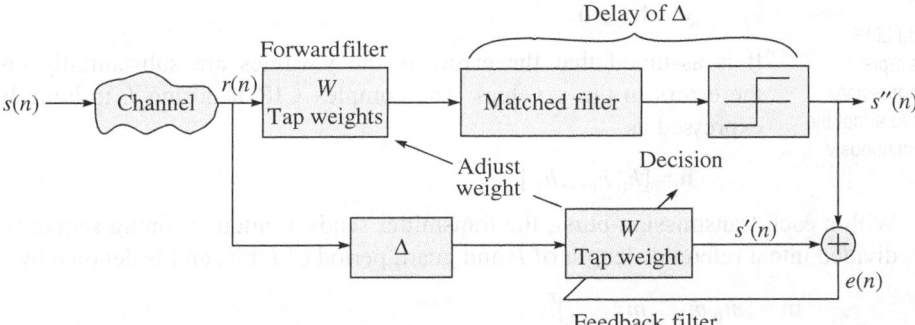

Fig. 8.31 Decision-directed feedback equalizer

BLIND EQUALIZATION

Blind equalization does not require known training sequences. To avoid transmission power and bandwidth consumption, it is preferable to have a blind equalization scheme that can operate without access to the channel input. It uses the Bayesian estimation algorithm. Furthermore, in some applications, such as blurred images, all that is available is the distorted signal and the only restoration method applicable is blind equalization. However, blind equalization is feasible only if some statistical knowledge of the channel input, and perhaps that of the channel, is available.

Blind equalization involves two stages: channel identification and deconvolution of the input signal and the channel response. Here, the channel identification or channel estimation is done by using some statistical or behavioural models.

---

transversal filter. This is achieved by having control over the decision feedback transversal filter by the control signal generating circuit using tap coefficients generated through correlation and time-average processing of a polarity signal included in the inputted matched signal and an error signal within the signal outputted after equalization. It eliminates from the inputted signal the ISI due to fading and outputs a signal after equalization.

As long as the error rate in $s(n)$ is not too high (say, 75%), this method works. Otherwise, $s'(n)$ is so inaccurate that the adaptive filter can never find the Wiener solution. This method is widely used in the telephone system and other digital communication networks.

8.9 LEAST SQUARES AND LEAST MEAN SQUARES ALGORITHMS

Two basic algorithms are introduced here for the understanding of the mathematics involved in the estimators and equalizers.

8.9.1 Least Squares Algorithms

Least squares (LS) means that the overall solution minimizes the sum of the squares of the errors made in the results of every single equation. The method of LS channel estimation is a standard approach to the approximate solution of overdetermined systems, that is, sets of equations in which there are more equations than unknowns.

The LS regression process can be understood with the help of the following mathematics. Consider a noise-corrupted communication system through a fading multipath channel **h**, after which the signal has memory of L symbols. Moreover, **n** is the white Gaussian noise, which is sampled at the symbol rate. The demodulation problem here is to detect the transmitted bits **x** from the received signal **y**, and hence, the estimation procedure is required. For a general linear equation of the received signal,

$$\mathbf{y} = \mathbf{hx} + \mathbf{n} \tag{8.22}$$

The LS and LMS algorithms represent the regression processes to adapt the system continuously.

It is assumed that the errors in the $\mathbf{y}$ values are substantially greater than the errors in the $\mathbf{x}$ values. The complex CIR $\mathbf{h}$ during L training bits can be expressed as

$$\mathbf{h} = [h_0 \, h_1 \ldots h_L]^T \tag{8.23}$$

Within each transmission burst, the transmitter sends a unique training sequence, which is divided into a reference length of P and guard period of L bits and is denoted by

$$\mathbf{m} = [m_0 \, m_1 \ldots m_{P+L-1}]^T \tag{8.24}$$

having bipolar elements $m_i \in \{-1, +1\}$. For estimation during the training period, the equation for the received signal can be rewritten as $\mathbf{y} = \mathbf{hm} + \mathbf{n}$. The circulant training sequence matrix can be formed as

$$\mathbf{M} = \begin{bmatrix} m_L & \cdots & m_1 & m_0 \\ m_{L+1} & \cdots & m_2 & m_1 \\ \vdots & \cdots & \vdots & \vdots \\ m_{L+P+1} & \cdots & m_P & m_{P-1} \end{bmatrix} \tag{8.25}$$

The vertical deviation can be calculated using the following formula:

$$\mathbf{d}_i = \mathbf{y}_i - \mathbf{y} = \mathbf{y}_i - (\mathbf{hm}_i + \mathbf{n}) \tag{8.26}$$

If the square of the deviations is minimized, the best line can be calculated as

$$\mathbf{d}_i^2 = (\mathbf{y}_i - \mathbf{y})^2 = (\mathbf{y}_i - \mathbf{hm}_i + \mathbf{n})^2 \tag{8.27}$$

In other way,

$$\hat{\mathbf{H}} = \arg_h \min \|y - Mh\|^2 \tag{8.28}$$

8.9.2 Least Mean Squares Algorithms

Least mean squares (LMS) algorithms are a class of adaptive filters used to mimic a desired filter by finding the filter coefficients that relate to producing the LMS of the error signal (difference between the desired and actual signals). It is a stochastic gradient descent method in which the filter is adapted based on the error at the current time. We have already explained this concept while discussing the adaptive equalization (Section 8.8.2). A simplified diagram explaining the algorithm is shown in Fig. 8.32. The following equations represent the formulation for the algorithm:

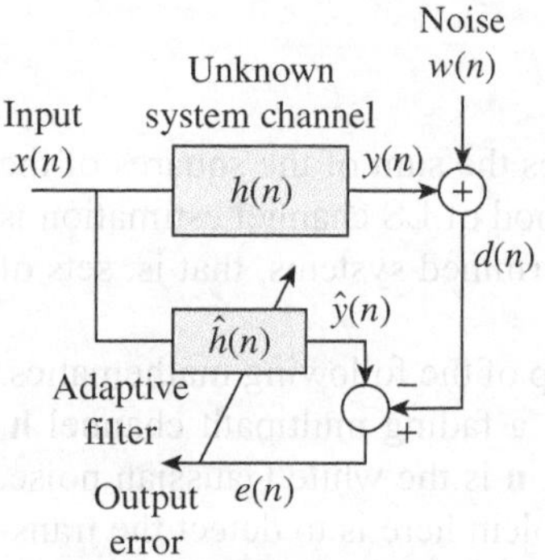

$$\mathbf{h}^H(n) = [h_0 * (n), h_1 * (n), \ldots, h_{p-1} * (n)] \tag{8.29}$$

(Hermitian transpose or conjugate transpose)

$$y(n) = \mathbf{h}^H(n) \cdot x(n) \tag{8.30}$$

$$d(n) = y(n) + w(n) \tag{8.31}$$

$$e(n) = d(n) - \hat{y}(n) = d(n) - \hat{\mathbf{h}}^H(n) \cdot x(n) \tag{8.32}$$

Fig. 8.32 LSM formulation diagram

The basic idea behind the LMS filter is to approach the optimum filter weights by updating the filter weights in a manner to converge to the optimum value. The algorithm starts by assuming a small weight (zero in most cases). At each step, the weights are updated by finding the gradient of the mean square error (MSE). That is, if the MSE gradient is positive, it implies that the error will keep increasing positively if the same weight is used for further iterations; hence, we need to reduce the weights. In the same way, if the gradient is negative, we need to increase the weights. Therefore, the basic weight update equation is

$$Wn + 1 = Wn - \mu \Delta \varepsilon(n) \tag{8.33}$$

where ε represents the MSE and μ is the step size or the gradient, W_n. A negative value indicates that we need to change the weights in a direction opposite to that of the gradient slope. The MSE as a function of filter weights is a quadratic function; this means that it has only one extreme value that minimizes the MSE, which is the optimum weight. Finally, we may write

$$\hat{h}(n + 1) = \hat{h}(n) + \mu e * (n) \cdot x(n) \tag{8.34}$$

A few other techniques, such as the MMSE estimator, also have multiple approaches. Correlation-based mathematical approaches are used in these techniques. Further theory and mathematics for estimators and equalizers are so wide that they are beyond the scope of this book.

MORE SOLVED EXAMPLES

Example 8.3 The CIR coefficients are given as $h(n) = [1, -2, 2, 1]$ (refer to Chapter 4 for channel modelling). Let us consider a small period discrete input signal $x(n) = [2, -1, 2]$. If the signal is transmitted through a channel, find the output signal $y(n)$. [If $x(n)$ is the training sequence known in advance from the received components, try to estimate the channel coefficients.]

Solution The signal passing through a channel undergoes convolution process. Hence,

$$y(n) = x(n)*h(n)$$

Now, using multiplication and shift method, if M represents the number of channel coefficients and N the samples in the period,

$x(n)$ \ $h(n)$	1	-2	2	1		
2	2	-4	4	2		
-1		-1	2	-2	-1	
2			2	-4	4	2
$y(n)$	2	-5	8	-4	3	2

Maximum length of $y(n) = M + N - 1 = 4 + 3 - 1 = 6$

These results are useful during simulation when channel coefficients represent the channel vector.

$h(n) = [h_0, h_1, h_2, ..., h_k]^T$, where, $h_0 ... h_k$ are the channel coefficients.

Here, $y(n) = [2, -5, 8, -4, 3, 2]$

[*Hint:* For estimating the channel, use $h(n) = y(n)/x(n)$ neglecting the noise part and also use discrete mathematics.]

Example 8.4 An eight-level pulse amplitude modulated pulses have a data rate of 9.6 kbps. Due to raised cosine filtering, the final bandwidth consumes 50% extra spectrum. Calculate the required bandwidth to fulfil this transmission. How will it affect the channel capacity?

Solution $W = \text{Symbol rate} \times (1 + \alpha)$

From Fig. 8.5, we can find that 50% extra bandwidth compared to the ideal requirement is consumed with the roll-off factor 0.5.

$$W = 9600 \times (1 + 0.5) = 9600 \times 1.5 = 14.4 \, \text{kHz}$$

The capacity will also increase by a factor of 1.5 (it can be tallied).

Example 8.5 A QPSK signal with 1 MHz bandwidth and 1 kW power is to be transmitted. Find the channel capacity of a simple SISO system and a 4 × 4 MIMO system. The noise density is 1×10^{-9} W/Hz.

Solution $SNR = 1 \times 10^3 / 1 \times 10^{-9} \times 1 \times 10^6 = 10^6 = 60$ dB

Channel capacity with SISO

$$C = W \log_2(1 + SNR)$$
$$\approx 1 \times 10^6 \log_2(10^6)$$
$$\approx 20 \, \text{Mbps}$$

Channel capacity with MIMO

$$C = W \log_2(1 + (4 \times 4)SNR)$$
$$\approx 1 \times 10^6 \log_2(16 \times 10^6)$$
$$\approx 24 \, \text{Mbps}$$

Example 8.6 A pure selection combiner with four branches of independent Rayleigh fading reception has an average SNR of just an acceptable range of 12 dB. With the combining, find the SNR improvement. With this improved SNR, find the probability of the drop of SNR below the acceptable SNR. If all the branches with the just acceptable SNRs are combined with a maximum ratio combiner, what is the SNR improvement for the maximum SNR?

Solution Let us define the terminology for selection combining.

SNR with just acceptable value = SNR_j
Improved average SNR = SNR_i
Number of branches $k = 1, ..., m$ (here, $m = 4$)
$SNR_j = 12$ dB = 15.85
For SNR improvement, it can be shown that

$$SNR_i = SNR_j \sum_{k=1}^{m} \frac{1}{k}$$

$$SNR_i = 15.85(1 + 1/2 + 1/3 + 1/4)$$
$$= 15.85(1 + 0.5 + 0.33 + 0.25)$$
$$= 15.85(2.08) = 32.97$$
$$SNR_i = 15.18 \, \text{dB}$$

Probability of dropping SNR is given by

$$P(12 \, dB) = [1 - e^{-(SNR_j/SNR_i)}]^m = [1 - e^{-(2.08)}]^4$$
$$= [1 - 0.125]^4 = [0.875]^4 = 0.58$$

With the MRC technique, after combining the just acceptable SNRs of all the four branches, the maximum SNR becomes $4 \times 15.85 = 63.4 = 18$ dB. Hence, 6 dB improvement is achieved.

SUMMARY

- All channels will act as low-pass filters (except a few cases in wireless) and spread a pulse due to some reason or the other, thereby causing ISI.
- Nyquist pulse is a sinc-shaped signal.
- Pulse shaping can be employed to remove spectral spreading or to improve spectral efficiency.
- Windowing can improve spectral efficiency at the cost of BER.
- A signal over an AWGN linear time-invariant channel can be detected optimally using a matched filter (simple equalization concept). If there are complicated channel assumptions, then better equalization techniques are required.
- There are two types of diversity techniques: macro and micro.
- Diversity techniques are used to combat the fading effects over a channel. They may be based on space, time, frequency, angle, polarization, site, and so on.
- Combining is the process to extract the main transmitted signal with minimum channel effects. This process is based on digital signal processing techniques in most of the cases.
- MRC gives the best performance but it requires complex hardware.

- MIMO is a multi-antenna system both at the transmitter and at the receiver side.
- MIMO exploits space diversity.
- MIMO based on spatial diversity improves channel capacity along with diversity gain, whereas MIMO based on spatial multiplexing improves spectral efficiency as well as speed of data communication.
- Channel estimation means estimation of CIR. This response is used for the equalization or compensation of the channel effects.
- Channel estimation can be done by using training sequences or using blind methods based on channel statistics.
- Equalization techniques are required for phase compensations, that is, indirectly channel compensations.
- Adaptive equalizers and decision feedback equalizers are more useful in fast fading channels where continuous corrections are required for the equalizations. However, they must be fast enough.
- The LS technique is basically for estimation whereas LMS is basically for adaptive filters.

EXERCISES

Multiple-choice Questions

8.1 The processes that is used to identify how the spectrums of two signals interact is
  (a) correlation
  (b) discrete Fourier transform
  (c) convolution
  (d) short-time Fourier transform

8.2 The windowing process will
  (a) increase the spectrum efficiency
  (b) decrease the out-of-band components
  (c) shape the spectrum
  (d) all of these

8.3 Which of the following processes can be used to detect the required signal from all the coexisting ones?
  (a) Cross correlation   (c) Auto correlation
  (b) Convolution   (d) Multiplexing

8.4 Which of the following processes can be used to reject an undesired signal from the channel?
  (a) Cross correlation   (c) Auto correlation
  (b) Convolution   (d) Windowing

8.5 In a space diversity reception system, at the output of which stage do we combine the outputs of several receivers?
  (a) RF amplifier
  (b) Detector
  (c) Intermediate frequency amplifier
  (d) Audio frequency amplifier

8.6 The most efficient combining method is
  (a) maximum ratio   (c) pure selection
  (b) equal gain   (d) threshold based

8.7 MIMO is based on
  (a) space diversity   (c) both (a) and (b)
  (b) spatial multiplexing (d) OFDM

8.8 MIMO channel capacity can be increased by
  (a) space diversity
  (b) STBC coding
  (c) increasing the number of antennas
  (d) more data

8.9 MIMO represents a _____ signal.
  (a) multidimensional   (c) two-dimensional
  (b) single-dimensional (d) none of these

8.10 The most important block in the equalizer design is the
  (a) integrator   (c) FIR filter
  (b) comparator   (d) matched filter

8.11 Which of the following method is hard to implement over real-time systems?

  (a) Pilot-based channel estimation
  (b) adaptive equalization
  (c) Blind channel estimation
  (d) none of these

8.12 Spread spectrum systems employ a form of diversity called
  (a) frequency diversity (c) spatial diversity
  (b) radio diversity   (d) bandwidth diversity

Review Questions

8.1 Why is the whole range of α not suitable for designing a raised cosine filter? What are the practical limitations?

8.2 Think about the windowing stage in the overall wireless link. Where should it be placed? Why?

8.3 Why does a channel act as a filter? Why is convolution process incorporated to find the output of a wireless channel? Will the channel always act as a filter?

8.4 Justify that diversity techniques definitely help combat the multipath channel.

8.5 Why do we require combining methods in space diversity? Do we require combining methods in the case of frequency, time, or polarization diversity? Why?

8.6 What are the types of diversity techniques used in cellular communication?

8.7 How are the various types of communication systems classified on the basis of the number of antennas used? Describe their mathematical representations for CIRs.

8.8 How does MIMO exploit multipath? Give the concept of spatial multiplexing.

8.9 How can we say that 'MIMO increases speed, range, and reliability'?

8.10 How does the smart antenna system differ from MIMO technology?

8.11 What is the difference between channel equalization and channel estimation?

8.12 What do you mean by weights in the case of adaptive equalizers? On what basis are they varied?

8.13 How is the training sequence useful in channel equalization as well as estimation?

8.14 Explain the working of a matched filter and justify its suitability for equalizers.

8.15 What is the complexity in the blind estimation method in comparison with the methods based on training sequence? How is the training sequence method disadvantageous?

Numerical Problems

8.1 Compare the channel capacities of the SISO, SIMO, MISO, and MIMO systems when a single-channel SNR is 12 dB and 1 MHz signal is to be transmitted. For MIMO, the transmitting antennas are four and the receiving antennas are also four.

8.2 Design a three-tap linear transversal equalizer for the received pulse $r(t)$.

$r(0) = 1, r(1) = 0.3, r(-1) = -0.3, r(2) = 0.1,$
$r(-2) = 0.2, r(3) = -0.03, r(-3) = -0.02$

Find the setting of the coefficient values.

9 Multiplexing and Multi-user Access

Theme of the Chapter

A communication system that is shared by many users should be designed efficiently because all users have to play within the limited allocated spectrum. Hence, a wireless channel is commonly shared by many applications. The efficiency of a system may be determined in terms of efficient utilization of the available spectrum, and all users must be treated equally. Multiplexing and multiple access schemes provide such efficient ways to design communication systems. The spectrum can be shared on a time-sharing or frequency-sharing basis, or any other basis described in this chapter, of which any one or a combination can be selected. For example, global system for mobile communication uses a combination of time division multiple access and frequency division multiple access schemes over space division multiple access.

Key Topics

- Multiplexing and multiple access
- FDM, TDM, CDM, and SDM
- Fixed allocation access: FDMA, TDMA, CDMA, and SDMA
- OFDMA multicarrier scheme
- Random access: ALOHA, Slotted ALOHA, CSMA/CD, ISMA, and DAMA
- Reservation-based access: PRMA, polling, token passing

9.1 MULTIPLEXING AND MULTIPLE ACCESS

When the multiple user informations are to be transmitted together, special techniques are required. This issue is resolved by the following schemes. The schemes may have their own pros and cons, and they may be selected as per the application requirement.

9.1.1 Multiplexing Schemes

In a multiplexing technique, allocation of frequency and time is fixed; hence, if the user does not use them, they will be a waste.

Multiplexing schemes are used when signals from multiple users are to be combined and sent on a single channel as a single input stream. Multiplexing is used to enable several users share a medium with minimum or no interference. All the incoming signals are upconverted or downconverted simultaneously and transmitted together as one signal. For wireless communication, multiplexing can be carried out in four dimensions: frequency, time, code, and space. Thus, there are

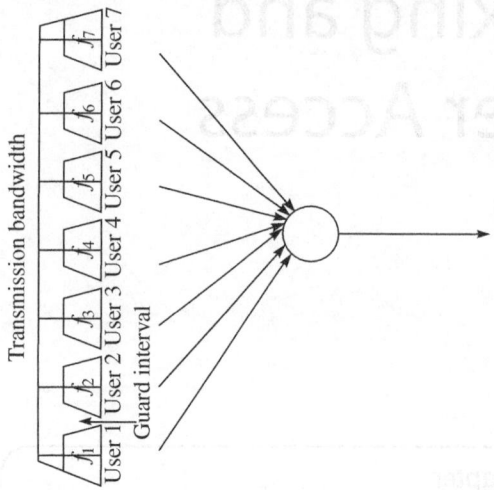

Fig. 9.1 FDM

In a synchronous TDM technique, the signal stream results in a frame once a cycle is over.

mainly four schemes for multiplexing based on the four possibilities:

(a) Frequency division multiplexing (FDM)
(b) Time division multiplexing (TDM)
(c) Code division multiplexing (CDM)
(d) Space division multiplexing (SDM)

The task of multiplexing is to assign frequency, time, code, or space to each user with minimum interference and maximum medium utilization. However, here the association of senders and receivers over the communication channel is the key feature. The concept of multiplexing indirectly gives birth to the concept of an individual's independent medium access.

Frequency division multiplexing In FDM, individual users are provided individual channels, which will in combination make the whole transmission bandwidth. For example, in Fig. 9.1, the seven user bandwidths make one transmission bandwidth.

Time division multiplexing In TDM, each individual user is pre-assigned a time slot in which he or she can send the information, and once that slot is over, the slot for the next user will start. The scenario for two users is given in Fig. 9.2. All users are scanned in a cyclic manner to collect the data. For n users, the bit rate of the TDM stream will increase n-fold. Time synchronization is a very important issue in TDM. There are two methods of TDM: *synchronous* and *asynchronous*.

Code division multiplexing The CDM scheme is shown in Fig. 9.3. It can be combined with orthogonal frequency division multiplexing (OFDM) to create hybrid configurations like orthogonal frequency and code division multiplexing (OFCDM). In CDM, separation is achieved by assigning each user channel its own code. Guard spaces are realized by using codes with the necessary distance in code space. Good protection against unauthorized reception is the main advantage of CDM.

Space division multiplexing In SDM, signals can be transmitted by different directional antennas, or the signals received by a multidimensional antenna can be combined to get

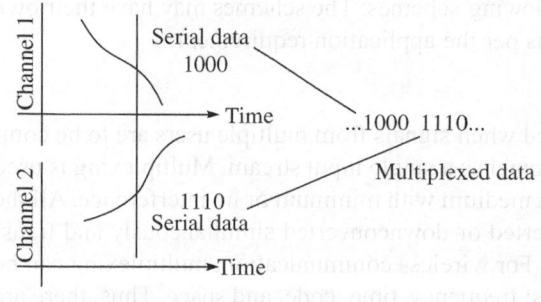

Fig. 9.2 TDM of two channels

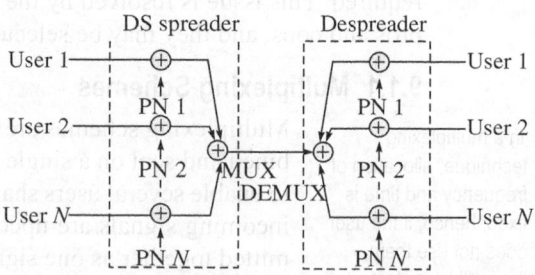

Fig. 9.3 CDM for *N* Users

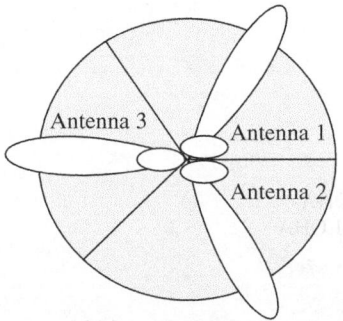

Fig. 9.4 SDM

An efficient sharing of spectrum schemes is required to increase the user capacity of a wireless network.

all of them back. Three directional antennas along with their lobes are shown in Fig. 9.4. Sometimes, differently polarized signals can be transmitted simultaneously. SDM is similar to a highway with a provision of different lanes.

9.1.2 Multiple Access Schemes

Multiple access schemes allow many simultaneous users to share the same available channel bandwidth or radio spectrum on an individual basis. In situations where channel-sharing or channel-access issues might arise, multiple access schemes are to be used. The available bandwidth in a radio system is always limited. For example, in mobile phone systems, the total bandwidth is typically 50 MHz, which is split in half to provide the forward and reverse links of the system. Further, the forward and reverse links must be established for thousands of subscribers.

Multiple access may be achieved by four different ways as follows:

(a) Fixed assignment of resources in terms of carrier allotment, time slot allocation, code allocation, or area allocation to specific users

(b) Demand assignment

(c) Random access, that is, a dynamic assignment of spectrum resources in time or bandwidth to the users, according to their needs or on the basis of demand

(d) Reservation-based access, where prior reservations intimate other users about the request of a particular user

Multiple access requirements can be visualized by Fig. 9.5(a) and a detailed classification of the important multiple access schemes is given in Fig. 9.5(b).

Frequency division multiple access (FDMA), time division multiple access (TDMA), code division multiple access (CDMA), and space division multiple access (SDMA) are the four major methods of multiple access by fixed assignment of resources to the users. These methods follow the same concepts of FDM, TDM, CDM, and SDM, correspondingly, with the only difference being that multiplexed information is transmitted on a combined basis, whereas multiple access is done on an individual basis.

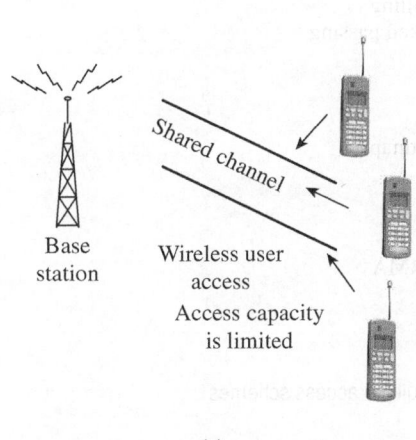

Base station

Wireless user access

Access capacity is limited

(a)

Fig. 9.5 Multiple access schemes (a) User access over shared channel

In Figs 9.6, 9.7, and 9.8, FDMA, TDMA, and CDMA are compared on the basis of bandwidth sharing, considering three users share the available channel bandwidth. This concept can be extended for *n* users. SDMA is a special case and is explained in Section 9.5.

The following points need to be noted:

- In FDMA, it is not possible for each user to use the entire bandwidth and only limited bandwidth is allocated.

Note: FDMA supports transmission of direct analog or digital data such as amplitude modulation (AM), frequency modulation (FM), and frequency shift keying (FSK) because FDMA requires no buffering.

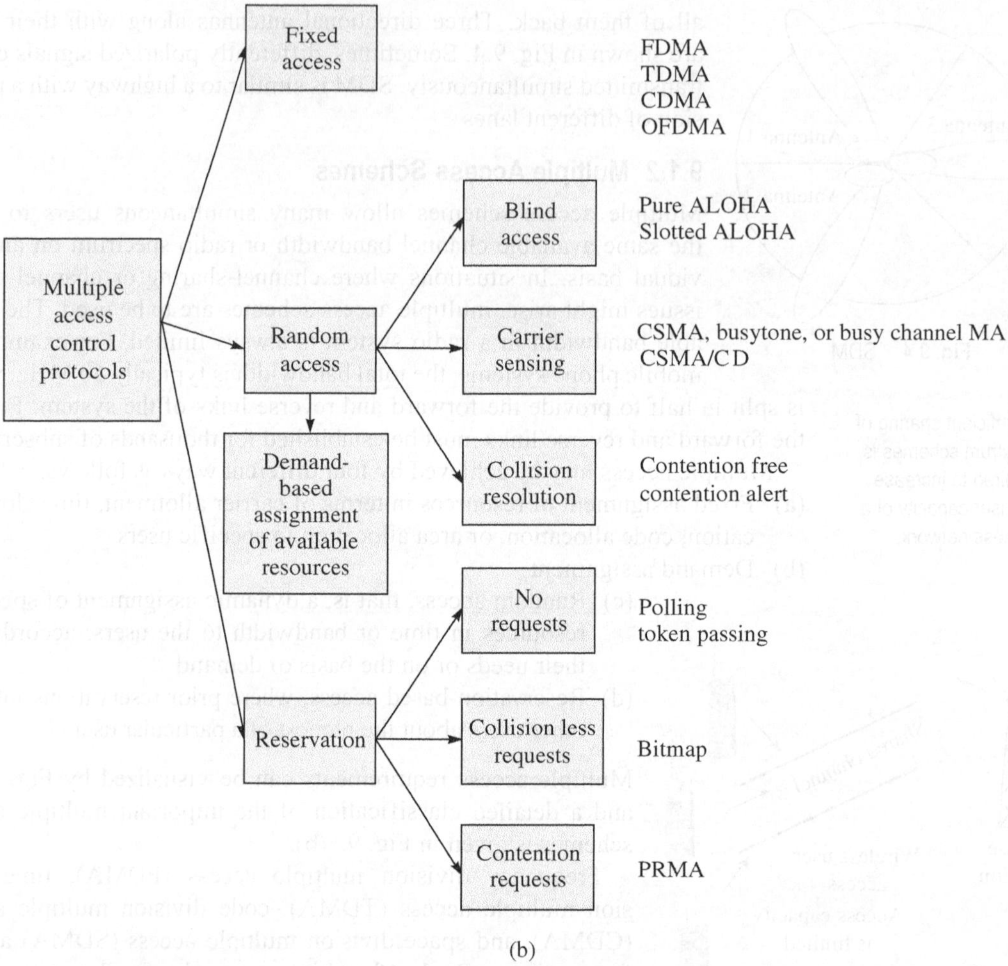

Fixed access
FDMA
TDMA
CDMA
OFDMA
Blind access
Pure ALOHA
Slotted ALOHA
Multiple access control protocols
Random access
Carrier sensing
CSMA, busytone, or busy channel MA
CSMA/CD
Demand-based assignment of available resources
Collision resolution
Contention free
contention alert
No requests
Polling
token passing
Reservation
Collision less requests
Bitmap
Contention requests
PRMA
(b)

Fig. 9.5 Multiple access schemes (b) Classification of multiple access schemes

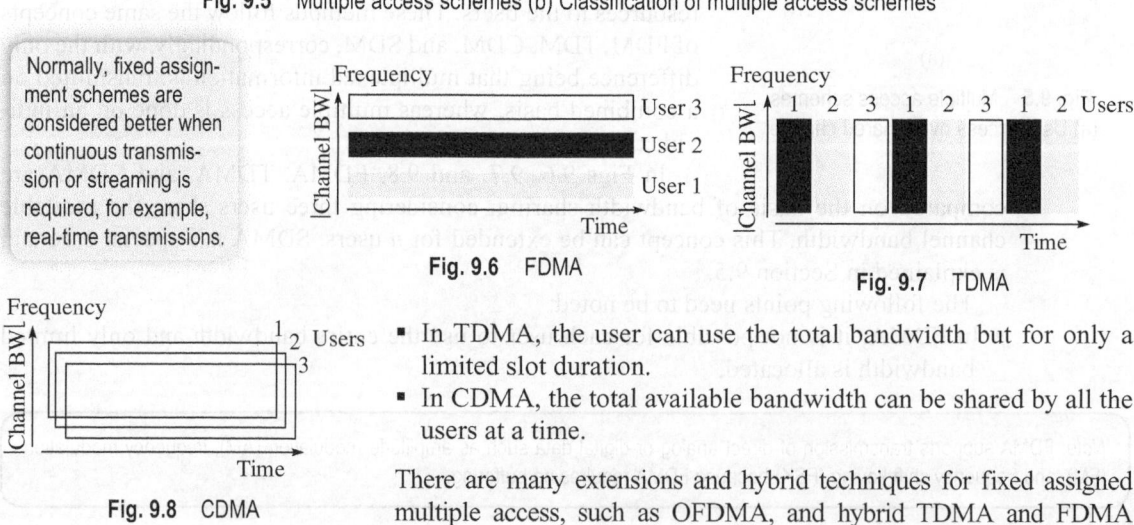

> Normally, fixed assignment schemes are considered better when continuous transmission or streaming is required, for example, real-time transmissions.

Frequency
Channel BW
User 3
User 2
User 1
Time

Fig. 9.6 FDMA

Frequency
Channel BW
1 2 3 1 2 3 1 2 Users
Time

Fig. 9.7 TDMA

Frequency
Channel BW
1
2 Users
3
Time

Fig. 9.8 CDMA

- In TDMA, the user can use the total bandwidth but for only a limited slot duration.
- In CDMA, the total available bandwidth can be shared by all the users at a time.

There are many extensions and hybrid techniques for fixed assigned multiple access, such as OFDMA, and hybrid TDMA and FDMA systems along with multiplexed data. However, an understanding of the three major methods is required for understanding any extension to these methods.

Random access mainly deals with packet radio and mostly storage data is utilized. Reservation-based access deals with channel reservation in advance whenever data transmission is needed. Demand-based channel assignment deals with the allocation of free channel at the time of request. All these techniques are discussed in this chapter.

9.2 FREQUENCY DIVISION MULTIPLE ACCESS

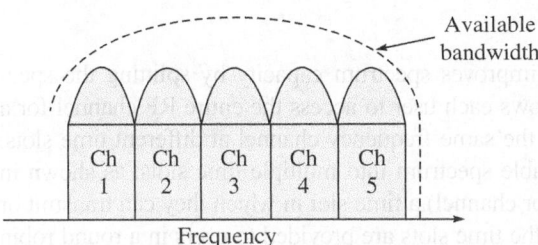

Fig. 9.9 Another way to represent FDMA compared to Fig. 9.6 (guard interval is implied between two channels)

Frequency division multiple access is the most common multiple access scheme that deals with radio frequency (RF) carriers. It is a technique whereby the spectrum is equally divided into frequencies and then assigned to different users, as shown in Fig. 9.9. Thus, the available bandwidth is subdivided into a number of narrower band channels. In FDMA, at any given time, only a single subscriber is assigned to a channel. The channel therefore is closed for other conversations (or until the conversation is handed off to a different channel in the case of the cellular system).

To avoid interference, FDMA transmission requires two one-way channels, one for transmitting and the other for receiving. A *full-duplex* or frequency division duplex (FDD) FDMA has been used since first-generation analog systems. Each user is allocated a unique frequency band in which to transmit and receive. No other user can use the same frequency band. The channel bandwidth used in most FDMA systems is typically low, as each channel needs to support only a single user and the associated baseband.

Limitations   The FDMA scheme has a few limitations. Carrier synchronization is required in FDMA because transmission is continuous, but with less overhead compared to TDMA. It also requires expensive filters to reduce adjacent channel interference. *Intermodulation* (IM) is another problem faced. The following are the other limitations of FDMA systems:

- Non-linearity in power amplifiers causes signal spreading in the frequency domain.
- There is undefined RF radiation that leaks into other channels.
- There is generation of undesirable harmonics that cause interference to other users in the mobile system or other systems in adjacent spectrum bands.

Capacity

Suppose the number of channels is N_c, total bandwidth is $W_{channel}$, and guard band is W_{guard}, and each channel has a bandwidth W_{signal}, then the FDMA capacity is given by

$$N_c = \frac{W_{channel} - (N_c - 1)W_{guard}}{W_{signal}} \tag{9.1}$$

More specifically,

$$N_u = N_c - N_{cch} \tag{9.2}$$

where N_u is the number of users supported and N_{cch} is the number of control channels.

Example 9.1 Consider an analog mobile phone system (AMPS) uplink (mobile to base station). The total available bandwidth is 12.5 MHz, per channel bandwidth is 30 kHz, and guard band is 10 kHz. 21 channels are used for control (setting up voice calls). Find the number of user channels.

Solution $W_{channel} = 12.5$ MHz, $W_{guard} = 10$ kHz,

$W_{signal} = 30$ kHz

So, $N_c = \dfrac{12.5 \times 10^6 - (N_c - 1) \times 10^4}{30,000}$

or $N_c \approx 312$ channels

$N_u = 312 - 21 = 291$ channels can be assigned to the users.

9.3 TIME DIVISION MULTIPLE ACCESS

> TDMA can suffer from multipath effects, as the transmission rate is normally high, resulting in significant intersymbol interference (ISI).

Time division multiple access improves spectrum capacity by splitting the spectrum's use into time slots. It allows each user to access the entire RF channel for a short period. Other users share the same frequency channel at different time slots. Thus, TDMA divides the available spectrum into multiple time slots, as shown in Fig. 9.10, by giving each user (or channel) a time slot in which they can transmit or receive. The figure shows how the time slots are provided to users in a round robin manner making time frames, with each user being allotted one time slot per frame.

The base station continually switches users on the channel. TDMA may be asynchronous or synchronous, just like TDM. It is the dominant technology in the second-generation mobile cellular networks.

Time division multiple access systems transmit data in a buffer and hence this is a bursty communication method. Thus, the transmission of each channel is non-continuous. The input data to be transmitted is buffered over the previous frame and burst transmitted at a higher rate during the time slot for the channel. As TDMA cannot send analog signals directly due to buffering requirement, it is used for transmitting digital data. Table 9.1 lists the advantages and disadvantages of a TDMA system.

Capacity

Suppose the total number of slots is N_s and the number of users per channel is n, then the TDMA capacity is given by

$$N_s = \frac{n[W_{channel} - (N_c - 1)W_{guard}]}{W_{signal}} \tag{9.3}$$

This is rather an FDMA–TDMA case in general.

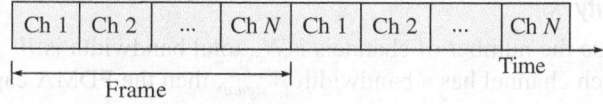

Fig. 9.10 TDMA frame generation

Example 9.2 Find the total number of slots for a GSM having 25 MHz forward link. Per channel bandwidth is 200 kHz and eight speech channels are supported per radio channel. No guard band is assumed.

Solution Given $W_{channel} = 25$ MHz, $W_{guard} = 0$, and $W_{signal} = 200$ kHz. Hence,

$$N_s = 8 \times 25 \times 10^6 / 200 \times 10^3 = 1000 \text{ slots}$$

Table 9.1 Advantages and disadvantages of TDMA

Advantages	Disadvantages
It is cheaper than FDMA as narrowband filters are not required.	It requires guard time between time slots to separate the users and accommodate time inaccuracies due to clock instability, delay spread of transmitted symbols, and transmission time delay.
Mobile devices can save battery power by turning off the transmitter and receiver during the slots when they are not active.	
It does not require a duplexer or multiple antennas for transmitting and receiving even when using FDD.	It requires signal processing techniques and high overhead for synchronization due to burst transmissions.
It can allocate multiple time slots to a user to provide increased data rate.	

For GSM, the total bandwidth of 25 MHz is divided into 125 channels, each of 200 kHz, using FDMA. These channels are subdivided further by using TDMA so that each 200 kHz channel allows 8-16 users.

Generally, TDMA is used in conjunction with FDMA to subdivide the total available bandwidth into several channels. This is done to reduce the number of users per channel, thereby allowing a lower data rate to be used. This helps reduce the effect of delay spread on the transmission. Figure 9.11 shows the use of TDMA with FDMA. Each channel based on FDMA is further subdivided using TDMA so that several users can transmit in the single channel. This type of transmission technique is used by most digital second-generation mobile phone systems.

So far, we have explained the fixed TDMA pattern. If a bidirectional link is to be handled or if a duplex channel between a base station and a mobile station is to be established, another approach is utilized in which different slots are assigned for uplink and downlink using the same frequency. This method is called *time division duplex* (TDD) and is used in digital enhanced cordless telecommunication (DECT) systems and universal mobile telecommunication systems (UMTS).

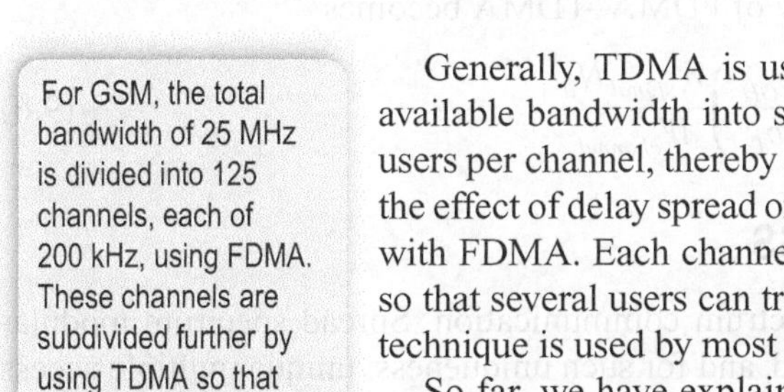

Spectral Efficiency

Frequency division multiple access requires guard bands in frequency to avoid adjacent channel interference; it reduces the overall channel bandwidth for data transmission. TDMA requires guard bands and synchronization sequences to reduce the time for data transmission. Spectral efficiency is measured by the ratio of the total time–frequency domain dedicated for voice or data transmission to the total time frequency domain available to the system. The spectral efficiency of an FDMA system is given by

Fig. 9.11 TDMA–FDMA hybrid (the bandwidth is split into frequency channels and time slot)

$$\eta_{fdma} = \frac{W_{signal} N_u}{W_{channel}} \leq 1 \tag{9.4}$$

For a TDMA system, if
N_r = Number of reference bursts per frame
b_r = Number of overhead bits per reference burst

N_t = Number of data slots per frame
b_p = Number of bits in each slot preamble
b_g = Number of bits in each guard interval
T_f = Frame time
R = Channel bit rate

and if overhead bits $b_{OH} = N_r b_r + N_t b_p + N_t b_g + N_r b_g$ and total bits $b_T = T_f R$, then the spectral efficiency is given by

$$\eta_{tdma} = \left(1 - \frac{b_{OH}}{b_T}\right) \leq 1 \tag{9.5}$$

Total available bandwidth transmits smaller channels. Each of the N_u channels that are used for data transfer by FDMA are broken into time frames. b_{OH} is the overhead of the total bits b_T. Then, the combined spectral efficiency of FDMA–TDMA becomes

$$\eta_{fdma-tdma} = \eta_{fdma} \times \eta_{tdma} = \left(1 - \frac{b_{OH}}{b_T}\right) \frac{W_{signal} N_u}{W_{channel}} \tag{9.6}$$

9.4 SPREAD SPECTRUM MULTIPLE ACCESS

There are various methods of spread spectrum communication. Spread spectrum modulation (SSM) is a unique modulation method, and for such uniqueness, unique multiple access methods are required. Depending upon the method of modulation selected, different multiple access methods can be derived.

Notes: There are mainly two multiple access schemes that are very popular with SSM techniques:
1. *Direct sequence SSM:* It enables CDMA with sufficient fading rejection and security.
2. *Frequency hopping SSM:* It enables frequency hopped multiple access (FHMA).

9.4.1 Code Division Multiple Access

Code division multiple access is used along with SSM, which uses neither frequency channels nor time slots. In SSM, spreading is achieved through the pseudo-noise (PN) code. If a unique code is assigned to each individual user, demodulation will be possible only if the code matches at the receiver end. This enables multiple access. The users in a CDMA system use the same frequency band and transmit simultaneously. They can use the entire available bandwidth all the time. The transmitted signal is recovered by correlating the received signal with the PN code used by the transmitter. Figure 9.12 shows the general use of the spectrum using CDMA. At one frequency, four users with different codes are shown sharing the same bandwidth.

The following are some of the properties that have made CDMA useful:

- Signal hiding and non-interference with the existing systems
  - Anti-jam and interference rejection
  - Information security
  - Accurate ranging
  - Multipath tolerance

CDMA suffers from near-far problem, which is solved by adjusting the transmitter power using a closed loop system.

For many years, spread spectrum technology was considered solely for military applications. However, with rapid development in large scale integration (LSI)

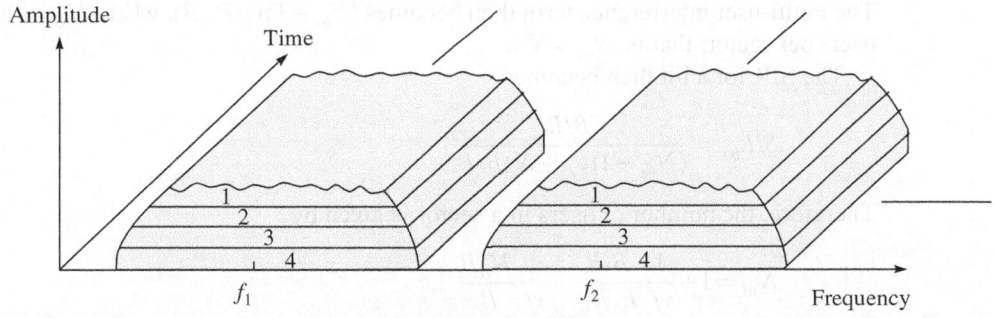

Fig. 9.12 CDMA

> The dominant radio interface for the third-generation mobile, or IMT-2000, is the wideband version of CDMA.

and very large scale integration (VLSI) designs, commercial systems have also developed. CDMA increases spectrum capacity by allowing all the users to occupy all the channels at the same time. Transmissions are spread over the entire radio band, and each voice or data call is assigned a unique code to differentiate it from the other calls carried over the same spectrum. CDMA allows for a *soft hand-off*, which means that terminals can communicate with several base stations at the same time.

Capacity

If all N users received power P_r and there are $N-1$ interferers and no channel noise, the signal-to-interference ratio (SIR) is calculated as

$$SIR = \frac{P_r}{(N-1)P_r} = 1/(N-1) \tag{9.7}$$

$$SIR_{bit} = \frac{E_b}{I_o} = \frac{P_r/R}{(N-1)(P_r/B)} = \frac{B/R}{(N-1)} \tag{9.8}$$

Including channel noise, SIR is calculated as

$$SIR_{bit} = \frac{E_b}{I_o} = \frac{P_r/R}{(N-1)(P_r/B)+N_o} = \frac{B/R}{(N-1)+N_oB/P_r} \tag{9.9}$$

Hence, the number of users in a cell

$$N = 1 + \frac{B/R}{E_b/I_o} - \frac{N_oB}{P_r} \tag{9.10}$$

We can increase the capacity of a direct sequence CDMA (DS-CDMA) system as follows:

(a) *Cell sectorization*: If a cell is split into s sectors, then the capacity increases by a factor s. (This is applied to other multiple access schemes too.)

(b) *Voice activity factor*: Users do not speak continuously, so the transmitter should be turned off when silence is detected. The percentage of time a user is speaking is equal to the voice activity factor v_f. (FDMA and TDMA cannot take advantage of voice activity factor easily.)

The multi-user interference term then becomes $(N_{us} - 1)v_f(P_r/B)$, where N_{us} is the number of users per sector; that is, $N_{us} = N/s$.

The SIR for a bit then becomes

$$SIR_{bit} = \frac{B/R}{(N_{us} - 1)v_f + N_o B/P_r} \tag{9.11}$$

Therefore, the number of users in a sector is given by

$$N_{us} = 1 + \frac{1}{vf}\frac{B/R}{E_b/I_o} - \frac{1}{vf}\frac{N_o B}{P_r} \tag{9.12}$$

Example 9.3 The parameters for an IS-95 system are as follows: total available bandwidth = 1.25 MHz, bit rate = 9600 bps, SIR_{bit} = 10 dB, s = 3 sectors, and v_f = 3/8. Find the number of users supported, avoiding interference from one another. Ignore channel noise.

Solution Total number of users = Number of sectors × Number of users per sector

Hence,

$$N = 3\left(1 + \frac{1}{3/8}\frac{1,250,000/9600}{10}\right)$$

$$= 107\,\text{users}$$

9.4.2 Frequency Hopped Multiple Access

Frequency hopping spread spectrum (FHSS) is a method of transmitting radio signals by rapidly switching a carrier among many frequency channels using a pseudo-random sequence known to both the transmitter and the receiver. Such spread spectrum signals are difficult to intercept. An FHSS signal simply sounds like an increase in the background noise to a narrowband receiver. In FHSS, frequency is constant in each time chip but changes from chip to chip. There may be two types of systems: *slow hopping* and *fast hopping*.

Different users are allocated different codes, that generates a frequency hopping pattern and hence, the carrier sequence generated will also be different (Fig. 9.13). There will be less possibility that the same carrier transmitted from different users interferes with each other. Thus, multiple access on the basis of different hopping patterns will be possible.

> Like FHMA, there can be a time hopping multiple access system.

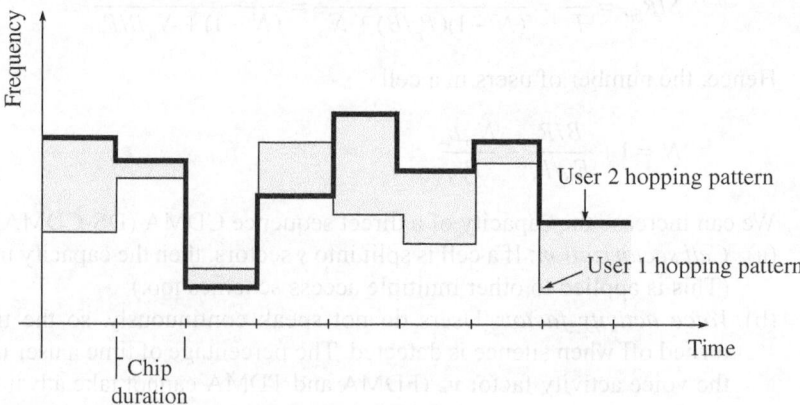

Fig. 9.13 Hopping patterns for an FHMA scheme

9.5 SPACE DIVISION MULTIPLE ACCESS

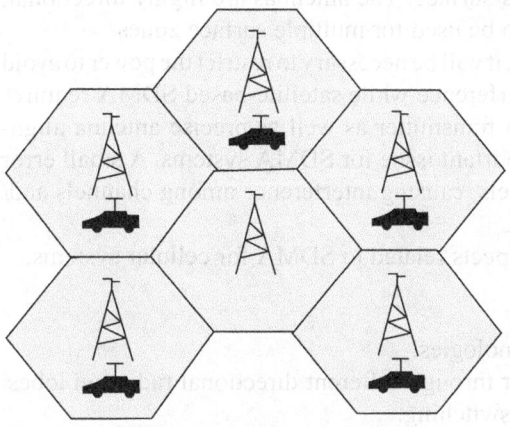

Fig. 9.14   SDMA scenario

Space division multiple access is geographical or cellular, as shown in Fig. 9.14. The idea behind the concept is that if two transmitter–receiver pairs are far enough to interfere, they can operate on the same frequency (by reusing the carrier) without interfering with each other. Chapter 2 explains the fundamentals of cellular technology, which utilizes the geographical division to accommodate all the users belonging to different areas. Cell site design is a very important factor in mobile communication networks design if SDMA is to be realized.

An SDMA system uses a directional antenna for splitting the coverage, as explained in Section 2.9 of Chapter 2. It is also a satellite communication mode that optimizes the use of radio spectrum and minimizes system cost by taking advantage of the directional

FREQUENCY REUSE IN SDMA FOR MOBILE SATELLITE COMMUNICATION

Consider a scenario in which signals must be transmitted simultaneously by a satellite to various portable wireless receivers in 20 different surface zones. In a conventional system, 20 channels and 20 antennas would be necessary to maintain channel separation. In contrast, SDMA can have fewer channels than zones. If duplicate channel zones are sufficiently separated, then 20 signals can be transmitted to earth using three or four channels. The narrow signal beams from the satellite antennas ensure that interference will not occur between the zones using the same frequency.

In Fig. 9.15, it is shown that the fourth zone is using the frequency of the first zone as they are far enough to interfere. Using this concept of frequency reuse, SDMA can support a large number of zones with a few channels.

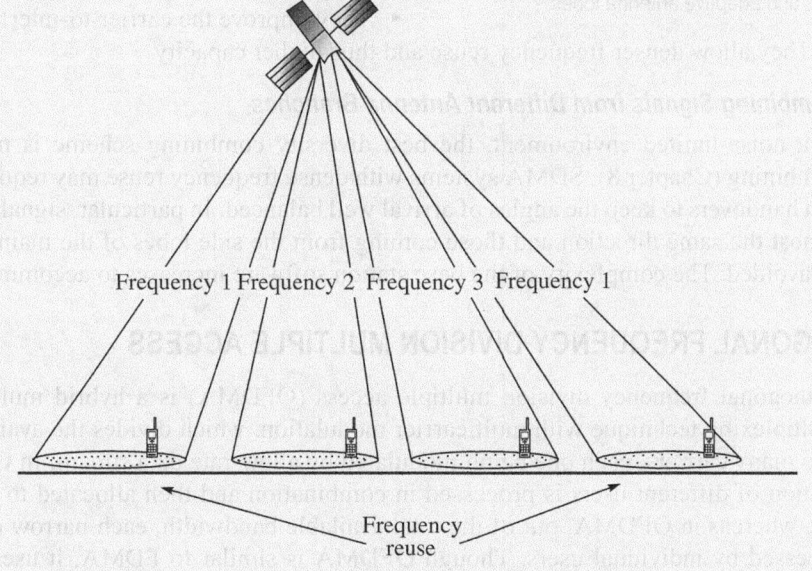

Frequency 1 Frequency 2   Frequency 3   Frequency 1

Frequency reuse

Fig. 9.15   Frequency reuse concept in SDMA for a satellite system

SDMA is very much dependent upon division of zones, frequency reuse, and antenna technologies.

properties of dish antennas. In SDMA, satellite dish antennas transmit signals to numerous zones on the earth's surface. The antennas are highly directional, allowing duplicate frequencies to be used for multiple surface zones.

In the case of cellular structure, it will be necessary to restrict the power to avoid co-channel and interchannel interference while satellite-based SDMA requires careful choice of zones for each transmitter as well as precise antenna alignment. Hence, antenna design becomes an important issue for SDMA systems. A small error can result in the failure of one or more channels, causing interference among channels and/or confusion between surface coverage zones.

The following are the various important aspects related to SDMA for cellular systems.

Antenna Technologies

The following are the important antenna technologies:

1. Beam switching is used to switch the user through different directional radiation lobes. A selector circuit is used to control beam switching.
2. Adaptive antennas at the base station estimate the location of the mobile terminal, and they receive and transmit only from and to this desired direction. An adaptive antenna system is also called beam steering.
3. Sectorization is done by using multiple highly directional antennas placed orthogonally to radiate in their corresponding limited time zone.

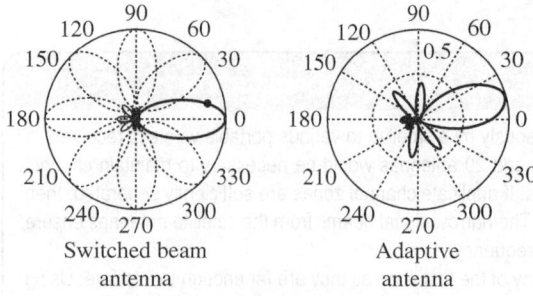

Fig. 9.16 Difference between switched beam antenna lobes and adaptive antenna lobes

The scenario of beam switching and adaptive antennas is shown in Fig. 9.16. In addition to achieving a multiple access scenario, these technologies help in the following:

- They improve range.
- They improve the carrier-to-interference ratio.
- They allow denser frequency reuse and thus higher capacity.

Combining Signals from Different Antenna Branches

In a noise-limited environment, the best diversity combining scheme is maximum ratio combining (Chapter 8). SDMA systems with dense frequency reuse may require many intra-cell handovers to keep the angles of arrival well balanced. In particular, signals arriving from almost the same direction and those coming from the side lobes of the main beams should be avoided. The complexity of the base station software increases to accommodate SDMA.

9.6 ORTHOGONAL FREQUENCY DIVISION MULTIPLE ACCESS

Orthogonal frequency division multiple access (OFDMA) is a hybrid multiple access or multiplexing technique with multicarrier modulation, which divides the available spectrum into many carriers, each one being modulated by a low-rate data stream. In OFDMA, information of different users is processed in combination and then allocated to multiple carriers, whereas in OFDMA, out of the total available bandwidth, each narrow channel can be accessed by individual users. Though OFDMA is similar to FDMA, it uses the spectrum much more efficiently by spacing the channels much closer together. This is achieved by making all the carriers orthogonal to one another, thereby preventing interference between the closely spaced carriers.

9.6.1 Comparison of Different Multiple Access Techniques

The FDMA, TDMA, and OFDMA techniques are compared in this section. The comparison shows how OFDMA is superior to the other techniques.

FDMA In FDMA, each user is allocated a single channel, which is used to transmit all the user information. The bandwidth of each channel is typically 10–30 kHz for voice communications. However, the minimum required bandwidth for speech is only 3.1 kHz. The allocated bandwidth is made wider to prevent channels from interfering with one another. This extra bandwidth is to allow for signals from neighbouring channels to be filtered out and to allow for any drift in the centre frequency of the transmitter or receiver. In a typical system, up to 50 per cent of the total spectrum is wasted due to extra spacing between different frequency bands. Most digital telephone systems use vocoders to compress digitized speech. They increase the system capacity due to a reduction in the bandwidth required for each user. Current vocoders require a data rate approximately within 4–13 kbps, so each user requires a minimum bandwidth of only about 2–7 kHz, using quadrature phase shift keying (QPSK) modulation. However, simple FDMA does not handle such narrow bandwidths very efficiently.

TDMA TDMA partly overcomes the problem of wasted bandwidth by using wider channels, which are used by several users. Multiple users access the same channel by transmitting their data in different time slots. Thus, many low data rate users can be combined together to transmit in a single channel that has sufficient bandwidth so that the spectrum can be used efficiently. There are however, two main problems in TDMA. Firstly, there is an overhead associated with the changeover between users due to time slotting on the channel. A changeover time must be allocated to allow for any tolerance in the start time of each user due to propagation delay variations and synchronization errors. It limits the number of users that can be sent efficiently in each channel. Secondly, the symbol rate of each channel is high (as the channel handles the information from multiple users), resulting in problems with multipath delay spread.

OFDMA OFDMA overcomes most of the problems encountered in both FDMA and TDMA systems. It splits the available bandwidth into many narrowband channels (typically 100–8000 depending upon the application). The carriers for each channel are made orthogonal to one another, allowing them to be spaced very close together, with almost no overhead (unlike FDMA). Thus, there is no great need for the users to be time multiplexed as in TDMA and consequently no overhead is associated with switching between the users. Each carrier in an OFDMA signal has a very narrow bandwidth and therefore the resulting symbol rate is low. This results in the signal having a high tolerance to multipath delay spread, as the delay spread must be very long to cause significant ISI (e.g., greater than 100 µs).

> OFDM deals with subcarriers, whereas OFDMA deals with subchannels.

In short, information of multiple users can be transmitted, jointly or on an individual basis using multiple orthogonal carriers in OFDM or OFDMA. Figure 9.17 shows that better spectral efficiency can be obtained with more number of carriers for more number of users in the case of OFDMA, where all the carriers are orthogonal to one another. Guard interval is omitted without loss of generality.

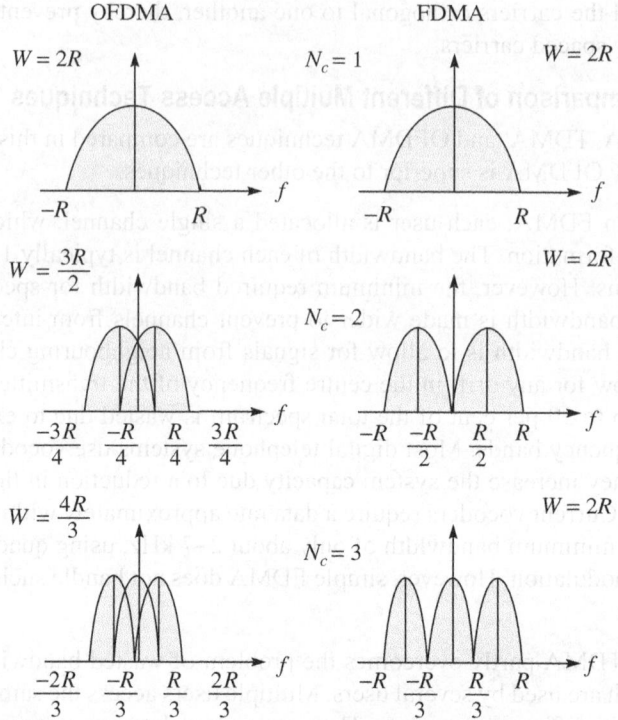

Fig. 9.17 Comparison between OFDMA and FDMA in terms of spectral consumption for equal number of users (W = available bandwidth, R = maximum bit rate, and N_c = number of carriers)

9.7 HYBRID METHODS OF MULTIPLE ACCESS

The following are a few combinations of multiple access techniques:

FDMA–CDMA The spectrum is divided into channels and each channel is a narrowband CDMA system with processing gain lower than the original CDMA system.

DSSS–FHSS The direct sequence modulates the signal and hops centre frequency using a pseudo-random hopping pattern. The method avoids near–far effect.

TDMA–CDMA Different spreading codes are assigned to different cells. One user per cell is allotted a particular time slot. Only one CDMA user transmits in each cell at any given time. The method avoids near–far effect.

TDMA–FHSS It involves a hop to a new frequency at the start of a new TDMA frame. The method avoids severe fades on the channel. Hopping sequences are predefined and unique per cell. It avoids co-channel interference if other base stations transmit on different frequencies at different times.

9.8 MULTIPLE ACCESS FOR PACKET RADIO SYSTEMS

When wireless networks are established, they especially follow the layered protocol stack. Chapter 10 will explain the networking aspects of wireless networks. However, an elementary introduction is given here.

- The lowermost layer is the physical layer, which takes care of the actual physical transmission of the packets by choosing the appropriate modulation scheme, bit rate, synchronization methods, and so on.
- Issues concerning packet access control (or user access control) are handled by the data link layer according to open system interconnect (OSI) reference model.
- Data link control layer is subdivided into two layers:
  - □ The lower sublayer is the medium access control (MAC) layer.
  - □ The upper sublayer is the logical link control (LLC) layer.

> Packet radio systems deal with random access and dynamic channel allocations.

In networks, whenever a request, acknowledgement, or data packet is to be transmitted, the channel will be utilized. This will not be a continuous process but will be bursty in nature. At the same time, for the datagram approach, all packets will be routed independently towards the destination. In this situation, the task of the data link layer is to establish a reliable point-to-point or point-to-multipoint connection between different devices. The same is true for wired and wireless media. Here, we shall discuss only the MAC sublayer.

Note: Standard MAC schemes from wired networks often fail in the wireless domain, so wireless networks require special MACs. In contrast to a wired network, wireless networks face serious problems such as hidden and exposed terminals and near and far terminals. Fixed assignment type of schemes can accommodate only low number of users. In addition, most of the time, active terminals will release their packets continuously, whereas idle terminals may release packets with a long gap of time. This will result in a severe wastage of the allocated fixed resources. Hence, some sort of dynamism is required for sharing the channel.

Three types of methods are discussed here. Pure ALOHA and slotted ALOHA are blind access schemes, whereas carrier sense multiple access (CSMA) and CSMA with collision detection (CSMA/CD) are carrier sensing based random access schemes. There are two more schemes, which are collision-free methods, called *collision-avoidance protocols* and *contention-free protocols*.

9.8.1 Pure ALOHA

The ALOHA protocol provides the fundamental solution for wireless access to computer systems. The main advantage of the ALOHA random access scheme is its simplicity. According to ALOHA, terminals can transmit their data regardless of the activity of other terminals. A terminal is allowed to transmit without considering whether the channel is idle or busy. If a packet is received correctly, the base station transmits an acknowledgement. If no acknowledgement is received by the mobile computer, the following takes place:

- The transmitter assumes the packet to be lost.
- It retransmits the packet after waiting for a random time.

The delay or random time is mainly determined by the probability that a packet is not received (because of interference from another transmission, called a *collision*) and the average value of the random waiting time before a retransmission is made. The basic flow of ALOHA is illustrated in Fig. 9.18.

ALOHA needs some adaptive control of the retransmission scheme. Otherwise, the system will become unstable. A popular method is to increase the mean waiting time if too many collisions occur. Shorter delay and higher throughput as compared to conventional ALOHA can

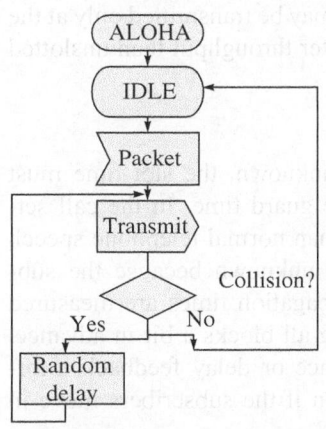

Fig. 9.18 Terminal behaviour in ALOHA random access network

> Pure ALOHA is also known as unslotted ALOHA and is unreliable due to randomness, whereas slotted ALOHA reduces randomness due to slot-based attempts.

be achieved using collision resolution schemes that exploit feedback about the collisions.

ALOHA in mobile radio sets The call set-up procedure of almost any (analog or digital) cellular telephone system uses ALOHA-type random access. Nevertheless, the performance differs from what one would expect in a wireline network.

In a radio channel, packets may be lost because of signal fading even if no other contending signal is present. On the other hand, packets may be received successfully despite interference from competing terminals. This is called *receiver capture*, which has a significant influence on the throughput.

> *Note*: The frequency reuse for ALOHA random access networks in a distributed scenario is based on the freely available channel or frequency and differs from the frequency reuse for cellular telephony, which exhibits a centralized scenario. The best reuse pattern for an ALOHA system is to use the same frequency in all cells.

Collision resolution For an infinite population of users and under certain channel conditions, the ALOHA system is unstable. Packets lost in a collision are retransmitted, but the retransmission again experiences collision. This may set off an avalanche of retransmission attempts. The number of previously unsuccessful packets that need to be retransmitted grows beyond any finite bound. One method to mitigate instability is to dynamically adapt the random waiting times of all terminals if the base station notices that many collisions occur. There are two solutions to ensure stability:

(a) *Dynamic frame length (DFL) ALOHA by Frits Schout*: It uses a centralized control mastered by the base station.

(b) *The Stack Algorithm by Boris Tsybakov, et al.*: This is a decentralized method.

9.8.2 Slotted ALOHA

In pure ALOHA, a transmission may start at any time. In slotted ALOHA, the time axis is divided into slots. All terminals are assumed to know the times at which a new slot begins; that is, time synchronization is needed in the network. Packets may be transmitted only at the beginning of a new slot. Slotted ALOHA has significantly better throughput than unslotted ALOHA.

GSM Call Set-up: Example of Slotted ALOHA

If the transmitter-to-receiver propagation time is large and unknown, the slot time must be equal to the sum of packet length and a sufficiently large guard time. In the call set-up of GSM, random access packets are substantially shorter than normal telephone speech blocks. During the call set-up, the propagation time is still unknown because the subscriber can be anywhere in the cell. During the call, the propagation times are measured and the terminal transmitter will compensate for it by sending all blocks a bit in advance. A closed loop control circuit, sending adaptive timing advance or delay feedback information, is used to ensure that the timing remains correct even if the subscribers move in the cell.

Performance analysis is done by measuring the throughput, drift, stability, and delay for both pure ALOHA and slotted ALOHA.

9.8.3 Carrier Sense Multiple Access

By sensing the channel empty condition, the reliability of medium access increases by a large amount, and this is what is done in various versions of CSMA.

For high traffic loads, collisions occur frequently in the most basic random access system, the ALOHA system. It reduces the throughput and may lead to instability if collided packets collide again during their retransmission. A number of other protocols have been proposed to mitigate this problem.

Busy channel multiple access (BCMA) is the class of multiple access schemes in which no new packet transmissions are allowed when the inbound channel is busy. Various strategies have been proposed to acquire information on the channel state. One of them is CSMA, in which carrier sensing is used to identify whether the channel is busy before transmission.

In its basic form, CSMA does not employ feedback other than positive acknowledgements of correct reception of a data packet. All terminals listen to the inbound (terminal-to-base) channel. No new packet transmission is initiated when the inbound channel is sensed busy by the mobile terminal. This requires that all mobile terminals should receive all others' signals on the inbound frequency.

There are two versions of CSMA.

CSMA with collision detection   In CSMA/CD, the receiver continuously informs all terminals about the incoming signal. If a collision is detected, the transmissions are aborted by all the terminals. Otherwise, the channel would remain occupied without any useful signal. This enhances the throughput compared to a system that acknowledges reception only after the transmission of the full message or packet.

CSMA with collision avoidance   In CSMA with collision avoidance (CSMA/CA), an attempt is made to improve the performance of CSMA. CSMA/CA by default uses the carrier sensing mechanism with exponential back-off.

In mobile radio networks with fading channels, a mobile terminal might not be able to sense a transmission by another (remote) terminal. This makes CSMA/CD unreliable. This effect is known as the *hidden terminal* problem.

Hidden terminal problem   To understand this problem, consider Fig. 9.19 in which three mobiles with their coverage are shown. The following observations can be made:

- *A* sends to *B*; *C* cannot receive *A* as it is out of range.
- *C* wants to send to *B*; *C* senses a *free* medium (carrier sensing fails).
- Collision occurs at *B*; *A* cannot receive the collision (collision detection fails).
- *A* is *hidden* for *C*.

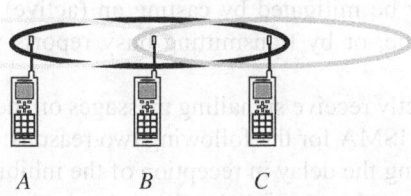

Fig. 9.19   Scenario for a hidden terminal problem

Exposed terminal problem   To understand this problem, consider Fig. 9.20 in which three mobiles with their coverage are shown. The following can be observed:

- *B* sends to *A*; *C* wants to send to another terminal (not *A* or *B*).
- *C* has to wait; carrier sensing signals a medium in use but *A* is outside the radio range of *C*, and therefore, waiting is not necessary.
- *C* is *exposed* to *B*.

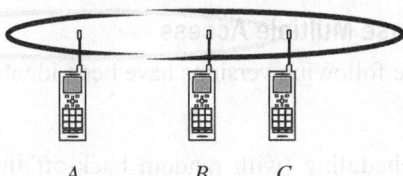

Fig. 9.20   Scenario for an exposed terminal problem

ANALYSING STATES OF CSMA

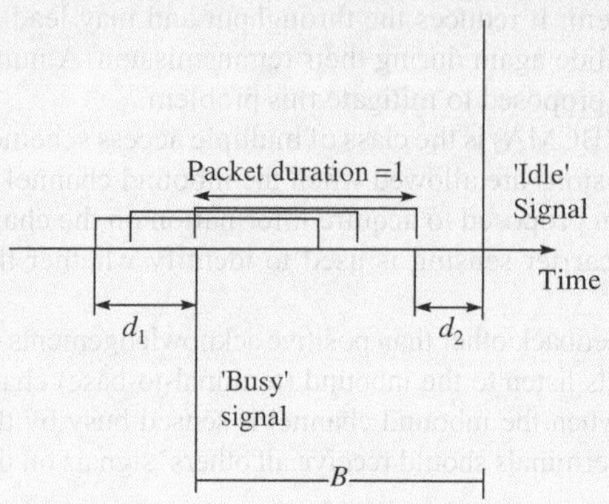

Fig. 9.21   CSMA/ISMA states

Carrier sense multiple access has two states of channel: *busy* and *idle*. When the base station receives an inbound packet, a busy signal is broadcast to all mobiles to inhibit them from transmitting. In a practical system, this occurs only after a short processing delay d_1. The effect of this delay depends on its magnitude relative to the duration of the data packet. After termination of (all contending) transmissions, the base station starts transmitting an idle signal after a delay of duration d_2. The whole scenario is represented in Fig. 9.21.

The *busy period* is defined as the sum of the period during which the base station broadcasts a busy signal and the preceding signalling delay d_1. For memoryless Poisson arrivals, the duration of the idle period—that is, the time interval between the release of the channel and the arrival of the first packet—is exponentially distributed with mean value $I = 1/G$, where G is the traffic.

The busy period is the time interval between the first arrival of a packet and the moment that the channel becomes idle. During the initial period d_1 of the busy period, the outbound channel thus still reports an idle inbound channel. If packet duration is of 1 unit time, the duration of the busy period is at least $1 + d_2$ but may be longer if a collision is caused by a packet arrival during the inhibit delay. Moreover, persistent terminals that sense a busy signal may start to transmit immediately after the channel becomes idle. In such cases, the busy period has a longer duration than two (or more) units of time.

Hence, the average duration of the busy period B depends on the signalling delays d_1 and d_2 and on the persistency p in scheduling (rescheduling) inhibited packets.

The same states can be observed for ISMA as well.

In inhibit sense multiple access (ISMA), the base station transmits a busy signal on an outbound channel to inhibit all other mobile terminals from transmitting as soon as an inbound packet is being received. A disadvantage of ISMA is the necessity of a real-time (continuous) feedback channel.

> CSMA/CA and ISMA do not suffer from hidden or exposed terminal problems.

Like ISMA, *busy tone multiple access* (BTMA) has been proposed. If the feedback channel contains only a narrowband *busy tone*, the mobile terminal may erroneously miss the presence of this tone if the outbound signal happens to be in a deep fade. This may be mitigated by casting an (active) idle tone, rather than an (active) busy tone, or by transmitting busy reports with error control coding.

Even if all mobile terminals always correctly receive signalling messages on the feedback channel, collisions can nonetheless occur in ISMA for the following two reasons:

(a) New packet transmissions can start during the delay in reception of the inhibit signal.

(b) Packets from two or more persistent terminals, awaiting the channel to become idle, can collide immediately after the termination of the previous packet transmission.

9.8.4 Versions of Carrier and Inhibit Sense Multiple Access

Based on the states of CSMA and ISMA, the following versions have been identified.

Non-persistent

For non-persistent CSMA and ISMA, rescheduling (with random back-off time) always occurs if the channel is busy at the instant of sensing. Hence, if a packet arrives at a non-persistent terminal when the base station transmits a busy signal, the attempt is considered to have failed. If the feedback channel (or the channel sensing mechanism in the case of CSMA) is imperfect, a transmission may erroneously be started in the period. The packet is rescheduled for later transmission.

1-persistent

Terminal behaviour in a 1-persistent CSMA random access network is shown in Fig. 9.22.

p-persistent

Terminal behaviour in a p-persistent CSMA random access network is shown in Fig. 9.23.

There is a probability $I/(I + B)$ that a test packet starts at an instant when the channel is idle. A collision can occur if one or more terminals start transmitting during the time delay d_1 of the inhibit signal. This allows us to compute the conditional probability of n transmissions overlapping with the test packet that initiated the busy period.

Alternatively, the test packet itself starts during a period of duration d_1 when the channel is busy because of a transmission by another terminal but seems idle since the inhibit signal is not yet being broadcast. This event occurs with a probability $d_1/(B + I)$. The test packet thus experiences interference from the packet that initiated the busy period but possibly also from other arriving packets. The additional contending signals occur with a Poisson arrival rate during the interval d_1.

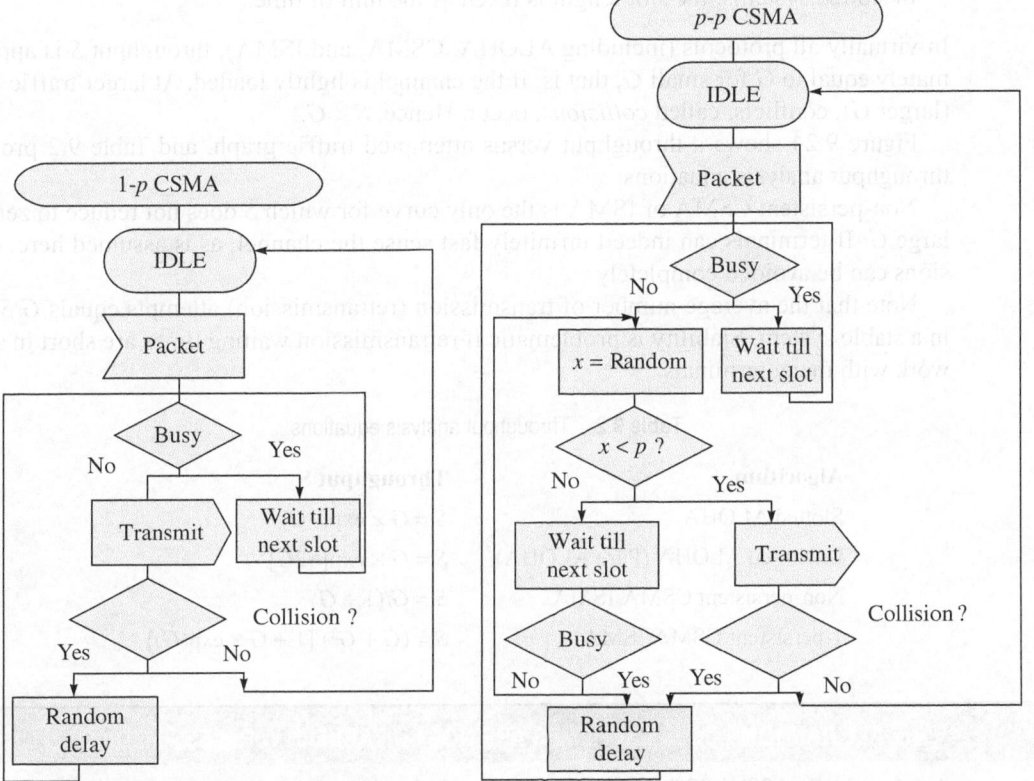

Fig. 9.22 Description of terminal behaviour in a 1-persistent CSMA random access network

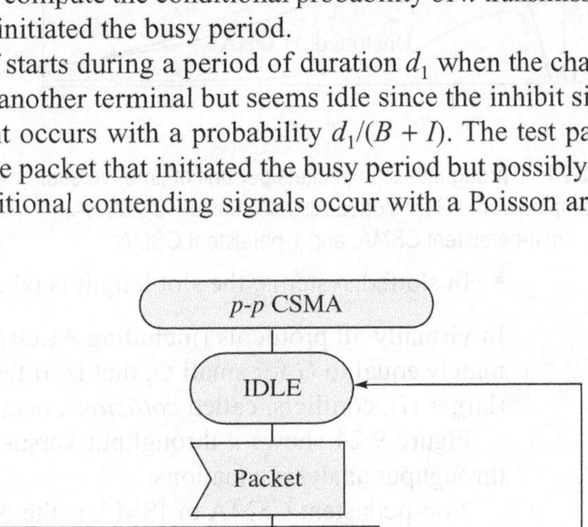

Fig. 9.23 Description of terminal behaviour in a p-persistent CSMA random access network

Taking account of these three possible events, the unconditional probability of successful transmission can be derived. The derivation differs from techniques typically used for wireline local area networks (LANs), because in radio systems we mostly want to be able to consider expressions for capture probabilities, depending on the location of one particular terminal.

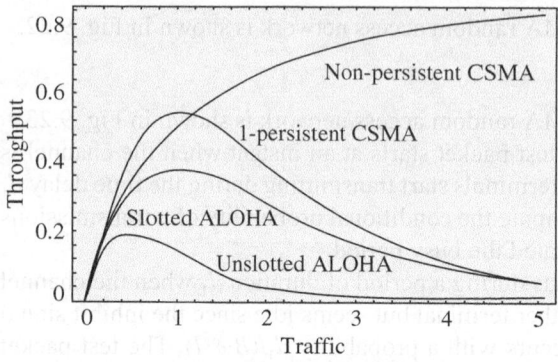

Fig. 9.24 Throughput S (in packets per unit of time) versus attempted traffic G for slotted ALOHA, unslotted ALOHA, non-persistent CSMA, and 1-persistent CSMA

9.8.5 Throughput of Random Access Schemes

Throughput is an important measure of performance. The following are the important factors considered in throughput analysis:

- The *offered* or *attempted* traffic, denoted as G, is the number of transmission attempts made on the channel per unit of time. In non-persistent CSMA, attempts made when the channel is busy are inhibited. These attempts contribute to the offered traffic.
- Throughput, denoted as S, is the average fraction of time that the radio channel takes to successfully carry packets.

- In slotted systems, the slot length is taken as the unit of time.

In virtually all protocols (including ALOHA, CSMA, and ISMA), throughput S is approximately equal to G for small G, that is, if the channel is lightly loaded. At larger traffic loads (larger G), conflicts, called *collisions*, occur. Hence, $S < G$.

Figure 9.24 shows a throughput versus attempted traffic graph, and Table 9.2 provides throughput analysis equations.

Non-persistent CSMA or ISMA is the only curve for which S does not reduce to zero for large G. If terminals can indeed infinitely fast sense the channel, as is assumed here, collisions can be avoided completely.

Note that the average number of transmission (retransmission) attempts equals G/S only in a stable system. Stability is problematic if retransmission waiting times are short in a network with many terminals.

Table 9.2 Throughput analysis equations

Algorithm	Throughput S
Slotted ALOHA	$S = G \times \exp(-G)$
Unslotted ALOHA (Pure ALOHA)	$S = G \times \exp[-2G]$
Non-persistent CSMA/ISMA	$S = G/(1 + G)$
1-persistent CSMA/ISMA	$S = (G + G^2)/[1 + G \times \exp(G)]$

WHY ALL RANDOM ACCESS SCHEMES ARE NOT SUITABLE IN WIRELESS MEDIA?

Let us consider a typical case of CSMA/CD that works as follows over a wired line:
- A sender senses the medium to see if it is free.
- In case the medium is busy, the sender waits until it becomes free.

- As soon as the medium is free, the sender starts transmitting data.
- During transmission, if the sender finds a collision, it stops transmitting and sends a jamming signal.

This scheme fails in case of wireless networks because of the following reason. In CSMA/CD, a signal must reach its receiver without any collisions. A collision might occur close to the sender or at a far distance. This is not a major problem in a wired network because everyone gets to know if a collision occurs anywhere in a wire.

In contrast, the situation is entirely different in wireless networks, which work as follows:

- The strength of a signal is inversely proportional to its distance from the sender.
- The sender applies carrier sense and finds an idle medium.
- The sender starts sending but a collision occurs at the receiver due to a second sender. This is the hidden terminal problem.
- If the receiver is in the middle of the first and second senders, the second sender will be at a far distance from the first one. The sender detects no collision and assumes that the data has been transmitted without errors, but a collision might have destroyed the data at the receiver.

Hence, all random access schemes are not suitable in wireless media.

9.9 RESERVATION-BASED MULTIPLE ACCESS SCHEMES

In multiple access schemes based on reservation, there is a provision for users to reserve their slot or resources in advance and then the transmission will be followed in that order. Such reservation times are separately provided.

9.9.1 Packet Reservation Multiple Access

Packet reservation multiple access (PRMA) is an implicit reservation scheme. Here, slots can be reserved implicitly as follows. A certain number of slots form a frame, which is repeated in time just like a TDM pattern. A base station broadcasts the status of each slot to all mobile stations. All stations receiving this vector will then know which slot is occupied and which slot is currently free. If all the slots in a vector are reserved except one, then more than one station attempts to access this free slot. Hence, a collision occurs, and a new status will be generated with the same free slot, indicating that one slot is still available for reservation. Again, stations can compete for this slot and the procedure continues. The actual transmission of packets in the network will follow the order in the final reservation status.

9.9.2 Polling and Token Passing

Polling is a strictly centralized scheme with one master station and several slave stations. The master can poll the slaves according to many schemes: round robin (efficient only if traffic patterns are similar over all stations), randomly according to the reservations, and so on. The master can also establish a list of stations wishing to transmit during a contention phase. After this phase, the station polls each station on the list. Similar schemes are used, for example, in the Bluetooth and as one possible access function in IEEE 802.11 systems.

Token passing is mainly suitable for wired networks. IEEE 802.4 is a token bus protocol on the LAN whereas IEEE 802.5 is a token ring protocol. Here, the key part is the token, which is a small bit pattern. The station that is transmitting captures the token during the transmission; after reception of the complete data, it releases the token, again circulating it among various users. Thus, a logical or physical ring configuration is necessary for the token passing type of scheme.

DEMAND ASSIGNED MULTIPLE (RANDOM) ACCESS

Demand assigned multiple access (DAMA) or random access schemes dynamically assign radio resources to users as per availability of the resources. This scheme is suitable in a mobile scenario and in a satellite system. The request for communication from the user will be transferred to the central authority that can assign the available or free carrier or time slot to the user on the basis of the current situation. The SPADE system for satellite communication and GSM are examples of systems using such a method of multiple access in support.

MORE SOLVED EXAMPLES

Example 9.4 Twenty-four voice signals are sampled and time division multiplexed. Narrow sampling pulses occur with 100 ms duration and the maximum frequency content of each signal is 4 kHz. One synchronization channel is added by some extra pulses with the same pulse duration as the samples. Assuming Nyquist sampling frequency criteria, calculate the spacing between successive pulses of the multiplexed signal.

Solution As per Nyquist criteria, the sampling frequency must be 8 kHz.

So, the sampling duration $T_s = 1/8000 = 125$ μs

Now, 24 voice channels plus 1 synchronization channel makes 25 channels.

Effective time slot for each channel is 125/25 = 5 μs

Now, 100 ms = 0.1 μs is the pulse duration.

Therefore, spacing between successive pulses = 5 − 0.1 μs = 4.9 μs

Example 9.5 An FDMA system is allocated a bandwidth of 12.5 MHz, where there are 360 channels of 30 kHz each. What will be the guard interval? What will be the spectral efficiency?

Solution Total bandwidth occupied = 360 channels × 30 kHz = 10,800 kHz

Bandwidth for guard interval = 12,500 − 10,800 = 1700 kHz

Hence, guard band = 1700/(360 − 1) = 4.735 kHz

Spectral efficiency = $(30 \times 360)/(12.5 \times 1000) = 0.864$

Hence, spectral efficiency is 86.4%, as more spectrum bands are wasted as guard bands.

Example 9.6 A DS-CDMA system has one-way bandwidth of 12.5 MHz and the other parameters are as follows:

Voice activity factor = 0.6
Data bit rate = 16 kbps
Bandwidth efficiency = 0.9
Frequency reuse factor = 0.5

$E_b/I_o = 8$ dB

Find the number of users and spectral efficiency for this system with an omnidirectional cell.

Solution $E_b/I_o = 8$ dB = 6.3
Neglecting noise part,

Number of users = Frequency reuse factor ×

Bandwidth efficiency $\left(\dfrac{1}{v_f} \dfrac{B/R}{E_b/I_o} \right) = 0.5 \times 0.9$ [(12.5 $\times 10^6$)/(0.6 $\times 16 \times 10^3 \times 6.3$)] ≈ 93

Spectral efficiency = Number of users × Bit rate/ Bandwidth = $93 \times (16 \times 10^3)/(12.5 \times 10^6) = 0.119$ bits/s/Hz

---

SUMMARY

- Multiplexing is required to accommodate the information of multiple users in a single information base, transmitted together on a single channel.
- Multiple access schemes are necessary in a multiple-user environment to share the available channels, bandwidth, and resources efficiently, and to accommodate a high number of users in the system.
- Multiple access schemes are mainly of three types: fixed assignment based access, random access, and

reservation-based access. An additional category is demand assigned multiple access.
- Fixed assignment of resources can be done by choosing one of the four parameters: frequency, time, code, and space.
- A wireless channel is more susceptible to noise and interference, and hence, multiple access schemes may not give the exact performance as in a wired line.

- FDMA does not use the spectrum efficiently whereas TDMA critically requires synchronization.
- CDMA schemes use both time and frequency efficiently but there is a limit on the number of users due to interference noise.
- SDMA is mainly suitable for cellular systems.
- ALOHA, slotted ALOHA, CSMA, CSMA/CD, and ISMA are some of the random access MAC schemes.
- DAMA is a special type of random multiple access scheme

and is a combination of random and reservation-based access.
- Performance of non-persistent CSMA gives better throughput with rise in offered traffic.
- PRMA is a method based on packet transmission slot reservation.
- Polling and token passing are different types of reservation-based medium access schemes.

EXERCISES

Multiple-choice Questions

9.1 Time division multiplexing
   (a) can be used only with PCM
   (b) interleaves pulses belonging to different transmissions
   (c) combines fine groups into a supergroup
   (d) stacks channels in adjacent frequency slots

9.2 Synchronous TDM is called synchronous because
   (a) synchronous transmission is used
   (b) time slots are pre-assigned to sources and fixed
   (c) preamble is transmitted
   (d) none of these

9.3 DAMA stands for
   (a) data accessibility master aerial
   (b) demand assigned multiple access
   (c) digital attenuator microwave antenna
   (d) dual accessibility mode antenna

9.4 A communication satellite is a repeater between
   (a) a transmitting station and a receiving station
   (b) a transmitting station and many receiving stations
   (c) many transmitting stations and one receiving station
   (d) many transmitting stations and many receiving stations

9.5 Each channel in a TDM input has a bit rate of 16 kbps. If there are four channels, the output bit rate will be
   (a) 16 kbps   (c) 32 kbps
   (b) 4 kbps   (d) 64 kbps

9.6 Which of the following multiple access schemes are used in combination by the SPADE system?
   (a) SDMA, TDMA, FDMA
   (b) DAMA, FDMA
   (c) SDMA, DAMA
   (d) DAMA, FDMA, TDMA

9.7 GSM uses TDMA–FDMA together
   (a) to accommodate many low bit rate voice channels
   (b) to accommodate higher number of users

   (c) to eliminate low bandwidth problems
   (d) for all these reasons

9.8 Which of the following is a random access method?
   (a) PRMA   (c) ISMA
   (b) FDMA   (d) Token passing

9.9 Which of the following multiple access methods allows all users to simultaneously use the whole bandwidth?
   (a) TDMA (b) FDMA (c) CDMA (d) SDMA

9.10 Which of the following multiple access schemes allows a user to use only a part of the total bandwidth?
   (a) TDMA (b) FDMA (c) CDMA (d) SDMA

9.11 Which of the following multiple access schemes allows the utilization of the frequency reuse concept?
   (a) DAMA (b) FDMA (c) CDMA (d) SDMA

9.12 Which of the following random access methods provides the best performance?
   (a) ALOHA
   (b) Slotted ALOHA
   (c) 1-persistent CSMA
   (d) Non-persistent CSMA

9.13 For real-time voice transmissions, the best suitable multiple access type is
   (a) random access
   (b) fixed assignment based access
   (c) reservation-based access
   (d) none of these

9.14 Which of the following techniques requires fast and accurate power control because each cell must be tightly synchronized?
   (a) TDMA   (c) OFDMA
   (b) CDMA   (d) FDMA

9.15 Which of the following statements is true?
   (a) OFDMA provides the benefits of frequency diversity and out-of-cell interference.

(b) OFDMA provides code sequences that are orthogonal.

(c) OFDMA is wideband and hence less spectrally efficient.

(d) OFDMA is nothing but FDMA.

9.16 Which of the following statements is true?

(a) CDMA uses frequency reuse concept in cellular infrastructure.

(b) A CDMA system has code sequences that are truly orthogonal in the presence of multipath delay.

(c) Due to voice activity factor, inclusion of the number of users in a CDMA system decreases.

(d) None of these is true.

Review Questions

9.1 Why do the MAC schemes from wired networks sometimes fail when used in wireless networks?

9.2 Give the detailed classification of various medium access control schemes.

9.3 Explain the problem of hidden and exposed terminals.

9.4 How can you use combinations of multiple access schemes? Explain with typical examples.

9.5 Justify that fixed assignment type of multiple access schemes are more suitable for real-time transmissions.

9.6 Why is the near–far problem more serious in CDMA-based schemes?

9.7 Assume all stations can hear all other stations. One station wants to transmit and senses that the carrier is idle. Why does collision still occur after the start of the transmission?

9.8 Explain the term *interference* in the frequency, time, code, and space domains. What are the countermeasures to avoid such problems in FDMA, TDMA, CDMA, and SDMA?

9.9 What is the difference between random access and reservation-based access?

9.10 What are the advantages of reservation schemes? How are collisions avoided during data transmission? Why is the probability of collisions lower in reservation schemes than in classical ALOHA? What are the disadvantages of reservation schemes?

9.11 Compare the performance of the various random access schemes and identify the scheme best suitable in the wireless environment.

9.12 Prove that OFDMA is more spectrally efficient than FDMA.

Numerical Problems

9.1 Eight pulse code modulation (PCM) channels, each band limited to 4 kHz, are time division multiplexed. Each sample is coded into an eight-bit word. Find the output rate and the required bandwidth.

9.2 How many broadcast stations can be accommodated in a 100 MHz bandwidth if the highest modulation bandwidth is 2 MHz? The guard band is 50 KHz.

9.3 Suppose the bandwidth is 1.75 MHz, bit rate is 9.6 kbps, and minimum acceptable E_b/I_o due to interference signals from other users is found to be 9 dB. Ignore channel noise. Determine the maximum number of users that can be supported in a single-cell CDMA system with following specifications: (a) an omnidirectional BS antenna without voice activity detection and (b) three sectors at the base station along with voice activity detection with $v_f = 3/8$.

9.4 The forward link of an FDD–FDMA–TDMA system is having 25 MHz allotted bandwidth with each radio channel of 200 kHz. If each radio channel supports eight speech channels, how many simultaneous users can be accommodated at a time in the forward link?

9.5 If the offered traffic in a network is 1000 packets and the medium access is based on slotted ALOHA, find the throughput of the network.

9.6 Sixteen delta modulation channels with 32 kbps bit rate are time division multiplexed. Find the data rate of the multiplexed signal. If 16 channels create one frame, find the period of a frame.

9.7 A satellite system employs QPSK modulation with around 30% extra bandwidth due to roll-off and 10% guard bandwidth. The PCM voice channels use a rate of 64 kbps. What will be the number of channels supported if the transponder allows total of 35 MHz bandwidth?

PART 4
Wireless Networks

Wireless networks provide voice, video, and data services to the subscribers. The last part of the book has three chapters that throw light on various networking fundamentals, cellular networks, and ad-hoc wireless networks. Networking fundamentals give the requirement of wireless networking by the overall protocol structure, security, and computing aspects. The other two chapters deal with a specific infrastructure/architecture part of the standard networks, their physical layer specifications, and MAC layer.

If a network deals with data services, it has to link up with the standard TCP/IP protocol stack of the Internet, and for voice services, higher layers of the protocol stack are minor. This means that the physical and MAC layer aspects are distinct and of prime importance for various wireless systems.

10 Networking Fundamentals

Theme of the Chapter

The chapter summarizes the networking fundamentals in terms of conventional computer networks and enhancements for mobile networks. Networking is the interconnection of various communication devices such as computers, mobiles, smartphones, and laptops. The interconnection is established in such a way that the resources available with each device can be shared among each other.

Networks follow various protocols and standards. Open systems interconnection (OSI) is the reference model for designing network protocols and the transmission control protocol/internet protocol (TCP/IP) suite is adopted in the existing Internet. Networks are based on wired or wireless media. In the present scenario, various networks are converged, and interworking among various types of standards and systems is possible.

Wireless networks are more critical to handle than wired networks due to uncertainty of the wireless channel. Medium access control protocols, routing protocols, and transport protocols of wired networks need many modifications in the wireless environment. Security aspects and mobile computing are other important issues in wireless networks.

Key Topics

- Wireless or mobile networks
- OSI reference model and TCP/IP stack
- IP and TCP headers
- Medium access control in wireless networks
- Routing algorithms
- Transport control mechanisms
- Security aspects
- Application protocols
- Mobile computing, mobile IP, and mobile TCP

10.1 WIRELESS NETWORKS

Networking of devices is required to share resources and exchange available information among various users. Wired computer networks are all well established. Today, there is a demand for anywhere–anytime communication, hence the need for wireless networking. The main difference in wireless networking is the communication medium, which is a wireless channel. As studied earlier, a wireless channel is unreliable, and many protocols that are designed for wired networks are not suitable for the wireless environment.

Topology-wise, there are two possibilities for networking in general:

Ad hoc networks are distributed ones and need processing at each individual node; hence, the battery life of the nodes needs high attention.

Fully meshed connections Here, if N is the number of users, then the total number of transmission links required are $N(N-1)/2$, which increases with an increase in the number of users.

Centralized switch   Here, a complex switch is required to handle N users and hence N transmission links. The centralized topology is often converted into a hierarchy, which is more convenient in the case of telephony.

Based on another point of view, networks can be of the following two types:

Infrastructure-based networks   These networks can have a centralized view of the problem. Hence, a centralized, hierarchy-based topology is very much suitable.

Ad hoc networks   These networks require a distributed approach and hence mesh topology may work. Still, the topological scenario may vary with the type of network being dealt with.

The following can be summarized to be the limitations of wireless networks against wired networks.

Complex systems   Users of a wireless network are mobile, whereas those of wired networks are static. Hence, the air interface between the user and the base station must be maintained at all times. There is a wide range of propagation conditions as well as any possible user locations. Heavy computing support is required. To assure adequate area coverage and handover with seamless connectivity, deployment of hundreds of base stations is required in the case of infrastructure-based networks. Collision avoidance type of multiple access mechanism and dynamic routing protocols are required. This makes a complex set of protocols. Security aspects also become critical.

Computations and handling capacity   As users are mobile, they may go out of coverage in a fraction of time. Hence, fast processors and fast computation algorithms are required to handle the heavy mass of subscribers, especially in the case of a vehicular mobility scenario, with a fast fading channel. Location management, channel allocation, call handling, and other such features require fast computations. Network protocols are complex and add extra overhead due to extra signalling or control packets. The throughput of packet-based networks is dependent upon the number of users or nodes, channel condition, and available resources.

Networks can also be classified in terms of their size. Networks of a room size are called personal area networks (PAN), networks of the size of a building or a campus are called local area networks (LAN), networks of the size of a small city are called metropolitan area network (MAN), and networks beyond that size are called wide area networks (WAN). Internetworking is possible in each of these categories.

10.2 OPEN SYSTEMS INTERCONNECTION REFERENCE MODEL

The fundamentals explained in this section are applicable to both wired and wireless networks. Certain rules are required to regulate the activities in a network. In networking parlance, these rules and terms are called *protocols*.

Open systems interconnection (OSI), which is the reference model suggested by the International Standards Organization (ISO), was developed to cover all the issues of a network design in a proper order. The OSI model suggests all the possible protocols but there is no standardization of any protocol. This model is not adopted by any system directly but any other standard protocol suite can be mapped in terms of the OSI model to establish equivalence. The presently adopted transmission control protocol/internet protocol (TCP/IP) suite is defined with five layers, but still equivalence can be established with the OSI model.

> Wired network protocols are sometimes not applicable to the wireless network scenario due to its dynamism and randomness.

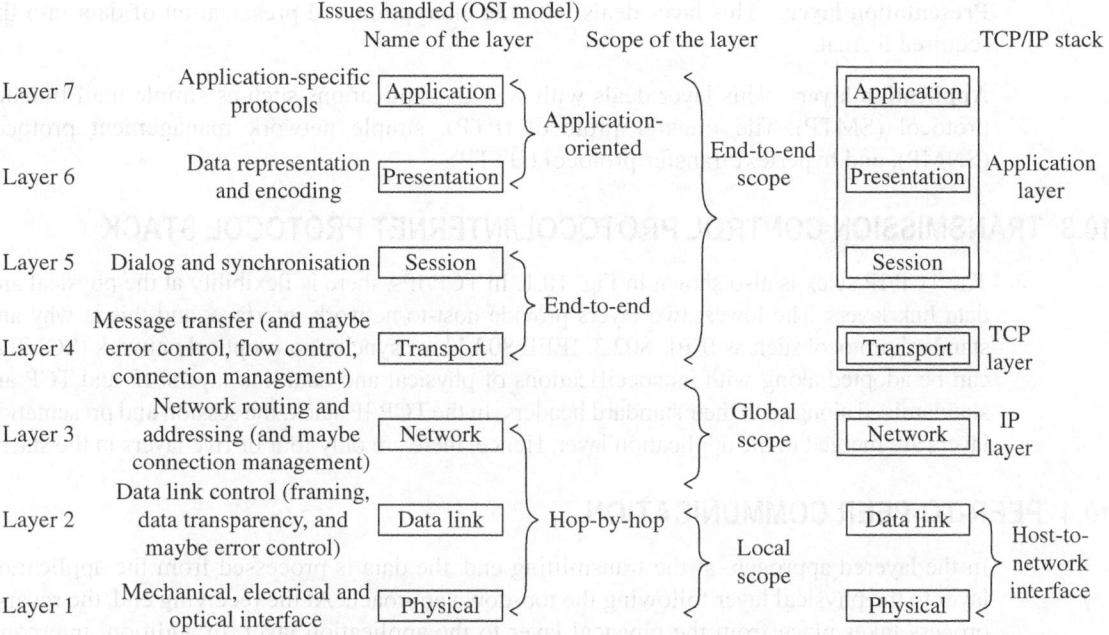

Fig. 10.1 OSI reference model with its layered architecture and TCP/IP stack with OSI equivalence

The OSI reference model is shown in Fig. 10.1. The following points provide a small rundown of its layers and the functions they perform.

Physical layer The lowermost layer is the physical layer. It deals with wireless communication techniques such as source and channel coding, media of transmission, modulation techniques, synchronization techniques, and signalling. All device-level and system-level hardware specifications are as per the physical layer designed for a network.

Data link layer A transmitter and a receiver are linked through the channel. In a network, many such point-to-point links are to be handled in terms of data flow control, error control by allowing acknowledgement and retransmissions, medium access control (MAC), and so on. The data link layer helps in handling all these issues. This layer is subdivided into two layers—MAC and logical link control (LLC).

Network layer Each point-to-point data link traversed by the packet is also called a hop. In packet transmissions, there are multiple hops in a route between the sender and a far distance receiver that is followed by a packet. For determining the optimum route for a packet, some algorithms are required. The network layer helps in finding the best route as well as deals with connection management (even in the case of internetworking) and congestion control.

Transport layer This layer deals with the handing over of the data to a particular session or application. It provides quality of service as well. Segmentation and reassembling of data are done at the transport layer.

Session layer This layer normally maintains the various sessions opened by multiple applications.

Presentation layer This layer deals with the encryption and presentation of data into the required format.

Application layer This layer deals with various applications such as simple mail transfer protocol (SMTP), file transfer protocol (FTP), simple network management protocol (SNMP), and hypertext transfer protocol (HTTP).

10.3 TRANSMISSION CONTROL PROTOCOL/INTERNET PROTOCOL STACK

The TCP/IP stack is also shown in Fig. 10.1. In TCP/IPs, there is flexibility at the physical and data link layers. The lowest two layers provide host-to-network interface, and this is why any standard protocol such as IEEE 802.3, IEEE 802.11, or synchronous optical network (SONET) can be adopted along with its specifications of physical and data link layers. IP and TCP are standardized along with their standard headers. In the TCP/IP stack, the session and presentation layers are merged in the application layer. Hence, there are only four or five layers in the stack.

10.4 PEER-TO-PEER COMMUNICATION

In the layered approach, at the transmitting end, the data is processed from the application layer to the physical layer following the top-down approach. At the receiving end, the reverse process takes place from the physical layer to the application layer. In addition, intercommunication between two peer layers can also be visualized due to the stack-natured representation and the possibility of identifying individual reverse operations. These layer-to-layer processes are called *peer processes* and are represented as shown in Fig. 10.2.

> Logical and port addresses together are known as the socket address.

Figure 10.2(a) represents the peer processes through one relay and Fig. 10.2(b) represents them through multiple nodes. Peer-to-peer communication is a logical communication between each corresponding layer. Figure 10.2 can be thought of for bidirectional scenarios or actual end-to-end communication.

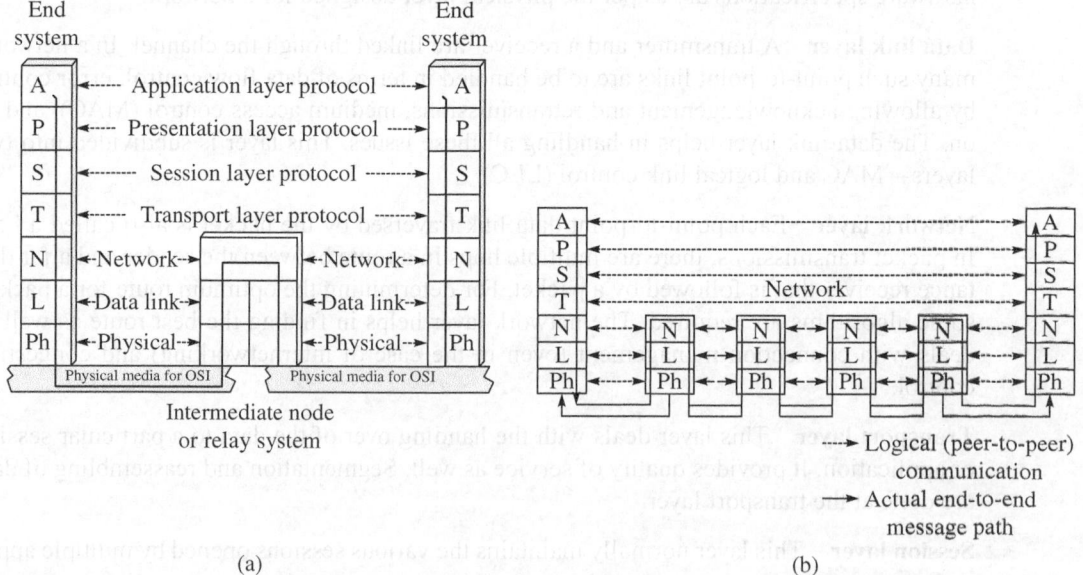

Fig. 10.2 Peer processes (a) Through one relay (b) Through multiple nodes

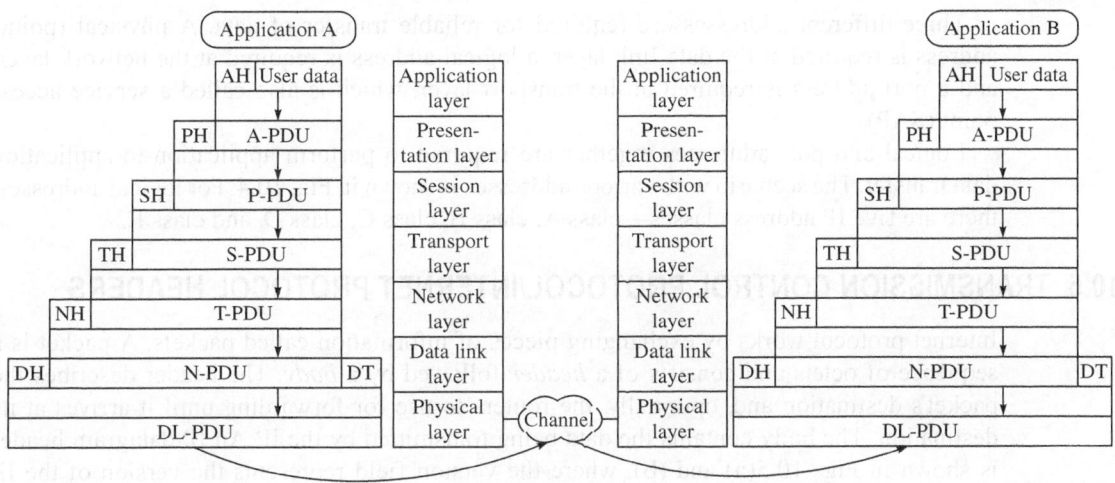

PDU — Protocol data unit

AH — Application header, similarly, other layer headers are tagged PH, SH, TH, NH, and DH

DT — Data link trailer

Fig. 10.3 Protocol data units with header tags at different layers

During processing, each layer attaches its header tag to the data packets or segments so that it is possible to identify the level at which the data undergoes processing. The data under processing is called the *protocol data unit* (PDU). Figure 10.3 shows how the PDUs are operated at different layers and how the headers of different layers are attached. The reverse operation is performed at the receiving end, where the header tags are removed step by step.

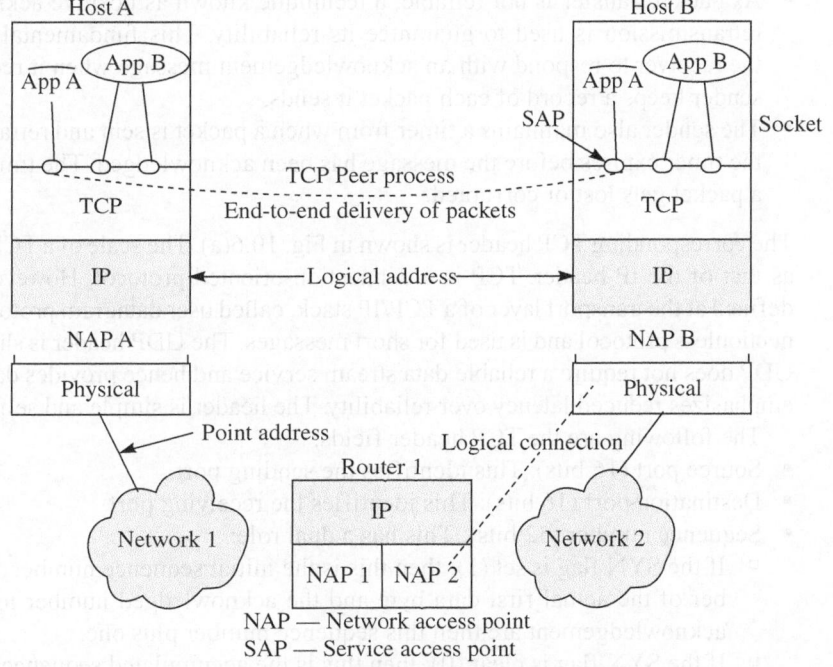

NAP — Network access point

SAP — Service access point

Fig. 10.4 Scenario with various addresses in TCP/IP

Three different addresses are required for reliable transfer of data. A physical (point) address is required at the data link layer, a logical address is required at the network layer, and a port address is required at the transport layer, which is also called a service access point (SAP).

Logical and port addresses together are required to perform application-to-application data transfer. The scenario with various addresses is shown in Fig. 10.4. For logical addresses, there are five IP address classes—class A, class B, class C, class D, and class E.

10.5 TRANSMISSION CONTROL PROTOCOL/INTERNET PROTOCOL HEADERS

Internet protocol works by exchanging pieces of information called packets. A packet is a sequence of octets and consists of a *header* followed by a *body*. The header describes the packet's destination and, optionally, the routers to use for forwarding until it arrives at its destination. The body contains the data being transmitted by the IP. An IP datagram header is shown in Figs 10.5(a) and (b), where the version field represents the version of the IP datagram. The scales of both the figures are the same.

There are two versions of IP headers: IPv4 and IPv6. The format of Fig. 10.5(a) represents only version 4 whereas that of Fig. 10.5(b) represents only version 6. The header checksum calculation in IPv4 is the checksum of the entire IP header. The details of the other fields are represented along with the header formats. There is no direct compatibility between IPv4 and IPv6 headers. The address field is larger in the IPv6 format (128 bits) as it accommodates more number of user addresses.

The TCP is a reliable stream delivery service that guarantees that all bytes received will be identical to the bytes sent, and in the correct order. However, the following additional support is required due to unreliable lower layers:

- As packet transfer is not reliable, a technique known as positive acknowledgement with retransmission is used to guarantee its reliability. This fundamental technique requires the receiver to respond with an acknowledgement message when it receives the data. The sender keeps a record of each packet it sends.
- The sender also maintains a timer from when a packet is sent and retransmits the packet if the timer expires before the message has been acknowledged. The timer is needed in case a packet gets lost or corrupted.

The corresponding TCP header is shown in Fig. 10.6(a). The scale of a TCP header is the same as that of the IP header. TCP is a connection-oriented protocol. However, another protocol defined at the transport layer of a TCP/IP stack, called user datagram protocol (UDP), is a connectionless protocol and is used for short messages. The UDP header is shown in Fig. 10.6(b). UDP does not require a reliable data stream service and hence provides datagram service that emphasizes reduced latency over reliability. The header is simple and self-explanatory.

The following are the TCP header fields:

- Source port (16 bits): This identifies the sending port.
- Destination port (16 bits): This identifies the receiving port.
- Sequence number (32 bits): This has a dual role:
  - If the SYN flag is set (1), then this is the initial sequence number. The sequence number of the actual first data byte and the acknowledged number in the corresponding acknowledgement are then this sequence number plus one.
  - If the SYN flag is clear (0), then this is the accumulated sequence number of the first data byte of this segment for the current session.

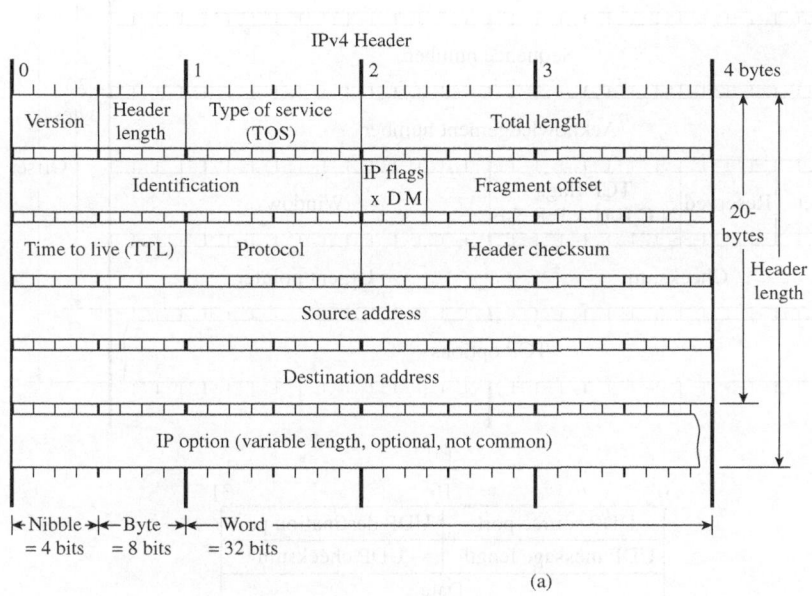

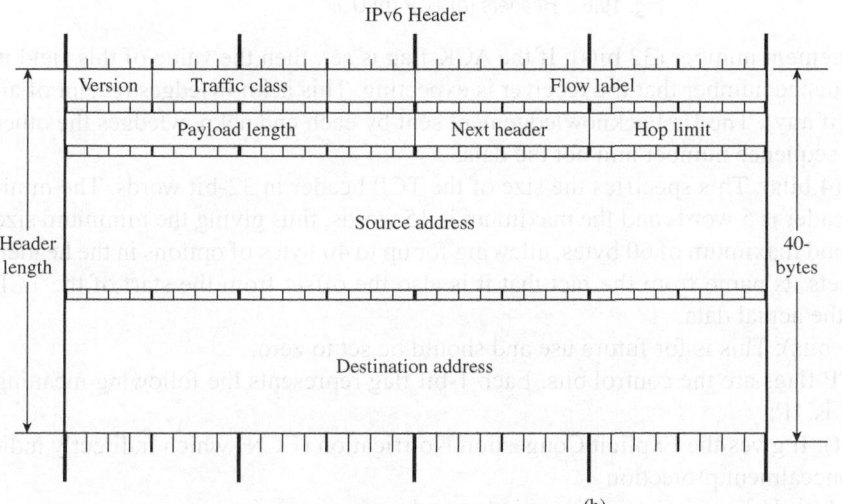

Fig. 10.5 Header of IP datagram (a) IPv4 (b) IPv6

Header length
Number of 32-bit words
in TCP header, minimum
value of 5. Multiply by
4 to get byte count.

Total length
Total length of IP datagram,
or IP fragment if fragmented.
Measured in bytes.

Fragment offset
Fragment offset from start of
IP datagram. Measured in
8-byte (2 words, 64 bits)
increments. If IP datagram
is fragmented, fragment size
must be a multiple of 8 bytes.

IP flags
x 0x80 reserved
D 0x40 do not fragment
M 0x20 more fragments
follow

Protocol
IP protocol ID. Including
(but not limited to):
1 ICMP 17 UDP 57 SKIP
2 IGMP 47 GRE 88 EIGRP
6 TCP   50 ESP 89 OSPF
9 IGRP 51 AH 115 L2TP

Traffic class
8-bit traffic class field.

Flow label
20-bit flow label.

Payload length
16-bit unsigned integer.
Length of the IPv6 payload, i.e.,
the rest of the packet following
this IPv6 header, in ocets.
Any extension headers are
considered part of the payload.

Next header
8-bit selector. Identifies the
type of header immediately
following the IPv6 header.
Uses the same values as the
IPv4 protocol field.

Hop limit
8-bit unsigned integer.
Decremented by 1 by each
node that forwards the packet.
The packet is discarded in hop
limit is decremented to zero.

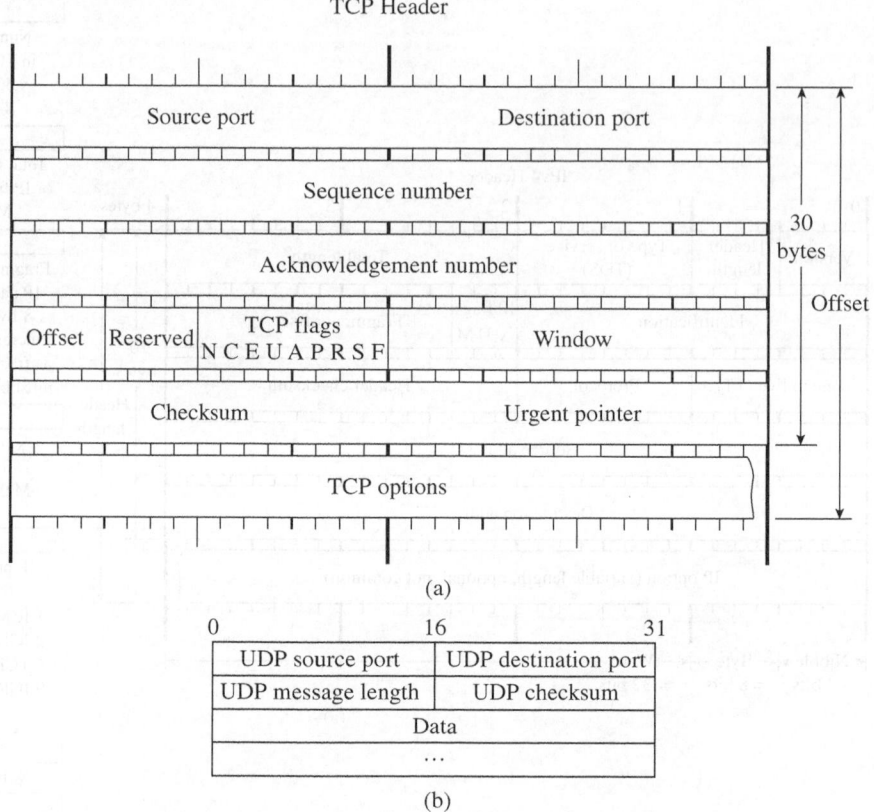

Fig. 10.6 Headers (a) TCP (b) UDP

- Acknowledgement number (32 bits): If the ACK flag is set, then the value of this field is the next sequence number that the receiver is expecting. This acknowledges receipt of all prior bytes (if any). The first acknowledgement sent by each end acknowledges the other end's initial sequence number and not the data.
- Data offset (4 bits): This specifies the size of the TCP header in 32-bit words. The minimum size header is 5 words and the maximum is 15 words, thus giving the minimum size of 20 bytes and maximum of 60 bytes, allowing for up to 40 bytes of options in the header. This field gets its name from the fact that it is also the offset from the start of the TCP segment to the actual data.
- Reserved (3 bits): This is for future use and should be set to zero.
- 9 bits of TCP flags are the control bits. Each 1-bit flag represents the following meaning if it set, that is, '1':
  □ NS (1 bit): It gives the Explicit Congestion Notification (ECN) which indirectly indicates concealment protection
  □ CWR (1 bit): Indicates Congestion window reduced
  □ ECE (1 bit): Indicates ECN echo
  □ URG (1 bit): Indicates the urgent pointer field
  □ ACK (1 bit): Indicates the acknowledgement field
  □ PSH (1 bit): Push function that asks to push the buffered data to the receiving application

        □ RST (1 bit): Resets the connection

        □ SYN (1 bit): Synchronizes sequence numbers; only the first packet sent from each end should have this flag set. Some other flags change meaning based on this flag—some are valid only when it is set and others only when it is clear.

        □ FIN (1 bit): No more data from sender

- Window size (16 bits): This gives the size of the *receive window*, which specifies the number of window size units (by default, bytes) (beyond the sequence number in the acknowledgement field) that the sender of this segment is currently willing to receive.

- Checksum (16 bits): The 16-bit checksum field is used for error-checking if error has occurred in the header and data.

- Urgent pointer (16 bits): If the URG flag is set, then this 16-bit field is an offset from the sequence number, indicating the last urgent data byte.

- Options (variable 0–320 bits, divisible by 32): The length of this field is determined by the data offset field. Options have up to three fields: option-kind (1 byte), option-length (1 byte), and option-data (variable).

- Padding: The TCP header padding is used to ensure that the TCP header ends and data begins on a 32-bit boundary. The padding is composed of zeros.

10.6 MEDIUM ACCESS CONTROL

Medium access control is very reliable in case of wired networks. In case of wireless networks, it is concerned with controlling each terminal's access to the radio resource. The main issues that MAC has to deal with are (a) prevention of collisions, (b) prevention of blocking, and (c) fairness in allocation of resources. The quality of service parameters with which it deals are latency (delay), jitter, losses, and so on.

Three broad classes of MAC were already described in Chapter 9:

> In infrastructure-based networks, the infrastructure is responsible for solving MAC-related problems. In ad hoc networks, each terminal has to be responsible for its own MAC.

    (a) Channel partitioning (divides the channel into small pieces of time slots, frequencies, or codes)

    (b) Fixed allocation of resources, random access (allows and recovers from collisions)

    (c) Reservation-based access (tightly coordinated shared access to avoid collisions)

Apart from these, roll call polling-based MAC is a scheme that is equally responsible for link management.

Various multiple access with collision avoidance (MACA) protocols are available, one of which is described here.

REQUEST TO SEND/CLEAR TO SEND SIGNALLING

A simple and elegant solution to the hidden node problem is MACA protocol with RTS/CTS (request to send/clear to send). RTS/CTS is signalling with *handshaking*.

- Before sending a packet, the transmitter sends an RTS and waits for a CTS from the receiver (see Fig. 10.7).
- The reception of a CTS indicates that the receiver is able to receive the RTS, so the packet may then be transmitted (the channel is clear in its area).
- At the same time, every node in the range of the receiver hears the CTS (even if it does not hear the RTS), so understands that a transmission is going on.
- The nodes hearing the CTS are the nodes that could potentially create collisions in the receiver (assuming a symmetric channel).

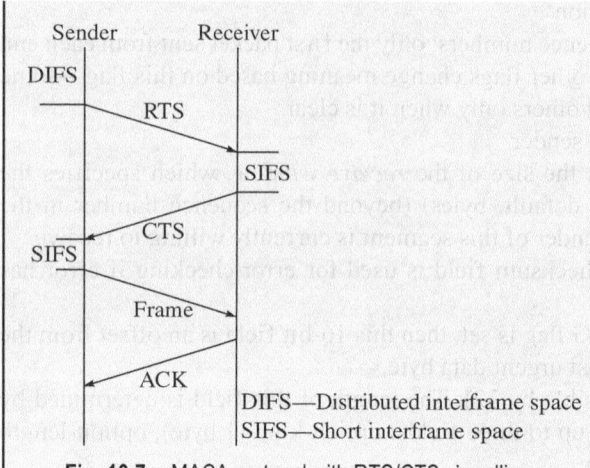

Fig. 10.7 MACA protocol with RTS/CTS signalling

DIFS—Distributed interframe space
SIFS—Short interframe space

- Since these nodes may not hear the data transmission, the RTS and CTS messages contain the size of the expected transmission (to know how long the transmission will last).
- This is the collision avoidance feature of the RTS/CTS mechanism (also called virtual carrier sense); all nodes avoid accessing the channel after hearing the CTS even if their carrier sensing indicates that the medium is free.

RTS/CTS has another advantage. It lowers the overhead of a collision on the medium (collisions are much shorter in time). If two nodes attempt to transmit in the same slot, their RTSs collide and they do not receive any CTS; hence, they lose only an RTS. In a normal scenario, they would have lost a whole packet.

RTS/CTS handshaking is not used for small packets or lightly loaded networks as it adds a significant overhead on the network. The unicast method is represented on the timing diagram as shown in Fig. 10.7.

10.7 ROUTING ALGORITHMS

> The router does the job of switching among the various nodes forming the route of packet transmission between the end users.

The throughput of a network is dependent upon the (a) type of traffic carried, (b) type of access mechanism, (c) real or non-real time communication, (d) control and signalling techniques, and (e) routing techniques adopted by the network.

Network layer design issues play an important in improving the network throughput. Switching involves strategies for making and breaking the connections among multiple users as per the requirement. Two types of switching techniques are defined in general—*circuit switching* and *packet switching*. Circuit switching is normally applied in centralized networks, whereas packet switching is applied in distributed networks or mesh networks.

Circuit switching In voice call oriented networks, circuit switching is the preferred option. Here, during a call, the link between the two parties remains dedicatedly engaged until the conversation is over. A centralized switch provides such dedicated resource during the whole span of the conversation.

Packet switching In the case of Internet and data access type of applications, packet switching is the preferred option. There are two approaches for packet switching—*connection-oriented approach* and *connectionless approach*.

Connect-oriented approach Here, when a path for end-to-end connection is selected before transferring data, signals are exchanged and all the nodes on the path mutually agree for that particular transaction. All packets are transferred through a predefined path. The drawback of this approach is that if any node or hop is in failure, the whole task of establishing the connection or data transfer may be a waste. The whole procedure must be repeated in that case. The cost incurred due to such failures increases with the size of the network.

Note: There is very little scope for routing issues in the case of centralized architecture, because the central unit handles most of the data transfer task.

Connectionless approach In the case of connectionless approach, each node has to decide the next hop for a packet to be routed. Once this is done, the job of that node finishes, and the job of the recipient of that packet starts to decide the next best hop as per the destination address entered by the sender.

For transfer of data or control packets, three different approaches are used: (a) unicasting (one-to-one), (b) multicasting (one-to-many), and (c) broadcasting (one-to-all). The approach used is decided based on the structure and protocols adopted by that network.

> *Note:* The simplest routing algorithm is the *shortest path routing algorithm*, whereby the shortest path is calculated in terms of the number of hops, distance, delay, or any suitable metric. The information to all the nodes in a fraction of time is sent using flooding, in which each node copies the available packets and forwards it to all the neighbouring nodes except from which it is received. It rejects duplicate packets.

Routing in mobile ad hoc networks Routing in mobile ad hoc networks (MANETs) is difficult because mobility causes frequent network topology changes; hence, it requires robust mechanisms to search for and maintain routes. When the network nodes move, the established paths may break and the routing protocols must dynamically search for other feasible routes. With a changing topology, even maintaining connectivity is very difficult. In addition, keeping the routes free from loops is even more difficult when the users are mobile. Besides handling the topology changes, routing protocols in MANETs must deal with other constraints such as low bandwidth, limited energy consumption, and high error rates, all of which are inherent in any wireless environment. Asymmetric links with different power levels and different terrain conditions make the routing protocols even more complicated.

There are three types of routing protocols:

Proactive protocols These are mostly table-driven protocols. Proactive methods maintain routes to all nodes, including those to which no packets are sent. Such methods are adaptive to topology changes. Tables are exchanged periodically among the neighbouring nodes with the latest upgrades.

Reactive protocols These protocols are based on the demand for data transmission. Routes between hosts are determined only when they are explicitly needed to forward packets. These methods significantly reduce routing overhead when the traffic is light and the topology changes less dramatically.

Hybrid protocols These protocols combine the proactive and reactive methods to find efficient routes, without much control overhead.

A few basic routing protocols for ad hoc networks are explained in Sections 10.7.1–10.7.4.

10.7.1 Destination-sequenced Distance Vector Routing

A routing table may contain stale entries–those that have not been updated for a while. Such entries as well as the routes using those nodes as next hops are deleted.

Destination-sequenced distance vector (DSDV) is a table-driven routing scheme for MANETs based on the Bellman–Ford algorithm. In a distance vector algorithm, each router maintains a vector of the shortest path lengths to every other node (distances) and the first hop on each path (directions). Each router performs *n* separate computations, one for each potential destination node.

Bellman–Ford algorithm is an algorithm that computes the shortest paths from a single source vertex (node) to all the other nodes in a spanning tree. It initializes

Dest	Next hop	Metric	Seq.no.
H1	H4	3	406
H2	H4	2	128
H3	H4	3	564
H4	H4	1	710
H5	H7	3	392
H6	H6	0	076
H7	H7	1	128
H8	H7	2	050

Fig. 10.8 Typical routing table for DSDV protocol

Each entry in a routing table contains a sequence number. The sequence numbers are normally even numbers if a link is present; else, an odd number is used. A typical table is shown in Fig. 10.8. The sequence number is generated by the receiver or destination, and the sender needs to send the next update with this number. Routing information is distributed between nodes by sending *full dumps* infrequently and smaller incremental updates more frequently.

If the router receives new information, then it uses the latest sequence number. If the sequence number is the same as the one present in the table, the route with the better metric is used.

> WRP is the enhanced version of DSDV algorithm and performs so because of maintaining multiple tables.

the distance to the source vertex as zero and all other vertices to infinity. It then does V passes (V is the number of vertices), over all edges and communication links, updating the distance to the destination of each edge. Finally, it checks each edge again to detect negative weight cycles, in which case it returns false.

> *Note*: Bellman–Ford algorithm is slower than Dijkstra's shortest path algorithm. However, it is more versatile, as it can handle those graphs where some of the edge weights are negative numbers. Dijkstra's shortest path algorithm needs global information about the topology, whereas in Bellman–Ford, only local information about the neighbouring nodes is sufficient.

The main disadvantages of the Bellman–Ford algorithm in this setting are as follows:
- It does not scale well. Changes in network topology are not reflected quickly since updates are spread node by node.
- It leads to the count-to-infinity problem (if link or node failures render a node unreachable from some set of other nodes, then those nodes may spend forever, gradually increasing their estimates of the distance to it; in the meantime, this results in routing loops).

The main contribution of the DSDV algorithm is to solve the routing loop problem because of the inclusion of sequence numbers.

The DSDV protocol requires a regular update of its routing tables, which uses up battery power and a small amount of bandwidth even when the network is idle. Whenever there is a change in the topology, a new sequence number is necessary before the network re-converges. Thus, DSDV is not suitable for highly dynamic networks.

10.7.2 Wireless Routing Protocol

Wireless routing protocol (WRP) is a proactive unicast routing protocol for MANETs. It uses an enhanced version of the distance vector routing protocol, which uses the Bellman–Ford algorithm to calculate paths. Due to the mobile nature of the nodes within the MANET, the protocol introduces mechanisms that reduce route loops and ensure reliable message exchange. WRP inherits the following properties:
- To counter the count-to-infinity problem and to enable faster convergence, it employs a unique method of maintaining tabular information regarding the following:
  - The shortest distance to every destination node in the network

□ The distance to the penultimate hop node on the path to every destination node
- As WRP maintains an up-to-date view of the network, every node has a readily available route to every destination node in the network.
- It differs from DSDV in table maintenance and in update procedures. DSDV maintains only one topology table, whereas WRP uses a set of tables to maintain more accurate information. The following tables are maintained by a node:
  □ The *distance table* (DT) contains the network view of the neighbours of a node due to timely exchange of the tables. It contains a matrix where each element contains the distance and the penultimate node reported by a neighbour for a particular destination.
  □ The *routing table* (RT) contains the up-to-date view of the network for all known destinations. The table entries contain the distance to the destination node and the previous and next nodes along the route and are tagged to identify the route's state: whether it is a simple path, loop, or invalid route. Storing the previous and successive nodes assists in detecting loops and avoiding the count-to-infinity problem, a shortcoming of distance vector routing.

  Thus, the RT maintains the following information: (a) the shortest distance, (b) the predecessor node (penultimate node), (c) the successor node (the next node to reach the destination), and (d) a flag indicating the status of the path.

> *Note*: The path status may be a simple path (correct), a loop (error), or the destination node not marked (null).

  □ The *link cost table* (LCT) contains the cost (e.g., the number of hops to reach the destination) of relaying messages through each link. The cost of a broken link is infinity. It also contains the number of update periods (intervals between two successive periodic updates) passed since the last successful update was received from that link. This is done to detect link breaks.

In addition, each node implementing WRP maintains a *message retransmission list* (MRL). This list contains an entry for every update message to be retransmitted and maintains a counter for each entry. This counter is decremented after every retransmission of an update message. Each update message contains a list of updates. A node also marks each node in the RT that has to acknowledge the update message it transmitted. Once the counter reaches zero, the entries in the update message for which no acknowledgements have been received are to be retransmitted and the update message is deleted. Thus, a node detects a link break by the number of update periods missed since the last successful transmission. After receiving an update message, a node not only updates the distance for transmission neighbours but also checks the other neighbours' distance; hence, convergence is much faster than in DSDV.

Nodes periodically exchange routing tables with their neighbours via update messages or whenever the link state table changes. The MRL maintains a list of neighbours that are yet to acknowledge an update message so that they can be retransmitted if necessary. When there is no change in the routing table, a node is required to transmit a *hello* message to affirm its connectivity. When an update message is received, a node updates its distance table and reassesses the best route paths. It also carries out a consistency check with its neighbours to help eliminate loops and speed up convergence.

WRP has advantages similar to those of DSDV. Apart from that, WRP has faster convergence and involves fewer table updates.

WRP is not suitable for large MANETs as it suffers from limited scalability.

However, the complexity of maintenance of multiple tables demands a larger memory and greater processing power from nodes in the ad hoc wireless network.

At high mobility, the control overhead involved in updating table entries is almost the same as that of DSDV and hence is not suitable for highly dynamic and for very large ad hoc wireless networks. Hence, WRP requires large memory storage and resources in maintaining its tables.

10.7.3 Dynamic Source Routing

Dynamic source routing (DSR) protocol is a reactive, simple, and efficient routing protocol designed specifically for use in multihop wireless ad hoc networks of up to about two hundred mobile nodes. It is designed to work well with even very high rates of mobility. DSR allows the network to be completely self-organizing and self-configuring, without the need for any existing network infrastructure or administration. Networks using the DSR protocol have been connected to the Internet.

The following are a few properties of DSR:

▪ DSR can interoperate with mobile IP (MIP, discussed in Section 10.11), and nodes using MIP and DSR have seamlessly migrated between wireless LANs, cellular data services, and DSR MANETs.

▪ All aspects of the protocol operate entirely on demand, allowing the routing packet overhead of DSR to scale automatically to only what is needed to react to changes in the routes currently in use.

▪ It uses source routing instead of relying on the routing table at each intermediate device.

▪ It allows multiple routes to any destination and allows each sender to select and control the routes used in routing its packets.

The protocol is composed of two main mechanisms phases—(a) route discovery and (b) route maintenance. These two work together to allow the nodes to discover and maintain routes to arbitrary destinations in the ad hoc network.

> *Note:* There are two control packets—Route Request and Route Reply.

A *Route Request packet* is generated by the source node. The basic approach of the DSR protocol (and all other on-demand routing protocols) during the route construction phase is to establish a route by flooding Route Request packets in the network.

On receiving a Route Request packet, which carries the route traversed by the packet received, the destination node responds by sending back a *Route Reply packet* to the source. Route Reply will be generated only if the message has reached the intended destination node. To send the Route Reply, the destination node must have a route to the source node. If the route is in the destination node's route cache, it will be used. Otherwise, the node will reverse the route based on the route record in the Route Request message header. This requires that all links be symmetric.

In addition, a *Route Error packet* is generated at a node when the route maintenance phase is initiated in the event of fatal transmission. The erroneous hop will be removed from the node's route cache; all routes containing the hop are truncated at that point. Again, the route discovery phase is initiated to determine the most viable route.

The following are the various phases in a DSR algorithm.

Route Discovery Phase

The DSR protocol uses cache memory in an efficient way to reduce the delays.

This phase is required to find the packet transmission route. Due to the dynamic topology, some specific procedures are to be incorporated as follows:

Requirements Determining source routes requires accumulating the address of each device between the source and destination during route discovery. The accumulated path information is cached by the nodes processing the route discovery packets. The learned paths are used to route the packets. To accomplish source routing, the routed packets normally contain the address of each device the packet will traverse. This may result in high overhead for long paths or large addresses, like IPv6. To avoid this, DSR optionally defines a flow id option that allows packets to be forwarded on a hop-by-hop basis. The major difference between this and the other on-demand routing protocols is that this is beaconless and hence does not require periodic hello packet (beacon) transmissions, which are used by a node to inform its neighbours of its presence.

Steps The following are the steps involved in the route discovery phase:

- Consider a source node that does not have a route to the destination. When it has data packets to be sent to that destination, it initiates a Route Request packet.
- This Route Request is flooded throughout the network.
- Upon receiving the Route Request packet, each node rebroadcasts the packet to its neighbours if it has not forwarded it already, provided that the node is not the destination node and that the packet's time to live (TTL) counter has not been exceeded.
- Each Route Request carries a sequence number generated by the source node and the path it has traversed. A node, upon receiving the Route Request packet, checks the sequence number on the packet before forwarding it. The packet is forwarded only if it is not a duplicate Route Request. The sequence number on the packet is used to prevent loop formations and to avoid multiple transmissions of the same Route Request by an intermediate node that receives it through multiple paths.

Thus, all nodes except the destination node forward a Route Request packet during the route construction phase. The destination node, after receiving the first Route Request packet, replies to the source node through the reverse path the Route Request packet had traversed. Nodes can also learn about the neighbouring routes traversed by data packets if operated in the promiscuous mode (the mode of operation in which a node can receive the packets that are neither broadcast nor addressed to itself). This route cache is also used during the route construction phase. The scenario is presented in Fig. 10.9, in which the path decided for communication is A-D-G-H-J.

Route Maintenance Phase

The route maintenance process is triggered when a link breaks between two nodes along the path from the source to the destination. The node that discovers the link break in the route sends a Route Error packet back to the source to inform about the broken link. After receiving the Route Error packet, the source node initiates another route discovery process. Additionally, at

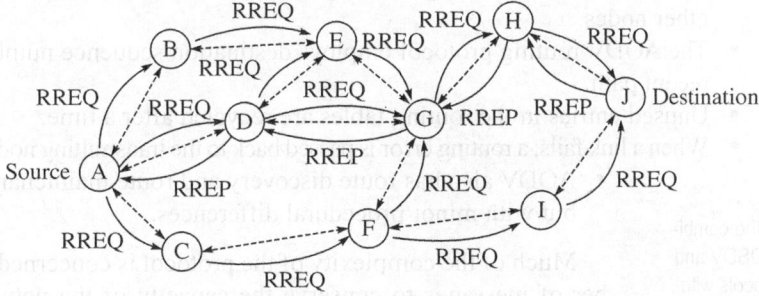

Fig. 10.9 DSR route discovery

the same time, the routes containing the broken link should be removed from the route caches of the intermediate nodes when the Route Error packet is transmitted to the source.

Advantages

This protocol eliminates the need to periodically flood the network with table update messages that are required in a table-driven approach. Hence, it restricts the bandwidth consumed by control packets. In a reactive approach such as this, a route is established only when it is required, and hence, the need to find routes to all other nodes in the network as required by the table-driven approach is eliminated. The intermediate nodes also utilize the route cache information efficiently to reduce the control overhead. Other advantages of the DSR protocol include easily guaranteed loop-free routing, support for use in networks containing unidirectional links, use of only *soft state* in routing, and very rapid recovery when routes in the network change.

Disadvantages

The main disadvantage of this protocol is that the route maintenance mechanism does not locally repair a broken link. Stale route cache information can also result in inconsistencies during the route reconstruction phase. The connection set-up delay is higher than in table-driven protocols. Even though the protocol performs well in static and low-mobility environments, the performance degrades rapidly with increasing mobility. In addition, considerable routing overhead is involved due to the source-routing mechanism employed in DSR. This routing overhead is directly proportional to the path length.

10.7.4 Ad hoc On Demand Vector Routing

Ad hoc on demand distance vector (AODV) routing is a routing protocol for MANETs and other wireless ad hoc networks. It is a reactive routing protocol, meaning that it establishes a route to a destination only on demand. AODV avoids the count-to-infinity problem of other distance vector protocols by using sequence numbers on route updates, a technique pioneered by DSDV. AODV is capable of both unicast and multicast routing.

The following are the features of AODV:

- The network is silent until a connection is needed.
- Whenever a network node needs a connection, it broadcasts a request for connection.
- Other AODV nodes forward this message and record the node that they heard it from, creating an explosion of temporary routes back to the needy node.
- When a node receives such a message and already has a route to the desired node, it sends a message back through a temporary route to the requesting node.
- The needy node then begins using the route that has the least number of hops through other nodes.
- The AODV routing protocol employs destination sequence numbers to identify the most recent path.
- Unused entries in the routing tables are recycled after a time.
- When a link fails, a routing error is passed back to the transmitting node, and the process repeats.
    - AODV also has route discovery and route maintenance phases just like DSR but with minor procedural differences.

> AODV is the combination of DSDV and DSR protocols with minor modifications.

Much of the complexity of the protocol is concerned with lowering the number of messages to conserve the capacity of the network. For example, each request for a route has a sequence number. Nodes use this sequence number so that they do not repeat route requests that they have already passed on. Another feature is that the route requests have a TTL number that limits the number of times they can be retransmitted. Moreover, if a route request fails, another route request may not be sent until twice as much time has passed as the timeout of the previous route request.

Note: The major difference between AODV and DSR stems from the fact that DSR uses source routing in which a data packet carries the complete path to be traversed. However, in AODV, the source node and the intermediate nodes store the next-hop information corresponding to each flow for data packet transmission.

Another difference is that it uses a *destination sequence number* (DestSeqNum) to determine an up-to-date path to the destination just like the sequence number used in DSDV. A node updates its path information only if the DestSeqNum of the current packet received is greater than or equal to the last DestSeqNum stored at the node with a smaller hop count.

In an on-demand routing protocol, the source node floods the Route Request packet in the network when a route is not available for the desired destination. It may obtain multiple routes to different destinations from a single Route Request.

Fields of Route Request Packet

A Route Request carries the *source identifier* (SrcID), *destination identifier* (DestID), *source sequence number* (SrcSeqNum), *destination sequence number* (DestSeqNum), *broadcast identifier* (BcastID), and TTL field.

DestSeqNum indicates the freshness of the route that is accepted by the source. When an intermediate node receives a Route Request, it either forwards it or prepares a Route Reply if it has a valid route to the destination. The validity of a route at the intermediate node is determined by comparing the sequence number at the intermediate node with the destination sequence number in the Route Request packet.

If a Route Request is received multiple times, which is indicated by the BcastID–SrcID pair, the duplicate copies are discarded. Any intermediate node having valid routes to the destination, or the destination node itself, is allowed to send Route Reply packets to the source. While forwarding a Route Request, every intermediate node enters the previous node address and its BcastID.

A timer is used to delete this entry in case a Route Reply is not received before the timer expires. This helps in storing an active path at the intermediate node because AODV does not employ source routing of data packets. When a node receives a Route Reply packet, information about the previous node from which the packet was received is also stored in order to forward the data packet to the next node as the next hop towards the destination.

Advantages

The main advantage of AODV protocol is that routes are established on demand and that destination sequence numbers are applied to find the latest route to the destination. The connection set-up delay is low. Another advantage is that it creates no extra traffic for communication along existing links. In addition, distance vector routing is simple and does not require much memory or calculation.

Disadvantages

The AODV protocol requires more time to establish a connection and the initial communication to establish a route is heavier than in some other approaches. Moreover, intermediate nodes can lead to inconsistent routes if the source sequence number is very old and the intermediate nodes have a higher but not the latest destination sequence number, thereby having stale entries. Multiple Route Reply packets in response to a single Route Request packet can lead to heavy control overhead. Another disadvantage of AODV is unnecessary bandwidth consumption due to periodic beaconing.

There are other routing techniques such as location-based routing (LBR) and zone-based routing (ZBR). LBR uses the global positioning system (GPS) for finding the tentative location of a user. ZBR is a hybrid technique. The references given at the end of the book may be used to study more routing techniques.

10.8 TRANSPORT CONTROL MECHANISMS

> TCP is optimized for accurate delivery rather than timely delivery of packets.

The transport layer and its control issues provide reliable and ordered delivery of a stream of octets from a program on one computer to another program on another computer. The main function of segmentation of the database and reassembling is shown in Fig. 10.10. This task is applied to mobile TCP as well.

Due to network congestion, traffic load balancing, or other unpredictable network behaviour, IP packets can be lost, duplicated, or delivered out of order. The transport layer detects these problems, requests retransmission of lost data, rearranges out-of-order data, and even helps minimize network congestion to reduce the occurrence of the other problems.

Once the transport layer of the receiver reassembles the sequence of octets originally transmitted, it passes them to the application program. Thus, transport control abstracts the application's communication from the underlying networking details. TCP sometimes incurs long delays (in the order of seconds) while waiting for out-of-order messages or retransmissions of lost messages. It is not particularly suitable for real-time applications such as Voice over IP. For such applications, protocols like real-time transport protocol (RTP) running over the UDP are usually recommended instead.

This section discusses the basic protocols required to have end-to-end flow control and congestion control, which are equally applicable on a link or hop basis. Indirectly, they also support error control on a link.

10.8.1 Stop and Wait Protocol

Stop and wait is the simplest technique and is adequate for a very simple communications protocol working on a half duplex basis. A stop and wait protocol transmits a protocol data unit (PDU) of information and then waits for a response. The receiver receives each PDU and

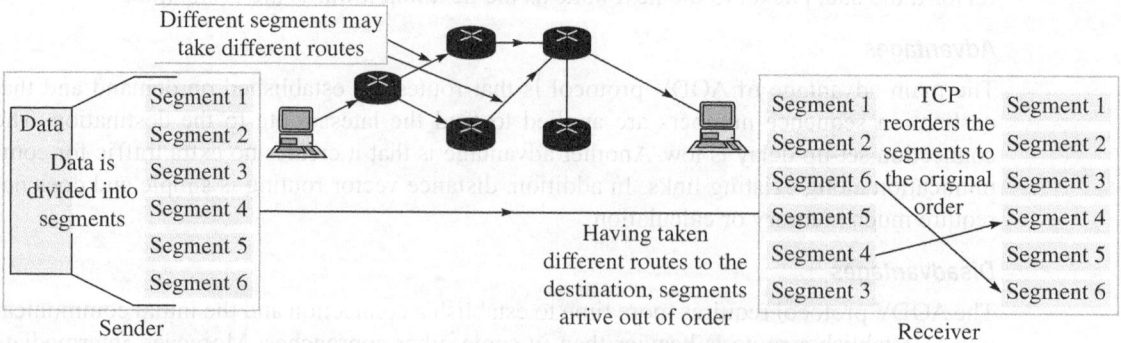

Fig. 10.10   Segmentation and reassembling of the data

sends an acknowledgement (ACK) PDU if the data PDU is received correctly, and a negative acknowledgement (NACK) PDU if the data is not received or received in error. In practice, the receiver may not be able to reliably identify whether a PDU has been received, and the transmitter will usually also need to implement a timer to recover from the condition where the receiver does not respond. Under normal transmission, the sender will receive an acknowledgement for the data and then commence transmission of the next data block. For a long delay link, the sender may have to wait an appreciable time for the response; hence, it is a slow method.

10.8.2 Sliding Window Protocol

If the stop and wait protocol is implemented on a bidirectional basis, we have two separate physical circuits, each with a *forward* channel (for data) and a *reverse* channel (for acknowledgements). In both cases, the bandwidth of the reverse channel is almost entirely wasted. In effect, the user is paying for two circuits but using only the capacity of one. The sliding window protocol is a protocol that allows simultaneous data transmission in both the directions on a full duplex basis. Figure 10.11 represents a frame-oriented protocol. The window size is of five frames. The figure shows the transmitting end scenario. The window slides on the right side as per the acknowledgement received on the reverse link unless all the frames are transmitted.

Cumulative acknowledgement is the method used by TCP. The other kinds of acknowledgement methods are selective ACK and selective NACK. Apart from this, various timers are maintained to identify the active and time out connections.

Transport protocol operations may be divided into three phases. Connections must be properly established in a multi-step handshake process (*connection establishment*) before entering the *data transfer* phase. After data transmission is completed, *connection termination* closes established virtual circuits and releases all allocated resources. A TCP connection is managed by an operating system through a programming interface that represents the local end-point for communications in the form of the *Internet socket*. During the lifetime of a connection-oriented transport (TCP as per the presently adopted protocol), the local end-point undergoes a series of state changes as follows:

> The stop and wait protocol and sliding window protocol are for flow and error control. Both may be bit oriented or byte oriented.

LISTEN (server) represents waiting for a connection request from any remote TCP and port.

SYN-SENT(client) represents waiting for a matching connection request after having sent a connection request.

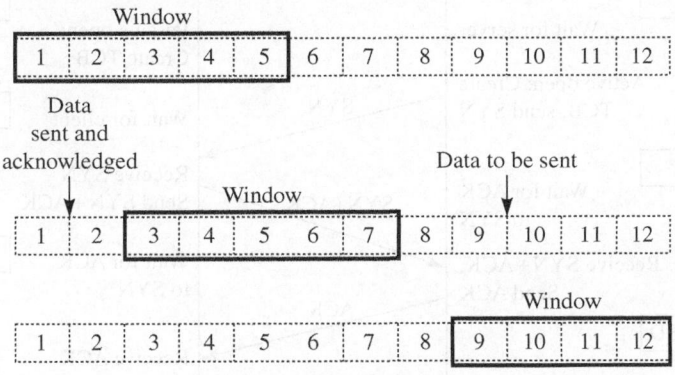

Fig. 10.11   Sliding window protocol

Transport control follows the client–server model for correspondence.

SYN-RECEIVED(server) represents waiting for a confirming connection request acknowledgement after having both received and sent a connection request.

ESTABLISHED(both server and client) represents an open connection; data received can be delivered to the user (the normal state for the data transfer phase of the connection).

FIN-WAIT-1(both server and client) represents waiting for a connection termination request from the remote TCP or an acknowledgement of the connection termination request previously sent.

FIN-WAIT-2(both server and client) represents waiting for a connection termination request from the remote TCP.

CLOSE-WAIT(both server and client) represents waiting for a connection termination request from the local user.

CLOSING(both server and client) represents waiting for a connection termination request acknowledgement from the remote TCP.

LAST-ACK(both server and client) represents waiting for an acknowledgement of the connection termination request previously sent to the remote TCP (which includes an acknowledgement of its connection termination request).

TIME-WAIT(either server or client) represents waiting for enough time to pass to be sure that the remote TCP received the acknowledgement of its connection termination request. [According to RFC 793, a connection can stay in TIME-WAIT for a maximum of 4 min, which is known as the maximum segment lifetime (MSL).]

> *Note:* RFC 793 is a transmission control protocol. Sometimes written as IETF RFC 793 (IETF is Internet Engineering Task force). It is specified for Defense Advanced Research Projects Agency (DARPA) internet program.

CLOSED(both server and client) represents no connection state at all.

The timing diagrams using these signals for connection establishment and termination phases are shown in the Figs 10.12 and 10.13.

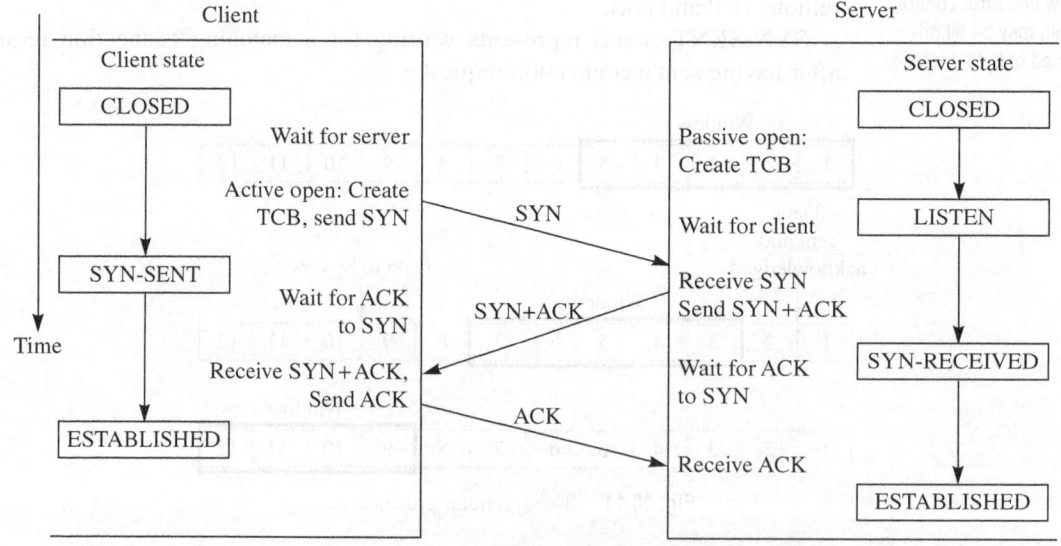

Fig. 10.12   Connection establishment phase

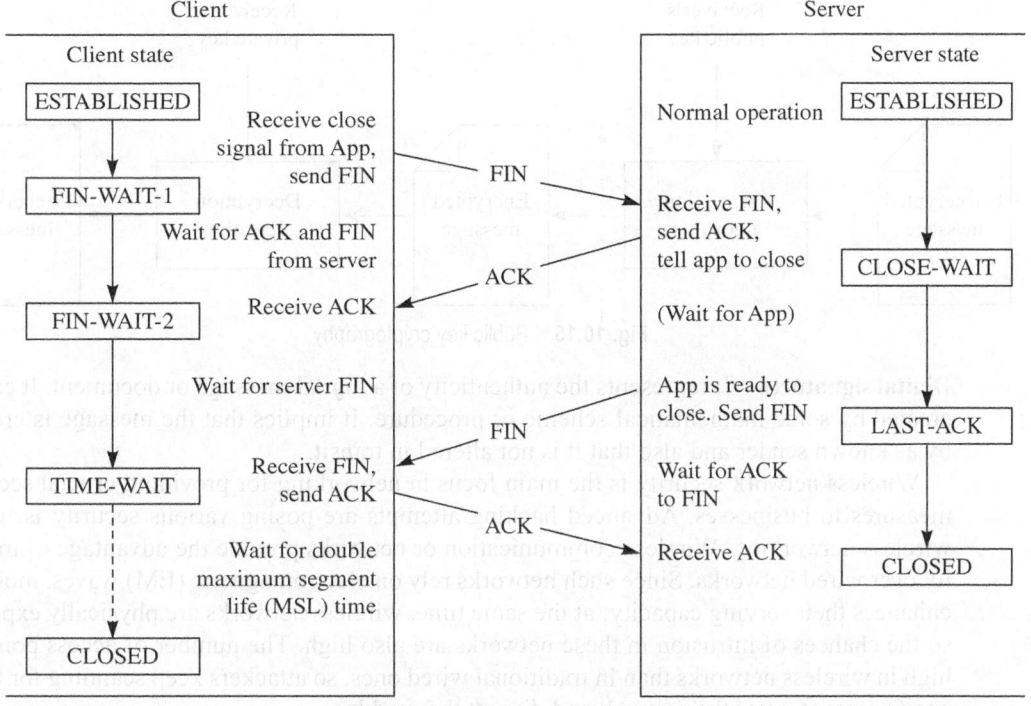

Fig. 10.13 Connection termination phase

10.9 SECURITY ASPECTS

Many solutions have been developed for wired security, such as the use of cryptography, authentication procedure, and digital signature. Cryptography is the encryption of the data at the transmitting end and decryption at the receiving end using a secret key (Fig. 10.14). When the key is private and known only to the user for both encryption and decryption, it is known as private key cryptography.

Another option is to encrypt the data with a public key that is known to everybody and to decrypt it with a private key. This approach is called public key cryptography (Fig. 10.15). These techniques are used in the wireless scenario as well. There are many problems in wireless networks, which are discussed here.

Authentication It is the process of identifying the correct user name and password. Authentication and authorization are two different processes. Authorization is the process of giving individuals access to system objects based on their identity.

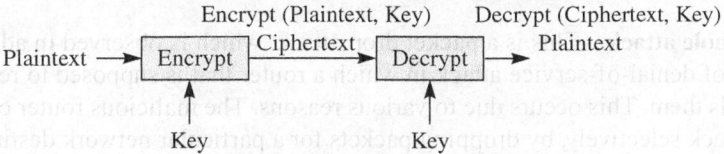

Fig. 10.14 Private key cryptography

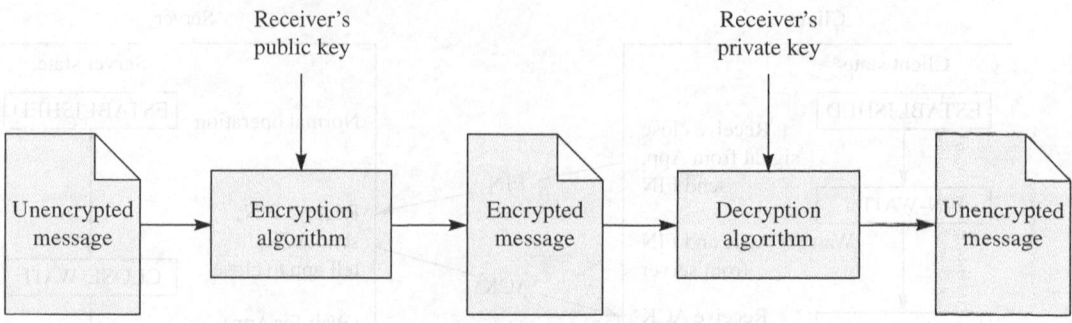

Fig. 10.15 Public key cryptography

Digital signatures It represents the authenticity of a digital message or document. It can be created by some mathematical scheme or procedure. It implies that the message is created by as known sender and also that it is not altered in transit.

Wireless network security is the main focus in networking for providing crucial security measures to businesses. Advanced hacking attempts are posing various security issues in wireless networking. Wireless communication or networks provide the advantage of mobility over wired networks. Since such networks rely on electromagnetic (EM) waves, mobility enhances their serving capacity; at the same time, wireless networks are physically exposed, so the chances of intrusion in these networks are also high. The number of access points is high in wireless networks than in traditional wired ones, so attackers keep scanning for those open points to enter the network and disrupt the working.

Carrier jamming is the most serious problem at the physical layer. Some other security issues or threats affecting wireless networks are as follows:

Creating rogue access points Rogue access points are created within the range of existing wireless LAN. These create an illusion to a node of the network that rogue point is a part of the network and associates with such point. These are short duration attacks, which are vulnerable for a short duration of time. Once the attacker associates itself with the physical port of a company's network, it can extend the period of vulnerability.

Sniffing Sniffer is a software application used by attackers or hackers to scan the traffic and different access points in a location having many wireless networks. It helps them find an open unsecured network, which hackers use to latch themselves with.

Denial of service This need not access any port or location in the server. The attacker simply overwhelms the network of a company by sending large packets of data so as to slow down the processing capabilities of its server. This forces the server to deny the service to be provided to the user. In wireless networks, this is achieved by interfering with the frequency of operation of any wireless network.

Black hole attack This is a packet drop attack, which is observed in ad hoc networks. It is a type of denial-of-service attack in which a router that is supposed to relay packets instead discards them. This occurs due to various reasons. The malicious router can also accomplish this attack selectively, by dropping packets for a particular network destination alone.

Bluesnarfing and Bluejacking Not only the conventional LANs but also Bluetooth can be used with the intention of data tampering and data theft. Bluejacking allows any unauthorized

Attacks are effective on various layers. Jamming is the most serious attack at the physical layer.

device to send to a device messages that could be malicious in nature. Bluesnarfers can steal data from a device. The biggest danger in this technique is that the bluesnarfer steals data from the device without the user's knowledge.

Elvin twin   This threat creates a replica of the authorized host at the attacker's point. The authorized access point is blocked by the attacker and the user is redirected via another access point under the authority of the attacker. This allows the attacker to control and analyse all the traffic from the user, including the keystrokes.

War driver   In this method, a wireless device such as a laptop or personal digital assistant (PDA) is required to attack a network. The device roams around the network and locks on with an unprotected wireless network. The hacker then records all the information IDs, the physical address of the network, and other such details and uploads them on an online database. This directs all intruders to the unsecured address and creates a chaos in the wireless network.

Wireless networks provide the attackers, the ease of being present in close proximity to the network, rather than being physically connected with the network. This enhances the chances of attacks and intrusion in a wireless network, thus compromising the privacy of the users.

10.10 APPLICATION LAYER

The following protocols are developed for the application layer of the Internet.

FTP   This is the standard way to transfer (send or fetch) files between different machines. In order to transfer files from the FTP server on the Unix or Telnet, an FTP client program is needed on the user's computer. The FTP program uses the TCP/IP to transfer files to and from remote nodes. Sometimes, it will read the host files to determine the IP address. FTP deals with authentication, listing a directory contents, ASCII or binary files, and so on.

Remote login (Telnet)   This is a remote terminal protocol that allows a user at one site to establish a TCP connection to another site and then pass keystrokes from the local host to the remote host. Here, pre-registration of the user along with password is required to create a secure login.

SMTP   This protocol allows a mail delivery agent on a local machine to connect to a mail delivery agent on a remote machine and deliver text-based email. Its multipurpose Internet mail extension (MIME) allows multimedia attachment with the email.

WIRELESS APPLICATION PROTOCOL

Wireless application protocol (WAP) is a standard protocol for the presentation, delivery, and access of wireless information over a mobile wireless network. A WAP browser is a Web browser for mobile devices. It offers interactive data services. The WAP standard is described as a protocol suite allowing the interoperability of WAP equipment and software with different network technologies (bearer standards or physical layers) such as global system for mobile communication (GSM) and IS-95.

The WAP forum has published a global wireless protocol specification, based on existing Internet standards such as Extensible Markup Language (XML) and IP, for all wireless networks. The WAP specification does not specify how data should be transmitted over the air interface. Instead, the WAP specification is intended to sit on top of existing bearer standards, so that any bearer standard can be used with the WAP to implement complete product solutions.

It defines a protocol stack that can operate on high-latency, low-bandwidth networks such as Short Message Service (SMS) and GSM Unstructured Supplementary Service Data (USSD) channels. In addition to being air interface independent, the WAP specification is also independent of any particular device.

The four WAP layers above bearers are as follows:
(a) Wireless Application Environment (WAE): Based on HTML JavaScript
(b) Wireless Session Layer (WSL): Based on the HTTP session
(c) Wireless Transport Layer Security (WTLS): Based on transport layer security (TLS) and secure socket layer (SSL) protocols that provide data encryption and authentication between applications and servers, where data is being sent across an insecure network
(d) Wireless Transport Layer (WTP): Based on TCP/IP and UDP/IP

HTTP This protocol is the base protocol for communication on the World Wide Web. It is an application protocol for a distributed, collaborative, and hypermedia information system. HTTP functions as a request–response protocol in the client–server model. The Web browser is treated as a client and the application running on a computer hosting a website is treated as a server.

10.11 MOBILE COMPUTING

Mobile computing is the task of near real time calculations for decision-making, when the user is mobile and its location is to be identified for communication purposes. It reflects the ability of fast computing of a mobile communication system. Mobile computing is human–computer interaction by which a computer is expected to be transported during normal usage. It allows portability. Wireless communication plays an important role in mobile computing. However, mobile computing does not necessarily mean wireless. It involves mobile communication, mobile hardware, and mobile software. Communication issues include ad hoc and infrastructure networks as well as communication properties, protocols, data formats, and concrete technologies. Hardware includes mobile devices or device components. Mobile software deals with the characteristics and requirements of mobile applications. The development scenario from computing aspects of mainframe computers to modern-day computing for mobile devices is summarized in Fig. 10.16.

> Mobile computing deals with real-time calculations for instant decision-making in wireless networks, for example, routing and location management, Mobile IP and TCP management, and resource allocations.

The following are the different kinds of mobility:
- *Terminal mobility*: movement of devices, caused by mobile devices
- *User mobility*: movement of users, caused by traveling people
- *Logical mobility*: movement of processes, caused by mobile codes

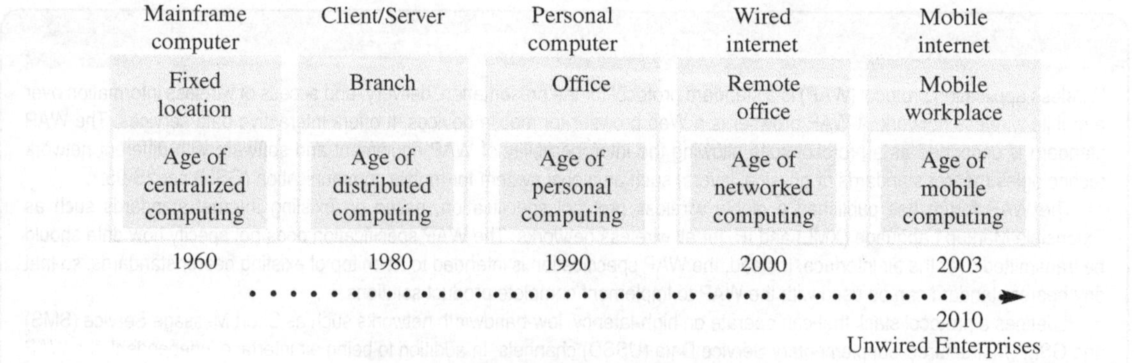

Mainframe computer	Client/Server	Personal computer	Wired internet	Mobile internet
Fixed location	Branch	Office	Remote office	Mobile workplace
Age of centralized computing	Age of distributed computing	Age of personal computing	Age of networked computing	Age of mobile computing
1960	1980	1990	2000	2003

2010
Unwired Enterprises

Fig. 10.16 Evolution of computing paradigm

The following are the different modes of computing:
- *Small office*: Take the service or application with users; computing is offline.
- *Home office*: Maintain access to the same service or application at different places; computing is online.
- *Mobile office*: Move to different places and the service or application follows; computing is moving.

MIP has been implemented vastly due to the new version of IP header IPv6. Tunnelling is possible with MIP.

Mobile computing and its complexity depend on the following key technical issues:
- *Architectures*: This technical issue deals with mobile network architectures and hosts, mobile agents and proxy architectures, mobility management algorithms, integrated wireline and wireless systems, and planning and standardization of the system.
- *Support services*: This issue includes mobility and roaming support, nomadic computing, multimedia operating system support, and power management.
- *Design and analysis of algorithms and protocols*: Suitable algorithms and protocols are required in the unreliable wireless channel environment with limited bandwidth and other resources. Algorithms are again dependent upon intermittent connectivity.
- *Mobile environments*: This issue involves data handling and knowledge management, performance modelling and characterization, security, scalability and reliability, operations, systems, and technologies.
- *Mobile communication systems*: This issue is related to wireless, cellular, and spread spectrum systems, multi-user and multi-access techniques and algorithms, multichannel processing, channel coding, data coding, and compression.
- *Applications*: Computing is not the same in all the cases and depends upon applications such as nomadic computing, wearable computers and body area networks, multimedia applications and multimedia signal processing, pervasive computing, and wireless sensor networks as well as whether the application is location dependent and sensitive.

Mobile IP

There are two versions of IP: MIP and cellular IP. MIP can efficiently provide mobility support on a global scale, for instance, migrations between local area networks. Cellular IP is optimized to support host mobility in a cellular wireless access network. Cellular IP is highly beneficial when the hosts migrate frequently. It can network with MIP to support migrations between cellular IP access networks. In short, cellular IP can be thought of as a protocol supporting micro-mobility whereas MIP is for macro-mobility.

In the context of future wireless networks, in order to make IP technology integrate all services and build true convergent networks, it is crucial to have an IP layer solution for the mobility management problem. Users will demand the same quality of service that they experience with fixed networks. Mechanisms are being developed at the IP layer to make handovers seamless for users, even when different access technologies are involved (inter-technology handovers). The use of IPv6 native mechanisms such as address auto-configuration and extension headers (such as the routing header and the destination option header) has increased the simplicity of mobility protocol implementation and, at the same time, its efficiency. MIP has the following features:
- Enables hosts to stay connected to the Internet regardless of their location
- Enables hosts to be tracked without a need to change their IP address
- Requires no changes to software of non-mobile hosts or routers
- Requires addition of some infrastructure
- Has no geographical limitations

- Requires no modifications to IP addresses or IP address format
- Supports security

There are five basic elements in an MIP scenario:

Mobile node The mobile node (MN) is the device that will travel through the different networks and also the entity that may change its point of attachment from network to network in the Internet. It registers with the *best* foreign agent (FA) when travelling in a network other than home. It is assigned a permanent IP, called its *home address*, to which other hosts send packets regardless of the MN's location. Since this IP does not change, it can be used by long-lived applications though the MN's location changes.

Home agent The home agent (HA) is a router with additional functionalities such as keeping track of the MN's location and intercepting the traffic sent to it. It is located in the home network of the MN. When the MN is away, the HA forwards packets to the appropriate network through encapsulation. It does mobility binding of the MN's IP address with its care-of-address (CoA). CoA is the address that identifies the MN's current location, which is sent by the FA to the HA when the MN attaches. It is usually the IP address of the FA.

Correspondent node The correspondent node (CN) is a regular host with whom the MN communicates.

Associated with these elements are the concepts of home network (HN) and foreign network (FN).

Home network This is the network where the HA is attached. While in this network, the MN is accessible through its home address and no MIPv6 process is necessary to the communication.

Foreign network This is the network to which the MN travels for a limited time. As it has to acquire a new IP address, a MIPv6 process is necessary to keep all the connections alive. It is another router with enhanced functionality. If the MN is away from the HA, then it uses an FA to send and receive data to and from the HA. The FA advertises itself periodically and forwards the MN's registration request. It decapsulates messages for delivery to the MN.

Let us consider a simple scenario in which we can describe and explain the MIPv6 process, shown in Fig. 10.17(a).

A MN listens for agent advertisement and then initiates registration. If the HA is the responding agent, then MIP is not necessary. On receiving the registration request from an MN, the HA acknowledges the receipt and the registration is complete. Registration happens as often as the MN changes networks. The HA intercepts all packets destined for the MN and then encapsulates all the packets addressed to the MN and forwards them to the FA; this is called IP tunnelling. The FA decapsulates all packets addressed to the MN and forwards them via the hardware address (learned as part of the registration process). The MN acquires an IP address from the dynamic host control protocol (DHCP). In general, when a mobile is in the home network, the HA manages the packet delivery; however, when a mobile is away from home, it sends information about its current location to the HA. Packets destined to a roaming MN are intercepted by its HA and, using a table, delivered via tunnelling to its present location CoA. The following points need to be noted:

- There is a specific lifetime for service before an MN must re-register.
- There is also a de-registration process with the HA if the MN returns home.

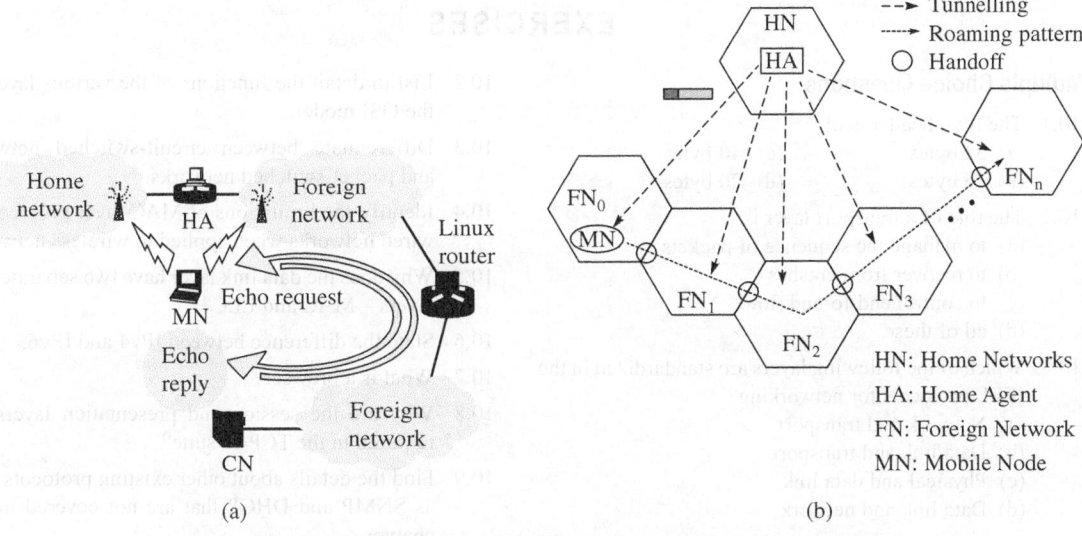

Fig. 10.17 MIPv6 processes (a) MIP elements (b) MIP Handover scenario

Changing IP address will cause on-going TCP sessions to break. Mobile TCPs must be able to restore the connection breaks.

A handoff scenario is shown in Fig. 10.17(b). A mobile node may roam across multiple subnets. At each boundary crossing, a handoff is initiated such that the CoA is updated and a new tunnel is established. Both basic MIP handoff and smooth handoff to the FA occurs at each subnet change. Reliable packet delivery in MIP networks can be modelled as a renewal process, because the retransmission over a new tunnel after each boundary crossing is independent of the previous history.

SUMMARY

- Wireless networks are more complicated than wired networks.
- Wired networks form the basis for the protocols but they are not applicable directly to wireless networks as the wireless environment is unpredictable.
- The OSI reference model forms the basis to develop networking.
- Presently, the TCP/IP suite is adopted for wired or wireless networking in the Internet—World Wide Web.
- The IP and TCP layers in the TCP/IP stack follow standard headers.
- MAC is handled by various techniques such as fixed access, random access, and reservation-based access. However, in the wireless environment, extra signalling adds more reliability due to collision avoidance, because hidden terminal and exposed terminal problems restrict the capabilities of the carrier sensing mechanism.
- Routing protocols are more complicated in a wireless scenario than in wired ones. Proactive, reactive, and hybrid protocols are designed to handle the dynamic topologies of wireless networks.

- A transmission control or transport control mechanism also requires extra concern due to uncertainty of the wireless environment and frequent breaks.
- Conventional security aspects such as authentication and cryptography are applicable to wireless networks too. Apart from that, the wireless networks should be protected from various attacks.
- Many applications such as FTP, SMTP, and HTTP that are defined for conventional networks are also applicable to the wireless scenario. A few more applications are specifically developed for the critical wireless environment.
- Mobile computing is the extra task required for tracking user location, dynamic assignment of the resources, mobility management, and so on. It needs faster processors and sophisticated algorithms to deal with near real-time calculations.
- MIP and mobile TCP are handled in wireless networks in a very specific manner with the help of computing aspects.

EXERCISES

Multiple Choice Questions

10.1 The IPv6 header is of
  (a) 32 bytes   (c) 40 bytes
  (b) 16 bytes   (d) 20 bytes

10.2 The role of a transport layer is
  (a) to manage the sequence of packets
  (b) to recover from crashes
  (c) to control end-to-end flow
  (d) all of these

10.3 Which of the following layers are standardized in the IEEE protocols for networking?
  (a) Network and transport
  (b) Data link and transport
  (c) Physical and data link
  (d) Data link and network

10.4 How many IP address classes are there?
  (a) Five   (b) Four   (c) Three   (d) Six

10.5 Which of the following components becomes important when an MN is roaming?
  (a) MN   (b) CN   (c) HA   (d) FA

10.6 Which of the following protocol is used for the short message routing?
  (a) UDP   (b) TCP   (c) MACA   (d) AODV

10.7 Which of the following routing protocol is table driven?
  (a) ZBR   (b) DSDV   (c) DSR   (d) AODV

Review Questions

10.1 Why is the OSI model not adopted as a standard but treated as a reference model?

10.2 List in detail the functions of the various layers in the OSI model.

10.3 Differentiate between circuit-switched networks and packet-switched networks.

10.4 Identify the limitations of MAC layer protocols of wired networks when applied to wireless networks.

10.5 Why does the data link layer have two separate sub-layers—MAC and LLC?

10.6 State the difference between IPv4 and IPv6.

10.7 What is a browser?

10.8 Why are the session and presentation layers not required in the TCP/IP suite?

10.9 Find the details about other existing protocols such as SNMP and DHCP that are not covered in this chapter.

10.10 Why is it necessary to study separately MIP and mobile TCP for wireless networks? What are the requirements of the Internet when the user is roaming?

10.11 Why do we need the concept of mobile computing?

10.12 How is it possible to have convergence in a network?

10.13 What is mobile Internet?

10.14 Explain the authentication procedure in general for security purpose and authentication in GSM.

10.15 How can we apply digital signatures?

10.16 What do you mean by a server? What is DNS?

11 Cellular Networks

Theme of the Chapter

Mobile networks are divided into two types: permanent infrastructure-based networks and ad hoc or personal area networks. These networks provide both voice and data services. Different protocols are designed for the two types, where provisions are made for voice and data transfer with reliability.

Only infrastructure-based networks are discussed in this chapter. All infrastructure-based networks are cellular networks, and the infrastructure is used in such a way that mobility is maintained; moreover, the management is mostly in a centralized way except in few cases like long-term evolution (LTE). The physical layer and medium access control (MAC) layer aspects of most of the systems are explained in this chapter.

Global system for mobile telecommunication (GSM), general packet radio service (GPRS), and enhanced data rate for GSM evolution (EDGE) systems, wireless local loop (WLL), Interim Standard 95 (IS-95) based on code division multiple access (CDMA), third-generation universal mobile telecommunication system (3G UMTS) based on wideband CDMA (WCDMA), and LTE—all these systems provide services for mobile phones, with voice and data applications, using the infrastructure based on the cellular theory discussed in Chapter 2. As these systems work in licensed bands, service providers must buy the required spectrum. Mobile networks can be combined with satellites, which is called mobile satellite communication. The scenario for 3G and beyond, which again gives rise to the concept of converged networks, is also provided in this chapter.

Key Topics

- GSM system
- GSM upgradations
- GPRS addition to GSM system
- EDGE technology

- CDMA systems
- WLL and CorDECT WLL
- UMTS and IMT-2000 system
- LTE networks

- Mobile satellite communication
- Converged networks and beyond 3G system

11.1 GLOBAL SYSTEM FOR MOBILE TELECOMMUNICATION

Since its deployment in 1991, the use of global system for mobile telecommunication (GSM) has grown steadily, and it is the most widely used mobile telecommunications system in the world today. GSM originally stood for Groupe Speciale Mobile, but it was changed later. GSM standards were developed for mobile telephony in digital form using a cellular structure. Though the development scenario was discussed in Chapter 1, it is recapitulated here for continuity. The following are the main aspects of wireless technology growth for mobile telephony:

- The analog mobile phones developed initially had a limited area of coverage (first generation or 1G). This system offered limited mobility and voice communication only within a group. No data communication was possible.
- Thereafter, analog cellular phones arrived with wide area coverage. This system offered complete mobility within the service area and connectivity to the public-switched telephone network (PSTN) and other networks but still without data communication facility.
- Next, digital cellular phone systems (beginning of GSM) of second generation (2G) arrived, and they still exist. This system offers complete mobility within the service area and connectivity to the PSTN and other networks. It also provides data communication facility along with other advanced features. The limitation of this system is low spectral efficiency and low data rates.
- Lastly came the digital cellular communication systems (third generation or 3G). This system offers complete mobility within the service area and connectivity to the PSTN and the Internet or intranet. It has very high data rates and other advanced features.

The following are a few points worth noting about GSM:

- The GSM was designed as a 2G cellular communication system.
- It is a fully digitized technology for better speech quality, greater capacity, and more users (within the available bandwidth) compared to previous 1G analog systems.
- GSM makes use of digital time division multiple access (TDMA) with frequency division multiple access (FDMA) and space division multiple access (SDMA), with the provision of channel allocations as per demand.
- In addition, ciphering of digitally encoded speech was adopted to retain privacy. Owing to its encryption feature, GSM calls cannot be captured by any unauthorized party.
- It can coexist with existing fixed-line networks.
- GSM allows worldwide roaming with a single number.
- Available versions of GSM are GSM 900, GSM 1800, and GSM 1900.

Table 11.1 compares the features of the three available versions of GSM.

Table 11.1 Comparison of the three available versions of GSM

Features	GSM 900	GSM 1800 (DCS 1800)	GSM 1900 (PCS 1900)
Allotted spectrum	50 MHz with FDD (i.e.,) 25 MHz uplink 25 MHz downlink	150 MHz with FDD (i.e.,) 75 MHz uplink 75 MHz downlink	120 MHz with FDD (i.e.,) 60 MHz uplink 60 MHz downlink
Frequency range—Uplink	890–915 MHz	1710–1785 MHz	1850–1910 MHz
Frequency range— Downlink	935–960 MHz	1805–1880 MHz	1930–1990 MHz
Access technology	FDMA/TDMA with SDMA	FDMA/TDMA with SDMA	FDMA/TDMA with SDMA
Number of channels	124	374 (512–885)	308 (512–810)
Channel bandwidth/spacing	200 kHz	200 kHz	200 kHz
Modulation technique	GMSK ($BT = 0.3$)	GMSK ($BT = 0.3$)	GMSK ($BT = 0.3$)
Time slot per RF channel	8 slots	8 slots	8 slots

(Contd)

Table 11.1 *(Contd)*

Features	GSM 900	GSM 1800 (DCS 1800)	GSM 1900 (PCS 1900)
Transmit—receive time slot spacing	3 slots	3 slots	3 slots
Transmit—receive frequency spacing or duplex spacing	45 MHz	95 MHz	—
Data rate	273.833 kbps	273.833 kbps	273.833 kbps
Frame duration	4.615 ms	4.615 ms	4.615 ms
Encoding	RPE—LTP, EFR	RPE—LTP, EFR	RPE—LTP, EFR
Compatibility	ISDN and PSPDN	ISDN and PSPDN	ISDN and PSPDN
Transmission power—handset	2 W	1 W	1 W

Note: BT—Bandwidth time product; DCS—Distributed control system; EFR—Enhanced full rate; FDD—Frequency division duplex; FDMA—frequency division multiple access; GMSK—Gaussian minimum shift keying; GSM—Global system for mobile communication; ISDN—Integrated services digital network; PCS—Personal communication system; PSPDN—Public switched packet data network; RF—Radio frequency; RPE—LTP—Regular pulse excitation—long-term prediction; SDMA—Space division multiple access; TDMA—Time division multiple access.

> GSM supports common channel signalling technique SS7. It uses paired radio channels with duplex spacing of 45 MHz.

In addition to voice services, primary 2G GSM supports many other data services. Though its performance is nowhere near that of 3G, it is still important and useful. Various data services are supported with user data rates up to 9.6 kbps. The services include group 3 facsimile, videotext and teletext, voicemail, and reverse message call alert (RMCA).

One service that has grown enormously is the short message service (SMS). It was developed as a part of the GSM specification, but now it has also been incorporated into other cellular systems. It can be thought of as being similar to the paging service but is far more comprehensive, allowing bidirectional messaging and store and forward facility. It also allows alphanumeric messages.

11.1.1 GSM Architecture

The components of a cellular system were explained in detail in Chapter 2. However, the GSM architecture is discussed here in brief. The GSM system diagram is shown in Fig. 11.1(a). The architecture of the GSM system with its hardware can broadly be grouped into three main areas: (a) mobile station (MS), (b) base station subsystem (BSS), and (c) network subsystem (NSS).

Mobile Station

An MS consists of two units: (a) mobile handset with battery and (b) subscriber identity module (SIM). The mobile handset is one of the most complicated GSM devices. It provides the user with access to the network. Each handset has a unique identity number known as the *international mobile equipment identity* (IMEI). This is installed in the phone at the time of manufacturing and it *cannot* be changed. It is accessed by the network during registration to check whether the equipment has been reported as stolen.

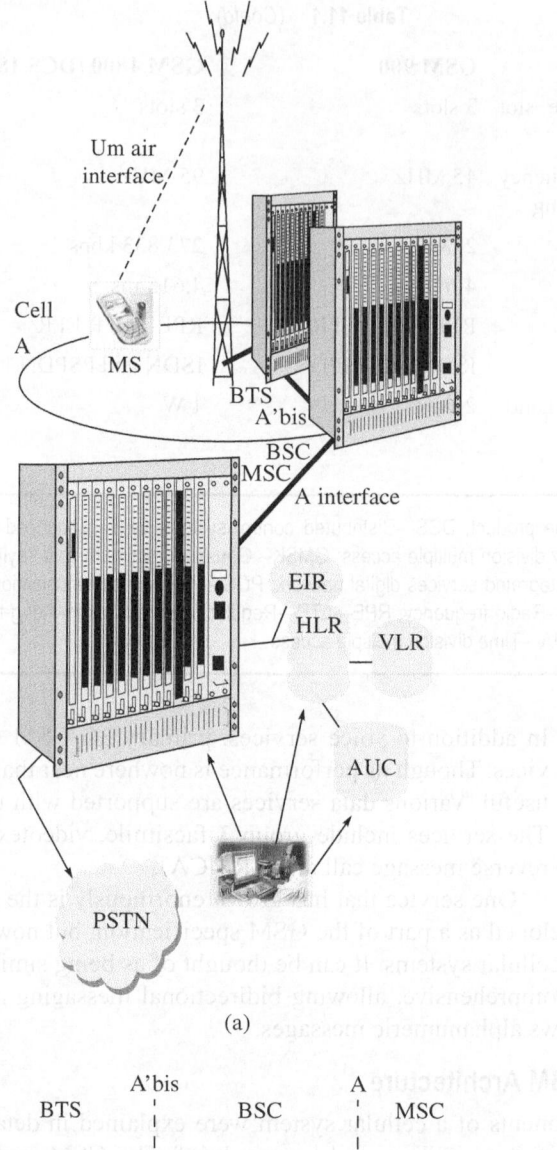

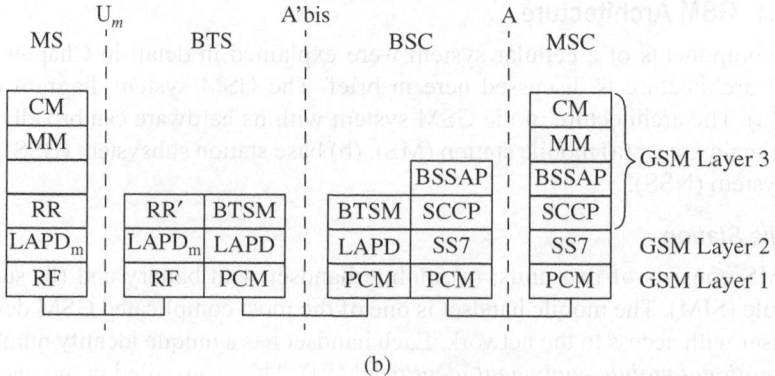

(a)

MS	U_m	BTS	A'bis	BSC	A	MSC	
CM						CM	
MM						MM	GSM Layer 3
					BSSAP	BSSAP	
RR		RR'	BTSM	BTSM	SCCP	SCCP	
LAPD$_m$		LAPD$_m$	LAPD	LAPD	SS7	SS7	GSM Layer 2
RF		RF	PCM	PCM	PCM	PCM	GSM Layer 1

(b)

Fig. 11.1 GSM (a) Complete GSM system scenario with interconnection between subsystems (b) Protocol architecture for GSM

Memory internal to a mobile handset is the limiting factor in accessing large web pages; however, the latest smartphones have overcome such issues.

The SIM is used to provide the identity of the user to the network. It contains a variety of information, including a number known as the *international mobile subscriber identity* (IMSI). It is a removable module that fits in GSM-supported mobile handsets. Each SIM has a unique IMSI. The SIM inserted in a mobile unit is self-configured. The phone number of a mobile unit is usually of 10–15 digits, out of which the first three digits are the country code, the next two digits are for the specific mobile switching centre (MSC), and the rest are the telephone's number. Mobile handsets have built-in microcomputer and memory.

Base Station Subsystem

The base station (BS) of a GSM network is fundamentally associated with communicating with the mobiles on the network. It consists of two elements, namely the *base transceiver station* (BTS) and the *base station controller* (BSC).

Base transceiver station   The BTS used in a GSM network comprises radio transceivers and their associated antennas that transmit and receive to directly communicate with the mobiles. The BTS is the defining element in each cell, and one BTS covers one or more cells. Its capacity depends on the number of transceivers present. The BTS communicates with the mobiles, and the interface between the two is known as the *Um* interface with its associated protocols. The BTS is connected to the BSC via the A'bis interface. The transmission rate on A'bis is 2 Mbps. The interface between the MS and BTS is called the *air interface*. The transmission rate on the air interface is typically 13 kbps. The BTS controls the radio frequency (RF) parameters of the MS. Each transceiver has eight TDMA channels to carry voice and signalling.

Base station controller   The BSC controls several BTSs. It manages channel allocation and handover of calls from one BTS to another. The BSC is connected to the MSC via the A interface. The transmission rate on the A interface is 2 Mbps. The BSC has the database for all of its BTS parameters. It provides the path from the MS to the MSC.

Network Subsystem

The network subsystem contains several elements and is often termed the *core network* (CN). It provides the main control and the interface for the entire mobile network. The elements included in it are described here (Fig. 11.1a).

Mobile switching centre   The MSC is the heart of the entire network connecting the fixed-line network to the mobile network. It manages all call-related functions and billing information. It is connected to the home location register (HLR) and visiting location register (VLR) for subscriber identification and for routing incoming calls. The MSC is connected to the BSC at one end through A interface and to the fixed-line network through the gateway mobile switching centre (GMSC) at the other end. A *call detail record* (CDR) is generated for each call in the MSC.

Home location register   The HLR has a permanent database of all the registered subscribers. It has a series of numbers for all subscribers. When a user switches on the phone, the phone registers with the network, and it is possible to determine the BTS with which it communicates so that incoming calls can be routed appropriately. Even when the phone

The capacity of a MSC is calculated in terms of the number of subscribers it can handle.

is not active (but switched on), it re-registers periodically to ensure that the network (HLR) is aware of its latest position. There is one HLR per network, although it may be distributed across various subcentres for operational reasons.

Visiting location register   An active subscriber is registered in the VLR, which has a temporary database of all the active subscribers used for their call routing. The HLR validates a subscriber before registration. The MSC asks the VLR before routing an incoming call.

Authentication centre (AuC)   The AuC is a protected database that contains the secret key, which is also included in the user's SIM card. It is used for authentication and for ciphering on the radio channel. Authentication is a process to verify the subscriber SIM. Secret data and the verification algorithm are stored in the AuC, which in combination with the HLR is used to authenticate the subscribers. Subscriber authentication can be done on every call, if required.

Equipment identity register (EIR)   The mobile handset data of all subscribers is stored in the EIR. It is the entity that decides whether the given mobile equipment may be allowed onto the network. Each mobile handset has an IMEI, which is installed in the equipment and is checked by the network during registration. The MSC asks the mobile to send its IMEI and then checks it with the data available in the EIR. The EIR has different classifications for mobile handsets such as white list, grey list, and black list. Depending upon the information held in the EIR, the mobile may be allocated one of three states—allowed onto the network, barred access, or monitored in case of problems. According to the categorization, the MS can make calls or can be stopped from making calls.

Operation and maintenance centre (OMC)   All network elements are connected to the OMC, which monitors the health of the network elements and carries out any maintenance operation, if required. The OMC links from MSC to the BTS are via the parent BSC. The OMC keeps records of all the faults occurred. It can also do traffic analysis and prepare the management information system (MIS) report for the network.

Gateway mobile switching centre   This is the point to which a mobile terminating call is initially routed, without any knowledge of the MS's location. The GMSC is thus in charge of obtaining the *mobile station roaming number* (MSRN) from the HLR and routing the call to the correct visited MSC.

Note: CM—Connection management; MM—Mobility management; RR—Radio resource management function; BTSM—BTS management; BSSAP—BSS (mobile) application; SCCP—Signalling connection control protocol; SS7—Signalling system 7; LAPD—Link access protocol of D channel; LAPD$_m$—modified LAPD; RF—radio air interface used for communication; PCM—pulse code modulation form of transmission—64 kbps line

---

The supported protocol standards for the architecture shown in Fig. 11.1(a) are given in Fig. 11.1(b) in terms of a layered model. The protocols are represented with equivalence to open systems interconnection (OSI) reference model and its layers.

ADDITIONAL SERVICE COMPONENTS

The SMS-G or SMS gateway (Fig. 11.2) is the term used to collectively describe the following two *short message services gateways*. These two gateways handle messages directed in different directions.

(a) The SMS-GMSC is for short messages being sent to a mobile. Its role is similar to that of the GMSC

(b) The SMS interworking mobile switching centre (SMS-IWMSC) is used for short messages originated in a mobile on the network. It provides a fixed access point to the SMS centre.

Similarly, voicemail services are also provided by voice-mail service gateway (VMS-G) (Fig. 11.2). It has the database of all voice mail service subscribers and also stores voice messages for them.

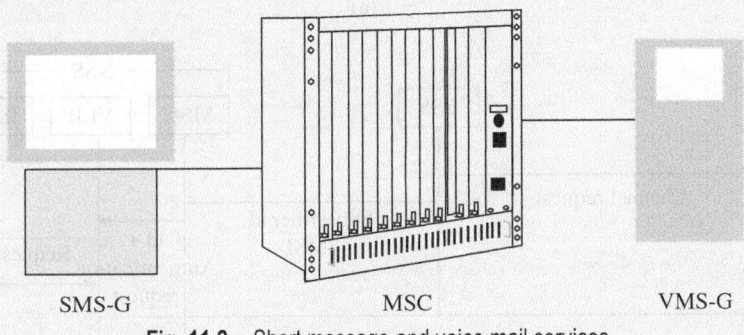

SMS-G   MSC   VMS-G

Fig. 11.2   Short message and voice mail services

11.1.2 Call Handling in GSM

This section discusses how mobile-originated and mobile-terminated calls are handled in GSM.

Mobile-originated Calls

The following steps describe the minimum procedure to handle mobile-originated calls (Fig. 11.3).

- Mobile-originated calls go to the BTS first and then to the BSC.
- The BSC forwards this call to the MSC.
- The MSC authenticates and routes the call as per the dialled digits.
- If a mobile originated call is for another mobile subscriber, then the process for the other mobile is the same as that for a mobile-terminated call, which is discussed in the next topic.

Figure 11.3(b) explains the timing sequences of a few operations for mobile-originated calls.

Mobile-terminated Calls

The following steps describe the minimum procedure to handle mobile-terminated calls (Fig. 11.4).

> A mobile sends a channel request to the BSS for initializing a call. After authentication procedures at the MSC, the request is granted in terms of channel assignment.

- Mobile-terminated calls come to the MSC first, where HLR/VLR inquiry is carried out; as per the information, the MS is paged in the suitable BSC.
- The BSC forwards this page to all children BTS where the actual paging is done.
- After BTS gets a response from the mobile, it allocates a channel for this call.
- On ending of the call, the BTS informs the BSC and the MSC.

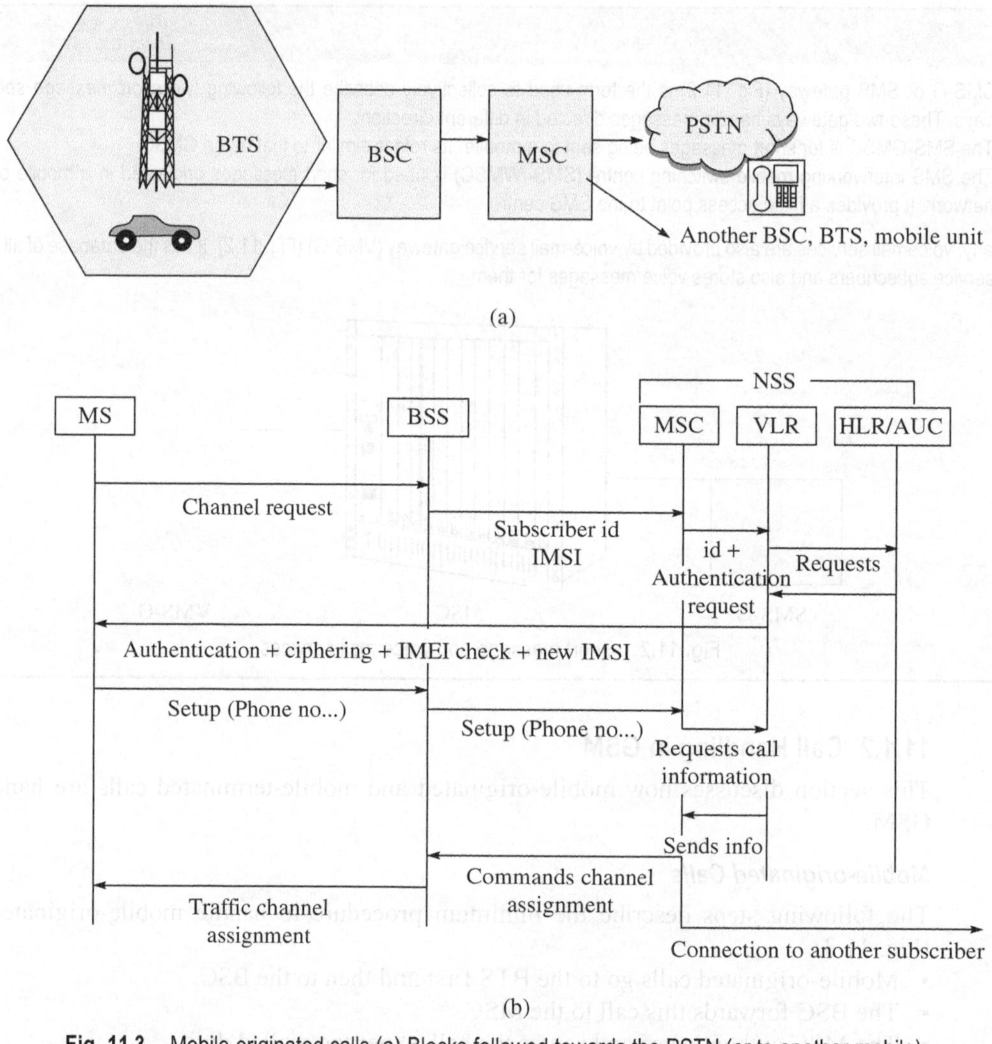

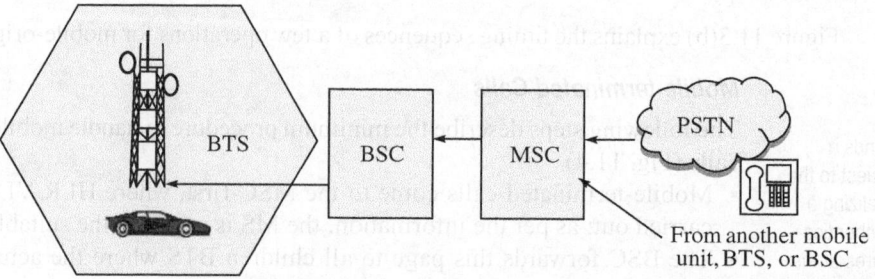

(a)

(b)

Fig. 11.3 Mobile-originated calls (a) Blocks followed towards the PSTN (or to another mobile)
(b) Sequences followed in simplified form

Fig. 11.4 Blocks followed towards BTS (with which the mobile unit is logged) for mobile-
terminated calls

GSM NETWORK MANAGEMENT

The following are a few points to be noted about GSM network management:
- Management of hard handovers was discussed in Chapter 2. This applies to the GSM system for continuity of calls.
- Different data channels are employed to carry data and exchange protocol or control messages that enable the radio sub-system to operate correctly.
- GSM system management can also be studied in terms of layered architecture. For example, air interface is a physical layer issue whereas multiple access is a data link layer issue. GSM channels are defined on both physical and data link layers. SS7 signalling and control data are conveyed through data link and network layers.

Apart from this, radio resource management, mobility management, and communication management are various issues in GSM. The control channels (CCHs) explained in Section 11.1.5 play a very important role in handling call set-up, termination, and handover.

11.1.3 GSM Radio Interface

The specifications of the physical layers of the various versions of GSM are given in Table 11.1. Some relevant information and calculations are presented in this section.

Modulation   The carrier is modulated using Gaussian minimum shift keying (GMSK). The details of the scheme were provided in Chapter 7. GMSK is resilient to noise when compared to some other forms of modulation. It occupies a relatively narrow bandwidth and has a constant power level. The data transported by the carrier serves up to eight different users under the basic system.

Voice coding   If digitized in a linear fashion by simple waveform coding, the speech would acco-modate a far greater bandwidth. To overcome this, voice coding systems or vocoders are used. These systems involve analysing the speech properties and then performing various actions upon the speech to reduce the data rate. At the receiving end, the reverse process is undertaken to re-constitute the speech data such that it can be understood. In GSM, a variety of vocoders are used, including regular pulse excitation—long-term prediction RPE-LTP) and enhanced full rate (EFR).

RPE-LTP   This vocoder was originally used in the GSM systems. It takes each 20 ms block of speech and then represents it using just 260 bits. This actually equates to a data rate of 13 kbps. In GSM, some bits are recognized to be more important than others, and the loss or corruption of such bits will have more effect on the voice quality. Accordingly, the different bits are classified as follows:

Class	No. of bits	Remarks on error sensitivity
Class Ia	50 bits	Most important and sensitive to bit errors
Class Ib	132 bits	Moderately sensitive to bit errors
Class II	78 bits	Least sensitive to bit errors
Total	260 bits	

BIT SETTINGS

The following points need to be noted about bit settings:
- The 50 bits of Class Ia are appended a three-bit cyclic redundancy code (CRC) such that errors can be detected. This makes a total length of 53 bits. If there are any errors, the frame is not used and is discarded. In its place, a version of the previously correctly received frame is used.

- These 53 bits, together with the 132 Class Ib bits with a four-bit tail sequence, are entered into a 1/2 rate convolutional encoder. The total length is now 189 bits. As a result, the output from the convolutional encoder consists of 378 bits.
- The remaining 78 Class II bits are considered the least sensitive to errors. They are not protected and are simply added to the data.
- In this way, every 20 ms speech sample generates a total of 456 bits. Accordingly, the overall bit rate is 22.8 kbps.
- Once in this format, the data is interleaved to add further protection against interference and noise. The 456 bits output by the convolutional encoder are divided into eight blocks of 57 bits; these blocks are transmitted in eight consecutive time slots, that is, a total of four bursts, as each burst takes two sets of data. The concept of GSM frames and superframes is represented in Fig. 11.5.

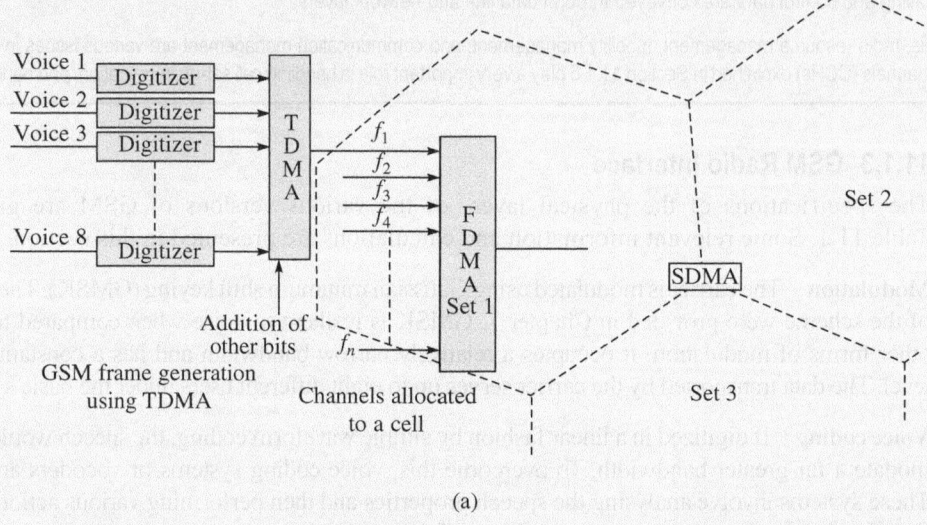

(a)

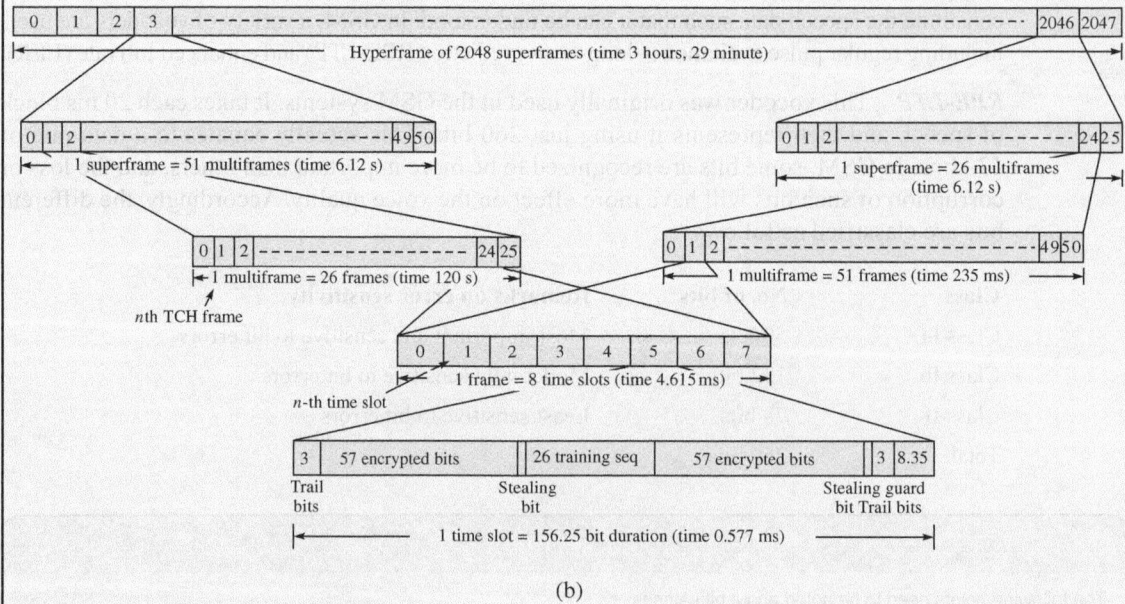

(b)

Fig. 11.5 Multiple access in GSM (a) Concept of TDMA and FDMA used along with SDMA using cellular structure (b) TDMA frame formats

EFR

This vocoder was later added in response to the poor quality perceived by users. Using the algebraic code excitation linear prediction (ACELP) compression technology, this vocoder gives a significant improvement in quality over the original RPE—LTP encoder. This became possible because of the increase in the processing power available in mobile phones combined with their low current consumption.

There is also a half-rate vocoder. Although this gives much inferior voice quality, it does allow for an increase in network capacity. It is used in some instances when the network loading is very high to accommodate all the calls.

Power levels in GSM Many power levels are allowed by the GSM standard, the lowest being only 800 mW (29 dBm). As mobiles may only transmit for one-eighth of the time, that is, for their allocated slot, which is one of eight, the average power is an eighth of the maximum. Additionally, to reduce the levels of transmitted power and hence the levels of interference, mobiles are able to step the power down in increments of 2 dB from the maximum to a minimum of 13 dBm (20 mW). The MS measures the signal strength or signal quality (based on the bit error rate) and passes the information to the BTS and hence to the BSC, which ultimately decides if, and when, the power level should be changed.

Discontinuous transmission (DTX) is a power-saving and interference-reducing technique. It is particularly useful because there are long pauses in speech. It is found that a person speaks for less than 40 per cent of the time during normal telephone conversations. The most important element of DTX is the voice activity detector. It must correctly distinguish between voice and noise inputs, a task that is not trivial. If a voice signal is misinterpreted as noise, the transmitter is turned off and an effect known as clipping results, which is particularly annoying to the person listening to the speech. However, if noise is too often misinterpreted as a voice signal, the efficiency of DTX is dramatically decreased.

11.1.4 Multiple Access in GSM

The GSM uses a combination of TDMA and FDMA techniques as shown in Fig. 11.5(a) along with the cellular structure (SDMA). A set of frequencies or channels is allocated to each cluster and the frequency reuse scenario is adopted. The set of frequencies is identified with absolute radio frequency channel numbers (ARFCN).

GSM 900 Calculations

Forward link (BS to mobile) $\rightarrow$ 960 MHz $-$ 935 MHz $=$ 25 MHz

Reverse link (mobile to BS) $\rightarrow$ 915 MHz $-$ 890 MHz $=$ 25 MHz

> A GSM frame structure uses a hierarchy of slots, frames, multi-frames, superframes, and hyperframes.

$$\text{Total number of channels} = \frac{\text{Total available bandwidth}}{\text{Per channel bandwidth}} = \frac{25\,\text{MHz}}{200\,\text{kHz}} = 125$$

ARFCN = 0 to 124

Eight time slots are assigned per channel and the channel data rate is 270.833 kbps.

So, effective channel transmission rate = 270.833/8 = 33.854 kbps

In GSM, with overhead, the user data is actually sent at 24.7 kbps instead of 33.854 kbps.

Note: The combination of a time slot number along with ARFCN constitutes the physical channels for both forward and reverse links.

> GSM TCHs are used to carry encoded speech or user data. The GSM forms combinations of speech channels and data channels.

Signalling bit duration = $1/270.833 = 3.692$ μs

The carriers are divided in time using a TDMA scheme.

- Starting from bottom to top, according to Fig. 11.5(b), the fundamental unit of time (one slot) is called a *burst period* and it lasts for approximately 0.577 ms (15/26 ms).
- Eight of these burst periods are grouped into a TDMA frame. This lasts for approximately 4.615 ms (i.e., 120/26 ms) and it forms the basic unit for the definition of logical channels.
- One physical channel is one burst period allocated in each TDMA frame.
- Different types of frames are transmitted to carry different data. The frames are organized into *multiframes* and *superframes* to provide overall synchronization.

Due to digital TDMA technology combined with a channel bandwidth of 200 kHz, the system is able to offer a high level of spectrum efficiency. As there are many carrier frequencies, one or more can be allocated to each BS. The system also operates using frequency division duplex (FDD); as a result, paired bands are needed for the uplink and downlink transmissions.

Due to cellular architecture, nearby cells use a different frequency set, and the whole set of frequencies in a cluster is repeated in another cluster to allow frequency reuse. Thus, TDMA—FDMA along with SDMA support can accommodate billions of users in the network.

11.1.5 GSM Channels

The GSM uses a variety of channels to carry data. These channels are separated into *physical channels* and *logical channels*. Physical channels are determined by their time slots, whereas logical channels are determined by the data carried by the physical channel. Several recurring time slots on a carrier constitute a physical channel. These slots are then used by different logical channels to transfer data. These channels may be used for either user data (payload) or signalling to enable the system to operate correctly (Fig. 11.6).

The following logical channels and their functions are defined in GSM:

(a) Traffic channels (TCHs)
(b) Control channels

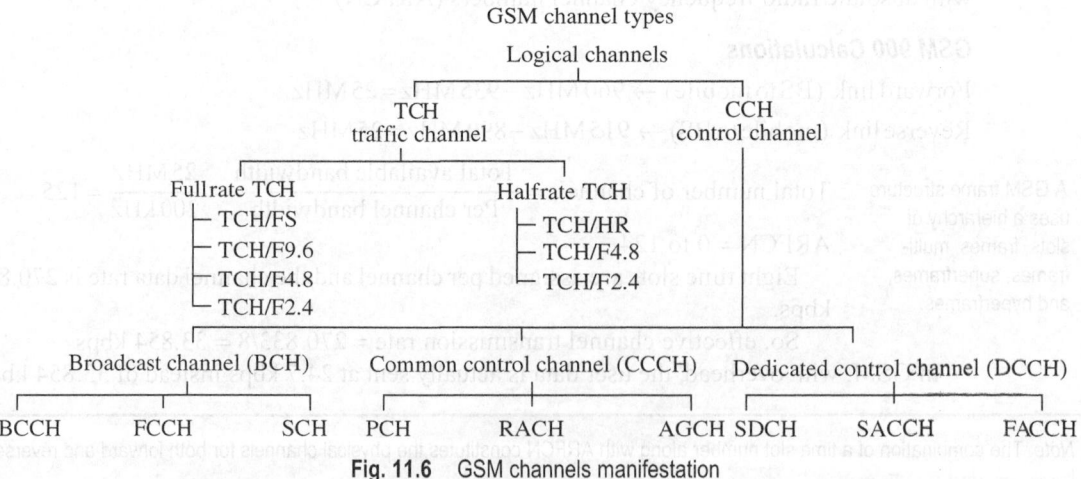

Fig. 11.6 GSM channels manifestation

Traffic Channels

Traffic channels carry digitally encoded user speech or user data and have identical functions and formats on both the forward and reverse links.

TCH/F This is a full-rate TCH. When transmitted at full rate, the user data is contained within one time slot per frame (Table 11.2).

TCH/H This is a half-rate TCH. When transmitted as half rate, the user data is mapped onto the same slot but is sent in alternate frames. Thus, two half-rate channel users can use the same time slot but alternately transmit during every other frame (Table 11.3).

For every 26 frames of TCH, the data is broken up at the 13th frame by either slow associated control channel (SACCH) or idle frames. The 26th frame contains idle bits when full-rate TCHs are used and SACCH data when half-rate TCHs are used.

Control Channels

> Control channels are intended to carry signalling and synchronization data between the BS and the MS.

Control channels carry signalling and synchronizing commands between the mobile and the BS. CCHs are categorized into three categories.

Broadcast channel (BCH) This channel operates on the forward link of a specific ARFCN within each cell and transmits data only in the first time slot of certain GSM frames. It provides synchronization for all mobiles within the cell and is occasionally monitored by the mobiles in the neighbouring cells so that received power and handover decisions may be made by out-of-cell users. The BCH is defined by the following three separate channels:

Broadcast control channel (BCCH) This channel is a point-to-multipoint channel (BS to MS). It broadcasts information such as cell and network identity and the operating characteristics of the cell such as the current CCH structure, channel availability, and congestion. It also broadcasts a list of channels that are currently in use within the cell.

Synchronization channel (SCH) This channel is used for the synchronization of the MSs. It is used to identify the serving BS while allowing each mobile to frame synchronize with the BS.

Frequency correction channel (FCCH-MS) This channel is used for frequency correction and allows each subscriber unit to synchronize its internal frequency standard or local oscillator to the exact frequency of the BS.

Table 11.2 TCH/F channel rates

	TCH/FS	**TCH/F9.6**	**TCH/F4.8**	**TCH/F2.4**
Speech channel raw data rate after digitization	13 kbps	9.6 kbps	4.8 kbps	2.4kbps
Data rate after channel coding	22.8 kbps	22.8 kbps	22.8 kbps	22.8 kbps

Table 11.3 TCH/H channel rates

	TCH/HS	**TCH/H4.8**	**TCH/H2.4**
Speech channel raw data rate after digitization	6.5 kbps	4.8 kbps	2.4 kbps
Data rate after channel coding	11.4 kbps	11.4 kbps	11.4 kbps

Common control channels (CCHs) These channels are of two main types—forward and return channels. *Forward common channels* are used for paging to inform a mobile of an incoming call, responding to channel requests, and broadcasting bulletin board information. *Return common channels* are random access channels (RACHs) used by the mobile to request channel resources before the timing information is conveyed by the BSs.

Access grant channel (AGCH) The BS acknowledges channel requests from the MS and allocates a stand-alone dedicated control channel (SDCCH). It uses the AGCH to provide forward link communication to the mobile. The AGCH carries data that instructs the mobile to operate in a particular physical channel (with a time slot and ARFCN), with a particular dedicated CCH. It is the final CCCH message sent by the BS before a subscriber is moved off the CCH and is also used by the BS to respond to an RACH sent by the MS in a previous CCCH frame.

Paging channel (PCH) The PCH is used for terminating call announcement by providing paging signals from the BSs to all mobiles in the cell and for notifying a specific mobile of an incoming call originating from the PSTN. The PCH transmits the IMSI of the target subscriber along with a request for acknowledgement from the mobile on a RACH. The PCH may also be used to provide *cell broadcast* ASCII text messages to all MSs in the form of SMS.

Random access channel (RACH) The main functions of the RACH-MS are access requests, response to call announcement, and location updates. This channel is used by the subscriber unit to acknowledge a page from the PCH and also to originate a call by mobile. Slotted ALOHA is the suitable access method. All mobiles must request access or respond to the PCH alert within zeroth time slot of a GSM frame. At the BTS, every frame, including the idle one, will accept RACH transmissions from mobiles during the zeroth slot. To establish service, the BS must respond to the RACH transmission by allocating a channel and assigning an SDCCH for signalling during a call. This connection is confirmed by the BS over the AGCH.

Dedicated channels (DCH) These channels are of two types: (a) those used for signalling and (b) those used for traffic. Signalling channels are used for maintaining the call, enabling call set-up, providing facilities such as handover when the call is in progress, and finally, terminating the call. TCHs handle the actual payload.

Fast associated control channel (FACCH) This channel is used for supervisory data transmissions between the MS and BS during a call. It is associated with the TCH and SDCCH. It is used for time critical signalling over the TCH (e.g., for handover signalling). Traffic burst is stolen for a full signalling burst. The FACCH carries urgent messages and essentially the same type of information as the SDCCH. It is assigned when the SDCCH has not been dedicated for a particular user and an urgent message comes in (handover).

Slow associated control channel (SACCH) This channel is associated with TCH and SDCCH in-band signalling (e.g., for link monitoring). In both the cases, the SACCH maps them onto some physical channel. Thus, each ARFCN systematically carries SACCH data for all of its current users. On the forward link, the SACCH is used to send slow but regularly changing control information to the mobile, for example, transmit power level instructions and specific timing advance instructions for each user on the ARFCN. On the reverse link, it carries information about the received signal strength and quality of the TCH and also the

BCH measurement resulting from the neighbouring cells. The SACCH is transmitted in the 13th frame (also in the 26th frame when half-rate traffic is used) of every speech or dedicated CCH multiframe. Within the SACCH frame, eight time slots are allocated to provide SACCH data to each of the eight full-rate (or 16 half rate) users on the ARFCN.

> With HSCSD, only circuit switching is possible and spectral efficiency is low.

Stand-alone dedicated control channel (SDCCH) The SDCCH is used for signalling exchanges, for example, during call set-up and registration or location updates. The SDCCH carries signalling data following the connection of the MS with the BS and just before a TCH assignment is issued by the BS. It ensures that the mobile and the BS remain connected while the BS and the MSC verify the subscriber unit and then allocates resources for the mobile. It can be considered as an intermediate temporary channel accepting a newly completed call from the BCH. It holds the traffic while waiting for the BS to allocate a TCH channel. The SDCCH is used to send authentication and alert messages.

11.1.6 GSM Enhancements and HSCSD

GSM mobile service started in India with 2G capabilities, which supported voice and low-speed circuit-switched data. To support an ever-increasing demand, many more advanced technologies are being developed on the GSM platform, such as general packet radio service (GPRS) and enhanced data rate for GSM evolution (EDGE). This brought about a revolution in the services provided by GSM operators, a few of which are listed as follows:

- High-speed packet-switched data connectivity
- Multimedia applications
- High-speed remote corporate local area network (LAN) access
- Web browsing, email, fax, and wireless imaging
- Video telephony and digital television reception on mobile phones

Existing GSM services with enhanced features are called GSM phase 2 and 2+.

GSM Phase 2+ Services

GSM phase 2 services are as follows:

- Voice services
- Fax and data services up to 9.6 kbps and SMS

High-speed circuit-switched data (HSCSD) is a straightforward improvement of GSM's data transmission capabilities becoming phase 2+. In this system, higher data rates are achieved by bundling several TCHs. Here, the MS requests one or more TCHs from the GSM network; that is, it allocates several TDMA slots within a TDMA frame. This allocation may be asymmetrical—more slots can be allocated on the downlink than on the uplink, which fits the typical user behaviour of downloading more data compared to uploading. Normal data rate is enhanced up to 57.6 kbps. Typical applications are in the Internet on mobile, mobile LAN, wireless real-time applications, and file transfer.

> GPRS and GSM are able to operate along-side each other on the same network using the same BS.

Thus, HSCSD may be an attractive interim solution for higher bandwidth and rather constant traffic, like file download. However, it does not make much sense for bursty Internet traffic as long as the user is charged for each channel allocated for communication. GPRS is the next step towards more flexible and powerful data transmission avoiding the problems of HSCSD, as it is fully packet oriented.

Note: The following are the limitations of HSCSD:
1. HSCSD still uses connection-oriented mechanisms of GSM. Hence, it is not efficient for bursty computer data traffic.
2. It supports only circuit switching and there can be no packet switching.
3. Service costs depend on the number of channels allocated. While downloading a heavy file, all the channels may get utilized; however, during normal browsing, most of the channels will remain idle.
4. For *n* channels, HSCSD requires *n* times signalling during handover, connection set-up, and release. Each channel is treated separately.
5. There may be service degradation during handovers because the BSC has to check the resources for all the *n* channels and not just one channel.

11.2 GENERAL PACKET RADIO SERVICE

The GSM technology was developed for voice services, but it did not have the capability to provide data services. To develop a higher data rate capability and to enhance the services, the GPRS protocol was developed on the GSM platform. The scenario and GPRS architecture are shown in Fig. 11.7. GPRS became the bridging gap between the 2G GSM system and the wideband code division multiple access (WCDMA) or universal mobile telecommunication system (UMTS). GPRS could offer data rates up to 115 kbps, which allowed Web browsing and other services requiring data transfer. Although GSM allows data services, the rate was too slow for real data applications.

GGSN—Gateway GPRS support node; SGSN—Serving GPRS support node; PDN—Packet data network

GPRS uses packet-switched data rather than circuit-switched data, which is much more efficient for using the available capacity. This is because data transfer takes place in a bursty manner. The transfer occurs in short peaks, followed by breaks when there is little or no activity. Packet switching permits sharing the overall capacity among several users.

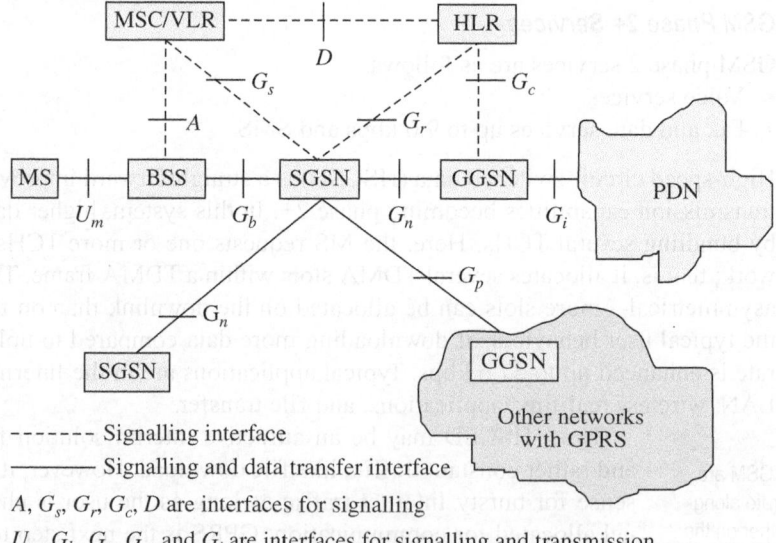

A, G_s, G_r, G_c, D are interfaces for signalling

U_m, G_b, G_n, G_p, and G_i are interfaces for signalling and transmission

Fig. 11.7 GPRS architecture with modifications in the GSM, including additional GPRS components

The CN structure of GSM is enhanced for GPRS and two new elements are added to support packet data.

The data is split into packets and then tags are inserted into each packet to mark its destination address. Packets from several sources can then be transmitted over the link. It is unlikely that the data burst for all the users will occur at the same time, so sharing the overall resource in this fashion makes the system efficient. Packet switching is at the core of many cellular data systems like GPRS.

Note: GSM frames cannot be called packets, as they are synchronous time division multiplexing (TDM) frames. However, GPRS packet format is defined. GPRS is TDMA based and follows Interim Standard (IS)-136 protocol.

In computer networks, the physical addresses of the network interface controller (NIC) and the Internet protocol (IP) are required to identify a destination device. For GPRS-based systems, the SIM and IMSI numbers can be used as the physical address. In GPRS, the IP address and the *domain name system* (DNS), which are developed for computer networks, are adopted directly. In order to accommodate packet data within GPRS, coding schemes were developed; these are discussed in Section 11.2.2. Additionally, the layers based on the OSI system have become very important. However, there are a few points that differentiate GPRS from the computer networks, some of which are listed as follows:

- A wireless channel is required for data transfer.
- Mobility management is necessary.
- GSM infrastructure, which incorporates HLR, VLR, AuC, MS, BSS, BTS, and MSC (Fig. 11.1a), is used.
- For routing of packets, GPRS requires supporting hardware and software, which are different from the computer networks.

For GPRS, the data from the BSC is routed through the *serving GPRS support node* (SGSN). This forms the gateway to the services within the network and then a *gateway GPRS support node* (GGSN), which forms the gateway to the outside world. SGSN and GGSN are two major additional elements along with basic GSM infrastructure to have GPRS services.

SGSN function It enables authentication and then tracks the location of a mobile within a network. It also ensures that the quality of service (QoS) is at the required level.

GGSN function It appears as a data gateway to the public packet network and provides identity to the mobiles for Internet access as follows. The network protocols used and supported by GPRS are X.25 and IP. In operation, the protocols assign addresses (packet data protocol or PDP addresses) to the devices in the network for routing the data through the system. A GPRS mobile must attach itself to the SGSN and activate its PDP address. This address is supplied by the GGSN, which is associated with the SGSN. As a result, a mobile can be attached to only one SGSN, although once its address is assigned, it can receive data from multiple GGSNs using multiple PDP addresses.

11.2.1 GPRS Functional Groups

The GPRS functional groups (defined according to the functions required for GPRS) are shown in Fig. 11.8.

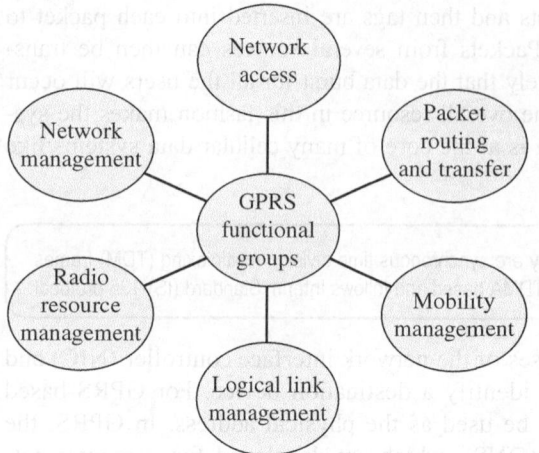

Fig. 11.8 GPRS functional groups

Network Access

Network access supports standard point-to-point data transfer and anonymous access (without authentication and ciphering). The functions include the following:

- Registration, which associates the MS identity with the PDPs
- Authentication and authorization are for security purposes to avoid anonymous access.
- Packet terminal adaptation, which adapts data transmission across the GPRS network
- Admission control, which determines the radio and network resources to be used for communication of MSs
- Message screening, which filters out unsolicited messages
- Charging information collection for packet transmission in GPRS and external networks

Packet Routing and Transfer

Packet routing and transfer is used to route the data between the MS and the destination through the SGSNs and GGSNs. Figure 11.7 shows the GPRS architecture in which SGSN and GGSN nodes are shown within the highlighted boxes. The functions include the following:

- Relay function, which is used by the BSS to forward packets between the MS and the SGSN and is also used by the SGSN to forward packets between the BSS and the SGSN or GGSN
- Routing, which determines the destinations of packets
- Address translation and mapping, which converts a GPRS network address to an external data network address and vice versa
- Encapsulation and tunnelling, which encapsulate packets at the source of a tunnel, deliver the packets through the tunnel, and decapsulate them at the destination
- Domain name service functions, which resolve logical GSN names to their IP addresses
- Compression and ciphering, which reduce and protect the database

Mobility Management

Mobility management keeps track of the current location of an MS. Three different scenarios can exist when the MS enters a new cell and, possibly, a new routing area:

- Cell update
- Routing area update
- Combined routing area and location area update

Logical Link Management

Logical link management maintains the communication channel between the MS and the GSM network across the radio interface, which includes the following:

- Logical link establishment
- Logical link maintenance
- Logical link release

Radio Resource Management

Radio resource management allocates and maintains radio communication paths, which includes the following:

- Um interface management, which determines the amount of radio resources to be allocated for GPRS usage
- Cell selection, which enables the MS to select the optimal cell for radio communication
- Um-tranx, which provides packet data transfer capability, such as medium access control (MAC), packet multiplexing, packet discrimination, error detection and correction, and flow control across the radio interface between the MS and the BSS
- Path management, which maintains the communication paths between the BSSs and SGSNs

Network Management

Network management provides the mechanisms to support operations, authentication, and maintenance (OAM) functions related to GPRS.

11.2.2 Coding Schemes in GPRS

Many coding schemes are offered by GPRS with different levels of error detection and correction. These are used depending upon the RF signal conditions and the requirements for the data being sent. These are given labels CS-1 to CS-4.

CS-1

CS-1 applies the highest level of error detection and correction. It is used in scenarios wherein interference levels are high or signal levels are low. Applying high levels of detection and correction prevents the need to re-send the data too often. Although it is acceptable for many types of data to be delayed, for others there is a more critical time element. This level of detection and coding results in a half code rate; that is, for every 12 bits that enter the coder, 24 bits result. It results in an actual throughput of 9.05 kbps data rate.

CS-2

The CS-2 error detection and coding scheme is for better channels. It effectively uses a 2/3 encoder and results in a real data throughput of 13.4 kbps, which includes the radio link control (RLC) or MAC header.

CS-3

CS-3 effectively uses a 3/4 coder and results in a data throughput of 15.6 kbps.

CS-4

CS-4 is used when the signal is high and interference levels are low. No correction is applied to the signal, allowing for a maximum throughput of 21.4 kbps. If all eight slots were used, then this would enable a data throughput of 171.2 kbps to be achieved.

In addition to the error detection and coding schemes, GPRS employs interleaving techniques to ensure that the effects of interference and spurious noise are reduced to a minimum. As blocks of 20 ms data are carried over four bursts, with a total of 456 bits of information, either 181, 268, 312, or 428 bits of payload data are carried, respectively, depending upon the error detection and coding scheme chosen, that is, CS-1 to CS-4.

11.2.3 GPRS Layers and Functions

The GPRS system is divided into three distinct layers based on its various functions so that the functional areas can be addressed separately.

Layers of GPRS	
LLC	Layer three
RLC	
	Layer two
MAC	
PL	
	Layer one
PRF	

Layer one concerns the physical link between the mobile and the BS. It is the lowermost layer. This is often subdivided into the following two sublayers:

Physical RF layer   It deals with the modulation and demodulation aspects.

Physical link layer   It manages the responses and controls required for the operation of the RF link. These include elements such as error correction, interleaving and correct assembly of data, and power control.

Layer two also concerns with two sublayers—RLC and MAC. They are above *layer one*. These organize the logical links between the mobile and the BS. They control the radio link access and organize the logical channels that route the data to and from the mobile.

Above all these layers is the *logical link layer* (LLC), which formats the data frames and is used to link the elements of the CN to the mobile. It can be considered as *layer three*.

MAC and Its Allocation Modes

As mentioned, the MAC layer is the central layer in the GPRS system. There are three MAC modes that control the transmissions, namely fixed allocation, dynamic allocation, and extended dynamic allocation.

Fixed allocation mode   This is required when a mobile requires a data to be sent at a consistent data rate. To achieve this, a set of packet data channels (PDCHs) (see Section 11.2.4) is allocated for a given amount of time. When this mode is used, there is no requirement to monitor for availability, and the mobile can send and receive data freely. This mode is used for applications like video conferencing.

Dynamic allocation mode   This is required by the network to allocate time slots as required. A mobile is allowed to transmit in the uplink when it sees an identifier flag, known as the uplink status flag (USF), that matches its own. The mobile then transmits its data in the allocated slot. This is required because up to eight mobiles can have potential access to a slot, but only one can transmit at any given time.

Extended dynamic allocation mode   This allows for much higher data rates to be achieved because it enables mobiles to transmit in more than one slot. When the USF indicates that a mobile can use this mode, it can transmit in the number allowed, thereby increasing the rate at which it can send data.

Like other cellular systems, GPRS uses various physical and logical channels to carry the data payload as well as the signalling required to control the calls.

11.2.4 GPRS Channels

General packet radio service builds on the basic structure of GSM. It uses the same modulation and frame structure employed by GSM, and in this way, it is an evolution of the GSM standard. Slots can be assigned dynamically by the BSC to GPRS calls depending on the demand, the remaining ones being used for GSM traffic.

A new data channel called PDCH is used for GPRS. It has the same power profile and timing advance attributes to overcome different signal travel times to the BS depending on the distance of the mobile from the BS. This enables the burst to fit in with the existing GSM structure.

In GPRS, each burst of information is 0.577 ms in length (the same as that of GSM). It also carries two blocks of 57 bits of information, giving a total of 114 bits per burst. It, therefore, requires four bursts to carry each 20 ms block of data, that is, $114 \times 4 = 456$ bits

> A PDCH channel is for packet data and uses the same slot structure as GSM.

of encoded data. The BSC assigns PDCHs to particular time slots; there will be times when the PDCH is inactive, allowing the mobile to check for other BSs and monitor their signal strengths to enable the network to judge when a handover is required. The BS may also use the GPRS slot to judge the time delay using a logical channel known as the *packet timing advance control channel* (PTCCH).

A variety of channels are used in GPRS, and these channels can be set into groups, depending on whether they are for common or dedicated use. The system uses the GSM control and BCHs for the initial set-up, but all the GPRS actions are carried out within the GPRS logical channels available within the PDCH.

Broadcast Channels

Packet broadcast central channel (PBCCH)   This is a downlink only channel that is used to broadcast information to mobiles and to inform them of incoming calls. It is very similar in operation to the BCCH used for GSM. In fact, the BCCH is still required in the initial stage to provide a time slot number for the PBCCH. In operation, the PBCCH broadcasts general information, such as power control parameters, access methods, operational modes, and network parameters, required to set up calls.

Common Control Channels

Packet paging channel (PPCH)   This is a downlink only channel and is used to alert the mobile to an incoming call or to be ready for receiving data. It is used for control signalling prior to the call set-up. Once the call is in progress, a dedicated channel referred to as the packet associated control channel (PACCH) takes over.

Packet random access channel (PRACH)   This is an uplink channel that enables the mobile to initiate a burst of data in the uplink. There are two types of PRACH burst; one is an 8-bit standard burst and the other, using an 11-bit burst, adds data to allow for a priority setting. Both types of bursts allow for timing advance setting.

Packet access grant channel (PAGCH)   This is also a downlink channel and it sends information to the mobile regarding the TCH that has been assigned to it. It occurs after the mobile has been informed by the PPCH of an incoming call.

Packet notification channel (PNCH)   This is another downlink only channel that is used to alert mobiles when there is broadcast traffic intended for a large number of mobiles. It is typically used in point-to-point multicasting.

Dedicated Control Channels

Packet associated control channel   This channel is present in both uplink and downlink directions and is used for control signalling while a call is in progress. It takes over from the PPCH once the call is set up, and it carries information such as channel assignments, power control messages, and acknowledgements of received data.

Packet timing advance control channel (PTCCH)   This channel is present in both uplink and downlink directions and is used to adjust the timing advance. This is required to ensure that messages arrive at the correct time at the BS regardless of the distance of the mobile from the BS. As timing is critical in a TDMA system and signals take a small but finite time to travel, this aspect is very important if long guard bands are not to be transmitted.

The SIM of a GSM system may be GPRS aware or non-aware. If non-aware, the GPRS service pack must be installed in the mobile phone.

Dedicated Traffic Channel

Packet data traffic channel (PDTCH) This channel is present in both uplink and downlink directions and is used to send the traffic. Up to eight PDTCHs can be allocated to a mobile to provide high-speed data.

11.2.5 GPRS Device Categories and Modes

It is possible to use GSM mobiles for GSM speech on a network that also carries GPRS; however, a GSM mobile cannot be upgraded for use as a GPRS mobile. Therefore, new GPRS mobiles are required. To utilize GPRS, new modes are required to enable the mobile device to transmit the data in the required format. Not all GPRS mobiles are designed to offer the same levels of service. As a result, they are split into three basic categories according to their capability to connect to GSM and GPRS facilities.

Class A Mobile phones in this class can be connected to both GPRS and GSM services at the same time.

Class B Mobile phones in this class can be attached to both GPRS and GSM services, but they can be used on only one service at a time. A class B mobile can make or receive a voice call, or send and/or receive text messages through SMS, during a GPRS connection. During voice calls or text, the GPRS service is suspended but it is re-established when the voice call or SMS session is complete.

Class C Mobile phones in this class can be attached to either GPRS or GSM services, but the user needs to switch manually between the two different types.

GPRS mobiles are also categorized by the data rates they can support. In GSM, there are eight time slots that can be used to provide TDMA, allowing multiple mobiles onto a single RF signal carrier. In GPRS, it is possible to use more than one slot to support much higher data rates. The different speed classes of the mobiles are dependent upon the number of slots that can be used in either direction. There are 29 speed classes. Class 1 mobiles are able to send and receive in one slot in either direction, that is, uplink and downlink, and class 29 mobiles are able to send and receive in all eight slots. The class within these two limits support sending and receiving in different combinations of uplink and downlink slots.

Modes

It can be seen from the way in which GPRS operates that there are three basic modes of operation. These are initialization or idle, standby, and ready.

Initialization or idle When a mobile is turned on, it must register with the network and update the location register. This operation is very similar to that performed in a GSM mobile, but it is referred to as *location update*. It first locates a suitable cell and transmits a radio burst on the RACH using a shortened burst, because it does not know what timing advance is required. The data contained within this burst temporarily identifies the mobile and indicates that the reason for the update is to perform a location update.

When the mobile performs its location update, the network also performs an authentication to ensure that it is allowed to access the network. Like GSM, it accesses the HLR and VLR as necessary for the location update and the AuC for authentication. The network detects at registration whether the mobile has GPRS capability. The SGSN also maintains a record of the location of the mobile so that data can be sent there if required.

GPRS OPERATIONS AND FUNCTIONING

Uplink Set-up

When a mobile is attached to the network, it is prepared for a call or data transfer:

- To transmit data, the mobile attempts a packet channel request using the PRACH uplink channel.
- As this may be busy, the mobile monitors the PBCCH, which contains a status bit indicating the status of the BS receiver, that is, whether it is busy or idle and capable of receiving data. When the status bit indicates that the receiver is idle, the mobile sees this and it sends its packet channel request message.
- If accepted, the BS will respond by sending an assignment message on the PAGCH on the downlink. This will indicate the channel to be used by the mobile for its packet data transfer as well as other details required for the data transfer.

Downlink Set-up

If data needs to be transferred in the downlink direction, then a separate assignment is performed for the downlink channel. When data is transferred, this is controlled by the action of the MAC layer. In most instances, it will operate in an acknowledge mode whereby the BS acknowledges each block of data. The acknowledgement may be contained within the data packets being sent in the downlink or the BS may send data packets down purely to acknowledge the data.

Disconnection

While disconnecting, the mobile will send a packet temporary block flow message, and this is acknowledged. Once this has taken place, the USF assigned to the mobile becomes redundant and can be assigned to another mobile wanting access. With this, the mobile effectively becomes disconnected, and although it is still attached to the network, no more data transfer takes place unless it is re-initiated. Separate messages are needed to detach the mobile from the network.

The following points are worth noting:

- Packet reservation protocol is used in GPRS for multiple accesses.
- It is possible to achieve 9 kbps to 150 kbps bit rate per user in GPRS.
- All securities of the GSM system and ciphering are utilized in GPRS.

GPRS is a relatively inexpensive mobile data service compared to SMS and circuit-switched data.

Standby The mobile then enters a standby mode, periodically updating its position as required. It monitors to ensure that it has not changed BSs and also looks for stronger BS CCHs. The mobile will also monitor the PPCH in case of an incoming alert indicating that data is ready to be sent. Like GSM, most BSs set up a schedule for paging alerts based on the last figures of the mobile number. In this way, it does not have to monitor all the available alert slots and can instead monitor only a reduced number where it knows alerts can be sent for it. In addition, the receiver can be turned off for a long period and hence the battery life can be extended.

Ready In the ready mode, the mobile is attached to the system and a virtual connection is made with the SGSN and GGSN. This connection enables the network to know where to route the packets when they are sent and received. In addition to this, the mobile is likely to use the PTCCH to ensure that its timing is correctly set so that it is ready for a data transfer should one be needed.

11.3 EDGE TECHNOLOGY

Enhanced data for GSM evolution is an enhancement to the GSM mobile cellular phone system. It enables data to be sent over a GSM TDMA system at speeds up to 384 kbps. It is

EDGE systems may also be known as EGPRS, or enhanced general packet radio service systems.

regarded as a 2.5G system. Operators who have not been able to secure full 3G licences use EDGE to provide data services. It is a highly spectrally efficient and high-speed system.

EDGE System Requirements

The EDGE technology is applied to GSM networks where the enhancements provided by GPRS have already been added. GSM provides voice services based on circuit switching whereas GPRS provides data services based on packet switching, as discussed earlier. Hence, the infrastructure for both is to be adopted by EDGE.

In terms of implementation, EDGE systems require an EDGE transceiver unit to be added to each cell along with software upgrades to allow its use.

As mentioned, EDGE is intended to build on the enhancements provided by the addition of GPRS where packet switching is applied to a network. It then enables a threefold increase in the speed at which data can be transferred by adopting a different form of modulation. EDGE changes the modulation to 8PSK, which is a form of phase shift keying (PSK) where eight phase states are used. The advantage is that it can transmit high data rates, although it is not as immune to interference and noise. By using 8PSK, data can be transferred at 48 kbps per channel rather than 9.6 kbps, the rate that is possible using GMSK. By allowing the use of multiple channels, the technology allows the transfer of data at rates up to 384 kbps. However, it should be remembered that these data transfer rates are possible only when the network is not highly loaded, as access to all channels would not be allowed.

EDGE services have the following specifications:

- They offer data services up to 400 kbps.
- They are used on voice over packet-switched network.
- They support IP-based application.
- Typical applications are Internet on mobile, mobile LAN, videophone, wireless real-time applications, and file transfer.

Table 11.4 gives the summarized comparison between GSM and EDGE technologies.

Table 11.4 Comparison between GSM and EDGE

Parameter	GSM	EDGE
Modulation	GMSK	8PSK
Bit rate	270.833 kbps	812.499 kbps
Channel bandwidth	200 kHz	200 kHz
Pulse shaping	Gaussian prefilter $BT = 0.3$	Linearized GMSK pulse
Modulation type	Non-linear, constant envelope	Linear

11.4 CDMA-BASED STANDARDS: IS-95 TO CDMA2000

The IS-95 system is a standard for cellular telephone system based on direct sequence CDMA (DS-CDMA). Multiple users can simultaneously share the same (wideband) channel. This was the first CDMA mobile phone system to gain widespread use. Gradually, it evolved and different versions of CDMA came up, like CDMA2000.

The first CDMA systems under the standard IS-95 offered voice as well as data services up to a speed of 14.4 kbps.

Note: Earlier CDMA mobiles did not have SIM cards, although recently it has changed. The subscriber data is stored in the internal mobile memory with a method of over-the-air programming of this data being available.

Before CDMA2000, the scenario was different. The GSM system used 200 kHz bandwidths whereas the CDMA system, IS-95A, used a 1.25 MHz bandwidth, which was much wider than anything that had been used before. CDMA operates well with a wide bandwidth, but it was limited to 1.25 MHz to remain compatible with the spectrum allocations that were available. The maximum speech encoded user data rate was 9.6 kbps. However, with the market moving towards data applications, the IS-95 specification was upgraded to IS-95A to cater to the needs of the operators. This new specification allowed data transmission up to a speed of 64 kbps.

Often, IS-95 A and B versions are marketed under the brand name *cdmaOne*. This is a registered trademark of the *CDMA development group* (CDG). Apart from voice, the mobile phone system is also able to carry data at rates up to 14.4 kbps for IS-95A and 115 kbps for IS-95B.

11.4.1 IS-95 System

The CDMA technology used for IS-95 is significantly different from conventional systems. CDMA is discussed in detail in Chapter 9 and partly in Chapter 7. To recapitulate, the basic principle of CDMA is that different orthogonal codes are used to distinguish between different users.

A CDMA system uses a form of modulation known as *direct sequence spread spectrum* (DSSS). Here, a signal that spreads out over a wide bandwidth is generated using a code known as *spreading code*. By using orthogonal codes for spreading, it is possible to pick out a signal with a given code in the presence of many other signals with different orthogonal codes. In fact, many different baseband signals with different spreading codes can be modulated onto the same carrier to enable many different users to be supported. By using different orthogonal codes, interference between the signals is minimal. Conversely, when signals are received from several MSs, the BS is able to isolate each one.

Note: CDMA has an edge over FDMA and TDMA. It supports a greater number of users. However, the improvement in efficiency is hard to define as it depends on many factors including the size of the cells and the level of interference between cells.

Unlike traditional cellular systems where neighbouring cells use different sets of channels, a CDMA system reuses the same channels. Signals from other cells will appear as interference, but the system is able to extract the required signal by using the correct code in the demodulation and signal extraction process. More than one channel is often used in each cell, and this provides additional capacity because there is a limit to the amount of traffic that can be supported on each channel.

Forward Link: Base to Mobile

The IS-95 system specifies 869–894 MHz band for the forward link. The downlink transmission (i.e., BS to the mobile) in IS-95 consists of a number of elements. There are logical channels for pilot, paging, sync, and traffic. The pilot channel corresponds to the CCH in GSM and enables the mobile to estimate the path loss and set its power level accordingly. In addition to this, there are other channels for paging, speech, data, and so on.

Note: In IS-95, the forward link has 64 channels whereas the reverse link has 94 channels. Both directions are operated differently.

The forward link consists of 64 channels and supports four types of channels.

Pilot channel (channel 0)   It is possible to have a continuous signal on this single channel, allowing the mobile unit to acquire timing information, provide phase reference for demodulation, and provide a means for signal strength comparison for determining the handover. This channel transmits all zeros.

Synchronization channel (channel 32)   This channel obtains identification information about the cellular system (e.g., system time, long code state, and protocol revision). It is a 1200 bps channel used by the MS.

Traffic channel (channels 8–31 and 33–63)   These channels originally supported data rates of up to 9600 bps. A subsequent revision added a second set of rates up to 14,400 bps.

Paging channel (channels 1–7)   These channels contain messages for one or more MSs.

All these channels use the same bandwidth. The channels are distinguished using the chip code, which is derived from a 64×64 Walsh matrix. Speech is encoded using a voice encoder. Error correction is then applied to this data to enable it to be carried even under poor conditions. This brings the data rate up to 19.2 kbps. Data is then interleaved in blocks to reduce the effect of errors by spreading them out.

In the IS-95 system, all signals originate at the same transmitter. Thus, it is fairly simple to reduce mutual interference from users within the same cell by assigning orthogonal Walsh—Hadamard codes to all the channels. As this is a 64-bit Walsh code, the data rate is multiplied by 64 to bring the overall data rate to 1.228 Mbps. This signal is then transmitted. The long pseudo-noise (PN) code provides a measure of voice privacy and improves time synchronization. The short PN code in the forward link has a limited resolution but makes synchronization easier. The following is a summary of the forward link parameters:

- Chip rate: 1.2288 Mchip/s = 128 times 9600 bit/s (128 spreading factor)
- Codes: combines 64 Walsh—Hadamard (for orthogonality among users) and a maximum length code sequence (for effective spreading and multipath resistance)
- Transmit bandwidth: 1.25 MHz
- Convolutional coding with rate 1/2 for error correction

Final carrier modulation scheme: quadrature phase shift keying (QPSK)

Reverse Link: Mobile to Base

The IS-95 system specifies 824–849 MHz band for the reverse channel. On the reverse link, every user uses the same set of short sequences for modulation. The length of these sequences is 2^{15}; that is, it is a modified 15-bit linear feedback shift register maximum length sequence ($2^{15} + 1 - 1$).

The reverse link consists of 94 logical CDMA channels, each occupying the same 1228 kHz bandwidth. The link supports up to 32 access channels and up to 62 TCHs. The following is a summary of the reverse link specifications:

- Chip rate: 1.228 Mchip/s
- Codes: Walsh codes, 64×64 matrix
- Transmit bandwidth: 1.25 MHz
- Convolutional coding with rate 1/3 for error correction
- Final carrier modulation scheme: orthogonal QPSK (OQPSK)

The uplink signal for IS–95 is generated in a different way, as follows:

- The reverse link uses rate 1/3 convolutional coding, and the resulting data has a greater degree of error correction or protection applied than in the forward link. Accordingly, the resulting data rate is brought up to 28.8 kbps.
- The data is then block interleaved.
- The next step is a spreading of the data using the Walsh matrix. The way in which the matrix is used and its purpose are different from those of the forward channel. In the reverse channel, the data coming out of the block interleaver is grouped in units of six bits. Each six-bit unit serves as an index to select the row of the Walsh matrix. Walsh code ($2^6 = 64$ and 64×64 matrix) is used for spreading. However, this results in 307.2 kbps data stream. The purpose of this encoding is to improve reception at the BS. As the 64 possible codings are orthogonal, block coding enhances the decision-making algorithm at the receiver and is also computationally efficient. We can view this Walsh modulation as a form of block error-correcting code with $(n, k) = (64, 6)$ and $d_{min} = 32$. The data burst randomizer is implemented to help reduce interference from other MSs. The operation involves the use of a long code mask to smooth the data out over each 20 ms frame.
- The next step is the DSSS function. Further spreading is also required. This is provided by using a different form of orthogonal spreading code known as the PN code. Each access channel as well as each TCH gets a different long PN sequence. Access channels use a distinct long code whereas the TCH uses a user-specific long code. The long sequences are used to separate the signals from different users on the reverse link. This is multiplied with the signal to increase its data rate by four to bring it up to the final data rate of 1.228 Mbps, the same as the downlink signal. The bit stream is then modulated onto the carrier using an OQPSK modulation scheme. This differs from the forward channel, as it uses the delay element in the modulator to produce orthogonality.

Notes:

1. The reason the modulators in the reverse channel are different from those in the forward channel is that in the forward channel, spreading codes are orthogonal, all coming from the Walsh matrix, whereas in the reverse channel, orthogonality of the spreading codes is not guaranteed.

2. The uplink and downlink transmissions for IS-95 are generated in a different way because of the fact that it is difficult to synchronize mobile handsets. Each one is at a different distance from the BS and the time delays will be different. As a result, synchronization is not possible. For the Walsh codes to maintain their orthogonality and to operate correctly, they must be properly synchronized. Thus, operations in forward direction and reverse direction are to be handled separately.

3. The PN codes do not require synchronization and can be used more successfully under these circumstances.

For Walsh code orthogonality in forward link, all users operating in the forward direction must be synchronized and vice versa.

In IS-95, all BSs use the same channel (frequency reuse factor = 1). Thus, there are very few critical issues related to frequency reuse. The following factors enumerate the interference between the cells:

- It highly depends on path loss law.
- It would theoretically diverge to infinity for free space loss with '20 log d'.

POWER CONTROL IN IS-95

Power control in IS-95 is achieved as follows:
- CDMA performance is optimized if all signals are received with the same power.
- Update is needed every 1 ms.
- Performance is sensitive to imperfections of only 1 dB.
- For flat (frequency non-selective) Rayleigh and Rician fading, perfect power control is impossible. Fades are so deep that the average gain needed to compensate is unbounded.

- According to Qualcomm, the surrounding cells contribute to the total interference as follows:
  - 1st tier: 6 cells, 6 per cent per cell
  - 2nd tier: 12 cells, 0.2 per cent per cell
  - 3rd tier: 18 cells, 0.03 per cent per cell
  - 4th tier: 24 cells, 0.01 per cent per cell

where the aforementioned percentages of interference draw power from each cell.

11.4.2 Soft Handover in IS-95

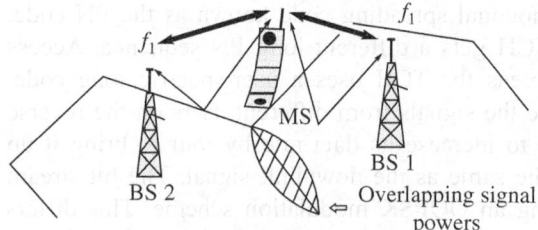

Fig. 11.9 DS-CDMA system with a soft handover (two BSs transmit at the same frequency)

Figure 11.9 illustrates the scenario for a soft handover.

The rake receiver detects a DS-CDMA signal transmitted over a dispersive multipath channel and capable of performing the required signal processing during a soft handover from one BS to another. The soft handover takes place as follows:

- When a mobile is about to cross a cell boundary, it starts receiving CDMA signals from different nearby BSs overlapping each other, with slightly different delays.

- Transmissions from the BSs in the adjacent cells may be made on the same frequency,
- As per Fig. 11.9, considering a scenario involving two BSs, signal power overlaps in the transition range and both BSs participate in maintaining the link with the user.
- If a unit enters a region in which the transmissions from the two BSs are comparable within some threshold of each other, the mobile enters the soft handover state in which it is connected to the two BSs.
- During the period of the handover, the two BSs transmit the same signal enabling the mobile to receive the signal via two routes at the same time. This means that during this handover phase, the mobile should not lose signal.
- The mobile remains in this state until one BS clearly predominates, at which time it is assigned exclusively to that cell.
- Then, as the mobile moves further into the second cell and the signal is firm, it can rely on one station alone and the handover is complete (the mobile would reject the signal from the second BS if it moves towards the first one).

> The soft handover between sectors is called softer handover.

Role of MSC In the soft handover state, the transmissions from the mobile reaching the two BSs are both sent on to the MSC, which estimates the quality of both the signals and selects one. The switch sends data or digitized speech

> While moving from one cell to another in hard handover, the system arranges for a new channel to be used. The mobile then changes the channel and hopes to receive the signal on the new channel. Here, the risk of loss is more.

signals to both the BSs, which transmit them to the mobile. The mobile unit then combines the two incoming signals to recover the information. The properties of the CDMA signalling scheme have made this possible. Soft handover results in a diversity gain (called soft handover gain).

Note: In CDMA, soft handover can be made to be easy and reliable. Soft handover considerably reduces the risk of loss of connection during handover. It also minimizes the risk of short breaks in speech during the handover. However, it is not free and there are associated costs, as follows:

- The mobile needs two decoders to monitor and decode the two signals, and this increases the complexity of the mobile.
- On the network side, it means that two channels are used instead of one, which can reduce the overall capacity. This is dependent upon the speed of handover and the degree of overlap in the cells.

CDMA2000—A 3G TECHNOLOGY FAMILY

The IS-95 system has an evolutionary migration path to 3G with CDMA2000 to provide higher data rates needed for video streaming and data transfer while retaining compatibility with existing networks. The CDMA2000 mobile phone system is designed to be used in the existing cellular telecommunications frequency allocations in addition to those mobile phone bands assigned to International Mobile Telecommunications-2000 (IMT-2000) (3G). The actual band allocations used for mobile phones vary from one country to another, dependent upon the frequency allocations available. It is easy for service providers to upgrade the system for adopting CDMA2000. The following are the major points related to CDMA2000 systems:

- CDMA2000–1X is the basic 3G standard; in fact, some people consider it only as a 2.75G system. It works on 1.25 MHz channel width.
- CDMA2000–1xEV is further developed to bring it in line with the UMTS or WCDMA system.
  - CDMA2000–1xEV-DO (Evolution, Data Only) is something of a sideline from the main evolutionary development of the standard. It carries only data but at speeds of up to 2.4 Mbps in the forward direction and the same as 1X in the reverse direction. The forward channel forms a dedicated variable-rate PDCH with signalling and control time multiplexed into it. The channel is time-divided and allocated to each user on a demand and opportunity driven basis. A data-only format was adopted so that the system could be optimized for data applications; if voice is required, then a dual mode phone using a separate 1X channel for the voice call is required. In fact, phones used for data-only applications are referred to as access terminals (ATs).
  - CDMA2000–1xEV-DV (Evolution, Data and Voice) is the mostly superseded one. It combines the 1xEV-DO with the widely deployed 1X (also called 1xRTT), providing full backward compatibility and simultaneous voice and data.

11.5 WIRELESS LOCAL LOOP

Local loops can be thought of as the last mile of the telecommunication network. They reside between the central office and the individual homes and business sectors in close proximity to the central office. The following are the *last-mile technologies*:

- *Wireline*: PSTN, digital subscriber line (DSL), cable modem (modulator—demodulator)
- *Wireless*: Fixed wireless such as local multipoint distribution service (LMDS), multichannel multipoint distribution service (MMDS), wireless LAN, and personal communication system (PCS)

The systems based on optical fibres as well as microwave links (line of sight or LOS) and satellite links (very small aperture terminal or VSAT) are categorized under *long-haul communication*, where maximum coverage with respect to the central unit is concerned.

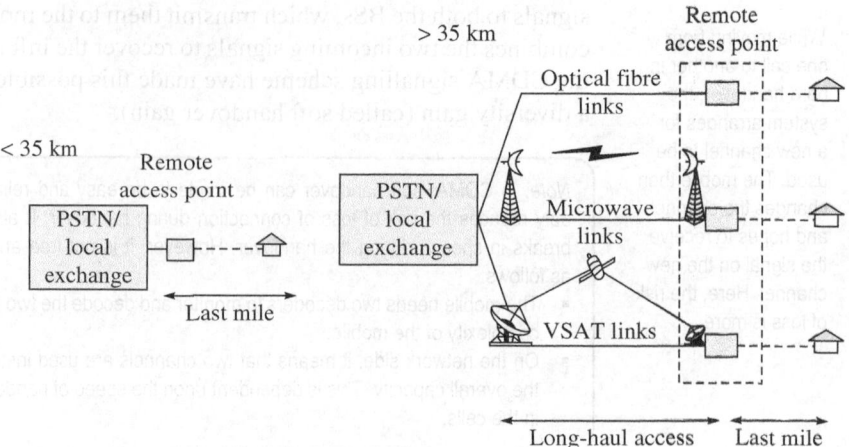

Long-haul communication offers high frequencies of the order of 28 GHz and more, which are greater than 10 times the carrier frequency of a terrestrial cellular network.

Fig. 11.10 Long-haul and last-mile systems

When long-haul and last-mile systems are combined, the complete wireless local loop (WLL) architecture can be formed, covering urban and suburban or rural areas, as in Fig. 11.10. Wireless equipment can be easily deployed in just a couple of hours, and once the equipment is purchased, there is no additional cost except monthly bills.

A WLL is defined in the last-mile system category. WLL services may be defined as fixed wireless services intended to provide access to the telephone network. In general, local loop means exchange-to-home-to-exchange closed loop. Conventionally, telephone local loops are unshielded twisted pair (UTP) or shielded twisted pair (STP) based cables. If we remove the wires used for communication, that is, establish an RF link, it becomes a wireless local loop, of course with supporting hardware.

Wireless local loop systems will generally divide a geographical region into many similar-sized cells (just like cellular telephony or maybe like PCS). Each cell will be serviced by a BS

LOCAL MULTIPOINT DISTRIBUTION SERVICE

The concept of LMDS led to the WLL concept. LMDS has vast bandwidth capabilities for WLL applications. It exhibits LOS fixed wireless applications. One of the most promising applications for LMDS is in a local exchange carrier (LEC) network. LEC may own wide bandwidth asynchronous transfer mode (ATM) or synchronous optical network (SONET) backbone switch capable of connecting huge traffic with the Internet, the PSTN, or to its own private network. Due to LOS, LMDS allows LECs to install wireless equipment in the customers' premises for rapid broadband connectivity without having to lease or install its own cable to the customers. LMDS architecture is shown in Fig. 11.11. The remote places may be Ethernet, large or small business houses, or residences connected through LOS by antennas.

Fig. 11.11 Structure of LMDS

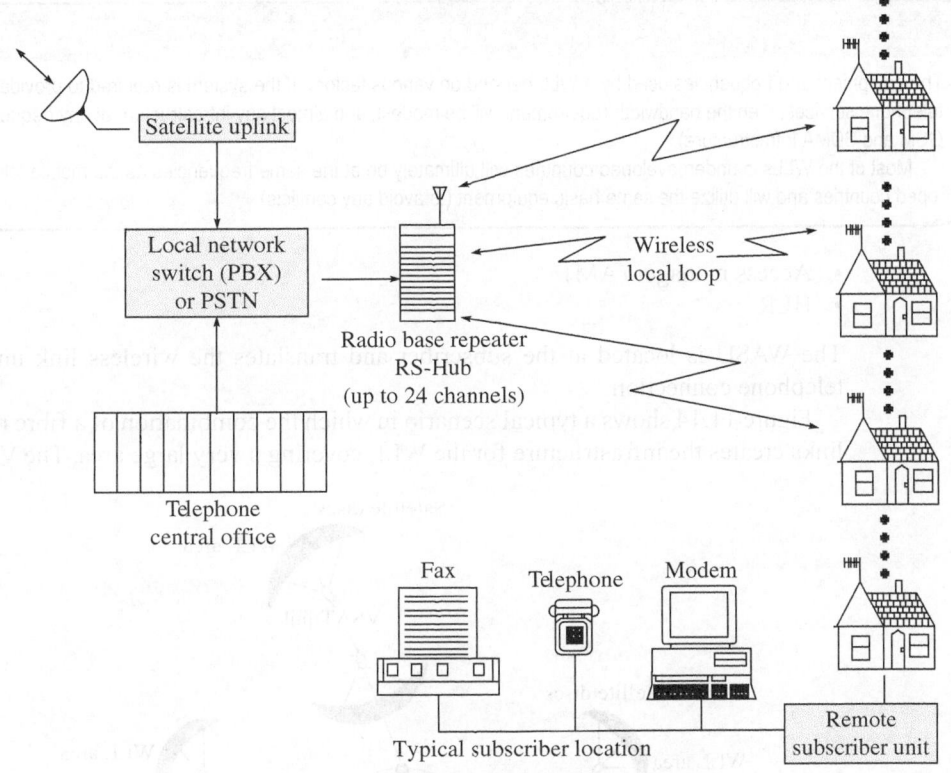

Fig. 11.12 WLL system

(wireless access network unit or WANU), which will communicate with all the WLL customers (wireless access subscriber units or WASUs) within the cell, as shown in Figs 11.12 and 11.13.

The BS may be as simple as a small omnidirectional antenna and control box hanging from the overhead electrical lines. Each customer will be equipped with a transceiver and a small patch antenna. The transceiver may have several outputs, one for a telephone, one for a modem, and maybe even one for a television. The small antenna, which may be inside or outside, will be positioned to communicate with the BS.

The WANU, the interface between the underlying telephone network and the wireless link, consists of the following:

- BTS
- Radio port control unit (RPCU)

WLLs may be based on either analog (e.g., AMPS) or digital (e.g., GSM, DECT, or CDMA) standards, and the architecture of the WLL may be as per the standard adopted.

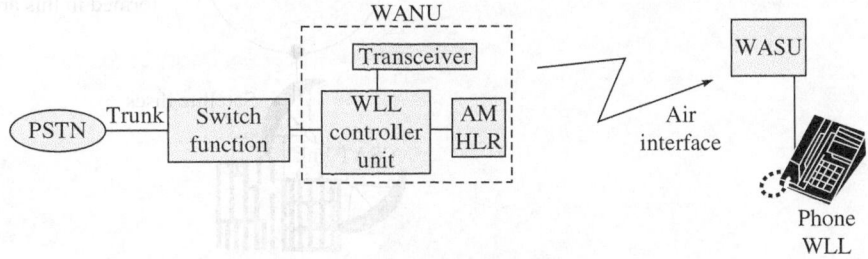

Fig. 11.13 WLL access units and their functions

EQUIPMENT AND FREQUENCIES USED BY WLL

The equipment and frequencies used by a WLL depend on various factors. If the system is required to provide only basic local telephone services, then the bandwidth requirement will be modest, and almost any infrastructure at any frequency will do (e.g., GSM and CDMA infrastructure).

Most of the WLLs in underdeveloped countries will ultimately be at the same frequencies as the mobile telephony in developed countries and will utilize the same basic equipment (to avoid any conflicts).

- Access manager (AM)
- HLR

The WASU is located at the subscriber and translates the wireless link into a traditional telephone connection.

Figure 11.14 shows a typical scenario in which the combination of a fibre ring with VSAT links creates the infrastructure for the WLL covering a very large area. The VSAT links may

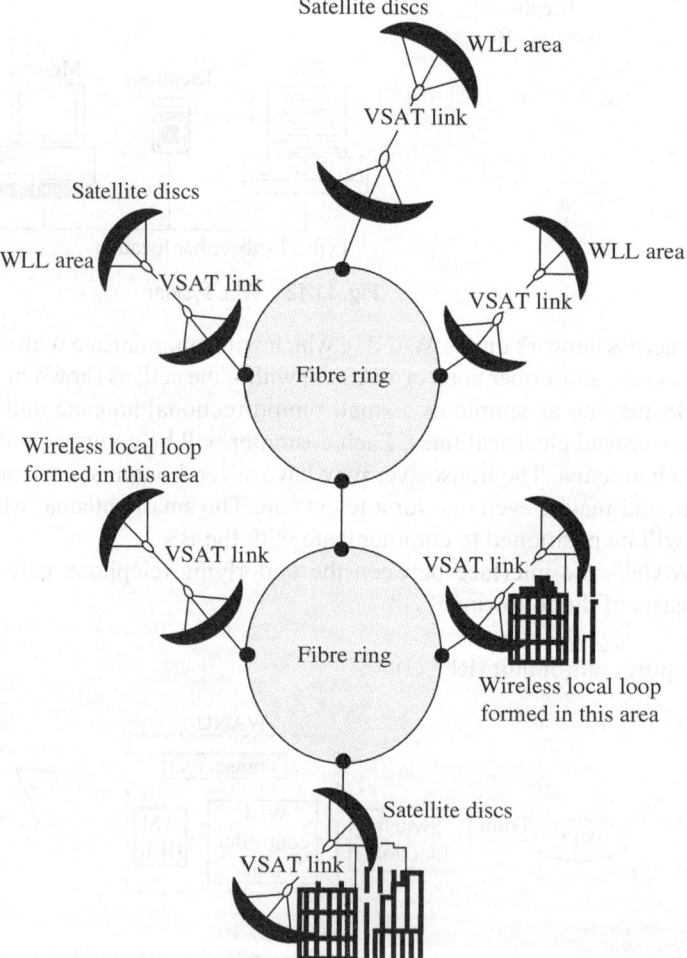

Fig. 11.14   WLL in fibre ring topology connecting satellite links

be replaced by microwave LOS links. In the WLL area, the BS as well as subscriber units will communicate as usual. The BSs are linked with satellite disks or microwave antennas to complete the scenario.

11.5.1 Digital Enhanced Cordless Telecommunication

The digital enhanced cordless telecommunications (DECT) system is developed based on the specifications of the European Telecommunications Standards Institute (ETSI). It replaces older analog cordless phone systems such as CT1 and CT1+. These analog systems ensured security only to a limited extent, because they did not use encryption for data transmission, and offered only low capacity. However, DECT is a more powerful alternative to the digital system CT2.

The DECT standard defines the specifications for the system and the way in which it operates. It specifies both the simplex (half slot) and duplex (full slot) operations. Table 11.5 provides the salient features of a DECT system.

The basic DECT system has 10 possible carrier frequencies between 1880 MHz and 1900 MHz. In addition to this, the time dimensions for each carrier are divided to provide time frames repeating every 10 ms. Each frame consists of 24 time slots (Fig. 11.15), each of which is individually accessible and may be used for either transmission or reception.

Table 11.5 Specifications of a DECT system

Parameter	Specification
Frequency band	1880 MHz to 1980 MHz and 2010 MHz to 2025 MHz (new)
Access technique	FDMA/TDMA/TDD
Channel bit rate	1152 kb/s
Carrier spacing	1728 kHz
Frame duration	10 ms
Slots in a frame	24
Access channels/RF carrier	120 duplex 32 kbit/s channels
TCH assignment	Instant dynamic
Control carriers	Not required
Modulation	GFSK/GMSK ($BT = 0.5$) and optional higher-level modulation schemes possible
Portable average RF power	10 mW
Portable peak RF power	240 mW 24 dBm
Speech codec	32 kbit/s ADPCM
BS sensitivity at 0.1% BER	−86 dBm (for GAP) (typically −90 dBm to −94 dBm)
Basic link budget	110 dBm (typically 114 dBm to 118 dBm)
Protected 64 kbit/s bearer service	Yes
BS antenna	Diversity switched; post-detection selection optional; dual antennas in handset optional
Tolerance to time dispersion with selection antenna diversity	200 ns (500 ns possible with low-cost non-coherent equalizer)

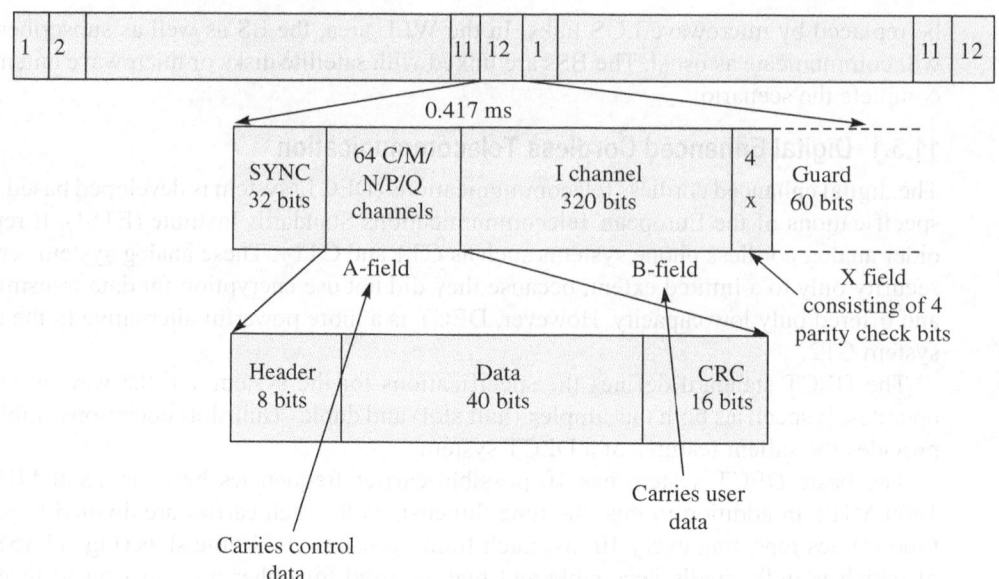

Fig. 11.15 DECT—a hybrid TDMA—FDMA system that uses TDD

DECT is a hybrid
TDMA-FDMA system
that uses time division
duplex (TDD).

The first 12 time slots are used for downlink transmissions and the remaining 12 are used for the uplink. This reduces the level of complexity, and as this is not needed for basic implementations, it can reduce the cost involved. The DECT TDMA structure enables up to 12 simultaneous basic voice connections per transceiver.

The system uses dynamic channel allocation and is thereby able to reduce the levels of interference and ensure that links are set up on the channels with the least interference. The DECT equipment scans the frequency allocation at least every 30 s as a background activity. This produces a list of free and occupied channels along with the available time slots to be used for the channel selection, should this be required.

For a basic DECT speech service, two time slots, with 5 ms separation, are paired to provide the bearer capacity for typically 32 kbps (adaptive differential pulse code modulation or ADPCM G.726 coded speech) full duplex connections. It consists of 10 carriers. The carrier bit rate is 1.152 MHz. As mentioned, each carrier is time slotted. A slot consists of 420 bits plus 60 bits guard time. The structure is shown in Figs 11.15 and 11.16.

A DECT system has the following features for signalling and control:
- DECT specifies two logical channels for *user data* transmission and five logical channels for *network control*.
- The data channel (B-field) features are as follows:
  □ *Unprotected I_N data* uses 320 bits in every slot and the rate is 32 kbps.
  □ *Protected I_p data* uses 256 bits. The error control uses 64 bits and the rate is 25.6 kbps.
- The CCH (A-field) features are as follows:
  □ Five logical channels are muxed over this field in a slot.
  □ An eight-bit header is used to identify the type of channel and nature of message.
  □ Forty bits are developed to transmit the CCHs.

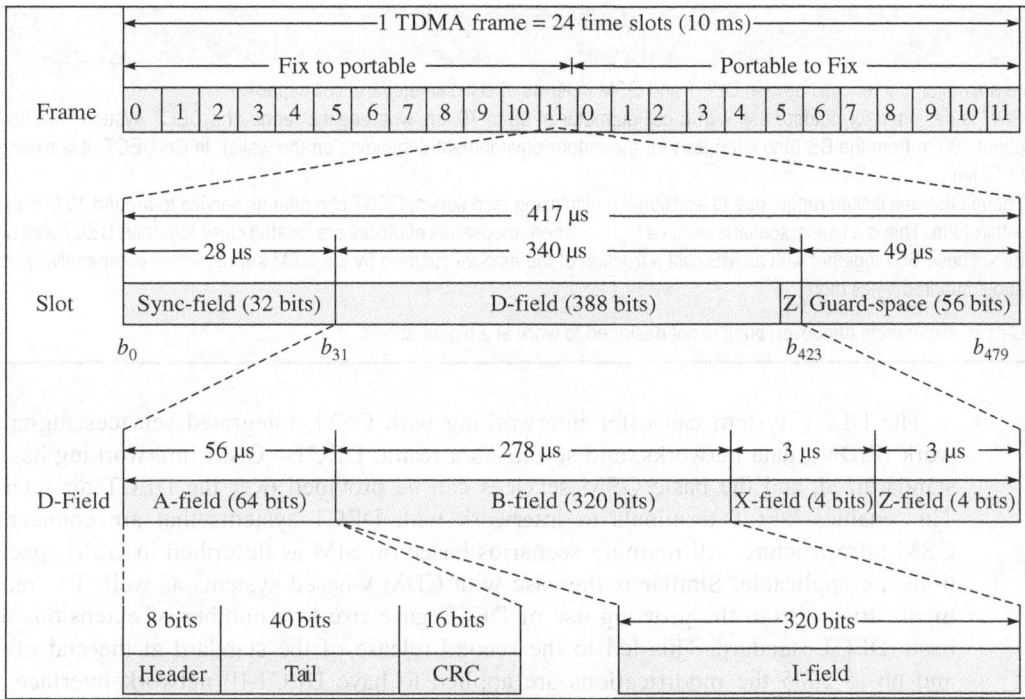

Fig. 11.16 DECT frame structure of Fig. 11.15 in detail

- The terminals are required to do most of the work in DECT. If a terminal wants to originate a call, it searches for an *idle channel* and starts to communicate with the BS.
- DECT supports *soft handoff*. The mobile terminal can communicate with two BSs when necessary.
- DECT supports authentication and encryption.

Additionally, the DECT portable continuously analyses the signals to ensure that they originate from the BS to which it is connected and has access rights. The portable locks onto the strongest BS and checks whether it can access the BS, as detailed in the DECT standard. The channels with the best signal strength (receive signal strength indication or RSSI) are used for the radio link as required. This dynamic channel selection and allocation mechanism guarantees that radio links are always set up on the channel available with the least interference and hence the best performance is obtained.

All DECT systems are based on a main standard, the *common interface* (CI), which is often used in association with the *generic access profile* (GAP). The GAP ensures interoperability of equipment from different providers for voice applications. It defines the minimum interoperability requirements including mobility management and security features. It has different requirements on public and private systems. This means that the GAP is effectively the industry standard for a basic fallback speech service with mobility management. This basic service is not always used, but instead it forms the fallback, which is always available, especially when requested by a roaming phone and other such instances.

DIFFERENCES BETWEEN DECT AND GSM

There are major differences between DECT and GSM in terms of cell diameter and cell capacity.

- GSM is designed for outdoor use with a cell diameter of up to 70 km, whereas the range of a DECT system is limited to about 300 m from the BS (and even less for the indoor environment depending on the walls). In CorDECT, it is extended to 10 km.
- Though this is a limited range, due to additional multiplexing techniques, DECT can offer its service to around 10,000 users within 1 km. This is a typical scenario within a big city where thousands of offices are located close together. DECT also uses BSs. These BSs together with an MS cost a fraction of the amount incurred by the GSM set-up, which comparatively costs about hundred times more.

DECT can also handle handover, but it is not designed to work at a higher speed.

The DECT system can offer interworking with GSM, integrated services digital network (ISDN), data networks, and so on. As a result, DECT—GSM interworking has been standardized, and the basic GSM services can be provided over the DECT air interface. This enables DECT terminals to interwork with DECT systems that are connected to GSM infrastructure. All roaming scenarios based on SIM as described in GSM specifications are applicable. Similar is the case with CDMA-based systems as well. The requirements arising from the growing use of DECT gave rise to a number of extensions to the basic DECT standard. This led to the second release of the standard at the end of 1995 and up to 2005 the modifications are applied to have DECT-IP network interface. This included facilities such as emergency call procedures, definition of the wireless relay station (WRS), an optional direct portable-to-portable communication feature, and the Internet services.

In a DECT system, access points to the PSTN can be established within, say, railway stations, large government buildings, and hospitals, offering a much cheaper telephone service compared to a GSM system. DECT could also be used to bridge the last few hundred metres between a new network operator and its customers. Using these small-range local loops, new companies can offer their service without having their own lines installed in the streets.

11.5.2 CorDECT WLL

This section discusses the architecture and deployments of a WLL system based on the DECT standard identified as CorDECT WLL, the technology adopted by India. CorDECT technology allows telephone and Internet services together on the same line. It is a European standard. This cost-effective technology can be a great boon to those who have no access to such facilities today, especially in rural areas.

In a PSTN, individual copper lines are wired to each subscriber from the backbone network (and linked to the PSTN central office), whereas WLL connects end-users to the backbone network wirelessly through an access unit.

Figure 11.17 provides the visual diagram of PSTN and dial-up network access over the telephone line (wired scenario) as a reference for comparing the difference with CorDECT WLL.

A simplified version of the CorDECT WLL system is shown in Fig. 11.18. In this figure, the CorDECT WLL system is composed of the following:

- DECT interface unit (DIU)
- Remote access switch (RAS)
- Compact base station (CBS)
- Subscriber unit at the receiving end

The following are the features of this CorDECT system:

- In Fig. 11.18, the subscriber terminal is a Wallset with Internet port (WS-IP) as a standard interface (RJ-11) and a serial port interface (RS-232).
- It supports a standard telephone, a modem, and a fax through the RJ-11 port and provides a direct connection to a personal computer (PC) through the RS-232 port.
- The WS-IP provides high-speed Internet access at 40–70 kbps as well as normal telephone service for subscribers in the CorDECT system.
- The WS-IP is connected to the CBS using a wireless link. The CBS provides the radio interface between the subscribers and the access unit, which consists of a DIU and RAS.

Fig. 11.17 PSTN dial-up wired network scenario

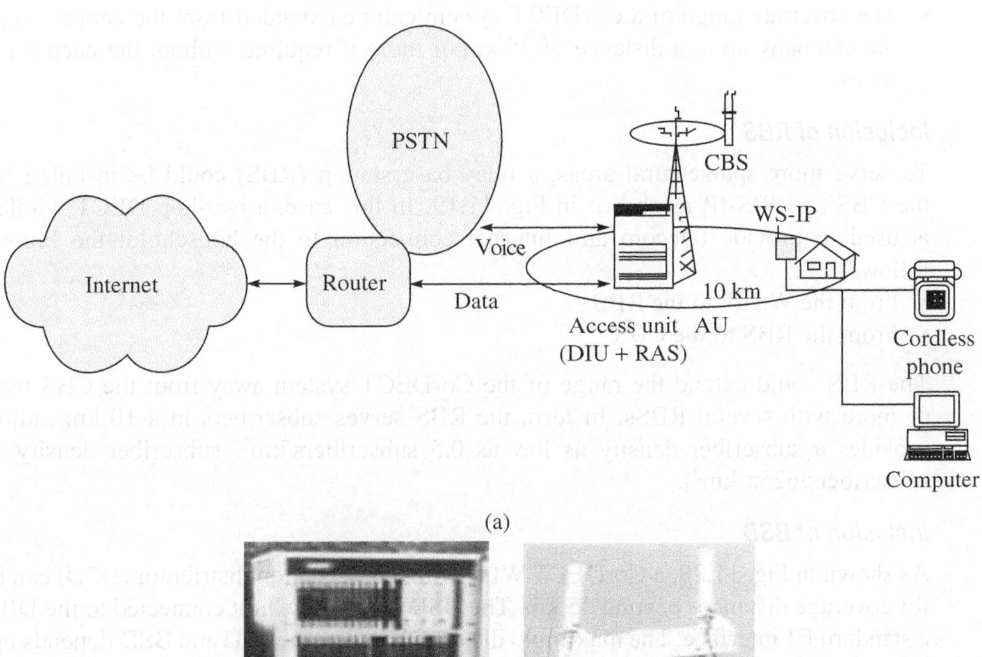

(a)

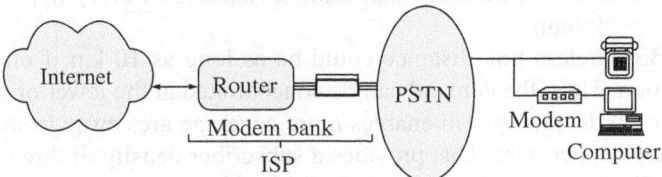

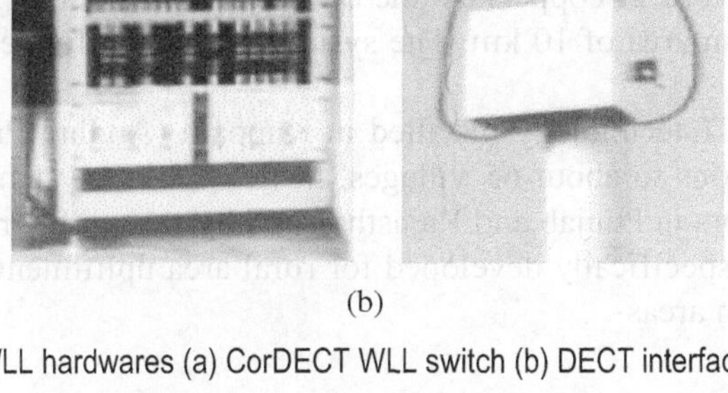

(b)

Fig. 11.18 WLL hardwares (a) CorDECT WLL switch (b) DECT interface unit and BS

Advantages of CorDECT WLL are fast deployment, immediate realization of revenues, low construction costs, low network maintenance, low management and operating costs, and high flexibility to increasing users.

- The DIU separates the voice traffic and directs it to the telecom network as well as switches the Internet calls to a built-in RAS.
- The RAS then routes the traffic to the Internet network. Since Internet traffic does not have to pass through the telecom network, CorDECT WLL gets rid of the Internet tangle associated with plain old telephone service (POTS).
- The DIU can be co-located with the PSTN through the backbone network at, say, any big company premises along with its township by utilizing the optical fibre network.
- It should be noted that LOS is required in every hop in the CorDECT WLL system.
- The CBS wireless link distance could be as long as 10 km if measured using an LOS connection. Thus, the demand can be concentrated at the tower or within a 10 km radius. In other words, this system enables more coverage area away from the company without using any copper wire. This provides a subscriber density of three subscribers/km^2 (subscriber density = 1000 subscribers/100π km^2). The subscriber density, however, can be increased by using more DIUs at the central premise.
- The coverage range of a CorDECT system can be extended from the central location in the company up to a distance of 35 km or more if required without the need for copper wires.

Inclusion of RBS

To serve more sparse rural areas, a relay base station (RBS) could be installed between the CBS and WS-IP, as shown in Fig. 11.19. In this case, a two-hop DECT wireless link is used to provide telecom and Internet connection to the household; the links are as follows:

- From the WS-IP to the RBS
- From the RBS to the CBS

The RBS could extend the range of the CorDECT system away from the CBS by 25 km or more with several RBSs. In turn, the RBS serves subscribers in a 10 km radius. This provides a subscriber density as low as 0.5 subscribers/km^2 (subscriber density = 1000 subscribers/625π km^2).

Inclusion of BSD

As shown in Fig. 11.20, a CorDECT WLL with a base station distributor (BSD) can be used for coverage distances beyond 35 km. The BSD is a remote unit connected to the DIU using a standard E1 interface. The maximum distance between the DIU and BSD depends upon the E1 link (radio, fibre, or copper). At the BSD site, a cluster of CBSs is mounted on a rooftop tower to serve an area of 10 km. The system is suitable for serving in a remote mid-sized town or city.

The CorDECT technology installed in Kuppam (Andhra Pradesh, India) has been providing connections to about 65 villages. Moreover, the system is being installed by basic services operators in Punjab and Rajasthan and by several Internet service providers (ISPs). CorDECT was specifically developed for rural area upliftment but has now been commercialized in urban areas

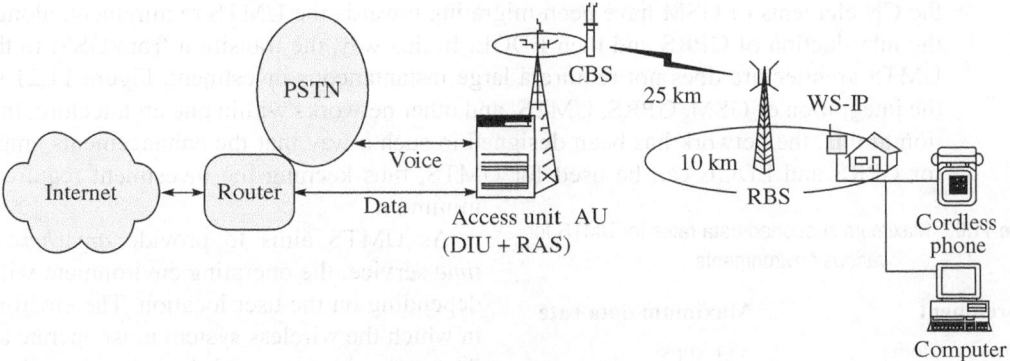

Fig. 11.19 Deployment with RBS

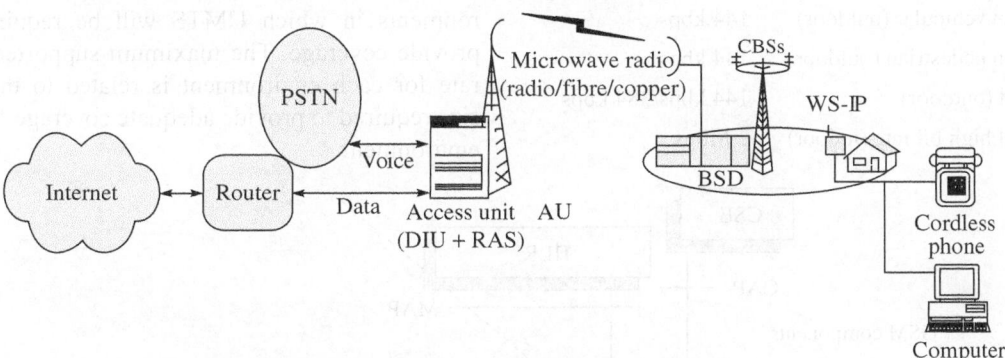

Fig. 11.20 Deployment with BSD

11.6 IMT-2000 AND UMTS

The global standard for 3G wireless communications is IMT-2000, defined by a set of interdependent International Telecommunication Union (ITU) recommendations. IMT-2000 provides a framework for worldwide wireless access by linking the diverse systems of terrestrial and/or satellite networks. In IMT-2000, 2000 stands for the year it was introduced and also the spectrum used (around 2000 MHz).

Conceptually, UMTS represents an evolution from 2G to 3G. It uses WCDMA to carry the radio transmissions, and the system is often referred to by the name WCDMA. In order to create and manage a system as complicated as UMTS or WCDMA, it is necessary to develop and maintain a large number of documents and specifications. For UMTS, these are now managed by a group known as the 3GPP—third-generation partnership programme.

11.6.1 UMTS or WCDMA Architecture

> Many of the ideas that were incorporated in GSM have been carried over and enhanced for UMTS.

The scope of the 3GPP was to produce globally applicable technical specifications and technical reports for a 3G mobile telecommunications system. This would be based upon the GSM CNs and the radio access technologies that they support. This is why UMTS is also called by another name—3GSM, because it is a 3G migration for GSM. Although UMTS uses a totally different air interface, the CN elements of GSM have been migrating towards the UMTS requirement, along with the introduction of GPRS and then EDGE. In this way, the transition from GSM to the 3G UMTS architecture does not require a large instantaneous investment. Figure 11.21 shows the integration of GSM, GPRS, UMTS, and other networks within one architecture. In addition to this, the network has been designed in such a way that the enhancements employed for GPRS and EDGE can be used for UMTS, thus keeping the investment required to a minimum.

As UMTS aims to provide *anywhere, anytime* service, the operating environment will vary depending on the user location. The environment in which the wireless system must operate affects the system capacity and the type of services that can be provided. Table 11.6 lists some of the environments in which UMTS will be required to provide coverage. The maximum supported data rate for each environment is related to the cell size required to provide adequate coverage for the environment.

Table 11.6 Maximum supported data rates for UMTS for various environments

Environment	Maximum data rate
Business (indoor)	384 kbps
Suburban (indoor/outdoor)	144 kbps
Urban vehicular (outdoor)	144 kbps
Urban pedestrian (outdoor)	144 kbps
Fixed (outdoor)	144 kbps/384 kbps
Local high bit rate (indoor)	2 Mbps

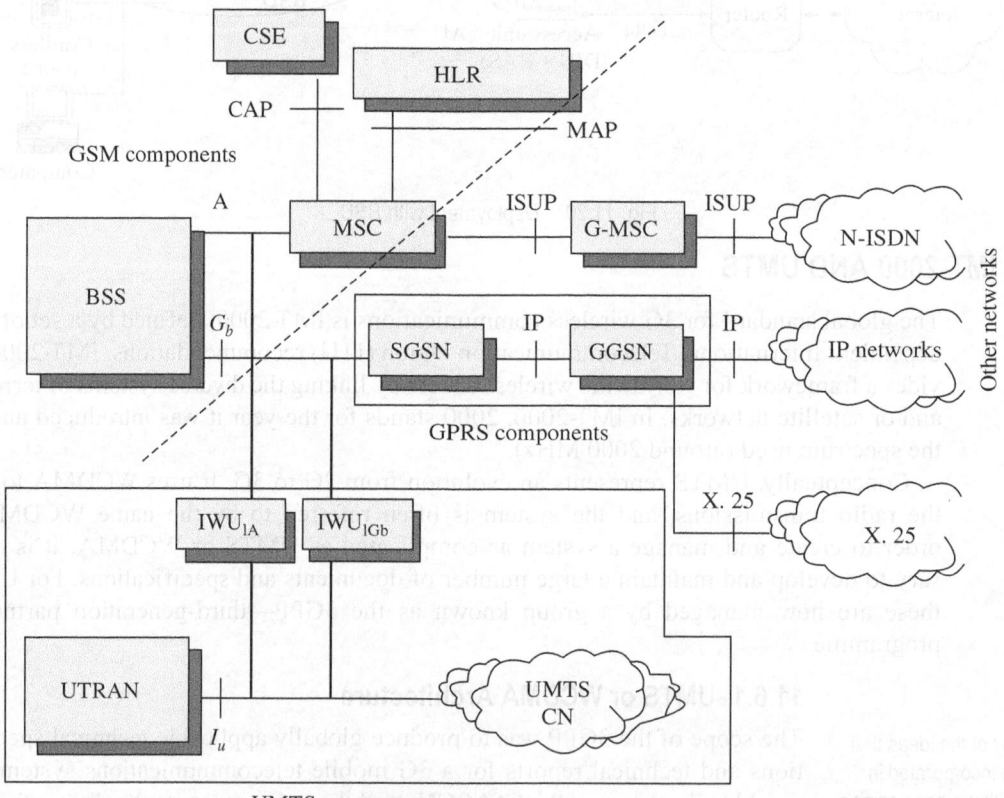

Fig. 11.21 Evolution of GSM towards UMTS—introduction to UMTS CN (2005) (various standard interfaces are shown over the connecting lines)

GLOBAL COVERAGE WITH SATELLITE SYSTEMS

The aim of UMTS aim is *anywhere, anytime*; however, cellular networks can cover only a limited area due to high infrastructure costs. Hence, satellite systems form an integral part of the UMTS network. Satellites will be able to provide an extended wireless coverage to remote areas and to aeronautical and maritime mobiles.

Cellular Structure

A cellular network is required to ensure that UMTS can provide a high-capacity network. As with any cellular system, the total capacity of the network is dependent on the size of the cells used. However, the cell size is limited by the amount of infrastructure. The cell size determines the maximum channel capacity for each cell, as propagation effects, multipath delay spread, and high path loss force large cells to have a lower data rate. Large cells also have to service a large number of users, and as the cell capacity is approximately fixed, each user can have only a reduced data rate with respect to a smaller cell.

In order to optimize the cellular network, three cell types are used. These are the picocell, microcell, and macrocell. These cell types trade off cell size with total capacity and services. The size and type of coverage of each cell type affects the radio propagation problems encountered. This will determine the most suitable radio transmission technique to be used. Table 11.7 shows some of the cell characteristics of the three cell types used in the UMTS system.

11.6.2 Elements of UMTS or WCDMA

The network for UMTS can be split into three main constituents (Fig. 11.22):
(a) MS, called the *user equipment* (UE)
(b) BSS, known as the radio (access) network subsystem (RNS)
(c) CN

Table 11.7 Cell types and cell characteristics for UMTS

Characteristic	Picocell	Microcell	Macrocell
Cell radius	<100 m	<1000 m	<20 km
Antenna	Ceiling or wall mounted	Below roof top height	Roof top mounting
Maximum multipath delay spread	1 µs	5 µs	20 µs
Applications and environments	Indoor or outdoor, within buildings, city centres	High-density outdoor, business (indoor), fixed (outdoor), inner city areas	Low-density areas, suburban areas, urban areas, fixed (outdoor)
Services and data rate supported	All services (up to 2 Mbps)	Up to 384 kbps	Limited subset (up to 144 kpbs)

User Equipment

The UE for UMTS or WCDMA is equivalent to the mobile equipment used on GSM networks. It is essentially the handset, although it can be much more versatile, containing many more applications, because of its access to much higher speed data communications.

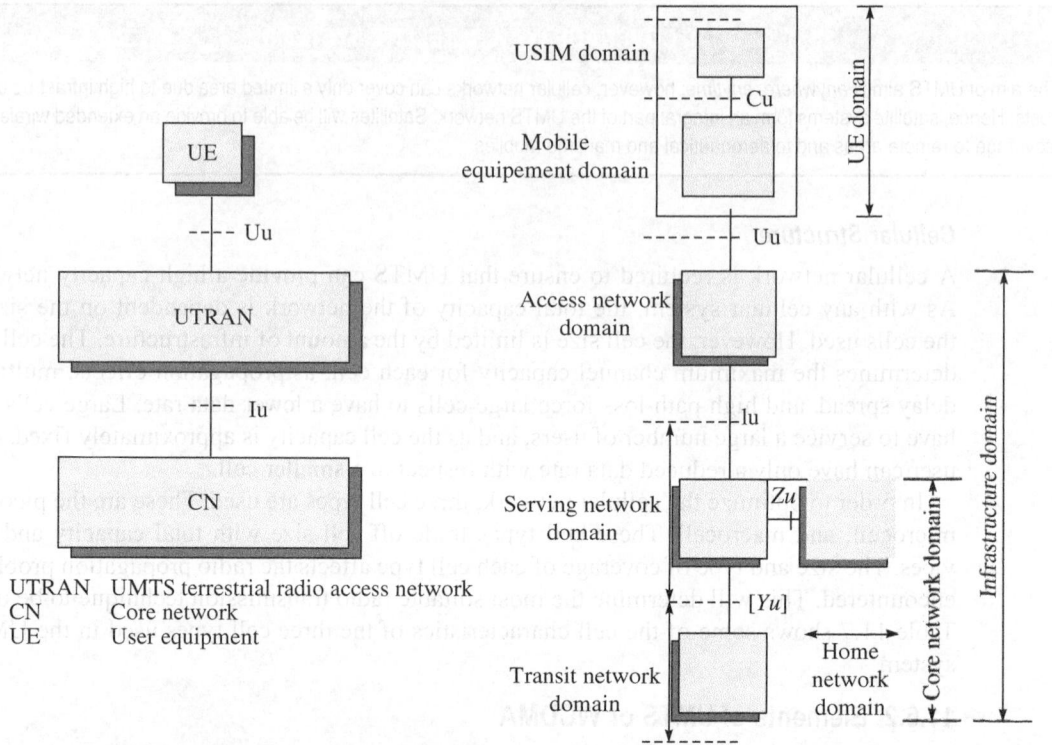

Fig. 11.22 UMTS architecture and domains along with defined interfaces

It consists of a variety of elements including RF circuitry, processing, antenna, and battery. The circuitry used within the UE can be broadly split into RF and baseband processing areas.

- The *RF* areas handle all elements of the signal, for both the receiver and the transmitter. One of the major challenges for the RF power amplifier was to reduce power consumption. The form of modulation used, WCDMA requires the use of a linear amplifier. Linear amplifiers inherently take more current than non-linear amplifiers, which are used for modulation in GSM. Accordingly, to maintain battery life, measures were introduced into many of the designs to ensure optimum efficiency.

- *Baseband signal processing* consists mainly of digital circuitry. This is considerably more complicated than that used in phones of previous generations. Again, this has been optimized to reduce the current consumption as far as possible; however, there has been an increase in the current drain on the battery. With users expecting the same lifetime between charging batteries as experienced on the previous-generation phones, this has necessitated the use of new and improved battery technology. Now, lithium ion (Li-ion) batteries are used in phones. These batteries are small and relatively light while still retaining or even improving the overall life between charges.

The UE contains a SIM card, although in the case of UMTS, it is termed as the universal subscriber identity module (USIM). This is a more advanced version of the SIM card used in GSM and other systems but embodies the same types of information. It contains the IMSI and mobile station international subscriber directory number (MSISDN). It also holds information

In UMTS, the SIM is transformed into a far more powerful USIM.

GSM—UMTS equivalent elements
SIM—USIM
MS—UE
BTS—NodeB
BSS—RNS
NSS—CN
ARFCN—UARFCN

regarding the preferred language to enable the correct language information to be displayed, especially when roaming, and a list of preferred and prohibited public land mobile networks (PLMN). The USIM further contains a short message storage area that allows messages to stay with the user even when the phone is changed. Similarly, *phone book* numbers and call information of the numbers of incoming and outgoing calls are stored.

Radio Access Network Subsystem

The RNS is the section of the UMTS or WCDMA network that acts as an interface between the UE and the CN. RNS is similar to the BTS and BSC. Under UMTS terminology, the radio transceiver or BTS equivalent (in GSM) is known as the Node B. This communicates with the various UE. It also communicates with the radio network controller (RNC). This is undertaken over an interface known as the Iub (Fig. 11.23). The overall radio access network is known as the *UMTS terrestrial radio access network* (UTRAN) (Fig. 11.24). The RNC component of the radio access network (RAN) connects to the CN.

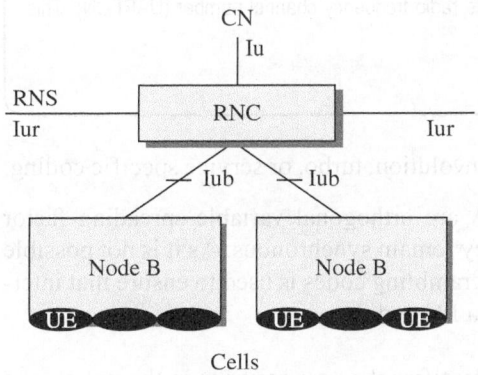

Fig. 11.23 UMTS RNC

Core Network

The CN used for UMTS is based upon a combination of the circuit-switched elements used for GSM and the packet-switched elements used for GPRS and EDGE. Thus, the CN is divided into circuit-switched and packet-switched domains. Some of the circuit-switched elements are MSC, VLR, and GMSC, and the packet-switched elements are SGSN and GGSN. Some network elements, such as EIR, HLR, VLR, and AuC mentioned earlier, are shared by both the domains and operate in the same manner as they do in GSM.

The ATM is specified for UMTS core transmission. The architecture of the CN may change when new services and features are introduced. *Number portability database* (NPDB) is used to enable a subscriber to change the network provider while retaining the old phone number. Gateway location register (GLR) may be used to optimize subscriber handling between network boundaries. The MSC, VLR, and SGSN can merge to become a UMTS MSC.

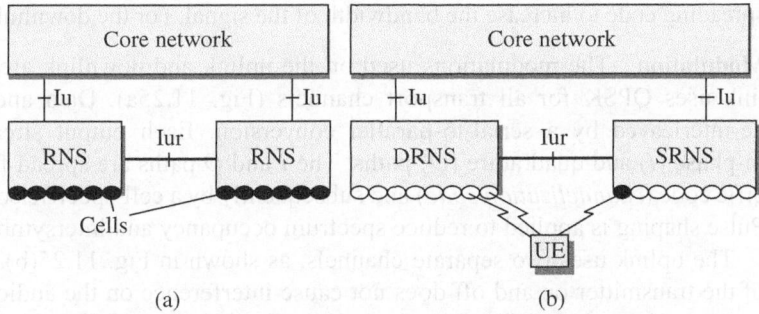

Fig. 11.24 UTRAN (a) Structure (b) Service and drift radio access networks

The codes required to spread the signal must be orthogonal if they are to enable multiple users and channels to operate without mutual interference.

11.6.3 UMTS or WCDMA Radio or Air Interface

The physical layer within UMTS or WCDMA is totally different from that employed by GSM. It employs a spread spectrum transmission in the form of CDMA. Additionally, it uses different frequencies to those allocated for GSM.

Frequency range Currently, *six bands* are specified for use in UMTS or WCDMA, although operation in other frequencies is not precluded. However, much of the focus for UMTS is currently on frequency allocations around 2 GHz. Within these bands, different portions have been reserved for different uses as follows:

- *1920–1980 MHz and 2110–2170 MHz (FDD, WCDMA)*: Uplink and downlink are paired, channel spacing is 5 MHz, and raster is 200 kHz. An operator needs three to four channels (2 × 15 MHz or 2 × 20 MHz) to be able to build a high-speed, high-capacity network.
- *1900–1920 MHz and 2010–2025 MHz (TDD, TD/CDMA)*: Unpaired, channel spacing is 5 MHz, and raster is 200 kHz. Transmit and receive transmissions are not separated in frequency.
- *1980–2010 MHz and 2170–2200 MHz*: Satellite uplink and downlink

> *Note*: Carrier frequencies are designated by UTRA absolute radio frequency channel number (UARFCN). This can be calculated as follows:
>
> UARFCN = 5 × (frequency in MHz)

Channel coding The channel coding may be convolution, turbo, or service specific coding.

Spreading codes The codes used in WCDMA are orthogonal variable spreading factor (OVSF) codes, which can be operated only if they remain synchronous. As it is not possible to retain exact synchronization, a second set of scrambling codes is used to ensure that interference does not result. This scrambling code is a PN code.

Thus, there are two stages of spreading:

(a) The first one uses the OVSF code, which identifies the user services in the uplink and user channels in the downlink (channelization codes).

(b) The second one uses a scrambling PN code, which identifies the individual Node B or UE (scrambling codes)

These codes are used to provide different levels of separation.

On the uplink, 512 different codes are available for use, one of which will be assigned to each Node B. The data of the spread spectrum modulation signal is multiplied by the chip or spreading code to increase the bandwidth of the signal. For the downlink, a short code is used.

Modulation The modulations used on the uplink and downlink are different. The downlink uses QPSK for all transport channels (Fig. 11.25a). Data and control streams are de-interleaved by a serial-to-parallel conversion. Each output stream is applied to the in-phase (I) and quadrature (Q) paths. The I and Q paths are spread first by a channel-specific code (*channelization code*) and subsequently by a cell-specific code (*scrambling code*). Pulse shaping is applied to reduce spectrum occupancy and intersymbol interference (ISI).

The uplink uses two separate channels, as shown in Fig. 11.25(b), such that the cycling of the transmitter on and off does not cause interference on the audio lines, a problem that was experienced on GSM. The dual channels are achieved by applying (a) the coded user data (data information) to the I input to the differential QPSK (DQPSK) modulator and

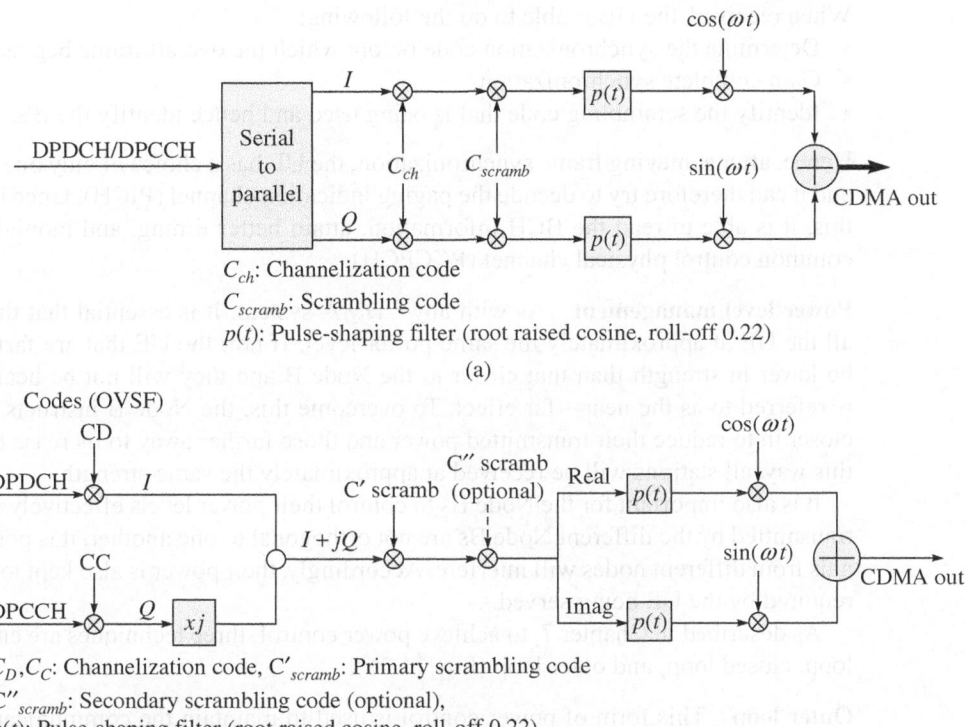

C_{ch}: Channelization code
C_{scramb}: Scrambling code
$p(t)$: Pulse-shaping filter (root raised cosine, roll-off 0.22)

(a)

C_D, C_C: Channelization code, C'_{scramb}: Primary scrambling code
C''_{scramb}: Secondary scrambling code (optional),
$p(t)$: Pulse-shaping filter (root raised cosine, roll-off 0.22)

(b)

Fig. 11.25 Spreading and modulation of dedicated channels (a) Downlink (b) Uplink

> The SCHs do not spread using OVSF and PN codes but other synchronization codes are used.

(b) the control data (control information), which has been encoded using a different (*channelization*) spreading code, to the Q input to the modulator. A UE-specific spreading code (*scrambling code*) is subsequently applied. The data to be transmitted is encoded using a spreading code particular to a given user; as a result, only the desired recipient is able to correlate and decode the signal, with all other signals appearing as noise. This allows the physical RF channel to be used by several users simultaneously.

Synchronization codes   The level of synchronization required for the WCDMA system to operate is provided by the primary synchronization channel (PSCH) and the secondary synchronization channel (SSCH) (described in Section 11.6.5). Two types of codes are used:
(a) Primary code is used on the PSCH.
(b) Secondary code is used on the SSCH.

The *primary synchronization code* is the same for all cells and is a 256-chip sequence that is transmitted during the first 256 chips of each time slot. This allows the UE to synchronize with the BS for the time slot. On gaining time slot synchronization, the UE knows only the start and stop of the time slot, but it does not have any information about the particular time slot or the frame. This information is gained using the secondary synchronization codes.

There are 16 different *secondary synchronization codes*. One code is sent at the beginning of each time slot, that is, the first 256 chips. It consists of numerous synchronization codes. There are various scrambling code groups, each having eight codes.

When received, the UE is able to do the following:
- Determine the synchronization code before which the overall frame begins.
- Gain complete synchronization.
- Identify the scrambling code that is being used and hence identify the BS.

Hence, after achieving frame synchronization, the UE has a choice of only one in eight codes and it can therefore try to decode the paging indication channel (PICH). Once it has achieved this, it is able to read the BCH information, attain better timing, and monitor the primary common control physical channel (PCCPCH).

Power level management As with any CDMA system, it is essential that the BS receives all the UE at approximately the same power level. If not, the UE that are farther away will be lower in strength than that closer to the Node B and they will not be heard. This effect is referred to as the near—far effect. To overcome this, the Node B instructs those stations closer in to reduce their transmitted power and those farther away to increase their power. In this way, all stations will be received at approximately the same strength.

It is also important for the Node Bs to control their power levels effectively. As the signals transmitted by the different Node Bs are not orthogonal to one another, it is possible that signals from different nodes will interfere. Accordingly, their power is also kept to the minimum required by the UE being served.

As described in Chapter 7, to achieve power control, three techniques are employed: open loop, closed loop, and outer loop (Fig. 11.26).

Outer loop This form of power control is used to maintain the communication quality at the level of bearer service quality, while using as low power as possible. The uplink outer loop power control is for setting a target SIR in the node B for each individual uplink inner loop power control. This target SIR is updated for each UL according to the estimated uplink quality, for each RRC correction. In downlink outer loop power control the UE receiver maintains the link quality set by RNC in the downlink.

Open loop techniques These techniques are used during the initial access before communication between the UE and the Node B has been fully established. It operates by making a measurement of the received signal strength and thereby estimating the transmitter power

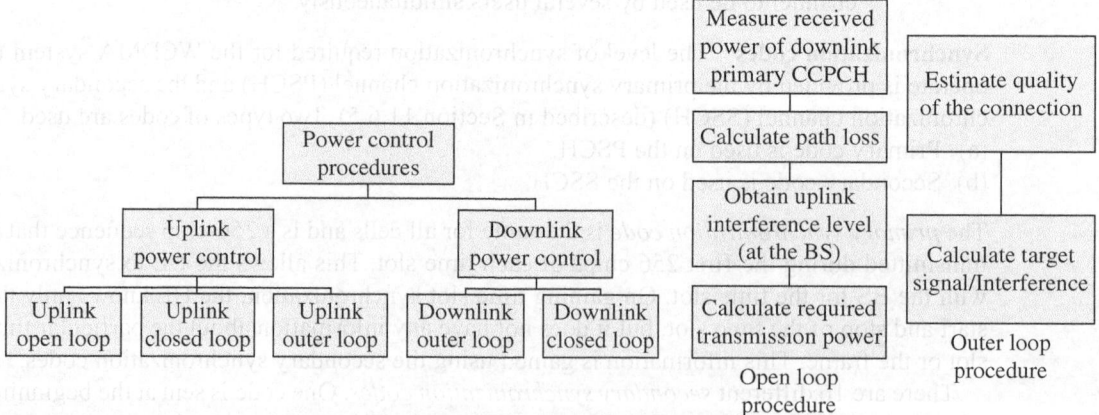

Fig. 11.26 Power control

required. As transmit and receive frequencies are different, path losses in either direction will be different, and therefore, this method cannot be any more than a good estimate.

Closed loop techniques The techniques are used once the UE has accessed the system and is in communication with the Node B. A measurement of the signal strength is taken in each time slot. As a result, a power control bit is sent requesting the power to be stepped up or down. This process is undertaken on both the uplink and the downlink. The fact that only one bit is assigned to power control means that the power will be continuously changing. Once it reaches approximately the right level, it will step up and then down by one level.

11.6.4 UMTS TDD and FDD

A communications system requires that communication be possible in both directions: to and from the BS to the remote station. The most obvious way is to transmit on one frequency and receive on another. The frequency difference between the two transmissions is such that the two signals do not interfere. This is known as FDD. It is one of the most commonly used schemes by cellular networks. Three major procedures are involved in the physical layer of FDD: cell search, power control, and random access.

It is also possible to use a single frequency and, rather than different frequency allocations, different time allocations. If the transmission times are split into slots, then transmissions in one direction take place in one time slot and those in the other direction take place in another slot. This scheme is known as TDD. Both the FDD and TDD concepts are shown in Fig. 11.27

> The data carried by the UMTS or WCDMA transmissions is organized into channels, frames, and slots; thus, all payload data as well as control data can be carried in an efficient manner.

11.6.5 UMTS Channels

Though UMTS uses CDMA techniques (WCDMA) as its multiple access technology, it additionally uses time division techniques with a slot and frame structure to provide the full channel structure.

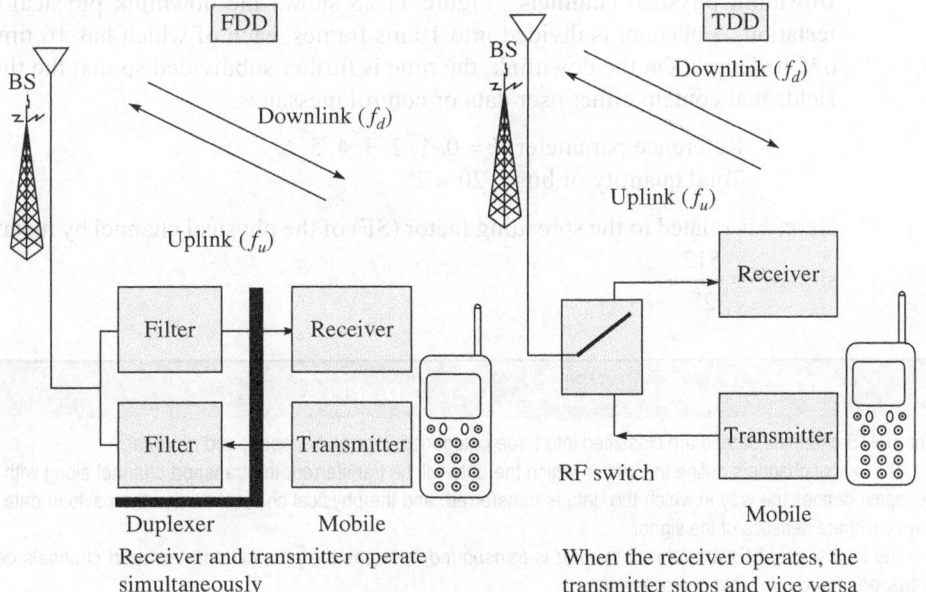

Fig. 11.27 FDD and TDD concepts

TDD IS BETTER THAN FDD

Both FDD and TDD modes are allowed in UMTS. In the FDD mode, the uplink and downlink are on different frequencies separated by a guard band. In the TDD mode, the uplink and downlink are split in time with the BSs and then the mobiles transmit alternately on the same frequency.

The following points show why TDD is better than FDD:

- TDD performs well with small cells and shorter channel delays.
- Typically, there is more traffic in the downlink (network to the mobile) than in the uplink (mobile to network). Therefore, it is better to allocate more bandwidth for the downlink and less for the uplink. TDD can support this kind of allocation, but FDD cannot. This is because FDD systems require the use of a paired spectrum, in which it is not possible to reallocate the use of the different bands.
- In TDD mode, it is possible to vary the capacity in either direction. By altering the proportion of the time allocated for transmission in each direction (downlink and uplink), it is possible to match the traffic load in each direction.
- UMTS TDD can support high peak data rates. Release 5 of the UMTS standard provides high-speed downlink packet access (HSDPA). The scheme allows the use of a higher order modulation scheme called 16QAM, which enables peak rates of 10 Mbps per sector in commercial deployments. The next release increases the modulation to 64QAM and introduces intercell interference cancellation and multiple input, multiple output (MIMO). In combination, these increase the peak rate to 31 Mbps per sector.
- UMTS TDD offers significant advantages for many applications. It is currently being used for mobile broadband; it can also serve to provide mobile television communication.
- TDD can achieve good overall performance by altering its transport format for each user.

The MAC layer provides data transfer services on logical channels. A set of logical channel types is defined for different kinds of data transfer services.

FDD Mode Channels

In UMTS, for both FDD and TDD modes, a separate set of channels are defined; For FDD, it is again required to define the channels for uplink and downlink separately.

Downlink physical channels Figure 11.28 shows the downlink physical channels manifestation. A channel is divided into 10 ms frames, each of which has 16 time slots, each of 625 μs length. On the downlink, the time is further subdivided so that the time slots contain fields that contain either user data or control messages.

Reference parameter: $k = 0, 1, 2, 3, 4, 5, 6$
Total quantity of bits $= 20 \times 2^k$

Here, k is related to the spreading factor (SF) of the physical channel by means of the expression $SF = \dfrac{512}{2^k}$.

UMTS CHANNELS

The UMTS channels carried are classified into three categories: *logical*, *transport*, and *physical*.

The logical channels define the way in which the data will be transferred, the transport channel along with the logical channel again defines the way in which the data is transferred, and the physical channel carries the payload data and governs the physical characteristics of the signal.

The logical channels are related to *what is transported*, whereas the physical and transport channels deal with *how it is transported and with what characteristics*.

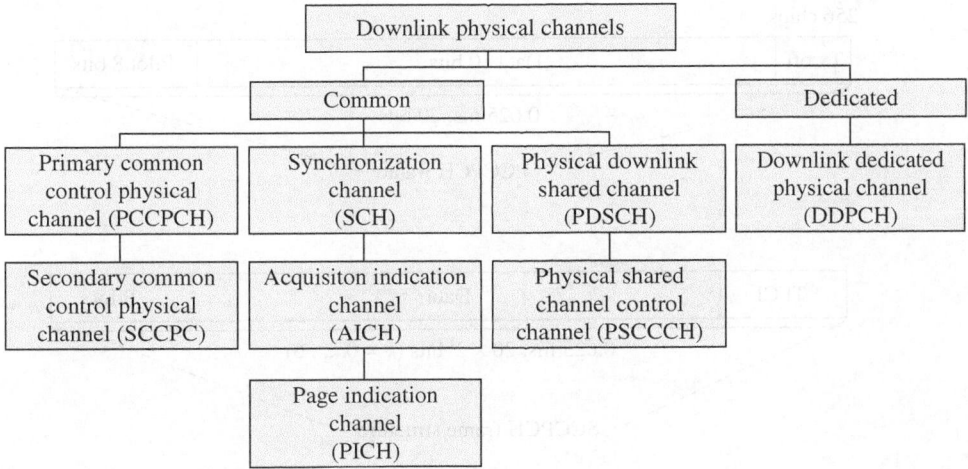

Fig. 11.28 UMTS downlink physical channels manifestation

The frame structures of some downlink physical channels are shown in Fig. 11.29.

> UTRAN determines whether the TFCI should be transmitted. It is mandatory for the UE to be able to support the TFCI.

Primary common control physical channel This channel continuously broadcasts system identification and access control information. It is used to carry the BCH. It is not transmitted during the first 256 chips of each time slot. This period is left to the PSCH and SSCH.

Secondary common control physical channel **(SCCPCH)** This channel carries the forward access channel (FACH) providing control information and PCH with messages for UE that are registered on the network.

> The number of SCCPCHs depends on cell traffic.

There are two types of SCCPCH: with transport format channel indicator (TFCI) and without TFCI. The SCCPCHs may be transmitted on narrow lobes pointed to the target UE channel (only valid for an SCCPCH carrying the FACH). A few differences are listed in Table 11.8.

Synchronization channel This channel consists of two subchannels: PSCH and SSCH. The PSCH is transmitted once per time slot. The contents are the same in each time slot and aligned in time with the BCCH. It is unmodulated and the spreading factor is $SF = 256$. The same code c_p is used for all bursts of the PSCH. The SSCH consists of repeatedly transmitting

Table 11.8 Comparison of PCCPCH and SCCPCH

PCCPCH	SCCPCH
Fixed rate (32 kbps) and fixed spreading factor (256)	Can support variable rate between 32 kbps and 2048 kbps, based on the value of the associated TFCI; $SF = 256/2^k$ ($k = 0 \ldots 6$)
Transmitted over the whole cell continuously (broadcast)	Transmitted only when there is data to be sent

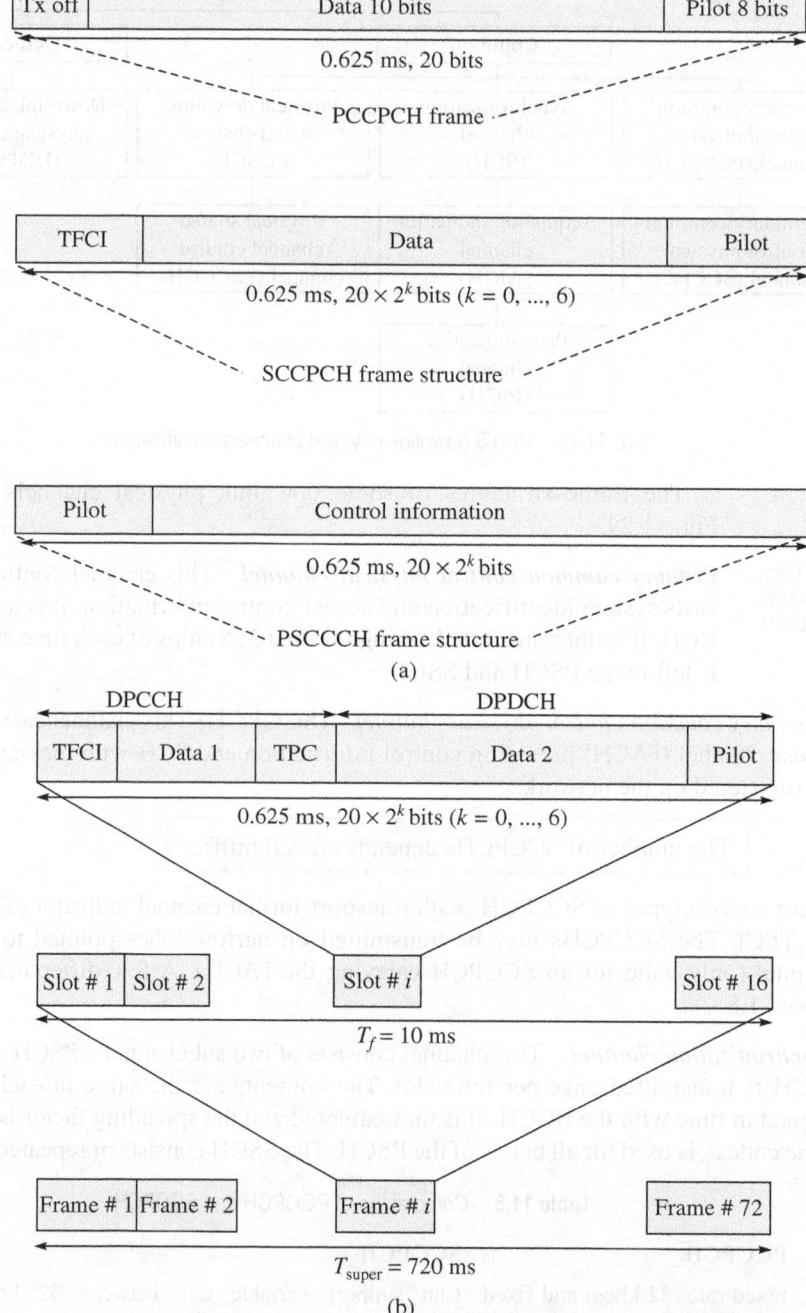

Fig. 11.29 Frame structures of FDD downlink physical channels (a) PCCPCH, SCCPCH, and PSCCCH (b) DDPCH (the number of frames and number of slots will remain the same and only the frame format for different channels will change)

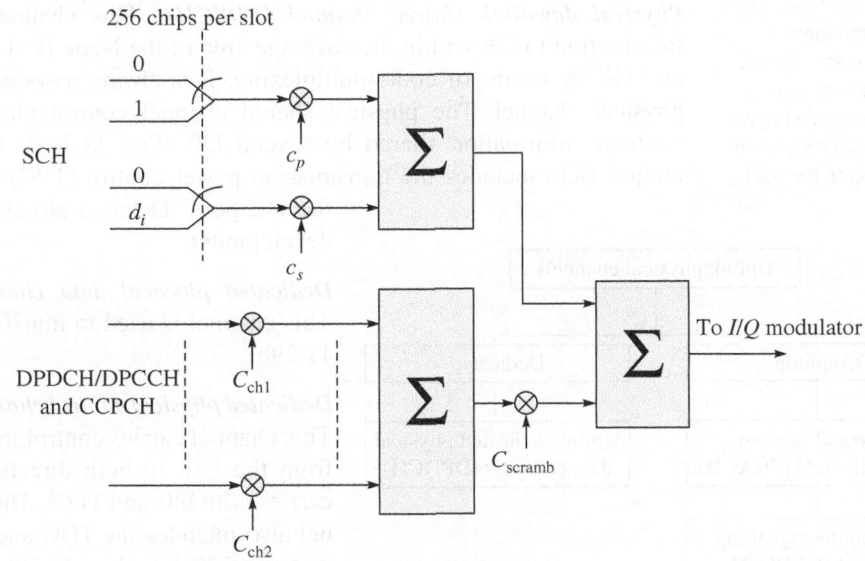

Fig. 11.30 Multiplexing of SCH

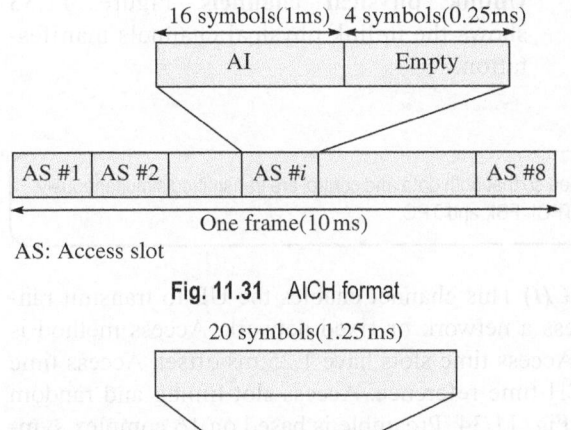

Fig. 11.31 AICH format

Fig. 11.32 PICH format

16 sequences of *unmodulated* codes c_s of length 256 chips. The SSCH sequence indicates which of the 32 different codes the cell's downlink scrambling code belongs to. The 32 sequences are used to encode 32 different code groups, each containing 16 scrambling codes. The SSCH sequence identifies the cell group. The multiplexing of PSCH and SSCH is shown in Fig. 11.30.

Acquisition indicator channel (AICH) This physical channel is used to carry acquisition indicators (AIs) corresponding to the signature of the RACH preamble. The AICH is used to inform the UE about the DCH it can use to communicate with the Node B. This channel assignment occurs as a result of a successful random access service request from the UE. One AICH frame is of 10 ms length. Each frame comprises eight access slots (Fig. 11.31).

Paging indication channel The PICH provides paging indicators (PIs) to the UE. As the UE needs to know when to monitor the PCH, data is provided on the PICH to assign the UE a paging repetition ratio to enable it to determine how often it needs to *wake up* and *listen* to the PCH. One PICH frame is of 10 ms length. Each frame comprises eight access slots, and 5, 10, or 20 PIs may be included in an access slot (Fig. 11.32).

The SCH is used for allowing the UE to synchronize with the network. It is used for cell search.

> PICH provides information to the UE to be able to operate its sleep mode to conserve its battery when listening on the PCH.

Physical downlink shared channel (PDSCH) This channel shares control information to UE within the coverage area of the Node B. It is shared by several UE by means of code multiplexing. It is always associated with another physical channel. The physical shared channel control channel (PSCCCH) contains information shared by several UE (Fig. 11.29a). Its control information field includes the transmission power control (TPC) to be applied by the UE pool. Detailed structure is still under development.

Dedicated physical data channel (DPDCH) This channel is used to transfer user data (Fig. 11.29b).

Dedicated physical control channel (DPCCH) This channel carries control information to and from the UE. In both directions, the channel carries pilot bits and TFCI. The downlink channel also includes the TOC and feedback information (FBI) bits (Fig. 11.29b)

Uplink physical channels Figure 11.33 shows the uplink physical channels manifestation.

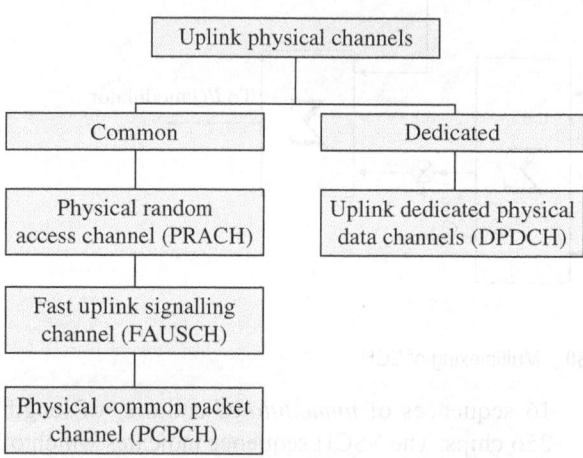

Fig. 11.33 Uplink physical channels manifestation

> *Note*: On the uplink, dual channel modulation is used so that both data and control are transmitted simultaneously. Here, the control elements contain a pilot signal, TFCI, FBI, and TPC.

Physical random access channel (PRACH) This channel enables the UE to transmit random access bursts in an attempt to access a network or fixed network. Access method is based on the slotted ALOHA principle. Access time slots have 1.25 ms offset. Access time slot boundaries are referred to the BCCH time reference. Access slot timing and random access bursts with its detail are given in Fig. 11.34. Preamble is based on 16 complex symbols (*signature*). Each symbol is spread by means of a 256-chip real orthogonal Gold code. There are 16 possible signatures.

Physical common packet channel (PCPCH) This channel is specifically intended to carry packet data. In operation, the UE monitors the system to check whether it is busy, and if not, it then transmits a brief access burst. This is retransmitted if no acknowledgement is obtained, with a slight increase in power each time. Once the Node B acknowledges the request, the data is transmitted on the channel.

Fast uplink signalling channel (FAUSCH) This channel deals with faster signalling in special conditions.

Dedicated physical data channel This channel is used to transfer user data (Fig. 11.35).

Dedicated physical control channel This channel carries control information to and from the UE. In both directions, the channel carries pilot bits and TFCI. The downlink

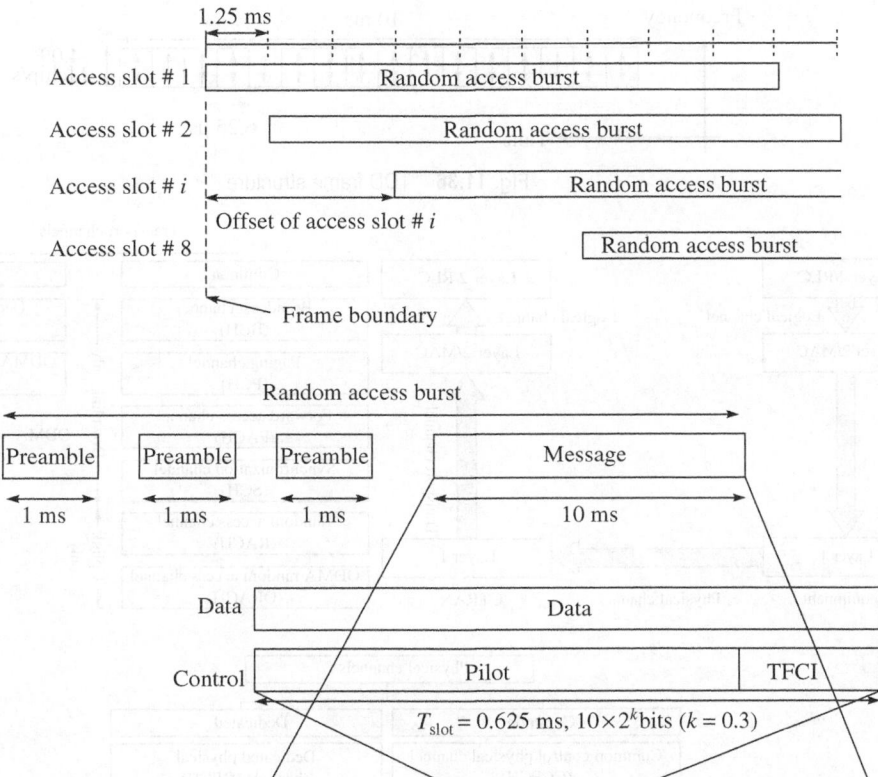

Fig. 11.34 Access slot timing and random access burst structure correlated with message timing

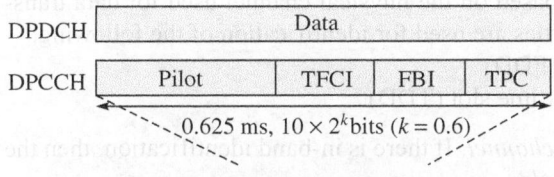

Fig. 11.35 DPDCH and DPCCH frames within the slot time within the superframe

channel also includes the TPC and FBI bits. The TFCI field is optional. UTRAN may request its presence to the UE. If it is present, it is represented by a 32-bit word transmitted in each frame. The TFCI value may be negotiated between the UTRAN and UE on a frame-to-frame basis (Fig. 11.35).

TDD Mode Channels

Figure 11.36 represents the TDD frame structure. (Here, each time slot may be used in the uplink or downlink. At least one time slot each must be used in the uplink and the downlink.) The services provided by layers 1 and 2 are called *transport channels* (Fig. 11.37). The part of layer 2 that directly interfaces with layer 1 is the MAC sublayer. The upper part of layer 2 that interfaces with the MAC sublayer is called the RLC sublayer. The services provided by the MAC sublayer to the RLC sublayers are called *logical channels*.

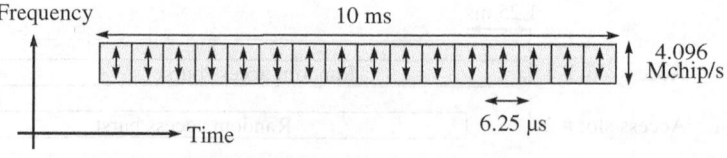

Fig. 11.36   TDD frame structure

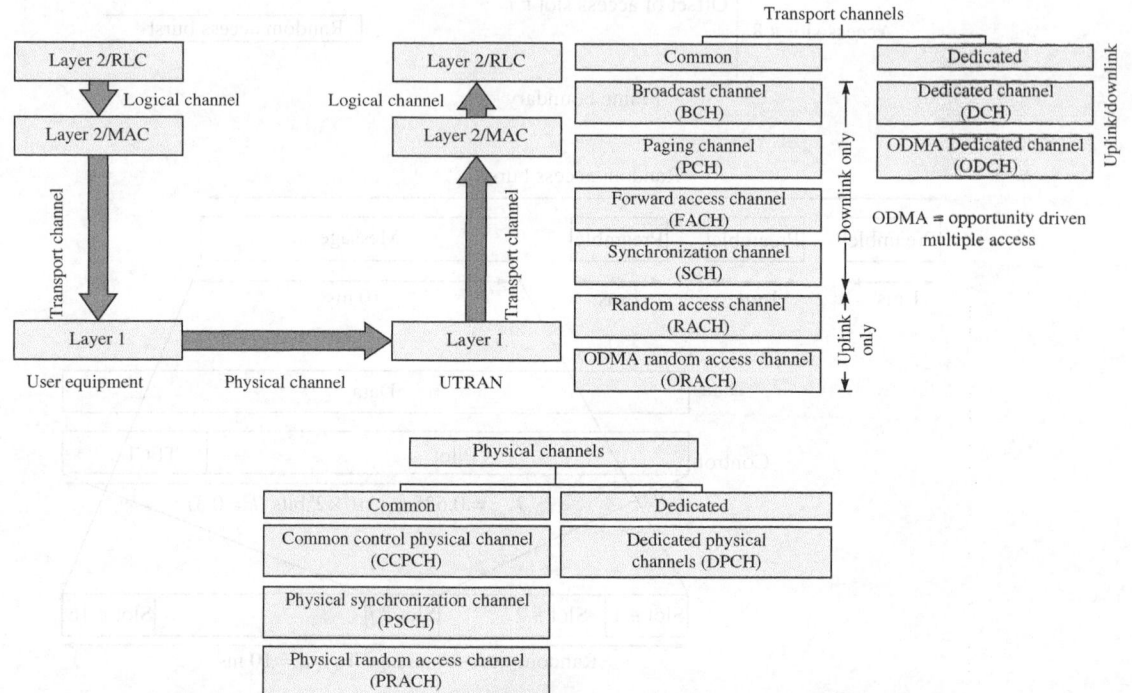

Fig. 11.37   Uplink and downlink channels for TDD mode correlated with layers

Transport channel identification may be based on the physical channel used for data transport. In this case, physical channel properties are used for identification of the following:

- Carrier frequency and spreading code (FDD)
- Carrier frequency, spreading code, and time slot (TDD)

This class of channels is called *dedicated channel*. If there is in-band identification, then the class of channels is called *common channels*.

Physical channels   Physical channels are for the procedures of actual data transfer to be handled.

Dedicated physical channel (DPCH)   The DPCH burst-type applications as shown in Fig. 11.38 are considered here. Burst type 1 has a longer midamble. Therefore, it allows a more accurate multi-user detection, as is required in the uplink. Burst type 2 has a shorter midamble and, therefore, higher user data throughput. It is more suitable for downlink, which will require more capacity, in, for example, wireless Internet applications.

Common control physical channel (CCPCH)   This channel is used to transport the BCH, PCH, and FACH. The burst type is the same as for the DPCHs.

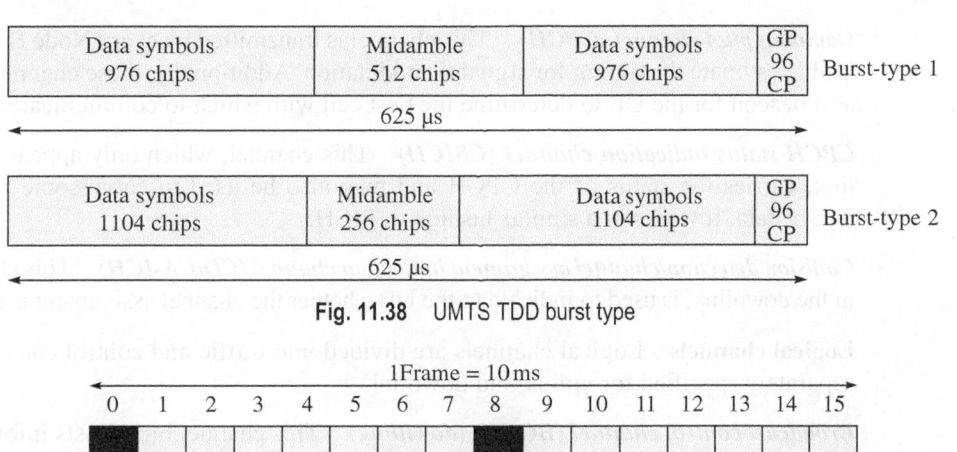

Fig. 11.38 UMTS TDD burst type

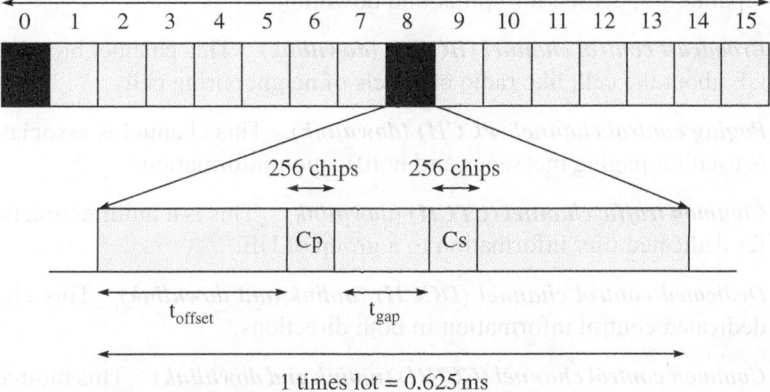

Fig. 11.39 PSCH internal burst structure

Physical synchronization channel (PSCH) The structure of this channel is similar structure to that in the FDD mode. Two code sequences are periodically broadcasted: primary synchronization code (c_p) and secondary synchronization code (c_s). In each frame, two time slots are allocated for the PSCH: TS0 and TS7 (Fig. 11.39).

Physical random access channel (PRACH) This channel supports the bursts sent by the UE in order to signal their presence to the BS of the cell where they intend to camp on (Fig. 11.40). As bursts are randomly sent, a certain risk of collision in the same time slot exists. However, as each access burst has its own spreading code (eight different options), the risk of collision is significantly reduced. Two types of random access bursts are specified. Each burst uses only 1/2 of the time slot. The UE may send both bursts in a single time slot, thereby increasing the probability of reaching the BS without collision.

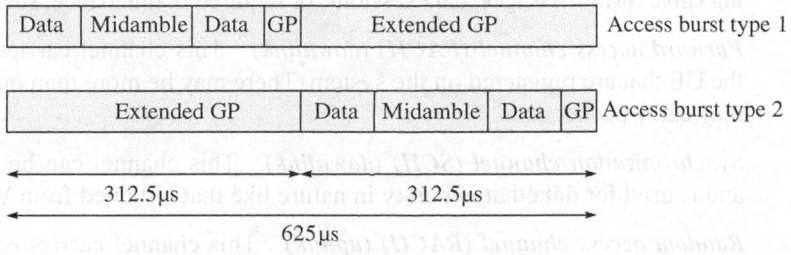

Fig. 11.40 PRACH access bursts

Common pilot channel (CPCH)   This channel is transmitted by every Node B so that the UE is able estimate the timing for signal demodulation. Additionally, these channels can be used as a beacon for the UE to determine the best cell with which to communicate.

CPCH status indication channel (CSICH)   This channel, which only appears in the downlink, carries the status of the CPCH and may also be used to carry some intermittent or *bursty* data. It works in a similar fashion to PICH.

Collision detection/channel assignment indication channel (CD/CA-ICH)   This channel, present in the downlink, is used to indicate to the UE whether the channel assignment is active or not.

Logical channels   Logical channels are divided into traffic and control channels. They are separately specified for uplink and downlink.

Broadcast control channel (BCCH) (downlink)   This channel broadcasts information to the UE about the cell, like radio channels of neighbouring cells.

Paging control channel (PCCH) (downlink)   This channel is associated with the PICH and is used for paging messages and notification information.

Common traffic channel (CTCH) (downlink)   This is a unidirectional channel used to transfer dedicated user information to a group of UE.

Dedicated control channel (DCCH) (uplink and downlink)   This channel is used to carry dedicated control information in both directions.

Common control channel (CCCH) (uplink and downlink)   This bidirectional channel is used to transfer control information.

Shared channel control channel (SHCCH)   This channel is bidirectional and is found only in the TDD form of WCDMA or UMTS, where it is used to transport shared channel control information.

Dedicated traffic channel (DTCH) (uplink and downlink)   This is a bidirectional channel used to carry user data or traffic.

Transport channels   Transport channels perform tasks similar to the functions of the transport layer in networking. There are a total of seven transport channels.

Broadcast channel (BCH) (downlink)   This channel broadcasts information to the UE in the cell to enable them to identify the network and the cell.

Paging channel (PCH) (downlink)   This channel carries messages that alert the UE to incoming calls, SMS messages, data sessions, or required maintenance, such as re-registration.

Forward access channel (FACH) (downlink)   This channel carries data or information to the UE that are registered on the system. There may be more than one FACH per cell as they may carry packet data.

Synchronization channel (SCH) (downlink)   This channel can be shared by several users and is used for data that is bursty in nature like that obtained from Web browsing.

Random access channel (RACH) (uplink)   This channel carries requests for service from the UE trying to access the system.

UMTS incorporates many vocoders including GSM EFR and adaptive multirate (AMR), and the rate is chosen as per capacity and system requirements.

Uplink common packet channel (CPCH) (uplink) This channel provides additional capability beyond that of the RACH and is used for fast power control.

Dedicated transport channel (DCH) (uplink and downlink) This channel is used to transfer data to a particular UE. Each UE has its own DCH in each direction.

11.6.6 Packet Handling, Power Saving, and Handover in UMTS or WCDMA

This section deals with three independent elements of the UMTS or WCDMA system, namely the way packet data is carried, the way power saving is accomplished, and handover, including hard, soft, and softer handover.

Packet Handling

Packet data is an increasingly important element in mobile phone applications. The WCDMA system is able to carry data in this format in the following two ways.

(a) The first is for short data packets to be appended directly to a random access burst. This method is called common channel packet transmission and is used for short infrequent packets. It is preferable to transmit short packets in this manner, because the link maintenance needed for a dedicated channel would lead to an unacceptable overhead. Additionally, the delay in setting up a PDCH and transferring the operational mode to this format is avoided.

(b) Larger or more frequent packets have to be transmitted on a dedicated channel. A large single packet is transmitted using a single-packet scheme, where the dedicated channel is released immediately after the packet has been transmitted. In a multipacket scheme, the dedicated channel is maintained by transmitting power control and synchronization information between subsequent packets.

Power Saving

One of the major issues with mobile phones in general is that of battery life. It is one of the key differentiators that people take into account when buying a phone, and this gives a measure of its importance. Considering this while developing the UMTS/WCDMA standard, a discontinuous reception or sleep mode was introduced. This mode allows several non-essential segments of the phone circuitry to power down during periods when paging messages will not be received. The presence of a voice activity detector also helps.

To enable this facility to be introduced into the UMTS UE circuitry, the PCH is divided into groups or subchannels. The actual number of paging subchannels to be used by a particular UE is assigned by the network. In this way, the UE has to listen only for part of the time. To achieve this, the PICH is split into 10 ms frames, each of which comprises 300 bits—288 for paging data and 12 idle bits. At the beginning of each PCH frame, there is a PI that identifies the paging group being transmitted. By synchronizing with the PCHs being transmitted, it is able to turn the receiver on only when it needs to monitor the PCH. As the receiver, with its RF circuitry, will consume power, it is possible to save power by switching it off.

Handover

Handover in UMTS follows many concepts that are similar to those used for other CDMA systems. There are three basic types of handover: hard, soft, and softer. Hard handover is interfrequency handover, whereas soft and softer handovers are intrafrequency. All three types are used but under different circumstances. Hard and soft handovers were explained earlier in Sections 2.6.5 and 11.4.2. Hence, we will be discussing softer handover here.

High-speed packet access with UMTS is identified as 3.5G scenario.

Softer handover is a special case of soft handover in which active set BSs are part of the same physical site; that is, if the overlapping sectors are from the same physical cell site, the soft handover is referred to as softer handover. It allows more efficient combining implementations than soft handover (e.g., to use maximal ratio combining instead of selection combining).

Decisions regarding handover are generally handled by the RNC. It continually monitors information regarding the signals being received by both the UE and node B, and when a particular link falls below a given level and another better radio channel is available, it initiates a handover. As part of this monitoring process, the UE measures the *received signal code power* (RSCP) and *received signal strength indicator* (RSSI) and the information is then returned to the Node B and hence to the RNC on the uplink CCH.

If a hard handover is required, then the RNC will instruct the UE to adopt a compressed mode, allowing for short time intervals in which the UE is able to measure the channel quality of other radio channels.

11.6.7 High-speed Uplink Packet Access

High-speed uplink packet access (HSUPA) is used to improve the data rates on the 3G WCDMA mobile phone system. Similarly, the standard is for downlink. Standards are now developed to enable the uplink from the mobile handset or UE to the BS (Node B) so that it is possible to handle data at similar speeds. HSUPA will allow the introduction of new features including full video conferencing.

For most applications, including Internet surfing, emails, video downloads, and the like, data flowing in the downlink is far greater than that in the uplink. However, for applications such as video conferencing, data flows equally in both directions. It is anticipated that video conferencing will become an increasing requirement and a significant revenue generator for the operators in the near future. To enable high-quality video to be passed, it is, therefore, essential to ensure that the uplink performs as fast as the downlink. Although it is very early days for the standards, work on HSUPA has already started under the auspices of the 3GPP, the body that controls the WCDMA standards.

Many of the techniques used in HSDPA are used in HSUPA, but these still need to be formalized. Accordingly, adaptive modulation, along with hybrid automatic repeat request (HARQ), is used. There are also improvements in the BS similar to those employed in HSDPA.

Originally, WCDMA had used only QPSK as the modulation scheme; however, under the new HSUPA system, 16QAM, which can carry a higher data rate but is less resilient to noise, is also used when the link is sufficiently robust. The robustness of the channel and its suitability to use 16QAM instead of QPSK are determined by analysing the information fed back about various parameters. These include details of the channel physical layer conditions, power control, QoS, and information specific to HSDPA.

Fast HARQ has also been implemented along with multicode operation, and this eliminates the need for a variable spreading factor. By using these approaches, all users, irrespective of their distance from the BS, are able to receive the optimum available data rate.

11.6.8 High-speed Downlink Packet Access

Improvements and enhancements are being made to the WCDMA or UMTS 3G telecommunications system. HSDPA, a new technology, promises to increase the download data rate fivefold. In addition to this, it also provides a twofold increase in BS capacity.

The introduction of HSDPA technology has come about as a result of the need to drive down costs as well as increase the data rates possible. Current trends show the volume of packet-switched data rising and overtaking the more traditional circuit-switched traffic. By adopting a packet-based approach to the delivery of digital content as well as IP-based person-to-person digitized voice, a single session can be used for multiple purposes and this can be used to drive revenues upwards. Thus, the use of HSDPA is a key element in providing the user with a better service as well as the service providers with increased capacity and usage.

Release 4 of the 3GPP WCDMA standard provided efficient IP support to enable the provision of services through an all-IP CN. Then, Release 5 included HSDPA, with support for packet-based multimedia services. Release 6 contained a further enhancement—MIMO. As HSDPA needs to work alongside the original Release 99 systems, the new technology is completely backwards compatible.

> *Note*: HSDPA also uses 16QAM modulation. Fast HARQ has been implemented along with multicode operation.

Further advances have been made in the area of *scheduling*. By moving more intelligence into the BS, data traffic scheduling can be achieved in a more dynamic fashion. This enables variations arising from fast fading to be accommodated, and the cell is even able to allocate much of the cell capacity for a short period of time to a particular user. In this way, the user is able to receive the data as fast as conditions allow.

Moreover, another channel, known as the *high-speed downlink shared channel* (HS-DSCH), has been introduced. The WCDMA normally carries data over DCHs, several of which are multiplexed onto one RF carrier. This approach has been adopted because it provides optimum performance with continuous user data. Under the new scheme, the bursty nature of the data has been accounted for and more efficient use of the available spectrum has been made.

Using the new HSDPA scheme, it will be possible to achieve peak data rates of 10 Mbps within the 5 MHz channel bandwidth offered under WCDMA. The new scheme has many benefits. It improves the overall network packet data capacity and the spectral efficiency and enables networks to achieve a low delivery cost per bit. Users will experience high data speeds as well as short service response times and better availability of services. However, new mobile designs should be able to handle increased data throughput rates. Reports indicate that handsets will need to have at least double the memory currently contained within handsets. Nevertheless, the advantages of HSDPA mean that it will be widely used as networks are upgraded and new phones are introduced.

The QoS for different applications in UMTS also needs separate attention and can be self-studied.

11.7 LONG-TERM EVOLUTION

The 3GPP worked on long-term-evolution (LTE), which is built on UMTS, as the industry looks beyond 3G and it is marketed as 4G technology.

The function of the eNodeB is equivalent to that of the Node B in UMTS and the BTS in GSM.

11.7.1 LTE Architecture

Long-term evolution transits from the existing UMTS circuit and packet switching combined network to an all-IP flat architecture system. Figure 11.41 shows the simplicity of the LTE architecture. The network side of the evolved UMTS

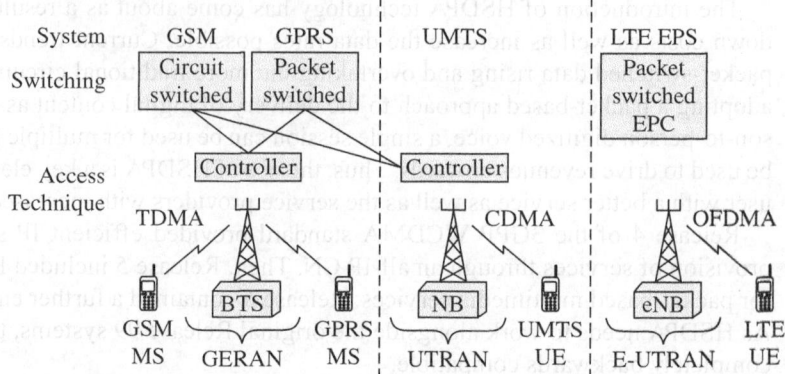

Fig. 11.41 Major differences in the architectures of GSM, GPRS, UMTS, and LTE

terrestrial radio access network (E-UTRAN) is composed only of evolved Node Bs (eNodeBs or eNBs), hence the simplified architecture.

Long-term evolution supports interoperation and coexistence with legacy standards (e.g., GSM or EDGE, UMTS, and CDMA2000). Users can start a call or transfer of data in an area using an LTE standard and, should coverage be unavailable, continue the operation without any action on their part using GSM, GPRS, WCDMA-based UMTS, or even 3GPP2 networks such as cdmaOne or CDMA2000.

> *Note*: Long-term evolution supports multicast-broadcast single-frequency network (MBSFN). This feature can deliver services like mobile television using the LTE infrastructure and is a competitor for television broadcast based on digital video broadcasting—handheld (DVB-H).

In addition, LTE supports cell sizes from tens of metres radius (femtocells and picocells) to 100 km radius macrocells. In the lower-frequency bands to be used in rural areas, 5 km is the optimal cell size, 30 km has reasonable performance, and up to 100 km cell sizes are supported with acceptable performance. In city and urban areas, higher-frequency bands (like 2.6 GHz) are used to support high-speed mobile broadband. In this case, cell sizes may be 1 km or even less.

The new blocks specific to evolved UMTS evolution, LTE, also known as the evolved packet system (EPS), are the E-UTRAN and the evolved packet core (EPC). The EPS (Fig. 11.41) is purely IP-based. Both real-time services and data communication services will be carried by the IP. The IP address is allocated when the mobile is switched on and released when switched off. The CN—EPC is prepared to work with other access technologies not developed by 3GPP, such as WiMAX and WiFi (described in Chapter 12). Access solutions not developed by 3GPP are divided into trusted and non-trusted categories. This division is based not on the technical solution but on the business relation or agreement between the operators.

> The reason for distributing the intelligence amongst the BSs in LTE is to speed up the connection set-up and reduce the time required for a handover.

The LTE access network—E-UTRAN—is simply a network of BSs and eNBs, generating a flat architecture (Fig. 11.42). There is no centralized intelligent controller, and the eNBs are normally interconnected by the X2 interface and towards the CN by the S1 interface. For an end-user, the connection set-up time for a real-time data session is in many cases crucial, especially in

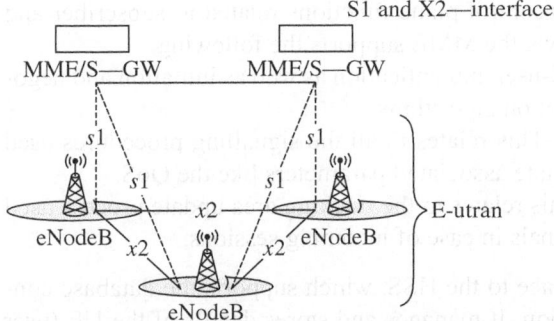

S1 and X2—interfaces

Fig. 11.42 Simplified LTE architecture—EPS

> The LTE eNodeB is similar to an access point of WiFi (explained in Chapter 12). There is no more intermediate controlling node as in the 2G BSC or 3G RNC.

on-line gaming. The time for handover is critical in real-time services where end-users tend to end calls if the handover takes too long.

11.7.2 Elements of LTE—EPS

Two major elements of LTE are E-UTRAN and within it eNode Bs. Major and subelements with various interfaces are described below.

E-UTRAN

The E-UTRAN is an orthogonal frequency division multiplexing (OFDM)-based structure and is quite simple compared to UTRAN. It is composed of only one network element, the eNodeB. The 3G RNC inherited from the 2G BSC has disappeared from the E-UTRAN, and the eNodeB is directly connected to the CN using the S1 interface. Consequently, the features supported by the RNC have been distributed between the eNodeB or the CN mobility management entity (MME) or serving gateway entities (Fig. 11.42).

eNode B   The eNB interfaces with the UE. The following points need to be noted by the ENB:

- It supports all physical and data link layer functions (MAC, RLC, and packet data control protocol) associated with the E-UTRAN OFDM physical interface, and eNBs are directly connected to network routers.
- It also hosts radio resource control (RRC) functionality corresponding to the control plane.
- This has the advantage of simple network architecture with few nodes of different types. This results in simplified network operation and allows good performance over the radio interface.
- The termination of layer 2 protocols in the eNodeB rather than in the RNC helps to decrease data transmission latency by saving the delay incurred by the transmission of packet repetitions over the Iub interface.
- From a functional perspective, the eNodeB supports a set of legacy features, all related to physical layer procedures for transmission and reception over the radio interface.

> *Note*: The eNodeB performs many functions including radio resource management, admission control, scheduling, enforcement of negotiated uplink QoS, cell information broadcast, ciphering or deciphering of user and control plane data, and compression or decompression of the downlink or uplink user plane packet headers.

X2 interface   A new interface known as X2 has been defined between eNodeBs, working in a meshed way to make them possible to be linked together. The main purpose of this interface is to minimize packet loss due to user mobility. As the terminal moves across the access network, unsent or unacknowledged packets stored in the old eNodeB queues can be forwarded or tunnelled to the new eNodeB by the X2 interface.

EPC   This is composed of several functional entities: the MME, the home subscriber server (HSS), the serving gateway (S-GW), the packet data network gateway (PDN-GW), and the policy and charging rules function (PCRF) server.

MME The MME is in charge of all the control plane functions related to subscriber and session management. From that perspective, the MME supports the following:

- *Security procedures*: This relates to end-user authentication as well as initiation and negotiation of ciphering and integrity protection algorithms.
- *Terminal-to-network session handling*: This relates to all the signalling procedures used to set up packet data context and negotiate associated parameters like the QoS.
- *Idle terminal location management*: This relates to the tracking area update process used for the network to be able to join terminals in case of incoming sessions.

The MME is linked through the S6 interface to the HSS, which supports the database containing all the user subscription information. It manages and stores details of the UE (user identities, mobility states, user security parameters, etc.) and generates temporary identities and allocates them to the UE. It also checks the authentication of the users.

HSS The HSS is the concatenation of the HLR and the AuC—two functions that are already present in GSM and UMTS networks. The HLR part of the HSS is in charge of storing and updating when necessary the database containing all the user subscription information, including the following:

- User identification and addressing, which corresponds to the IMSI and MSISDN or mobile telephone number
- User profile information, which includes service subscription states and user-subscribed QoS information (such as maximum allowed bit rate or allowed traffic class).

The AuC part of the HSS is in charge of generating security information from user identity keys. This information is provided to the HLR and further communicated to other entities in the network. The security information is mainly used for mutual network—terminal authentication, radio path ciphering, and integrity protection to ensure that data and signalling are transmitted between the network, and the terminal is neither eavesdropped nor altered.

S-GW From a functional perspective, the S-GW is the termination point of the packet data interface towards the E-UTRAN. The S-GW routes and forwards user data packets. When terminals move across the eNodeB in the E-UTRAN, the S-GW serves as a local mobility anchor for the user plane during handovers. This means that packets are routed through this point for intra-E-UTRAN mobility and mobility with other 3GPP technologies, such as GSM and UMTS (terminating S4 interface and relaying the traffic between the 2G or 3G systems and the PDN-GW).

PDN-GW Similar to the serving gateway, the PDN-GW is the termination point of the packet data interface towards the packet data network (PDN). The PDN-GW provides connectivity from the UE to external PDNs by being the point of exit and entry of traffic for the UE. The UE may have simultaneous connectivity with more than one PDN-GW for accessing multiple PDNs. As an anchor point for sessions towards the external PDNs, the PDN-GW also supports policy enforcement features (which apply operator-defined rules for resource allocation and usage), packet filtering (e.g., deep packet inspection for virus signature detection), evolved charging support (e.g. per uniform resource locator or URL charging), lawful interception, and packet screening.

PCRF Server The PCRF server manages the service policy and sends the QoS setting information for each user session and the accounting rule information. It combines the functionalities of the following two UMTS nodes: (a) the policy decision function (PDF) and (b) the charging rules function (CRF)

11.7.3 LTE Radio or Air Interface

Evolved UMTS terrestrial radio access (E-UTRA) is the air interface of LTE and is a packet-switched radio interface. The LTE standard can be used with many different frequency bands. The frequencies used are 700/800 MHz and 1700/1900 MHz in North America, 2500 MHz in South America, 800 MHz, 900 MHz, 1800 MHz, and 2600 MHz in Europe, 1800 MHz and 2600 MHz in Asia, and 1800 MHz in Australia. As a result, phones from one country may not work in other countries. Users will need a multiband-capable phone for roaming internationally.

Modulation

Orthogonal frequency division multiple access (OFDMA) with a subcarrier spacing of 15 kHz and single-carrier frequency division multiple access (SC-FDMA) have been chosen as the transmission schemes for the downlink and uplink, respectively, to conserve power. The physical layer at the eNB is responsible for protecting data against channel errors using adaptive modulation and coding (AMC) schemes based on channel conditions. QPSK, 16QAM, and 64QAM will be the downlink and uplink modulation schemes in E-UTRA. For the uplink, 64QAM is optional at the UE.

In addition to RF processing including modulation and demodulation, it processes measurement reports from the UE such as channel quality index (CQI) and provides indications to the upper layers. Each radio frame is 10 ms long and contains 10 subframes, with each subframe capable of carrying 14 OFDM symbols. LTE supports both FDD and TDD communication systems as well as half-duplex FDD with the same radio access technology. It also maintains frequency and time synchronization.

Other Features

The following are some of the other important features of LTE:

- The minimum unit of scheduling is a time—frequency block corresponding to one subframe (1 ms) and 12 subcarriers, which is called one resource block.
- The scheduling is not done at a subcarrier granularity in order to limit the control signalling.
- Multiple antennas at the UE are supported, with the two receive and one transmit antenna configuration being mandatory. MIMO is also supported at the eNB, with two transmit antennas being the baseline configuration.
- Peak download rates are up to 299.6 Mbps and upload rates are up to 75.4 Mbps, depending on the UE category (with 4×4 antennas using 20 MHz of spectrum).
- Five different terminal classes have been defined from a voice-centric class up to a high-end terminal that supports peak data rates. All terminals will be able to process 20 MHz bandwidth.

Radio resource management and scheduling aspects in LTE are very important. One can treat these aspects between the physical layer and the MAC layer.

Long-term evolution has low data transfer latencies (sub-5 ms latency for small IP packets in optimal conditions) and lower latencies for handover and connection set-up time than with previous radio access technologies. It also has improved support for mobility, exemplified by support for terminals moving at up to 350 km/h or 500 km/h depending on the frequency band. LTE has increased spectrum flexibility: 1.4 MHz, 3 MHz, 5 MHz, 10 MHz, 15 MHz, and 20 MHz wide cells are standardized. (WCDMA requires 5 MHz slices, leading to some problems with roll-outs of the technology in countries where 5 MHz is a commonly allocated amount of spectrum, and is frequently already in use with legacy standards such as 2G GSM and cdmaOne.)

11.7.4 LTE Channels

Long-term evolution channels are of three types: physical, logical, and transport.

Physical Channels

The LTE physical channels are again grouped into uplink (SC-FDMA based) and downlink (OFDMA based) channels as each has different requirements and operates in a different manner.

Downlink channels There are four downlink physical channels for correspondance from eNode B to UE.

Physical broadcast channel (PBCH) This channel carries system information for the UE requiring access to the network. It carries only the master information block (MIB) messages. The modulation scheme is always QPSK, and the information bits are coded and rate matched. The bits are then scrambled using a scrambling sequence specific to the cell to prevent confusion with data from other cells.

The MIB message on the PBCH is mapped onto the central 72 subcarriers or six central resource blocks regardless of the overall system bandwidth. A PBCH message is repeated every 40 ms; that is, one transmission time interval (TTI) of the PBCH includes four radio frames. The PBCH transmission has 14 information bits, 10 spare bits, and 16 CRC bits.

Physical control format indicator channel (PCFICH) This channel informs the UE about the format of the signal being received. It indicates the number of OFDM symbols used for the physical downlink control channels (PDCCHs), whether 1, 2, or 3. The information within the PCFICH is essential because the UE does not have prior information about the size of the control region. A PCFICH is transmitted on the first symbol of every subframe and carries a control format indicator (CFI) field. The CFI contains a 32-bit code word that represents 1, 2, or 3. CFI 4 is reserved for possible future use. The PCFICH uses (32, 2) block coding, which results in a 1/16 coding rate, and it always uses QPSK modulation to ensure robust reception.

Physical downlink control channel (PDCCH) The main purpose of this channel is to carry mainly scheduling information of different types, downlink resource scheduling, uplink power control instructions, uplink resource grant, and indication for paging or system information. The PDCCH contains a message known as the downlink control information (DCI), which carries the control information for a particular UE or group of UE. The DCI format has several different types, which are defined with different sizes. The different format types include Type 0, 1, 1A, 1B, 1C, 1D, 2, 2A, 2B, 2C, 3, 3A, and 4.

Physical hybrid automatic repeat request indicator channel (PHICH) This channel is used to report the HARQ status. It carries the HARQ ACK/NACK signal indicating whether a transport block has been correctly received. The HARQ indicator is one-bit long; that is, 0 indicates ACK and 1 indicates NACK. The PHICH is transmitted within the control region of the subframe and is typically transmitted only within the first of symbol. If the radio link is poor, then the PHICH is extended to a number of symbols for robustness.

Uplink channels The following three different channels are specified as uplink physical channels:

Physical uplink control channel (PUCCH)   This channel provides the various control signalling requirements. Many different formats have been defined to enable the channel to carry the required information in the most efficient format for the particular scenario encountered. It includes the ability to carry scheduling requests (SRs). For the different PUCCH formats, different modulation schemes and different bits per subframe are defined, as for the PDCCH.

Physical uplink shared channel (PUSCH)   This physical channel found on the LTE uplink is the uplink counterpart of the PDSCH.

Physical random access channel (PRACH)   This channel is used for random access functions. This is the only non-synchronized transmission that the UE can make within LTE. The downlink and uplink propagation delays are unknown when PRACH is used, and therefore, it cannot be synchronized. The PRACH is made up from two sequences—a cyclic prefix and a guard period. The preamble sequence may be repeated to enable the eNodeB to decode the preamble when link conditions are poor.

Logical Channels

The logical channels cover the data carried over the radio interface. The service access point between the MAC sublayer and the RLC sublayer provides the logical channel. Logical channels are again grouped into TCHs and CCHs, like in the GSM system.

Traffic channels   The LTE TCHs carry the user plane data.

Dedicated traffic channel   This channel is used for the transmission of user data.

Multicast traffic channel (MTCH)   This channel is used for the transmission of multicast data.

Control channels   The LTE CCHs carry the control plane information

Broadcast control channel (BCCH)   The channel provides system information to all mobile terminals connected to the eNodeB.

Paging control channel (PCCH)   This channel is used for paging information when searching a unit on a network.

Common control channel (CCCH)   This channel is used for random access information, for example, for actions like setting up a connection.

Multicast control channel (MCCH)   This control channel is used for information needed for multicast reception.

Dedicated control channel (DCCH)   This channel is used for carrying user-specific control information, for example, for controlling actions such as power control and handover.

LTE Transport Channels

The LTE transport channels are also defined for the uplink and the downlink, as each has different requirements and operates in a different manner. Physical layer transport channels offer information transfer to MAC and higher layers.

Downlink   Four downlink transport channels are specified as follows:
Broadcast channel   This channel maps to the BCCH.

Downlink shared channel (DL-SCH) This channel is the main channel for downlink data transfer. It is used by many logical channels.

Paging channel This channel is used to convey the PCCH information.

Multicast channel (MCH) This channel is used to transmit MCCH information to set up multicast transmissions.

Uplink The following two transport channels are specified for uplink:

Uplink shared channel (UL-SCH) This channel is the main channel for uplink data transfer. It is used by many logical channels.

Random access channel (RACH) This channel is used for random access requirements.

It can be seen that many of the LTE channels bear similarities to those used in previous generations of mobile telecommunications.

11.8 MOBILE SATELLITE COMMUNICATION

In mobile satellite communication, the satellite and mobile systems are combined, as shown in Fig. 11.43. Antenna spot beams can be used for frequency reuse. The beam projection areas are treated as cells. However, instead of the normal transmitter and receiver used in GSM- or CDMA-based mobile systems, small satellite earth stations are used to send and receive the signals from the satellite. These signals can be further routed to the MSC, which will then take care of further routing of the signals.

> Both low earth orbit (LEO) and medium earth orbit (MEO) satellites are used in mobile communication services, but LEO requires more handovers than MEO.

The system offers communication services to mobile users operating within a predefined service area. The users communicate with other mobiles or with fixed users through one of the visible satellites. Users in the fixed network are accessed through large fixed stations called *gateways*, which carry a large amount of traffic, whereas the mobiles are small portable units that can support only a few channels. Mobile terminals may be mounted on top of vehicles such as ships, aircrafts, and trucks or may be carried by individuals.

There may be one or more space segments, which may consist of one satellite or a group of interlinked satellites. Depending on the service area and

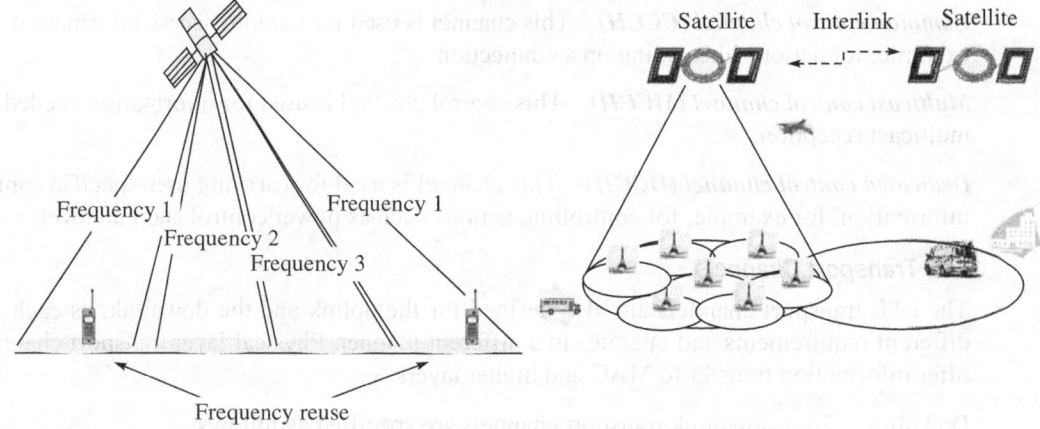

Fig. 11.43 Mobile satellite communication scenario

application, the space segments can be utilized. Telemetry and control ground stations, used for monitoring and controlling satellites, constitute a part of the space segment. To simplify mobile terminals, complexity is shifted to the space segment and hence satellites used for this system tend to be complex. A 3–3.5 kW geostationary satellite with 5–10 spot beams is typically used for 2G systems. A 5 kW geosatellite with 100–200 spot beams is a typical case of a 3G system.

Note: As a geosynchronous satellite is synchronized with the Earth in terms of angular velocity, it comprises one or more static footprints of the spot beam. Therefore, fixed stations can operate with a single antenna with minimal tracking, and the network topology of the satellites is simple.

According to the static spot beams, there are entries in HLR and VLR. Hence, identifying a mobile unit is not very difficult. The main problems in mobile satellite communication are difficult propagation environment and small mobile terminal size.

Iridium The Iridium system is a large group of satellites providing voice and data services through satellite phones. It cannot be a replacement for existing terrestrial cellular systems. However, it can be considered as an extension of existing wireless systems to provide mobile services to remote and populated areas, which are not covered by terrestrial cellular services. The Iridium system provides more channel capacity, which means a large number of channels and better QoS with shorter transmission delays to areas that currently receive mobile services from the geostationary satellites. It can provide emergency services in the event that terrestrial cellular services are disabled like in a disaster situation.

SOLVED EXAMPLES

Example 11.1 For a GSM system with 25 MHz spectrum and 200 kHz channel bandwidth, assume a cluster of four cells. If there are three CCHs per cell, find the number of voice channels per BTS.

Solution Total channels are divided into uplink and downlink; hence, considering one-way spectrum for available channels to establish a call, 12.5 MHz/200 kHz ≈ 62.5 total channels.

Now, as per TDMA—FDMA structure, eight time slots means there are eight voice channels per frame. Therefore, total number of voice channels effectively is 62.5 × 8 = 500.

These channels are divided into four cells, so each cell has 125 voice channels. Subtracting 3 CCHs, each cell has 122 voice channels.

Example 11.2 The following data is given for a GSM 1800 system in a particular area:
- Average usage per subscriber per month is 175 min.
- Busy hours per day is 7.
- Allotted spectrum is 4.8 MHz.
- Per channel bandwidth is 200 kHz.
- Total subscribers in the zone are 35,000 (assuming zero growth rate of subscribers).
- Total area is 10,000 m^2.
- Capacity of one BTS is 36 erlangs (one BTS per cell).
- Frequency reuse factor is 4.
- There are two CCHs per cell.

Estimate the average busy hour traffic per subscriber in that area and the number of BSs required to handle such traffic.

Solution Assuming 30 days for the calculations,
Average busy hour traffic per subscriber in erlangs = 175/30 × 7 × 60 = 0.0138 erlangs
Total number of channels = 4.8 × 10^6/200 × 10^3 = 24
Channels per cell with reuse factor of 4 = 24/4 = 6
TCHs per cell with two CCHs per cell = 6 × 8 − 2 = 46
This is assuming 2% grade of service and using Erlang B table, traffic capacity of a cell as 36.5 erlangs, which is nearly equal to the BTS capacity.

Hence, the maximum number of subscribers supported by that cell is 36.5/0.0138 ≈ 2644.

For 35,000 subscribers, 35,000/2644 = 13.23 ≈ 14 BTSs are required to handle such traffic.

Note: For designing the various cellular systems, the receiver sensitivity and link budget analysis are required. Calculations and mathematical formulations for this are beyond the scope of this book, but reader may study further from the appropriate references.

SUMMARY

- There are two types of networks: infrastructure networks based on cellular theory and ad hoc networks without requirement of infrastructure.
- Infrastructure-based networks use wired infrastructure partly along with wireless infrastructure; 100 per cent wireless systems are very difficult to handle.
- Infrastructure-based networks are GSM, GPRS, EDGE, WLL, IS-95 based mobile phone system, UMTS, and now LTE.
- Cell-wise BS is required in all these networks to receive the signals from mobile phones. The cells are formed on the basis of geographical situations and population density, and cell size dependent bit rates are achieved as radio propagation effects also differ with cell size.
- Handovers are necessary in such systems. GSM systems deal with hard handover as adjacent cells use different carriers, whereas CDMA systems use soft handover as the carrier is the same for all the cells but with different user codes. UMTS offers hard, soft, and softer handovers.
- GSM is a 2G system which is enhanced to a 2.5G system—EDGE. It is especially used for voice services whereas GPRS is specially used for data services, but both can use the same basic infrastructure. UMTS is a 3G system based on WCDMA for which the infrastructure is developed on the existing GSM—GPRS infrastructure. LTE, the 4G system, also adds with modifications in the existing infrastructure. Interoperation and coexistence among all these systems is possible.
- The important components of the GSM system are the mobile equipment, BSC, BTS, MSC, and so on. The MSC is supported by the HLR, VLR, EIR, AUC, and OMC.

- The EDGE technology is developed only for data rate enhancement.
- The IS-95 is a CDMA-based standard for mobile telephony, which has become popular after cdmaOne and its evolution resulted in CDMA2000.
- Spread spectrum modulation used in CDMA systems has made secure communication possible.
- A WLL is a modified telephone system in which wired local loops are replaced by wireless loops.
- A WLL can use different existing wireless technologies over cell as it is based on interworking.
- The WLL and GSM differ from each other in terms of cell size, coverage, power, and handovers.
- The CorDECT WLL is especially suitable for communication in rural areas.
- The infrastructure of UMTS is mostly similar to that of GSM—GPRS systems; LTE also uses similar infrastructure. There are basic differences only in the interfaces.
- TDD and FDD are both used in UMTS and LTE.
- Satellite systems are to be incorporated in UMTS for an *anywhere, anytime* scenario.
- Universal SIM is required for UMTS.
- LTE uplink is based on SC-FDMA whereas the downlink is based on OFDMA.
- For all infrastructure-based systems, a set of channels is defined for the various functions.
- Cellular systems when combined with satellite interlinks make mobile satellite communication system. Here, an earth station is connected with an MSC.

EXERCISE

Multiple-choice Questions

11.1 Which of the following multiple access technique is used in GSM?
   (a) FDMA
   (b) FDMA/SDMA
   (c) TDMA/FDMA/SDMA
   (d) TDMA/SDMA

11.2 The modulation method for GSM is

   (a) GMSK   (c) MSK
   (b) GFSK   (d) FSK

11.3 The uplink frequency of GSM is
   (a) 890–915 MHz   (c) 935–960 MHz
   (b) 890–915 kHz   (d) 1800–1900MHz

11.4 The channel width for GSM is
   (a) 124 kHz   (c) 270 kHz
   (b) 200 kHz   (d) 890 kHz

11.5 The permanent database of all registered subscribers is maintained in the
 (a) VLR   (b) AUC   (c) EIR   (d) HLR

11.6 Which of the following is a protected database with a secret key also contained in the user's SIM card?
 (a) VLR   (b) AUC   (c) EIR   (d) HLR

11.7 EDGE is a
 (a) 2G technology   (c) 2.5G technology
 (b) 3G technology   (d) 1.5G technology

11.8 GSM-2G uses
 (a) NB—TDMA   (c) NB—FDMA
 (b) WB—FDMA   (d) WB—TDMA

11.9 The GMSK modulation scheme used in GSM is a form of
 (a) ASK   (c) (a) and (b)
 (b) PSK   (d) none of these

11.10 FDMA is used in
 (a) radio   (c) television
 (b) GSM   (d) all of these

11.11 How many bit identification numbers known as the electronic serial number (ESD) are installed in the mobile phone at the time of its manufacture?
 (a) 40   (b) 32   (c) 15   (d) 4

11.12 Frequency reuse
 (a) maintains spectrum efficiency
 (b) increases spectrum efficiency
 (c) decreases spectrum efficiency
 (d) does not affect spectrum efficiency

11.13 'SDMA is used only in combination with TDMA, FDMA, and CDMA.' This statement is
 (a) true
 (b) false
 (c) true in case of CDMA only
 (d) true in case of FDMA only

11.14 A high-capacity switch that provides handover, cell configuration data, and control of RF power levels in base transceiver stations is known as
 (a) VLR   (c) EIR
 (b) BSC   (d) none of these

11.15 In 2G-GSM, the rate of voice transmission over TCH/F is
 (a) 11.8 kbps   (c) 13 kbps
 (b) 9.6 kbps   (d) none of these

11.16 Which of the following is the code used in IS-95 for spreading purpose?
 (a) PN   (c) PN and Walsh
 (b) Walsh   (d) Gold

11.17 Which of the following is not a UMTS interface?
 (a) Uu   (b) Iub   (c) Iuc   (d) Iur

11.18 Which of the following is not a characteristic of 3G systems?
 (a) Support to only data transmissions
 (b) Worldwide use
 (c) High data rates
 (d) Vehicular mobility

11.19 The UMTS increases the transmission speed to
 (a) 1 Mbps per mobile user
 (b) 2 Mbps per mobile user
 (c) 3 Mbps per mobile user
 (d) 5 Mbps per mobile user

11.20 The most important revolutionary step of GSM towards UMTS is
 (a) CAMEL   (b) GPRS
 (c) EDGE   (d) GGSN

11.21 Which of the following enables worldwide access to operator-specific applications such as prepaid, call screening, and supervision?
 (a) 3GPP   (b) Node B
 (c) CN   (d) CAMEL

11.22 Release 99 for UMTS
 (a) migrates the core voice network to voice over IP network
 (b) enables IP multimedia service
 (c) adds new radio access network, UTRAN
 (d) defines BTS as Node B

11.23 Which of the following statements is correct?
 (a) The WLL is based only on CDMA.
 (b) The WLL is the direct substitute of wireline local loop.
 (c) The WLL system is designed only for rural areas.
 (d) WLL coverage is limited than GSM coverage.

11.24 UMTS is a 3G technology
 (a) being deployed by a GSM or GPRS service provider
 (b) that is the successor to GSM
 (c) developed for fixed wireless access
 (d) that support both TDD and FDD

11.25 Which of the following channels are used for downlink in GSM or UMTS?
 (a) BCH, PCH   (b) FACH, SCH
 (c) DCH, PRACH   (d) RACH, CCPCH

11.26 The prevalent versions of UMTS releases are
 (a) 99 and 4   (b) 9 and 10
 (c) 5 and 6   (d)   none of these

11.27 GPRS supports
 (a) voice
 (b) data
 (c) voice and data
 (d) video

Review Questions

11.1 What is meant by infrastructure-based networks? How do they differ from ad hoc networks?

11.2 How does the 2G GSM system differ from the analog mobile and cellular phone systems?

11.3 What are the upgradations made in the 2G GSM system?

11.4 List the 2G GSM services and also the services after its enhancement.

11.5 How does the GPRS service enhance GSM services?

11.6 List the basic components of the GSM system and the additional component due to GPRS addition.

11.7 How many identity numbers are required in the GSM system? List the GSM service providers in your area.

11.8 Mention the steps for location updation in a GSM network.

11.9 How does the GPRS differ from computer networks?

11.10 Why is it necessary to define GSM and GPRS logical channels?

11.11 How does EDGE differ from 2G GSM?

11.12 Find the infrastructure required for IS-95 to CDMA2000-based systems. Draw the forward and reverse link diagrams for an IS-95 system.

11.13 What is the type of infrastructure required for a WLL?

11.14 Why are separate specifications not required for a WLL system?

11.15 Explain the interworking in DECT with other systems such as GSM and CDMA.

11.16 'The CorDECT WLL is the best technology for rural areas.' Explain.

11.17 What will be the access equipment used by a WLL?

11.18 State the solutions for long-haul transmissions in case of a WLL.

11.19 What is LMDS?

11.20 List the basic attributes of a WLL system.

11.21 Draw the CorDECT WLL architecture.

11.22 What is WCDMA?

11.23 What are the features of 3G systems that do not appear in the systems of previous generations?

11.24 How can you say that UMTS is a 3G GSM? List the similarities and differences between 2G GSM and 3G UMTS.

11.25 How is it possible to have coexistence of LTE with previous generation systems?

11.26 State the basic differences between 3G UMTS and 4G LTE.

11.27 Which channels are the same in 2G, 3G, and 4G systems?

11.28 Draw a detailed diagram of mobile satellite communication showing interlinks between the satellites.

11.29 Fill in the blanks:
   (a) In a cellular system, one cluster uses frequencies $f1, f2, f3, f4, f5, f6$, and $f7$. The frequency reuse factor for that system is _____.
   (b) If the total number of duplex channels available is 56 and the total number of cells in one cluster is 7, the total number of channels allocated to each cell is _____.
   (c) _____ uplink logical channel is used for mobile-originating calls.
   (d) In TCH/F 9.6, the speech signal is digitized at raw data rate_____.
   (e) In a GSM frame structure, one hyperframe equals _____TDMA frame.
   (f) For GSM security, _____ key is dynamic.
   (g) Each 20 ms segment of a digitized speech signal applied to an RPE—LPC speech encoder of GSM consists of _____samples.
   (h) The additional components for GPRS inclusion in the GSM system are _____ and _____.
   (i) The DECT standard uses _____ modulation technique.
   (j) The conventional frequency band of the DECT standard is _____.
   (k) DECT supports _____handoff.
   (l) The transmission bandwidth of IS-95 is _____.
   (m) The interface between RNC and MSC in UMTS is _____.
   (n) _____ protocol is used to transfer data packets between SGSN and MS.
   (o) The interface used between SGSN and GGSN is _____ when they are in different networks.
   (p) Downloading documents is _____ type of service of UMTS.
   (q) The ____kHz channel bandwidth is used for EDGE.
   (r) The agency responsible for developing UMTS and LTE is _____.
   (s) The UMTS and LTE equivalents of a GSM BTS are _____ and _____, respectively.
   (t) The transmission scheme on the LTE uplink is _____.
   (u) The size of a resource block in the downlink LTE is _____subcarriers.

11.30 Match the following in terms of GSM:

A	B
(a) A_3 algorithm	(a) Public number
(b) A_5 algorithm	(b) Authentication algorithm
(c) A_8 algorithm	(c) Used for generation of cipher key Kc
(d) MSISDN	(d) Not public number
(e) IMSI	(e) Timely upgrade number
(f) TMSI	(f) Used for encryption

11.31 Strike out the incorrect word:
  (a) In GSM, **AGCH/PCH** sends the final message to the CCCH before the subscriber is moved from the CCH.
  (b) For fixed cell size, if cluster size N reduces, the probability of co-channel interference **decreases/increases**.
  (c) **Transport signals/Data streams** are not one of the five major UTRAN protocol blocks.
  (d) **Iuc/Iub** is not one of the new interfaces introduced in UTRAN.
  (e) **USF/TBF** controls multiple accesses in GPRS.

11.32 State true or false:
  (a) One of the most important IMT-2000 proposals was to have UMTS as the successor to GSM.
  (b) Fixed wireless access is primarily a rural application.
  (c) Location management in GPRS depends on the MS state.
  (d) In class C, mobile terminal reattachment is required to switch between GSM and GPRS modes.
  (e) LTE supports circuit switching.

12 Ad hoc Networks

Theme of the Chapter

Ad hoc networks are voice and data networks that are established temporarily and on the go without having to set up an infrastructure. There may be limited number of users in such systems. Mostly, ad hoc networks are established for personal use or for use within a limited domain such as an office or a plant. These networks can be established anywhere, maybe on a temporary basis, and because of this, the carrier frequencies chosen are industrial, scientific, and medical (ISM) band frequencies.

Bluetooth can be used to make a wireless home network for personal devices. Wireless local area network (WLAN) in an ad hoc mode is an example of an ad hoc network when formed on short ranges using personal laptops. However, the convergence in WLAN and wireless metropolitan area network (WMAN) industry made it necessary to have some infrastructure (cellular) to allow for interoperability and long ranges.

Sensor networks are specially designed networks in which various physical changes are sensed by sensor-based nodes. These changes are noted in a remote computer connected in a wireless way; it may be connected to WLAN as well. We can consider sensor networks as ad hoc networks or under a special category. Similarly, ultra-wideband is also considered as a special type of network for very high speed personal area communication. In this chapter, the main focus is on the architecture, key aspects of the system, and physical and data link layers.

Key Topics

- Bluetooth
- UWB
- Wi-fi—IEEE 802.11 standards
- WiMAX—IEEE 802.16 standards

12.1 INTRODUCTION

An ad hoc (or spontaneous) network is one with wireless or temporary plug-in connection devices, where some of the network devices are a part of the network only for the duration of a communications session or, in the case of mobile or portable devices, are in some close proximity to the rest of the network. Ad hoc wireless networks do not need any infrastructure. In Latin, ad hoc literally means *for this*, further meaning *for this purpose only* and thus usually *temporary*. The term has been applied to future office or home networks in which new devices can be quickly added to a network or detached from a network as and when required and the devices communicate with each other using wireless transmission.

Ad hoc networks with or without the need for infrastructure and with limited coverage can be established for connecting wireless devices in close proximity without expensive wiring. An ad hoc network can be of the following types:

- It can be a personal area for personal devices. This kind of architecture is known as a *wireless personal area network* (WPAN). The concept of a personal area network (PAN) refers to a space of small coverage (less than 100 m) around a person where ad hoc communication occurs.
- It can be a *wireless local area network* (WLAN) with a coverage of around 500 m. Some configurations of WLANs need infrastructure that provides access to other networks. They follow centralized management. Others are pure ad hoc and distributed networks.

In ad hoc systems on a wide scale, mobile stations (MSs) may act as a relay station in a multihop transmission environment from distant mobiles to base stations (BSs). MSs will have the ability to support BS functionality in this case. The network organization will be based on interference measurements by all mobiles and BSs for automatic and dynamic network organization according to the actual interference and channel assignment situation for channel allocation of new connections and link optimization. The central challenge in the design of ad hoc networks is the development of dynamic routing protocols that can efficiently find routes between two communication nodes. A mobile ad hoc networks (MANETs) working group has been formed within the Internet engineering task force (IETF) to develop a routing framework in ad hoc networks that is based on internet protocol (IP) technologies. Another challenge is the design of proper medium access control (MAC) protocols for multihop ad hoc networks.

This chapter discusses examples of such networks. One observation made is that the range of frequencies is industrial, scientific, and medical (ISM) band in most of the ad hoc networks, and hence, the network can be established in any part of the world.

12.2 BLUETOOTH

Bluetooth represents a single-chip low-cost radio-based wireless network technology. It is a standard for wireless networking of small peripherals. Bluetooth technology aims at ad hoc piconets, with very limited coverage and without the need for infrastructure. It is for personal area communication, such as communication between personal Bluetooth-supported devices within maximum range up to 100 m, in the latest scenario. Bluetooth devices consist of a hardware and protocol stack support as shown in Fig. 12.1.

Bluetooth can connect peripheral devices and support their ad hoc networking. The technology is based on a set of specifications to have wireless communication of data, voice, image, and video clips. The Bluetooth radio technology can be built into both mobile phones and laptops and can connect a laptop and a mobile phone without the use of cables. Apart from this, printers, personal digital assistants (PDAs), desktops, fax machines, keyboards, joysticks, and virtually any other digital device can be a part of the Bluetooth system, making all of them wireless.

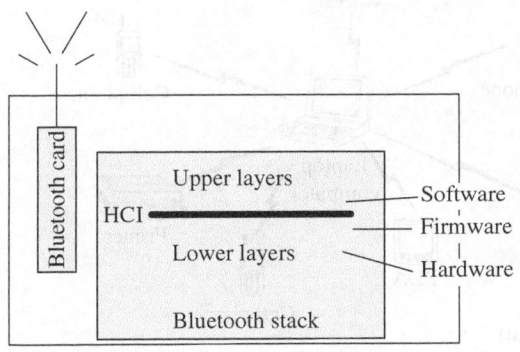

Fig. 12.1   Bluetooth device requirement

A piconet supports one master and maximum seven slave units

The founder members or five promoters of Bluetooth are IBM, Ericsson, Intel, Nokia, and Toshiba, and they jointly are known as the *Bluetooth Special Interest Group* (SIG).

12.2.1 Bluetooth Network Structure

The following are some important terms related to the Bluetooth network structure:

Piconet   It is a topology of devices connected via Bluetooth technology in an ad hoc fashion, as shown in Fig. 12.2. A Piconet starts with two connected devices, such as a portable personal computer (laptop) and mobile phone, and may grow to eight connected devices. All Bluetooth devices are peer units and have identical implementations. However, when establishing a piconet, one unit will act as a master and the other(s) as slave(s) for the duration of the piconet connection. Each piconet is identified by a different frequency hopping sequence and is determined by the Bluetooth device address (BD_ADDR) of the master. All users participating on the same piconet are synchronized to this hopping sequence, the phase of which is determined by the Bluetooth clock of the master. Parked and standby devices are not considered as a part of piconet.

Scatternet   Multiple independent and non-synchronized piconets form a scatternet, as shown in Fig. 12.3.

Note: Normally, maximum ten piconets are allowed in a scatternet.

Master unit   It is the device in a piconet whose clock and hopping sequences are used to synchronize all other devices in the piconet.

Slave units   All devices in a piconet that are not the master are called slave units.

Active member address (AM_ADDR)   It is a three-bit address to distinguish between the active units participating in a piconet.

Parked units   These are the devices in a piconet that are synchronized but do not have an AM_ADDR. They are identified with PM_ADDR, which is valid only as long as the slave is parked.

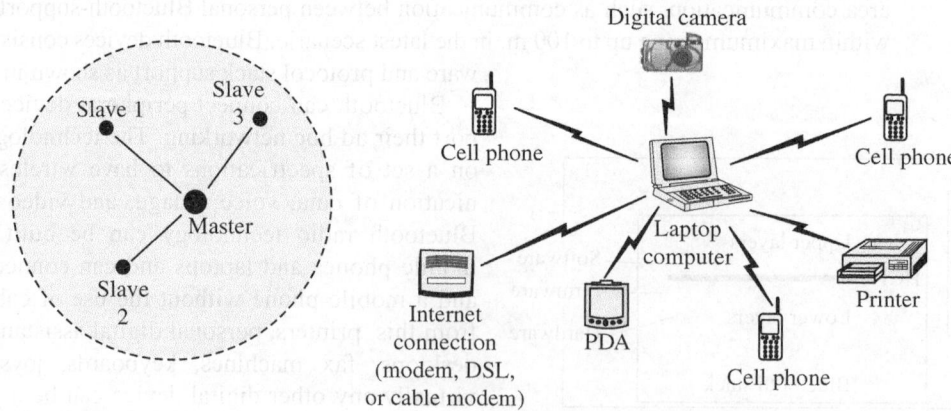

Fig. 12.2   Piconet

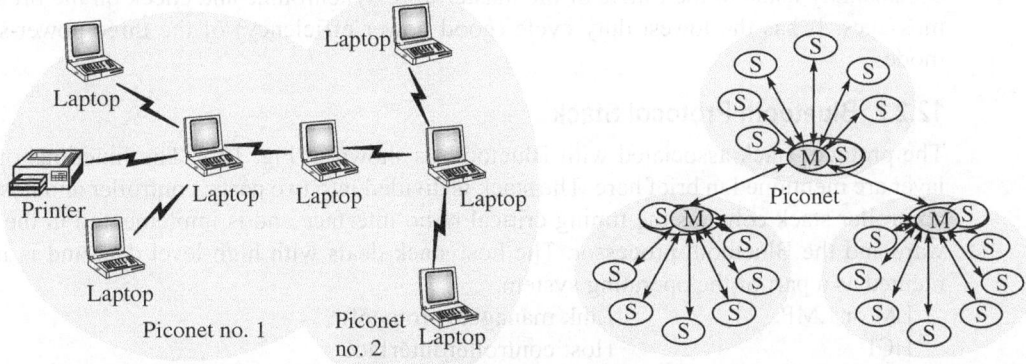

Fig. 12.3 Scatternet: (a) Concept (b) Master of one piconet acting as slave of another piconet

> All Bluetooth devices have a clock that runs at twice the hopping speed and this provides synchronization to the whole network. Clock synchronization is a mandatory requirement.

Access request address (AR_ADDR) This is used by a parked slave to determine the slave-to-master half slot in the access window it can use to send access request messages.

A Bluetooth device in the connection state can be in any of the four modes: active, hold, sniff, and park.

Active mode In the active mode, the Bluetooth unit actively participates on the channel. The master schedules transmissions based on traffic demands to and from the different slaves. In addition, it supports regular transmissions to keep the slaves synchronized to the channel. Active slaves listen in the master-to-slave slots for packets. If an active slave is not addressed, it may sleep until the next new master transmission.

Sniff mode The devices synchronized to a piconet can enter power-saving modes in which the device activity is lowered. In the sniff mode, a slave device listens to the piconet at reduced rate, thus reducing its duty cycle. The sniff interval is programmable and depends on the application. It has the highest duty cycle (least power efficient) of all three power-saving modes (sniff, hold, and park).

Hold mode This is another power-saving mode. The master unit can put the slave units into the hold mode, where only an internal timer is running. The slave units can also demand to be put into the hold mode. Data transfer restarts instantly when units transit out of the hold mode. It has an intermediate duty cycle (medium power efficient) of the three power-saving modes.

Park mode In the park mode, a device is still synchronized to the piconet but does not participate in the traffic. Parked devices have given up their MAC (AM_ADDR) address and

---

SETTING UP A BLUETOOTH NETWORK

To set up a network, the master transmits an enquiry message every 1.28 s to discover whether there are any devices within range. If a reply is received, then an invitation to join the net is transmitted to the specific device that has responded. This is achieved by a scanning procedure, followed by a one-time pairing procedure. After this, the master allocates each device a member address and then controls all the transmissions. The master transmits in the even-numbered time slots whereas the slaves transmit in the odd-numbered slots once they have been given permission to transmit. Clock synchronization is maintained.

occasionally listen to the traffic of the master to re-synchronize and check on the broadcast messages. It has the lowest duty cycle (good power efficiency) of the three power-saving modes.

12.2.2 Bluetooth Protocol Stack

The protocol stack associated with Bluetooth is shown in Fig. 12.4. The functions of each layer are mentioned in brief here. The stack is divided into two parts: controller and host. The controller stack contains the timing critical radio interface and is implemented in the hardware and the Bluetooth processor. The host stack deals with high-level data and is implemented as a part of the operating system.

LM or LMP	Link manager (protocol)
HCI	Host controller interface
L2CAP	Logical link control and adaptation protocol
RFCOMM	Radio frequency communication
OBEX	Generalized multi-transport object exchange protocol
BNEP	Bluetooth network encapsulation protocol
IP	Internet protocol
TCP/UDP	Transfer control protocol/user datagram protocol
AT	Attention sequence
TCS BIN	Telephony control protocol specification binary
SDP	Service discovery protocol

At the bottom of the stack is the hardware component that implements the Bluetooth *radio*, *baseband*, and *link manager* protocols. Neither an application nor even the host has access to this layer of the stack.

The *HCI* transmits data and commands from the layers above to the Bluetooth module below. Conversely, it receives events from the Bluetooth module and transmits them to the upper layers. The functions of the HCI are implemented in the kernel, which is introduced by various companies on their own. The applications can use objects in the user-level L2CAP and RFCOMM layers to access the corresponding in-kernel objects, although many applications will not need to do so directly.

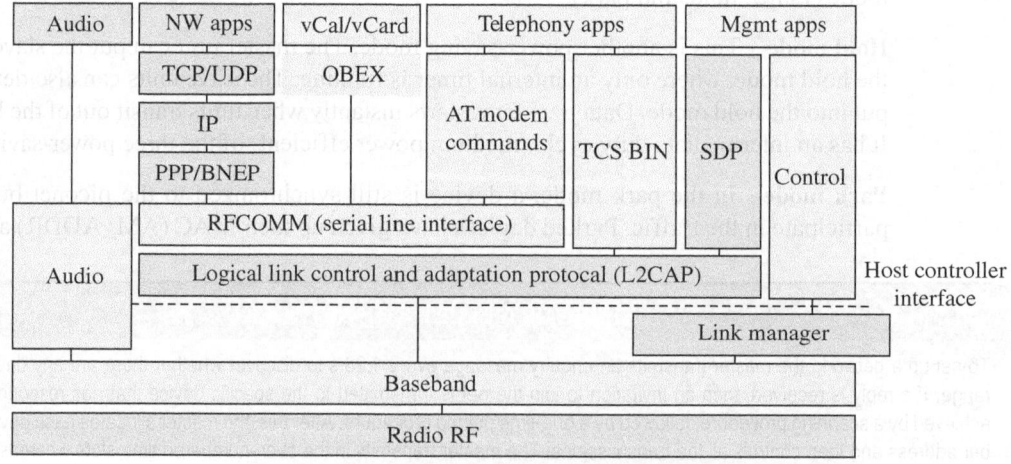

Fig. 12.4 Bluetooth protocol stack

The *L2CAP* layer provides the following:

- Multiplexing of data channels
- Segmentation and reassembly of data packets to conform to a device's maximum packet size
- Support for different channel types and channel IDs, such as RFCOMM

L2CAP functions are nearer to the functions of the transport layer. RFCOMM is a simple set of transport protocols on top of L2CAP.

The in-kernel L2CAP layer provides transport for higher-level protocols and profiles. Using the L2CAP layer's multiplexing feature, it is possible to send and receive data to and from the RFCOMM layer and the SDP layer at the same time.

The in-kernel *RFCOMM* protocol layer is a RS-232 serial port emulation protocol. Its primary mission is to make a data channel appear as a serial port. It supports up to 60 simultaneous connections to a Bluetooth device at a time. It also implements the ability to create and destroy RFCOMM channels and to control the speed of the channel as if it were a physical serial port cable.

The *SDP* layer is more of a service than a protocol. It is shown connected to the user level L2CAP layer, because it uses an L2CAP channel to communicate with remote Bluetooth devices to discover their available (location-based) services.

The *BNEP* layer is similar to the subnetwork access protocol in WLAN and is used for delivering network packets on top of L2CAP during personal area networking. BNEP is bound to L2CAP.

Above the user-level RFCOMM layer is the *OBEX* protocol layer. The OBEX protocol is an HTTP-like protocol that supports the transfer of simple objects, like files, between devices. It uses an RFCOMM channel for transport because of the similarities between IrDA (which defines the OBEX protocol) and serial port communication. The *TCP/UDP* and *IP* layers support normal Internet services.

12.2.3 Bluetooth Physical Layer

Bluetooth radios operate in the unlicensed ISM band at 2.4 GHz. There are a few restrictions while selecting this range of frequency, which are as follows:

- Spectrum spreading must be employed.
- The channel bandwidth is limited to 1 MHz.
- Multiple uncoordinated networks may exist and cause interference.
- Microwave ovens also use this band.
- The 2.4 GHz IC electronics must run at high current levels.
- The IEEE 802.15 standard was also designed to operate on the same frequency.

Bluetooth device implementations are found with the following features:

- Frequency hopping spread spectrum (FHSS) or time division duplex (TDD) with Gaussian frequency shift keying (GFSK) carrier modulation ($BT = 0.5$) is used.
- The 1 Mbps symbol rate exploits the maximum channel bandwidth of 1 MHz.
- Fast frequency hopping and short data packets are used.
- Continuous variable slope delta modulation (CVSD) voice coding enables operation at high bit error rates. (It is a 1 bit/sample vocoder and allows 8–16 kHz input sampling and yields 8–16 kbps compressed speech.)
- The air interface is tailored to minimize current consumption.
- The relaxed link budget supports low-cost single-chip integration.
- The standard uses a hopping rate of 1600 hops/s. These are spread over 79 fixed frequencies and they are chosen in a pseudo-random sequence. The fixed frequencies occur at $2400 + n$ MHz, where the value of n varies from 1 to 79. This gives frequencies of 2402, 2404, …, 2480 MHz.

> Bluetooth supports real-time audio streaming to headset from player using audio-video data and control transport protocol.

With many other users on the ISM band from microwave ovens to Wi-fi, the hopping carrier enables interference to be avoided by Bluetooth devices. A Bluetooth transmission remains on a given frequency for only a short time, and if there is any interference, the data will be resent later when the signal has changed to a different channel, which is likely to be clear of other interfering signals. In order to ensure effective communication in an environment where many devices may receive the signal, each device has its own identifier. This is provided by having a 48-bit hard-wired address identity, giving a total of 2.815×10^{14} unique identifiers.

Table 12. 1 Bluetooth transmitter power levels

Class	Distance covered	Power level
I	Long range up to 100 m	20 dBm
II	Ordinary range up to 10 m	4 dBm
III	Short range up to 10 cm	0 dBm

Note: Bluetooth standard adopts an FHSS rather than a direct sequence spread spectrum (DSSS) approach because it can operate over a greater dynamic range. If DSSS techniques were used, then the transmitters nearer to the receiver would block the required transmission if it is farther away and weaker.

The transmitter power levels for different distances are given in Table 12.1.

Data Transfer Links in Bluetooth

There are two ways in which data is transferred. The first is termed the *asynchronous connectionless* (ACL) communications link. This is the normal type of radio link and uses a polling time division multiple access (TDMA) scheme. It is used for file and general data transfers. The second method is termed the *synchronous connection-oriented* (SCO) *communications link*. This is used for applications such as digital audio, that is, voice data. Figure 12.5 represents the timing diagram of execution of some ACL and SCO links between a master and two slaves.

Asynchronous connectionless link The ACL link is a point-to-multipoint (P2MP) link between the master and all the slaves, as shown in Fig. 12.5. The ACL enables data to be transferred via a Bluetooth radio at speeds up to the maximum rate of 732.2 kbps. This occurs when it is operating in an asymmetric mode. This is commonly used, because for most applications, there is far more data transferred in one direction than in the other. When a symmetrical mode is needed, with data being transferred at the same rate in both directions, the data transfer rate falls to 433.9 kbps. It supports one, three, or five consecutive time slots for data transfer depending on the required payload size, forward error

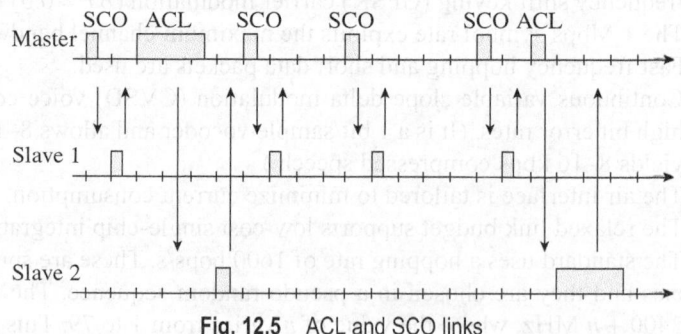

Fig. 12.5 ACL and SCO links

correction (FEC), and up to triple data rate by using a different radio frequency (RF) modulation for the payload.

A connection must be explicitly set up and accepted between two devices before packets can be transferred. ACL packets are retransmitted automatically, if unacknowledged, allowing for correction of a radio link that is subject to interference. ACL links are disconnected if there is nothing received for the supervision timeout period; the default timeout is 20 s, but this may be modified by the master.

Synchronous connection-oriented link   The SCO link is a symmetric point-to-point (P2P) link between a master and a slave in the piconet (circuit-switched type). The master can support up to three simultaneous SCO links, and the slaves can support two or three SCO links. The SCO packets are never retransmitted but FEC can be optionally applied. In an SCO link, the channel reserves two consecutive slots for bidirectional connections at a rate of 64 kbps between a master and its corresponding slave. The data rates are adequate for audio and most file transfers but insufficient for applications like high-rate DVDs that require 9.8 Mbps and for many other video applications, including games.

An SCO link is a set of reserved timeslots on an existing ACL link. Each device transmits encoded voice data in the reserved timeslot. Enhanced SCO (eSCO) links allow greater flexibility in setting up links: they may use retransmissions to achieve reliability and may allow a wider variety of packet types and greater intervals between packets than SCO, thus increasing radio availability for other links.

The data rates in Bluetooth for ACL and SCO are provided in Table 12.2.

Table 12.2   Data rates in Bluetooth (in kbps)

Packet type	Symmetric (SCO)	Asymmetric (ACL)	
		Downlink (master)	Uplink (slave)
DM1	108.8	108.8	108.8
DH1	172.8	172.8	172.8
DM3	258.1	387.2	54.4
DH3	390.4	585.6	86.4
DM5	286.7	477.8	36.3
DH5	433.9	723.2	57.6

Notes:
1. In packet type, M indicates with FEC coding and H indicates no FEC coding. The numbers 1, 3, and 5 represent the occupying slots.
2. Other packet types are NULL, POLL, FHS (common), AUX1 (ACL), HV1, HV2, HV3, and DV (SCO). All these occupy one slot.

12.2.4 Bluetooth MAC Layer

Just like in any other networks, the MAC layer in Bluetooth deals with the packet format in which the data transfer takes place. The format is given in Fig. 12.6(a) and the details of the packet header are given in Fig. 12.6(b).

The data is organized into packets to be sent across a Bluetooth link. The Bluetooth specification lists 17 different formats that can be used depending upon the requirements. The standard packet consists of a 72-bit access code field, a 54-bit header field, and then the data to be transmitted, which may be between 0 and 2745 bits. This data includes the 16-bit cyclic redundancy check (CRC) if it is needed for error detection.

Error handling is incorporated within the system, as it is likely that interference will cause errors. The following points needed to be noted:

The Bluetooth MAC format has options for elements such as FEC and automatic repeat request.

- For asynchronous links, packet sequence numbers are transmitted. If an error is detected in a packet, then the receiver can request it to be resent.
- For synchronous links, the packets cannot be resent, as there is unlikely to be sufficient bandwidth available to resend data and *catch up*. However, it is possible to include some forward error control.

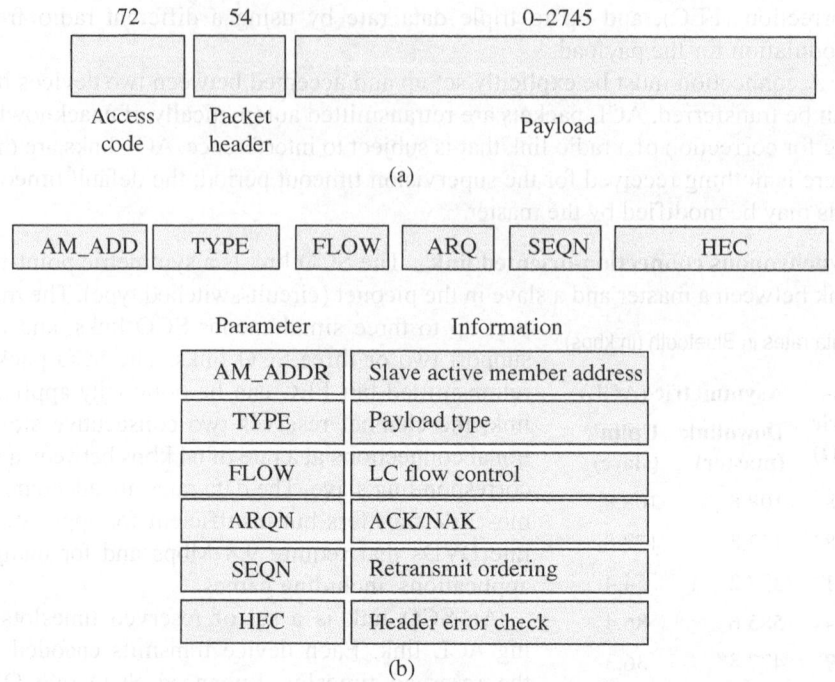

Fig. 12.6   Bluetooth (a) Packet format (b) Packet header

BLUETOOTH SECURITY

Security is becoming an important issue, especially where links to computers are concerned. However, secure communications are possible over Bluetooth, with the devices encrypting the data transmitted. A key up to 128 bits is used, and it is claimed that the level of security provided is sufficient for financial transactions. Bluetooth provides confidentiality and authentication. Bluetooth key generation is generally based on a Bluetooth PIN, which must be entered into both the pairing devices. Sometimes, one of the devices has a fixed PIN, such as in some headsets or similar devices. During pairing, a master key is generated using the E22 algorithm, which is used for encryption of the packets later on.

12.2.5 Modified Version of Bluetooth

Bluetooth is now well established as a wireless technology. It has many significant applications, particularly in areas like connecting mobile phones to hands-free headsets. One of the disadvantages of the original version of Bluetooth in some applications was that the data rate was not sufficiently high, especially when compared to other wireless technologies like 802.11. Hence, modified versions of Bluetooth have been introduced. Of all the features included in Bluetooth 2, the most important is the *enhanced data rate* (EDR) facility. In the new specification, the maximum data rate is able to reach 3 Mbps, a significant increase on what was available in the previous Bluetooth specifications.

Bluetooth 2 supports a much higher data throughput because it uses a different modulation scheme for the payload data. However, this is implemented with reverse compatibility with its previous versions.

Bluetooth data is transmitted as packets made up from a standard format. This consists of the following four elements:

(a) The *access code*, which is used by the receiving device to recognize the incoming transmission

(b) The *header*, which describes the packet type and its length

(c) The *payload*, which is the data required to be carried

(d) The *interpacket guard band*, which is required between the transmissions to ensure that the transmissions from two sources do not collide and to enable the receiver to re-tune

Modulation In the earlier versions of Bluetooth, all three elements of the transmission, namely access code, header, and payload, were transmitted using GFSK, where the carrier is shifted by ±160 kHz, indicating a one or a zero; in this way, one bit is encoded per symbol.

The Bluetooth 2 specification uses a variety of modulation schemes. The GFSK is still used for transmitting the access code and header, and thus, compatibility is maintained. However, other forms of modulation can be used for the payload. Two additional forms of modulation have been introduced. One of these is mandatory, whereas the other is optional.

- The first of the new modulation formats gives a twofold improvement in the data rate and thereby allows a maximum speed of 2 Mbps. This is achieved by using $\pi/4$ differential quaternary phase shift keying ($\pi/4$ DQPSK). This form of modulation is significantly different from the GFSK in that the new standard uses a form of phase modulation, whereas the previous one used frequency modulation. The use of quaternary phase shift modulation means that there are four possible phase positions for each symbol. Accordingly, this means that two bits can be encoded per symbol.

- To enable full threefold increase in data rate to be achieved, a further form of modulation is used. Eight phase differential phase shift keying (8DPSK) enables eight positions to be defined, with 45° between each of them. By using this form of modulation, eight positions are possible and three bits can be encoded per symbol. It can support data rates up to 3 Mbps. As the separation between the different phase positions is much smaller than it was with the quadrature phase shift keying (QPSK) used to provide twofold increase in speed, noise immunity has been reduced in favour of increased speed. Accordingly, this optional form of modulation is used only when a link is sufficiently robust.

A further small change in Bluetooth 2 is the addition of a small guard band between the header and the payload. In addition to this, a short synchronization word is inserted at the beginning of the payload.

New packet formats The Bluetooth 2 specification defines 10 new packet formats for use with the higher data rate modulation schemes, five each of the enhanced data rate schemes:

- Three of these are for the first, third, and fifth slot asynchronous packets used for transferring data.

- The remaining two are used for the third and fifth slot enhanced synchronous connection-oriented (eSCO) packets. These use bandwidth that is normally reserved for voice communications.

The new format for these packets does not incorporate FEC. If this is required, then the system switches back automatically to the standard rate packets. However, many of the links are over a very short range where the signal level is high and the link quality is good.

It is necessary for the packet type to be identified so that the receiver can decode them correctly, knowing also the type of modulation being used. An identifier is, therefore, included in the header, which is sent using GFSK. The packet header used for the earlier versions of Bluetooth used only four bits. This provided sufficient capability to the original system. However, the space was insufficient for the additional information that needed to be sent for Bluetooth 2.

The header format could not be changed because backward compatibility would not be possible. Instead, different link modes are defined. When two Bluetooth 2 or EDR devices communicate, the messages are used in a slightly different way, indicating the Bluetooth 2 or EDR modes. This retains compatibility while still being able to carry the required information.

12.3 WI-FI STANDARDS

The IEEE 802.11 protocols are designed and standardized by the Institute of Electrical and Electronics Engineers (IEEE) and such equipments are handled by the Wi-fi forum. Networking with 802.11 equipment is being treated as WLAN. Compatibility is maintained with previous IEEE standards for networking. Mainly, two layers are standardized: *physical* and *data link*. The data link layer consists of two sublayers: *logical link control* (LLC) and MAC. The 802.11 protocols use the same 802.2 LLC and 48-bit addressing as other 802 local area networks (LANs), allowing for very simple bridging from IEEE wireless to wired networks, but the MAC is unique to WLANs.

The Wi-fi is able to compete with many wired systems. Owing to the flexibility and performance of the system, many Wi-fi *hotspots* have been set up and more are following. These enable people to use their laptops as they wait in hotels, airport lounges, cafes, and many other places using a wireless link rather than a cable. Wi-fi applications are discussed in Section 12.3.5.

Earlier, the 802.11 protocols were designed for infrared communication. They include various standards, each with a letter suffix, specifically for security aspects, quality of service (QoS), and so on. The following are the various 802.11 standards along with their main features:

- 802.11a: Wireless network bearer operating in the 5 GHz ISM band with data rate up to 54 Mbps
- 802.11b : Wireless network bearer operating in the 2.4 GHz ISM band with data rates up to 11 Mbps
- 802.11e: QoS and prioritization (QoS extension)
- 802.11f: Handover
- 802.11g: Wireless network bearer operating in the 2.4 GHz ISM band with data rates up to 54 Mbps
- 802.11h: Power control
- 802.11i: Authentication and encryption (enhanced security)
- 802.11j: Interworking
- 802.11k: Measurement reporting
- 802.11n: Wireless network bearer operating in the 2.4 GHz and 5 GHz ISM bands with data rates up to 600 Mbps
- 802.11s: Mesh networking

802.11 products do not require licensing for carrier frequency. They use ISM band frequencies.

Of these, the standards that are most widely known are the network-bearer standard— 802.11a, 802.11b, 802.11g, and 802.11n. All the 802.11 Wi-fi standards operate within the ISM frequency bands. These are shared by a variety of other users, as no licence is required for operation within these frequencies. This makes them ideal for a general system for widespread use.

12.3.1 Wi-fi Architecture

The architecture of a Wi-fi network contains the following elements in general.

WLAN roaming happens in two ways: (a) internal—client moves from one AP to another AP of the home network and (b) externa—client moves in another service provider's zone, say hotspot, that is, home network to foreign network.

Access points (APs)   The AP is the WLAN transceiver or BS that can connect one or many wireless devices simultaneously to the Internet.

Wi-fi cards   These accept the wireless signal and relay information. They can be internal or external. Examples are PCMCIA card for laptop and PCI card for desktop computers.

Safeguards   These include firewalls or anti-virus software applications that are used to protect the network from uninvited users and keep information secure.

There are three basic topologies for Wi-fi: AP-based, peer-to-peer, and P2MP bridge.

AP-based Topology (Infrastructure Mode)

In the AP-based topology, the client communicates through an AP. RF coverage is provided by the AP. AP's are installed in such a way that their RF coverages overlap 10–15% of the total coverage area, and allow roaming too. (Fig. 12.7).

A Wi-fi hotspot is created by installing an AP to an Internet connection. The AP acts as a BS. When a Wi-fi enabled device encounters a hotspot, the device can connect to that network wirelessly. A single AP can support up to 30 users and can function within a range of 100–150 feet indoors and up to 300 feet outdoors. Many APs can be connected to each other via Ethernet cables to create a single large network.

Peer-to-peer Topology (Ad hoc Mode)

An AP is not required in the peer-to-peer topology. The client devices within a cell can communicate directly with each other (Fig. 12.8). It is useful for setting up of a wireless network quickly and easily.

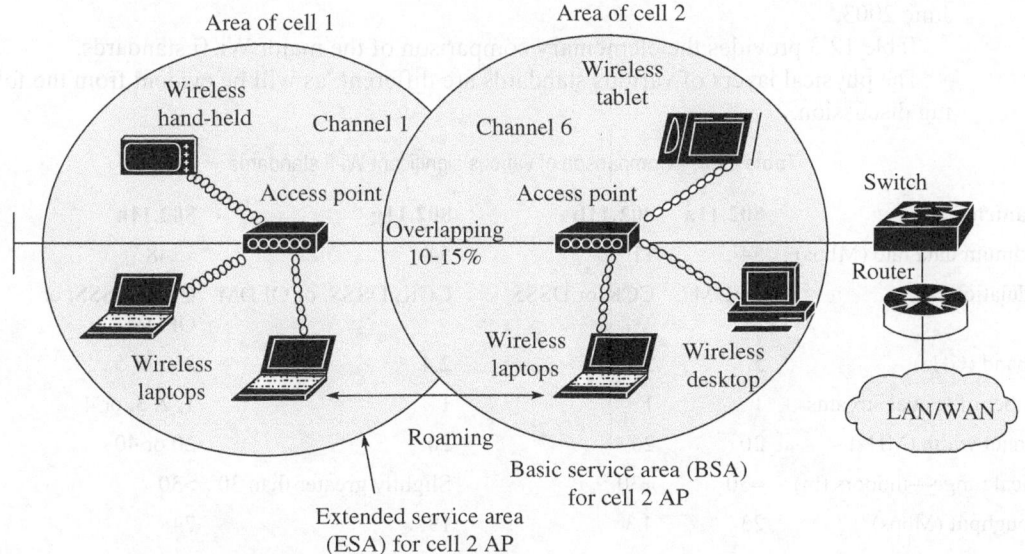

Fig. 12.7   Typical AP-based Wi-fi infrastructure (an ESA is larger than or equal to a BSA)

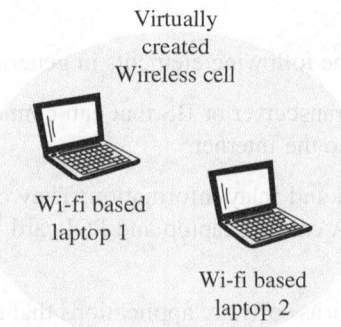

Fig. 12.8 Typical peer-to-peer topology in ad hoc mode

Point-to-multipoint Bridge Topology

The point-to-multipoint bridge topology is used to connect a LAN in one building to LANs in other buildings even if the buildings are miles apart (Fig. 12.9). These conditions receive a clear line of sight (LOS) between buildings. The LOS range varies based on the type of wireless bridge and antenna used as well as the environmental conditions.

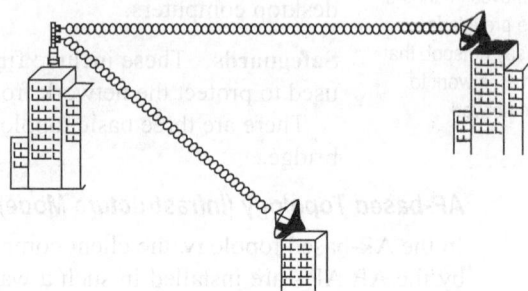

Fig. 12.9 P2MP bridge topology for Wi-fi

12.3.2 Wi-fi Physical Layer

Each of the different Wi-fi standards has different features and the standards were launched at different times. The first accepted 802.11 WLAN standard was 802.11b, supporting the DSSS technique. Almost in parallel with this, another standard was defined. This was 802.11a, which used a different modulation technique—orthogonal frequency division multiplexing (OFDM). Of the two standards, the 802.11b variant caught on primarily because the chips for the lower 2.4 GHz band support were easier and cheaper to manufacture compared to 5 GHz for 802.11a. The 802.11b standard remained the main Wi-fi standard for many years. Looking to increase speed, another standard, 802.11g, was introduced and ratified in June 2003.

Table 12.3 provides the elementary comparison of the major Wi-fi standards.

The physical layers of various standards are different, as will be evident from the following discussion.

Table 12.3 Comparison of various significant Wi-fi standards

Parameter	802.11a	802.11b	802.11g	802.11n
Maximum data rate (Mbps)	54	11	54	≈248
Modulation	OFDM	CCK or DSSS	CCK, DSSS, or OFDM	CCK, DSSS, or OFDM
RF band (GHz)	5	2.4	2.4	2.4, or 5
Number of spatial streams	1	1	1	1, 2, 3, or 4
Channel width (MHz)	20	20	20	20 or 40
Typical range—indoors (m)	≈30	≈30	Slightly greater than 30	>30
Throughput (Mbps)	23	4.3	19	74

IEEE 802.11a

The following are the main features of the 802.11a standard:

- The 802.11a standard uses basic 802.11 concepts as its base, and it operates within the 5 GHz ISM band.
- The modulation is OFDM to enable it to transfer raw data at the maximum rate of 54 Mbps, although a more realistic practical level is in the region of the mid-20 Mbps region. The data rate can be reduced to 48 Mbps, 36 Mbps, 24 Mbps, 18 Mbps, 12 Mbps, and 9 Mbps, and then to 6 Mbps if required.
- The 802.11a standard has 12 non-overlapping channels, 8 dedicated to indoors and 4 to P2P.
- The OFDM signal used for 802.11a comprises 52 subcarriers. Of these, 48 are used for data transmission and 4 are used as pilot subcarriers. The separation between the individual subcarriers is 0.3125 MHz. This results from the fact that the 20 MHz bandwidth is divided by 64 . Although only 52 subcarriers are used, occupying a total of 16.6 MHz, the remaining space is used as a guard band between the different channels. The symbol duration is of 4 μs in which 0.8 μs guard interval is used. For more details on the spectrum setting concept, readers can refer to OFDM explained in Chapter 7.
- Various forms of modulation can be used on each of the 802.11a subcarriers. Binary phase shift keying (BPSK), QPSK, 16QAM (quadrature amplitude modulation), and 64QAM can be used as the conditions permit (Table 12.4). For each set data rate, a corresponding form of modulation is used.

As with many data transmission systems, signal generation is performed using digital signal processing techniques and a baseband signal is generated. It is then upconverted to the final frequency. Similarly, for signal reception, the incoming 802.11a signal is downconverted to baseband and then converted to its digital format after which it can be processed digitally.

IEEE 802.11b

The IEEE 802.11b standard was the first WLAN standard to be widely adopted and built into many laptop computers and other forms of equipment.

Complementary code keying (CCK) The RF signal format used for 802.11b is CCK, which can be considered as a block code generalization of the lower-rate Barker code. This

Table 12.4 Details of OFDM-based modulation stage in 802.11a

Data rate (Mbps)	Modulation	Coding rate	Coded bits per subcarrier	Coded bits per OFDM symbol	Data bits per OFDM symbol
6	BPSK	1/2	1	48	24
9	BPSK	3/4	1	48	36
12	QPSK	1/2	2	96	48
18	QPSK	3/4	2	96	72
24	16QAM	1/2	4	192	96
36	16QAM	3/4	4	192	144
48	64QAM	1/2	6	288	192
54	64QAM	3/4	6	288	216

CCK works only in conjunction with the DSSS technology. It does not work with FHSS.

is a slight variation on code division multiple access (CDMA), which uses the basic DSSS as its basis. In view of the fact that the original 802.11 specification uses CDMA/DSSS, it was easy to upgrade any existing chipset and other investment to provide the new 802.11b standard. As a result, 802.11b chipsets appeared relatively quickly in the market.

The *complementary code* contains a set of finite bit sequences of equal length, such that the number of pairs of identical elements—1 or 0—with any given separation in one sequence is equal to the number of pairs of unlike elements having the same separation in another sequence. A set of 64 eight-bit code words is used to encode data for 5.5 Mbps and 11 Mbps data rates in the 2.4 GHz band of 802.11b wireless networking. The code words have unique mathematical properties that allow them to be correctly distinguished from one another by a receiver even in the presence of substantial noise and multipath interference.

Although 802.11b cards are specified to operate at a basic rate of 11 Mbps, the system monitors the signal quality. If the signal falls or interference levels rise, then it is possible for the system to adopt a slower data rate with increased error correction that is more resilient. Under these conditions, the system will first fall back to a rate of 5.5 Mbps, then 2 Mbps, and finally 1 Mbps. This scheme is known as *adaptive rate selection* (ARS).

Packet binary convolution code   The 22 Mbps version of IEEE 802.11b (sometimes called 802.11b+) technology uses the packet binary convolutional code (PBCC). Ordinary 802.11b uses a short block length for its 8QPSK data symbols. The PBCC, on the other hand, uses 64-state symbols. A PBCC symbol can carry more data than others, but it also requires a more powerful digital signal processor at the AP and the network interface controller (NIC) to make that data available.

Another advantage of PBCC over CCK is that its *convolutional coding* is a method of FEC that enables it to reduce the bit error rate without increasing transmission power. In real life, this means that one can get a higher data transmission rate and expand the range, all the while not using any more power than a conventional 802.11b device.

Though CCK–OFDM lags behind PBCC in development, it has the potential to deliver at speeds up to 55 Mpbs at 2.4 GHz, whereas PBCC currently would top out at 33 Mbps. Figure 12.10 shows the PBCC generation. Here, the cover sequence is used for keying purpose.

IEEE 802.11g

In order to provide higher speeds than what 802.11a provided while operating at the 2.4 GHz ISM band, a new standard known as 802.11g was introduced. It soon took over from the *b*

PRACTICAL DATA RATES OF WI-FI

The actual data rates achieved over a real-time network are significantly less even under reasonably good radio conditions, that is, good signal and low interference.

The drop in data rates results from a number of factors such as the following:

- One is the use of carrier sense multiple access with collision avoidance (CSMA/CA) where the system has to wait for clear times on a channel to transmit. Using this technique, when a node wants to make a transmission, it listens for a clear channel and then transmits. It then listens for an acknowledgement. If it does not receive one, it backs off a random amount of time, assuming that another transmission caused interference. It again listens for a clear channel and then retransmits the data.
- Another is associated with the use of TCP/IP and the additional overhead required. If UDP is used rather than TCP, there can be an increase in the data rate.

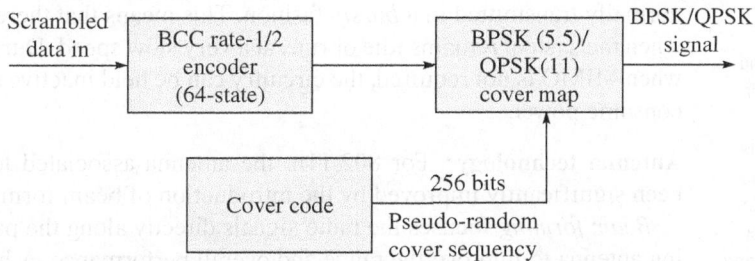

Fig. 12.10 Typical packet binary convolution code generation

MIMO is introduced in IEEE 802.11n for the first time and it operates in dual bands, 2.4 GHz and 5 GHz.

standard. Even before the standard was ratified, 802.11g products were available in the market and it became the dominant Wi-fi technology. The 802.11g standard provided a number of improvements over the 802.11b standard, which was its predecessor.

A variety of modulation schemes can be used by 802.11g. For speeds of 6 Mbps, 9 Mbps, 12 Mbps, 18 Mbps, 24 Mbps, 36 Mbps, 48 Mbps, and 54 Mbps, OFDM is used, but for 5.5 Mbps and 11 Mbps, it uses CCK, and then for 1 Mbps and 2 Mbps, it uses differential BPSK or differential QPSK–DSSS.

The maximum range that can be achieved by 802.11g devices is slightly greater than that achieved by 802.11b, but the range at which the full 54 Mbps can be achieved is much shorter than the maximum range of an 802.11 device. Only when signal levels and interference levels are low can the maximum specified performance be achieved.

IEEE 802.11n

The 802.11n standard was developed with the objective to provide much better performance than earlier standards and be able to keep pace with the rapidly growing speeds provided by technologies such as Ethernet. To achieve this, many new features have been incorporated. The following are the major innovations:

Changes in OFDM implementation One of the major changes to the physical layer of 802.11n is to improve the performance of the implementation of the OFDM modulation. By adapting the way it is set up, the data rate can be increased from the 54 Mbps data rate achieved for 802.11a and g to 300 Mbps and more.

Use of multiple input, multiple output (MIMO) This is a technique that exploits multipath propagation (refer to Chapter 8). Normally, when a signal is transmitted from *A* to *B*, the signal will reach the receiving antenna via multiple paths, causing interference. However, MIMO uses this multipath propagation to increase the data rate by using a technique known as spatial division multiplexing. The data is split into a number of spatial streams, and these are transmitted through separate antennas to corresponding antennas at the receiver. Doubling the number of spatial streams doubles the raw data rate, enabling a far greater utilization of the available bandwidth.

Power saving One of the problems with using MIMO is that it increases the power of the hardware circuitry. More transmitters and receivers need to be supported and this entails the use of more current. Though it is not possible to eliminate the power increase resulting from the use of MIMO in 802.11n, it is possible to make the most efficient use of it. Data is

There are also 802.11ac, ad, af, and ah protocols, supporting wider channels, more spatial streams, and higher data rates than the earlier protocols. The af protocol supports television and cognitive technologies, and ah supports large-scale sensor networks.

normally transmitted in a *bursty* fashion. This means that there are long periods when the system remains idle or runs at a very slow speed. During these periods, when MIMO is not required, the circuitry can be held inactive so that it does not consume power.

Antenna technology   For 802.11n, the antenna-associated technologies have been significantly improved by the introduction of beam forming and diversity.

Beam forming focuses the radio signals directly along the path for the receiving antenna to improve the range and overall performance. A higher signal level and better signal-to-noise ratio will mean that full use can be made of the channel.

Diversity uses the multiple antennas available and combines or selects the best subset from a larger number of antennas to obtain the optimum signal conditions. This can be achieved because there are often surplus antennas in MIMO system. As 802.11n supports any number of antennas between one and four, it is possible that one device has three antennas while another with which it is communicating will have only two. The supposedly surplus antenna can be used to provide diversity reception or transmission as appropriate.

Increased channel bandwidth   An optional mode for the new 802.11 chips is to run using a double-sized channel bandwidth. Previous systems used the 20 MHz bandwidth, but the new ones have the option of using 40 MHz. The main trade-off for this is that other devices have fewer channels for use. There is sufficient scope at 2.4 GHz for three 20 MHz channels, but only one 40 MHz channel can be accommodated. Thus, the choice of whether to use 20 MHz or 40 MHz has to be made dynamically by the devices in the network.

Removal of backward compatibility   Although 802.11n retains support for backward compatibility, this feature can be removed when all the devices operating in a network are 802.11n devices. This removes an overhead that is not required and enables the maintenance of maximum efficiency. When earlier devices enter the network, the backward compatibility overhead and features are re-introduced. As with 802.11g, when earlier devices enter a network, the operation of the whole net is considerably slowed. Therefore operating a network in only 802.11n offers considerable advantages.

12.3.3 Wi-fi MAC Layer

The following aspects are to be considered in the study of the 802.11 MAC layer protocol. Wi-fi MAC layer uses CSMA/CA protocol, which avoids collision by explicit acknowledgement (ACK). This will result in additional overhead of ACK packets and therefore slow performance. It also supports *request to send/clear to send* (RTS/CTS) protocol (Chapter 10), which is considered the solution for the *hidden node* problem. This protocol adds additional overhead by temporarily reserving the medium. Therefore, if used for large size packets, retransmission would be expensive.

The MAC layer supports power conservation to extend the battery life of portable devices. There are two power utilization modes:

(a) In the *continuous aware mode*, the radio is always on and continually drawing power.

(b) In the *power save polling mode*, the radio is *dozing* with the AP queuing any data for it. The client radio will wake up periodically in time to receive regular beacon signals broadcast from the AP. The beacon includes information regarding the stations that have traffic waiting for them.

Note: PCF supports time-bound delivery of data frames, but this approach is not widely adopted because the transmission times are not predictable.

The way in which data is transmitted and controlled has a major impact on the way that QoS is achieved. This is largely determined by the way the MAC layer operates. Within 802.11, there are two options for the MAC layer.

(a) The centralized control scheme, referred to as the *point coordination function* (PCF), for delay sensitive service

(b) The contention-based approach called *distributed coordination function* (DCF), for best effort delivery service

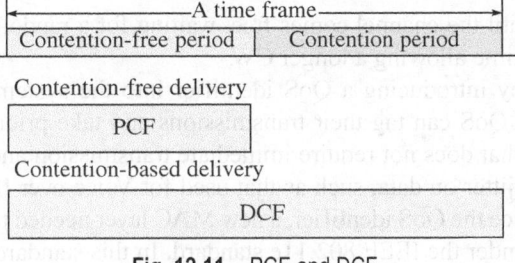

Fig. 12.11 PCF and DCF

Point Coordination Function

The PCF mode supports time-sensitive traffic flows to some degree. Wireless APs periodically send beacon frames to communicate network management and identification that is specific to that WLAN. Between the sending of these frames, PCF splits the time frame into a contention-free period (CFP) and a contention period (CP) (Fig. 12.11). If PCF is enabled on the remote station, it can transmit data during the contention-free polling periods.

Note: Unfortunately, PCF has limited support and numerous limitations; for example, it does not define classes of traffic.

Distributed Coordination Function

The DCF scheme works for mesh topology and uses CSMA/CA. The following are the steps involved:

- The MAC layer sends instructions to the receiver to look for other carriers transmitting.
- If it sees none, then it sends its packet after a given interval and waits for an acknowledgement.
- If no acknowledgement is received, then it infers that its packet was not successfully received.
- It then waits for a given time interval and again checks the channel before retrying to send its data packet.

The transmitter uses a variety of methods to determine whether the channel is in use. It does so by monitoring the activity looking for real signals and also determining whether any signals is expected. This can be achieved because every packet that is transmitted includes a value indicating the length of time the transmitting station is expected to occupy the channel. This is noted by the stations that receive the signal, and only when this time has expired may they consider transmitting.

DCF is the mandatory requirement of the Wi-fi MAC.

If the channel appears to be idle, the prospective transmitting station must wait for a period equal to the DCF interframe space (DIFS). If the channel has been active, it must first wait for a time consisting of the DIFS and a random number of back-off slot times, also known as the *contention window* (CW). This is to ensure that two stations waiting for transmission do not transmit together. If a transmitter senses that

---

LIMITATIONS OF DCF

The DCF option has several limitations, some of which are as follows:

- If many stations communicate at the same time, many collisions will occur, which will lower the available bandwidth [just like in Ethernet, which uses carrier sense multiple access with collision detection (CSMA/CD)].
- There is no notion of high or low priority traffic. While the system works well in preventing stations transmitting together, the result of using this access system is that if the network usage level is high, then the time that it takes for data to be successfully transferred increases. This results in the system appearing to become slower for the users.
- Once a station gains access to the medium, it may keep the medium for as long as it chooses. If a station has a low bit rate (1 Mbps, for example), then it will take a long time to send its packet, and all other stations will suffer.
- Generally, there are no QoS guarantees when real-time data transfer is required.

---

a channel has become active, it must wait until the channel comes free, waiting for a random period for the channel to come free, but this time allowing a longer CW.

The problem of QoS can be addressed by introducing a QoS identifier into the system. Then, those applications that require a high QoS can tag their transmissions and take priority over the transmissions carrying the data that does not require immediate transmission and response. In this way, the level of delay and jitter on data, such as that used for Voice over IP (VoIP) and video, may be reduced. To introduce the QoS identifier, a new MAC layer needed to be developed and this has been undertaken under the IEEE 802.11e standard. In this standard, the traffic is assigned a priority level prior to transmission. These are termed *user priority* (UP) levels and there are eight such levels. Having done this, the transmitter then prioritizes the data with it waiting to be sent by assigning it one of the four *access categories* (AC).

In order to achieve the required functions, the re-developed MAC layer takes on aspects of both the DCF and PCF from the previous MAC layer alternatives and is termed the *hybrid coordination function* (HCF). The modified elements of the DCF are termed the *enhanced distributed channel access* (EDCA), whereas the elements of the PCF are termed the *HCF controlled channel access* (HCCA). Both EDCA and HCCA define traffic classes.

EDCA

The following are the main features of EDCA:

- It uses a mechanism called *transmit opportunity* (TXOP), which is a bounded time interval during which a station can send as many frames as possible, but the transmission time must not extend beyond the maximum duration of the TXOP. Each priority level is assigned a TXOP, and this mechanism prevents low-speed stations from spending too much time using the media when other clients (including those with traffic in the higher priority queues) are waiting.
- It provides a mechanism whereby traffic can be prioritized, but it remains a contention-based system, and therefore, it cannot guarantee a given QoS. When using EDCA, a new class of interframe space (IFS) called an *arbitration interframe space* (AIFS) has been introduced. This is chosen such that the higher is the priority of the message, the shorter will be the AIFS and the CW. The transmitter then gains access to the channel in the normal way, but due to the shorter AIFS and CW, the possibility of gaining access to the channel will be high.

HCCA

The following are the main features of HCCA:

- It adopts a different technique, using a polling mechanism. Accordingly, it can provide guarantees about its level of service, thereby ensuring a true QoS level.

HCCA is more complex than EDCA—it acts like PCF in that it uses a CP and a CFP.

- Although it has a point coordination IFS, it is shorter than the DIFS mentioned earlier, and therefore, it will always gain control of the channel.
- Once it has taken control, it polls all the stations or transmitters in the network. To do this, it broadcasts a particular frame indicating the start of polling, and it will poll each station in turn to determine the highest priority.
  - It will then enable the transmitter with the highest priority data to transmit, although it will result in longer delays for traffic that has a lower priority.

The HCCA mode allows for a CFP to be initiated any time during a CP, which is called the controlled access phase (CAP). The control station, which is normally the AP, is known as the *hybrid coordinator* (HC), and it takes control of the channel. An AP can initiate a CAP whenever it needs to send a frame to a station; this allows the AP to make better decisions about how the wireless medium is used. During the CP, all wireless clients operate in the EDCA mode.

Another feature of HCCA is that *traffic classes* and *traffic streams* are defined, allowing the AP to provide a per-session service for QoS on top of the per-station service that EDCA enables. HCCA is a very powerful but complex coordination function that even allows clients to request specific transmission requirements such as jitter and data rate, allowing for very effective implementations of voice over Wi-fi and video over Wi-fi.

MAC Frame and Addressing

Figure 12.12 will help understand the MAC frame. For an ad hoc mode, the addressing is very simple and straightforward. The following short forms are used for representing the various components of the AP-based Wi-fi topology for which the typical use of the various addresses and conversion of the MAC addresses for Ethernet-based Internet access through AP are shown in Fig. 12.13.

AP: Access point
H1: Wireless host 1
R: Router

Here, for the frame of 802.3 protocol, the source address is the address of the AP and the destination address is that of the router.

The frame control field is further explored in Fig. 12.14. The type field represents the type of the frame—RTS, CTS, ACK, or data. The interpretation of the tags is self-explanatory. The last bit is reserved for future use.

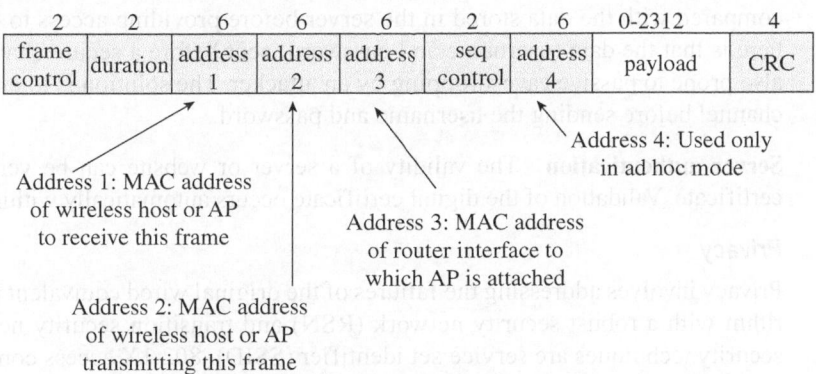

Fig. 12.12 MAC frame for IEEE 802.11

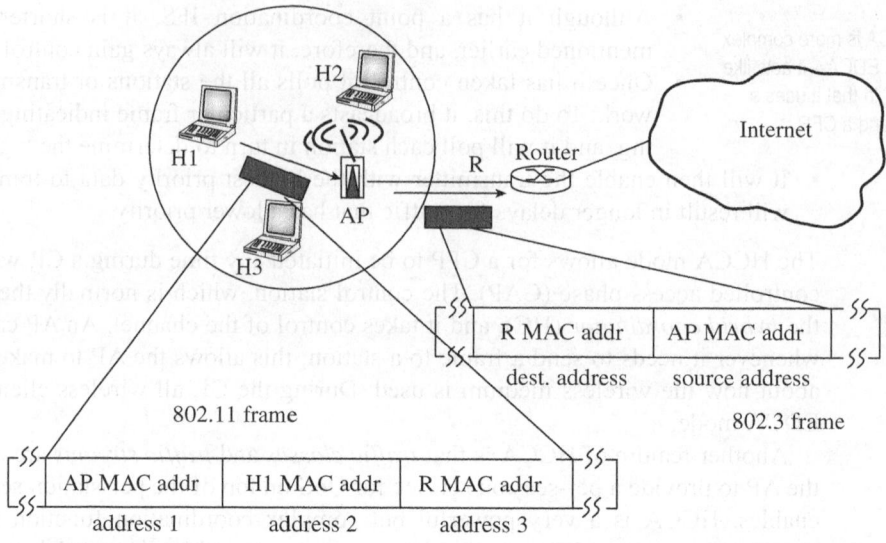

Fig. 12.13 Example for conversion of IEEE 802.11 MAC frame to 802.3 MAC frame

2	2	4	1	1	1	1	1	1	1	1
Protocol version	Type	Subtype	To AP	From AP	More frag	Retry	Power mgt	More data	WEP	Rsvd

Fig. 12.14 Frame control field of two bytes explored in terms of bits representing different tags

12.3.4 Wi-fi Security Aspects

Wireless technology does not remove any old security issues but introduces new ones such as eavesdropping, man-in-the-middle attacks, and denial of service. The requirements for Wi-fi network security can be broken down into two primary components: authentication and privacy.

Authentication

Authentication is used to keep unauthorized users off the network. This is of two types.

User authentication User authentication is a process of identifying the user and verifying that he/she is allowed to access some service. The username and password of the user are compared with the data stored in the server before providing access to the service. The risk here is that the data (username and password) sent before a secure channel is established is also prone to passive eavesdropping by an attacker. The solution is to establish an encrypted channel before sending the username and password.

Server authentication The validity of a server or website can be verified using a digital certificate. Validation of the digital certificate occurs automatically within the client software.

Privacy

Privacy involves addressing the failures of the original wired equivalent privacy (WEP) algorithm with a robust security network (RSN) and transition security network (TSN). Other security techniques are service set identifier (SSID), 802.1X access control, Wi-fi protected access (WPA), and IEEE 802.11i.

WEP   It provides the same level of security as that provided by a wired network. This is the original security solution offered for the IEEE 802.11 standard. It uses RC4 encryption with pre-shared keys and 24-bit *initialization vectors* (IVs). The key is generated by concatenating the shared secret key with a randomly generated 24-bit IV. There is total of 32 bit integrity check value (ICV). The number of bits in the key schedule is equal to the sum of the lengths of the plaintext and the ICV. The following are the two shared secret keys:

(a) 64-bit pre-shared key—WEP

(b) 128-bit pre-shared key—WEP2

These encrypt data only between 802.11 stations. Once the data enters the wired side of the network (between APs), WEP is no longer valid. WEP has a short IV and it is a static key. Therefore, it offers very little security.

> WEP is a weak security standard as the password can be cracked. WPA is more secure and WPA2 is the latest version and is incorporated in Wi-fi devices with firmware upgrade or replacement.

SSID   It is used to identify an 802.11 network and can be pre-configured or advertised in the beacon broadcast. It is transmitted in clear text and provides very little security.

802.1X access control   It is designed as a general-purpose network access control mechanism and is not Wi-fi specific, which authenticates each client connected to an AP (for WLAN) or a switch port (for Ethernet). Authentication is done with the RADIUS server, which informs the AP whether access to controlled ports should be allowed or not, through the following procedure:

- The AP forces the user into an unauthorized state.
- The user sends an extensible authentication protocol (EAP) start message.
- The AP returns an EAP message requesting the user's identity.
- The identity send by user is then forwarded to the authentication server by the AP.
- The authentication server verifies the user and returns an accept or a reject message to the AP.
- If accept message is returned, the AP changes the client's state to authorized and normal traffic flows.

The access control scenario is shown in Fig. 12.15.

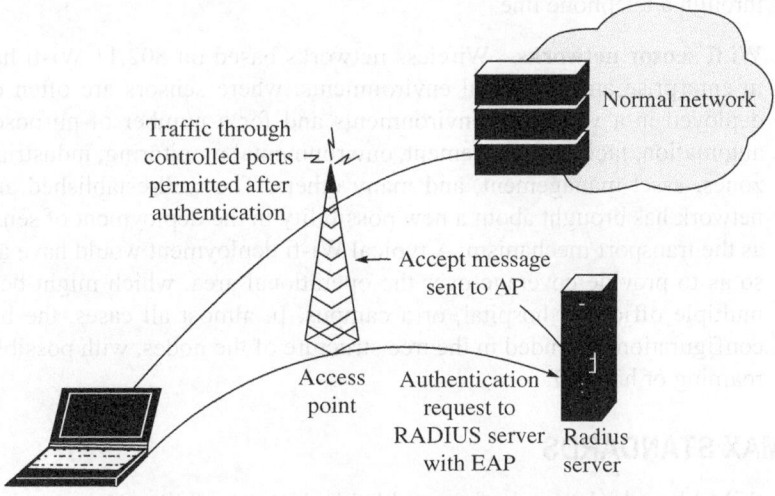

Fig. 12.15   IEEE802.1X access control scenario

> A set of sensors deployed scattered in a location and monitored at a central facility is commonly referred to as a *sensor network* being organized as a mesh of interconnected nodes.

WPA It is a specification of standard-based interoperable security enhancements that strongly increase the level of data protection and access control for existing and future WLAN systems. The following are its features:

- User authentication is done by 802.1x and EAP.
- Temporal key integrity protocol (TKIP) based encryption is used with RC4 (most widely used software stream cipher) dynamic encryption keys, in which 48-bit IV and per-packet key mixing functions are present.
- It fixes all issues found in WEP. It also uses message integrity code (MIC) to ensure data integrity.
- Old hardware should be upgradeable to WPA.

Wi-fi protected access is of two types:

(a) *WPA pre-shared keys*: In this, a single master key is used for all users. It is suitable for small office/home office (SOHO) environments.
(b) *WPA-Enterprise*: It is the most secure method. It has a unique key as well as a separate username and password for each user. It is suitable for large organizations.

IEEE 802.11i It provides the standard for WLAN security along with authentication. For data encryption, advanced encryption standard (AES) protocol is used. It supports secure and fast handoff, which allows roaming between APs without requiring the client to re-authenticate every AP. The protocol will require new hardware.

12.3.5 Wi-fi Applications

The following are some of the important applications of Wi-fi.

Internet access A Wi-fi enabled device can connect to the Internet when within the range of a wireless network that is configured to permit this. The coverage of one or more (interconnected) APs—called hotspots—can extend from an area as small as a few rooms to as large as many square miles. Coverage in large areas may require a group of APs with overlapping coverage. Outdoor public Wi-fi technology has been used successfully in wireless mesh networks in many countries. Wi-fi based modems can be used for Internet access through a telephone line.

Wi-fi sensor networks Wireless networks based on 802.11 Wi-fi have become pervasive in enterprise and industrial environments, where sensors are often deployed. Sensors are deployed in a variety of environments and for a number of purposes, including building automation, facility management, environmental monitoring, industrial automation, military zones, asset management, and many others. The well-established and standardized Wi-fi network has brought about a new possibility in the deployment of sensors—the use of Wi-fi as the transport mechanism. A typical Wi-fi deployment would have a set of APs positioned so as to provide coverage over the operational area, which might be a large factory floor, multiple offices, a hospital, or a campus. In almost all cases, the basic network is a star configuration, extended in the tree structure of the nodes, with possible provisions for client roaming or handoff.

12.4 WIMAX STANDARDS

WiMAX, which stands for *worldwide interoperability for microwave access*, is a new broadband wireless data communications technology or mobile Internet based around the IEEE 802.16 standard that provides high-speed data communications (70 Mbps) over a wide area. It is a technology for P2MP wireless networking. WiMAX is a standard for WMANs that has been developed by working group number 16 of IEEE 802, specializing in broadband wireless access. It is suitable for rural applications. The WiMAX Forum is a wireless industry consortium with over 100 members, including many industry leaders. It has been set up to support and develop WiMAX technology worldwide and to ensure common standards across the globe. One of the aims of the forum is to enable a standard to be adopted that will facilitate full interoperability between the products.

12.4.1 WiMAX Architecture

> IEEE 802.16j is a relay-based wireless access standard for multihop relay, and 802.16m is for higher bandwidths and higher data rates of 100 Mbps.

WiMAX has two important standards or usage models:

(a) A *fixed usage model* IEEE 802.16d-2004 for fixed wireless broadband access (FWBA)

(b) A *portable usage model* IEEE 802.16e-2005 for mobile wireless broadband access (MWBA)

IEEE 802.16–2004 WiMAX is through LOS communications, which incorporates a stationary transmitter and receiver, and is less complicated, but MWBA non-line-of-sight (NLOS) communication is much more complicated. A practical WiMAX scenario is given in Fig. 12.16 (the figure represents that Wi-fi can also

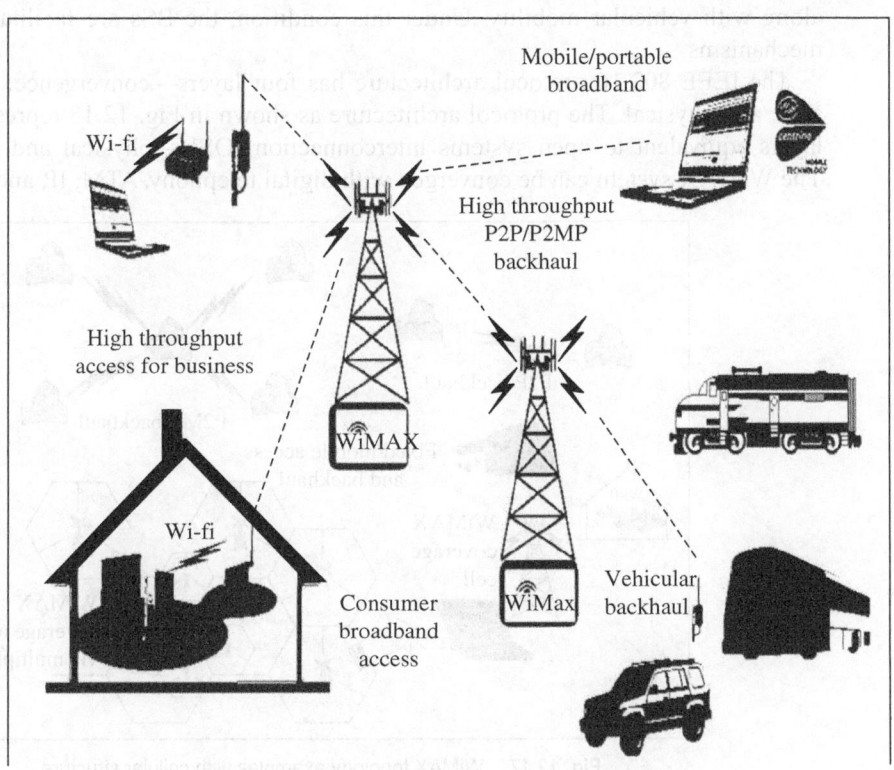

Fig. 12.16 WiMAX practical scenario

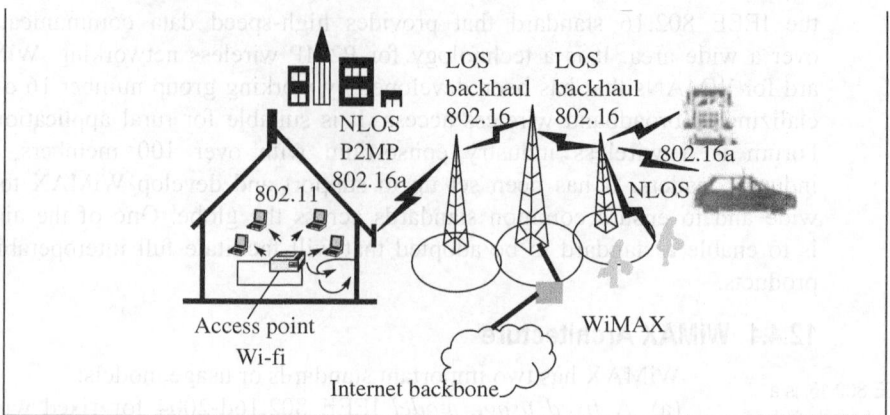

Fig. 12.16 *(Contd)*

be made to communicate with WiMAX devices). For LOS, 802.16 protocol is used between two BSs. For fixed wireless access (e.g., a Wi-fi connection), 802.16d protocol is utilized, whereas for mobile applications, 802.16e is more suitable. Normally, P2P or P2MP architecture is a basic one, and with increase in the number of connections, it may result in a mesh architecture. The basic topologies are given in Fig. 12.17. The use of cellular infrastructure and BSs are required to have continuous broadband access along with vehicular mobility. Under this condition, the BSs are facilitated by handover mechanisms.

The IEEE 802.16 protocol architecture has four layers—convergence, MAC, transmission, and physical. The protocol architecture as shown in Fig. 12.18 represents mainly two layers equivalent to open systems interconnection (OSI)—physical and data link layers. The WiMAX system can be converged with digital telephony, ATM, IP, and so on.

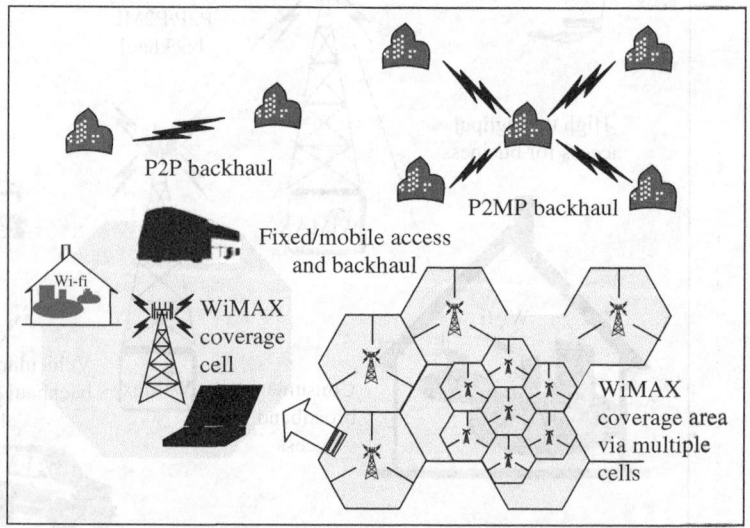

Fig. 12.17 WiMAX topology examples with cellular structure

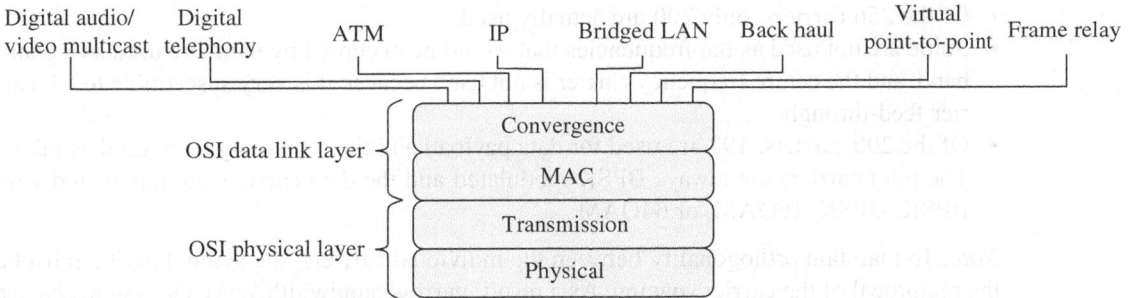

Fig. 12.18 WiMAX protocol architecture with various possible applications

12.4.2 WiMAX Physical Layer

The evolution of three IEEE 802.16 standards is given in Table 12.5 in terms of physical layer parameters. These standards are widely used in WiMAX architecture.

The WiMAX 802.16 standard describes four different RF or air interfaces depending upon the application envisaged. However, currently the most widely implemented interface is the one that is intended for NLOS applications up to 30 km and for frequencies below 11 GHz. As a result, it is often thought of as *the* WiMAX air interface.

IEEE 802.16d

The following are the spectrum and modulation features of the IEEE 802.16d standard:

- WiMAX 802.16d RF signals use OFDM techniques. The signal incorporates 256 carriers in the total signal bandwidth, which may range from 1.25 MHz to 20 MHz.

Table 12.5 Evolution in IEEE 802.16 standards

Parameter	802.16–2001	802.16d–2004	802.16e–2005
Spectrum or band-width	10–66 GHz	2–11 GHz	2–6 GHz
Propagation channel condition	LOS	NLOS (pedestrian)	NLOS (vehicular)
Bit rate	Up to 134 Mbps (28 MHz channelization	Up to 75 Mbps (20 MHz channelization	Up to 15 Mbps (5 MHz channelization
Modulation	QPSK, 16QAM (optional in UL), 64QAM (optional)	256 subcarriers OFDM, BPSK, QPSK, 16QAM, 64QAM,	Scalable OFDMA, QPSK, 16QAM, 64QAM, 256QAM(optional)
Mobility	Fixed	Fixed/nomadic	Portable/mobile

Note: Only significant changes are highlighted.

USE OF BASIC WIMAX STANDARD

The basic 802.16 WiMAX standard uses frequencies in the range 10–66 GHz, although extensions to the standard allow the use of other frequencies. At these frequencies, particularly those in the range between 10 GHz and 66 GHz, the transmission path is essentially LOS and multipath reflections are reduced. This accelerates the rate at which data can be sent. This is the reason 802.16 is used between BS-to-BS LOS links in the WiMAX architecture.

- Of the 256 carriers, only 200 are actually used.
- Some are not used as the frequencies that would be occupied by them are used as a guard band, and the centre frequency carrier is not used because it is very susceptible to RF carrier feed-through.
- Of the 200 carriers, 192 are used for data payload and the remaining 8 are used as pilots. The pilot carriers are always BPSK modulated and the data carriers are modulated with BPSK, QPSK, 16QAM, or 64QAM.

Note: To maintain orthogonality between the individual carriers, the symbol period must be the reciprocal of the carrier spacing. As a result, narrow bandwidth WiMAX systems have a longer symbol period, which helps overcome problems such as multipath interference that is prevalent on NLOS applications. This is a major advantage of WiMAX systems.

The IEEE 802.16 WiMax standard allows data transmission using multiple broadband frequency ranges. The fact that the technology can work in multiple frequency ranges allows it to avoid interference with other wireless applications, thereby being able to operate satisfactorily in the presence of other transmissions. The frequency bands that are chosen for a particular system affect a number of factors including the data rate that can be carried and the range that can be achieved. Thus, it is possible to choose the frequency band to be used according to the prevailing conditions and the requirements for the system.

IEEE 802.16e

> The 802.6 specification supports a mix of MSs with different MIMO capabilities; hence, MIMO configuration is auto-negotiated between the BS and the MS.

The following are the spectrum and modulation features of the IEEE 802.16e standard:
- The mobile WiMAX air interface adopts orthogonal frequency division multiple access (OFDMA) for improved multipath performance in NLOS environments.
- Scalable OFDMA (SOFDMA) is introduced in the IEEE 802.16e amendment to support scalable channel bandwidths from 1.25 MHz to 20 MHz. The scalability is supported by adjusting the fast Fourier transform (FFT) size while fixing the subcarrier frequency spacing at 10.94 kHz. 802.16e systems offer scalability in both radio access technology and network architecture, thus providing a great deal of flexibility in network deployment options and service offerings.
- Since the resource unit subcarrier bandwidth and symbol duration are fixed, the impact to higher layers is minimal when scaling the bandwidth.
- The 802.16e supports TDD and full and half duplex frequency division duplex (FDD) operations.
- The inclusion of MIMO antenna techniques along with flexible subchannelization schemes and advanced modulation and coding (AMC) together enable the 802.16e technology to support peak downlink data rates up to 63 Mbps per sector and peak uplink data rates up to 28 Mbps per sector in a 10 MHz channel.

Hybrid automatic repeat request (HARQ) error control protocol and fast channel feedback (CQICH) through channel quality index make the system more reliable.

Power Management in WiMAX

Mobile WiMAX supports sleep mode and idle mode for power saving and traffic control.
- Idle mode provides a mechanism for the MS to become periodically available. It removes the requirement for handoff and other normal operations. It eliminates air interface and network handoff traffic from essentially inactive MSs.

WiMAX coverage is measured in square kilometres, while that of Wi-fi is measured in square metres. Due to larger area, WiMAX has more security threats.

- In sleep mode, the MS conducts pre-negotiated periods of absence from the serving base station (SBS) air interface. These periods are interpreted as unavailability of the MS to the SBS concerning downlink or uplink traffic. Sleep mode minimizes MS power usage and the usage of SBS's air interface resources.

Handoff Management in WiMAX

Handoffs between BSs are available during sleep mode. Mobile WiMAX also supports seamless handoff to enable the MS to switch from one BS to another at vehicular speeds without interrupting the connection.

There are three handoff methods:
(a) Hard handover (HHO) (mandatory)
(b) Fast base station switching (FBSS) (optional)
(c) Macro diversity handover (MDHO) (optional)

In HHO, the MS communicates with only one BS at a time. The connection with the old BS is broken before a new connection is established. Signal strength is the main parameter for decision-making as usual.

When FBSS is supported, the MS and the BS keep a list of BSs that are involved in FBSS with the MS. This set is called an active set. The MS defines an anchor BS and communicates only with it. The transition between the anchor BSs is carried out without any handoff signalling. Handoff begins with a decision by an MS to receive or transmit data from the anchor BS that may change within the active set. An important requirement of FBSS is that the data be simultaneously transmitted to all members of an active set of BSs that are able to serve the MS.

When MDHO is supported, the MS and the BS maintain an active set of BSs that are involved in MDHO with the MS. Moreover, an anchor BS is defined. The MS communicates with all BSs in the active set of the uplink and downlink unicast messages and traffic. MDHO begins when an MS decides to transmit or receive unicast messages and traffic from multiple BSs in the same time interval.

Antenna Technologies for WiMAX

Smart antenna technologies typically involve complex vector or matrix operations on signals due to multiple antennas. OFDMA allows smart antenna operations to be performed on vector-flat subcarriers. Mobile WiMAX supports a full range of MIMO smart antenna technologies to enhance system performance. It supports transmit diversity MIMO and spatial multiplexing MIMO. It also supports advance techniques such as uplink collaborative MIMO, adaptive antenna steering, and cyclic delay diversity.

WiMAX with Wi-fi

The WiMAX 802.16 standard has a number of possible applications. The initial plan is to use WiMAX as a P2MP broadband technology to provide individual users with access to broadband data services. In this way, it would provide an attractive alternative to technologies like digital subscriber loop (DSL). Typically, providers would want to limit the number of subscribers to any one BS to around 500 to preserve the bandwidth needed by each user. In this application, it is unlikely that ranges will exceed 10 miles.

While it will no doubt be used in this role, with a maximum range up to 31 miles, WiMAX may also achieve widespread use and provide a backbone for other services when no wired service exists. It may be used as part of the service providers' infrastructure. It is anticipated that services will use directional antennas, and in this way, several services can share the same frequency bands with minimum levels of interference. As a result, the capacity of the system can be huge.

In some instances, there may be competition between WiMAX and Wi-fi 802.11. However, this is not the intention. Although WiMAX can be used to provide applications such as Internet or intranet access in a way similar to Wi-fi, the range of Wi-fi is limited up to around 50 m. In contrast, WiMAX can transmit data up to a distance of 50 km.

The aim of WiMAX is not to compete with Wi-fi but to address a different market and hence to coexist with it, as shown in Fig. 12.19. A general comparison of Wi-fi and WiMAX is given in Table 12.6.

WiMAX Modes

Although WiMAX can be deployed as TDD, FDD, and half duplex FDD, the most common arrangement is the TDD mode. This allows for a greater efficiency in spectrum usage than that obtained in FDD mode.

In TDD mode, the WiMAX BS and the end-users transmit on the same frequency, but to enable them not to interfere with each other, their transmissions are separated in time, as shown in Fig. 12.20 along with the MAC frame. The following are the steps involved:

> The timing of the uplink subframes needs to be accurately controlled and synchronized so that they do not overlap whatever be their distance from the BS.

- In order to achieve this, BS first transmits a subframe and this is followed by a short gap called the *transmit/receive transition gap* (TTG). After this gap, the users or remote stations are allowed to transmit their subframes.
- Once all the uplink subframes have been transmitted, another short gap known as the *receive/transmit transition gap* (RTG) is left before the BS transmits again.

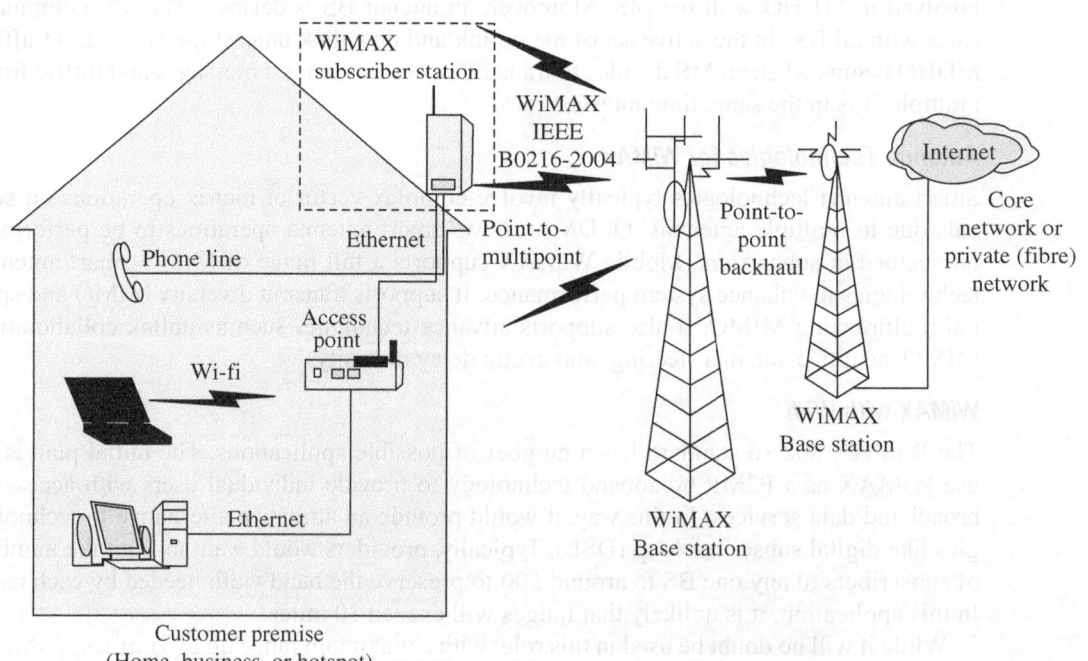

Fig. 12.19 P2MP architecture of WiMAX incorporating Wi-fi

Table 12.6   General comparison of Wi-fi and WiMAX

Wi-fi	WiMAX
1. Wide (20 MHz) frequency channels	1. Channel bandwidth can be chosen by operator (e.g., for sectorization); 1.5 MHz to 20 MHz wide channels; MAC designed for scalability independent of channel bandwidth
2. MAC designed to support tens of users	2. MAC designed to support thousands of users
3. Contention-based MAC (CSMA/CA); no guaranteed QoS	3. Grant request MAC; centrally enforced QoS
4. Currently no guarantee latency for voice and video	4. Designed to support voice and video from ground up
5. Does not allow for differentiated levels of service on a per-user basis	5. Supports differentiated service levels: for example, best effort for residential and T1 for business customers
6. Optimized for ~50+ m	6. Optimized for up to ~50 km
7. No near–far compensation	7. Designed to handle many users spread over kilometres
8. Optimized for indoor performance; designed to handle indoor multipath delay spread of 0.8 µs	8. Optimized for outdoor NLOS performance; designed to tolerate greater multipath delay spread up to 10 µs
9. No mesh topology support within ratified standards	9. Supports mesh network topology and advanced antenna techniques

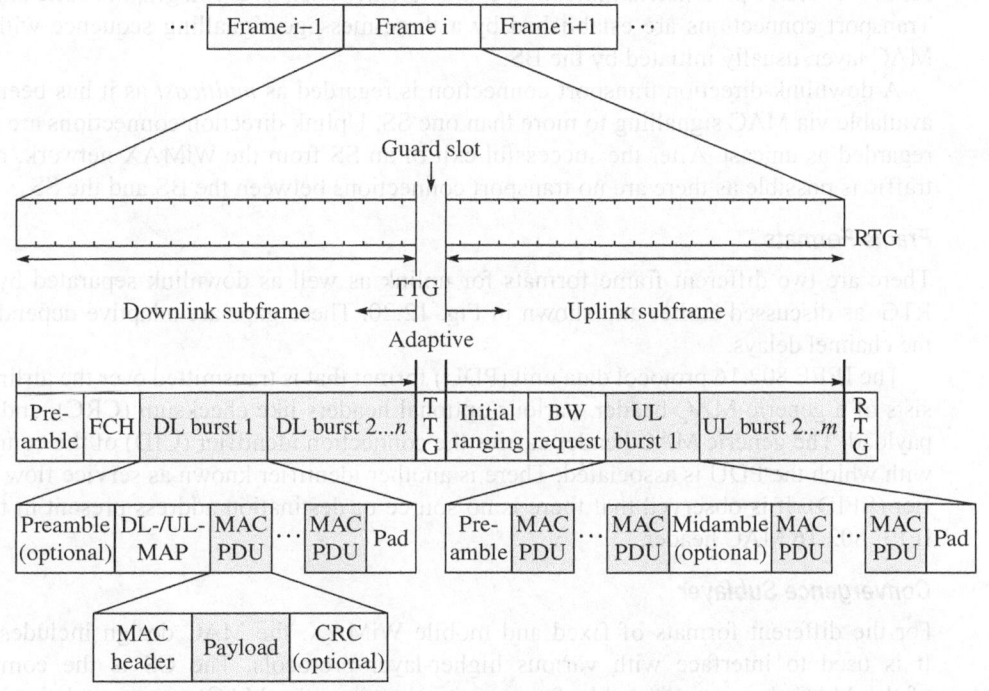

Fig. 12.20   TDD system with MAC frame

Note: There are slight differences between the WiMAX subframes transmitted on the uplink and the downlink. The downlink subframe begins with a preamble after which a header is transmitted, and this is followed by one or more bursts of data. The modulation within a subframe may change, but it remains the same within an individual burst. Nevertheless, it is possible for the modulation type to change from one burst to the next. The first bursts to be transmitted use the more-resilient forms of modulation such as BPSK and QPSK. Later bursts may use the less-resilient forms of modulation such as 16QAM and 64QAM, which enable more data to be carried.

12.4.3 WiMAX MAC Layer

The IEEE 802.16 standard is P2MP and connection oriented at the MAC, physically arranged in a P2MP structure with the BS terminating one end of each connection and an individual subscriber station (SS) terminating the other end. Each SS in the network possesses a 48-bit MAC address and each BS possesses a 48-bit unique identifier called *BSId*.

The BS and SS learn the MAC address or BSId of each other during the entry of SS into the network. Additionally, the BS may possess a 48-bit MAC address, but this is known to the SS only if it is using the Ethernet convergence sublayer (CS). The WiMAX MAC supports communication in all traffic conditions. It uses the spectrum efficiently and has various QoS options.

Transport Connections

The WiMAX MAC layer is primarily an adaptation layer between the physical and upper layers of the protocol stack. User data traffic in both the uplink and downlink directions is carried on unidirectional transport connections. Each transport connection has a particular set of associated parameters indicating characteristics such as cryptographic suite and QoS. Transport connections are established by a three-message signalling sequence within the MAC layer, usually initiated by the BS.

A downlink-direction transport connection is regarded as *multicast* as it has been made available via MAC signalling to more than one SS. Uplink direction connections are always regarded as unicast. After the successful exit of an SS from the WiMAX network, no data traffic is possible as there are no transport connections between the BS and the SS.

Frame Formats

There are two different frame formats for uplink as well as downlink separated by TTG/RTG, as discussed earlier and shown in Fig. 12.20. These gaps are adaptive depending on the channel delays.

The IEEE 802.16 protocol data unit (PDU) format that is transmitted over the airlink consists of a generic MAC header, various optional headers like checksum (CRC), and a data payload. The generic MAC header carries the connection identifier (CID) of the connection with which the PDU is associated. There is another identifier known as service flow identifier (SFID). It is observed that there is no source or destination address present in the raw IEEE 802.16 MAC header.

Convergence Sublayer

For the different formats of fixed and mobile WiMAX, the MAC design includes a CS. It is used to interface with various higher-layer protocols. The CS is the component of the MAC that is responsible for mapping between the MAC service and the internal connection-oriented service of the MAC common part sublayer (CPS) through classification and encapsulation.

The classification process assigns transmit direction service data units originating from a higher-layer application, for example, an IP stack at the BS or SS, to a specific outbound transport connection.

> *Note:* The CS maintains an ordered classifier table. Each entry in the classifier table includes a classifier and a target CID.

A classifier, in turn, consists of a conjunction of one or more subclassifiers, where each subclassifier specifies a packet field [e.g., the destination MAC address in an Ethernet frame or the type of service (TOS) field of an IP datagram contained in an Ethernet frame] together with a particular value or range of values for the field.

To perform classification on an outbound service data unit, the CS proceeds from the first entry of the classifier table to the last entry and evaluates the fields of the service data unit for a match with the table entry's classifier. When a match is found, the CS associates the service data unit with the target CID (for eventual transmission), and the remainder of the IEEE 802.16 MAC and physical layer processing can take place.

12.4.4 WiMAX Security Aspects

Security issues are equally important for WiMAX. The various issues along with the solutions are discussed in this section.

Physical Layer Security Threats

Two basic types of attacks can affect the physical layer of WiMAX. One is *jamming* and the other is *packet scrambling*.

Jamming   This is the commonly known type and is sometimes the result of interference rather than an attack. Jamming consists of a stronger signal than the WiMAX network signal overwhelming network data feeds either in intermittent bursts or with sustained carrier waves. *Constant jamming*, whether malicious or otherwise, can usually be found quickly using a spectrum analyser and directional antennas to triangulate the signal. It is more difficult to find the location of *intermittent jamming* or interference. However, intermittent jamming is less intrusive to the network, resulting in some packet retransmission and slowdowns but less often in blanket outages.

Packet scrambling   This attack occurs when control packets in the respective downlink and uplink subframes are sniffed, that is, identified about a browser or which links, information and terms they use, and then scrambled and returned to the network. This attack is much harder to mount than a jamming attack. Since most WiMAX networks today use TDD, wherein signals are sliced via timeslots, an attacker can parse this timing sequence and capture control data, the preamble and map, scramble them, and send them back with correct timing to interrupt the legitimate signal, resulting in slowdowns and effectively lowered bandwidth. Intercepted and scrambled packets are possible with FDD as well, which transmits both the uplink and downlink s imultaneously.

The WiMAX security model focuses primarily on *theft-of-service* and *theft-of-data* threat vectors.

Upper Layer Security Threats

WiMAX implements protections in the MAC layer, with a security sublayer that implements authentication and provides data encryption. Still, WiMAX offers a significant number of theoretical vulnerabilities. The various attacks due to these are discussed here.

Man-in-the-middle attack   WiMAX implements a unidirectional authentication scheme using X.509 from subscriber to BS, but there is no provision for BS to subscriber authentication in return. This opens a potential vulnerability for rogue BSs to attempt the impersonation of legitimate devices. Attackers can simply intercept subscriber initiation requests and spoof responses, authorizing them to use the rogue AP. Using strong mutual authentication via IEEE 802.1x and a RADIUS server would mitigate this risk significantly.

Denial of service attack   In addition to MAC-layer vulnerabilities, WiMAX could suffer from potential physical layer vulnerabilities as well. WiMAX implements cryptographic protections to prevent most of the spoofing attacks. However, it is vulnerable to replay-based attacks to flood a network with rogue management frames, effectively creating a denial of service. Many WiMAX adopters are also concerned about jamming or scrambling attacks, where an attacker attempts to manipulate RF signals in order to interfere with connections. Even electromagnetic interference is being watched closely to determine whether it has an adverse impact on reliability and availability.

Base station attack   WiMAX protocols and security models are well implemented, but attackers still attempt to exploit them, as some products have security flaws. This vulnerability allows an attacker to create a malicious request, which can allow administrative access to the BS. Although these vulnerabilities are typically found and patched rapidly, the challenge with worldwide WiMAX adoption is that not all consumers will monitor their equipment for intrusion, and even when the vulnerabilities are published, many do not patch them.

Application layer attack   While WiMAX is intended to run securely, making use of strong encryption, it is possible to deploy a solution without enabling either data encryption standard (DES) or AES. When data is transmitted in the clear, it is vulnerable to sniffing as well as attacks against applications, just as Wi-fi is. The key to prevent these is for administrators to take advantage of the built-in encryption capabilities of WiMAX.

SOLUTIONS FOR SECURITY

Authentication If a BS is not set up with adequate authentication measures, an attacker can capture control packets and pose as a legitimate subscriber. Therefore, it is important to have authentication of wireless transmission. The X.509 certificate from subscriber to BS makes it very hard for an intruder to impersonate a subscriber. This certificate is embedded in WiMAX subscriber units and incorporates a public key authentication encryption. This effectively means that a WiMAX BS can detect legitimate subscriber stations quickly and easily. Unfortunately, this is a one-way protocol. The MAC control headers are never encrypted in WiMAX; however, with EAP, carriers can choose to authenticate them. This approach, called *hashed message authentication code* (HMAC), uses a form of encrypted private key.

Encryption Older wireless iterations used the DES, which relied on a 56-bit key for encryption. This is largely considered obsolete. WiMAX 802.16e certainly supports DES (3DES) but it also adds support for the AES, which can work with 128-bit, 192-bit, or 256-bit encryption keys.

12.4.5 Quality of Service and Scheduling Aspects in WiMAX

Quality of services are generally requested by the user and the network protocols will try to satisfy them. The issues are generally associated with the transport layer.

Quality of Service

In WiMAX, a service flow is a MAC transport connection provided for transmission of uplink and downlink traffic, which forms the base of the QoS architecture.

Resource (bandwidth) allocation in the OFDMA-based physical layer is an important issue in WiMAX related to mobile computing. Apart from this, the primary purpose of the QoS feature is to define transmission ordering and scheduling on the air interface. These features often need to work in conjunction with mechanisms beyond the air interface in order to provide end-to-end QoS.

A service flow is a unidirectional flow of packets that provides a particular QoS. The various service flows admitted in a WiMAX network are usually grouped into service flow classes, each identified by a unique set of QoS requirements. This concept of service flow classes allows higher-layer entities at the SS and BS to request QoS parameters in globally consistent ways. Each data service or service flow class is associated with a unique set of QoS parameters, such as latency, jitter throughput, and packet error rate, that quantifies the behavioural aspects of the system.

The WiMAX networks is a connection-oriented MAC in that it assigns traffic to a service flow and maps it to the MAC connection using a CID, as discussed previously. In this way, even connectionless protocols, such as IP and UDP, are transformed into connection-oriented service flows. The connection can represent an individual application or a group of applications sending with the same CID. The SS and BS provide the QoS according to the QoS parameter set defined for that service flow.

These parameters are managed through a series of MAC management messages:

- The dynamic service addition (DSA) messages create a new service flow.
- The dynamic service change (DSC) messages change an existing service flow.
- The dynamic service deletion (DSD) messages delete an existing service flow.

The following five services are supported in the mobile WiMAX:
(a) Unsolicited grant service (UGS)
(b) Real-time polling service (rtPS)
(c) Extended real-time polling service (ErtPS)
(d) Non-real-time polling service (nrtPS)
(e) Best effort service (BE)

Each of these scheduling services has a mandatory set of QoS parameters that must be included in the service flow definition when the scheduling service is enabled for a service flow. These are summarized in Table 12.7.

Scheduling

Scheduling is the main component of the MAC layer that assures QoS to various service classes. The MAC scheduling services are adopted to determine which packet will be served first in a specific queue to guarantee its QoS requirement. In fact, the scheduler works as a distributor in order to allocate the available resources among the SSs. The scheduling

---

Note: The scheduling decision for the downlink traffic is relatively simple as only the BS transmits during the downlink subframe and the queue information is located in the BS, whereas an uplink scheduler at the BS must synchronize its decision with all the SSs.

Table 12.7 WiMAX QoS service classes

QoS Category	Applications	QoS Specifications
UGS	VoIP	-Maximum Sustained Rate -Maximum Latency Tolerance -Jitter Tolerance
rtPS	Streaming Audio or Video	-Minimum Reserved Rate -Maximum Sustained Rate -Maximum Latency Tolerance -Traffic Priority
ErtPS	Voice with Activity Detection (VoIP)	-Minimum Reserved Rate -Maximum Sustained Rate -Maximum Latency Tolerance -Jitter Tolerance -Traffic Priority
nrtPS	File Transfer Protocol (FTP)	-Minimum Reserved Rate -Maximum Sustained Rate -Traffic Priority
BE	Data Transfer, Browsing, Web, etc.	-Maximum Sustained Rate -Traffic Priority

architecture should ensure good use of bandwidth and fairness among users along with satisfying the requirements of QoS.

It is important to mention that scheduling algorithms can be implemented in the BS as well as in the SSs. Those implemented at the BS have to deal with both uplink and downlink traffics. Therefore, there are three different schedulers: two at the BS to schedule the packet transmission in the downlink and uplink subframes and one at the SS for the uplink to apportion the assigned bandwidth to its connections. In order to indicate the allocation of transmission intervals in both uplink and downlink, in each frame, the signalling messages UL-MAP and DL-MAP (Fig. 12.20) are broadcast at the beginning of the downlink subframe.

12.5 WIRELESS SENSOR NETWORKS

A wireless ad hoc sensor network consists of numerous sensor nodes spread across a geographical area. Each sensor node has wireless communication capability and some level of intelligence for signal processing and networking of the data. Some examples of wireless ad hoc sensor networks are as follow:

(a) Military sensor networks to detect and gain as much information as possible about enemy movements, explosions, and other phenomena of interest

(b) Sensor networks to detect and characterize chemical, biological, radiological, nuclear, and explosive (CBRNE) attacks and material

(c) Sensor networks to detect and monitor environmental changes in plains, forests, oceans, and so on

(d) Wireless traffic sensor networks to monitor vehicle traffic on highways or in congested parts of a city making it an intelligent transport system

(e) Wireless surveillance sensor networks for providing security in shopping malls, parking garages, and other facilities

(f) Wireless parking lot sensor networks to determine which spots are occupied and which are free

The examples suggest that wireless ad hoc sensor networks offer certain capabilities and enhancements in operational efficiency in civilian applications as well as assist in the national effort to increase alertness to potential terrorist threats.

Wireless ad hoc sensor networks can be classified based on the following two aspects:
(a) Whether the nodes are individually addressable
(b) Whether the data in the network is aggregated

The sensor nodes in a parking lot network should be individually addressable, so that it is possible to determine the locations of all the free spaces. This application shows that it may be necessary to broadcast a message to all the nodes in the network. If the temperature in a corner of a room needs to be determined, then addressability may not be so important. Any node in the given region can respond. The ability of a sensor network to aggregate the data collected can greatly reduce the number of messages that need to be transmitted across the network.

The basic goals of a wireless ad hoc sensor network generally depend upon the application, but the following tasks are common to many networks.

Determining the value of some parameter at a given location In an environmental network, the temperature, atmospheric pressure, amount of sunlight, and relative humidity at a number of locations can be determined. A given sensor node may be connected to different types of sensors, each with a different sampling rate and range of allowed values.

Detecting the occurrence of events of interest and estimating the parameters of the detected events In a traffic sensor network, it is possible to detect a vehicle moving through an intersection and estimate the speed and direction of the vehicle.

Classifying a detected object It is possible to identify the type of vehicle detected in a traffic sensor network such as a car, a minivan, a light truck, or a bus.

Tracking an object In a military sensor network, an enemy tank is tracked as it moves through the geographic area covered by the network.

In these four tasks, an important requirement of the sensor network is that the required data be disseminated to the proper end-users. In some cases, there are fairly strict time requirements on this communication. For example, the detection of an intruder in a surveillance network should be immediately communicated to the police so that action can be taken.

The requirements of a wireless ad hoc sensor network are as follows.

Large number of (mostly stationary) sensors Aside from the deployment of sensors on the ocean surface or the use of mobile, unmanned, and robotic sensors in military operations, most nodes in a smart sensor network are stationary. Networks of 10,000 or even 1,00,000 nodes are envisioned and so scalability is a major issue.

Low energy use As in many applications the sensor nodes will be placed in a remote area, service of a node may not be possible. In this case, the lifetime of a node may be determined by the battery life, thereby requiring minimization of energy expenditure.

A sensor node is a small transceiver with a low-capability processor, an analog-to-digital converter, a special operating system like TinyOS, and a battery.

Network self-organization Given the large number of nodes and their potential placement in hostile locations, it is essential that the network be able to self-organize; manual configuration is not feasible. Moreover, nodes may fail (either from lack of energy or from physical destruction) and new nodes may join the network. Therefore, the network must be able to periodically reconfigure itself so that it can

Sensor networks can be connected to the outside world or the Internet via any broadband access network like Wi-fi.

continue to function. Individual nodes may become disconnected from the rest of the network, but a high degree of connectivity must be maintained.

Collaborative signal processing Yet another factor that distinguishes the wireless ad hoc sensor networks from MANETs is that the end goal is detection or estimation of some events of interest, and not just communications. To improve the detection or estimation performance, it is often quite useful to fuse data from multiple sensors. This data fusion requires the transmission of data and control messages, and so it may put constraints on the network architecture.

Querying ability A user may want to query an individual node or a group of nodes for information collected in the region. Depending on the amount of data fusion performed, it may not be feasible to transmit a large amount of the data across the network. Instead, various local sink nodes will collect data from a given area and create summary messages. A query may be directed to the sink node nearest to the desired location.

Figure 12.21 shows the various possible topologies of a sensor network and Fig. 12.22 shows the protocol architecture of the sensor networks. Figure 12.23 shows the IEEE 802.15.4 standard frame format.

With the availability of low-cost short-range radios along with advances in wireless networking, it is expected that wireless ad hoc sensor networks will become commonly deployed. In these networks, each node may be equipped with a variety of sensors, such as acoustic, seismic,

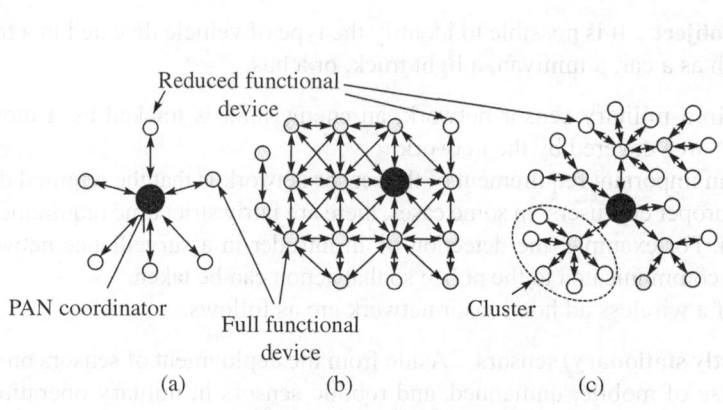

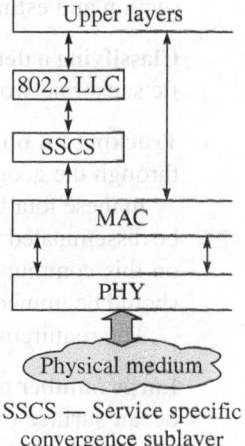

Fig. 12.21 Topologies supported by sensor networks (a) Star (b) Peer to peer (c) Cluster tree

SSCS — Service specific convergence sublayer

Fig. 12.22 WSN protocol stack

Synchronization header SHR		PHY header	PHY payload
Preamble sequence	Start of frame delimiter	Frame length	PSDU (physical service data unit)
4 octets	1 octet	1 octet	<= 127 bytes

Fig. 12.23 IEEE 802.15.4 standard frame format

infrared, and still or motion video camera. These nodes may be organized in clusters such that a locally occurring event can be detected by most of, if not all, the nodes in a cluster. Each node may have sufficient processing power to make a decision, and it will be able to broadcast this decision to the other nodes in the cluster. One node may act as the cluster master or head, and it may also contain a longer-range radio using a protocol such as IEEE 802.11 or Bluetooth.

12.6 ULTRA-WIDEBAND

Ultra-wideband (UWB) is for indoor and short-range outdoor communication. It is a rapidly emerging wireless technology that promises data rates well beyond those possible with currently deployed technologies such as 802.11a, b, g, and WiMAX. UWB is gaining considerable acceptance and is being proposed for use in a number of areas. Already Bluetooth, wireless universal serial bus (USB), and others are developing solutions, and in these areas alone, it should have numerous applications. UWB can be used in both commercial and military applications.

The method of transmission employed by UWB is totally different to that used by most other wireless technologies in use today. Rather than using a specified frequency with a carrier, the technique that is used by traditional transmissions, UWB uses what can be termed *time domain electromagnetics*.

The fact that UWB transmissions have such a wide bandwidth means that they will cross the boundaries of many of the currently licensed carrier-based transmissions. As such, one of the fears is that UWB transmission may cause interference. However, the very high bandwidth used also allows the power spectral density to be very low, and the power limits on UWB are being strictly limited by the regulatory bodies. In many instances, they are lower than the spurious emissions from electronic apparatus that has been certified. In view of this, it is anticipated that they will cause no noticeable interference to other carrier-based licensed users. Nevertheless, the regulatory bodies are moving forward cautiously so that users who already have spectrum allocations are not affected.

Despite the single name used for the UWB transmissions, there are two very different technologies being developed:

(a) One is based around a carrier free technology where a series of impulses is transmitted. In view of the very short duration of the pulses, the spectrum of the signal occupies a very wide bandwidth.

(b) The alternative technology uses a wide band or multiband OFDM (MBOFDM) signal, which is effectively a 500 MHz wide OFDM signal that is hopped in frequency to enable it to occupy a sufficiently high bandwidth.

Both these systems have their advantages and disadvantages, each one having its supporters and applications for which it is most suited. The impulse-based technology, also called direct sequence ultra-wideband (DS-UWB), is being used for many high data rate data transmissions, such as short-range video transmissions. The MBOFDM, on the other hand, is being adopted for wireless USB, where it performs well.

> UWB uses pulses that spread out over a wide bandwidth of the order of gigahertz rather than transmissions that are confined within a given channel; hence, it supports data rates in the order of gigabits per second.

UWB Air Interface

The UWB technology is a revolutionary wireless technology that enables data to be transmitted at speed well in excess of 100 Mbps. It is often referred to as an *impulse*, *baseband*, or *zero carrier* technology. The lowest frequencies for UWB have been set above 3 GHz to ensure they do not cut across bands currently used for global positioning system (GPS), cellular, and many other services.

UWB pulses It operates by sending low-power Gaussian-shaped pulses that are coherently received at the receiver. In view of the fact that the system operates using pulses, the transmissions are spread out over a wide bandwidth, typically many hundreds of megahertz or even several gigahertz. This means that it will overlay the bands and transmissions used by more traditional channel-based transmissions. Each of the UWB pulses has an extremely short duration. This is typically between 10 ps and 1000 ps, and as a result, it is shorter than the duration of a single bit of the data to be transmitted. The short pulse duration means that multipath effects can usually be ignored, giving rise to a large degree of resilience in the UWB transmissions when the signal path is within buildings.

Power level In view of the wide bandwidth over which the UWB transmissions are spread, the actual energy density is exceedingly low. In fact, many of the transmissions themselves are less that the unintentional or spurious radiation levels from a typical personal computer. Typically, a UWB transmitter might transmit less than 75 nW/MHz. When integrated over the total bandwidth of the transmission, it means that transmissions may only be about 0.25 mW. This is very small when compared to 802.11 transmissions, which may be between 25 mW and 100 mW, or Bluetooth transmissions, which may be anywhere between 1 mW and 1 W.

This very low spectral density means that the UWB transmissions do not cause harmful interference to other radio transmissions using traditional carrier-based techniques and operating in the existing bands. Even in the bands that are likely to be more sensitive to interference, such as the GPS, it is possible to reduce the UWB transmission power density levels even further to ensure that there is no noticeable interference.

Modulation There are a number of ways in which UWB transmissions can be modulated to enable data to be carried. The strict power density limits placed on any UWB transmissions by the federal communications commission (FCC) means that the form of modulation applied must be efficient. It must provide the optimum error performance for a given level of energy per bit. The choice of modulation also affects the UWB transmission spectrum, and this must be taken into account to ensure that the spectral density limits are not exceeded.

Two of the most popular forms of modulation used for UWB are pulse position modulation (PPM) and BPSK. These provide the best performance in terms of modulation efficiency and spectral performance. The PPM encodes the information by modifying the time interval and hence the position of the pulses. The BPSK reverses the phase of the pulse to signify the data to be transmitted. This is a 180° reversal. As the pulses consist of an initial upward or downward voltage, this is easy to reverse. Looking at a pulse on an oscilloscope, it would appear that a pulse is either the right way up or upside down.

Ultra-wideband is being recognized as a technology with a huge capability, and as such, it is being adopted in many new areas. As mentioned already, many silicon manufacturers have already developed solutions, which are being demonstrated, and more are being developed. With the growing level of wireless communications, UWB offers significant advantages in many areas. One of the main attractions for WAN and LAN applications is the very high data rates that can be supported. With computer technology requiring ever-increasing amounts of data to be transported, it is likely that standards such as 802.11 and others may not be able to support the data speeds required in some applications. The UWB may well become a major technology of the future in overcoming this problem.

DS-UWB receiver is a rake receiver with multiple fingers.

SOLVED EXAMPLES

Example 12.1 Following are the specifications of the IEEE 802.11a standard.

Total number of subchannels = 64
Subcarriers for useful information = 48
Reserved subcarriers = 12
Pilot subcarriers = 4
Total bandwidth allotted = 20 MHz
Symbol mapping scheme = 16QAM
Guard interval = 0.6 µs

Each subchannel supports 200 kilosymbols/s. The channel coin is rate ½ convolution coding. Find the subchannel bandwidth, user data rate, and time efficiency of the system.

Solution Bandwidth of the subchannel = $20 \times 10^6/$ $64 = 312.5$ kHz

User data rate = Information subchannels × Per-channel symbol rate = $48 \times 200 = 9.6$ Msps (Mega symbols per second)

For 16QAM 4 bits per symbol and rate ½ coding
User data rate = $½ \times 9.6 \times 4 = 19.2$ Mbps
Symbol duration is $1/200 \times 10^3 = 5000$ ns
Time utilization efficiency = $5000/5600 = 0.8928$

Example 12.2 On an SCO link of Bluetooth, HV1 voice packets of the size 240 bits are transferred. Find the packet transmission efficiency. How often should the HV1 packets be sent in each direction?

Solution The following data are as per the specifications of Bluetooth:

Hopping rate = 1600 hops/s
Bit rate support on each SCO digital voice link = 64 kbps
Total bits in a slot = 240
Packet transmission efficiency = $240/240 = 1$ (100%)
$64 \times 10^3/240 = 266$ times the HV1 packets should be sent.

SUMMARY

- Ad hoc networks are temporarily established.
- In all ad hoc networks, some hardware and software support is required because most of the networks are data networks.
- Bluetooth supports voice and data services whereas WLAN and WMAN are especially data networks.
- Bluetooth can be used to replace cables in homes.
- Bluetooth is based on piconet and scatternet structures in which communication takes place between the master and slave devices. These devices are identified on the basis of addresses.
- Bluetooth operates on 2.4 GHz and has its unique protocol stack.
- IEEE 802.11a, b, g, and n standards are significant Wi-fi standards for establishment of WLAN. All standards

have different physical layers (modulation schemes).
- IEEE 802.16 is a WMAN standard using which the range of communication can be extended compared to Wi-fi.
- Interworking of Wi-fi and WiMAX is possible.
- For all IEEE protocols for networking (802 series), physical and data link layers are standardized.
- Sensor networks require many sensor nodes to be deployed in large numbers.
- UWB can provide very high bit rate of communication for indoor applications.
- IEEE 802.20 and 21 are upcoming protocols that will have better features compared to Wi-fi and WiMAX.

EXERCISES

Multiple Choice Questions

12.1 What is the per-channel bandwidth in Bluetooth?
    (a) 10 MHz   (c) 79 MHz
    (b) 1 MHz   (d) 2.4 GHz

12.2 The total number of devices in a piconet
    (a) is not more than two

    (b) is not more than seven
    (c) is not more than eight
    (d) has no restriction

12.3 The RF carrier modulation scheme in Bluetooth is
    (a) FFH-SSM   (c) GFSK
    (b) GMSK   (d) FSK

12.4 The largest distance covered by Bluetooth devices is
(a) 10 feet (b) 30 feet (c) 60 feet (d) 300 feet

12.5 The hopping rate for Bluetooth is
(a) 1000 hops/s   (c) 1400 hops/s
(b) 1200 hops/s   (d) 1600 hops/s

12.6 Which of the following layers in the Bluetooth protocol stack is used to perform tasks equivalent to those of the transport layer?
(a) L2CAP   (c) SDP
(b) TCS   (d) Link manager

12.7 Which of the following devices is synchronized but may not have AM_ADDR in Bluetooth?
(a) Sniff (b) Hold (c) Park (d) Active

12.8 The size of a Bluetooth packet header is
(a) 68 bits   (c) 54 bits
(b) 72 bits   (d) 2745 bits

12.9 In Bluetooth, masters can start communication in the
(a) odd number of time slot
(b) any number of time slot
(c) even number of time slot
(d) first number of time slot

12.10 In Bluetooth devices, polling-based access is used in
(a) ACL links   (c) voice links
(b) SCO links   (d) active links

12.11 In which of the following Wi-fi standards was the MIMO concept introduced?
(a) 802.11a   (c) 802.11e
(b) 802.11n   (d) 802.11b

12.12 In which of the following networks does the physical layer combine the spread spectrum and OFDM modulation schemes?
(a) Wi-fi   (c) UWB
(b) Zigbee   (d) WiMAX

12.13 Which of the following modulation schemes is introduced in IEEE 802.16e?
(a) Scaleable OFDM   (c) CCK-DSSS
(b) MC-CDMA   (d) COFDM

12.14 Which of the following is considered as zero-carrier technology?
(a) Zigbee   (c) UWB
(b) Wi-fi   (d) Bluetooth

12.15 In IEEE 802.16d, the actual number of carriers used to carry data is
(a) 192   (c) 256
(b) 200   (d) none of these

12.16 In which of the following IEEE standards is QoS the main focus?
(a) 802.11n   (c) 802.11e
(b) 802.11b   (d) 802.11a

12.17 On which of the following layers of 802.11 are PCF and DCF implemented?
(a) Application   (c) Physical
(b) Link   (d) MAC

12.18 Which of the following modes is significant in WiMAX systems while communicating between the BS and the mobile receiver?
(a) Half duplex   (c) TDD
(b) TDM   (d) FDD

12.19 Which of the following statements is true?
(a) Bluetooth devices can communicate for only 10 m.
(b) Bluetooth is based on CDMA concept.
(c) The per-channel bandwidth in Bluetooth is 1 MHz.
(d) The master device in Bluetooth cannot work as slave at all.

12.20 Which of the following statements is/are true?
(a) The Wi-fi and WiMAX have independent protocols and interworking cannot be established.
(b) The IEEE 802.16d standard supports vehicular mobility.
(c) The IEEE 802.11n standard is designed only for power-saving mode.
(d) All of these are true.

12.21 Zigbee is the protocol for
(a) wireless sensor networks
(b) Wi-fi
(c) UWB
(d) data networks

12.22 The UWB physical layer is based on
(a) CDMA   (c) Multiband OFDM
(b) OFDM   (d) none of these

Review Questions

12.1 How can you say that Bluetooth is an ad hoc network?

12.2 Describe the concept of piconet and scatternet.

12.3 Why is it necessary to have a protocol stack along with Bluetooth chip-based hardware?

12.4 Differentiate the ACL and SCO modes in Bluetooth. Draw the timing diagrams for both. Justify the suitability of the type of data transferred using these modes.

12.5 Why is WLAN considered along with ad hoc networks? Is it a pure ad hoc network?

12.6 What are the major differences between Wi-fi and WiMAX systems?

12.7 What do you mean by air interface?

12.8 Give the data of all the Wi-fi standards and discuss their key features.

12.9 What are the upgradations still required in WiMAX? Find from the drafts of IEEE.

12.10 Study the Zigbee protocol of wireless sensor network. How does it differ from Bluetooth?

12.11 What are the different types of applications of sensor networks? Do you think that all the applications will be possible with a single standard?

12.12 How does UWB differ from other types of networks?

12.13 Fill in the blanks:
 (a) The lowest physical layer topology is identified as _____ in Bluetooth.
 (b) The final modulation scheme in Bluetooth is _____.
 (c) The number of hops per second used in Bluetooth devices is _____.
 (d) The maximum number of devices that can form a piconet is _____.
 (e) Security aspects are mainly covered in _____ Wi-fi standard.
 (f) _____ is the original security solution for 802.11.
 (g) The EDCA uses a mechanism called _____.

(h) A new modulation scheme called _____ is used along with DSSS in IEEE 802.11b.

(i) The _____ technique is used in mobile WiMAX in the physical layer.

12.14 Strike out the incorrect word:
 (a) Bluetooth is a wireless radio **standard/specification** for data and voice communication.
 (b) Bluetooth is a **personal/local** area network.
 (c) The link set-up and control in Bluetooth is managed by **LM/L2CA** protocol.
 (d) The physical layer security in Bluetooth is achieved due to **SSM/PIN**.
 (e) The IEEE 802.11a standard is based on **CDMA/OFDM** physical layer.
 (f) The IEEE 802.16e standard is a **fixed/mobile** wireless broadband access standard.
 (g) The LMDS exhibits **LOS/NLOS** links with **GHz/MHz** range frequencies.
 (h) The WiMAX is a **LAN/MAN** standard.

12.15 Define the following devices in terms of Bluetooth:
 (a) Master unit   (c) Hold mode
 (b) Park unit

12.16 Define the following terms with respect to WiMAX:
 (a) MAC PDU   (c) TTG
 (b) CID

Linear Systems Theory

A linear system is given by

$$y(t) = \int_{-\infty}^{\infty} x(\tau)h(t,\tau)d\tau \tag{A.1}$$

where $x(t)$ is the input and $h(t, \tau)$ is the system weighting function.

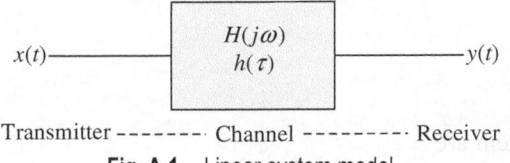

Transmitter ------- Channel -------- Receiver

Fig. A.1   Linear system model

A linear system model (Fig. A.1) is one with the input-output relationship represented by a transfer function as shown in the box. It is tried to correlate this model with the general diagram of a communication system in which the transmit-receive relationship is decided by the channel transfer function (or Impulse response). The whole theory with mathematical equations is described with the help of this model.

If the system is time invariant, then Eq. (A.1) becomes

$$y(t) = \int_{-\infty}^{\infty} h(\tau)x(t-\tau)d\tau \tag{A.2}$$

This type of integral is called *convolution integral*.

Differential equations involving random processes are called *stochastic differential*. A linear stochastic differential equation as a model of an RP (random process) with initial conditions has the general form

$$\overline{x}(t) = F(t)x(t) + G(t)w(t) + C(t)u(t) \tag{A.3}$$

$$y(t) = H(t)x(t) + v(t) + D(t)u(t) \tag{A.4}$$

where

$\overline{x}(t) = $ the next state estimation

$x(t) = n \times 1$, state vector (input)

$y(t) = l \times 1$, measurement vector (received output of the channel)

$u(t) = r \times 1$, deterministic input vector (known control input—not present in our

system

$F(t) = n \times n$, time-varying dynamic coefficient matrix

$C(t) = n \times r$, time-varying control input coupling matrix (not present in our system)

$H(t) = l \times n$, time-varying measurement sensitivity matrix (i.e. transfer function—here channel)

$D(t) = l \times r$, time-varying output coupling matrix (for further control input—not in our system)

$G(t) = n \times r$, time-varying process noise coupling matrix (related to input—omitted in our system)

$w(t) = r \times 1$, zero mean uncorrelated 'plant nose' process (to input—omitted in our system)

$v(t) = l \times 1$, zero mean uncorrelated 'measurement noise' process (here channel noise or interference)

The expected values (mean) are as follows:

$$E[w(t)] = 0, \text{ zero mean Gaussian noise at input}$$

$$E[v(t)] = 0, \text{ zero mean Gaussian noise at output}$$

$$E[w(t_1)w^T(t_2)] = Q(t_1)\delta(t_2 - t_1)$$

$$E[v(t_1)v^T(t_2)] = R(t_1)\delta(t_2 - t_1)$$

$$E[w(t_1)v^T(t_2)] = M(t_1)\delta(t_2 - t_1)$$

The simplified suitable equations for the system are

$$\bar{x}(t) = F(t)x(t) \tag{A.5}$$

$$y(t) = H(t)x(t) + v(t) \tag{A.6}$$

The symbols Q, R, and M represent $r \times r$, $l \times l$, and $r \times l$ matrices, respectively, and δ represents the Dirac delta 'function' (a measure).

The values over time of variable $x(t)$ in the differential equation model define vector-valued Markov processes. This model is a fairly accurate and useful representation for many real-world processes, including stationary Gaussian and non-stationary Gaussian processes, depending on the statistical properties of the random variables (say, input/output) and the temporal properties of the deterministic variables (say, training sequence).

Various filters can be derived on the basis of different equations mentioned in Equations (A.5) and (A.6).

Discrete model of a random sequence or frame-based data without initial conditions or control input can be given in the form

$$x_k = \Phi_{k-1}x_{k-1} \tag{A.7}$$

$$y_k = H_k x_k + v_k \tag{A.8}$$

Here

$x_k = n \times 1$, state vector (input)

$y_k = l \times 1$, measurement vector (received output of the channel)

$\Phi_{k-1} = n \times n$, time-varying dynamic coefficient matrix

$H_k = l \times n$, time-varying measurement sensitivity matrix (i.e. transfer—here channel)

$E[v_k] = 0$, zero mean Gaussian noise at output

$$E[w_{k1}w^T_{k2}] = Q_{k1} \, \Delta(t_2 - t_1)$$

$$E[v_{k1}v^T_{k2}] = R_{k1} \, \Delta(t_2 - t_1)$$

$$E[w_{k1}v^T_{k2}] = M_{k1} \, \Delta(t_2 - t_1)$$

Algebra for the Linear System

Many times, it is convenient to use vector and matrix notation for the representation of the discrete signals and operations performed on them. This representation simplifies many of the mathematical expressions and their solutions. Such representation is also useful while studying mathematics of MIMO systems.

A vector is an array of real-valued or complex-valued numbers or functions. Vectors can be denoted by lower-case bold letters and, in all cases, these vectors will be assumed to be column vectors. For example,

$$\mathbf{x} = \begin{bmatrix} x_1 \\ x_2 \\ \vdots \\ x_N \end{bmatrix} \tag{B.1}$$

is a column vector containing N scalars. If the elements are real, then it is a real vector, otherwise a complex vector of N dimensions. The transpose of a vector, $\mathbf{x}^T$ is a row vector and Hermitian transpose $\mathbf{x}^H$ is the complex conjugate of the transpose of $\mathbf{x}$. The vector elements can be represented with time index values in case of finite length sequence $x(n)$ as

$$\mathbf{x} = \begin{bmatrix} x(0) \\ x(1) \\ \vdots \\ x(N-1) \end{bmatrix} \quad \text{or} \quad \mathbf{x}(n) = \begin{bmatrix} x(n) \\ x(n-1) \\ \vdots \\ x(x-N+1) \end{bmatrix} \tag{B.2}$$

The convolution between $h(n)$ and $x(n)$, where $h(n)$ can be represented in the same vector form as $x(n)$, results in $y(n)$. Then $y(n)$ may be written as the inner product

$$y(n) = \mathbf{h}^T \mathbf{x}(n) \tag{B.3}$$

A set of linear equations can be better represented in matrix form. An $n \times m$ matrix is an array of numbers (real or complex) or functions having n rows and m columns. For example,

$$\mathbf{A} = \{a_{ij}\} = \begin{bmatrix} a_{11} & a_{12} & a_{13} & \cdots & a_{1m} \\ a_{21} & a_{22} & a_{23} & \cdots & a_{2m} \\ a_{31} & a_{32} & a_{33} & \cdots & a_{3m} \\ \vdots & \vdots & \vdots & \vdots & \vdots \\ a_{n1} & a_{n2} & a_{n3} & \cdots & a_{nm} \end{bmatrix} \tag{B.4}$$

The rank of $\mathbf{A}$ is defined to be the number of linearly independent columns in $\mathbf{A}$. Special matrix forms are square matrix, diagonal matrix, and identity matrix (where the diagonal elements are 1's). Other important terms, such as Toeplitz matrix, eigenvalues, and eigenvectors, can be studied from appropriate references.

Probability Theory

The following concepts are important to study the randomly fading channel behavioural mathematics and signal detection process mathematics.

Discrete Probability Distributions

Definition C.1

Suppose we have an experiment whose outcome depends on the chance. We represent the outcome of the experiment by a letter, such as X, called a *random variable*. The sample space of the experiment is the set of all possible outcomes. If the sample space is either finite or countably infinite, the random variable is said to be *discrete*. We generally denote a sample space by the capital Greek letter Ω. There are two additional definitions. These are subsidiary to the definition of sample space and serve to make precise some of the common terminology used in conjunction with sample spaces. First, we define the elements of a sample space to be outcomes. Second, each subset of a sample space is defined to be an event. Normally, we shall denote outcomes by lower-case letters and events by capital letters.

If a sample space has an infinite number of points, then the way a distribution function is defined depends upon whether or not the sample space is countable. A sample space is countably infinite if the elements can be counted, that is, can be put in one-to-one correspondence with the positive integers, and uncountably infinite otherwise.

Definition C.2

Let X be a random variable, which denotes the value of the outcome of a certain experiment, assuming that this experiment has only finitely many possible outcomes. Let Ω be the sample space of the experiment (i.e. the set of all possible values of X or equivalently, the set of all possible outcomes of the experiment). A distribution function for X is a real-valued function m whose domain is Ω and which satisfies the following conditions:

1. $m(\omega) \geq 0$ for all $\omega \in \Omega$
2. $\sum_{\omega \in \Omega} m(\omega) = 1$

For any subset E of Ω, we define the probability of E to be the number $P(E)$ given by

$$P(E) = \sum_{\omega \in E} m(\omega) \tag{C.1}$$

Note that an immediate consequence of the aforementioned definitions is that for every $\omega \, \varepsilon \, \Omega$,

$$P(\{\omega\}) = m(\omega) \tag{C.2}$$

That is, the probability of the elementary event $\{\omega\}$, consisting of a single outcome ω, is equal to the value $m(\omega)$ assigned to the outcome ω by the distribution function.

It is important to consider ways in which probability distributions are determined in practice. One way is by symmetry. In general, considerations of symmetry often suggest the uniform distribution function. Care must be taken here. We should not always assume that just because we do not know any reason to suggest that one outcome is more likely than another, it is appropriate to assign equal probabilities. Statistical estimates for probabilities are fine if the experiment under consideration can be repeated a number of times under similar circumstances.

Definition C.3

The uniform distribution on a sample space Ω containing n elements is the function m defined by

$$m(\omega) = 1/n \quad \text{for every } \omega \in \Omega \tag{C.3}$$

Definition C.4

If $P(E) = p$, the odds in favour of the event E occurring are $r : s$ (r to s), where $r/s = p/(1-p)$. If r and s are given, then p can be found by using the equation $p = r/(r + s)$.

Set Theory

In many cases, events can be described in terms of other events through the use of the standard constructions of set theory. We will briefly review the definitions of these constructions. The Venn diagrams shown in Fig. C.1 illustrate these constructions.

Let A and B be two sets.

Then the union of A and B is the set

$$A \cup B = \{x \mid x \in A \text{ or } x \in B\} \tag{C.4}$$

The intersection of A and B is the set

$$A \cap B = \{x \mid x \in A \text{ and } x \in B\} \tag{C.5}$$

The difference of A and B is the set

$$A - B = \{x \mid x \in A \text{ and } x \notin B\} \tag{C.6}$$

The set A is a subset of B, written as $A \subset B$, if every element of A is also an element of B.

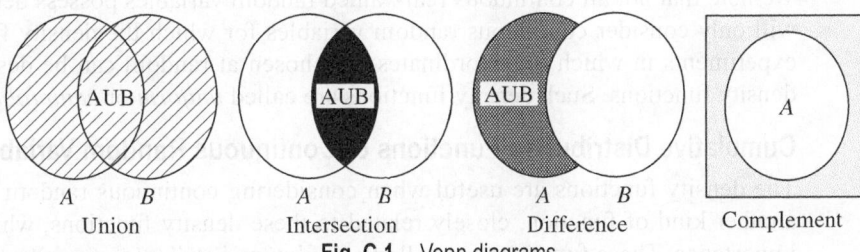

| A B | A B | A B | A |
| Union | Intersection | Difference | Complement |

Fig. C.1 Venn diagrams

Finally, the complement of A is the set

$$A = \{x \mid x \in \Omega \text{ and } x \notin A\} \tag{C.7}$$

For any two events A and B,

$$P(A) = P(A \cap B) + (A \cap \bar{B}) \tag{C.8}$$

Properties

Theorem C.1 The probabilities assigned to events by a distribution function on a sample space Ω satisfy the following properties:

1. $P(E) \geq 0$ for every $E \subset \Omega$
2. $P(\Omega) = 1$
3. If $E \subset F \subset \Omega$, then $P(E) \leq P(E)$
4. If A and B are disjoint subsets of Ω, then $P(A \cup B) = P(A) + P(B)$
5. $P(\bar{A}) = 1 - P(A)$ for every $A \subset \Omega$

Theorem C.2 If $A_1, ..., A_n$ are pairwise disjoint subsets of Ω (i.e. no two of the A_i's have an element in common), then

$$P(A_1 \cup ... \cup A_n) = \sum_{i=1}^{n} P(A_i) \tag{C.9}$$

Theorem C.3 Let $A_1, ..., A_n$ be pairwise disjoint events with $\Omega = A_1 \cup ... \cup A_n$ and E be any event. Then

$$P(E) = \sum_{i=1}^{n} P(E \cap A_i) \tag{C.10}$$

Theorem C.4 If A and B are subsets of Ω, then

$$P(A \cup B) = P(A) + P(B) - P(A \cap B) \tag{C.11}$$

Density Functions of Continuous Random Variables

Definition C.5

Let X be a continuous real-valued random variable. A density function for X is a real-valued function f that satisfies the following condition:

$$P(a \leq X \leq b) = \int_a^b f(x)dx \quad \text{for all } a, b \in R(\text{set of real values}) \tag{C.12}$$

We note that not all continuous real-valued random variables possess density functions. We will only consider continuous random variables for which the density functions exist. The experiments in which the coordinates are chosen at random can be described by constant density functions. Such density functions are called *uniform* or *equiprobable*.

Cumulative Distribution Functions of Continuous Random Variables

The density functions are useful when considering continuous random variables. There is another kind of function, closely related to these density functions, which is also of great importance. These functions are called *cumulative distribution functions*.

Definition C.6

Let X be a continuous real-valued random variable. Then the cumulative distribution function of X is defined by the equation

$$F_X(x) = P(X \leq x) \tag{C.13}$$

If X is a continuous real-valued random variable that possesses a density function, then it also has a cumulative distribution function. Theorem C.5 shows that the two functions are related in a very nice way.

Theorem C.5   Let X be a continuous real-valued random variable with density function $f(x)$. Then the function defined by

$$F(x) = \int_{-\infty}^{x} f(t)\,dt \tag{C.14}$$

is the cumulative distribution function of X. Furthermore, we have

$$\frac{d}{dx}F(x) = f(x) \tag{C.15}$$

In many experiments, the density function of the relevant random variable is easy to write down. However, it is quite often the case that the cumulative distribution function is easier to obtain than the density function (Fig. C.2). (Of course, once we have the cumulative distribution function, the density function can easily be obtained by differentiation, as the aforementioned theorem shows.)

A fundamental question in practice is: How shall we choose the probability density function in describing any given experiment? The answer depends to a great extent on the amount and kind of information available to us about the experiment. In some cases, we can see that the outcomes are equally likely. In some cases, we can see that the experiment resembles another experiment already described by a known density. In some cases, we can run the experiment a large number of times and make a reasonable guess at the density on the basis of the observed distribution of outcomes. In general, the problem of choosing the right density function for a given experiment is a central problem for the experimenter and is not always easy to solve.

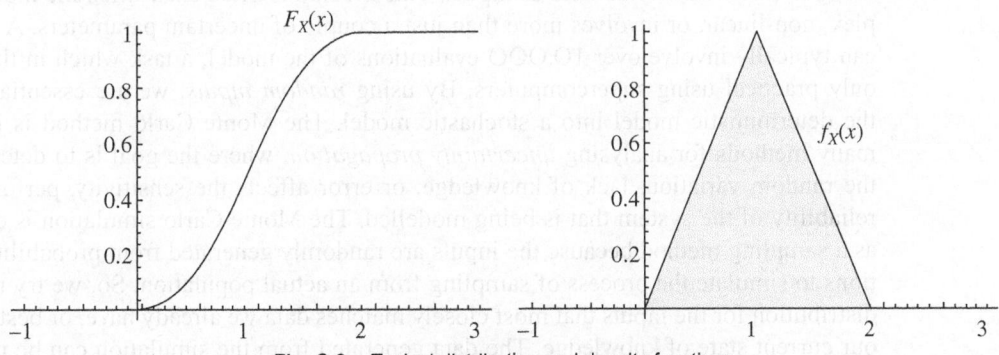

Fig. C.2   Typical distribution and density functions

Definition C.7

Let $X_1, X_2, ..., X_n$ be random variables associated with an experiment. Suppose the sample space (i.e. the set of possible outcomes) of X_i is the set R_i. Then the joint random variable $\overline{X} = (X_1, X_2, ..., X_n)$ is defined to be the random variable whose outcomes consist of ordered n-tuples of outcomes, with the ith coordinate lying in the set R_i. The sample space Ω of X is the Cartesian product of the R_i's

$$\Omega = R_1 \times R_2 \times ... \times R_n \tag{C.16}$$

The joint distribution function of $\overline{X}$ is the function that gives the probability of each of the outcomes of $\overline{X}$

Definition C.8

The random variables $X_1, X_2, ..., X_n$ are mutually independent if

$$P(X_1 = r_1, X_2 = r_2, ..., X_n = r_n)$$
$$= P(X_1 = r_1)P(X_2 = r_2)...P(X_n = r_n) \tag{C.17}$$

for any choice of $r_1, r_2, ..., r_n$. Thus, if $X_1, X_2, ..., X_n$ are mutually independent, then the joint distribution function of the random variable

$$\overline{X} = (X_1, X_2, ..., X_n) \tag{C.18}$$

is just the product of the individual distribution functions. When two random variables are mutually independent, we can say briefly that they are independent.

Continuous Probability Densities Simulation

It is sometimes desirable to estimate quantities whose exact values are difficult or impossible to calculate exactly. In some of these cases, a procedure involving chance, called a *Monte Carlo procedure,* can be used to provide such an estimate. The Monte Carlo method is a technique that involves using random numbers and probability to solve problems. The term *Monte Carlo method* was coined by S. Ulam, von Neumann, and Nicholas Metropolis in reference to games of chance, a popular attraction in Monte Carlo, Monaco (Metropolis, N. and S. Ulam, 'The Monte Carlo Method', *J. Amer. Stat. Assoc.*, 44, 335–41, 1949).

The *Monte Carlo simulation* is a method for iteratively evaluating a deterministic model using sets of random numbers as inputs. This method is often used when the model is complex, non-linear, or involves more than just a couple of uncertain parameters. A simulation can typically involve over 10,000 evaluations of the model, a task which in the past was only practical using supercomputers. By using *random inputs*, we are essentially turning the deterministic model into a stochastic model. The Monte Carlo method is just one of many methods for analysing *uncertainty propagation*, where the goal is to determine how the random variation, lack of knowledge, or error affects the sensitivity, performance, or reliability of the system that is being modelled. The Monte Carlo simulation is categorized as a sampling method because the inputs are randomly generated from probability distributions to simulate the process of sampling from an actual population. So, we try to choose a distribution for the inputs that most closely matches data we already have, or best represents our current state of knowledge. The data generated from the simulation can be represented as probability distributions (or histograms) or converted to error bars, reliability predictions, tolerance zones, and confidence intervals.

The steps in the Monte Carlo simulation correspond to the uncertainty propagation. All we need to do is follow the five simple steps listed below:

Step 1–Create a parametric model, $y = f(x_1, x_2, \ldots, x_p)$.
Step 2–Generate a set of random inputs, $x_{i1}, x_{i2}, \ldots, x_{ip}$.
Step 3–Evaluate the model and store the results as y_i.
Step 4–Repeat Steps 2 and 3 for $i = 1$ to n.
Step 5–Analyse the results using histograms, summary statistics, confidence intervals, etc.

Readers may refer to appropriate references for further reading on the following topics:
1. **Random processes:**
   • Bayes formula
   • Bernoulli processes
   • Markov processes
   • Poisson processes
2. **Distributions:**
   • Binomial distribution
   • Geometric distribution
   • Negative binomial distribution
   • Poisson distribution
3. **Densities:**
   • Continuous uniform density
   • Exponential and gamma densities
   • Maxwell and Rayleigh densities

Normal Density

The normal density is the most important density function for communication systems.

A very important theorem in probability theory, called the central limit theorem, states that under very general conditions, if we sum a large number of mutually independent random variables, then the distribution of the sum can be closely approximated by a certain specific continuous density, called the normal density. The normal density function with parameters μ and σ is defined as follows:

$$f_x(x) = \frac{1}{\sqrt{2\pi}\sigma} e^{-(x-\mu)^2/2\sigma^2} \tag{C.19}$$

The parameter μ represents the 'centre' of the density (it is the average or expected value of the density). The parameter σ is a measure of the 'spread' of the density and thus it is assumed to be positive (σ is the standard deviation of the density). We note that it is not at all obvious that the above function is a density, i.e. that its integral over the real line equals 1. The cumulative distribution function is given by the formula

$$F_x(x) = \int_{-\infty}^{x} \frac{1}{\sqrt{2\pi}\sigma} e^{-(u-\mu)^2/2\sigma^2} \, du \tag{C.20}$$

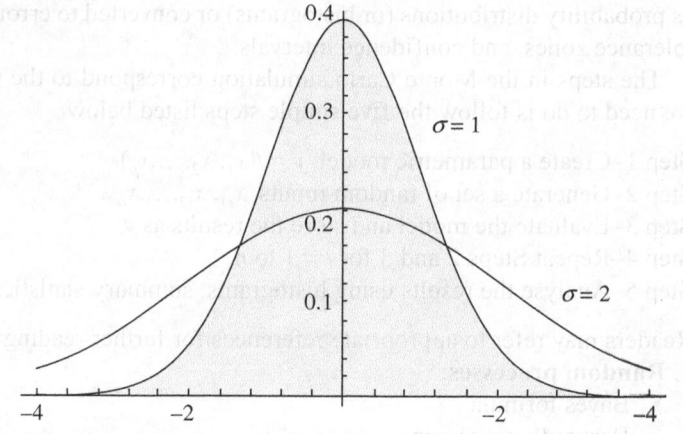

Fig. C.3   Normal density for the cases (a) $\mu = 0$ and $\sigma = 1$ (b) $\mu = 0$ and $\sigma = 2$

Definition C.9

Let X be a numerically valued discrete random variable with sample space Ω and distribution function $m(x)$. The expected value $E(X)$ is defined by

$$E(X) = \sum_{x \in \Omega} xm(x) \tag{C.21}$$

provided this sum converges absolutely. We often refer the expected value as the *mean* and denote $E(X)$ by μ for short. If the aforementioned sum does not converge absolutely, then we say that X does not have an expected value.

It is easy to prove by mathematical induction that the expected value of the sum of any finite number of random variables is the sum of the expected values of the individual random variables.

Definition C.10

Let X be a numerically valued random variable with expected value $\mu = E(X)$. Then the variance of X, denoted by $V(X)$, is

$$V(X) = E[(X - \mu)^2] \tag{C.22}$$

The standard deviation of X, denoted by $D(X)$, is

$$D(X) = \text{sqrt}[V(X)]$$

We often write σ for $D(X)$ and σ^2 for $V(X)$.

DSP Fundamentals Applied to OFDM Processing

The orthogonal frequency division multiplexing (OFDM) system is based on the fundamentals of digital signal processing. The subcarrier setting in frequency domain is considered as a discrete one. Hence, it is necessary to deal with discrete time signal and discrete time systems (the reason behind it will be clear with the mathematics in which the whole system is represented). Since a discrete time signal is an indexed sequence of real or complex numbers, a discrete time signal is a function of an integer valued index n, denoted by $x(n)$. A discrete time system is a mathematical operator or mapping that transforms one signal (the input) into another signal (the output) by means of a fixed set of rules or functions

$$y(n) = T[x(n)]$$ (D.1)

where T is the transformation. The relation between input and output can be represented by difference equation.

The OFDM is based on time and frequency domain transformations. The discrete time Fourier transform (DTFT) of a signal $x(n)$ is the complex-valued function with continuous (frequency) variable ω defined by

$$X(e^{j\omega}) = \sum_{n=-\infty}^{\infty} x(n)e^{-jn\omega}$$ (D.2)

The DTFT is, in general, complex-valued function of ω. Therefore, it is normally represented in polar form in terms of its magnitude and phase $X(e^{j\omega}) = |X(e^{j\omega})| e^{-j\phi_x \omega}$. The DTFT of the unit sample (impulse) response of a linear shift-invariant stable system (in our case, a channel) is

$$H(e^{j\omega}) = \sum_{n=-\infty}^{\infty} h(n)e^{-jn\omega}$$ (D.3)

This is called the frequency response of the filter. It defines how a complex exponential is changed in amplitude and phase by the system. The channel can be represented by a filter. The DTFT is invertible transformation and the reverse transformation is achieved by inverse DTFT (IDTFT).

$$x(n) = \frac{1}{2\pi} \int_{-\pi}^{\pi} X(e^{j\omega})e^{jn\omega} d\omega$$ (D.4)

The DTFT of a convolution of two signals is equal to the product of their transforms:

$$y(n) = x(n) * y(n) \tag{D.5}$$

or $\quad Y(e^{j\omega}) = X(e^{j\omega})H(e^{j\omega}) \tag{D.6}$

Since the DTFT is a function of a continuous variable ω, it is not directly amenable to digital computation. For finite length sequences, there is another representation, called discrete Fourier transform (DFT), which is a function of an integer variable k. For a finite length sequence $x(n)$ of length N, which is equal to zero outside the interval $[0, N-1]$, the N-point DFT is

$$X(k) = \sum_{n=0}^{N-1} x(n)e^{-j2\pi kn/N} \tag{D.7}$$

The IDFT is

$$x(n) = \frac{1}{N}\sum_{k=0}^{N-1} X(k)e^{-j2\pi kn/N} \tag{D.8}$$

The fast algorithm (due to special structures) of DFT is called Fast Fourier Transform (FFT) and its inverse is IFFT. The signal transmission over the channel exhibits the convolution process. The product of the DTFT of two signals corresponds, in time domain, to linear convolution of the two signals. For the DFT, however, if $H(k)$ and $X(k)$ are the N-point DFTs of $h(n)$ and $x(n)$, respectively, and if $Y(k) = X(k)H(k)$, then

$$y(n) = \sum_{k=0}^{N} x((k))_N \, h((n-k))_N \tag{D.9}$$

which is N-point circular convolution of $x(n)$ and $h(n)$. In general, circular convolution of two sequences is not the same as linear convolution. However, there is a special and important case in which the two are the same. Specifically, if $x(n)$ is a finite length sequence of length N_1 and $h(n)$ is a finite length sequence of length N_2, then the linear convolution of $x(n)$ and $h(n)$ is of length $L = N_1 + N_2 - 1$. In this case, the N-point circular convolution of $x(n)$ and $h(n)$ will be equal to the linear convolution provided $N \geq L$.

Satellite Communication Aspects

A satellite communication link consists of a transmitting earth station, a transponder, and a receiving earth station. Uplink for one earth station becomes the downlink for another and vice versa. The free space propagation model can be directly applied to a satellite link and can also be used for its budgetary analysis. Satellite communication is of two types: geo-stationary and orbital. Again, orbits may be low, medium, or high. Satellite communication systems are developed based on Kepler's three laws. Satellites are used for a wide variety of applications from satellite television broadcasting and navigation in the case of global positioning system (GPS) to photography, weather monitoring, and many others.

The various blocks of an earth station are shown in Fig. A.1. As the earth stations are capable of transmitting signals at a high power level, they can withstand path losses. Therefore, uplinks are at a higher frequency than downlinks. The transponder reproduces the signal and sends it back to the earth with a downconversion. At the transponder, solar cells are used for power generation, and hence, the downlink is at a lower frequency in order to reduce frequency-dependent path losses.

The circulator is used to make sure that the transmit signals go out through the dish and not back into the receive chain. It also makes sure that the received signals come from the dish into the receiver chain and not into the transmitter chain. It works much like a rounda-bout in principle. This is often referred to as an orthomode transducer (OMT) and is built into the feed assembly.

The Rx or receive filter is usually a waveguide filter that tightly controls the frequencies allowed into the receiver chain. This has the effect of reducing the unwanted noise from space and preventing interference from outside the receiver band of frequencies.

The Tx or transmit filter is usually a waveguide filter that tightly controls the frequencies allowed into the antenna. This has the effect of reducing the unwanted signals being accidentally transmitted onto the satellite and preventing interference from outside the transmit band of frequencies.

The low noise amplifier (LNA), sometimes known as an low noise block converter (LNB) on the receive-only terminals, is an amplifier that amplifies the small signals picked up by the antenna without amplifying the noise. Various kinds of LNAs exist but they all perform the same function—provide enough signal level to demodulate the data from the carrier.

The high power amplifier (HPA), otherwise known as a travelling wave tube amplifier (TWTA) or a solid-state high power amplifier (SSHPA), amplifies a specific band of frequencies by an amount that is sufficiently large to enable the antenna to beam them up to

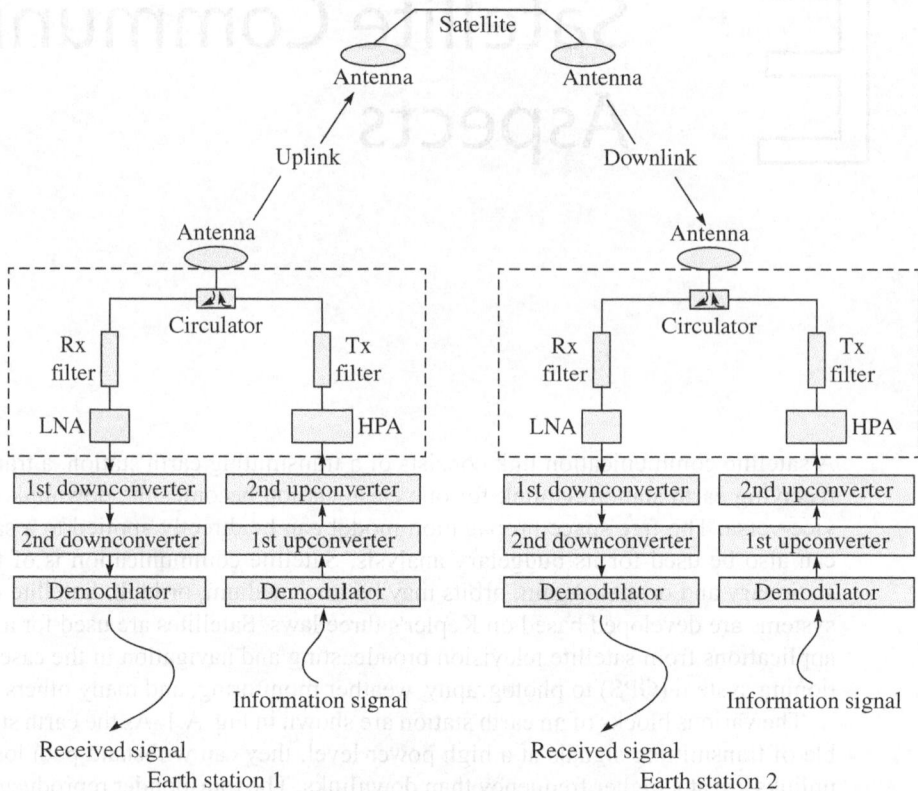

Fig. E.1 A typical bidirectional satellite link for video communication

the satellite. These can range in power from a few watts up to over 1000 watts. Large HPAs have to be cooled using liquid nitrogen and resemble electron microscopes. Small HPAs look more like a lump of metal bolted to a small heat sink.

Downconverters convert a band of frequencies from a higher frequency to a lower frequency. Signals arrive at the dish at frequencies between 10 GHz and 40 GHz and are then filtered and amplified. They now need to be moved down the frequency spectrum so that the equipment can be made cheaper and easier. The first downconverter mixes the signals with another frequency, and the result is both the sum and difference of the signals. The sum frequencies are then filtered out, retaining only the difference frequencies, which are lower in the frequency spectrum. An example would be the downconversion of 10 GHz to 1 GHz, which is from Ku band to L band. The second downconverter then downconverts the L band signals to an intermediate frequency (IF) of around 70 MHz. This is then ready for the demodulator.

Upconverters convert a band of frequencies from a lower frequency to a higher frequency. Signals are sent to the upconverters at around 70 MHz. They now need to be moved up the frequency spectrum so that the HPA can amplify them and transmit them through the antenna. The first upconverter mixes the signals with another frequency and the result is both the sum and difference of the signals. The difference frequencies are then filtered out, retaining only the sum frequencies, which are higher in the frequency spectrum. An example would be the upconversion of 70 MHz to 1 GHz, which is from the IF to L band. The second upconverter then upconverts the L band signals to a radio frequency (RF) of around 10 GHz. This is then ready for the HPA to transmit through the antenna.

A modulator and a demodulator are often combined as one, known as a modem. Like the computer modem, these units take digital data and modulate it onto a carrier and demodulate the digital data from a carrier. Source and channel coding may be the inclusive part of the modem. Voice, image, or video can be communicated.

For the transmission and reception of signals, satellites normally use frequencies above 500 MHz and often much higher. At these frequencies, the transmissions to and from the satellites might be assumed to be immune to the effects of the atmosphere. However, this is not the case, and the effects are introduced primarily by two layers of the atmosphere—the troposphere and the ionosphere.

Atmosphere for Satellite Link

The atmosphere can be divided into several layers. It is found that the temperature of these layers varies according to the height. Initially, the temperature falls until altitudes of about 10 km are reached. At this point, the temperature is around −50°C or −60°C. It is around this point that the temperature starts rising again. The region below this inflexion point is the troposphere, which is the first layer affecting the transmission and reception of signals.

The second layer that affects radio signals is the ionosphere. This region of the atmosphere starts at altitudes of around 50 km and extends to more than 400 km. Beyond the ionosphere, the signals can be considered to be in free space. The region between the upper reaches of the troposphere and the ionosphere is often termed inner free space. This region too has a little effect. The troposphere and ionosphere have refractive indices that differ from unity. The refractive index of the troposphere is greater than unity and that of the ionosphere is less than unity. As a result, refraction and absorption occur.

A further effect that is introduced by the ionosphere is known as Faraday rotation, which results from the magneto-ionic nature of the region. The Faraday rotation of a signal causes different elements of a signal to travel in different ways, particularly rotating the plane of polarization. This can create some problems during reception.

Another effect introduced by the ionosphere (already described previously) is ionospheric scintillations. These scintillations manifest themselves as a variety of variations of the amplitude, phase, and polarization angle. They can also change the angle of arrival of the signals. These variations change over a period of between 1 s and 15 s, and they can affect signals well into the microwave region. The variations are caused primarily by the variations in electron density arising in the E region, often because of sporadic E, as well as in the F layer, where a spreading effect is the cause. The level of scintillation is dependent upon numerous factors, including the location of the earth station and the state of the ionosphere because of the location, sunspot cycle, level of geomagnetic activity, latitude, and local time of day. Scintillations are more intense in equatorial regions, falling with increasing latitude away from the equator but then rising at high latitudes. The effects are also found to decrease with increasing frequency and generally not noticeable above frequencies of 1–2 GHz. As such, they are not applicable to many direct broadcast television signals, although they may affect GPS and some communication satellites.

Troposphere introduces various effects, including signal bending caused by refraction, scintillation, and attenuation. Signal refraction in the troposphere is in the opposite sense to that in the ionosphere. This is because the refractive index in the troposphere is greater than unity and is frequency independent. The signal refraction gives them a greater range than would be expected as a result of the direct geometric line of sight. Tropospheric ducting and extended range effects that are experienced by terrestrial very high frequency (VHF) and ultra-high frequency (UHF) communications may also be experienced when low angles of elevation are used.

The scintillations induced by the troposphere are often greater than those induced by the ionosphere. They are caused by the turbulence in the atmosphere where areas of differing refractive indices move around because of the wind or convection currents. The degree to which the scintillations occur is dependent upon the angle of inclination, and above angles of about 15°, the effect can normally be ignored. At angles within 5°–10°, the changes can often be about 6 dB at frequencies of 5–6 GHz.

Frequency changes caused by the Doppler shift principle (explained in Section 5.9.2) may be in evidence with signals from some satellites. Satellites in low earth orbits move very quickly, and as a result, a Doppler frequency shift is apparent in many cases. As a satellite moves towards the earth station, the frequency appears higher than nominal, and then as it moves away, the apparent frequency falls. The degree of shift is dependent upon various factors, including the speed of the satellite (more correctly, its speed relative to the earth station) and the frequencies in use. Shifts of the order of 10 kHz may be experienced. As most satellites operate in a cross-mode configuration, the Doppler shift is applicable not just to the band on which the signal is received but also to the cumulative effect of the uplink and downlink transmissions. In many instances, the effects will subtract because of the way the satellite mixing process is configured. Such effects can be observed even in mobile satellite communication.

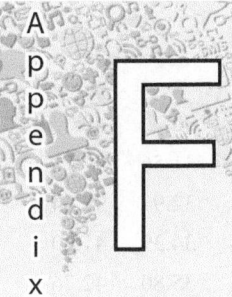

Erlang and Poisson Traffi Tables

ERLANG B TRAFFIC TABLE

Maximum Offered Load versus *B* and *N*

N/B	0.01	0.05	0.1	0.5	1.0	2	5	10	15	20	30	40
1	.0001	.0005	.0010	.0050	.0101	.0204	.0526	.1111	.1765	.2500	.4286	.6667
2	.0142	.0321	.0458	.1054	.1526	.2235	.3813	.5954	.7962	1.000	1.449	2.000
3	.0868	.1517	.1938	.3490	.4555	.6022	.8994	1.271	1.603	1.930	2.633	3.480
4	.2347	.3624	.4393	.7012	.8694	1.092	1.525	2.045	2.501	2.945	3.891	5.021
5	.4520	.6486	.7621	1.132	1.361	1.657	2.219	2.881	3.454	4.010	5.189	6.596
6	.7282	.9957	1.146	1.622	1.909	2.276	2.960	3.758	4.445	5.109	6.514	8.191
7	1.054	1.392	1.579	2.158	2.501	2.935	3.738	4.666	5.461	6.230	7.856	9.800
8	1.422	1.830	2.051	2.730	3.128	3.627	4.543	5.597	6.498	7.369	9.213	11.42
9	1.826	2.302	2.558	3.333	3.783	4.345	5.370	6.546	7.551	8.522	10.58	13.05
10	2.260	2.803	3.092	3.961	4.461	5.084	6.216	7.511	8.616	9.685	11.95	14.68
11	2.722	3.329	3.651	4.610	5.160	5.842	7.076	8.487	9.691	10.86	13.33	16.31
12	3.207	3.878	4.231	5.279	5.876	6.615	7.950	9.474	10.78	12.04	14.72	17.95
13	3.713	4.447	4.831	5.964	6.607	7.402	8.835	10.47	11.87	13.22	16.11	19.60
14	4.239	5.032	5.446	6.663	7.352	8.200	9.730	11.47	12.97	14.41	17.50	21.24
15	4.781	5.634	6.077	7.376	8.108	9.010	10.63	12.48	14.07	15.61	18.90	22.89
16	5.339	6.250	6.722	8.100	8.875	9.828	11.54	13.50	15.18	16.81	20.30	24.54
17	5.911	6.878	7.378	8.834	9.652	10.66	12.46	14.52	16.29	18.01	21.70	26.19
18	6.496	7.519	8.046	9.578	10.44	11.49	13.39	15.55	17.41	19.22	23.10	27.84
19	7.093	8.170	8.724	10.33	11.23	12.33	14.32	16.58	18.53	20.42	24.51	29.50
20	7.701	8.831	9.412	11.09	12.03	13.18	15.25	17.61	19.65	21.64	25.92	31.15
21	8.319	9.501	10.11	11.86	12.84	14.04	16.19	18.65	20.77	22.85	27.33	32.81
22	8.946	10.18	10.81	12.64	13.65	14.90	17.13	19.69	21.90	24.06	28.74	34.46
23	9.583	10.87	11.52	13.42	14.47	15.76	18.08	20.74	23.03	25.28	30.15	36.12

(Contd)

N/B	0.01	0.05	0.1	0.5	1.0	2	5	10	15	20	30	40
24	10.23	11.56	12.24	14.20	15.30	16.63	19.03	21.78	24.16	26.50	31.56	37.78
25	10.88	12.26	12.97	15.00	16.13	17.51	19.99	22.83	25.30	27.72	32.97	39.44
26	11.54	12.97	13.70	15.80	16.96	18.38	20.94	23.89	26.43	28.94	34.39	41.10
27	12.21	13.69	14.44	16.60	17.80	19.27	21.90	24.94	27.57	30.16	35.80	42.76
28	12.88	14.41	15.18	17.41	18.64	20.15	22.87	26.00	28.71	31.39	37.21	44.41
29	13.56	15.13	15.93	18.22	19.49	21.04	23.83	27.05	29.85	32.61	38.63	46.07
30	14.25	15.86	16.68	19.03	20.34	21.93	24.80	28.11	31.00	33.84	40.05	47.74
31	14.94	16.60	17.44	19.85	21.19	22.83	25.77	29.17	32.14	35.07	41.46	49.40
32	15.63	17.34	18.21	20.68	22.05	23.73	26.75	30.24	33.28	36.30	42.88	51.06
33	16.34	18.09	18.97	21.51	22.91	24.63	27.72	31.30	34.43	37.52	44.30	52.72
34	17.04	18.84	19.74	22.34	23.77	25.53	28.70	32.37	35.58	38.75	45.72	54.38
35	17.75	19.59	20.52	23.17	24.64	26.44	29.68	33.43	36.72	39.99	47.14	56.04
36	18.47	20.35	21.30	24.01	25.51	27.34	30.66	34.50	37.87	41.22	48.56	57.70
37	19.19	21.11	22.08	24.85	26.38	28.25	31.64	35.57	39.02	42.45	49.98	59.37
38	19.91	21.87	22.86	25.69	27.25	29.17	32.62	36.64	40.17	43.68	51.40	61.03
39	20.64	22.64	23.65	26.53	28.13	30.08	33.61	37.72	41.32	44.91	52.82	62.69
40	21.37	23.41	24.44	27.38	29.01	31.00	34.60	38.79	42.48	46.15	54.24	64.35
41	22.11	24.19	25.24	28.23	29.89	31.92	35.58	39.86	43.63	47.38	55.66	66.02
42	22.85	24.97	26.04	29.09	30.77	32.84	36.57	40.94	44.78	48.62	57.08	67.68
43	23.59	25.75	26.84	29.94	31.66	33.76	37.57	42.01	45.94	49.85	58.50	69.34
44	24.33	26.53	27.64	30.80	32.54	34.68	38.56	43.09	47.09	51.09	59.92	71.01
45	25.08	27.32	28.45	31.66	33.43	35.61	39.55	44.17	48.25	52.32	61.35	72.67
46	25.83	28.11	29.26	32.52	34.32	36.53	40.55	45.24	49.40	53.56	62.77	74.33
47	26.59	28.90	30.07	33.38	35.22	37.46	41.54	46.32	50.56	54.80	64.19	76.00
48	27.34	29.70	30.88	34.25	36.11	38.39	42.54	47.40	51.71	56.03	65.61	77.66
49	28.10	30.49	31.69	35.11	37.00	39.32	43.53	48.48	52.87	57.27	67.04	79.32
50	28.87	31.29	32.51	35.98	37.90	40.26	44.53	49.56	54.03	58.51	68.46	80.99
51	29.63	32.09	33.33	36.85	38.80	41.19	45.53	50.64	55.19	59.75	69.88	82.65
52	30.40	32.90	34.15	37.72	39.70	42.12	46.53	51.73	56.35	60.99	71.31	84.32
53	31.17	33.70	34.98	38.60	40.60	43.06	47.53	52.81	57.50	62.22	72.73	85.98
54	31.94	34.51	35.80	39.47	41.51	44.00	48.54	53.89	58.66	63.46	74.15	87.65
55	32.72	35.32	36.63	40.35	42.41	44.94	49.54	54.98	59.82	64.70	75.58	89.31
56	33.49	36.13	37.46	41.23	43.32	45.88	50.54	56.06	60.98	65.94	77.00	90.97
57	34.27	36.95	38.29	42.11	44.22	46.82	51.55	57.14	62.14	67.18	78.43	92.64
58	35.05	37.76	39.12	42.99	45.13	47.76	52.55	58.23	63.31	68.42	79.85	94.30
59	35.84	38.58	39.96	43.87	46.04	48.70	53.56	59.32	64.47	69.66	81.27	95.97
60	36.62	39.40	40.80	44.76	46.95	49.64	54.57	60.40	65.63	70.90	82.70	97.63

N/B	0.01	0.05	0.1	0.5	1.0	2	5	10	15	20	30	40
61	37.41	40.22	41.63	45.64	47.86	50.59	55.57	61.49	66.79	72.14	84.12	99.30
62	38.20	41.05	42.47	46.53	48.77	51.53	56.58	62.58	67.95	73.38	85.55	101.0
63	38.99	41.87	43.31	47.42	49.69	52.48	57.59	63.66	69.11	74.63	86.97	102.6
64	39.78	42.70	44.16	48.31	50.60	53.43	58.60	64.75	70.28	75.87	88.40	104.3
65	40.58	43.52	45.00	49.20	51.52	54.38	59.61	65.84	71.44	77.11	89.82	106.0
66	41.38	44.35	45.85	50.09	52.44	55.33	60.62	66.93	72.60	78.35	91.25	107.6
67	42.17	45.18	46.69	50.98	53.35	56.28	61.63	68.02	73.77	79.59	92.67	109.3
68	42.97	46.02	47.54	51.87	54.27	57.23	62.64	69.11	74.93	80.83	94.10	111.0
69	43.77	46.85	48.39	52.77	55.19	58.18	63.65	70.20	76.09	82.08	95.52	112.6
70	44.58	47.68	49.24	53.66	56.11	59.13	64.67	71.29	77.26	83.32	96.95	114.3
71	45.38	48.52	50.09	54.56	57.03	60.08	65.68	72.38	78.42	84.56	98.37	116.0
72	46.19	49.36	50.94	55.46	57.96	61.04	66.69	73.47	79.59	85.80	99.80	117.6
73	47.00	50.20	51.80	56.35	58.88	61.99	67.71	74.56	80.75	87.05	101.2	119.3
74	47.81	51.04	52.65	57.25	59.80	62.95	68.72	75.65	81.92	88.29	102.7	120.9
75	48.62	51.88	53.51	58.15	60.73	63.90	69.74	76.74	83.08	89.53	104.1	122.6
76	49.43	52.72	54.37	59.05	61.65	64.86	70.75	77.83	84.25	90.78	105.5	124.3
77	50.24	53.56	55.23	59.96	62.58	65.81	71.77	78.93	85.41	92.02	106.9	125.9
78	51.05	54.41	56.09	60.86	63.51	66.77	72.79	80.02	86.58	93.26	108.4	127.6
79	51.87	55.25	56.95	61.76	64.43	67.73	73.80	81.11	87.74	94.51	109.8	129.3
80	52.69	56.10	57.81	62.67	65.36	68.69	74.82	82.20	88.91	95.75	111.2	130.9
81	53.51	56.95	58.67	63.57	66.29	69.65	75.84	83.30	90.08	96.99	112.6	132.6
82	54.33	57.80	59.54	64.48	67.22	70.61	76.86	84.39	91.24	98.24	114.1	134.3
83	55.15	58.65	60.40	65.39	68.15	71.57	77.87	85.48	92.41	99.48	115.5	135.9
84	55.97	59.50	61.27	66.29	69.08	72.53	78.89	86.58	93.58	100.7	116.9	137.6
85	56.79	60.35	62.14	67.20	70.02	73.49	79.91	87.67	94.74	102.0	118.3	139.3
86	57.62	61.21	63.00	68.11	70.95	74.45	80.93	88.77	95.91	103.2	119.8	140.9
87	58.44	62.06	63.87	69.02	71.88	75.42	81.95	89.86	97.08	104.5	121.2	142.6
88	59.27	62.92	64.74	69.93	72.82	76.38	82.97	90.96	98.25	105.7	122.6	144.3
89	60.10	63.77	65.61	70.84	73.75	77.34	83.99	92.05	99.41	107.0	124.0	145.9
90	60.92	64.63	66.48	71.76	74.68	78.31	85.01	93.15	100.6	108.2	125.5	147.6
91	61.75	65.49	67.36	72.67	75.62	79.27	86.04	94.24	101.8	109.4	126.9	149.3
92	62.58	66.35	68.23	73.58	76.56	80.24	87.06	95.34	102.9	110.7	128.3	150.9
93	63.42	67.21	69.10	74.50	77.49	81.20	88.08	96.43	104.1	111.9	129.8	152.6
94	64.25	68.07	69.98	75.41	78.43	82.17	89.10	97.53	105.3	113.2	131.2	154.3
95	65.08	68.93	70.85	76.33	79.37	83.13	90.12	98.63	106.4	114.4	132.6	155.9
96	65.92	69.79	71.73	77.24	80.31	84.10	91.15	99.72	107.6	115.7	134.0	157.6
97	66.75	70.65	72.61	78.16	81.25	85.07	92.17	100.8	108.8	116.9	135.5	159.3

(Contd)

N/B	0.01	0.05	0.1	0.5	1.0	2	5	10	15	20	30	40
98	67.59	71.52	73.48	79.07	82.18	86.04	93.19	101.9	109.9	118.2	136.9	160.9
99	68.43	72.38	74.36	79.99	83.12	87.00	94.22	103.0	111.1	119.4	138.3	162.6
100	69.27	73.25	75.24	80.91	84.06	87.97	95.24	104.1	112.3	120.6	139.7	164.3

(N is the number of servers. The numerical column headings indicate blocking probability B in percentage.)

ERLANG C TRAFFIC TABLE

Maximum Offered Load versus B and N

N/B	0.01	0.05	0.1	0.5	1.0	2	5	10	15	20	30	40
1	.0001	.0005	.0010	.0050	.0100	.0200	.0500	.1000	.1500	.2000	.3000	.4000
2	.0142	.0319	.0452	.1025	.1465	.2103	.3422	.5000	.6278	.7403	.9390	1.117
3	.0860	.1490	.1894	.3339	.4291	.5545	.7876	1.040	1.231	1.393	1.667	1.903
4	.2310	.3533	.4257	.6641	.8100	.9939	1.319	1.653	1.899	2.102	2.440	2.725
5	.4428	.6289	.7342	1.065	1.259	1.497	1.905	2.313	2.607	2.847	3.241	3.569
6	.7110	.9616	1.099	1.519	1.758	2.047	2.532	3.007	3.344	3.617	4.062	4.428
7	1.026	1.341	1.510	2.014	2.297	2.633	3.188	3.725	4.103	4.406	4.897	5.298
8	1.382	1.758	1.958	2.543	2.866	3.246	3.869	4.463	4.878	5.210	5.744	6.178
9	1.771	2.208	2.436	3.100	3.460	3.883	4.569	5.218	5.668	6.027	6.600	7.065
10	2.189	2.685	2.942	3.679	4.077	4.540	5.285	5.986	6.469	6.853	7.465	7.959
11	2.634	3.186	3.470	4.279	4.712	5.213	6.015	6.765	7.280	7.688	8.336	8.857
12	3.100	3.708	4.018	4.896	5.363	5.901	6.758	7.554	8.099	8.530	9.212	9.761
13	3.587	4.248	4.584	5.529	6.028	6.602	7.511	8.352	8.926	9.379	10.09	10.67
14	4.092	4.805	5.166	6.175	6.705	7.313	8.273	9.158	9.760	10.23	10.98	11.58
15	4.614	5.377	5.762	6.833	7.394	8.035	9.044	9.970	10.60	11.09	11.87	12.49
16	5.150	5.962	6.371	7.502	8.093	8.766	9.822	10.79	11.44	11.96	12.77	13.41
17	5.699	6.560	6.991	8.182	8.801	9.505	10.61	11.61	12.29	12.83	13.66	14.33
18	6.261	7.169	7.622	8.871	9.518	10.25	11.40	12.44	13.15	13.70	14.56	15.25
19	6.835	7.788	8.263	9.568	10.24	11.01	12.20	13.28	14.01	14.58	15.47	16.18
20	7.419	8.417	8.914	10.27	10.97	11.77	13.00	14.12	14.87	15.45	16.37	17.10
21	8.013	9.055	9.572	10.99	11.71	12.53	13.81	14.96	15.73	16.34	17.28	18.03
22	8.616	9.702	10.24	11.70	12.46	13.30	14.62	15.81	16.60	17.22	18.19	18.96
23	9.228	10.36	10.91	12.43	13.21	14.08	15.43	16.65	17.47	18.11	19.10	19.89
24	9.848	11.02	11.59	13.16	13.96	14.86	16.25	17.51	18.35	19.00	20.02	20.82
25	10.48	11.69	12.28	13.90	14.72	15.65	17.08	18.36	19.22	19.89	20.93	21.76
26	11.11	12.36	12.97	14.64	15.49	16.44	17.91	19.22	20.10	20.79	21.85	22.69
27	11.75	13.04	13.67	15.38	16.26	17.23	18.74	20.08	20.98	21.68	22.77	23.63

N/B	0.01	0.05	0.1	0.5	1.0	2	5	10	15	20	30	40
28	12.40	13.73	14.38	16.14	17.03	18.03	19.57	20.95	21.87	22.58	23.69	24.57
29	13.05	14.42	15.09	16.89	17.81	18.83	20.41	21.82	22.75	23.48	24.61	25.50
30	13.71	15.12	15.80	17.65	18.59	19.64	21.25	22.68	23.64	24.38	25.54	26.44
31	14.38	15.82	16.52	18.42	19.37	20.45	22.09	23.56	24.53	25.29	26.46	27.38
32	15.05	16.53	17.25	19.18	20.16	21.26	22.93	24.43	25.42	26.19	27.39	28.33
33	15.72	17.24	17.97	19.95	20.95	22.07	23.78	25.30	26.32	27.10	28.31	29.27
34	16.40	17.95	18.71	20.73	21.75	22.89	24.63	26.18	27.21	28.01	29.24	30.21
35	17.09	18.67	19.44	21.51	22.55	23.71	25.48	27.06	28.11	28.92	30.17	31.16
36	17.78	19.39	20.18	22.29	23.35	24.53	26.34	27.94	29.00	29.83	31.10	32.10
37	18.47	20.12	20.92	23.07	24.15	25.36	27.19	28.82	29.90	30.74	32.03	33.05
38	19.17	20.85	21.67	23.86	24.96	26.18	28.05	29.71	30.80	31.65	32.97	34.00
39	19.87	21.59	22.42	24.65	25.77	27.01	28.91	30.59	31.71	32.57	33.90	34.94
40	20.58	22.33	23.17	25.44	26.58	27.84	29.77	31.48	32.61	33.48	34.83	35.89
41	21.28	23.07	23.93	26.23	27.39	28.68	30.63	32.37	33.51	34.40	35.77	36.84
42	22.00	23.81	24.69	27.03	28.21	29.51	31.50	33.26	34.42	35.32	36.70	37.79
43	22.71	24.56	25.45	27.83	29.02	30.35	32.36	34.15	35.33	36.23	37.64	38.74
44	23.43	25.31	26.22	28.63	29.84	31.19	33.23	35.04	36.23	37.15	38.58	39.69
45	24.15	26.06	26.98	29.44	30.67	32.03	34.10	35.93	37.14	38.07	39.51	40.64
46	24.88	26.82	27.75	30.24	31.49	32.87	34.97	36.83	38.05	39.00	40.45	41.59
47	25.60	27.57	28.52	31.05	32.32	33.72	35.84	37.72	38.96	39.92	41.39	42.54
48	26.34	28.33	29.30	31.86	33.14	34.56	36.72	38.62	39.87	40.84	42.33	43.50
49	27.07	29.10	30.08	32.68	33.97	35.41	37.59	39.52	40.79	41.76	43.27	44.45
50	27.80	29.86	30.86	33.49	34.80	36.26	38.47	40.42	41.70	42.69	44.21	45.40
51	28.54	30.63	31.64	34.31	35.64	37.11	39.35	41.32	42.61	43.61	45.15	46.36
52	29.28	31.40	32.42	35.12	36.47	37.97	40.23	42.22	43.53	44.54	46.10	47.31
53	30.03	32.17	33.21	35.94	37.31	38.82	41.10	43.12	44.44	45.47	47.04	48.27
54	30.77	32.95	33.99	36.76	38.15	39.67	41.99	44.02	45.36	46.39	47.98	49.22
55	31.52	33.72	34.78	37.59	38.99	40.53	42.87	44.93	46.28	47.32	48.93	50.18
56	32.27	34.50	35.57	38.41	39.83	41.39	43.75	45.83	47.20	48.25	49.87	51.13
57	33.03	35.28	36.37	39.24	40.67	42.25	44.64	46.74	48.12	49.18	50.82	52.09
58	33.78	36.06	37.16	40.07	41.51	43.11	45.52	47.64	49.04	50.11	51.76	53.05
59	34.54	36.85	37.96	40.90	42.36	43.97	46.41	48.55	49.96	51.04	52.71	54.01
60	35.30	37.63	38.76	41.73	43.20	44.83	47.29	49.46	50.88	51.97	53.65	54.96
61	36.06	38.42	39.56	42.56	44.05	45.70	48.18	50.37	51.80	52.90	54.60	55.92
62	36.82	39.21	40.36	43.39	44.90	46.56	49.07	51.27	52.72	53.83	55.55	56.88
63	37.59	40.00	41.16	44.23	45.75	47.43	49.96	52.18	53.64	54.77	56.49	57.84
64	38.35	40.80	41.97	45.06	46.60	48.30	50.85	53.10	54.57	55.70	57.44	58.80

(Contd)

N/B	0.01	0.05	0.1	0.5	1.0	2	5	10	15	20	30	40
65	39.12	41.59	42.78	45.90	47.45	49.16	51.74	54.01	55.49	56.63	58.39	59.76
66	39.89	42.39	43.58	46.74	48.30	50.03	52.64	54.92	56.42	57.57	59.34	60.72
67	40.66	43.18	44.39	47.58	49.16	50.90	53.53	55.83	57.34	58.50	60.29	61.68
68	41.44	43.98	45.20	48.42	50.01	51.77	54.42	56.75	58.27	59.44	61.24	62.64
69	42.21	44.78	46.02	49.26	50.87	52.65	55.32	57.66	59.20	60.37	62.19	63.60
70	42.99	45.58	46.83	50.10	51.73	53.52	56.21	58.57	60.12	61.31	63.14	64.56
71	43.77	46.39	47.64	50.95	52.59	54.39	57.11	59.49	61.05	62.25	64.09	65.52
72	44.55	47.19	48.46	51.79	53.45	55.27	58.01	60.41	61.98	63.18	65.04	66.48
73	45.33	48.00	49.28	52.64	54.31	56.14	58.90	61.32	62.91	64.12	65.99	67.44
74	46.11	48.81	50.10	53.49	55.17	57.02	59.80	62.24	63.84	65.06	66.94	68.40
75	46.90	49.61	50.92	54.34	56.03	57.90	60.70	63.16	64.76	66.00	67.89	69.37
76	47.68	50.42	51.74	55.19	56.89	58.78	61.60	64.07	65.69	66.94	68.85	70.33
77	48.47	51.23	52.56	56.04	57.76	59.65	62.50	64.99	66.63	67.88	69.80	71.29
78	49.26	52.05	53.38	56.89	58.62	60.53	63.40	65.91	67.56	68.82	70.75	72.25
79	50.05	52.86	54.21	57.74	59.49	61.41	64.30	66.83	68.49	69.76	71.70	73.22
80	50.84	53.68	55.03	58.60	60.36	62.30	65.21	67.75	69.42	70.70	72.66	74.18
81	51.63	54.49	55.86	59.45	61.22	63.18	66.11	68.67	70.35	71.64	73.61	75.14
82	52.43	55.31	56.69	60.30	62.09	64.06	67.01	69.59	71.28	72.58	74.57	76.11
83	53.22	56.13	57.52	61.16	62.96	64.94	67.92	70.52	72.22	73.52	75.52	77.07
84	54.02	56.95	58.35	62.02	63.83	65.83	68.82	71.44	73.15	74.46	76.47	78.04
85	54.81	57.77	59.18	62.88	64.70	66.71	69.73	72.36	74.08	75.40	77.43	79.00
86	55.61	58.59	60.01	63.73	65.57	67.60	70.63	73.28	75.02	76.35	78.38	79.97
87	56.41	59.41	60.84	64.59	66.45	68.48	71.54	74.21	75.95	77.29	79.34	80.93
88	57.21	60.23	61.67	65.45	67.32	69.37	72.45	75.13	76.89	78.23	80.30	81.90
89	58.02	61.06	62.51	66.32	68.19	70.26	73.35	76.06	77.82	79.18	81.25	82.86
90	58.82	61.88	63.34	67.18	69.07	71.15	74.26	76.98	78.76	80.12	82.21	83.83
91	59.62	62.71	64.18	68.04	69.94	72.04	75.17	77.91	79.69	81.06	83.16	84.79
92	60.43	63.54	65.02	68.90	70.82	72.92	76.08	78.83	80.63	82.01	84.12	85.76
93	61.23	64.36	65.86	69.77	71.70	73.81	76.99	79.76	81.57	82.95	85.08	86.73
94	62.04	65.19	66.70	70.63	72.57	74.71	77.90	80.69	82.50	83.90	86.03	87.69
95	62.85	66.02	67.54	71.50	73.45	75.60	78.81	81.61	83.44	84.84	86.99	88.66
96	63.66	66.85	68.38	72.36	74.33	76.49	79.72	82.54	84.38	85.79	87.95	89.62
97	64.47	67.69	69.22	73.23	75.21	77.38	80.63	83.47	85.32	86.74	88.91	90.59
98	65.28	68.52	70.06	74.10	76.09	78.27	81.54	84.39	86.26	87.68	89.87	91.56
99	66.09	69.35	70.90	74.97	76.97	79.17	82.46	85.32	87.20	88.63	90.82	92.53
100	66.91	70.19	71.75	75.84	77.85	80.06	83.37	86.25	88.13	89.58	91.78	93.49

(*N* is the number of servers. The numerical column headings indicate blocking probability *B* in percentage.)

POISSON TRAFFIC TABLE

Maximum Offered Load versus *B* and *N*

N/B	0.01	0.05	0.1	0.5	1.0	2	5	10	15	20	30	40
1	.0001	.0005	.0010	.0050	.0101	.0202	.0513	.1054	.1625	.2231	.3567	.5108
2	.0142	.0320	.0454	.1035	.1486	.2147	.3554	.5318	.6832	.8244	1.097	1.376
3	.0862	.1497	.1905	.3379	.4360	.5672	.8177	1.102	1.331	1.535	1.914	2.285
4	.2318	.3552	.4286	.6722	.8232	1.016	1.366	1.745	2.039	2.297	2.764	3.211
5	1.078	1.279	1.530	1.970	2.433	2.785	3.090	3.634	4.148			
6	.7137	.9672	1.107	1.537	1.785	2.089	2.613	3.152	3.557	3.904	4.517	5.091
7	1.030	1.348	1.520	2.037	2.330	2.684	3.285	3.895	4.348	4.734	5.411	6.039
8	1.387	1.768	1.971	2.571	2.906	3.307	3.981	4.656	5.155	5.576	6.312	6.991
9	1.778	2.220	2.452	3.132	3.508	3.953	4.695	5.433	5.973	6.429	7.220	7.947
10	2.198	2.699	2.961	3.717	4.130	4.618	5.425	6.221	6.802	7.289	8.133	8.904
11	2.643	3.202	3.492	4.321	4.771	5.300	6.169	7.021	7.639	8.157	9.050	9.864
12	3.112	3.726	4.042	4.943	5.428	5.996	6.924	7.829	8.484	9.031	9.972	10.83
13	3.600	4.269	4.611	5.580	6.099	6.704	7.690	8.646	9.336	9.910	10.90	11.79
14	4.106	4.828	5.195	6.231	6.782	7.424	8.464	9.470	10.19	10.79	11.82	12.76
15	4.629	5.402	5.794	6.893	7.477	8.153	9.246	10.30	11.06	11.68	12.75	13.72
16	5.167	5.990	6.405	7.567	8.181	8.891	10.04	11.14	11.92	12.57	13.69	14.69
17	5.718	6.590	7.028	8.251	8.895	9.638	10.83	11.98	12.79	13.47	14.62	15.66
18	6.281	7.201	7.662	8.943	9.616	10.39	11.63	12.82	13.67	14.37	15.56	16.63
19	6.856	7.822	8.306	9.645	10.35	11.15	12.44	13.67	14.55	15.27	16.50	17.60
20	7.442	8.453	8.958	10.35	11.08	11.92	13.26	14.53	15.43	16.17	17.44	18.57
21	8.037	9.093	9.619	11.07	11.83	12.69	14.07	15.38	16.31	17.08	18.38	19.54
22	8.642	9.741	10.29	11.79	12.57	13.47	14.89	16.24	17.20	17.99	19.32	20.51
23	9.255	10.40	10.96	12.52	13.33	14.25	15.72	17.11	18.09	18.90	20.27	21.48
24	9.876	11.06	11.65	13.26	14.09	15.04	16.55	17.98	18.98	19.81	21.21	22.46
25	10.50	11.73	12.34	14.00	14.85	15.83	17.38	18.84	19.88	20.73	22.16	23.43
26	11.14	12.41	13.03	14.74	15.62	16.63	18.22	19.72	20.77	21.64	23.10	24.41
27	11.78	13.09	13.73	15.49	16.40	17.43	19.06	20.59	21.67	22.56	24.05	25.38
28	12.43	13.78	14.44	16.25	17.18	18.23	19.90	21.47	22.57	23.48	25.00	26.36
29	13.09	14.47	15.15	17.00	17.96	19.04	20.75	22.35	23.48	24.40	25.95	27.33
30	13.75	15.17	15.87	17.77	18.74	19.85	21.59	23.23	24.38	25.32	26.91	28.31
31	14.42	15.87	16.59	18.53	19.53	20.66	22.45	24.11	25.29	26.24	27.86	29.29
32	15.09	16.58	17.32	19.31	20.32	21.48	23.30	25.00	26.19	27.17	28.81	30.26
33	15.76	17.30	18.05	20.08	21.12	22.30	24.15	25.89	27.10	28.09	29.76	31.24
34	16.44	18.01	18.78	20.86	21.92	23.12	25.01	26.77	28.01	29.02	30.72	32.22

(Contd)

N/B	0.01	0.05	0.1	0.5	1.0	2	5	10	15	20	30	40
35	17.13	18.73	19.52	21.64	22.72	23.95	25.87	27.66	28.92	29.95	31.67	33.20
36	17.82	19.46	20.26	22.42	23.53	24.77	26.73	28.56	29.84	30.88	32.63	34.18
37	18.52	20.19	21.01	23.21	24.33	25.60	27.60	29.45	30.75	31.81	33.59	35.16
38	19.21	20.92	21.75	24.00	25.14	26.44	28.46	30.35	31.66	32.74	34.54	36.14
39	19.92	21.66	22.51	24.79	25.96	27.27	29.33	31.24	32.58	33.67	35.50	37.11
40	20.62	22.40	23.26	25.59	26.77	28.11	30.20	32.14	33.50	34.60	36.46	38.09
41	21.33	23.14	24.02	26.38	27.59	28.95	31.07	33.04	34.42	35.54	37.42	39.07
42	22.05	23.88	24.78	27.18	28.41	29.79	31.94	33.94	35.33	36.47	38.38	40.05
43	22.76	24.63	25.54	27.99	29.23	30.63	32.81	34.84	36.26	37.41	39.34	41.04
44	23.48	25.38	26.31	28.79	30.05	31.47	33.69	35.74	37.18	38.34	40.30	42.02
45	24.20	26.14	27.08	29.60	30.88	32.32	34.56	36.65	38.10	39.28	41.26	43.00
46	24.93	26.90	27.85	30.41	31.71	33.17	35.44	37.55	39.02	40.22	42.22	43.98
47	25.66	27.65	28.62	31.22	32.53	34.01	36.32	38.46	39.94	41.16	43.18	44.96
48	26.39	28.42	29.40	32.03	33.37	34.87	37.20	39.36	40.87	42.09	44.14	45.94
49	27.13	29.18	30.18	32.85	34.20	35.72	38.08	40.27	41.79	43.03	45.10	46.92
50	27.86	29.95	30.96	33.66	35.03	36.57	38.97	41.18	42.72	43.97	46.06	47.90
51	28.60	30.72	31.74	34.48	35.87	37.43	39.85	42.09	43.65	44.91	47.03	48.89
52	29.34	31.49	32.53	35.30	36.71	38.28	40.73	43.00	44.58	45.85	47.99	49.87
53	30.09	32.26	33.31	36.13	37.55	39.14	41.62	43.91	45.50	46.80	48.95	50.85
54	30.84	33.04	34.10	36.95	38.39	40.00	42.51	44.82	46.43	47.74	49.92	51.83
55	31.59	33.82	34.90	37.78	39.23	40.86	43.40	45.74	47.36	48.68	50.88	52.82
56	32.34	34.60	35.69	38.60	40.07	41.72	44.29	46.65	48.29	49.63	51.85	53.80
57	33.09	35.38	36.48	39.43	40.92	42.59	45.18	47.56	49.22	50.57	52.81	54.78
58	33.85	36.16	37.28	40.26	41.77	43.45	46.07	48.48	50.15	51.51	53.78	55.77
59	34.60	36.95	38.08	41.09	42.61	44.32	46.96	49.40	51.09	52.46	54.74	56.75
60	35.36	37.73	38.88	41.93	43.46	45.18	47.85	50.31	52.02	53.40	55.71	57.73
61	36.13	38.52	39.68	42.76	44.31	46.05	48.75	51.23	52.95	54.35	56.68	58.72
62	36.89	39.31	40.48	43.60	45.16	46.92	49.64	52.15	53.89	55.30	57.64	59.70
63	37.66	40.11	41.29	44.43	46.02	47.79	50.54	53.07	54.82	56.24	58.61	60.68
64	38.42	40.90	42.09	45.27	46.87	48.66	51.43	53.99	55.76	57.19	59.58	61.67
65	39.19	41.70	42.90	46.11	47.73	49.53	52.33	54.91	56.69	58.14	60.54	62.65
66	39.96	42.49	43.71	46.95	48.58	50.41	53.23	55.83	57.63	59.08	61.51	63.64
67	40.74	43.29	44.52	47.79	49.44	51.28	54.13	56.75	58.56	60.03	62.48	64.62
68	41.51	44.09	45.33	48.64	50.30	52.16	55.03	57.67	59.50	60.98	63.45	65.61
69	42.29	44.89	46.15	49.48	51.16	53.03	55.93	58.59	60.44	61.93	64.41	66.59
70	43.07	45.70	46.96	50.33	52.02	53.91	56.83	59.52	61.37	62.88	65.38	67.58
71	43.84	46.50	47.78	51.17	52.88	54.79	57.73	60.44	62.31	63.83	66.35	68.56

N/B	0.01	0.05	0.1	0.5	1.0	2	5	10	15	20	30	40
72	44.63	47.31	48.60	52.02	53.74	55.66	58.63	61.36	63.25	64.78	67.32	69.54
73	45.41	48.11	49.42	52.87	54.60	56.54	59.54	62.29	64.19	65.73	68.29	70.53
74	46.19	48.92	50.24	53.72	55.47	57.42	60.44	63.21	65.13	66.68	69.26	71.52
75	46.98	49.73	51.06	54.57	56.33	58.30	61.35	64.14	66.07	67.63	70.23	72.50
76	47.76	50.54	51.88	55.42	57.20	59.19	62.25	65.06	67.01	68.58	71.20	73.49
77	48.55	51.36	52.70	56.28	58.07	60.07	63.16	65.99	67.95	69.54	72.17	74.47
78	49.34	52.17	53.53	57.13	58.94	60.95	64.06	66.92	68.89	70.49	73.14	75.46
79	50.13	52.98	54.35	57.98	59.80	61.84	64.97	67.85	69.83	71.44	74.11	76.44
80	50.92	53.80	55.18	58.84	60.67	62.72	65.88	68.77	70.77	72.39	75.08	77.43
81	51.72	54.62	56.01	59.70	61.54	63.61	66.79	69.70	71.72	73.35	76.05	78.41
82	52.51	55.43	56.84	60.55	62.41	64.49	67.70	70.63	72.66	74.30	77.02	79.40
83	53.31	56.25	57.67	61.41	63.29	65.38	68.60	71.56	73.60	75.25	77.99	80.39
84	54.10	57.07	58.50	62.27	64.16	66.27	69.51	72.49	74.54	76.21	78.96	81.37
85	54.90	57.89	59.33	63.13	65.03	67.15	70.43	73.42	75.49	77.16	79.93	82.36
86	55.70	58.72	60.16	63.99	65.91	68.04	71.34	74.35	76.43	78.11	80.91	83.34
87	56.50	59.54	61.00	64.85	66.78	68.93	72.25	75.28	77.38	79.07	81.88	84.33
88	57.31	60.37	61.83	65.72	67.66	69.82	73.16	76.21	78.32	80.02	82.85	85.32
89	58.11	61.19	62.67	66.58	68.53	70.71	74.07	77.14	79.27	80.98	83.82	86.30
90	58.91	62.02	63.51	67.44	69.41	71.61	74.98	78.08	80.21	81.93	84.79	87.29
91	59.72	62.84	64.34	68.31	70.29	72.50	75.90	79.01	81.16	82.89	85.77	88.28
92	60.52	63.67	65.18	69.17	71.17	73.39	76.81	79.94	82.10	83.85	86.74	89.26
93	61.33	64.50	66.02	70.04	72.05	74.28	77.73	80.88	83.05	84.80	87.71	90.25
94	62.14	65.33	66.86	70.91	72.93	75.18	78.64	81.81	83.99	85.76	88.68	91.24
95	62.95	66.16	67.70	71.77	73.81	76.07	79.56	82.74	84.94	86.72	89.66	92.22
96	63.76	66.99	68.55	72.64	74.69	76.97	80.47	83.68	85.89	87.67	90.63	93.21
97	64.57	67.83	69.39	73.51	75.57	77.86	81.39	84.61	86.83	88.63	91.60	94.20
98	65.38	68.66	70.23	74.38	76.45	78.76	82.31	85.55	87.78	89.59	92.58	95.19
99	66.19	69.50	71.08	75.25	77.33	79.65	83.22	86.48	88.73	90.54	93.55	96.17
100	67.01	70.33	71.92	76.12	78.22	80.55	84.14	87.42	89.68	91.50	94.52	97.16

(*N* is the number of servers. The numerical column headings indicate blocking probability *B* in percentage.)

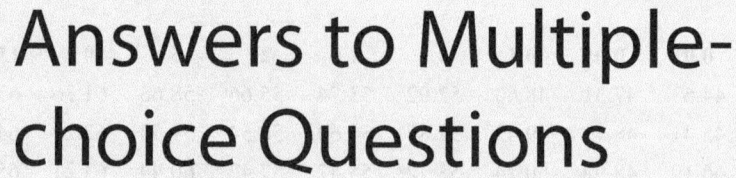

Answers to Multiple-choice Questions

CHAPTER - 1

1.1 (b)	1.2 (c)	1.3 (a)	1.4 (d)	1.5 (b)	1.6 (a)	1.7 (b)	1.8 (d)
1.9 (c)	1.10 (c)	1.11 (c)	1.12 (d)	1.13 (a)	1.14 (d)	1.15 (d)	1.16 (c)
1.17 (a)	1.18 (b)						

CHAPTER - 2

2.1 (d)	2.2 (d)	2.3 (c)	2.4 (d)	2.5 (c)	2.6 (b)	2.7 (a)	2.8 (c)
2.9 (c)	2.10 (b)	2.11 (d)	2.12 (c)	2.13 (b)	2.14 (b)	2.15 (b)	2.16 (d)

CHAPTER - 3

3.1 (b)	3.2 (c)	3.3 (c)	3.4 (c)	3.5 (b)	3.6 (b)	3.7 (a)	3.8 (c)
3.9 (d)	3.10 (d)	3.11 (a)	3.12 (b)	3.13 (a)	3.14 (b)	3.15 (b)	3.16 (b)
3.17 (a)	3.18 (d)	3.19 (a)	3.20 (c)	3.21 (b)	3.22 (c)	3.23 (d)	

CHAPTER - 4

4.1 (a)	4.2 (b)	4.3 (c)	4.4 (a)	4.5 (c)	4.6 (c)	4.7 (c)	4.8 (b)

CHAPTER - 5

5.1 (c)	5.2 (a)	5.3 (b)	5.4 (b)	5.5 (b)	5.6 (d)	5.7 (b)	5.8 (b)
5.9 (d)	5.10 (c)	5.11 (c)	5.12 (c)	5.13 (b)	5.14 (d)	5.15 (c)	5.16 (d)
5.17 (a)	5.18 (b)	5.19 (c)	5.20 (c)	5.21 (c)	5.22 (d)	5.23 (d)	5.24 (d)
5.25 (a)	5.26 (d)	5.27 (b)					

CHAPTER - 6

6.1 (b)	6.2 (b)	6.3 (d)	6.4 (b)	6.5 (c)	6.6 (c)	6.7 (a)	6.8 (c)
6.9 (b)	6.10 (a)	6.11 (c)					

CHAPTER - 7

7.1 (a)	7.2 (a)	7.3 (b)	7.4 (d)	7.5 (d)	7.6 (d)	7.7 (b)	7.8 (a)
7.9 (c)	7.10 (d)	7.11 (a)	7.12 (d)	7.13 (b)	7.14 (b)	7.15 (d)	7.16 (b)
7.17 (b)	7.18 (d)	7.19 (c)	7.20 (b)	7.21 (d)	7.22 (c)	7.23 (a)	7.24 (c)
7.25 (a)	7.26 (d)	7.27 (d)	7.28 (d)	7.29 (b)	7.30 (b)		

CHAPTER - 8

8.1 (c)	8.2 (d)	8.3 (c)	8.4 (a)	8.5 (d)	8.6 (a)	8.7 (c)	8.8 (c)
8.9 (a)	8.10 (c)	8.11 (c)	8.12 (c)				

CHAPTER - 9

9.1 (b)	9.2 (b)	9.3 (b)	9.4 (d)	9.5 (d)	9.6 (d)	9.7 (d)	9.8 (c)
9.9 (c)	9.10 (b)	9.11 (d)	9.12 (d)	9.13 (b)	9.14 (c)	9.15 (a)	9.16 (b)

CHAPTER - 10

10.1 (c)	10.2 (d)	10.3 (c)	10.4 (a)	10.5 (d)	10.6 (a)	10.7 (b)

CHAPTER - 11

11.1 (c)	11.2 (a)	11.3 (a)	11.4 (b)	11.5 (d)	11.6 (b)	11.7 (c)	11.8 (c)
11.9 (d)	11.10 (d)	11.11 (b)	11.12 (b)	11.13 (b)	11.14 (b)	11.15 (b)	11.16 (c)
11.17 (c)	11.18 (a)	11.19 (b)	11.20 (b)	11.21 (d)	11.22 (c)	11.23 (d)	11.24 (c)
11.25 (a)	11.26 (a)	11.27 (b)					

CHAPTER - 12

12.1 (b)	12.2 (c)	12.3 (c)	12.4 (d)	12.5 (d)	12.6 (a)	12.7 (c)	12.8 (c)
12.9 (c)	12.10 (a)	12.11 (b)	12.12 (c)	12.13 (a)	12.14 (c)	12.15 (a)	12.16 (c)
12.17 (d)	12.18 (c)	12.19 (c)	12.20 (d)	12.21 (a)	12.22 (c)		

Index

About the Author

Dr Upena Dalal is presently working as Associate Professor and Head of the Department of Electronics Engineering in Sardar Vallabhbhai National Institute of Technology (SVNIT), Surat, India. She obtained her Bachelor's degree from SVRCET, Surat, in 1991, completed her post-graduation from DDIT, Nadiad, with a Gold Medal in 2005, and went on to obtain a PhD from SVNIT in 2009. She has more than 23 years of academic experience.

She is also the author of the book *Wireless Communication*, published by Oxford University Press in 2009. She has published about 100 conference and reputed journal papers, at both national and international levels. She has guided many UG and PG projects, dissertations, and seminars pertaining to advance communication systems. Under her guidance, two scholars have received PhDs and 12 research scholars are pursuing their PhD. Her doctoral thesis has been published as a book by VDM Publishing, Germany.

Dr Dalal was awarded the 5th N V Gadadhar memorial award by IETE, the Rashtriya Gaurav Award, as well as the International Goldstar award by the India International Friendship Society. She has received the Divya Bhaskar woman of the year award in 2010. She has organized 11 national level training programs and one IEEE International conference. She is a member of IEEE, and a lifetime member of IETE, IE, ISTE, and IJERIA. She is also the chairperson of IETE Surat subcenter. She has chaired many reputed conferences and is an editorial board member of IJCT.

Related Titles

Communication Systems (9780198078050)

V. Chandra Sekar, SASTRA University, Kumbakonam

This textbook is specifically designed for students of CSE, IT, and ECE, and provides a balance between the theory and related applications with the help of figures, solved problems, numerical exercises, and review questions.

Key Features
- Provides discussion on topics such as waveform distortion, distortionless line, waveguide impedance matching, and microstrip antennas
- Includes practical circuits for topics such as FM detection and generation, transmitters and receivers, and phase-locked loop (PLL)
- Provides application case studies on PLL frequency synthesizer, satellite instructional television experiment (SITE), and digital microwave radio

Digital Image Processing (9780198070788)

S Sridhar, College of Engineering Guindy, Anna University, Chennai

The book aims to provide an understanding of the principles and processing techniques of digital images.

Key Features
- Provides simple explanations of topics such as Shannon–Fano coding, morphological gradient, polygonal approximations, shape number, and component labelling
- Includes solved numerical examples interspersed throughout the book
- Contains appendices that discuss the basics of MATLAB programming and ImageJ, and provides information on other public domain image processing software

Microprocessors and Microcontrollers (9780198066477)

N Senthil Kumar, Mepco Schlenk Engineering College, Sivakasi; **M Saravanan**, Thiagarajar College of Engineering, Madurai; **S Jeevananthan**, Pondicherry Engineering College, Puducherry

This textbook is designed as a comprehensive textbook for undergraduate engineering students to build a strong foundation in the basic principles, functioning, and applications of microprocessors and microcontrollers.

Key Features
- Includes case studies on traffic light control, washing machine control, and elevator control to enable students appreciate the applications of processors
- Includes discussions on advanced processors, such as 80186, 80286, 80386, 80486, Pentium, PowerPC, and PIC 16F877
- Contains a section on the advent of high-level language programming in 8051 processors, with examples illustrated using the most popular language C

Intelligent Systems and Control (9780198063155)

Laxmidhar Behera, IIT Kanpur; **Indrani Kar**, IIT Guwahati

The book covers the fundamentals of neural networks, fuzzy logic, and nonlinear control so that the readers can easily follow intelligent control techniques.

Key Features
- Comprehensively covers fuzzy and neural control schemes and introduces quantum neural networks using a novel paradigm
- Incorporates experiments on cart-pole inverted pendulum system and seven DOF robot manipulator
- Provides illustrative examples with MATLAB codes

Other Related Titles

9780195669305   **RP Khare**: Fiber Optics and Optoelectronics

9780198067665   **Debaprasad Das**: VLSI Design

9780198063575   **Satish Shah**: 8051 Microcontrollers

9780195671544   **N.P. Padhy**: Artificial Intelligence and Intelligent Systems

9780195686661   **A.R. Harish and M. Sachidananda**: Antennas & Wave Propagation

9780198066774   **Bhushan Trivedi**: Computer Networks